MW00586525

HANDBOOK OF SIMULATION

HANDBOOK OF SIMULATION

Principles, Methodology, Advances, Applications, and Practice

edited by

Jerry Banks
Georgia Institute of Technology
Atlanta, Georgia

A WILEY-INTERSCIENCE PUBLICATION

JOHN WILEY & SONS, INC.

New York • Chichester • Weinheim • Brisbane • Singapore • Toronto

This book is printed on acid-free paper. ♾

Published simultaneously in Canada.

Library of Congress Cataloging-in-Publication Data:

Handbook of simulation/edited by Jerry Banks.
p. cm.
"A Wiley-Interscience publication."
Includes index.
ISBN 0-471-13403-1 (cloth: alk. paper)
1. Simulation methods—Handbooks, manuals, etc. 2. Discrete-time systems—Handbooks, manuals, etc. I. Banks, Jerry.
T57.62.H37 1998
003′.83—dc21 97-51533

Printed in the United States of America.

10 9 8 7 6 5 4 3 2 1

To Nancy for all her patience, understanding, and love.

CONTENTS

PART III RECENT ADVANCES

PART IV APPLICATION AREAS

PART V PRACTICE OF SIMULATION

CONTRIBUTORS

Christos Alexopoulos, School of Industrial and Systems Engineering, Georgia Institute of Technology, Atlanta, Georgia

Sigrún Andradóttir, School of Industrial and Systems Engineering, Georgia Institute of Technology, Atlanta, Georgia

Osman Balci, Department of Computer Science, Virginia Polytechnic Institute and State University, Blacksburg, Virginia

Jerry Banks, School of Industrial and Systems Engineering, Georgia Institute of Technology, Atlanta, Georgia

Daniel T. Brunner, Systemflow Simulations, Inc., Indianapolis, Indiana

Russell C. H. Cheng, Institute of Mathematics and Statistics, The University of Kent at Canterbury, Canterbury, England

Wayne J. Davis, Department of General Engineering, University of Illinois, Urbana, Illinois

Richard M. Fujimoto, College of Computing, Georgia Institute of Technology, Atlanta, Georgia

David Goldsman, School of Industrial and Systems Engineering, Georgia Institute of Technology, Atlanta, Georgia

Ali Gunal, Production Modeling Corporation, Dearborn, Michigan

Alfred Hartmann, Mesquite Software, Inc., Austin, Texas

Jeffrey A. Joines, Department of Industrial Engineering, North Carolina State University, Raleigh, North Carolina

Keebom Kang, Department of Systems Management, Naval Postgraduate School, Monterey, California

Ali S. Kiran, Kiran and Associates, San Diego, California

Jack P. C. Kleijnen, Center for Economic Research, Tilburg University, Tilburg, Netherlands

Ron Laughery, Micro Analysis and Design, Inc., Boulder, Colorado

Pierre L'Ecuyer, Department of Computer Science and Operations Research, University of Montreal, Montreal, Canada

Mani S. Manivannan, CNF Transportation, Inc., Portland, Oregon

Frank McGuire, Premier, Inc., Charlotte, North Carolina

Ken Musselman, Pritsker Corporation, West Lafayette, Indiana

Barry Nelson, McCormick School of Engineering and Applied Science, Northwestern University, Evanston, Illinois

Van Norman, AutoSimulations, Inc., Bountiful, Utah

Beth Plott, Micro Analysis and Design, Inc., Boulder, Colorado

A. Alan B. Pritsker, Pritsker Corporation, Indianapolis, Indiana

Stephen D. Roberts, Department of Industrial Engineering, North Carolina State University, Raleigh, North Carolina

Matthew W. Rohrer, AutoSimulations, Inc., Bountiful, Utah

Ronald J. Roland, Rolands and Associates Corporation, Monterey, California

Thomas J. Schriber, School of Business Administration, University of Michigan, Ann Arbor, Michigan

Herb Schwetman, Mesquite Software, Inc., Austin, Texas

Shelly Scott-Nash, Micro Analysis and Design, Inc., Boulder, Colorado

Andrew F. Seila, Program in Management Science, University of Georgia, Athens, Georgia

Onur Ulgen, Production Modeling Corporation, Dearborn, Michigan

Stephen Vincent, Compuware Corporation, Milwaukee, Wisconsin

HANDBOOK OF SIMULATION

PART I

PRINCIPLES

CHAPTER 1

Principles of Simulation

JERRY BANKS
Georgia Institute of Technology

1.1 INTRODUCTION

The purpose of this handbook is to provide a reference to important topics that pertain to discrete-event simulation. All the contributors to this volume, who are a mix from academia, industry, and software developers, are highly qualified. The book is intended for those who want to apply simulation to important problems. If you are new to simulation, reading this chapter will provide an overview to the remainder of the book. If you studied simulation several years ago, reading this chapter will provide a useful review and update. [Much of this introductory chapter is from Banks et al. (1995).]

Chapter 1 is essentially in three parts. The first part begins with a definition and an example of simulation. Then modeling concepts introduced in the example are presented. Four modeling structures for simulation are then presented. The second part of the chapter concerns subjective topics. First, the advantages and disadvantages of simulation are discussed. Then some of the areas of application are mentioned. Last, the steps in the simulation process are described. The third part of the chapter has four sections. Each of these sections introduces operational aspects of discrete-event simulation. The chapter concludes with a summary.

1.2 DEFINITION OF SIMULATION

Simulation is the imitation of the operation of a real-world process or system over time. Simulation involves the generation of an artificial history of the system and the observation of that artificial history to draw inferences concerning the operating characteristics of the real system that is represented. Simulation is an indispensable problem-solving methodology for the solution of many real-world problems. Simulation is used to describe and analyze the behavior of a system, ask what-if questions about the real

Handbook of Simulation, Edited by Jerry Banks.
ISBN 0-471-13403-1

system, and aid in the design of real systems. Both existing and conceptual systems can be modeled with simulation.

Example 1 (Ad Hoc Simulation) Consider the operation of a one-teller bank where customers arrive for service between 1 and 10 minutes apart in time, integer values only, each value equally likely. The customers are served in between 1 and 6 minutes, also integer valued, and equally likely. Restricting the times to integer values is an abstraction of reality since time is continuous, but this aids in presenting the example. The objective is to simulate the bank operation, by hand, until 20 customers are served, and to compute measures of performance such as the percentage of idle time, the average waiting time per customer, and so on. Admittedly, 20 customers are far too few to draw conclusions about the operation of the system for the long run, but by following this example, the stage is set for further discussion in this chapter and subsequent discussion about using the computer for performing simulation.

To simulate the process, random interarrival and service times need to be generated. Assume that the interarrival times are generated using a spinner that has possibilities for the values 1 through 10. Further assume that the service times are generated using a die that has possibilities for the values 1 through 6.

Table 1.1 is called an *ad hoc simulation table*. The setup of the simulation table is for

Table 1.1 Ad Hoc Simulation

(1)	(2)	(3)	(4)	(5)	(6)	(7)	(8)	(9)
Customer	Time Between Arrivals	Arrival Time	Service Time	Service Begins	Time Service Ends	Time in System	Idle Time	Time in Queue
1	—	0	2	0	2	2	0	0
2	5	5	2	5	7	2	3	0
3	1	6	6	7	13	7	0	1
4	10	16	5	16	21	5	3	0
5	6	22	6	22	28	6	1	0
6	2	24	4	28	32	8	0	4
7	9	33	3	33	36	3	1	0
8	1	34	4	36	40	6	0	2
9	10	44	1	44	45	1	4	0
10	3	47	3	47	50	3	2	0
11	5	52	1	52	53	1	2	0
12	2	54	2	54	56	2	1	0
13	3	57	3	57	60	3	1	0
14	5	62	6	62	68	6	2	0
15	4	66	2	68	70	4	0	2
16	3	69	6	70	76	7	0	1
17	7	76	4	76	80	4	0	0
18	8	84	5	84	89	5	4	0
19	7	91	3	91	94	3	2	0
20	7	98	1	98	99	1	4	0
						79	30	10

the purpose of this problem but does not pertain to all problems. Column (1), Customer, lists the 20 customers who arrive at the system. It is assumed that customer 1 arrives at time zero; thus a dash is indicated in row 1 of column (2), Time Between Arrivals. Rows 2 through 20 of column (2) were generated using the spinner. Column (3), Arrival Time, shows the simulated arrival times. Since customer 1 is assumed to arrive at time 0 and there is a 5-minute interarrival time, customer 2 arrives at time 5. There is a 1-minute interarrival time for customer 3; thus the arrival occurs at time 6. This process of adding the interarrival time to the previous arrival time is called *bootstrapping*. By continuing this process, the arrival times of all 20 customers are determined. Column (4), Service Time, contains the simulated service times for all 20 customers. These were generated by rolling the die.

Now simulation of the service process begins. At time 0, customer 1 arrived and immediately began service. The service time was 2 minutes, so the service period ended at time 2. The total time in the system for customer 1 was 2 minutes. The bank teller was not idle since simulation began with the arrival of a customer. The customer did not have to wait for the teller.

At time 5, customer 2 arrived and began service immediately, as shown in column (6). The service time was 2 minutes, so the service period ended at time 7, as shown in column (6). The bank teller was idle from time 2 until time 5, so 3 minutes of idle time occurred. Customer 2 spent no time in the queue.

Customer 3 arrived at time 6, but service could not begin until time 7, as customer 2 was being served until time 7. The service time was 6 minutes, so service was completed at time 13. Customer 3 was in the system from time 6 until time 13, or for 7 minutes, as indicated in column (7), Time in System. Although there was no idle time, customer 3 had to wait in the queue for 1 minute for service to begin.

This process continues for all 20 customers, and the totals shown in columns (7), (8) (Idle Time), and (9) (Time in Queue) are entered. Some performance measures can now be calculated as follows:

Average time in system = 79/20 = 3.95 minutes.

Percent idle time = (30/99)(100) = 30%.

Average waiting time per customer = 10/20 = 0.5 minute.

Fraction having to wait = 5/20 = 0.25.

Average waiting time of those who waited = 10/3 = 3.33 minutes.

This very limited simulation indicates that the system is functioning well. Only 25% of the customers had to wait. About 30% of the time the teller is idle. Whether a slower teller should replace the current teller depends on the cost of having to wait versus any savings from having a slower server.

This small simulation can be accomplished by hand, but there is a limit to the complexity of problems that can be solved in this manner. Also, the number of customers that must be simulated could be much larger than 20 and the number of times that the simulation must be run for statistical purposes could be large. Hence, using the computer to solve real simulation problems is almost always appropriate.

Example 1 raises some issues that are addressed in this chapter and explored more fully in the balance of the book. The issues include the following:

1. How is the form of the input data determined?
2. How are random variates generated if they follow statistical distributions other than the discrete uniform?
3. How does the user know that the simulation imitates reality?
4. What other kinds of problems can be solved by simulation?
5. How long does the simulation need to run?
6. How many different simulation runs should be conducted?
7. What statistical techniques should be used to analyze the output?

Each of these questions raises a host of issues about which many textbooks and thousands of technical papers have been written. Although an introductory chapter cannot treat all of these questions in the greatest detail, enough can be said to give the reader some insight that will be useful in understanding the framework of the remainder of the book.

1.3 MODELING CONCEPTS

There are several concepts underlying simulation. These include system and model, system state variables, entities and attributes, list processing, activities and delays, and the definition of discrete-event simulation. Additional information on these topics is available from Banks et al. (1996) and Law and Kelton (1991). The discussion in this section follows that of Carson (1993). Chapter 2 provides an extensive discussion of the topic.

1.3.1 System, Model, and Events

A *model* is a representation of an actual system. Immediately, there is a concern about the limits or boundaries of the model that supposedly represent the system. The model should be complex enough to answer the questions raised, but not too complex. Consider an *event* as an occurrence that changes the state of the system. In Example 1, events include the arrival of a customer for service at a bank and the completion of a service. There are both internal and external events, also called *endogenous events* and *exogenous events*, respectively. For example, an endogenous event in Example 1 is the beginning of service of the customer since that is within the system being simulated. An exogenous event is the arrival of a customer for service since that occurrence is outside the simulation. However, the arrival of a customer for service impinges on the system and must be taken into consideration.

In this book we consider *discrete-event simulation models*. (Chapter 2 describes continuous and combined discrete-continuous models.) These are contrasted with other types of models, such as mathematical models, descriptive models, statistical models, and input–output models. A discrete-event model attempts to represent the components of a system and their interactions to such an extent that the objectives of the study are met. Most mathematical, statistical, and input–output models represent a system's inputs and outputs explicitly but represent the internals of the model with mathematical or statistical relationships. An example is the mathematical model from physics,

$$\text{force} = \text{mass} \times \text{acceleration}$$

based on theory. Discrete-event simulation models include a detailed representation of the actual internals.

Discrete-event models are *dynamic*; that is, the passage of time plays a crucial role. Most mathematical and statistical models are *static*, in that they represent a system at a fixed point in time. Consider the annual budget of a firm. The budget resides in a spreadsheet. Changes can be made in the budget and the spreadsheet can be recalculated, but the passage of time is usually not a critical issue. Further comments will be made about discrete-event models after several additional concepts are presented.

1.3.2 System State Variables

The *system state variables* are the collection of all information needed to define what is happening within a system to a sufficient level (i.e., to attain the desired output) at a given point in time. The determination of system state variables is a function of the purposes of the investigation, so what may be the system state variables in one case may not be the same in another case, even though the physical system is the same. Determining the system state variables is as much an art as a science. However, during the modeling process, any omissions will readily come to light. (On the other hand, unnecessary state variables may be eliminated.)

Having defined system state variables, a contrast can be made between discrete-event models and continuous models based on the variables needed to track the system state. The system state variables in a discrete-event model remain constant over intervals of time and change value only at certain well-defined points called *event times*. *Continuous models* have system state variables defined by differential or difference equations, giving rise to variables that may change continuously over time.

Some models are mixed discrete-event and continuous. There are also continuous models that are treated as discrete-event models after some reinterpretation of system state variables, and vice versa. The modeling of continuous systems is not treated in this book.

1.3.3 Entities and Attributes

An *entity* represents an object that requires explicit definition. An entity can be dynamic in that it "moves" through the system, or it can be static in that it serves other entities. In Example 1 the customer is a dynamic entity, whereas the bank teller is a static entity.

An entity may have *attributes* that pertain to that entity alone. Thus attributes should be considered as local values. In Example 1, an attribute of the entity could be the time of arrival. Attributes of interest in one investigation may not be of interest in another investigation. Thus, if red parts and blue parts are being manufactured, the color could be an attribute. However, if the time in the system for all parts is of concern, the attribute of color may not be of importance. From this example it can be seen that many entities can have the same attribute or attributes (i.e., more than one part may have the attribute "red").

1.3.4 Resources

A *resource* is an entity that provides service to dynamic entities. The resource can serve one or more than one dynamic entity at the same time (i.e., operate as a parallel server). A dynamic entity can request one or more units of a resource. If denied, the requesting entity joins a queue or takes some other action (i.e., is diverted to another resource, is ejected from the system). (Other terms for queues are *files*, *chains*, *buffers*, and *waiting lines*.) If permitted to capture the resource, the entity remains for a time, then releases the resource. There are many possible states of a resource. Minimally, these states are idle and busy. But other possibilities exist, including failed, blocked, or starved.

1.3.5 List Processing

Entities are managed by allocating them to resources that provide service; by attaching them to event notices, thereby suspending their activity into the future; or by placing them into an ordered list. Lists are used to represent queues.

Lists are often processed according to FIFO (first in, first out), but there are many other possibilities. For example, the list could be processed by LIFO (last in, first out), according to the value of an attribute, or randomly, to mention a few. An example where the value of an attribute may be important is in SPT (shortest process time) scheduling. In this case the processing time may be stored as an attribute of each entity. The entities are ordered according to the value of that attribute, with the lowest value at the head or front of the queue.

1.3.6 Activities and Delays

An *activity* is a period of time whose duration is known prior to commencement of the activity. Thus, when the duration begins, its end can be scheduled. The duration can be a constant, a random value from a statistical distribution, the result of an equation, input from a file, or computed based on the event state. For example, a service time may be a constant 10 minutes for each entity; it may be a random value from an exponential distribution with a mean of 10 minutes; it could be 0.9 times a constant value from clock time 0 to clock time 4 hours, and 1.1 times the standard value after clock time 4 hours; or it could be 10 minutes when the preceding queue contains at most four entities and 8 minutes when there are five or more in the preceding queue.

A *delay* is an indefinite duration that is caused by some combination of system conditions. When an entity joins a queue for a resource, the time that it will remain in the queue may be unknown initially since that time may depend on other events that may occur. An example of another event would be the arrival of a rush order that preempts the resource. When the preempt occurs, the entity using the resource relinquishes its control instantaneously. Another example is a failure necessitating repair of the resource.

Discrete-event simulations contain activities that cause time to advance. Most discrete-event simulations also contain delays as entities wait. The beginning and ending of an activity or delay are events.

1.3.7 Discrete-Event Simulation Model

Sufficient modeling concepts have been defined so that a discrete-event simulation model can be defined as one in which the state variables change only at those discrete points in time at which events occur. Events occur as a consequence of activity times and delays. Entities may compete for system resources, possibly joining queues while waiting for an available resource. Activity and delay times may "hold" entities for durations of time.

A discrete-event simulation model is conducted over time ("run") by a mechanism that moves simulated time forward. The system state is updated at each event, along with capturing and freeing of resources that may occur at that time.

1.4 MODELING STRUCTURES

There are four modeling structures taken by the simulation community. They are known as the process-interaction method, event-scheduling method, activity scanning, and the three-phase method. The descriptions are rather concise; readers requiring greater explanation are referred to Balci (1988) or Pidd (1992). The first two of these modeling structure topics are discussed in Chapter 2. We describe all four of them briefly here.

1.4.1 Process-Interaction Method

The simulation structure that has the greatest intuitive appeal is the process-interaction method. The notion is that the computer program should emulate the flow of an object through the system. The entity moves as far as possible in the system until it is delayed, enters an activity, or exits from the system. When the entity's movement is halted, the clock advances to the time of the next movement of any entity.

This flow, or movement, describes in sequence all the states that the object can attain in the system. For example, in a model of a self-service laundry a customer may enter the system, wait for a washing machine to become available, wash his or her clothes in the washing machine, wait for a basket to become available to unload the washing machine, transport the clothes in the basket to a drier, wait for a drier to become available, unload the clothes into a drier, dry the clothes, and leave the laundry.

1.4.2 Event-Scheduling Method

The basic concept of the event-scheduling method is to advance time to when something next happens. This usually releases a resource (i.e., a scarce entity such as a machine or transporter). The event then reallocates available objects or entities by scheduling activities where they can now participate. For example, in the self-service laundry, if a customer's washing is finished and there is a basket available, the basket could be allocated immediately to the customer and unloading of the washing machine could begin. Time is advanced to the next scheduled event (usually the end of an activity) and activities are examined to see if any can now start as a consequence.

1.4.3 Activity Scanning

The third simulation modeling structure is activity scanning. It is also known as the two-phase approach. Activity scanning is similar to rule-based programming. (If a specified condition is met, a rule is *fired*, meaning that an action is taken.) Activity scanning produces a simulation program composed of independent modules waiting to be executed. Scanning takes place at fixed time increments at which a determination is made concerning whether or not an event occurs at that time. If an event occurs, the system state is updated.

1.4.4 Three-Phase Method

The fourth simulation modeling structure is known as the three-phase method. Time is advanced until there is a state change in the system or until something next happens. The system is examined to determine all of the events that take place at this time (i.e., all the activity completions that occur). Only when all resources that are due to be released at this time have been released is reallocation of these resources into new activities started in the third phase of the simulation. In summary, the first phase is time advance. The second phase is the release of those resources scheduled to end their activities at this time. The third phase is to start activities given the global picture about resource availability.

Possible modeling inaccuracies may occur with the last two methods, as discrete time slices must be specified. With computing power growing so rapidly, high-precision simulation will be utilized increasingly, and the error due to discretizing time may become an important consideration.

1.5 ADVANTAGES AND DISADVANTAGES OF SIMULATION*

Competition in the computer industry has led to technological breakthroughs that are allowing hardware companies to produce better products continually. It seems that every week another company announces its latest release, each with more options, memory, graphics capability, and power.

What is unique about new developments in the computer industry is that they often act as a springboard for related industries to follow. One industry in particular is the simulation-software industry. As computer hardware becomes more powerful, more accurate, faster, and easier to use, simulation software does, too.

The number of businesses using simulation is increasing rapidly. Many managers are realizing the benefits of utilizing simulation for more than just the one-time remodeling of a facility. Rather, due to advances in software, managers are incorporating simulation in their daily operations on an increasingly regular basis.

For most companies, the benefits of using simulation go beyond simply providing a look into the future. These benefits are mentioned by many authors (Banks et al., 1996; Law and Kelton, 1991; Pegden et al., 1995; Schriber, 1991) and are included in the following.

1.5.1 Advantages

1. *Choose correctly.* Simulation lets you test every aspect of a proposed change or addition without committing resources to their acquisition. This is critical, because once the hard decisions have been made, the bricks have been laid, or the material handling systems have been installed, changes and corrections can be extremely expensive. Simulation allows you to test your designs without committing resources to acquisition.

2. *Compress and expand time.* By compressing or expanding time, simulation allows you to speed up or slow down phenomena so that you can investigate them thoroughly. You can examine an entire shift in a matter of minutes if you desire, or you can spend 2 hours examining all the events that occurred during 1 minute of simulated activity.

*Reprinted with the permission of the Institute of Industrial Engineers, 25 Technology Park, Norcross, GA 30092, 770-449-0161. Copyright ©1998.

3. *Understand why.* Managers often want to know why certain phenomena occur in a real system. With simulation, you determine the answer to the "why" questions by reconstructing the scene and taking a microscopic examination of the system to determine why the phenomenon occurs. You cannot accomplish this with a real system because you cannot see or control it in its entirety.

4. *Explore possibilities.* One of the greatest advantages of using simulation software is that once you have developed a valid simulation model, you can explore new policies, operating procedures, or methods without the expense and disruption of experimenting with the real system. Modifications are incorporated in the model, and you observe the effects of those changes on the computer rather than on the real system.

5. *Diagnose problems.* The modern factory floor or service organization is very complex, so complex that it is impossible to consider all the interactions taking place in a given moment. Simulation allows you to better understand the interactions among the variables that make up such complex systems. Diagnosing problems and gaining insight into the importance of these variables increases your understanding of their important effects on the performance of the overall system.

The last three claims can be made for virtually all modeling activities, queueing, linear programming, and so on. However, with simulation the models can become very complex and thus have a higher fidelity [i.e., they are valid representations of reality (as discussed in Chapter 10)].

6. *Identify constraints.* Production bottlenecks give manufacturers headaches. It is easy to forget that bottlenecks are an effect rather than a cause. However, by using simulation to perform bottleneck analysis, you can discover the cause of the delays in work in process, information, materials, or other processes.

7. *Develop understanding.* Many people operate with the philosophy that talking loudly, using computerized layouts, and writing complex reports convinces others that a manufacturing or service system design is valid. In many cases these designs are based on someone's thoughts about the way the system operates rather than on analysis. Simulation studies aid in providing understanding about how a system really operates rather than indicating someone's predictions about how a system will operate.

8. *Visualize the plan.* Taking your designs beyond CAD drwings by using the animation features offered by many simulation packages allows you to see your facility or organization actually running. Depending on the software used, you may be able to view your operations from various angles and levels of magnification, even in three dimensions. This allows you to detect design flaws that appear credible when seen just on paper on in a two-dimensional CAD drawing.

9. *Build consensus.* Using simulation to present design changes creates an objective opinion. You avoid having inferences made when you approve or disapprove of designs because you simply select the designs and modifications that provided the most desirable results, whether it be increased production or reducing the waiting time for service. In addition, it is much easier to accept reliable simulation results, which have been modeled, tested, validated, and visually represented, instead of one person's opinion of the results that will occur from a proposed design.

10. *Prepare for change.* We all know that the future will bring change. Answering all of the what-if questions is useful for both designing new systems and redesigning

existing systems. Interacting with all those involved in a project during the problem-formulation stage gives you an idea of the scenarios that are of interest. Then you construct the model so that it answers questions pertaining to those scenarios. What if an AGV is removed from service for an extended period of time? What if demand for service increases by 10%? What if ... ? The options are unlimited.

11. *Invest wisely.* The typical cost of a simulation study is substantially less than 1% of the total amount being expended for the implementation of a design or redesign. Since the cost of a change or modification to a system after installation is so great, simulation is a wise investment.

12. *Train the team.* Simulation models can provide excellent training when designed for that purpose. Used in this manner, the team provides decision inputs to the simulation model as it progresses. The team, and individual members of the team, can learn by their mistakes and learn to operate better. This is much less expensive and less disruptive than on-the-job learning.

13. *Specify requirements.* Simulation can be used to specify requirements for a system design. For example, the specifications for a particular type of machine in a complex system to achieve a desired goal may be unknown. By simulating different capabilities for the machine, the requirements can be established.

1.5.2 Disadvantages

The disadvantages of simulation include the following:

1. *Model building requires special training.* It is an art that is learned over time and through experience. Furthermore, if two models of the same system are constructed by two competent individuals, they may have similarities, but it is highly unlikely that they will be the same.

2. *Simulation results may be difficult to interpret.* Since most simulation outputs are essentially random variables (they are usually based on random inputs), it may be hard to determine whether an observation is a result of system interrelationships or randomness.

3. *Simulation modeling and analysis can be time consuming and expensive.* Skimping on resources for modeling and analysis may result in a simulation model and/or analysis that is not sufficient to the task.

4. *Simulation may be used inappropriately.* Simulation is used in some cases when an analytical solution is possible, or even preferable. This is particularly true in the simulation of some waiting lines where closed-form queueing models are available, at least for long-run evaluation.

In defense of simulation, these four disadvantages, respectively, can be offset as follows:

1. *Simulators.* Vendors of simulation software have been actively developing packages that contain models that only need input data for their operation. Such models have the generic tag "simulators" or templates.

2. *Output analysis.* Most simulation-software vendors have developed output-analysis capabilities within their packages or, as add on features for performing very extensive analysis. This reduces the computational requirements on the part of the user, although they still must understand the analysis procedure.

3. *Faster and faster.* Simulation can be performed faster today than yesterday, and even faster tomorrow. This is attributable to the advances in hardware that permit rapid running of scenarios. It is also attributable to the advances in many simulation packages. For example, many simulation software products contain constructs for modeling material handling using transporters such as conveyors and automated guided vehicles.

4. *Limitations of closed-form models.* Closed-form models are not able to analyze most of the complex systems that are encountered in practice. In nearly 8 years of consulting practice, not one problem has been encountered that could have been solved by a closed-form solution.

1.6 AREAS OF APPLICATION

The applications of simulation are vast. Recent presentations at the Winter Simulation Conference (WSC) can be divided into manufacturing, public systems, and service systems. WSC is an excellent way to learn more about the latest in simulation applications and theory. There are also numerous tutorials at both the beginning and advanced levels. WSC is sponsored by eight technical societies and the National Institute of Standards and Technology (NIST). The technical societies are the American Statistical Association (ASA), Association for Computing Machinery/Special Interest Group on Simulation (ACM/SIGSIM), Institute of Electrical and Electronics Engineers: Computer Society (IEEE/CS), Institute of Electrical and Electronics Engineers: Systems, Man and Cybernetics Society (IEEE/SMCS), Institute of Industrial Engineers (IIE), Institute for Operations Research and the Management Sciences, College on Simulation (INFORMS/CS), and Society for Computer Simulation (SCS). The societies can provide information about upcoming WSCs, usually held Monday through Wednesday in early December. Applications in the remainder of this section were presented at recent WSCs. (Chapter 25, in particular, contains references to recent Winter Simulation Conference Proceedings.) The major application areas of discrete-event simulation are discussed in Chapters 14 through 21.

1.6.1 Manufacturing and Material Handling Applications

Presentations included the following, among many others:

- Minimizing synchronization delays of prefabricated parts before assembly
- Evaluation of AGV routing strategies
- Flexible modeling and analysis of large-scale AS/RS-AGV systems
- Design and analysis of large-scale material handling systems
- Material flow analysis of automotive assembly plants
- Analysis of the effects of work-in-process levels on customer satisfaction
- Assessing the cost of quality

1.6.2 Public Systems Applications

Presentations included the following, among many others:

Health Systems

- Screening for abdominal aortic aneurysms
- Lymphocite development in immune-compromized patients
- Asthma dynamics and medical amelioration
- Timing of liver transplants
- Diabetic retinopathy
- Evaluation of nurse-staffing and patient-population scenarios
- Evaluation of automated equipment for a clinical processing laboratory
- Evaluation of hospital surgical suite and critical care area

Military Systems

- Air Force support equipment use
- Analysis of material handling equipment for prepositioning ships
- Development and implementation of measures of effectiveness
- Reengineering traditional stovepiped Army staffs for information operations
- Evaluation of theater airlift system productivity
- Evaluation of C-141 depot maintenance
- Evaluation of air mobility command channel cargo system

Natural Resources

- Nonpoint-source pollution analysis
- Weed scouting and weed control decision making
- Evaluation of surface water quality data

Public Services

- Emergency ambulance system analysis
- Evaluation of flow of civil lawsuits
- Evaluation of field offices within a government agency

1.6.3 Service System Applications

Presentations included the following, among many others:

Transportation

- Analysis of intelligent vehicle highway systems
- Evaluation of traffic control procedures at highway work zones
- Evaluation of taxi management and route control
- Animation of a toll plaza
- Port traffic planning model analysis
- Evaluation of rapid transit modeling with automatic and manual controls

Computer Systems Performance

- User transaction processing behavior analysis
- Evaluation of database transaction management protocols
- Evaluation of analytic models of memory queueing

Air Transportation

- Evaluation of human behavior in aircraft evacuations
- Analysis of airport/airline operations
- Evaluation of combination carrier air cargo hub

Communications Systems

- Trunked radio network analysis
- Evaluation of telephone service provisioning process
- Picture archiving and communications system analysis
- Evaluation of modeling of broadband telecommunication networks
- Analysis of virtual reality for telecommunication networks

1.7 STEPS IN A SIMULATION STUDY

Figure 1.1 shows a set of steps to guide a model builder in a thorough and sound simulation study. Similar figures and their interpretation can be found in other sources, such as Pegden et al. (1995) and Law and Kelton (1991). This presentation is built on that of Banks et al. (1996).

1. *Problem formulation.* Every simulation study begins with a statement of the problem. If the statement is provided by those that have the problem (client), the simulation analyst must take extreme care to ensure that the problem is clearly understood. If a problem statement is prepared by the simulation analyst, it is important that the client understand and agree with the formulation. It is suggested that a set of assumptions be prepared by the simulation analyst and agreed to by the client. Even with all of these precautions, it is possible that the problem will need to be reformulated as the simulation study progresses. This step is discussed further in Chapters 22 and 23.

2. *Setting of objectives and overall project plan.* Another way to state this step is "prepare a proposal." This step should be accomplished regardless of location of the analyst and client (i.e., as an external or internal consultant). The objectives indicate the questions that are to be answered by the simulation study. The project plan should include a statement of the various scenarios that will be investigated. The plans for the study should be indicated in terms of time that will be required, personnel that will be used, hardware and software requirements if the client wants to run the model and conduct the analysis, stages in the investigation, output at each stage, cost of the study and billing procedures, if any. This step is discussed further in Chapters 22 and 23.

3. *Model conceptualization.* The real-world system under investigation is abstracted by a conceptual model, a series of mathematical and logical relationships concerning

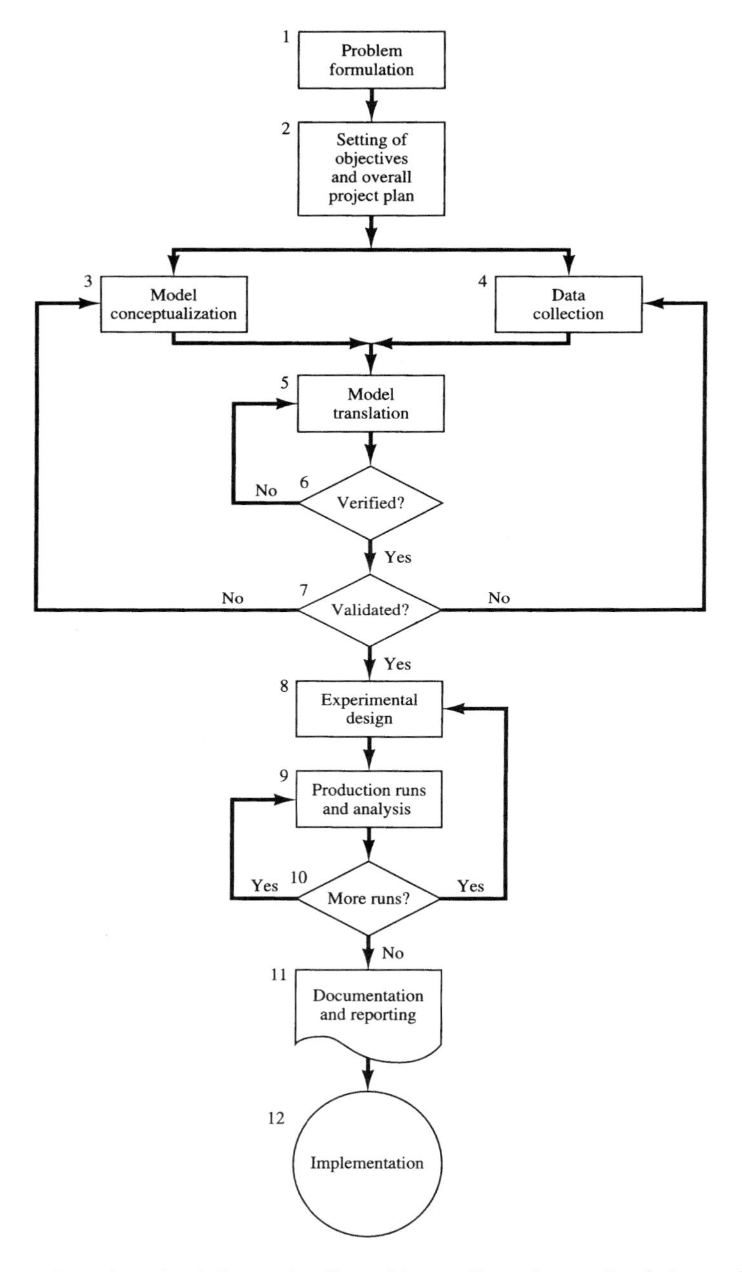

Figure 1.1 Steps in a simulation study. (From *Discrete-Event System Simulation*, 2nd ed., by Banks/Carson/Nelson, @ 1996. Reprinted by permission of Prentice Hall, Upper Saddle River, N.J.

the components and the structure of the system. It is recommended that modeling begin simply and that the model grow until a model of appropriate complexity has been developed. For example, consider the model of a manufacturing and material handling system. The basic model with the arrivals, queues, and servers is constructed. Then add the failures and shift schedules. Next, add the material-handling capabilities. Finally, add the special features. It is not necessary to construct an unduly complex model. This will add to the cost of the study and the time for its completion without increasing the quality of the output. The client should be involved throughout the model construction process. This will enhance the quality of the resulting model and increase the client's confidence in its use. This step is discussed further discussed in Chapters 2, 22, and 23.

4. *Data collection.* Shortly after the proposal is "accepted," a schedule of data requirements should be submitted to the client. In the best of circumstances, the client has been collecting the kind of data needed in the format required and can submit these data to the simulation analyst in electronic format. Often, the client indicates that the required data are indeed available. However, when the data are delivered they are found to be quite different than anticipated. For example, in the simulation of an airline-reservation system, the simulation analyst was told "we have every bit of data that you want over the last five years." When the study began the data delivered were the *average* "talk time" of the reservationist for each of the years. Individual values were needed, not summary measures. Model building and data collection are shown as contemporaneous in Figure 1.1. This is to indicate that the simulation analyst can readily construct the model while the data collection is progressing. This step is discussed further in Chapter 3.

5. *Model translation.* The conceptual model constructed in step 3 is coded into a computer-recognizable form, an operational model. This step is discussed further in Chapters 11, 12, 13, and 24.

6. *Verified?* Verification concerns the operational model. Is it performing properly? Even with small textbook-sized models, it is quite possible that they have verification difficulties. These models are orders of magnitude smaller than real models (say, 50 lines of computer code versus 2000 lines of computer code). It is highly advisable that verification take place as a continuing process. It is ill advised for the simulation analyst to wait until the entire model is complete to begin the verification process. Also, use of an interactive run controller, or debugger, is highly encouraged as an aid to the verification process. Verification is extremely important and is discussed further in this chapter. Additionally, this step is discussed extensively in Chapter 10.

7. *Validated?* Validation is the determination that the conceptual model is an accurate representation of the real system. Can the model be substituted for the real system for the purposes of experimentation? If there is an existing system, call it the base system, an ideal way to validate the model is to compare its output to that of the base system. Unfortunately, there is not always a base system (such as in the design of a new system). There are many methods for performing validation, and some of these are discussed further in this chapter. Additionally, this step is discussed extensively in Chapter 10.

8. *Experimental design.* For each scenario that is to be simulated, decisions need to be made concerning the length of the simulation run, the number of runs (also called *replications*), and the manner of initialization, as required. This step is discussed further in Chapter 6.

9. *Production runs and analysis.* Production runs, and their subsequent analysis, are used to estimate measures of performance for the scenarios that are being simulated. This step is discussed extensively in Chapters 7 to 9.

10. *More runs?* Based on the analysis of runs that have been completed, the simulation analyst determines if additional runs are needed and if any additional scenarios need to be simulated. This step is discussed extensively in Chapters 7 to 9.

11. *Documentation and reporting.* Documentation is necessary for numerous reasons. If the simulation model is going to be used again by the same or different analysts, it may be necessary to understand how the simulation model operates. This will stimulate confidence in the simulation model so that the client can make decisions based on the analysis. Also, if the model is to be modified, this can be greatly facilitated by adequate documentation. One experience with an inadequately documented model is usually enough to convince a simulation analyst of the necessity of this important step. The result of all the analysis should be reported clearly and concisely. This will enable the client to review the final formulation, the alternatives that were addressed, the criterion by which the alternative systems were compared, the results of the experiments, and analyst recommendations, if any. This step is discussed further in Chapters 22 and 23.

12. *Implementation.* The simulation analyst acts as a reporter rather than an advocate. The report prepared in step 11 stands on its merits and is just additional information that the client uses to make a decision. If the client has been involved throughout the study period, and the simulation analyst has followed all the steps rigorously, the likelihood of a successful implementation is increased. See Chapters 22 and 23 for more about implementation.

1.8 RANDOM NUMBER AND RANDOM VARIATE GENERATION

Example 1 used input values that were generated by a spinner and a die. Almost all simulation models are constructed within a computer, so spinners and dice are not devices that will be used. Instead, the computer will generate independent random numbers that are distributed continuously and uniformly between 0 and 1 [i.e., $U(0, 1)$]. These random numbers can then be converted to the desired statistical distribution, or random variate, using one of several methods. Random variates are used to represent interarrival times, batch sizes, processing times, repair times, and time until failure, among others. Many researchers have written on the two topics in this section. These topics are discussed further in Chapters 4 and 5.

Simulation software products have a built-in random number generator (RNG) that produces a sequence of random numbers. Most of these generators are based on the linear congruential method (LCM), documented by Knuth (1969). A RNG is defined by its parameters, and some of them have been tested extensively. Chapter 4 introduces the topic of RNG.

The numbers generated by a RNG are actually pseudorandom. They are deterministic since they can be reproduced. Knowing the starting value, the values that follow it can be predicted, totally determining the sequence. There is no reason for concern since the length of the sequence prior to repeating itself is very, very long. On a 32-bit computer, this sequence can be longer than 2 billion. As reported in Chapter 4, even-longer-period RNGs are available.

The importance of a good source of random numbers is that all procedures for generating nonuniformly distributed random variates involve a mathematical transformation of uniform random numbers. For example, suppose that R_i is the ith random number generated from $U(0, 1)$. Suppose further that the desired random variate is exponentially distributed with rate λ. These values can be generated from

$$X_i = \frac{1}{\lambda} \ln(1 - R_i) \qquad (1)$$

where X_i is the ith random variate generated [e.g., the time between the arrival of the ith and the $(i+1)$st entities]. Suppose that $\lambda = \frac{1}{10}$ arrival per minute. Using equation (1) [called the random variate generator (RVG)], if $R_1 = 0.3067$, then $X_1 = 3.66$ minutes. The RVG was developed using what is called the inverse-transform technique. Other techniques include convolution, acceptance–rejection, and composition. Techniques for RVG are discussed in Chapter 5.

Most simulation software products have built-in RVGs for the most widely used distributions and several that are not so widely utilized. The simulation software usually provides a facility for generating a sample from an empirical distribution (a distribution of the raw input data) that is either discrete or continuous. It is important that the simulation analyst know how to use RVGs, but it is not usually important to be concerned with their generation.

1.9 INPUT DATA

For each element in a system being modeled, the simulation analyst must decide on a way to represent the associated random variables. The presentation of the subject that follows is based on Banks et al. (1998). This topic is discussed in much more detail in Chapter 3.

The techniques used may vary depending on:

1. The amount of available data
2. Whether the data are "hard" or someone's best guess
3. Whether each variable is independent of other input random variables, or related in some way to other outputs

In the case of a variable that is independent of other variables, the choices are as follows:

1. Assume that the variable is deterministic.
2. Fit a probability distribution to the data.
3. Use the empirical distribution of the data.

These three choices are discussed in the next three subsections.

1.9.1 Assuming Randomness Away

Some simulation analysts may be tempted to assume that a variable is deterministic, or constant. This value could have been obtained by averaging historic information. The value may even be a guess. If there is randomness in the model, this technique can surely invalidate the results.

Suppose that a machine manufactures parts in exactly 1.5 minutes. The machine requires a tool change according to an exponential distribution with a mean of 12 minutes between occurrences. The tool change time is also exponentially distributed with a mean of 3 minutes. An inappropriate simplification would be to assume that the machine operates in a constant time of 1.875 minutes, and ignore the randomness. The consequences of these two interpretations are very great on such measures as the average number in the system or time waiting before the machine.

1.9.2 Fitting a Distribution to Data

If there are sufficient data points, say 50 or more, it may be appropriate to fit a probability distribution to the data using conventional methods. [Advanced methods for distribution fitting, such as that described by Wagner and Wilson (1993), are available to the interested reader.] When there are few data, the tests for goodness of fit offer little guidance in selecting one distribution form over another.

There are also underlying processes that give rise to distributions in a rather predictable manner. For example, if arrivals (1) occur one at a time, (2) are completely at random without rush or slack periods, and (3) are completely independent of one another, a Poisson process occurs. In such a case it can be shown that the number of arrivals in a given time period follows a Poisson distribution and the time between arrivals follows an exponential distribution.

Several vendors provide software to accomplish input data analysis. However, if a goodness-of-fit test is being conducted without the aid of input data analysis software, the following three-step procedure is recommended:

1. *Hypothesize a candidate distribution.* First, ascertain whether the underlying process is discrete or continuous. Discrete data arise from counting processes. Examples include the number of customers that arrive at a bank each hour, the number of tool changes in an 8-hour day, and so on. Continuous data arise from measurement (time, distance, weight, etc.). Examples include the time to produce each part and the time to failure of a machine. Discrete distributions frequently used in simulation include the Poisson, binomial, and geometric. Continuous distributions frequently used in simulation include the uniform, exponential, normal, triangular, lognormal, gamma, and Weibull. These distributions are described in virtually every engineering statistics text.

2. *Estimate the parameters of the hypothesized distribution.* For example, if the hypothesis is that the underlying data are normal, the parameters to be estimated from the data are the mean and the variance.

3. *Perform a goodness-of-fit test such as the chi-squared test.* If the test rejects the hypothesis, that is a strong indication that the hypothesis is not true. In that case, return to step 1, or use the empirical distribution of the data following the process described below.

The three-step procedure is described in engineering statistics texts and in many simulation texts, such as Banks et al. (1996) and Law and Kelton (1991). Even if software is being used to aid in the development of an underlying distribution, understanding the three-step procedure is recommended.

1.9.3 Empirical Distribution of the Data

When all possibilities have been exhausted for fitting a distribution using conventional techniques, the empirical distribution can be used. The empirical distribution uses the data as generated.

An example will help to clarify the discussion. The times to repair a conveyor system after a failure, denoted by x, for the previous 100 occurrences are given as follows:

Interval (hours)	Frequency of Occurrence
$0 < x \leq 1.0$	27
$1.0 < x \leq 2.0$	13
$2.0 < x \leq 3.0$	31
$3.0 < x \leq 4.0$	18
$4.0 < x \leq 8.0$	11

No distribution could be fit acceptably to the data using conventional techniques. It was decided to use the data as generated for the simulation. That is, samples were drawn, at random, from the continuous distribution shown above. This required linear interpolation so that simulated values might be in the form 2.89 hours, 1.63 hours, and so on.

1.9.4 When No Data Are Available

There are many cases where no data are available. This is particularly true in the early stages of a study, when the data are missing, when the data are too expensive to gather, or when the system being modeled is not in existence. One possibility in such a case is to obtain a subjective estimate, some call it a *guesstimate*, concerning the system. Thus if the estimate that the time to repair a machine is between 3 and 8 minutes, a crude assumption is that the data follow a uniform distribution with a minimum value of 3 minutes and a maximum value of 8 minutes. The uniform distribution is referred to as the *distribution of maximum ignorance* since it assumes that every value is equally likely. A better "guess" occurs if the "most likely" value can also be estimated. This would take the form "the time to repair the machine is between 3 and 8 minutes with a most likely time of 5 minutes." Now, a triangular distribution can be used with a minimum of 3 minutes, a maximum of 8 minutes, and a most likely value (mode) of 5 minutes.

As indicated previously, there are naturally occurring processes that give rise to distributions. For example, if the time to failure follows the (reasonable) assumptions of the Poisson process indicated previously, and the machine operator says that the machine fails about once every 2 hours of operation, an exponential distribution for time to failure could be assumed initially with a mean of 2 hours.

Estimates made on the basis of guesses and assumptions are strictly tentative. If, and when, data, or more data, become available, both the parameters and the distributional forms should be updated.

1.10 VERIFICATION AND VALIDATION

In the application of simulation, the real-world system under investigation is abstracted by a conceptual model. The conceptual model is then coded into the operational model. Hopefully, the operational model is an accurate representation of the real-world system. However, more than hope is required to ensure that the representation is accurate. There is a checking process that consists of two components:

1. *Verification:* a determination of whether the computer implementation of the conceptual model is correct. Does the operational model represent the conceptual model?
2. *Validation:* a determination of whether the conceptual model can be substituted for the real system for the purposes of experimentation.

The checking process is iterative. If there are discrepancies among the operational and conceptual models and the real-world system, the relevant operational model must be examined for errors, or the conceptual model must be modified to represent the real-world system better (with subsequent changes in the operational model). The verification and validation process should then be repeated. These two important topics are discussed extensively in Chapter 10.

1.10.1 Verification

The verification process involves examination of the simulation program to ensure that the operational model accurately reflects the conceptual model. There are many commonsense ways to perform verification.

1. *Follow the principles of structured programming.* The first principle is *top-down design* (i.e., construct a detailed plan of the simulation model before coding). The second principle is *program modularity* (i.e., break the simulation model into submodels). Write the simulation model in a logical, well-ordered manner. It is highly advisable (we would say mandatory if we could mandate such) to prepare a detailed flowchart indicating the macro activities that are to be accomplished. This is particularly true for real-world-sized problems. It is quite possible to think through all the computer code needed to solve problems at chapter ends of an academic text on discrete-event simulation. However, that computer code is minuscule compared to that of real-world problems.

2. *Make the operational model as self-documenting as possible.* This requires comments on virtually every line and sometimes between lines of code for those software products that allow programming. Imagine that one of your colleagues is trying to understand the computer code that you have written, but that you are not available to offer any explanation. For graphical software, on-screen documentation is suggested. In some cases, the text associated with documentation can be hidden from view when it is inappropriate to show it.

3. *Have the computer code checked by more than one person.* Several techniques have been used for this purpose. One of these can be called *code inspection*. There are four parties as follows: the moderator or leader of the inspection team, the designer or person who prepared the conceptual model, the coder or person who prepared the operational model, and the tester or the person given the verification responsibility. An

inspection meeting is held where a narration of the design is provided and the operational model is discussed, line by line, along with the documentation. Errors detected are documented and classified. There is then a rework phase, followed by another inspection. Alternatives to code inspection include the review, except that the interest is not line by line but in design deficiencies. Another alternative is the audit that verifies that the development of the computer code is proceeding logically. It verifies that the stated requirements are being met.

4. *Check to see that the values of the input data are being used appropriately.* For example, if the time unit is minutes, all of the data should be in terms of minutes, not hours or seconds.

5. *For a variety of input-data values, ensure that the outputs are reasonable.* Many simulation analysts are satisfied when they receive output. But that is far from enough. If there are 100 entities in a waiting line when 10 would be rather high, there is probably something wrong. For example, the resource actually has a capacity of two, but was modeled with a capacity of one.

6. *Use an interactive run controller (IRC) or debugger to check that the program operates as intended.* The IRC is a very important verification tool that should be used for all real-system models. An example of one of the capabilities of the IRC is the trace that permits following the execution of the model step by step.

7. *Animation is a very useful verification tool.* Using animation, the simulation analyst can detect actions that are illogical. For example, it may be observed that a resource is supposed to fail as indicated by turning red on the screen. While watching the animation, the resource never turned red. This could signal a logical error.

1.10.2 Validation

A variety of subjective and objective techniques can be used to validate the conceptual model. Sargent (1992) offers many suggestions for validation. Subjective techniques include the following:

1. *Face Validation.* A conceptual model of a real-world system must appear reasonable "on its face" to those who are knowledgeable (the "experts") about the real-world system. For example, the experts can validate that the model assumptions are correct. Such a critique by experts would aid in identifying deficiencies or errors in the conceptual model. The credibility of the conceptual model would be enhanced as these deficiencies or errors are eliminated.

2. *Sensitivity Analysis.* As model input is changed, the output should change in a predictable direction. For example, if the arrival rate increases, the time in queues should increase, subject to some exceptions. (An example of an exception is as follows: If a queue increases, it may be the case that resources are added within the model, negating the prediction.)

3. *Extreme-Condition Tests.* Does the model behave properly when input data are at the extremes? If the arrival rate is set extremely high, does the output reflect this change with increased numbers in the queues, increased time in the system, and so on?

4. *Validation of Conceptual Model Assumptions.* There are two types of conceptual model assumptions. They are structural assumptions (concerning the operation of the real-world system) and data assumptions. Structural assumptions can be validated by

observing the real-world system and by discussing the system with the appropriate personnel. No one person knows everything about the entire system. Many people need to be consulted to validate conceptual model assumptions.

Information from intermediaries should be questioned. A simulation consulting firm often works through other consulting firms. An extremely large model of a distant port operation was constructed. It was only after a visit by the simulation consulting firm to the port that it was discovered that one of the major model assumptions concerning how piles of iron ore are formed was in error.

Assumptions about data should also be validated. Suppose it is assumed that times between arrivals of customers to a bank during peak periods are independent and in accordance with an exponential distribution. To validate conceptual model assumptions, the following would be in order:

(a) Consult with appropriate personnel to determine when peak periods occur.
(b) Collect interarrival data from these periods.
(c) Conduct statistical tests to ensure that the assumption of independence is reasonable.
(d) Estimate the parameter of the assumed exponential distribution.
(e) Conduct a goodness-of-fit test to ensure that the exponential distribution is reasonable.

5. *Consistency Checks.* Continue to examine the operational model over time. An example explains this validation procedure. A simulation model is used annually. Before using this model, make sure that there are no changes in the real system that must be reflected in the structural model. Similarly, the data should be validated. For example, a faster machine may have been installed in the interim period, but it was not included in the information provided.

6. *Turing Tests.* Persons knowledgeable about system behavior can be used to compare model output to system output. For example, suppose that five reports of actual system performance over five different days are prepared and five simulated outputs are generated. These 10 reports should be in the same format. The 10 reports are randomly shuffled and given to a person, say an engineer, who has seen this type of information. The engineer is asked to distinguish between the two kinds of reports, actual and simulated. If the engineer identifies a substantial number of simulated reports, the model is inadequate. If the engineer cannot distinguish between the two, there is less reason to doubt the adequacy of the model.

Objective techniques include the following:

7. *Validating Input–Output Transformations.* The basic principle of this technique is the comparison of output from the operational model to data from the real system. Input–output validation requires that the real system currently exist. One method of comparison uses the familiar t-test, discussed in most statistics texts.

8. *Validation Using Historical Input Data.* Instead of running the operational model with artificial input data, we could drive the operational model with the actual historical record. It is reasonable to expect the simulation to yield output results within acceptable statistical error of those observed from the real-world system. The paired t-test,

discussed in most statistics texts, is one method for conducting this type of validation.

1.11 EXPERIMENTATION AND OUTPUT ANALYSIS

The analysis of simulation output begins with the selection of performance measures. Performance measures can be time weighted, based on counting of occurrences, or arise from the tabulation of expressions including means, variances, and so on.

An example of a time-weighted statistic is the average number in system over a time period of length T. Figure 1.2 shows the number in system, $L(t)$, at time t, from $t = 0$ to $t = 60$. The time-weighted average number in the system, $\overline{L}$, at $T = 60$ is given by the sum of the areas of the rectangles divided by T. Thus

$$\overline{L} = \frac{(0 \times 10) + (1 \times 10) + (2 \times 15) + (1 \times 10) + (0 \times 5) + (1 \times 6) + (2 \times 4)}{60}$$

$$= 1.07$$

An example of a statistic based on counting of occurrences is the number of acceptable units completed in 24 hours of simulated time. A statistic based on the tabulation of expressions is the patent royalties from three different part types, each with a different contribution per unit, for a 24-hour period.

The simulation of a stochastic system results in performance measures that contain random variation. Proper analysis of the output is required to obtain sound statistical results from these replications. Specific questions that must be addressed when conducting output analysis are:

1. What is the appropriate run length of the simulation (unless the system dictates a value)?
2. How do we interpret the simulated results?
3. How do we analyze the differences between different model configurations?

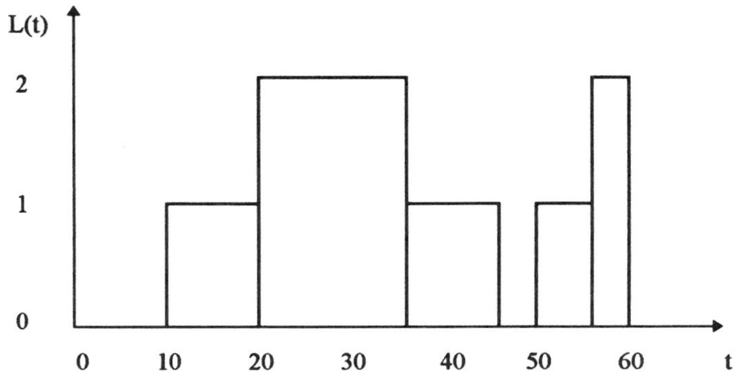

Figure 1.2 Number in system, $L(t)$, at time t.

These topics are introduced in the next section. They are discussed extensively in Chapters 6 to 8.

1.11.1 Statistical Confidence and Run Length

A confidence interval for the performance measure being estimated by the simulation model is a basic component of output analysis. A confidence interval is a numerical range that has a probability of $1 - \alpha$ of including the true value of the performance measure, where $1 - \alpha$ is the confidence level for the interval. For example, let us say that the performance measure of interest is the mean time in the queue, μ, and a $100(1 - \alpha)\%$ confidence interval for μ is desired. If many replications are performed and independent confidence intervals on μ are constructed from those replications, approximately $100(1 - \alpha)\%$ of those intervals will contain μ. Consider the following example.

Example 2 (Confidence Intervals) Given the data in Table 1.2, both a 95% ($\alpha = 0.05$) and a 99% ($\alpha = 0.01$) two-sided confidence interval are desired. Assuming that the values for X are normally distributed, a $1 - \alpha$ confidence interval for the mean, μ, is given by $(\overline{X} - h, \overline{X} + h)$, where $\overline{X}$ is the sample mean and h is the half-width. The equation for $\overline{X}$ is given by

$$\overline{X} = \sum_{i=1}^{n} \frac{X_i}{n} \tag{2}$$

The half-width h of the confidence interval is computed as follows:

$$h = t_{n-1,1-\alpha/2} \frac{S}{\sqrt{n}} \tag{3}$$

where $t_{n-1,1-\alpha/2}$ is the upper $1 - \alpha/2$ critical value of the t-distribution with $n - 1$ degrees of freedom, and S is the sample standard deviation. To compute S, first use equation (4) to compute S^2 as follows:

$$S^2 = \frac{\sum_{i=1}^{n} X_i^2 - n\overline{X}^2}{n - 1} \tag{4}$$

Table 1.2 Data for Example 2

Replication Number, i	Average Time in Queue, X_i
1	63.2
2	69.7
3	67.3
4	64.8
5	72.0

Taking the square root of S^2 yields S.

Since a two-sided confidence interval is desired, we use $\alpha/2$ to compute the half-width. Using equations (2) and (4), we obtain $\overline{X} = 67.4$ and $S = 3.57$. In addition,

$$t_{4,.975} = 2.78 \quad \text{(95\% confidence)}$$
$$t_{4,.995} = 4.60 \quad \text{(99\% confidence)}$$

resulting in

$$h = \begin{cases} 4.44 & \text{(95\% confidence)} \\ 7.34 & \text{(99\% confidence)} \end{cases}$$

The confidence interval is given by $(\overline{X}-h, \overline{X}+h)$. Therefore, the 95% confidence interval is (62.96, 71.84), and the 99% confidence interval is (60.06, 74.74).

As demonstrated in Example 2, the size of the interval depends on the confidence level desired, the sample size, and the inherent variation (measured by S). The higher level of confidence (99%) requires a larger interval than the lower confidence level (95%). In addition, the number of replications, n, and their standard deviation, S, are used in calculating the confidence interval. In simulation, each replication is considered one data point. Therefore, the three factors that influence the width of the confidence interval are:

1. Number of replications (n)
2. Level of confidence ($1 - \alpha$)
3. Variation of performance measure (S)

The relationship between these factors and the confidence interval is:

1. As the number of replications increases, the width of the confidence interval decreases.
2. As the level of confidence increases, the width of the interval increases. In other words, a 99% confidence interval is larger than the corresponding 95% confidence interval.
3. As the variation increases, the width of the interval increases.

1.11.2 Terminating Versus Nonterminating Systems

The procedure for output analysis differs based on whether the system is terminating or nonterminating. In a terminating system, the duration of the simulation is fixed as a natural consequence of the model and its assumptions. The duration can be fixed by specifying a finite length of time to simulate or by limiting the number of entities created or disposed. An example of a terminating system is a bank that opens at 9:00 A.M. and closes at 4:00 P.M. Some other examples of terminating systems include a check-processing facility that operates from 8:00 P.M. until all checks are processed, a ticket booth that remains open until all the tickets are sold or the event begins, and a manufacturing facility that processes a fixed number of jobs each day and then shuts down.

By definition, a terminating system is one that has a fixed starting condition and an event definition that marks the end of the simulation. The system returns to the fixed initial condition, usually "empty and idle," before the system begins operation again. The objective of the simulation of terminating systems is to understand system behavior for a "typical" fixed duration. Since the initial starting conditions and the length of the simulation are fixed, the only controllable factor is the number of replications.

One analysis procedure for terminating systems is to simulate a number of replications, compute the sample variance of the selected estimator measure, and determine if the width of the resulting confidence interval is within acceptable limits. For example, if the average number of parts in the queue is of interest, the first step is to conduct a pilot run of n replications. Next, compute the confidence interval for the expected average number of parts in the queue using the observations recorded from each replication. Then if the confidence interval is too large, determine the number of additional replications required to bring it within limits. Finally, conduct the approximate additional replications and recompute the new confidence interval using all the data. Iterate the last two steps until the confidence interval is of satisfactory size.

In a nonterminating system, the duration is not finite; the system is in perpetual operation. An example of a nonterminating system is an assembly line that operates 24 hours a day, 7 days a week. Another example of this type of system is the manufacture of glass fiber insulation for attics. If operation of the system is stopped, the molten glass will solidify in the furnace, requiring that it be chipped away tediously before restarting the system. The objective in simulating a nonterminating system is to understand the long-run, or *steady-state*, behavior. To study steady-state behavior accurately, the effects of the initial conditions, or transient phase, must be removed from the simulation results. This can be accomplished by swamping, preloading, or deletion.

The first method, *swamping*, suppresses the initial-condition effects by conducting a very long simulation run, so long that any initial conditions have only a minuscule effect on the long-run value of the performance measure. For example, if the initial conditions last for 100 hours, simulate for 10,000 hours. A problem with the swamping technique is that the bias from starting empty and idle will always exist, even if it is small.

The second method, *preloading*, primes the system before simulation starts by placing entities in delay blocks and queues. In other words, an attempt is made to have the initial conditions match the steady-state conditions. This requires some rough knowledge of how the system looks in steady state. Thus, if we are simulating a bank that has one line forming before three tellers, we need to observe the bank in operation to obtain information about the usual situation. For example, we may find that the three tellers are usually busy and that there are about four people in line. This is how the simulation would begin when using the preloading technique. The bank is a very simple system to observe. However, for more complex systems, this initialization procedure becomes somewhat difficult, especially if the system is still in the design phase.

The third method, *deletion*, excludes the initial transient phase that is influenced by the initial conditions. Data are collected from the simulation only after the transient (warm-up) phase has ended. This idea is demonstrated in Figure 1.3. The difficulty with the deletion method is the determination of the length of the transient phase. Although elegant statistical techniques have been developed, a satisfactory method is to plot the

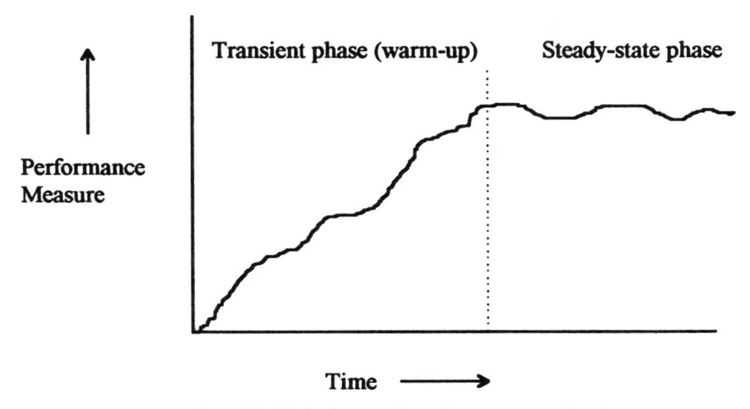

Fig. 1.3 Deletion of initial observations for a nonterminating system.

output of interest over time and visually observe when steady state is reached. Welch (1983) provides a formal description of this method.

1.12 SUMMARY

The chapter began with a definition of simulation, including an example. Underlying concepts were presented, including the system and model, system state variables, entities and attributes, list processing, activities and delays, and the definition of discrete-event simulation. Next, four modeling structures were discussed, including process interaction, event scheduling, activity scanning, and the three-phase method. The advantages and disadvantages of simulation were presented, with amelioration of the disadvantages. Next, areas of application from presentations at the Winter Simulation Conference were shown. The steps in a simulation study were given with a brief discussion of each. The manner in which random numbers and random variates are generated was presented next. Three ways that might be used for generating input data were described. However, the first method, assuming randomness away, is discouraged. The extremely important concepts of verification and validation were then discussed. The all-important topic of experimentation and output analysis was introduced. The topics introduced in this chapter are discussed much more extensively in the remaining chapters.

REFERENCES

Balci, O. (1988). The implementation of four conceptual frameworks for simulation modeling in high-level languages, in *Proceedings of the 1988 Winter Simulation Conference*, M. Abrams, P. Haigh, and J. Comfort, eds., IEEE, Piscataway, N.J.

Banks, J., and V. Norman (1995). Justifying simulation in today's manufacturing environment, *IIE Solutions*, November.

Banks, J., B. Burnette, H. Kozloski, and J. D. Rose (1995). *Introduction to SIMAN V and CINEMA V*, Wiley, New York.

Banks, J., J. S. Carson II, and B. L. Nelson (1996). *Discrete-Event System Simulation*, 2nd ed., Prentice Hall, Upper Saddle River, N.J.

Banks, J., J. S. Carson II, and D. Goldsman (1998). Discrete-event computer simulation, in *Handbook of Statistical Methods for Engineers and Scientists*, 2nd ed., H. M. Wadsworth, ed., McGraw-Hill, New York.

Carson, J. S. (1993). Modeling and simulation world views, in *Proceedings of the 1993 Winter Simulation Conference*, G. W. Evans, M. Mollaghasemi, E. C. Russell, and W. E. Biles, eds., IEEE, Piscataway, N.J., pp. 18–23.

Knuth, D. W. (1969). *The Art of Computer Programming*, Vol. 2: *Semi-numerical Algorithms*, Addison-Wesley, Reading, Mass.

Law, A. M., and W. D. Kelton (1991). *Simulation Modeling and Analysis*, 2nd ed., McGraw-Hill, New York.

Pegden, C. D., R. E. Shannon, and R. P. Sadowski (1995). *Introduction to Simulation Using SIMAN*, 2nd ed., McGraw-Hill, New York.

Pidd, M. (1992). *Computer Modelling for Discrete Simulation*, Wiley, Chichester, West Sussex, England.

Sargent, R. G. (1992). Validation and verification of simulation models, in *Proceedings of the 1992 Winter Simulation Conference*, J. J. Swain, D. Goldsman, R. C. Crain, and J. R. Wilson, eds., IEEE, Piscataway, N.J., pp. 104–114.

Schriber, T. J. (1991). *An Introduction to Simulation Using GPSS/H*, Wiley, New York.

Wagner, M. A. F., and J. R. Wilson (1993). Using univariate Bézier distributions to model simulation input processes, in *Proceedings of the 1993 Winter Simulation Conference*, G. W. Evans, M. Mollaghasemi, E. C. Russell, and W. E. Biles, eds., IEEE, Piscataway, N.J., pp. 365–373.

Welch, P. D. (1983). The statistical analysis of simulation results, in *The Computer Performance Modeling Handbook*, S. Lavenberg, ed., Academic Press, San Diego, Calif.

CHAPTER 2

Principles of Simulation Modeling

A. ALAN B. PRITSKER
Pritsker Corporation and Purdue University

2.1 INTRODUCTION

Simon [1] captures the essence of *modeling* in the following quote: "Modeling is a principal—perhaps the primary—tool for studying the behavior of large complex systems. . . . When we model systems, we are usually (not always) interested in their dynamic behavior. Typically, we place our model at some initial point in phase space and watch it mark out a path through the future." *Simulation* embodies this concept because it involves playing out a model of a system by starting with the system status at an initial point in time and evaluating the variables in a model over time to ascertain the dynamic performance of the model of the system. When the model is a valid representation of the system, meaningful information is obtained about the dynamic performance of the system. In this chapter, principles for building and using models that are analyzed using simulation, referred to as *simulation models*, are presented.

Models are descriptions of systems. In the physical sciences, models are usually developed based on theoretical laws and principles. The models may be scaled physical objects (iconic models), mathematical equations and relations (abstract models), or graphical representations (visual models). The usefulness of models has been demonstrated in describing, designing, and analyzing systems. Model building is a complex process and in most fields involves both inductive and deductive reasoning. The modeling of a system is made easier if (1) physical laws are available that pertain to the system, (2) a pictorial or graphical representation can be made of the system, and (3) the uncertainty in system inputs, components, and outputs is quantifiable.

Modeling a complex, large-scale system is usually more difficult than modeling a strictly physical system, for one or more of the following reasons: (1) few fundamental laws are available, (2) many procedural elements are involved which are difficult to describe and represent, (3) policy inputs are required which are hard to quantify, (4) random components are significant elements, and (5) human decision making is an integral part of such systems.

Handbook of Simulation, Edited by Jerry Banks.
ISBN 0-471-13403-1

Since a model is a description of a system, it is also an abstraction of a system. To develop an abstraction, a model builder must decide on the elements of the system to include in the model. To make such decisions, a purpose for the model building must be established. Thus the first step in model building is the development of a purpose for modeling that is based on a stated problem or project goal. Based on this purpose, system boundaries and modeling details are established. This abstraction results in a model that does not include all the rough, ill-defined edges of the actual system. Typically, the assessment process requires redefinitions and redesigns that cause the entire model building process to be performed iteratively.

Simulation models are ideally suited for carrying out the problem-solving approach described above. Simulation provides the flexibility to build either aggregate or detailed models. It directly supports iterative model building by allowing models to be embellished through simple and direct additions. Surveys indicate that simulation is one of the most widely used tools of industrial engineers and management scientists. In 1989, the U.S. Departments of Defense and Energy specified that simulation and modeling technology is one of the top 22 critical technologies in the United States [2].

2.2 BASIC PRINCIPLES

There are no established, published principles for simulation modeling. Modeling is considered an art [3] and a creative activity [4]. In this chapter an attempt is made to provide modeling principles, based on the author's experience and interaction with colleagues.

Modeling Principle 1 Conceptualizing a model requires system knowledge, engineering judgment, and model-building tools.

A modeler must understand the structure and operating rules of a system and be able to extract the essence of the system without including unnecessary detail. Usable models tend to be easily understood, yet have sufficient detail to reflect realistically the important characteristics of the system. The crucial questions in model building focus on what simplifying assumptions are reasonable to make, what components should be included in the model, and what interactions occur among the components. The amount of detail included in the model should be based on the modeling objectives established. Only those components that could cause significant differences in decision making, including confidence building, need to be considered.

A modeling project is normally an interdisciplinary activity and should include the decision maker as part of the team. Close interaction among project personnel is required when formulating a problem and building a model. This interaction causes inaccuracies to be discovered quickly and corrected efficiently. Most important is the fact that interactions induce confidence in both the modeler and the decision maker and help to achieve a successful implementation of results.

By conceptualizing the model in terms of the structural components of the system and product flows through the system, a good understanding of the detailed data requirements can be projected. From the structural components, the schedules, algorithms, and controls required for the model can be determined. These decision components are typically the most difficult aspect of a modeling effort.

Modeling Principle 2 The secret to being a good modeler is the ability to remodel.

Model building should be interactive and graphical because a model is not only defined and developed but is continually refined, updated, modified, and extended. An up-to-date model provides the basis for future models. The following five model-building themes support this approach and should be used where feasible:

1. Develop tailorable model input procedures and interfaces.
2. Divide the model into relatively small logical elements.
3. Separate physical and logical elements of the model.
4. Develop and maintain clear documentation directly in the model.
5. Leave hooks in the model to insert extensions or more detail; that is, build an open-ended model.

Models developed for analysis by simulation are easily changed, which facilitates iterations between model specification and model building. This is not usually the case for other widely used model analysis techniques. Examples of the types of changes that are easily made in simulation models are:

1. Setting arrival patterns and activity times to be constant, as samples from a theoretical distribution, or derived from a file of historical values
2. Setting due dates based on historical records, manufacturing resource planning (MRPII) procedures, or sales information
3. Setting decision variables based on a heuristic procedure or calling a decision-making subprogram that uses an optimization technique
4. Including fixed rules or expert-system-based rules directly in the model

Modeling Principle 3 The modeling process is evolutionary because the act of modeling reveals important information piecemeal.

Information obtained during the modeling process supports actions that make the model and its output measures more relevant and accurate. The modeling process continues until additional detail or information is no longer necessary for problem resolution or a deadline is encountered. During this evolutionary process, relationships between the system under study and the model are continually defined and redefined. Simulations of the model provide insights into the behavior of the model, and hence the system, and lead to a further evolution of the model. The resulting correspondence between the model and the system not only establishes the model as a tool for problem solving but provides system familiarity for the modelers and a training vehicle for future users.

2.3 MODEL-BASED PROBLEM SOLVING

Figure 2.1 presents the components in the problem-solving environment when models are used to support the making of decisions or the setting of policies.

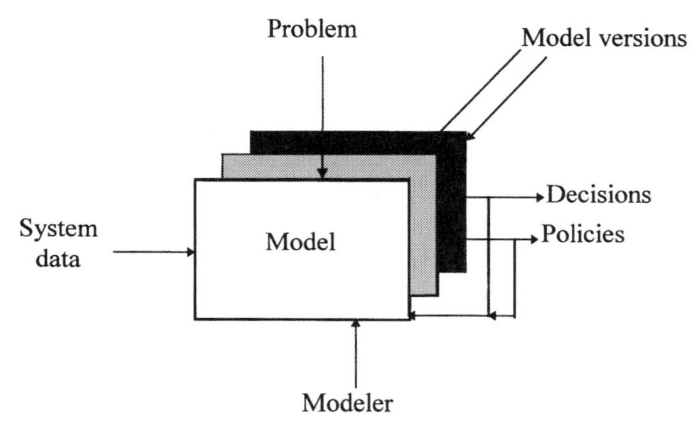

Figure 2.1 Model-based problem-solving process.

Modeling Principle 4 The problem or problem statement is the primary controlling element in model-based problem solving.

A problem or objective drives the development of the model. Problem statements are defined from system needs and requirements. Data from the system provide the input to the model. The availability and form of the data help to specify the model boundaries and details. The modeler is the resource used to build the model in accordance with the problem statement and the available system data. The outputs from the model support decisions to be made to solve the problem or the setting of policies that allow decisions to be made in accordance with established rules and procedures. These components are described in the following paragraphs with the aid of Figures 2.2 to 2.5.

The first step in model-based problem solving is to formulate the problem by understanding its context, identifying project goals, specifying system performance measures,

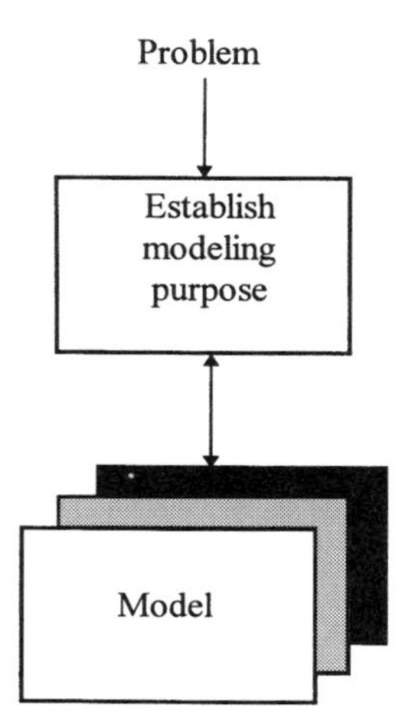

Figure 2.2 Model and its control.

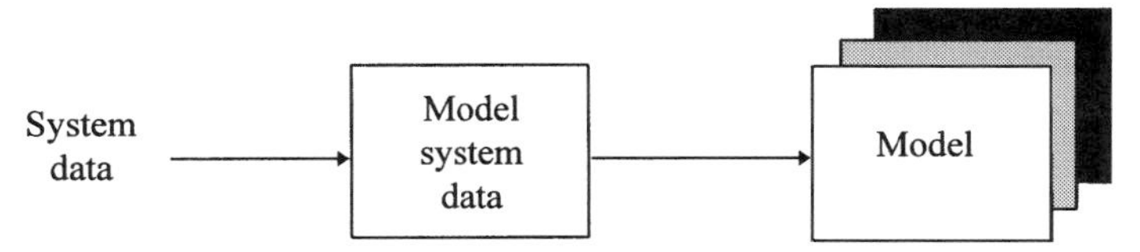

Figure 2.3 Model and its control.

setting specific modeling objectives, and in general, defining the system to be modeled. Figure 2.2 shows that from the problem, a purpose for modeling is developed that guides the modeling effort toward solving the problem formulated. The double arrow between model and modeling purpose indicates that during the modeling effort, the purpose for modeling can change. Great care is needed here to ensure that the modeling effort does not stray from its goal of solving the original problem.

Figure 2.3 shows that there is a process between obtaining system data and using those data in the model. This process is called *input modeling*. Input modeling involves the determination of whether the system data should be used directly to drive the model, whether the system data should be summarized in the form of a histogram or distribution function, or whether a cause-and-effect model (e.g., a regression model) should be developed to characterize the system data. The input modeling activity can involve a large effort. It is very useful in gaining an understanding of a system as well as refining and for generalizing the system data for input to a model.

The types of data that need to be collected to support model-based problem solving include data that describe the elements and logic of the system, data that measure the actual performance of the system, and data that describe the alternatives to be evaluated. Data that describe the logic of the system is concerned with system structure, individual components of the system, component interactions, and system operations. The possible states of the system are established from this information. Performance measures are functions of these states and relate to profitability, costs, productivity, and quality.

Data collection may involve performing detailed time studies, getting information from equipment vendors, and talking to system operators. Actual system performance histories are collected whenever possible to support the validation of model outputs. Data describing proposed solutions are used to modify the basic model for each alternative to be evaluated. Each alternative is typically evaluated individually, but performance data across alternatives are displayed together.

Figure 2.4 shows that a modeler may use a simulation system in developing the model. A simulation system provides an environment to build, debug, test, verify, run, and analyze simulation models. They provide capabilities for graphical input and output, interactive execution of the simulation, data and distribution fitting, statistical analysis, storing and maintaining data in databases, reporting outputs using application programs, and most important, a simulation language. A simulation language contains modeling concepts and procedures that facilitate the development of a model. Languages are composed of symbols. Defining a modeling language means defining the semantics and syntax of the symbols. *Semantics* specify the meaning of each symbol and the relationships among symbols and take into account the human comprehensibility of the symbols; *syntax* defines the formal expression of symbols and their relationships in human- and computer-readable form.

For many years, models have been built using the language of mathematics and

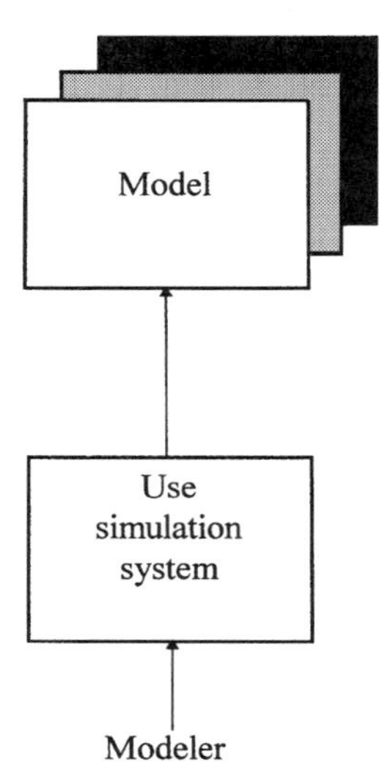

Figure 2.4 Role of a simulation system.

general-purpose computer languages. General-purpose computer languages provide a great deal of flexibility in modeling but do not contain a structure or set of concepts that facilitate the modeling task. Specializing such languages for use has simplified modeling tasks. In Chapter 24 we discuss how simulation software works and in Chapter 25, provide a survey of the software.

The final stage in the problem-solving process is to support decision making and policy setting as shown in Figure 2.5. *For a simulation analyst, no project should be considered complete until its results are used.* The use of the model involves both an interpretation of outputs and a presentation of results. Planning for the use of model outputs entails both strategic and tactical considerations. These considerations must include continual interaction between the model builder and the decision maker to ensure that the decision maker understands the model, its outputs, and its uses. If this is done, it is more likely that the results of the project will be implemented with vigor. The feedback from output analysis to the model provides information as to how the model can be adapted to satisfy better the problem statement. It is not uncommon that such feedback also influences the problem statement. When this occurs, communications to the decision maker are even more important.

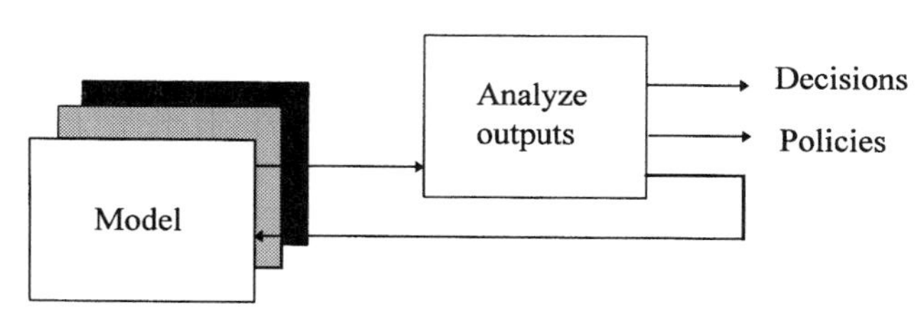

Figure 2.5 Model and its outputs.

2.4 SIMULATION MODELING WORLD VIEWS [5]

Simulation models of systems can be classified as discrete-change, continuous-change, or combined models. In most simulations, time is the major independent variable. Other variables included in the simulation, such as machine status and number of parts in inventory, are functions of time and are the dependent variables. The values of the dependent variables are used to calculate operational performance measures. In a manufacturing system, typical performance measures are throughput, probability of meeting deadlines, resource utilization, and in-process inventory. Profits and return on investments, when possible, are estimated from these operational performance measures.

A discrete model has dependent variables that change only at distinct points in simulated time, referred to as *event times*. For example, event times in a manufacturing system correspond to the times at which orders are placed in the system; material handling equipment arrives and departs from machines; and machines change status (e.g., from busy to either idle, broken, or blocked).

A continuous model has dependent variables that are continuous functions of time. For example, the time required to unload an oil tanker or the position of a crane. In some cases it is useful to model a discrete variable with a continuous representation by considering the entities in the system in the aggregate rather than individually. For example, the number of bottles on a conveyor may be modeled more efficiently using a continuous representation, even though the bottles are washed and filled individually.

In a combined model, the dependent variables of a model may change discretely, continuously, or continuously with discrete jumps superimposed. The most important aspect of combined simulation arises from the interaction between discretely and continuously changing variables. For example, when a crane reaches a prescribed location, unloading is initiated.

In the following sections, the terminology of simulation modeling is presented and examples of the use of modeling world views are given.

2.4.1 Discrete Simulation Modeling

The components that flow in a discrete system, such as people, equipment, orders, and raw materials, are called *entities*. There are many types of entities, and each has a set of characteristics or attributes. In simulation modeling, groupings of entities are called files, sets, lists, or chains. The goal of a discrete simulation model is to portray the activities in which the entities engage and thereby learn something about the system's dynamic behavior. Simulation accomplishes this by defining the states of the system and constructing activities that move it from state to state. The beginning and ending of each activity are events. The state of the model remains constant between consecutive event times, and a complete dynamic portrayal of the state of the model is obtained by advancing simulated time from one event to the next. This timing mechanism, referred to as the *next-event approach*, is used in many discrete simulation languages.

There are many ways to formulate a discrete simulation model. Four formulation possibilities are:

1. Defining the changes in state that occur at each event time
2. Describing the process (network) through which the entities in the model flow
3. Describing the activities in which the entities engage

4. Describing the objects (entities) and the conditions that change the state of the objects

In this chapter the first two of the formulations above are presented. In Chapter 11, object-oriented modeling and its implementation are discussed.

2.4.2 Example of Discrete Simulation Modeling

An example of the concept of simulation was presented in Chapter 1, where service is given to customers by a bank teller. The purpose for this model was to estimate the percent of time the teller is idle and the average time a customer spends in the system. Table 1.1 assumes that the time of arrival of each customer and the processing time by the teller for each customer are known. An ad hoc simulation was used to analyze the model.

To understand the model, we first define the state of the system, which for this example is the status of the teller (busy or idle) and the number of customers in the system. The state of the system is changed by (1) a customer arriving at the system, and (2) the completion of service by the teller and the subsequent departure of the customer. Note that for these two events, the status of the teller can be determined from the number of customers in the system (i.e., idle if number of customers in the system is zero, and busy otherwise). The art of modeling and the evolutionary nature of modeling lead to the definition of state above to accommodate any future need to model customer departures or teller breaks. A possible status variable not included in the definitions of the state of the system is the remaining processing time for a teller on a customer. This variable was omitted because a discrete-event model involving only two events was perceived. If a continuous model or a more elaborate discrete model is required to satisfy the purpose for modeling, the model might require this embellishment to the state of the system description. To illustrate a simulation, the state of the system over time is obtained by processing the events corresponding to the arrival and departure of customers in a time-ordered sequence, as shown in Table 1.1.

In Table 2.1, columns (1) to (4) are system data. It is assumed that initially there are no customers in the system, the teller is idle, and the first customer is to arrive at time 0.0. The start of service time given in column (5) depends on whether service on the preceding customer has been completed. It is taken as the larger value of the arrival time of the customer and the departure time of the preceding customer. Column (6), the time when service ends, is the sum of the column (5) value and the service time for the customer, column (4). Values for time-in-queue and time-in-system for each customer are computed in columns (7) and (8) as shown in Table 2.1. Average values per customer for these variables are 7/20 or 0.35 minute and 72/20 or 3.6 minutes, respectively. Table 2.1 presents a summary of information concerning each customer but does not provide information about the teller and the queue size of customers. To obtain such information, it is convenient to examine the events associated with the situation.

The logic associated with processing the arrival and departure events depends on the state of the system at the time of the event. In the case of the arrival event, the disposition of the arriving customer is based on the status of the teller. If the teller is idle, the status of the teller is changed to busy and the departure event is scheduled for the customer by adding a service time to the current time. However, if the teller is busy at the time of an arrival, the customer cannot begin service at the current time and therefore enters the queue (the queue length is increased by 1). At a departure event, the

Table 2.1 Ad Hoc Simulation of Customer–Teller Banking System

(1) Customer Number	(2) Time Between Arrivals	(3) Arrival Time	(4) Service Time	(5) Service Begins	(6) Time Service Ends	(7) Time in Queue	(8) Time in System
1	—	0	2	0	2	0	2
2	5	5	2	5	7	0	2
3	1	6	6	7	13	1	7
4	10	16	5	16	21	0	5
5	6	22	6	22	28	0	6
6	2	24	4	28	32	4	8
7	9	33	3	33	36	0	3
8	1	34	4	36	40	2	6
9	10	44	1	44	45	0	1
10	3	47	3	47	50	0	3
11	5	52	1	52	53	0	1
12	2	54	2	54	56	0	2
13	3	57	3	57	60	0	3
14	5	62	6	62	68	0	6
15	4	66	2	68	70	2	4
16	3	69	6	70	76	1	7
17	7	76	4	76	80	0	4
18	8	84	5	84	89	0	5
19	7	91	3	91	94	0	3
20	7	98	1	98	99	0	1
						10	79

status of the teller depends on whether a customer is waiting. If a customer is waiting in the queue, the teller's status remains busy, the queue length is reduced by 1, and a departure event for the customer removed from the queue is scheduled. If, however, the queue is empty, the status of the teller is set to idle. In this description, the initiation of service can occur at an arrival event or a departure event. If the initiation of service could occur at some other time, it would be necessary to define initiation of service as a separate event. This is also the case for completion of service, which, for this illustration, only happens when a departure event occurs.

An event-oriented description of customer status and number of customers in the system is given in Table 2.2. In the table the events are listed in chronological order. The average number of customers in the system is computed as a time-weighted average, that is, the sum of the product of the number in the system and the fraction of time that the number in the system existed. For this example there were 0 customers in the system for 30 minutes, 1 customer in the system for 59 minutes, and 2 customers in the system for 10 minutes. The weighted-average number of customers in the systems is $0(30/99) + 1(59/99) + 2(10/99)$, or 0.798. The fraction of time the teller is idle is the total idle time divided by the total simulation time, which for this example is $30/99$, or 0.303.

To place the arrival and departure events in their proper chronological order, it is necessary to maintain a record or calendar of future events to be processed. This is done by maintaining the times of the next arrival event and next departure event. The next event to be processed is then selected by comparing these event times. For situations

Table 2.2 Event-Oriented Description of Customer–Teller Simulation.

Event Time	Customer Number	Event Type	Number in Queue	Number in System	Teller Status	Teller Idle Time
0	—	Start	0	0	Idle	—
0	1	Arrival	0	1	Busy	
2	1	Departure	0	0	Idle	
5	2	Arrival	0	1	Busy	3
6	3	Arrival	1	2	Busy	
7	2	Departure	0	1	Busy	
13	3	Departure	0	0	Idle	
16	4	Arrival	0	1	Busy	3
21	4	Departure	0	0	Idle	
22	5	Arrival	0	1	Busy	1
24	6	Arrival	1	2	Busy	
28	5	Departure	0	1	Busy	
32	6	Departure	0	0	Idle	
33	7	Arrival	0	1	Busy	1
34	8	Arrival	1	2	Busy	
36	7	Departure	0	1	Busy	
40	8	Depature	0	0	Idle	
44	9	Arrival	0	1	Busy	4
45	9	Departure	0	0	Idle	
47	10	Arrival	0	1	Busy	2
50	10	Departure	0	0	Idle	
52	11	Arrival	0	1	Busy	2
53	11	Departure	0	0	Idle	
54	12	Arrival	0	1	Busy	1
56	12	Departure	0	0	Idle	
57	13	Arrival	0	1	Busy	1
60	13	Departure	0	0	Idle	
62	14	Arrival	0	1	Busy	2
66	15	Arrival	1	2	Busy	
68	14	Departure	0	1	Busy	
69	16	Arrival	1	2	Busy	
70	15	Departure	0	1	Busy	
76	16	Departure	0	0	Idle	
76	17	Arrival	0	1	Busy	
80	17	Departure	0	0	Idle	
84	18	Arrival	0	1	Busy	4
89	18	Departure	0	0	Idle	
91	19	Arrival	0	1	Busy	2
94	19	Departure	0	0	Idle	
98	20	Arrival	0	1	Busy	4
99	20	Departure	0	0	Idle	

with many events, an ordered list of events is maintained, which is referred to as an *event calendar.*

Several important concepts are illustrated by this example. We observe that at any instant in simulated time, the model is in a particular state. As events occur, the state of the model may change as prescribed by the logical-mathematical relationships associated with the events. Thus events define potential changes. Given the starting state, the logic for processing each event, and a method for specifying the sample values, the problem is largely one of bookkeeping. An essential element in the bookkeeping scheme is the event calendar, which provides a mechanism for recording and sequencing future events. Another point to observe is that state changes can be viewed from two perspectives: (1) the process that the entity (customer) encounters as it seeks service, or (2) the events that cause the state to change. These views are illustrated in a network model and in an event-based flowchart model in the next two sections.

2.4.3 Network Model of the Banking System

A *network* is a form of process model that depicts the flow of entities through nodes and branches. This view of dynamic systems modeling using activity networks was developed by Pritsker in the early 1960s [6]. Many variants of network models for analysis by simulation have been built on this theme.

A network model for the customer–teller bank system will now be developed. On a network, the passage of time is represented by a branch. A *branch* is a graphical representation of an activity. Clearly, teller processing is an activity and hence is modeled by a branch. If the teller activity is ongoing, arriving entities (customers) must wait. Waiting occurs in a QUEUE node. Thus a one-server, single-queue operation could be depicted as QUEUE node, Q, followed by an activity, Processing activity →, that is,

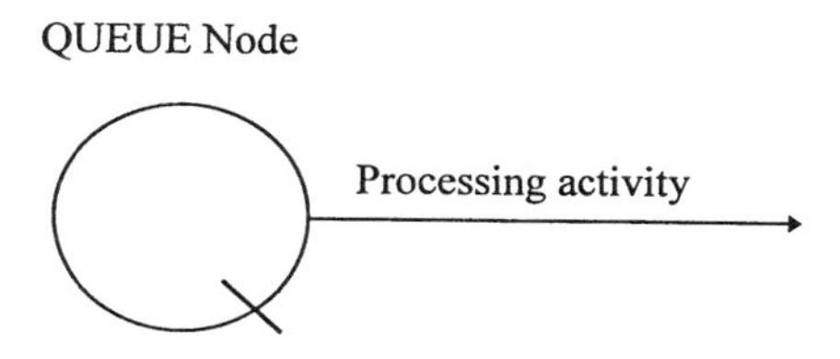

In this example, customers wait for service at the QUEUE node. When the teller is free, a customer is removed from the queue and a service activity is initiated. Many procedures exist for specifying the time to perform the activities.

A complete network model is shown in Figure 2.6. Customer entities are inserted into the network at the CREATE node. There is a zero time for the customer entity to travel to the QUEUE node, so that the customers arrive at the same time as they are

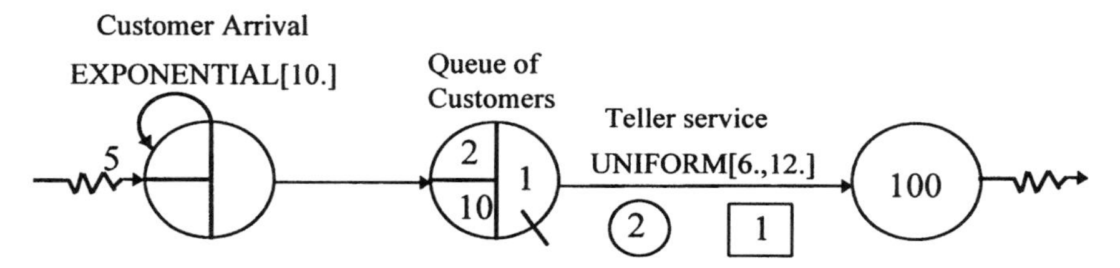

Figure 2.6 Network model of banking system.

Figure 2.7 Diagram of a banking system.

created. The customers either wait or are processed. The time spent in the bank system by a customer is then collected at the COLLECT node. As can be seen, the network model of Figure 2.6 resembles the diagram of operations presented in Figure 2.7. The symbol ⁓W→ is used to indicate the start or end of the path for an entity and provides a means to see easily an entity's starting and ending locations in the network.

2.4.4 Discrete-Event Model of the Banking System

The states of the bank system are measured by the number of customers in the system and the status of the teller. With the following two events, the changes in model state can be made: (1) at a customer-arrival event, and (2) at an end-of-service event. A modeler must determine the significant events to include in the model. Here all changes in status are assumed to occur at either the arrival time of a customer or at the time that service by the teller ends. Thus the state of the model will not change except at these event times.

The initialization logic for this example is depicted in Figure 2.8. The teller is initially set to idle. The arrival event corresponding to the first customer arrival is scheduled to occur. By *schedule* is meant the placing of an event on the event calendar to occur at a future time. This initialization establishes the initial state of the model as empty and idle.

The logic for the customer-arrival event is depicted in Figure 2.9. The first action that is performed is the scheduling of the next arrival. Thus each arrival causes another arrival to occur at a later time. In this way, only one arrival event is scheduled to occur at any one time, but a complete sequence of arrivals is generated. The arrival time of the customer is recorded. A test is then made on the status of the teller. If processing can begin, the teller is made busy and an end-of-processing event for the arriving customer is scheduled. Otherwise, the customer is placed in the queue representing waiting customers.

At each end-of-processing event, a value is collected on the time the customer spent in the queue and in processing. Next, the first waiting customer, if any, is removed from the queue and its processing is initiated. The logic for the end-of-processing event is depicted in Figure 2.10.

This simple example illustrates the basic concepts of discrete-event simulation modeling. Variables, entities, and file memberships make up the static structure of a simulation model. They describe the state of the model but not how it operates. The events specify the logic that controls the changes that occur at specific instants of time. The dynamic behavior is then obtained by sequentially processing events and recording status values at event times.

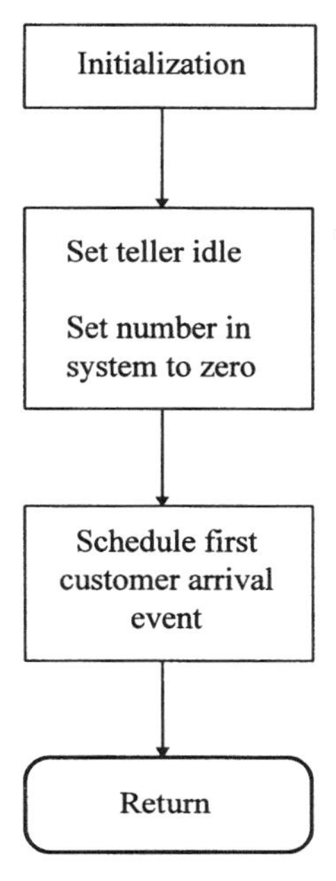

Figure 2.8 Initialization for banking system problem.

2.4.5 Continuous Simulation Modeling

In a continuous simulation model, the state of the system is represented by dependent variables that change continuously over time. To distinguish continuous-change variables from discrete-change variables, the former are, for communication convenience, referred to as state variables. A continuous simulation model is constructed by defining equations for a set of state variables.

State variables in a continuous model can be represented by one or more of the following forms:

- A system of explicit functional forms [e.g., $y = f(x, t)$]
- A system of difference equations (e.g., $y_{n+1} = ay_n + bu_n$)
- A system of differential equations [e.g., $dy/dt = f(x, t)$]

Typically, the independent variable is time which, in the examples above, is represented by t and n. Simulation solutions are obtained by specifying values of the variables in the equations at an initial (or specific) point in time and using these values as inputs to obtain solutions at a next point in time. The new values then become the starting values for the next evaluation. This evaluation-step-evaluation procedure for obtaining values for the state variables is referred to as *continuous simulation analysis*. It is used directly when variables are described by explicit functional forms or by a set of difference equations. When simultaneous equations are involved, numerical analysis procedures for obtaining variable values to satisfy the set of equations are required at each step.

Models of continuous systems are frequently written in terms of the derivatives of

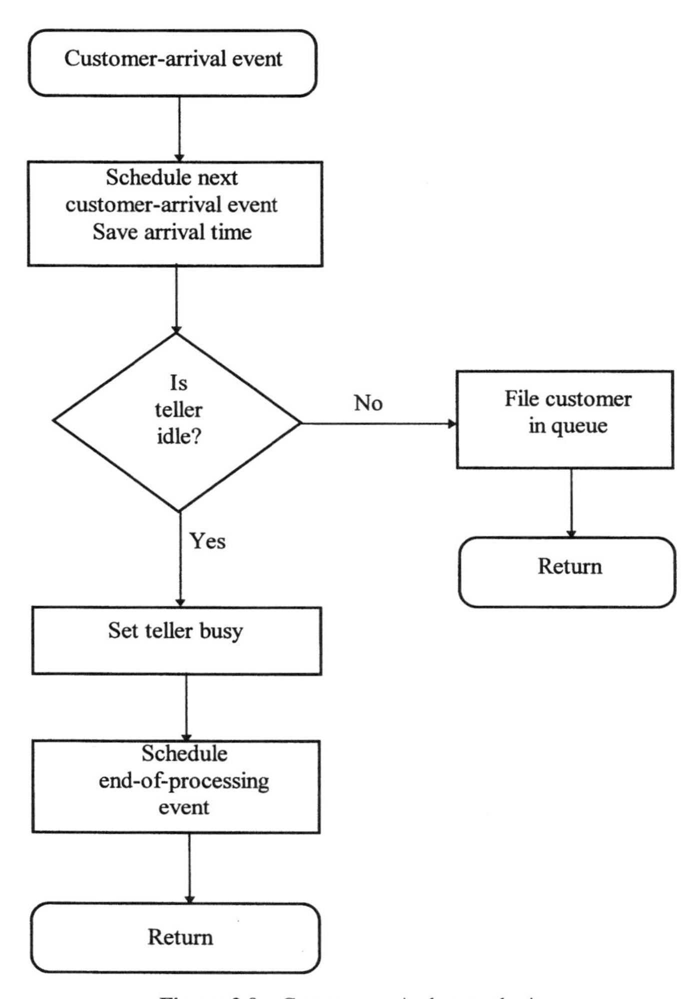

Figure 2.9 Customer-arrival event logic.

the state variables, that is, differential equations. The reason for this is that it is often easier to construct a relationship for the rate of change of a state variable than to devise a relationship for the state variable directly. If there is a differential equation in the model, $dy(t)/dt$, the values of $y(t)$ are obtained using numerical integration as follows:

$$y(t_2) = y(t_1) + \int_{t_1}^{t_2} \frac{dy}{dt}\, dt$$

Many methods exist for performing the integration indicated in the equation above.

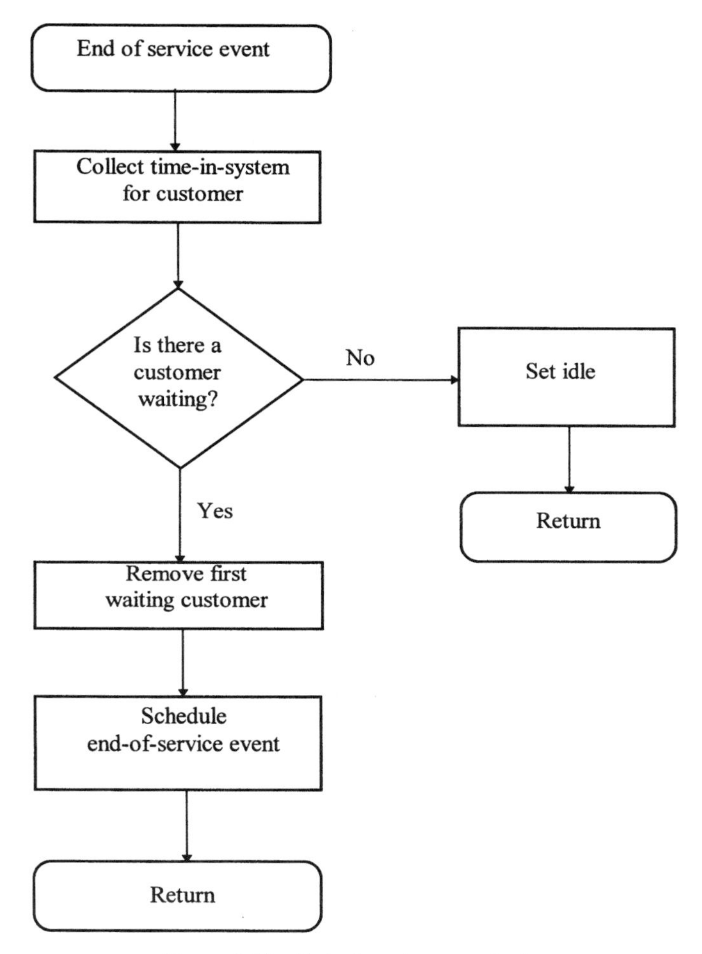

Figure 2.10 End-of-service event logic.

2.4.6 Building a Continuous Model

The tasks of a modeler when developing a continuous model are:

- Identify the state variables whose behavior is to be portrayed. A typical purpose for building continuous models is to describe behavior or to evaluate or optimize proposed designs for controlling behavior.
- Develop the equations describing the behavior of the state variables.
- Identify threshold conditions where status changes may occur. These are referred to as *state events* to distinguish them from scheduled or *time events*.
- Determine the value of the state variables in accordance with the defining equations subject to the changes possible when state events occur.

Complexity in continuous models occurs for the following reasons:

- Changes occur in the defining equations. These changes can be to the coefficients or in the form of the equations and may occur at a time event or a state event.
- Discrete (discontinuous) changes in the state variables can occur.
- Simultaneous sets of equations may exist because there are interactions among state variables.
- Random variables are included in the defining equations.

Because some or all of these complexities usually occur in problem-solving situations, a simulation approach is common when analyzing continuous models.

2.4.7 Combined Discrete–Continuous Models

The world view of a combined model specifies that the system can be described in terms of entities, global or model variables, and state variables. The behavior of the model is simulated by computing the values of the state variables at small time steps and by computing the values of attributes of entities and global variables at event times. The breakthrough in modeling combined systems occurred when the definition of an event was challenged [7].

Three fundamental interactions can occur between discretely and continuously changing variables. First, a discrete change in value may be made to a continuous variable. Examples of this type of interaction are the completion of a maintenance operation that instantaneously increases the rate of processing by machines within a system, and the investment of capital that instantaneously increases the dollars available for raw material purchase. Second, a continuous state variable achieving a threshold value may cause an event to occur or to be scheduled (e.g., the arrival of a material handler to a prescribed position initiates an unloading process). In general, events could be based on the relative value of two or more state variables. Third, the functional description of continuous variables may be changed at discrete time instants. An example of this is the change in the equations governing acceleration of a crane when a human being is in the vicinity of the crane.

The following principle describes a convenient initial approach to the combined modeling of a system. The evolutionary modeling principle stated previously applies to combined modeling so that any initial order to the modeling sequence will be superseded.

Modeling Principle 5 In modeling combined systems, the continuous aspects of the problem should be considered first. The discrete aspects of the model—including events, networks, algorithms, control procedures, and advanced logic capabilities—should then be developed. The interfaces between discrete and continuous variable should then be approached.

Combined discrete-event and continuous modeling constitutes a significant advance in the field of simulation. There are distinct groups within the simulation field for discrete-event simulation and continuous simulation. The disciplines associated with discrete-event simulation are industrial engineering, computer science, management science, operations research, and business administration. People who use continuous simulation are more typically electrical engineers, mechanical engineers, chemical engineers, agricultural engineers, and physicists. The overlap between the two groups has not been large. For years, the Summer Computer Simulation Conference was for continuous modelers and the Winter Simulation Conference was for discrete-event modelers. Only recently have the two groups started to mix. A large number of problems are in reality a combination of discrete and continuous phenomena and should be modeled using a combined discrete-event/continuous approach. However, due to group division, either a continuous or a discrete modeling approach is normally employed.

2.4.8 Combined Modeling Descriptions

An area where combined discrete continuous modeling has proven to be very powerful is in the development of procedures for analyzing the use of material handling equipment. The modeling of conveyors, cranes, and automated guided vehicles involve the continuous concepts of acceleration and velocity changes and also the discrete requirements associated with loading, unloading, picking up material, moving over network segments, and control strategies associated with movements. For packaging-line models, it is often advantageous to model the items on the conveyor using continuous concepts to maintain state variables of the number of items on the conveyor, the number of items in staging areas, and the number of items being processed at a machine. Advanced packaging lines employ variable-speed machines and variable-speed conveyors so that any linearization of the continuous variables to allow discrete-event scheduling is not practical.

For large crane systems, the acceleration and deceleration characteristics of the crane can make a large difference in its ability to process loads efficiently. Assumptions of instantaneous startups and stops are not appropriate. In addition, multiple cranes are typically on a single runway that requires modeling the interference among the cranes. This involves monitoring multiple state variables and the detection of state events when two variables are within a prescribed distance.

For automated guided vehicles (AGVs), a two-dimensional grid is typically required with the intersections being potential control points for directing the AGVs. Continuous variables are used to represent the position of the vehicle and also the amount of energy available to power the vehicle. Discrete events relate to the requests for the AGV, the loading and unloading of material from the AGV, and the decision logic associated with the disposition of an available AGV. Movement of the AGV, either loaded or unloaded, through the grid network is then described by the equations of motion governing the AGV system.

2.5 SIMULATION MODEL PURPOSE

Throughout this chapter, emphasis has been placed on the use of modeling and simulation to solve problems. The model-based problem solving process was presented as being driven by a system-specific purpose. Table 2.3 provides illustrations of systems and areas and the types of design, planning, and operational issues that can be addressed using modeling and simulation. The purpose for modeling can also have a functional level. The following list illustrates functional levels to which simulation modeling has been applied:

- As *explanatory devices* to understand a system or problem
- As a *communication vehicle* to describe system operation

Table 2.3 Modeling and Simulation Application Areas

Type of System	Design, Planning, and Operational Issues
Manufacturing systems	Plant design and layout Continuous improvement Capacity management Agile manufacturing evaluation Scheduling and control Materials handling
Transportation systems	Railroad system performance Truck scheduling and routing Air traffic control Terminal and depot operations
Computer and communication systems	Performance evaluation Work-flow generation and analysis Reliability assessment
Project planning and control	Product planning Marketing analysis Research and development performance Construction activity planning Scheduling project activities
Financial planning	Capital investment decision making Cash flow analysis Risk assessment Balance sheet projections
Environmental and ecological studies	Flood control Pollution control Energy flows and utilization Farm management Pest control Reactor maintainability
Health care systems	Supply management Operating room scheduling Manpower planning Organ transplantation policy evaluation

- As an *analysis tool* to determine critical elements, components, and issues and to estimate performance measures
- As a *design assessor* to evaluate proposed solutions and to synthesize new alternative solutions
- As a *scheduler* to develop on-line operational schedules for jobs, tasks, and resources
- As a *control mechanism* for the distribution and routing of materials and resources
- As a *training tool* to assist operators in understanding system operations
- As a *part of the system* to provide on-line information, status projections, and decision support

Table 2.4 Primary Outputs for Use in Applications by Functional Level

Functional Level	Primary Outputs
Explanatory devices	Animations
Communication vehicle	Animations, plots, pie charts
Analysis tool	Tabulations, statistical estimators, statistical graphs
Design assessor	Statistical estimators, summary statistics, ranking and selection procedures
Scheduler	Tabular schedules, Gantt charts, resource plots
Control mechanism	Tabular outputs, animations, resource plots
Training tool	Animations, event traces, statistical estimators, summary statistics
Embedded system element	Status information, projections

Since simulation modeling can be used at each of these levels and across a wide spectrum of systems, many types of outputs and analysis capabilities are associated with simulation models. For any given level, any output capability could be used. In an attempt to bring focus to this issue, the high-level primary outputs associated with each of the levels are listed in Table 2.4. With regard to the different purposes for which models are used, the following principle is presented:

Modeling Principle 6 A model should be evaluated according to its usefulness. From an absolute perspective, a model is neither good or bad, nor is it neutral.

As a summary to this section, modeling principle 7 is offered.

Modeling Principle 7 The purpose of simulation modeling is knowledge and understanding, not models.

Simulation modeling is performed to induce change. To achieve change, the results of the modeling and simulation effort require implementation. For the simulationist, implementation based on results is a rewarding experience.

2.6 RELATED MODELING RESEARCH

Research on modeling is extremely difficult. Basic research on understanding models and the modeling process are reported by Fishwick [8], Little [9], Polya [10], Wymore [11], and Zeigler [12–14]. Henriksen [15–18] has produced excellent papers that highlight the significant questions on specialized simulation modeling efforts. Geoffrion [19, 20] has written several basic papers on the fundamentals of structured modeling. The impact of these efforts on modeling practice remains to be seen.

ACKNOWLEDGMENTS

This material is based on work supported by the National Science Foundation under Grant DMS-8717799. The government has certain rights in this material. The author thanks the following persons for helping to improve this chapter by providing review comments: Barry Nelson of Northwestern University, Bob Schwab of Caterpillar Corporation, Bruce Schmeiser of Purdue University, and Steve Duket, Mary Grant, and Ken Musselman of Pritsker Corporation.

REFERENCES

1. Simon, H. A. (1990). Prediction and prescription in systems modeling, *Operations Research*, Vol. 38, No. 1, pp. 7–14.
2. Council on Competitiveness (1989). *Challenges*, Vol. 1, No. 6.
3. Morris, W. T. (1967). On the art of modeling, *Management Science*, Vol. 13, No. 12, pp. 707–717.
4. Evans, J. R. (1991). *Creative Thinking in the Decision and Management Sciences*, South-Western Publishing, Cincinnati, Ohio.
5. Pritsker, A. A. B. (1995). *Introduction to Simulation and SLAM II*, 4th ed., Systems Publishing Corporation, West Lafayette, Ind., and Wiley, New York, pp. 51–66.
6. Pritsker, A. A. B. (1990). *Papers · Experiences · Perspectives*, Wadsworth/ITP, Belmont, Calif., pp. 240–245.
7. Pritsker, A. A. B. (1990). *Papers · Experiences · Perspectives*, Wadsworth/ITP, Belmont, Calif., p. 253.
8. Fishwick, P. A. (1994). *Simulation Model Design and Execution: Building Digital Worlds*, Prentice Hall, Upper Saddle River, N.J.
9. Little, J. D. C. (1992). Tautologies, models and theories: can we find laws of manufacturing? *IIE Transactions*, July, pp. 7–13.
10. Polya, G. (1973). *How to Solve It*, 2nd ed., Princeton University Press, Princeton, N.J.
11. Wymore, A. W. (1976). *A Mathematical Theory of Modelling and Simulation*, Wiley, New York.
12. Ziegler, B. P. (1976). *Theory of Modelling and Simulation*, Wiley, New York.
13. Ziegler, B. P. (1984). *Multi-facetted Modelling and Discrete Event Simulation*, Academic Press, San Diego, Calif.
14. Ziegler, B. P. (1990). *Object-Oriented Simulation with Hierarchical, Modular Models*, Academic Press, San Diego, Calif.
15. Henriksen, J. O. (1986). You can't beat the clock: studies in problem solving, in *Proceedings of the 1986 Winter Simulation Conference*, Washington, D.C., December, J. R. Wilson, J. O. Henriksen, and S. D. Roberts, eds., IEEE, Piscataway, N.J., pp. 713–726.
16. Henriksen, J. O. (1987). Alternatives for modeling of preemptive scheduling, in *Proceedings of the 1987 Winter Simulation Conference*, A. Thesen, H. Grant, and W. D. Kelton, eds., Atlanta, Ga., December, IEEE, Piscataway, N.J., pp. 575–581.
17. Henriksen, J. O. (1988). One system, several perspectives, many models, in *Proceedings of the 1988 Winter Simulation Conference*, M. A. Abrams, P. L. Haigh, and J. C. Comfort, eds., San Diego, Calif., December, IEEE, Piscataway, N.J., pp. 352–356.
18. Henriksen, J. O. (1989). Alternative modeling perspectives: finding the creative spark, in *Proceedings of the 1989 Winter Simulation Conference*, Washington D.C., December,

E. A. MacNair, K. J. Musselman, and P. Heidelberger, eds., IEEE, Piscataway, N.J., pp. 648–652.

19. Geoffrion, A. M. (1987). An introduction to structured modeling, *Management Science*, Vol. 33, pp. 547–588.
20. Geoffrion, A. M. (1986). Integrated modeling systems, in *Proceedings of the Conference on Integrated Modeling Systems*, University of Texas, Austin, Texas, October.

PART II

METHODOLOGY

CHAPTER 3

Input Data Analysis

STEPHEN VINCENT
Compuware Corporation

3.1 NATURE OF SIMULATION INPUT

Developing a validated simulation model (Figure 3.1) involves three basic entities: the *real-world system* under consideration, a theoretical *model of the system*, and a computer-based representation of the model, the *simulation program*. The activity of deriving the theoretical model from the real-world system can be referred to as *simulation modeling*, and the activity whereby the computer-based representation is derived from the model can be referred to as *simulation programming*. Figure 3.1 also shows the basic checks of verification and validation that are applied in the development of a simulation model; these concepts are discussed in Chapter 10.

One of the primary reasons for using simulation is that the model of the real-world system is too complicated to study using the stochastic processes models described in Section 3.2. Major sources of complexity are the components of the model that "drive," or are inputs to, the logic of the model. Examples of such random inputs include arrivals of orders to a job shop, times between arrivals to a service facility, times between machine breakdowns, and so on. Each simulation model input has a correspondent both in the real-world system and in the simulation program, as shown in Figure 3.2. The activity of representing a random input model as a random variate generator is discussed in Chapter 5. The focus of this chapter is simulation input modeling, wherein probabilistic models of real-world processes or phenomena are derived. Often, the goal of simulation input modeling is to provide a model that is *reasonable*, given the goals of the simulation; in contrast to standard statistical applications, often we are not really interested in determining whether the model is *perfect*, as discussed in Section 3.5.3.

The impact of the choice of a random input model on the overall simulation validity may range from crucial to virtual irrelevance (depending on the system under consideration). There is no definitive manner to ascertain what the impact will be, without applying formal validation methods. On occasion it has been suggested that a rough *sensitivity analysis* can be used to judge the severity of the impact. The logic of this

Handbook of Simulation, Edited by Jerry Banks.
ISBN 0-471-13403-1

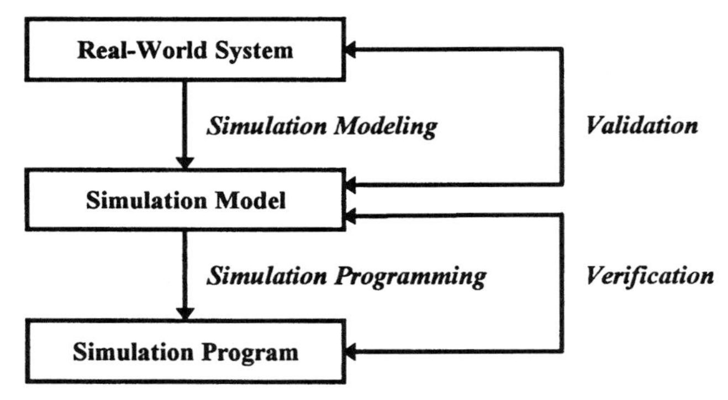

Figure 3.1 Overview of simulation model development.

analysis can be paraphrased as "try a number of representations; if the simulation results do not vary significantly, then the choice is not important." The fatal flaw in this logic is that the lack of variation does not establish the validity or even the reasonableness of *any* of the alternatives. Because, in general, the magnitude of the impact of the choice of the simulation input model on the overall simulation validity is not known a priori, we recommend the conservative approach of assuming that the impact is large. This implies that we should never ignore or short-change the input modeling process, which can be summarized by Figure 3.3: combine prior experience, relevant applicable theory (e.g., from stochastic processes or related to the system under consideration), and data (when available). Part of the input modeling process itself is an assessment of the

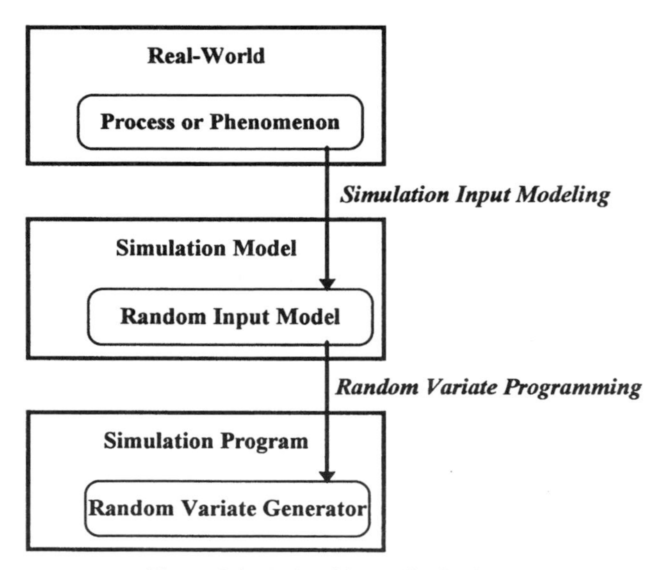

Figure 3.2 Role of input distributions.

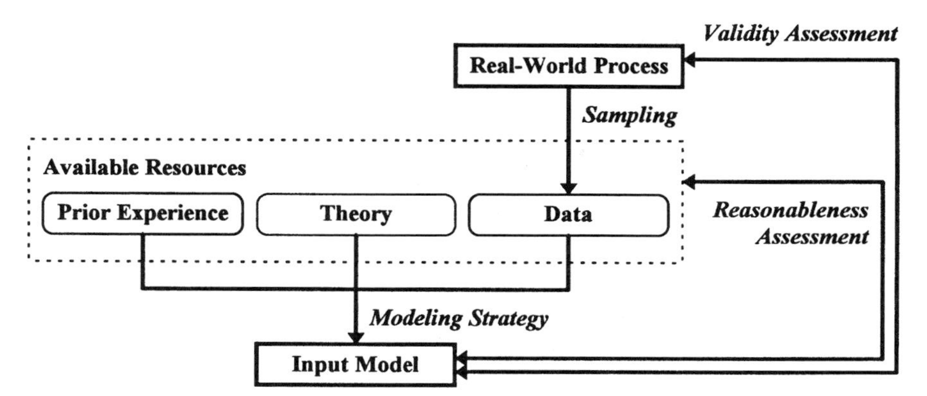

Figure 3.3 Modeling simulation inputs.

reasonableness of the resulting model, which concerns just the process under consideration. At a higher level is the validity check for the input model, which is possible only when considering the broader context in which the process occurs.

An effective job of simulation input modeling requires the use of a reasonable input modeling software package. It is not possible to perform most of the calculations "by hand," and spreadsheets and general-purpose statistical packages lack the capabilities to do even a limited analysis. In this chapter we provide an overview of simulation input modeling that will help to assess the qualities of alternative simulation input processors and to make effective use of the one selected. This chapter does not contain a rigorous development of all the statistical methodologies underlying simulation input modeling (we defer this task to texts intended for graduate-level study of simulation issues). For additional information and extensive references on the broad statistical issues related to simulation input modeling, we refer to the latest editions of the classic series on distributions by Johnson and Kotz [1–3] and the thorough collection of articles concerning goodness-of-fit techniques by D'Agostino and Stephens [4].

3.2 ROOTS OF SIMULATION INPUT MODELING

A random input variable to a simulation model can be viewed as a *stochastic process*. (Nelson [5] is a good introductory text with a simulation perspective.) A stochastic process is often defined as a collection of random variables $\{X(t), t \text{ in } T\}$, where T is called the *index set* of the process and t usually represents time. In the discrete-event simulation context the index set is typically taken to be the nonnegative integers, so the stochastic process itself is referred to as being *discrete-time*. For such a process let us use the simpler notation $\{X_k, k = 1, 2, \ldots\}$. Notice that the subscript k dictates the order of the variables but not the specific time of occurrence (i.e., X_2 occurs some time after X_1 and some time before X_3, not necessarily at time equal to 2). Here each X_k is a distinct occurrence of the same general random phenomenon X with probability distribution function $F_k(x) = \Pr\{X_k \leq x\}$. But what exactly is an X, how is X_k related to another member of the set X_j, and what do the $F_k(\cdot)$ functions "look" like? In the academic discipline of stochastic processes, various sets of answers are assumed and the characteristics of the resulting processes are studied, typically using analytic methods. Our current interest in stochastic processes is threefold:

1. It introduces terminology that will be of interest in later chapters.
2. There is one specific stochastic process, the IID (independent and identically distributed) process introduced below, that we would like to apply to the analysis of random simulation input.
3. We must be aware of the assumptions of the preferred IID process so that we can assess their reasonableness in a particular context, and if appropriate, consider alternative assumptions.

A good background in stochastic processes is always useful to a simulationist, but not essential for the remaining discussion.

Perhaps the most fundamental assumption to be made about a process is the dimensionality of X: univariate versus multivariate. If each random variable X represents a single quantity such as the service time of a customer, the process is called *univariate*; X_k would be the service time of the kth customer to arrive to the system. If, instead, X represents a number of quantities such as the amounts for different items within a single order, the process is called *multivariate*; here $X_k = \{A_k, B_k, \ldots\}$ could represent the amounts of items A, B, and so on, on the kth order submitted to an inventory system. In general, whenever a multivariate process is considered, the assumptions concerning the interrelationships of random variables become more complicated due to the increased dimensionality (e.g., we must consider how A_k is related to B_j, etc.). Due to these complexities, multivariate modeling is beyond the scope of this introductory chapter; we refer the interested reader to, for example, Johnson [6].

The strongest interrelationship assumptions that we can make are:

1. All of the X_k random variables are probabilistically independent of one another.
2. All of the X_k random variables follow the same probability distribution and thus are said to be *identically distributed*; that is, $F_k(\cdot)$ has a common form $F(\cdot)$ for all k.

This IID process is frequently encountered in classical statistics, in both its univariate and multivariate forms. The alternatives to these assumptions are endless, but we mention two general alternatives that are particularly germane to simulation:

1. In *nonstationary* processes the probability distribution of X_k varies as a function of k (time) in such a way that the basic mathematical form of the probability distributions $F_k(\cdot)$ is stable, but it has a parameterization that depends on time—$F_k(x) = F(x; g(k))$. [In practice, $g(\cdot)$ is typically parameterized directly in terms of t, time, rather than in terms of occurrence k.] Two examples are: (a) service systems, such as restaurants and airports, may have time-varying arrival rates of customers, and (b) physical devices that are subject to failure may have time-varying failure rates. In a particular simulation context, depending on the purposes of the study, the nonstationary nature of the process might be ignored intentionally, particularly if the variations in the process are small during the period to be simulated or if a worst-case scenario is being simulated with the most demanding rate for a process being used over the entire simulation.

2. In *correlated* processes the subsequent values in the process are not independent

of each other. It should be noted that in statistics the term *correlation* is solely a measure of linear dependence between two variables. (Consider the variable $Y = \sin X$, where X is uniformly distributed over the range 0 to 2π. Here Y is determined completely by X, yet the statistical correlation between X and Y is zero.) We therefore use the term "*correlated*" loosely in our present context. As shown in Chapter 7, output processes exhibiting positive correlations are common in simulations. In some systems order quantities from a single source, arriving sequentially over time, can be positively or negatively correlated; that is, some sources may produce similar values over time, whereas other sources may alternate between high and low amounts.

We defer discussion of particular forms of the distribution $F_k(\cdot)$ or $F(\cdot)$ until Section 3.5, but note now that in *Poisson* processes, $F(\cdot)$ is the exponential distribution, and much is known about different types of Poisson processes because their study is mathematically tractable. The use of this important process can often be justified as a model of arrivals to a system on the basis of three realistic assumptions:

1. Arrivals occur one at a time (if multiple arrivals occur simultaneously, we can still account for this phenomenon/complication when the number arriving can be justified as independent of the time of arrival).
2. The number arriving within a time interval starting at a point in time t is independent of the number that arrived by any point in time before t.
3. The distribution of the number arriving within a time interval starting at a point in time t is independent of the time t.

We refer the reader to an advanced text on simulation or stochastic processes for additional discussion of these points.

3.3 DATA COLLECTION

Arguably, the most difficult aspect of simulation input modeling is gathering data of sufficient quality, quantity, and variety to perform a reasonable analysis. In some contexts it may be impossible or infeasible to collect data; such is clearly the case when modeling proposed systems (e.g., existing systems with significant proposed modifications or nonexistent systems), but this can also be the case when labor agreements preclude ad hoc collection of performance data. For other studies there may be insufficient time or personnel to carry out extensive data collection. Existing data sets offer their own set of challenges. In contrast to classical statistics, wherein data are collected in a planned and systematic manner *after* the analysis method has been chosen, it is quite common to "mine" available sources of data that were collected for purposes other than the specification of a simulation model. Figure 3.4 presents a model of how observed data result from a real-world process. With precollected data a number of annoyances can occur:

- Data can be recorded in an order other than that in which it was observed, which precludes the checking of important assumptions (e.g., on autocorrelation).
- Data can be grouped into intervals (e.g., reported as histogram frequencies).

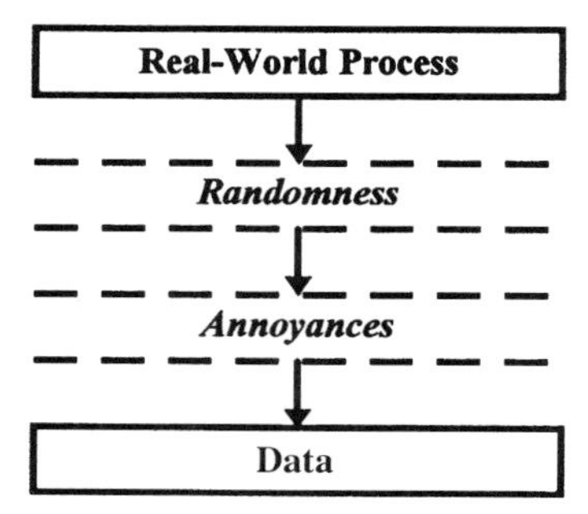

Figure 3.4 Reality of sampling.

- Data can be recorded with insufficient precision, perhaps even rounded to the closest integer, even though the observations were real-valued.
- Samples can hold obviously erroneous values, simply because the recorder did not anticipate the need for highly accurate data.
- Tables can hold values from more than one process, without documentation that would allow discrimination (e.g., repair times for a machine may vary with type of repair).
- Data values might be representative of a different real-world process (e.g., the historical data reflect operating conditions that are unlike those under consideration in the simulation study).

Although not all of these annoyances will occur with each data set in a simulation study, they occur so often that we are suspicious of the validity of any data set that is derived from historical records. Our skepticism concerning the accuracy and representativeness of data leads us to a paradoxical position: although a data sample is the best evidence available for use in specifying and evaluating a proposed model, it should not be taken *too* seriously. The impact of this conclusion is to mistrust model-selection strategies that are overly reliant on extremely "clean" samples: We want *robust* strategies.

When data will be collected specifically for a simulation study, we make the following practical suggestions:

- Whenever possible, collect between 100 and 200 observations. The decrease in quality of analyses performed with smaller samples is noticeable, whereas not much is gained by increasing the sample size.
- For real-valued observations, attempt to record them using at least two to three accurate significant digits for a nominal value such as the mean or median (50th percentile). For example, when collecting values that range between 0 and 100 with an anticipated mean of about 10, attempt to record values with at least one-decimal-place accuracy (e.g., 10.2). Standard mechanisms for assessing quality of fit react badly to the "value clumping" that occurs with limited recording accuracy, which can mislead an analyst about the quality of fit afforded by a model.
- When interested in interevent times, record event times and later calculate the interevent times manually or with the assistance of data analysis software.

- If there is any suspicion that real-world behavior depends on time of day or day of week, collect a number of samples from different time periods (this is discussed further in the next section).

3.4 PRACTICAL METHODS FOR TESTING ASSUMPTIONS

As indicated in Section 3.2, the easiest processes to deal with are IID. Therefore, we consider practical methods to detect gross deviations from these assumptions in our data. It should be noted that a deviation from a preferred assumption can manifest itself in more than one way, as will be seen below. *If a data set is available only in an order other than the values were observed (e.g., sorted), the important assumptions of IID cannot be tested, nor can IID alternatives be considered.*

To demonstrate the methods described in this section, we use six artificial sample data sets that can be thought of as observations over 200 time units of arrival processes with an (average) rate of arrival equal to 5. The sets differ in the way they were generated:

- Set 1: generated as an IID process with exponential interarrival times with a mean of 0.2 time unit (i.e., a rate of $1/0.2 = 5$).
- Set 2: generated as an IID process with gamma interarrival times, where the shape parameter of the gamma was 5.
- Set 3: generated as an IID process with lognormal interarrival times, where the shape parameter was 1.0.
- Set 4: generated as a nonstationary Poisson process (i.e., exponential interarrival times where the rate of arrival varies with time), where the rate function rises linearly from 1 at time 0 to 9 at time 200 (so an average of 5).
- Set 5: generated as a nonstationary Poisson process, where the rate of arrival for the first 100 time units is 1 and thereafter is 9.
- Set 6: generated as a correlated process, where an interarrival time is the weighted sum of the previous interarrival time and a random (uniformly distributed) value.

3.4.1 Methods for Assessing Independence

Two simple yet effective heuristic procedures that are readily available in almost all statistical software packages (intended for simulation input modeling or not) are (1) tables and/or plots of estimated lag (linear) correlations, and (2) scatter diagrams. Our general approach is to apply these heuristic procedures to a sample and to react only to the most disconfirming evidence. We do not recommend the application of formal tests to answer the question of whether sample values are independent of each other. Because we cannot make any assumptions concerning the distribution of the values, we can apply only nonparametric tests, which in our experience are of limited value.

First consider a *lag k correlation*, which is the correlation between observations k values apart. For example, a lag 2 correlation applies to the observation pairs with indexes (1,3), (2,4), (3,5), and so on. Formulas for estimated correlations appear in Chapter 7. For a sample of size n, the estimated lag k correlation is calculated using $n - k$ sample pairs, where clearly we need n larger than k. Lag correlations calculated from a very small number of pairs cannot be taken too seriously, due to the inherent variability of the estimate itself. In general, we do not recommend using a correlation estimate based

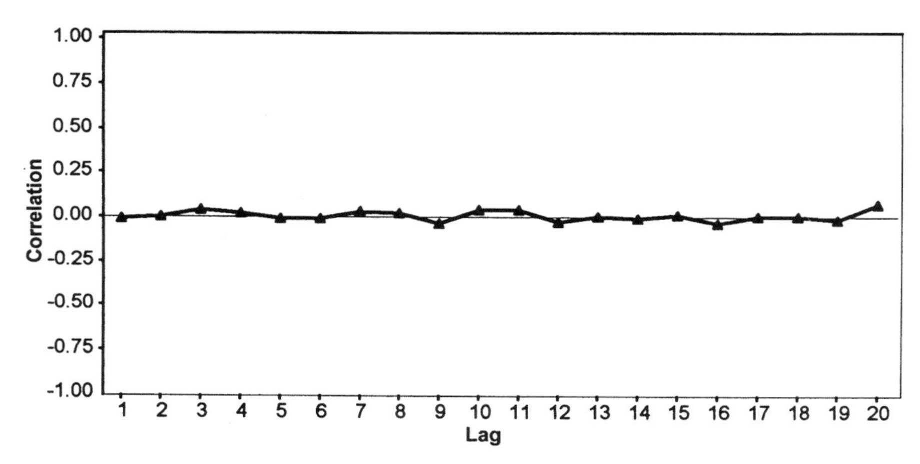

Figure 3.5 Estimated lag correlations for set 1 (IID exponential data).

on fewer than four pairs of observations. Further, we find that lags of size 1 through 10 are the most informative about a sample, whereas lags over 20 are noninformational. For independent samples, we expect the estimated lag correlations to be small in magnitude and clustered about zero, with both negative and positive estimated values. In Figures 3.5 to 3.7 we see examples of lag-correlation plots for sets 1, 5, and 6, where our concern over the validity of the independence assumption should be least for the IID sample and most for the correlated sample. (The example statistical analyses presented in this chapter were produced by the software described in Vincent and Law [7].) A rationale for the low-level positive correlations in the nonstationary plot is that small and large interarrival times tend to occur in the same time periods, due to the shift in arrival rate at time 100; despite the differences in assumptions, the plot for set 4 is very similar.

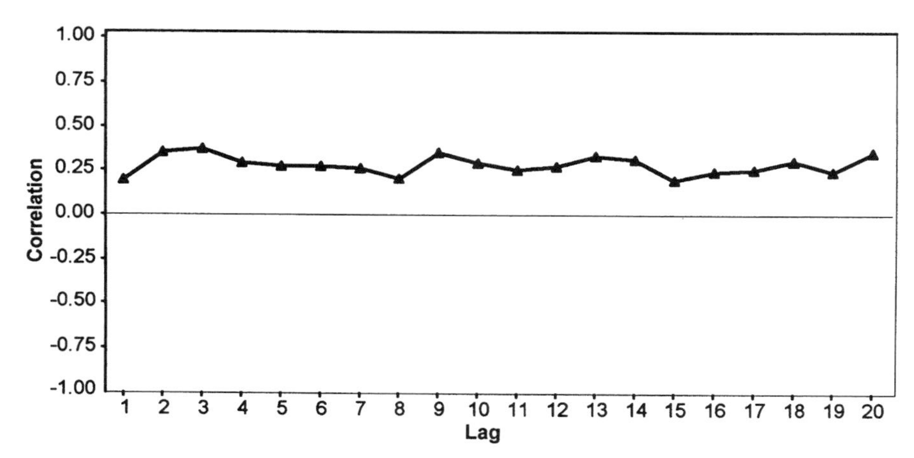

Figure 3.6 Estimated lag correlations plot for set 5 (dual-value nonstationary Poisson process sample).

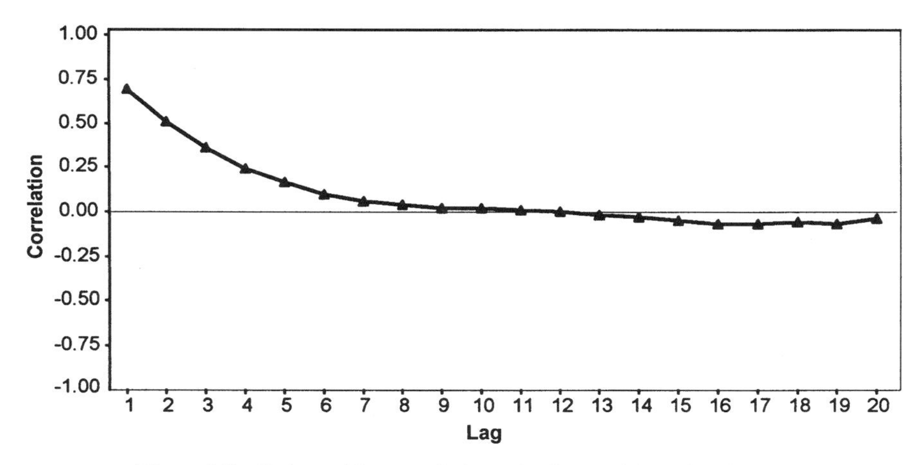

Figure 3.7 Estimated lag correlations plot for set 6 (correlated data).

Next we construct a *scatter diagram* by plotting the sequential pairs of observations with indexes (1,2), (2,3), (3,4), ... , $(n - 1, n)$. The x- and y-axes typically have the same endpoints, corresponding to the minimum and maximum observation. Correlated samples tend to produce plots in which values cluster tightly about one of the diagonals (for positive correlations the lower-left to upper-right diagonal). Independent samples tend to produce plots with points distributed over portions of the plot in accordance with the underlying distribution. In Figures 3.8 to 3.10 we compare scatter diagrams from sets 1 to 3, which show that IID samples from different distributions can produce different distributions of points in the plot region. In Figure 3.11 we show a scatter diagram for the data from set 6, which highlights the positive correlations between the values.

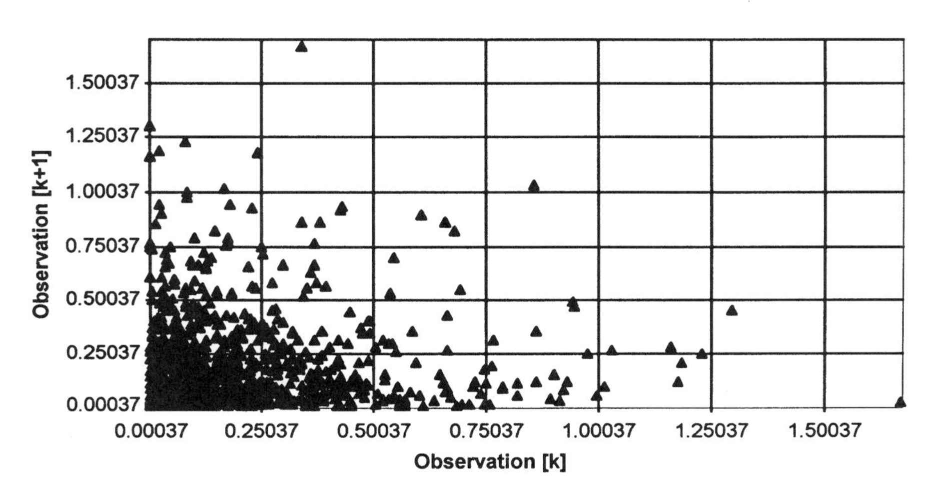

Figure 3.8 Scatter diagram for set 1 (exponential data).

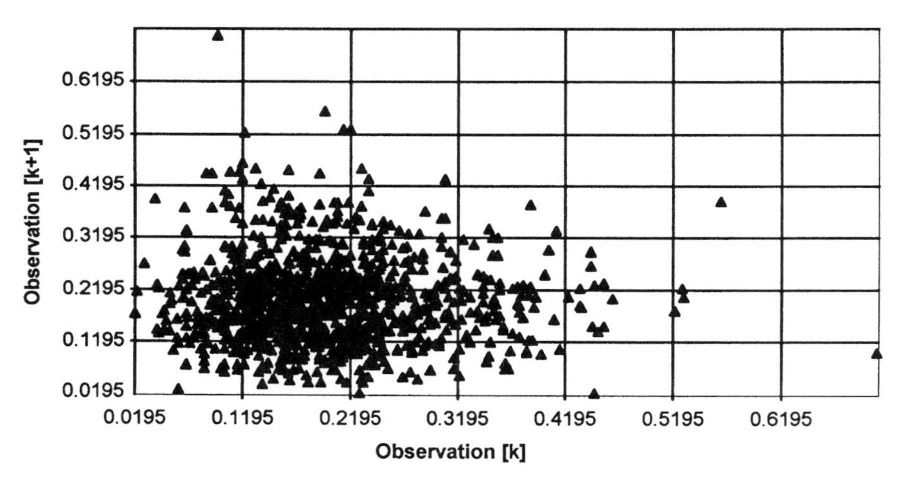

Figure 3.9 Scatter diagram for set 2 (gamma data).

3.4.2 Methods for Assessing Stability of Distribution

Stability of distribution is perhaps best discussed in terms of a real-valued timeline rather than an integer-valued order of occurrence. The essential question is whether or not, for any *relevant* pair of points in time, the probability distribution of the process under consideration is essentially the same. Many processes related to human activity are not stable for all points in time, even within relatively small time horizons. For example, arrival rates at airports and restaurants vary considerably during the day and from day to day. However, this lack of stability may not be relevant to a simulation study:

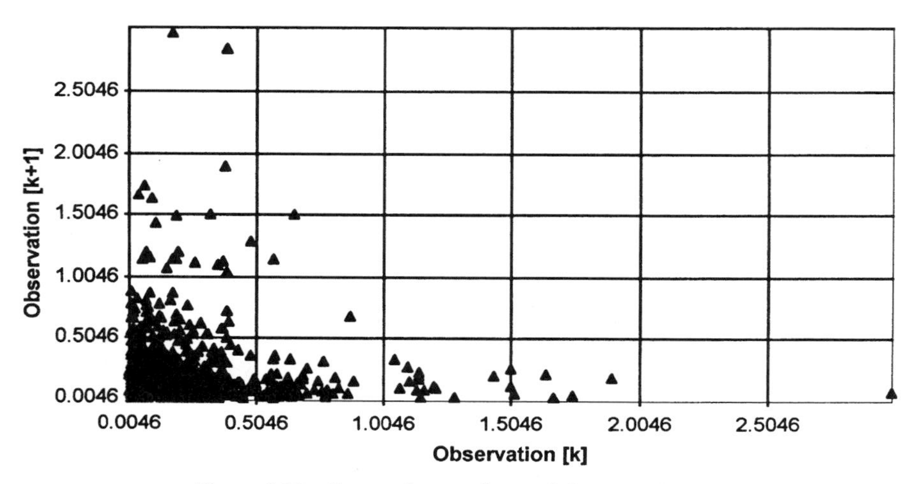

Figure 3.10 Scatter diagram for set 3 (lognormal data).

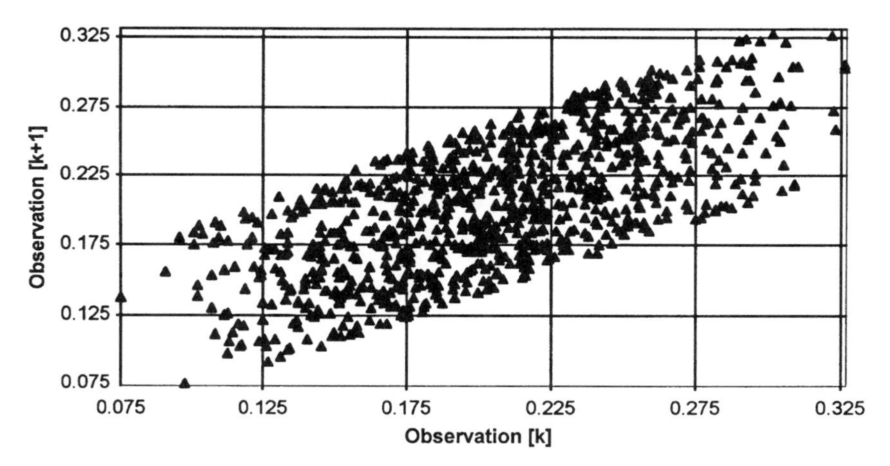

Figure 3.11 Scatter diagram for set 6 (correlated data).

- We may wish to simulate only a portion of the time period, in which the real-world phenomenon is relatively stable.
- We may wish to simulate a worst-case scenario in which peak demands on a system are sustained for long periods of time. In such a simulation study we use input models representative of specific intervals of time for all time intervals in order to place the most "stress" on the system.

Detecting even gross deviations from stability of distribution can sometimes be difficult. For arrival processes that will be modeled in terms of the interarrival times, we recommend plotting the proportion of arrivals up until a time T as a function of T. If the rate of arrival is relatively constant over the time period under consideration, the plot will be fairly linear, with few drastic deviations. In Figures 3.12 to 3.14 we show

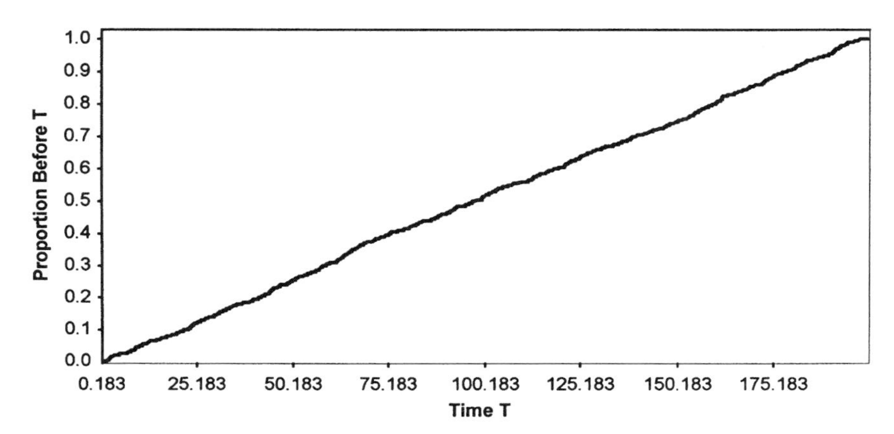

Figure 3.12 Arrival time pattern for set 1 (exponential data).

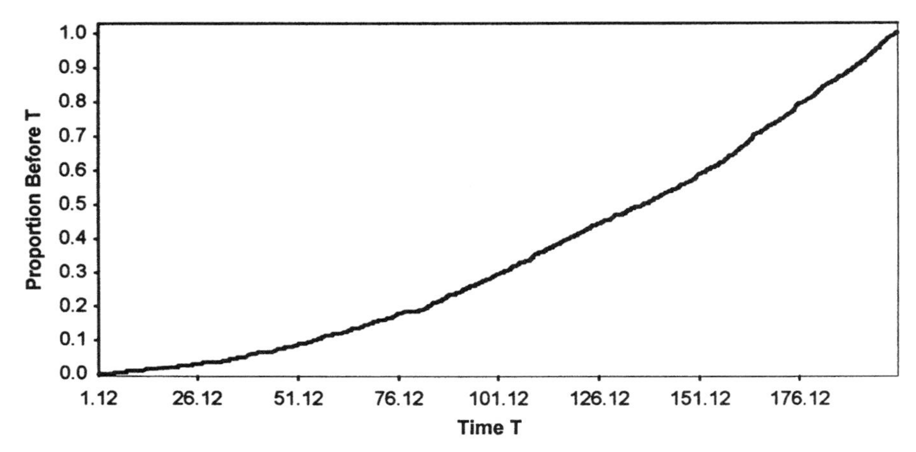

Figure 3.13 Arrival time pattern for set 4 (increasing nonstationary Poisson process sample).

the arrival patterns for sets 1, 4, and 5. The nonstationary aspects of sets 4 and 5 are very clear from these pictures. As an alternative or companion, the time period can be divided into intervals, and the number of arrivals that occur in each interval can be plotted, effectively estimating the arrival rate as a function of time. The resulting plot is dependent on the chosen interval width (this deficiency is closely related to the standard problems with histograms discussed later in the chapter). For this reason it may be difficult to assess whether variations from interval to interval are due to randomness possibly compounded by the interval-width decision, or whether there is a true difference in the rate across time. In Figures 3.15 to 3.17 we show the histogramlike plots of number of arrivals in 5-minute intervals for sets 1, 4, and 5. The nonstationary aspects of sets 4 and 5 are again very clear from these pictures.

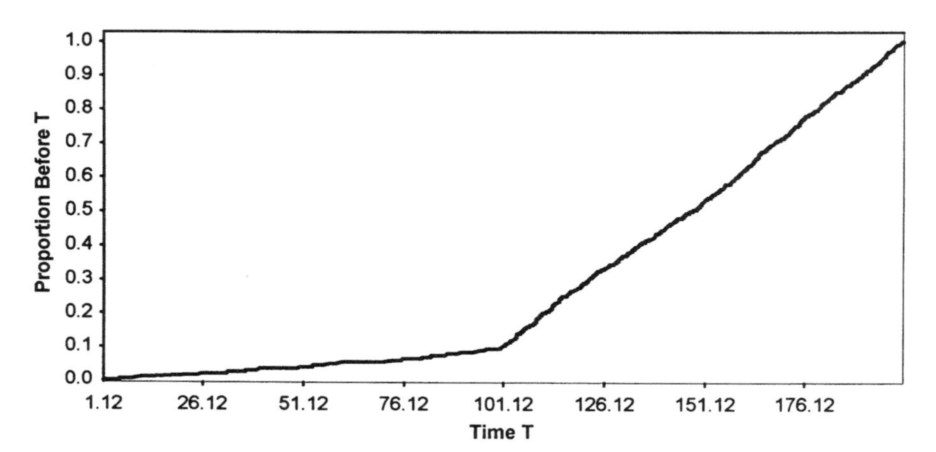

Figure 3.14 Arrival time pattern for set 5 (dual-value nonstationary Poisson process sample).

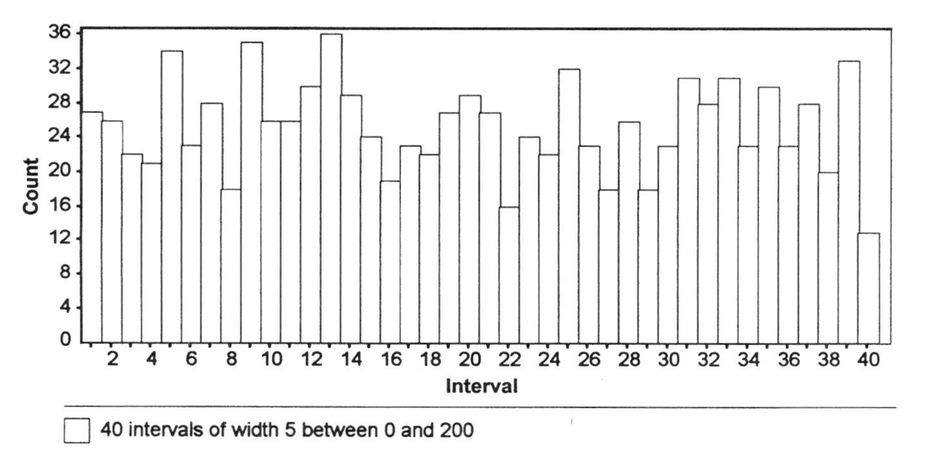

Figure 3.15 Arrival counts for set 1 (exponential data).

Other aspects of stability can be evaluated by considering a simple plot of the observed values against their integer order of occurrence. We expect the values to vary as we scan the plot, with no discernible overall patterns. In particular, we would expect the range of plotted points to be roughly the same across the plot. A deviation of this characteristic might occur, for example, in repair times for a system that is wearing out—as the state of the system deteriorates, it takes longer to complete a repair. A "typical" value should not change dramatically across the plot. If a process features stability of distribution, the mean of the distribution appropriate to time T will not vary with T; stability of distribution implies stability of mean value (as well as all other distribution moments). We can use this simple observation to provide a crude, yet effective detector of instability of distribution. If we aggregate values in the form of a moving average, a plot of the moving average should be roughly constant. Care must

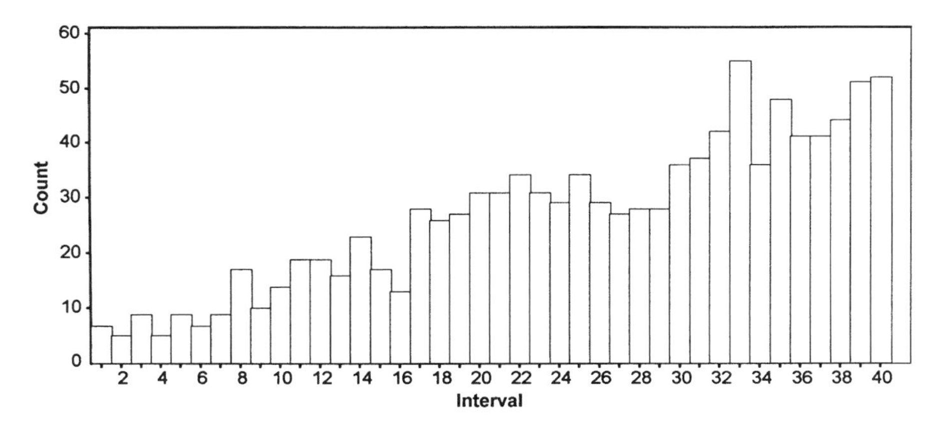

Figure 3.16 Arrival counts for set 4 (increasing nonstationary Poisson process sample).

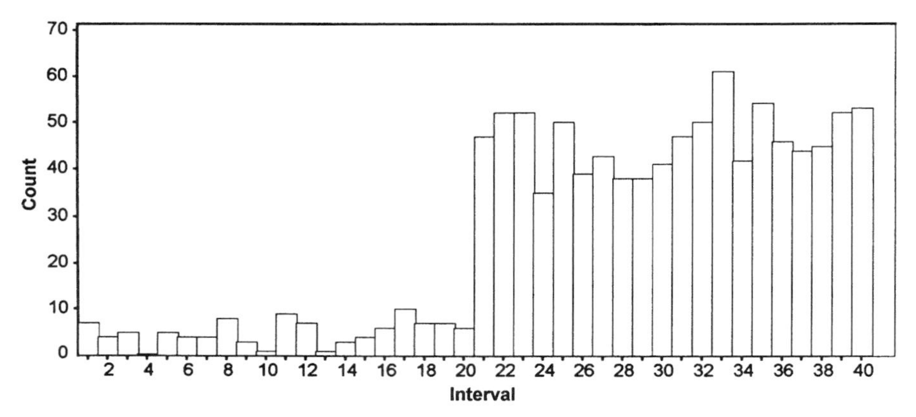

Figure 3.17 Arrival counts for set 5 (dual-value nonstationary Poisson process sample).

be taken when constructing and interpreting such plots, because the character of the plots can vary substantially with the choice of the moving-average "window" W. For most of the points in the plot the moving average is calculated using $2W + 1$ observations: if x_k is an observation and y_k is a calculated moving average, then generally $y_k = (x_{k-w} + \cdots + x_{k-1} + x_k + x_{k+1} + \cdots + x_{k+w})/(2W + 1)$. For the first and last W moving averages, fewer points are available for the moving average, and thus these estimates should be given less credence. In Figures 3.18 to 3.20, we show $W = 100$ moving-average plots for sets 1, 4, and 5. (In these figures, the sample average is shown as a solid horizontal line and the horizontal bound lines located 3 "standardized distances" from the sample average define a range in which we would expect the moving-average plot to lie if it were indeed stable.) The nonstationary aspects of sets 4 and 5 are very clear from these pictures.

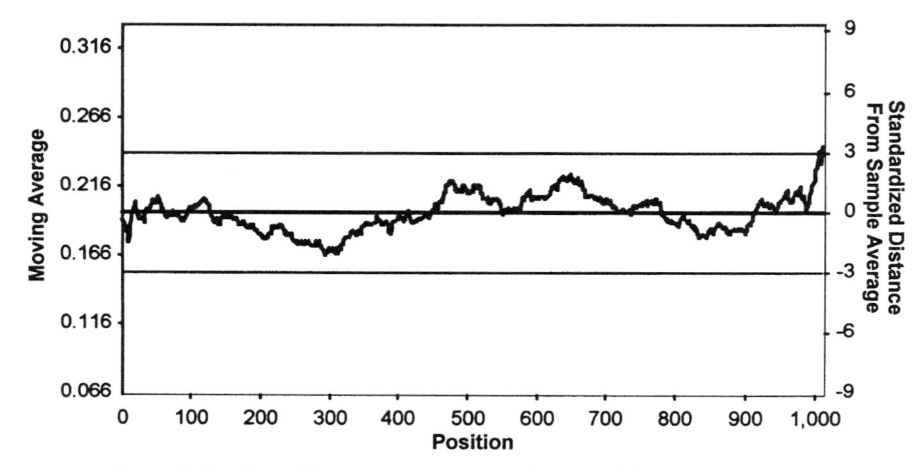

Figure 3.18 $W = 100$ moving-average plot for set 1 (HD exponential data).

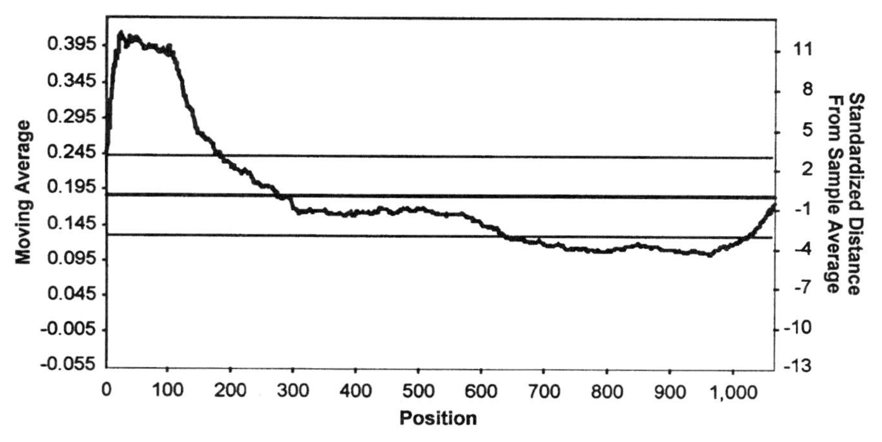

Figure 3.19 $W = 100$ moving-average plot for set 4 (increasing nonstationary Poisson process sample).

3.5 UNIVARIATE IID CASE WITH DATA

3.5.1 Alternative Distribution Forms

The following forms of probability distributions are commonly used in simulation input modeling:

- Standard distributions (e.g., normal or exponential)
- Flexible families (e.g., Johnson or Pearson)
- Empirical distributions

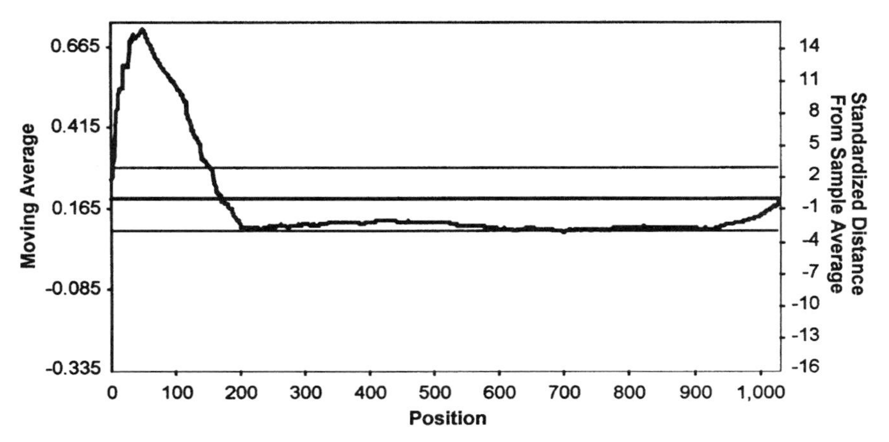

Figure 3.20 $W = 100$ moving-average plot for set 5 (dual-value nonstationary Poisson process sample).

Distributions are classified as being *discrete* when they can produce only a finite or countable number of different values (e.g., 0, 1, 2, ...), and as being *continuous* when they can produce an uncountable number of different values (e.g., all real numbers between 0 and 1). Due to the special attention paid to continuous distributions, they are further classified according to the range of values that they can produce:

- *Nonnegative continuous distributions* take on values in the range (γ, ∞), where γ is typically 0 or some positive value.
- *Bounded continuous distributions* take on values in the range (a, b), where $a < b$ are typically positive values.
- *Unbounded continuous distributions* take on values in the range $(-\infty, \infty)$ (i.e., they are not restricted).

Note that in some references finite endpoints are made inclusive (e.g., "$[a, b]$" for a bounded continuous distribution), which is unimportant since the theoretical probability of obtaining such an endpoint is zero. Table 3.1 lists distributions of each of the types mentioned.

For our purposes, we assume that distributions are defined by their distribution functions, or equivalently, by their related density (continuous) or mass (discrete) functions. All but the most trivial distribution functions are parameterized so that they are more flexible. For example, the normal distribution features two parameters corresponding to the distribution's mean (expected value) and standard deviation (or equivalently, the variance). Such parameters fulfill different roles. *Location* and *scale* parameters allow for any desired units of measurement. For example, if X is a random variable, then the *location and scale transformation* $Y = \gamma + \beta X$ introduces a location parameter γ and a scale parameter β. Such a transformation is useful for translating between different measurement systems, such as metric and English. The aforementioned normal distribution's mean parameter is a location parameter, and its standard deviation parameter is a

Table 3.1 Common Standard Simulation Input Probability Distributions

Nonnegative continuous	Unbounded continuous	Bounded continuous
chi-square	Cauchy	beta
Erlang	error	Johnson S_B
exponential	exponential power	power function
F	extreme value—maximum	triangular
gamma	extreme value—minimum	uniform
inverse Gaussian	Johnson S_U	**Nonnegative discrete**
inverted Weibull	Laplace	geometric
log-Laplace	logistic	logarithmic series
log-logistic	normal	negative binomial
lognormal	Pareto	Poisson
Pearson type 5	Student's t	**Bounded discrete**
Pearson type 6		Bernoulli
random walk		binomial
Rayleigh		uniform
Wald		
Weibull		

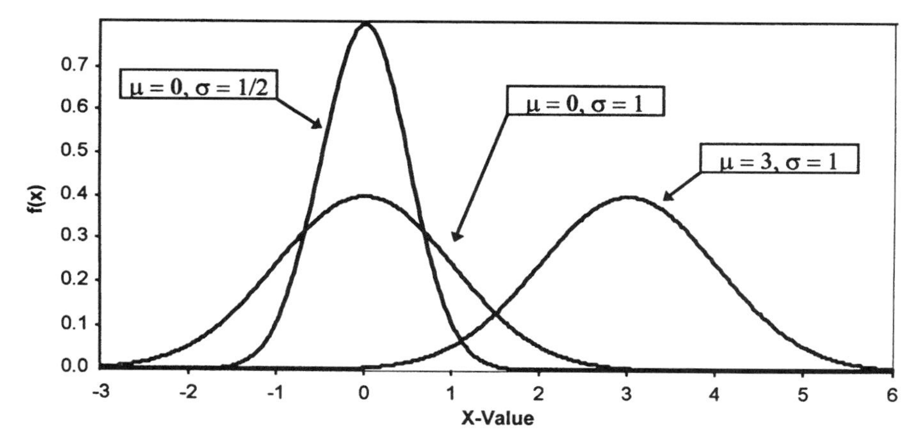

Figure 3.21 Three normal densities.

scale parameter. In Figure 3.21 we display three normal distributions that demonstrate the effects of location and scale parameters. *Location parameters* typically represent the center point (e.g., mean or median) of unbounded distributions and the lower endpoint or threshold of bounded and nonnegative distributions. *Scale parameters* serve the scaling role in the location and scale transformation, except in the case of bounded distributions, where they typically serve as upper endpoint parameters. Notice that in Figure 3.21, regardless of the parameter values chosen, the density always has a bell shape. Shape parameters permit distributions to take on different shapes (i.e., beyond simple changes in location and scale). The most frequently mentioned shape parameter in introductory statistics is the degrees of freedom parameter for a Student t-distribution, which unlike most shape parameters is typically assumed to be a positive integer. Figure 3.22 gives density plots for the Student t-distribution with four different values for the degrees-of-freedom parameter. It should be noted that as the degrees-of-freedom parameter approaches infinity, the density converges to that of a so-called *standard normal distribution*, which has a mean of zero and a standard deviation of 1. In Figure 3.23 we show density functions for four gamma distributions, all with the same mean value, yet with dramatically different shapes.

Standard distributions typically have location and scale parameters, and zero, one, or two shape parameters. "Flexible families" can be viewed as bounded, nonnegative, and unbounded continuous distribution forms that are mathematically related (i.e., one distribution can be derived from another through application of mathematical transformations and/or taking of limits). For example, the Johnson family is based on transformations of the normal distribution (unbounded) to both bounded and nonnegative ranges (see ref. 2). The typical method for applying Johnson distributions is to calculate the value of a special function, which based on its value indicates the "most appropriate" family member; notice that the application-dependent reasonableness of the family member's range (e.g., bounded) is not considered. Although a specific flexible family will work well in certain circumstances, it is not universally optimal. This is clear from the work of Johnson and others, who considered alternatives such as the logistic and Laplace distributions as substitutes for the normal when deriving bounded and nonnegative family

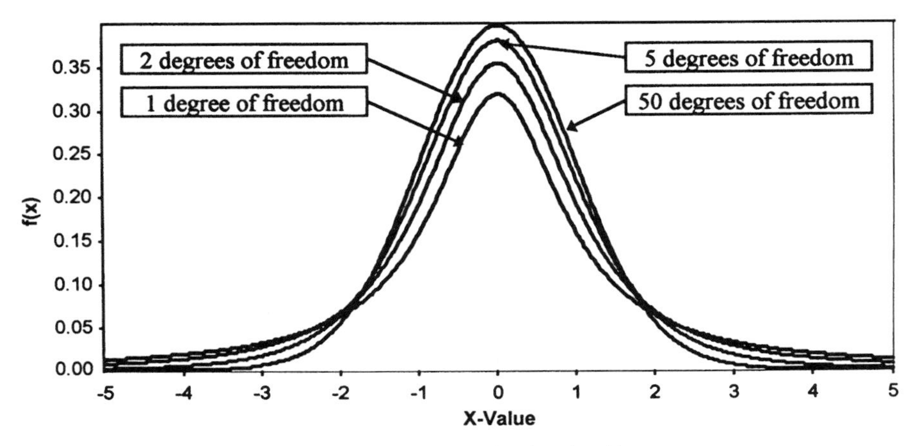

Figure 3.22 Four student's t densities.

members. Rather than limiting oneself to a specific flexible family, we recommend considering each family member to be another standard distribution.

Standard distributions have a number of intuitive positive characteristics. For many of the distributions, their mathematical form can be shown to be the direct consequence of a natural physical phenomenon. For example, in Section 3.2 it was stated that based on minimal assumptions, it can be shown that times between two successive arrivals for many systems is exponentially distributed. In some industries it is common practice to assume specific distributional forms for industry-specific probabilistic analyses.

A standard alternative to fitted parametric models are empirical distributions, for which there are a number of alternative forms. The general form for empirical distribution functions used in standard statistics is

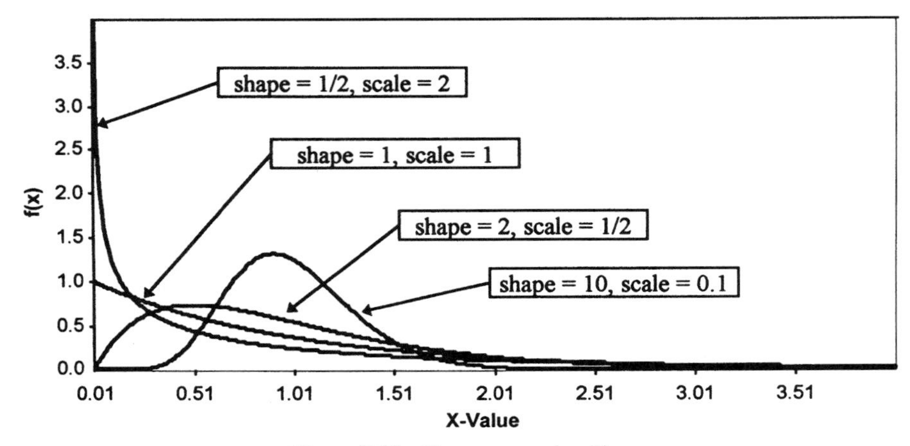

Figure 3.23 Four gamma densities.

$$\text{probability of a value less than or equal to } x = F_E(x;a,b) = \frac{N(x) - a}{n + b} \tag{1}$$

where $N(x)$ is the number of observations no larger than x and n is the number of observations; there may be an additional special case for x smaller than the smallest observation [e.g., $F_E(x;a,b) = 0$]. An intuitive choice for the constants a and b is $a = b = 0$, so that the empirical probability of getting a value less than or equal to the kth largest of n is k/n:

$$F_S(x) = F_E(x;0,0) = \frac{N(x)}{n} \tag{2}$$

$F_S(\cdot)$ is the basis for empirical distributions used to directly represent a data set in a simulation, as described in Section 3.5.4. However, because the empirical probability is 1 for the largest observation, $F_S(\cdot)$ is undesirable in some contexts, such as the heuristic plots described later. Common alternatives involve having one of the constants being nonzero, such as setting $a = \frac{1}{2}$ or $b = \frac{1}{2}$ (see ref. 8). For purposes of explication, we employ the former alternative as our standard empirical distribution function:

$$F_n(x) = F_E\left(x;\tfrac{1}{2},0\right) = \frac{N(x) - \frac{1}{2}}{n} \tag{3}$$

for x at least as large as the smallest observation and $F_n(x) = 0$ otherwise. Note that empirical distribution functions are step functions that rise at each unique observed value proportionally to the number of such values. As a practical note, despite the fact that the theoretical probability of ties is zero, ties occur very frequently in real samples due to the limits of recording accuracy; thus the step size can be larger than $1/(n+b)$. As we shall see in Section 3.5.4, an alternative formalization of the empirical distribution function for continuous distributions replaces the step with a linear interpolation between subsequent points.

Empirical distributions have a number of obvious drawbacks:

- In their standard form, empirical distribution functions can represent only bounded distributions. Further, unless we adjust them (as discussed in Section 3.5.4), they represent only the observed sample range. [Note that although it is possible to add an infinite upper tail to an empirical distribution function, making it a nonnegative distribution, this procedure works well only when we know the general form for the addition (e.g., adding an exponential distribution upper tail to an empirical works well only when the upper tail actually follows an exponential distribution, which we typically do not know).]
- The quality of representation afforded by an empirical distribution is completely dependent on the quality of the sample available. Although empirical distributions are classified as nonparametric models, each unique sample value does serve as a defining parameter of the empirical distribution function; thus the quality of the estimated distribution function relies on the quality of the available sample.
- For small to moderate sample sizes, the upper tail defined by the empirical distribution function can be unreliable due to the small number of values used to define

it; this is particularly true for distributions with long upper tails, which are common in simulation contexts.

- The probability that history repeats itself "exactly" is zero.

For these reasons we prefer standard distributions (or those from flexible families) over empirical distributions in a simulation context when they provide comparable or better representations.

3.5.2 Recommended Strategy for Selecting the Distribution Form

The following discussion draws heavily on the conclusion from Section 3.3 that although collected data provide our best evidence concerning a source of randomness, it may not be a *reliable* set of evidence. Our first step in any data analysis is thus to examine the available data in detail. We should be concerned with detecting the problems discussed in Section 3.3 and with detecting any inconsistencies with an IID assumption as discussed in Section 3.4. It is always a good idea to take time to create a good histogram of the data in order to familiarize ourselves with the "shape" of the data, as well as to establish a reference for any histogram-based comparisons that might be used to evaluate the quality of fit provided by a fitted distribution (as discussed below).

In the remainder of this section we will be using as the basis of our examples a set of 100 repair times recorded as decimal minutes. The repair times range from 3.79 to 125.75 minutes, with a sample average of 31.506 minutes. Examination of the plots similar to those found in Section 3.4 (not shown here) revealed nothing to question the assumptions of an IID sample.

Making histograms is an art and not a science. Although various rules or heuristics have been suggested for determining one or more of the histogram parameters (e.g., starting point, interval width, number of intervals), none will produce an "optimal" (or even a *good*) histogram in *all* cases. (In fact, researchers have to limit carefully the range of applicability before suggesting in print that any rule can be considered optimal in some sense.) Most, if not all statistical packages used for analyzing input data provide a default histogram configuration for a set of data. Use this as a *first approximation* to a good histogram, not the final choice. We suggest iterating through adjusting the starting point, adjusting the interval width, and then setting the number of intervals to cover all the data. (Some data may be well away from the rest, on either end; then they might be ignored during histogram construction.) The most difficult step is choosing an appropriate interval width. Too small a width will produce a ragged histogram, whereas too large a width will produce an overaggregated blocklike histogram, as shown in Figures 3.24 to 3.26. A "good" histogram is a relative concept but generally indicates that the histogram suggests a smooth underlying density shape, such as one of a common density function. However, there is no guarantee that a reasonable histogram can be constructed for any particular data set. If it is difficult to produce a reasonable histogram when there are at least 50 points, it is likely that no standard distribution will provide an outstanding fit to the data.

Once you have become familiar with the data, we suggest the following steps, many of which are detailed further in subsequent sections:

1. Use knowledge of the source of randomness to determine any definite limits on the values it can produce. Be clear about which values, if any, are absolutely impos-

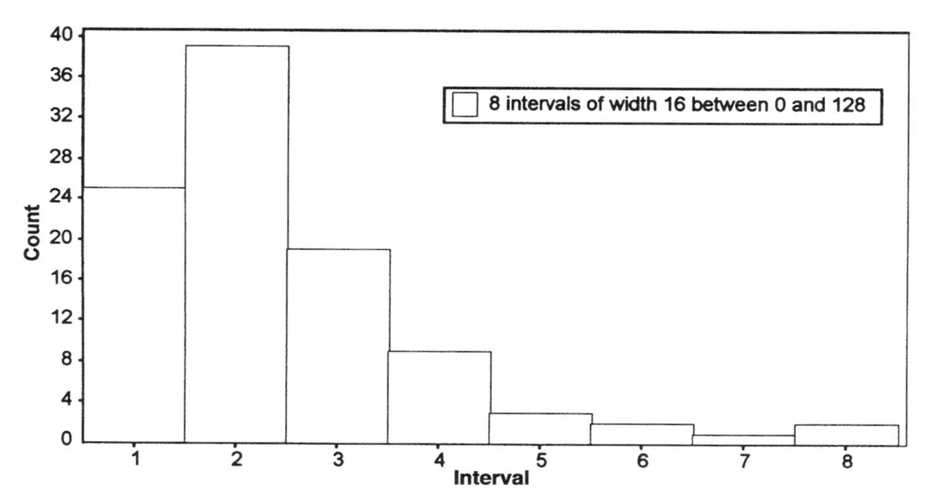

Figure 3.24 "Blocklike" histogram.

sible and thus undesirable in a simulation representation. When there are lower and/or upper bounds on permissible values, be clear on whether some flexibility can be allowed when choosing a probability model for the simulation; for example, whereas in the real world it is impossible to accomplish a task in less than one time unit, it might be perfectly acceptable to use a probability distribution in the simulation that produced values as small as 0.95 time unit. Note that we are suggesting flexibility in terms of larger ranges of possible values—we can't consider ranges that are subsets of that demonstrated by the data.

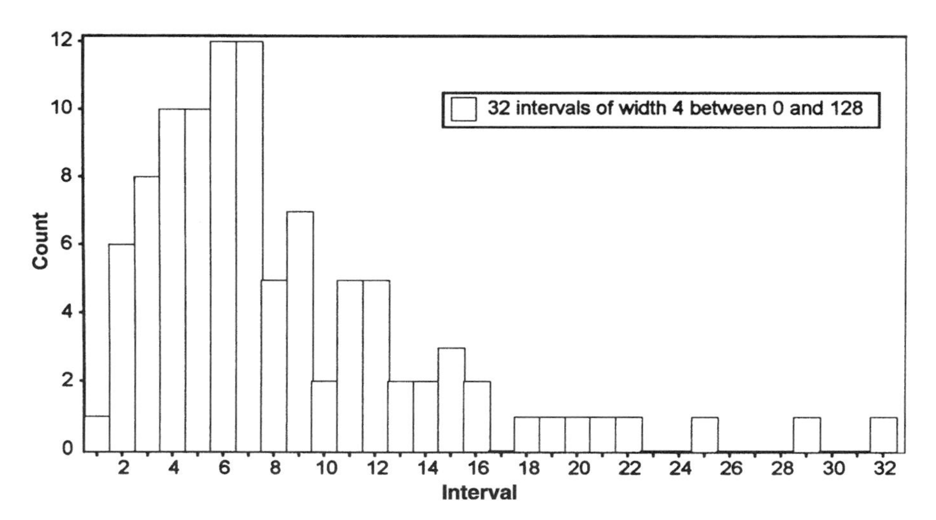

Figure 3.25 "Ragged" histogram.

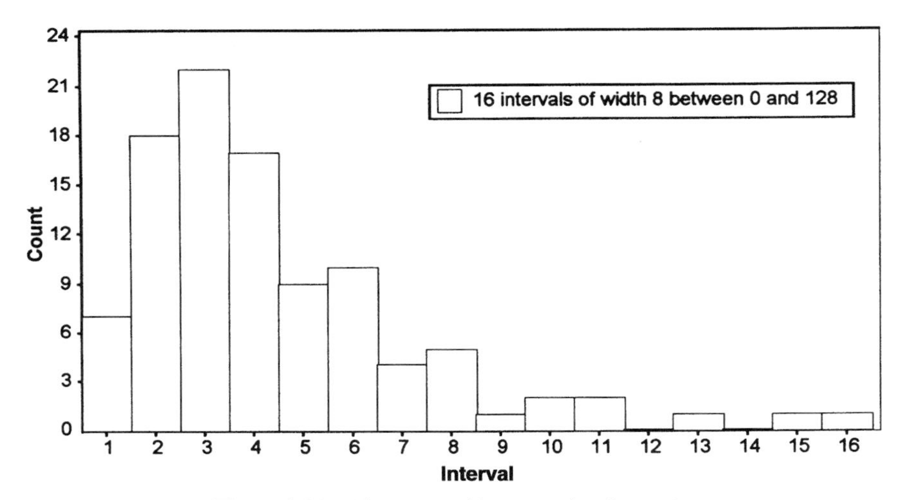

Figure 3.26 Histogram with appropriate intervals.

2. "Fit" as many standard distributions (and members of flexible families) to the data as possible, using ranges not completely inconsistent with those determined in step 1. In general, there is no harm in fitting distributions with broader ranges than we need; we just need to be careful about the final representation.
3. Use a reasonable set of criteria to rank the goodness of fit of the fitted distributions to the data observed.
4. If any of the top-ranked models are terribly inconsistent with the assumed range for source of randomness, rule them out.
5. Use a reasonable set of criteria to determine if the best of the fitted distributions is a reasonable representation for the data.
6. If the best of the fitted distributions provides a reasonable representation of the data, use it in the simulation. Otherwise, use an empirical distribution to represent the data directly.

Each authority on simulation input modeling might have a different "reasonable" set of criteria to apply in the steps outlined above, and the criteria applied might well depend on the type, quantity, and quality of data available. The following general rules of thumb should guide use of the goodness-of-fit options found in simulation input software:

1. Consider a number of measures of goodness of fit rather than a single one, since each goodness-of-fit measure will be unreliable in some cases.
2. Do not depend on goodness-of-fit measures that rely overly on "clean" data samples (e.g., those relatively free of the problems mentioned in Section 3.3) or on user-supplied parameters (such as histogram configurations), since such measures can provide inconsistent results.

3.5.3 Practical Application of Standard Distributions and Flexible Families

Distributions are "fit" to data sets by specifying values for their parameters with the intent of making the distribution "resemble" the data. As an introductory example, consider an exponential distribution that takes on values larger than zero [e.g., the location (threshold) parameter is zero] and has a scale parameter β that represents the mean of the distribution. If we collect a sample of values, an intuitive estimator (formula) for β is that of the sample average. (This estimator is also appropriate if we apply more sophisticated arguments than intuition.) Statisticians have long considered desirable methods for estimating parameters of distributions, and we can use their results, although the context in which statisticians typically consider the problem differs from ours in one important respect: they take the liberty of assuming that they *know* the form of the underlying distribution. The names attributed by statisticians to parameter estimators for particular distributions at times reflect the method used to derive the estimator (e.g., maximum likelihood, method of moments, or quantile), or one of the properties associated with the derived estimator (e.g., best linear unbiased estimator, uniformly minimum variance unbiased estimator) (see ref. 9). In general, we recommend use of maximum likelihood estimators (MLEs). When for technical reasons they are unavailable or have undesirable properties, quantile estimators are often used. For many standard distributions MLEs are not available in explicit forms (e.g., the estimate is the sample average), so iterative solution methods must be employed to find appropriate numerical solutions.

Distributions were fitted to the repair-time data, using a simulation input processing package with the assumption that the range of fitted models should be consistent with positive repair times. A total of 21 models were fitted to the data, and according to the software's internal algorithm for ranking the fitted models, a lognormal distribution provided the best fit. To provide a contrast to this model, an exponential distribution with a much lower ranking will also be shown in the example plots. In all cases the lognormal distribution produces prototypical "good-looking" plots, whereas the exponential produces prototypical "bad-looking" plots.

Once all of the parameters of a distribution have been specified, the distribution can be compared with the sample observed. Many readers may be familiar with goodness-of-fit tests in the form discussed in statistics courses (e.g., application of the chi-squared goodness-of-fit test) and thus may be predisposed to make exclusive use of such methods to compare fitted models with the sample. For the following reasons, in the context of simulation input modeling, these classical goodness-of-fit methods are not completely appropriate for and/or definitive in assessing quality of fit; therefore, more heuristic graphical methods will be emphasized:

1. Classical statistical theories and methods have different premises than those (commonly) appropriate to simulation. As shown in Figure 3.27, classical goodness-of-fit testing assumes that an appropriate (single) probability distribution is hypothesized and then data are collected in a manner appropriate to the hypothesis. In simulation input modeling, however, we almost always select multiple candidate distributions on the basis of a sample that we collected previously. (Some statisticians have considered a situation that lies between the procedural extremes stated. For example, they might assume it is *known* that data are produced by one of k listed distributions and investigate statistical procedures for inferring which of the distributions is the true parent population.) Classical goodness-of-fit testing may or may not hypothesize (before data collection)

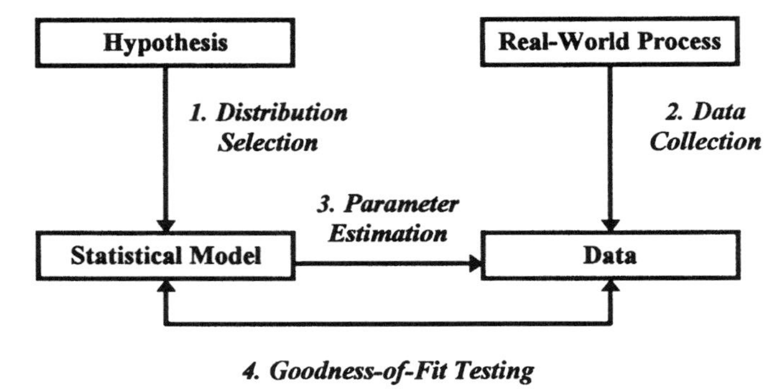

Figure 3.27 Steps in classical goodness-of-fit testing.

some known parameter values, with the remaining being nuisance parameters that must be estimated from the data. In contrast, in simulation input modeling, most if not all parameters are typically estimated from the data (we might assume values for location parameters for nonnegative continuous distributions, for example). Because we make much "looser" assumptions than in classical goodness-of-fit testing, we must assume that goodness-of-fit test results are only approximate. This approximate nature is compounded whenever we consider distribution and/or parameter estimation combinations for which there are no specific classical goodness-of-fit results; this topic is considered more extensively below.

2. When performing classical goodness-of-fit testing we are attempting to ascertain the truthfulness of the distributional hypothesis—we will either reject or fail to reject the hypothesis that the (single, specified) distribution is a "true" representation of how nature behaves. In simulation input modeling, the specific distribution (e.g., gamma) typically is not especially important to us; what is essential is that we have *some* reasonable representation of the underlying source of randomness—we cannot simply state that all input models were rejected.

3. The operational characteristics of classical goodness-of-fit tests do not include a provision for detecting "good-enough" fits, which might be appropriate in a simulation context. As the sample size increases, the sensitivity of a classical goodness-of-fit test to even minute differences increases, to the point that for large sample sizes (e.g., 1000 values) a goodness-of-fit test might reject a candidate model that by all other measures appears to provide a good representation of the data.

Due to the highlighted problems with goodness-of-fit tests, we often rely on more heuristic methods to assess the quality of fit, both in a relative sense (e.g., which is best) and an absolute sense (e.g., whether it is good enough). We now briefly consider a variety of heuristics that can be used to assess quality of fit and discuss their particular benefits and difficulties. We then consider formal goodness-of-fit tests that might be implemented in input analysis software. We consider only continuous random variables in the following discussion because they are much more frequently encountered in simulation practice; not all of these heuristics are available for discrete random variables.

A histogram is an estimate of the underlying probability distribution's density function. It is therefore reasonable to compare the estimate to the density function of a fitted model by plotting the density over the histogram. (As a practical matter, although his-

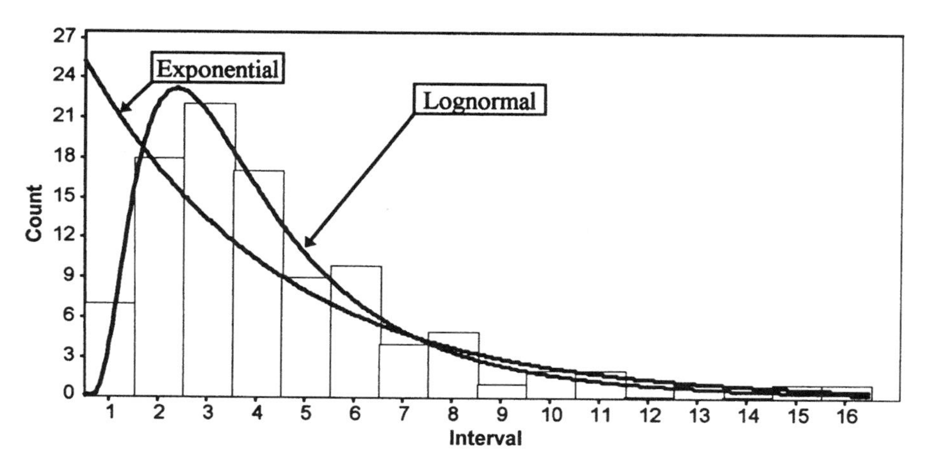

Figure 3.28 Lognormal (good fit) and exponential (bad fit) densities plotted over histogram.

tograms usually represent sample counts, proportions, or percentages, they can easily be converted to density units, and vice versa.) In Figure 3.28 we see densities for two distributions plotted over a histogram. As an alternative, one can determine the frequency attributed to each histogram interval by the fitted model and compare the theoretical (expected) probabilities (or counts) with those observed in the sample. Figure 3.29 shows a comparison between the interval counts from the sample and the expected counts from the lognormal and the exponential distribution. These are intuitive graphs that are immediately appealing. However, they suffer from the shortcoming that simple changes in the histogram configuration can drastically change the plot's appearance and subsequent interpretation of goodness of fit. (This is especially true for small samples and those samples where values have discrete attributes due to limited accuracy in value recording.) Single-valued composites of the quality of fit can be constructed for

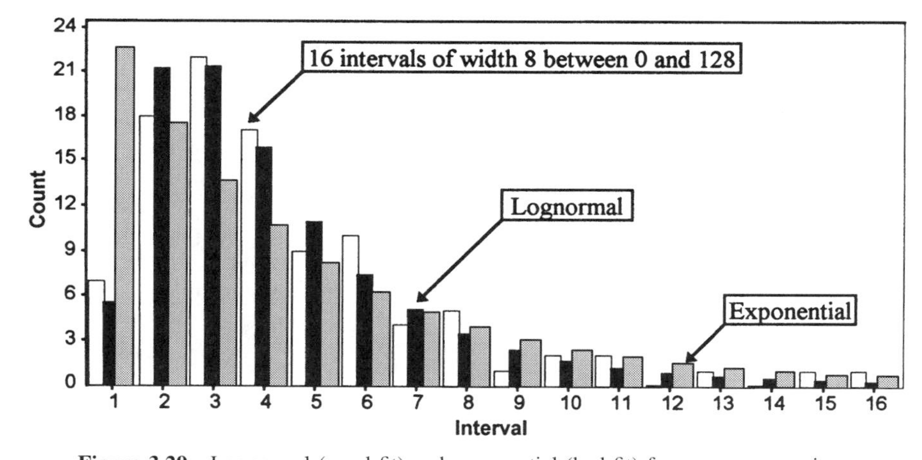

Figure 3.29 Lognormal (good fit) and exponential (bad fit) frequency comparison.

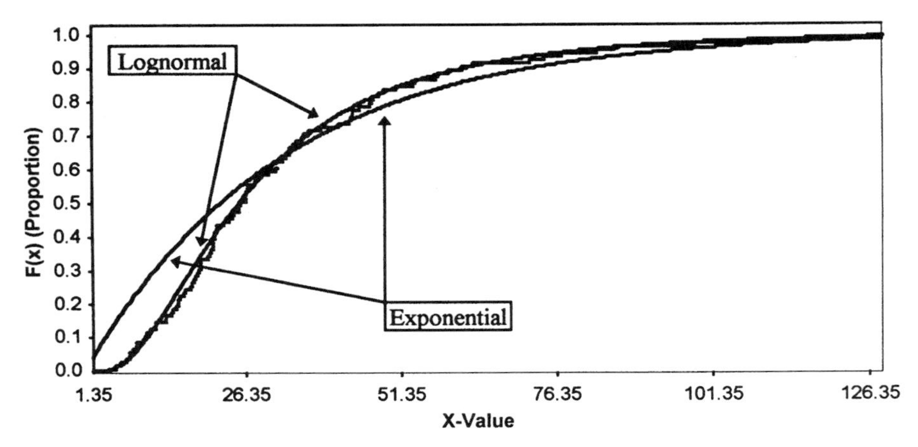

Figure 3.30 Lognormal (good fit) and exponential (bad fit) distribution function plots.

frequency plots, although their use should be limited due to the extreme variability in results with different histogram configurations.

Many heuristics are derived from consideration of the sample and fitted model distribution functions. Figure 3.30 shows the empirical cumulative distribution function $F_n(\cdot)$ [equation (3)] compared with the cumulative distribution functions of the two models. In general, cumulative distribution functions from standard distributions will have an S-shape, consistent with a single-mode density function [it might be noted that the exponential cumulative distribution function looks like the top half only of an S (as shown in Figure 3.30) and the uniform cumulative distribution function is a flattened S]. A number of goodness-of-fit tests are derived from comparisons of the empirical and model distribution functions; the most familiar might be the Kolmogorov–Smirnov test discussed below. Such tests provide simple numerical representations of the graphical comparison that can serve as the basis for ranking alternative candidate distributions. A number of alternative heuristics can be derived from the simple distribution function plot.

The simplest heuristic plots the differences ("errors") between the empirical and model cumulative probabilities at each of the unique observations. The x-axis remains the same as in the original plot, but the y-axis is changed to reflect positive and negative differences that necessarily must be smaller in absolute value than 1. A reference line corresponding to a difference of zero is very useful. In general, the closer the plot is to the reference line, the better the quality of fit. However, no plot will be perfect, so some standard should be derived that can be used to determine whether a particular difference is "large." Reasonable limit values can be derived from a number of goodness-of-fit theoretical results; these limits will, in general, narrow as the sample size increases. In Figure 3.31 we show the differences between the empirical distribution function and the distribution functions for the fitted distributions; the lines at approximately ± 0.13 are limit values provided by the analysis software. It is possible to derive single-valued composites of the graphed results, such as the largest absolute difference or the average absolute difference, that are suitable for use in ranking candidate models. This is one of the author's favorite plots for assessing both relative and absolute aspects of quality of fit.

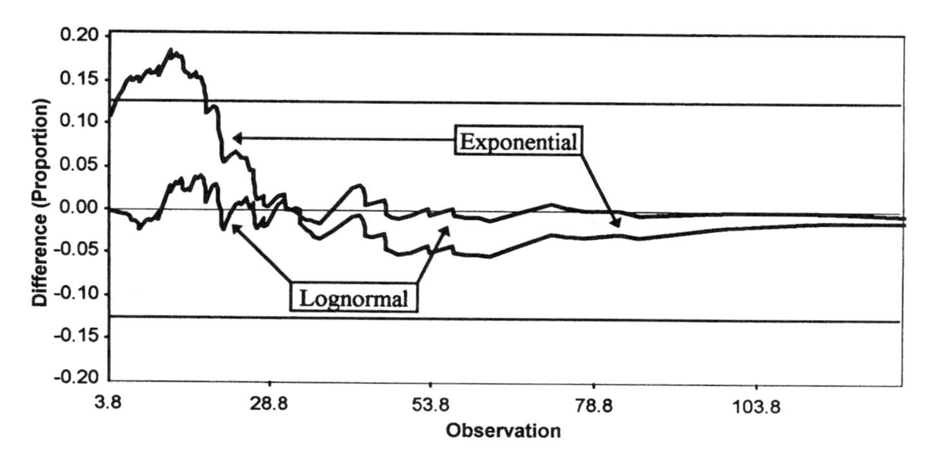

Figure 3.31 Distribution function differences for lognormal (good fit) and exponential (bad fit).

A P–P (probability–probability) plot compares the probabilities derived from the empirical distribution and the fitted model; for each unique and ordered observation x_k we plot

$$\text{abscissa} = F_n(x_k) \qquad \text{ordinate} = F_M(x_k) \tag{4}$$

where $F_M(\cdot)$ is the distribution function of the model under consideration. It is natural to plot these points on a grid with both axes ranging from zero to 1 and to include a reference diagonal line from (0,0) to (1,1), or, as will be shown in Figure 3.32, that corresponds to the sample probability range $\left(1 - \frac{1}{2}\right)/n$ to $\left(n - \frac{1}{2}\right)/n$. When the model and sample correspond closely, the x-values will be close together and thus close to the diagonal line. In general, the closer the resulting plot to the reference diagonal line, the better the quality of fit. Due to their construction, P–P plots are most sensitive to differences in the "centers" of the distributions (e.g., not the extremes or "tails"). P–P plots for the fitted lognormal and exponential distributions are shown in Figure 3.32. It is possible to derive numerical measures of the closeness of the plotted points to the reference line, either from a regression framework or from a simple spatial interpretation of the graph. In fact, goodness-of-fit tests can thus be derived from these plots.

A Q–Q (quantile–quantile) plot compares x-values (quantiles) derived from the empirical distribution and the fitted model; for each unique and ordered observation x_k we plot

$$\text{abscissa} = x_k \qquad \text{ordinate} = F_M^{-1}[F_n(x_k)] \tag{5}$$

where $F_M^{-1}(\cdot)$ is the inverse distribution function of the model under consideration:

$$\text{for } 0 < p < 1 \qquad \text{if } x = F_M^{-1}(p), \text{then } p = F_M(x) \tag{6}$$

It is natural to plot these points on a grid with both axes having the same range, cor-

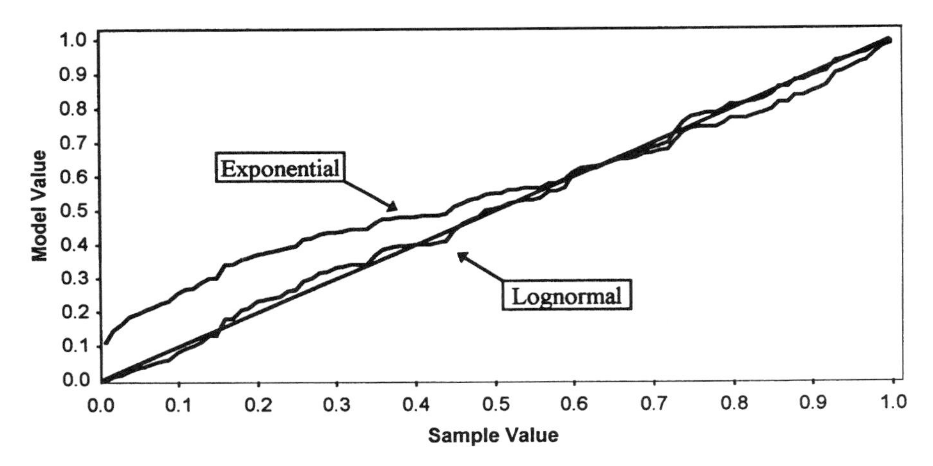

Figure 3.32 *P–P* plots for the lognormal (good fit) and exponential (bad fit).

responding to the minima/maxima of the extreme sample and model x-values, and to include a reference diagonal line. When the model and sample correspond closely, the quantities will be close together and thus close to the diagonal line. In general, the closer the resulting plot to the reference diagonal line, the better the quality of fit. Due to their construction, Q–Q plots are most sensitive to differences in the "tails" of the distributions. In general, the upper tail corresponding to large values will be the source of the largest problems for these plots. It is common in simulation to have distributions with relatively "thick" upper tails (e.g., one where there is significant probability in the upper tail, which contributes significantly to the expected value). Random samples from such distributions can produce very bad Q–Q plots, due to a few of the largest values. In general, we recommend being cautious about Q–Q plots: If they look "good," take that as confirming evidence of good quality of fit; if they look "bad," do not deem the model to be a bad model on the basis of this plot *alone*, particularly if the problems arise from just a few upper tail points. In Figure 3.33 we see Q–Q plots for the two distributions. Note that in general the plot for the lognormal is very close to the reference line, which extends over the range of the observations and deviates only at the very upper tail. This is a remarkably good-looking Q–Q plot for a thick-tailed distribution like the lognormal. Although single-valued numerical composites of the plot can be derived for the purposes of ranking, these typically are not converted into goodness-of-fit tests.

As indicated earlier in the section, the author values heuristics over goodness-of-fit tests when performing simulation input modeling. Yet these tests have their uses when properly applied and interpreted. Unfortunately, goodness-of-fit tests are not given reasonable treatment in many statistical texts and are sometimes misimplemented by software developers. For these reasons, a rather detailed discussion of goodness-of-fit tests is included for the interested reader.

Goodness-of-fit tests are specialized forms of hypothesis tests in which the null hypothesis specifies the distribution that produced the data (e.g., "the data were produced by a normal distribution"). The strongest null hypothesis used in goodness-of-fit testing, and unfortunately the only one typically covered in most introductory statistics courses, states that we not only know the distribution that produced the data but also the param-

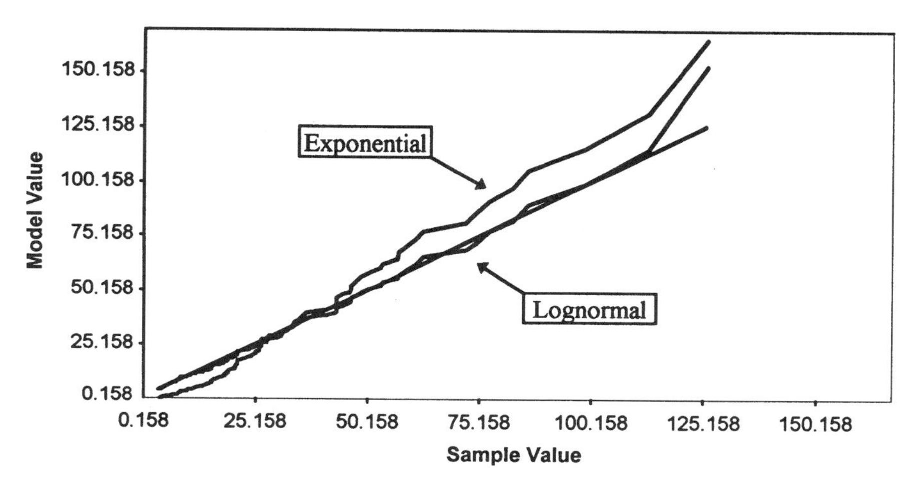

Figure 3.33 *Q–Q* plots for the lognormal (good fit) and exponential (bad fit).

eters of the distribution (e.g., "the data were produced by a normal distribution with known mean μ_0 and known standard deviation σ_0"). We will consider this strong hypothesis in detail before considering more realistic weaker alternatives.

The operational procedure of a goodness-of-fit test is to calculate a function of the data observed and the specified model, $T(x)$, called a *test statistic*, and then to compare the value of the test statistic to what we call a *critical value*: If the statistic exceeds the critical value, there is sufficient evidence to reject the stated null hypothesis. As in most statistical testing, a goodness-of-fit test cannot prove that a sample was produced by a specified distribution; we can only state that there was insufficient evidence to reject such a hypothesis. This process can be rationalized as follows: Goodness-of-fit test statistics are usually defined to measure errors between sample and model, such that the larger the statistic value, the worse the error and correspondingly, the worse the fit. Because $T(x)$ is a function of random variables (e.g., the sample values), it too is a random variable. Even when the hypothesis is true, there will be measurable errors between the observations and the parent population, and therefore in a test we must make some allowance for the natural occurrence of such errors: We should dismiss the reasonableness of a hypothesis only when faced with extreme errors. Since the distribution of the original sample values is completely known, it is often possible to derive an exact distribution for $T(x)$, which typically depends on the sample size only; however, it may also depend on the hypothesized distribution. We can therefore set probabilistic limits on the values of $T(x)$ that are acceptable—when a "level α test" is performed, we allow $T(x)$ to range over $100(1 - \alpha)\%$ of its range before rejecting the hypothesis. Here α is the probability that we reject the hypothesis even though it is true, called *type I error*. Since in goodness-of-fit testing $T(x)$ is typically a nonnegative function, there is single critical value (upper limit) on $T(x)$ that is employed [the convention for unbounded functions is to distribute $100(\alpha/2)\%$ into each tail].

We can weaken the null hypothesis to make it more reasonable by removing the assumption that we know the distribution's parameter values. This creates the immediate operational dilemma of how to calculate $T(x)$, since the parameters of a specified distribution are required. For the test results discussed in this chapter, statisticians have assumed that maximum likelihood estimators (or at least estimators with similar properties) replace the unknown parameters in the statistic, which we now denote $T(x,\theta)$ where θ represents the estimated parameters. The introduction of parameter estimators severely complicates the underlying problem for statisticians, because the parameter estimators themselves are random variables that must be accounted for when considering the distribution of $T(x,\theta)$. In general, the distribution of $T(x,\theta)$ is *not* the same as that of $T(x)$; it typically depends on the distribution being considered as well as values of the unknown parameters (we shall see one major exception). The aforementioned complications have proved to be difficult obstacles for statisticians, so there are useful results for relatively few distributions.

Before continuing, it should be emphasized that in simulation input modeling we typically violate even the weak null hypothesis if we do not have a fixed single hypothesized distribution choice before data collection. Even when we ignore this difficulty, an additional problem arises in the normal course of analyzing data: the *multiple comparisons* problem. Test results are not independent of each other if we perform multiple goodness-of-fit tests, regardless of whether we perform different tests with the same distributional null hypothesis or the same test with different distributional null hypotheses. The lack of independence violates the desired test level on each of the tests performed.

We now consider three general goodness-of-fit tests and summarize the availability of appropriate critical values. The most commonly encountered goodness-of-fit tests are based on histograms intervals (e.g., chi-square) or on the empirical distribution function (e.g., Kolmogorov–Smirnov). In a chi-squared goodness-of-fit test we divide the range of the random variable under consideration into m intervals:

$$(C_0,C_1), [C_1,C_2), [C_2,C_3), \ldots, [C_{n-1},C_m) \tag{7}$$

(note that the interval number is taken to be the upper bound index, so we have m intervals numbered from 1 to m). The simplest way to create such intervals is to start with a set of histogram intervals and to extend the first and last intervals to cover the entire range of the random variable. Here all but possibly the first and last intervals will share a common width $W = C_k - C_{k-1}$; be aware that the first and last intervals might be infinite in size in order to match the overall range of the distribution under consideration. For each interval we can count the number of observations falling into the interval, which we can denote O_k. The number of values that we might "expect" from the random variable is the number of sample values times the probability that the random variable places into the interval, which we can denote E_k. The chi-squared goodness-of-fit test statistic is then calculated as the sum of the error terms:

$$\sum_{k=1}^{m} \frac{(O_k - E_k)^2}{E_k} \tag{8}$$

We can consider the numerator to be a squared-error term and the denominator to be a weight that is inversely proportional to the number of "expected" points; the most weight is placed on intervals corresponding to "rare" events! Clearly, changing the intervals will alter both the O_k and E_k values, and thus the overall statistic can vary considerably and unpredictably. Under the strong null hypothesis, when certain conditions (especially

IID) are met, the test statistic follows a chi-squared distribution with $(m-1)$ degrees of freedom. This holds true for any choice of distribution, which makes this test universally applicable.

The distributional result for the test statistic under the weaker null hypothesis is not quite as straightforward. Here we know that the distribution of the test statistic, when certain conditions are met, is bounded between two chi-squared distributions, one with $(m-1)$ degrees of freedom and the other with $(m-p-1)$ degrees of freedom, where p is the number of estimated parameters. This result also holds regardless of the choice of distribution, which makes the test uniformly applicable.

For continuous distributions it is possible to choose the intervals in such a way as to make each of the intervals hold exactly the same number of "expected" points by using the interval bounds:

$$C_k = F_M^{-1}\left(\frac{k}{m}\right) \tag{9}$$

(e.g., for $m = 10$ intervals, we use C_k values that correspond to the 0th, 10th, 20th, ... , 90th, and 100th percentiles of the distribution under consideration). Although use of such "equal-probable" intervals can help to prevent the undue influence of intervals corresponding to "rare" events, it cannot change the inherent variability of the test—the test statistic can vary considerably with minor changes in the number of equally probable intervals. In the author's experience, it is generally possible to produce conflicting test results with any data sample and model, simply by changing interval configurations (even with equally probable intervals), and thus any strategy that attempts to compare the chi-squared goodness-of-fit test results of different models should not be trusted implicitly.

There has been serious consideration in the statistical literature of the conditions under which chi-squared goodness-of-fit tests are reliable. These results are summarized by Moore [10], who proposes a number of guidelines for application of the test. In general, equal-probable intervals should be used, and the "expected" number of model points per interval should be reasonable (say, 3 to 5). Even when these conditions are met, the test suffers from some well-known deficiencies. Since we are aggregating data into intervals, we are losing information; so other goodness-of-fit tests that do not use intervals are preferable (statistically more powerful). As mentioned in the preceding paragraph, any result that depends on histogram intervals can produce vastly different results when the interval configuration changes. The strongest advantage to the test is its universal applicability.

The *Kolmogorov–Smirnov* (K-S) and *Anderson–Darling* (A-D) tests both directly compare a hypothesized model's distribution function to the discontinuous empirical distribution function $F_S(\cdot)$ [Equation (2)]. Despite their disparate origins, both tests can be calculated by considering n sample points. We will follow the notation of Stephens [11], who defines

$$z_i = F_M(x_i) \tag{10}$$

and then uses a parenthetical subscript, $z_{(k)}$, to indicate the values in sorted order.

The K-S test looks for the supremum of the differences between the two distribution functions. As shown in Figure 3.34, at each sample value the empirical distribution has

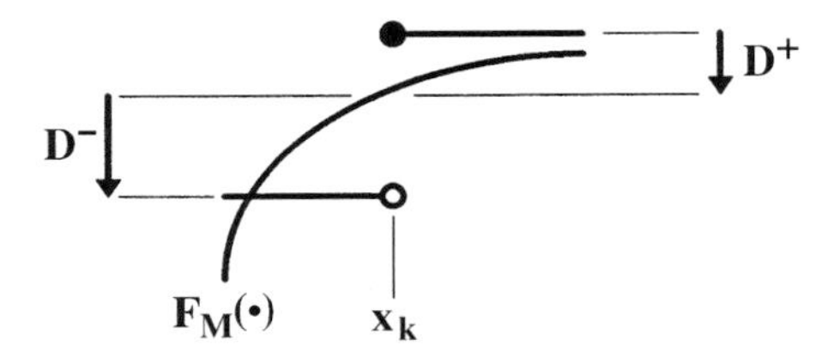

Figure 3.34 Kolmogorov–Smirnov test statistic components.

a discontinuity, and therefore there is both a "from the left" difference and a "from the right" difference at each point to consider. The test statistic can be calculated as

$$T_{\text{KS}}(x) = D = \max(D^+, D^-) \tag{11}$$

where

$$D^+ = \max_k \left\{ \frac{k}{n} - z_{(k)} \right\} \quad \text{and} \quad D^- = \max_k \left\{ z_{(k)} - \frac{k-1}{n} \right\} \tag{12}$$

Under the strong null hypothesis there is a single distributional result for $T_{\text{KS}}(x)$ that applies to all hypothesized distributions; Stephens provides a handy approximation method that depends only on the sample size.

This test result cannot be extended to the case where parameters are estimated, so each distributional choice in a weak hypothesis must be considered separately (although some results can be applied to distributions that have functional relationships, such as the normal/lognormal or extreme value/Weibull). Stephens presents a thorough review of the critical value tables available and provides a number of handy approximation methods.

The A-D test statistic is defined as a weighted average of the squared difference between the distribution functions:

$$n \int_{-\infty}^{\infty} \frac{[F_S(x) - F_M(x)]^2}{F_M(x)[1 - F_M(x)]} f_M(x)\, dx \tag{13}$$

The A-D statistic places greatest emphasis on differences found in the tails. Fortunately, the test statistic can be calculated using the much simpler formula:

$$T_{\text{AD}}(x) = A^2 = -\left[n + \frac{1}{n} \sum_k [(2k-1)\log\, z_{(k)} + (2n+1-2k)\log(1 - z_{(k)})] \right] \tag{14}$$

Under the strong null hypothesis the distribution of the test statistic varies little for samples of size at least 5. As with the K-S goodness of fit test, critical values must be determined for each distributional assumption under the weak null hypothesis. Stephens summarizes the available results for the A-D test. It is generally accepted that the A-D test provides better results than the K-S when appropriate critical values for both tests are available.

It is possible to use goodness-of-fit test statistic values as heuristic indicators of fit, since the statistics are designed to measure the errors between fitted models and the data available; the author relies heavily on test statistic values when comparing the relative quality of fit afforded by alternative distributions. Note that simulation and classical statistical practices diverge in this case. A ranking procedure based on goodness-of-fit statistic values is statistically improper unless the critical values applicable to each distribution are identical; a statistician would prefer to compare the observed levels of the tests (the largest level at which the test does not reject the null hypothesis).

There are a number of nongraphical heuristics that can be used to compare candidate distributions and to help determine whether a particular fitted distribution provides a reasonable fit. Log-likelihood functions, which are central to the maximum likelihood parameter estimation method, are defined for continuous distributions:

$$L(x) = \log\left[\prod_k f(x_k)\right] = \sum_k \log[f(x_k)] \tag{15}$$

These values can be calculated for each candidate distribution and used as a ranking mechanism. The log-likelihood function comparisons have natural appeal to a statistician and work very well in a simulation context. Characteristics of fitted distributions such as quantiles, probabilities of exceeding specific values, means, and variances can be compared with sample values or with reference values. The author believes that it is essential to check such values when assessing distributional reasonableness.

3.5.4 Practical Application of Empirical Distributions

The empirical distribution function $F_S(\cdot)$ defined by equation (2) is applicable only to the discrete case in a simulation context. For continuous random variables the distribution function must be modified so that a linear interpolation of distribution values occurs between the unique observations. (Note that this is actually intuitive if one takes the sample histogram to be the shape of the empirical density function; constant density in an interval implies a linear rise in the distribution across the interval.) The simulation-compatible empirical distribution function for the sample repair times is shown in Figure 3.35.

It is possible to extend one or both tails of an empirical distribution function to overcome the deficiency of having the limits of the distribution correspond exactly to the observed sample range. To do so, one or two artificial sample points are added to the list of unique sample values, which increases the number of points used in the definition of $F_S(\cdot)$. This is a reasonable course of action in simulation where nonnegative continuous models are most common; we would choose as the new lower bound either zero or a known limit. In Figure 3.36 we show an empirical distribution function extended to cover the range [0,150].

The larger the number of unique sample values used in the definition of $F_S(\cdot)$, the less time efficient will be the corresponding generator used in the running simulation.

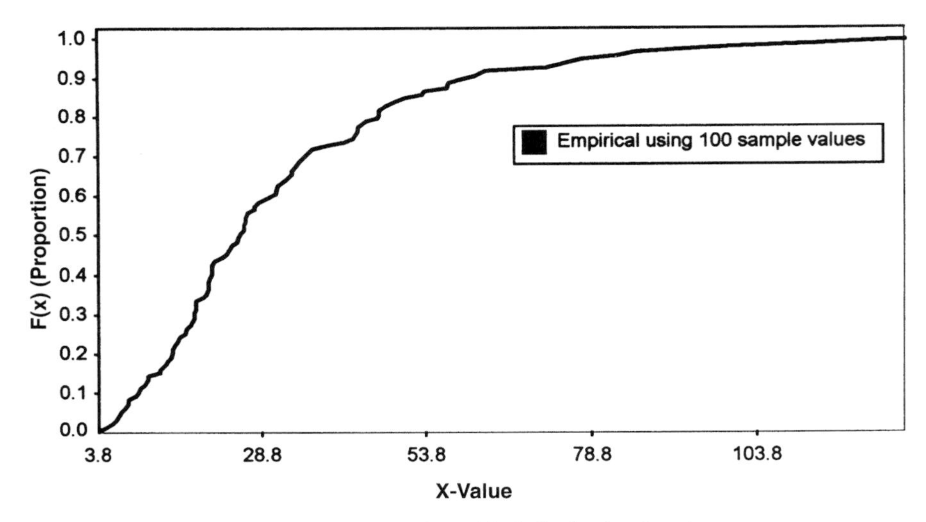

Figure 3.35 Standard empirical distribution function.

To minimize the impact of large sample sizes and to provide a degree of smoothing of the distribution, one can define alternative empirical distributions that aggregate sample values into intervals just as with the chi-squared test, either with equal width or with (approximately) equal probability.

One consideration when using continuous empirical distribution functions is that the expected value of the empirical will not necessarily match the sample mean (due to the linear interpolation in the intervals). Since the mean value may play an important role in the simulation, the empirical mean should be assessed for reasonableness.

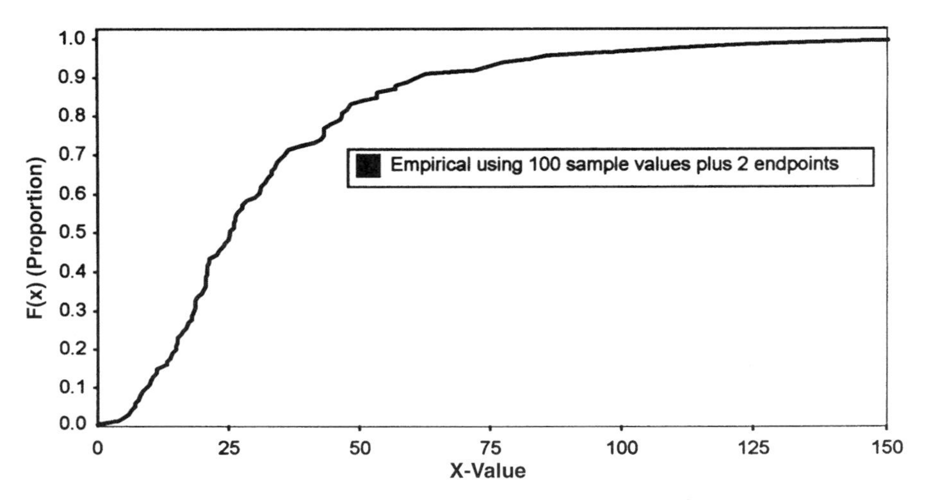

Figure 3.36 Extended empirical distribution function.

3.6 APPROACHES FOR HANDLING THE NON-IID CASE

Once the strong assumptions of IID are abandoned, there is very little definitive guidance available for the simulationist. In fact, for the most part only partial methodologies are available, and the supporting literature often serves as a demonstration of the method's usefulness for a specific data set or class of generated data sets. Due to the difficulty of the underlying statistical issues, few alternative methodologies exist and so comparative performance reviews are not available.

One of the most promising areas of research is the derivation of methods for estimating and assessing the goodness of fit of models applicable to certain forms of nonstationary processes. Kuhl et al. [12] report on methods for fitting a Poisson process (i.e., exponential interevent times) where the exponential rate function can contain trend and/or periodicity components. Such components are particularly germane to the simulation of arrival processes to many systems. Once practical methods exist for handling this case, there is the promise that methods could be developed to handle more complicated interevent models such as distributions with shape parameters (e.g., gamma, Weibull). It would then be possible to consider comparing the qualities of fit provided by alternative models for the nonstationary case.

3.7 WHAT TO DO WHEN NO DATA ARE AVAILABLE

Simulation textbooks have very little to say on the topic of what to do when no data are available. The cause is the vast majority of research in the mainstream statistical and simulation communities and subsequent software development by vendors concerns the cases where data *are* available. Although unfortunate, the truth is that we do not have adequate methodologies to handle reliably the case where data are not available. In this section we consider a few of the general approaches to this situation and discuss their applicability to simulation practice.

If a simulationist is *very* fortunate, he or she will be able to apply confidently a model chosen for a different context or a generally applicable model. The model could be standard for an industry or company, deduced from stochastic processes results (e.g., Poisson process arrivals), based on previous simulation input analyses or derived from an analysis of data collected from a different system. The parameters of the model might need to be adjusted to reflect the differing circumstances (e.g., increase the mean by 10%). The key is that there is sufficient reason to be *confident* that the selected distribution is reasonable for the new context.

If a simulationist is *less* fortunate, he or she will have access to one or more people with some *expert knowledge* of actual or expected behavior. In this situation the simulationist must elicit a number of subjective probability assessments from the expert(s), which can be difficult to accomplish and often produces unreliable results [13]. There are two opposing tendencies that must be balanced by the simulationist. The larger the number of probability assessments required, the more difficult and error-prone the elicitation process, whereas the larger the number of probability assessments available, the more refined the resulting probability model. The crudest model applicable to this situation is a uniform distribution, for which subjective assessments of a minimum and maximum value are required, where the range reflects *close* to 100% of occurrences. The uniform distribution represents a noninformative assessment of the likelihood of different values [14]. If a "central" value such as the mode (most likely) or mean can be elicited, it is possible to derive a triangular distribution, which supplies a more reasonable distributional shape. Although deriving an estimate of the central value typically proves most difficult for experts, finding a reasonable maximum value is often as difficult and no less critical. Specification of triangular distributions in this manner is almost universally recommended by authors of simulation textbooks, despite the fact that the resulting distribution will be grossly in error unless the true underlying probability distribution features the triangular shape and lacks a significant upper tail (an unlikely event). If two percentiles or both "central" values (e.g., mode and mean) can be assessed, an appropriate beta distribution can often be derived (there are some cases where the beta is infeasible). The resulting probability distribution is intuitively more appealing than the triangular, since the shape of a beta distribution can be dramatically different than that of a triangle. However, because this method involves use of only four subjectively assessed values, there is no guarantee that the resulting model would be reasonable, even were accurate values available. Practical experience suggests that many experts have difficulty producing the required information; further, there is often significant variation in answers produced by multiple experts. In the rare situation where a reliable and cooperative expert is available, other distributions can be "fitted" visually using simulation input software. This might involve manipulation of a particular distribution's parameters or deriving a special form of empirical distribution function from assessments of multiple percentiles or probabilities (e.g., a Bezier distribution [15]).

REFERENCES

1. Johnson, N. L., S. Kotz, and A. W. Kemp (1992). *Univariate Discrete Distributions*, 2nd ed., Houghton Mifflin, Boston.
2. Johnson, N. L., S. Kotz, and N. Balakrishnan (1994). *Continuous Univariate Distributions*, Vol. 1, 2nd ed., Houghton Mifflin, Boston.
3. Johnson, N. L., S. Kotz, and N. Balakrishnan (1995). *Continuous Univariate Distributions*, Vol. 2, 2nd ed., Houghton Mifflin, Boston.
4. D'Agostino, R. B., and M. A. Stephens (1986). *Goodness-of-Fit Techniques*, Marcel Dekker, New York.
5. Nelson, B. L. (1995). *Stochastic Modeling: Analysis and Simulation*, McGraw-Hill, New York.
6. Johnson, M. E. (1987). *Multivariate Statistical Simulation*, Wiley, New York.
7. Vincent, S., and A. M. Law (1995). ExpertFit: total support for simulation input modeling, in *Proceedings of the 1995 Winter Simulation Conference*, C. Alexopoulos, K. Kang, W. R. Lilegdon, and D. Goldsman, eds. Association for Computing Machinery, New York, pp. 395–401.
8. D'Agostino, R. B. (1986). Graphical analysis, in *Goodness-of-Fit Techniques*, R. B. D'Agostino and M. A. Stephens, eds., Marcel Dekker, New York, pp. 7–62.
9. Lehmann, E. L. (1983). *Theory of Point Estimation*, Wiley, New York.
10. Moore, D. S. (1986). Tests of the chi-squared type, in *Goodness-of-Fit Techniques*, R. B. D'Agostino and M. A. Stephens, eds. Marcel Dekker, New York, pp. 63–96.
11. Stephens, M. A. (1986). Tests based on EDF statistics, in *Goodness-of-Fit Techniques*, R. B. D'Agostino and M. A. Stephens, eds. Marcel Dekker, New York, pp. 97–194.
12. Kuhl, M. E., J. R. Wilson, and M. A. Johnson (1997). Estimating and simulating Poisson processes having trends or multiple periodicities, *IEE Transactions*, Vol. 29, No. 3, pp. 201–211.
13. Hogarth, R. M. (1980). *Judgement and Choice*, Wiley, New York.

14. Banks, J., J. S. Carson II, and B. L. Nelson (1996). *Discrete-Event System Simulation*, 2nd ed., Prentice Hall, Upper Saddle River, N.J.

15. Wagner, M. A. F., and J. R. Wilson (1996). Recent developments in input modeling with Bezier distributions, in *Proceedings of the 1996 Winter Simulation Conference*, J. M. Charnes, D. J. Morrice, D. T. Brunner, and J. J. Swain, eds., Association for Computing Machinery, New York, pp. 1448–1456.

CHAPTER 4

Random Number Generation

PIERRE L'ECUYER
Université de Montréal

4.1 INTRODUCTION

Random numbers are the nuts and bolts of simulation. Typically, all the randomness required by the model is *simulated* by a random number generator whose output is *assumed* to be a sequence of independent and identically distributed (IID) $U(0, 1)$ random variables [i.e., continuous random variables distributed uniformly over the interval (0,1)]. These *random numbers* are then transformed as needed to simulate random variables from different probability distributions, such as the normal, exponential, Poisson, binomial, geometric, discrete uniform, etc., as well as multivariate distributions and more complicated random objects. In general, the validity of the transformation methods depends strongly on the IID $U(0, 1)$ assumption. But this assumption is *false*, since the random number generators are actually simple deterministic programs trying to fool the user by producing a deterministic sequence that *looks* random.

What could be the impact of this on the simulation results? Despite this problem, are there "safe" generators? What about the generators commonly available in system libraries and simulation packages? If they are not satisfactory, how can we build better ones? Which ones should be used, and where is the code? These are some of the topics addressed in this chapter.

4.1.1 Pseudorandom Numbers

To draw the winning number for several million dollars in a lottery, people would generally not trust a computer. They would rather prefer a simple physical system that they understand well, such as drawing balls from one or more container(s) to select the successive digits of the number (as done, for example, by Loto Quebec each week in Montreal). Even this requires many precautions: The balls must have identical weights and sizes, be well mixed, and be changed regularly to reduce the chances that some numbers come out more frequently than others in the long run. Such a procedure is

Handbook of Simulation, Edited by Jerry Banks.
ISBN 0-471-13403-1

clearly not practical for computer simulations, which often require millions and millions of random numbers.

Several other physical devices to produce random noise have been proposed and experiments have been conducted using these generators. These devices include gamma ray counters, noise diodes, and so on [47, 62]. Some of these devices have been commercialized and can be purchased to produce random numbers on a computer. But they are cumbersome and they may produce *unsatisfactory* outputs, as there may be significant correlation between the successive numbers. Marsaglia [90] applied a battery of statistical tests to three such commercial devices recently and he reports that all three failed the tests spectacularly.

As of today, the most convenient and most reliable way of generating the random numbers for stochastic simulations appears to be via deterministic algorithms with a solid mathematical basis. These algorithms produce a sequence of numbers which are in fact not random at all, but seem to behave like independent random numbers; that is, like a realization of a sequence of IID $U(0, 1)$ random variables. Such a sequence is called *pseudorandom* and the program that produces it is called a *pseudorandom number generator*. In simulation contexts, the term *random* is used instead of *pseudorandom* (a slight abuse of language, for simplification) and we do so in this chapter. The following definition is taken from L'Ecuyer [62, 64].

Definition 1 A *(pseudo)random number generator* is a structure $\mathcal{G} = (S, s_0, T, U, G)$, where S is a finite set of *states*, $s_0 \in S$ is the *initial state* (or *seed*), the mapping $T : S \to S$ is the *transition function*, U is a finite set of *output* symbols, and $G : S \to U$ is the *output function*.

The state of the generator is initially s_0 and evolves according to the recurrence $s_n = T(s_{n-1})$, for $n = 1, 2, 3, \ldots$. At step n, the generator outputs the number $u_n = G(s_n)$. The u_n, $n \geq 0$, are the *observations*, and are also called the *random numbers* produced by the generator. Clearly, the sequence of states s_n is eventually periodic, since the state space S is finite. Indeed, the generator must eventually revisit a state previously seen; that is, $s_j = s_i$ for some $j > i \geq 0$. From then on, one must have $s_{j+n} = s_{i+n}$ and $u_{j+n} = u_{i+n}$ for all $n \geq 0$. The *period* length is the smallest integer $\rho > 0$ such that for some integer $\tau \geq 0$ and for all $n \geq \tau$, $s_{\rho+n} = s_n$. The smallest τ with this property is called the *transient*. Often, $\tau = 0$ and the sequence is then called *purely periodic*. Note that the period length cannot exceed $|S|$, the cardinality of the state space. Good generators typically have their ρ very close to $|S|$ (otherwise, there is a waste of computer memory).

4.1.2 A Linear Congruential Generator

Example 1 The best-known and (still) most widely used types of generators are the simple linear congruential generators (LCGs) [41, 57, 60, 82]. The state at step n is an integer x_n and the transition function T is defined by the recurrence

$$x_n = (ax_{n-1} + c) \bmod m \tag{1}$$

where $m > 0$, $a > 0$, and c are integers called the *modulus*, the *multiplier*, and the *additive constant*, respectively. Here, "mod m" denotes the operation of taking the least nonnegative residue modulo m. In other words, multiply x_{n-1} by a, add c, divide the result by m, and put x_n equal to the remainder of the division. One can identify s_n with x_n and the state space S is the set $\{0, \ldots, m-1\}$. To produce values in the interval [0,1], one can simply define the output function G by $u_n = G(x_n) = x_n/m$.

When $c = 0$, this generator is called a *multiplicative linear congruential generator* (MLCG). The maximal period length for the LCG is m in general. For the MLCG it cannot exceed $m-1$, since $x_n = 0$ is an absorbing state that must be avoided. Two popular values of m are $m = 2^{31} - 1$ and $m = 2^{32}$. But as discussed later, these values are too small for the requirements of today's simulations. LCGs with such small moduli are still in widespread use, mainly because of their simplicity and ease of implementation, but we believe that they should be discarded and replaced by more robust generators.

For a concrete illustration, let $m = 2^{31} - 1 = 2147483647$, $c = 0$, and $a = 16807$. These parameters were originally proposed in [83]. Take $x_0 = 12345$. Then

$$x_1 = 16,807 \times 12,345 \bmod m = 207,482,415$$

$$u_1 = \frac{207,482,415}{m} = 0.0966165285$$

$$x_2 = 16,807 \times 207,482,415 \bmod m = 1,790,989,824$$

$$u_2 = \frac{1,790,989,824}{m} = 0.8339946274$$

$$x_3 = 16,807 \times 1,790,989,824 \bmod m = 2,035,175,616$$

$$u_3 = \frac{2,035,175,616}{m} = 0.9477024977$$

and so on.

4.1.3 Seasoning the Sequence with External Randomness

In certain circumstances one may want to combine the deterministic sequence with external physical noise. The simplest and most frequently used way of doing this in simulation contexts is to select the seed s_0 randomly. If s_0 is drawn uniformly from S, say by picking balls randomly from a container or by tossing fair coins, the generator can be viewed as an extensor of randomness: It stretches a short, truly random seed into a longer sequence of random-looking numbers. Definition 1 can easily be generalized to accommodate this possibility: Add to the structure a probability distribution μ defined on S and say that s_0 is selected from μ.

In some contexts, one may want to rerandomize the state s_n of the generator every now and then, or to jump ahead from s_n to $s_{n+\nu}$ for some random integer ν. For example, certain types of slot machines in casinos use a simple deterministic random number generator, which keeps running at full speed (i.e., computing its successive states) even when there is nobody playing with the machine. Whenever a player hits the appropriate button and some random numbers are needed to determine the winning combination (e.g., in the game of Keno) or to draw a hand of cards (e.g., for poker machines), the generator provides the output corresponding to its current state. Each time the player hits the button, he or she selects a ν, as just mentioned. This ν is random (although not uniformly distributed). Since typical generators can advance by more than 1 million states per second, hitting the button at the right time to get a specific state or predicting the next output value from the previous ones is almost impossible.

One could go further and select not only the seed, but also some parameters of the generator at random. For example, for a MLCG, one may select the multiplier a at random from a given set of values (for a fixed m) or select the pairs (a, m) at random from a given set. Certain classes of generators for cryptographic applications are defined in a way that the parameters of the recurrence (e.g., the modulus) are viewed as part of the seed and must be generated randomly for the generator to be safe (in the sense of unpredictability).

After observing that physical phenomena by themselves are bad sources of random numbers and that the deterministic generators may produce sequences with too much structure, Marsaglia [90] decided to combine the output of some random number generators with various sources of white and black noise, such as music, pictures, or noise produced by physical devices. The combination was done by addition modulo 2 (bitwise exclusive-or) between the successive bits of the generator's output and of the binary files containing the noise. The result was used to produce a CD-ROM containing 4.8 billion random bits, which appear to behave as independent bits distributed uniformly over the set $\{0, 1\}$. Such a CD-ROM may be interesting but is no universal solution: Its use cannot match the speed and convenience of a good generator, and some applications require much more random numbers than provided on this disk.

4.1.4 Design of Good Generators

How can one build a deterministic generator whose output looks totally random? Perhaps a first idea is to write a computer program more or less at random that can also modify its own code in an unpredictable way. However, experience shows that random number generators should not be built at random (see Knuth [57] for more discussion on this). Building a good random number generator may look easy on the surface, but it is not. It requires a good understanding of heavy mathematics.

The techniques used to evaluate the quality of random number generators can be partitioned into two main classes: The *structural analysis* methods (sometimes called *theoretical tests*) and the *statistical* methods (also called *empirical tests*). An empirical test views the generator as a black box. It observes the output and applies a statistical test of hypothesis to catch significant statistical defects. An unlimited number of such tests can be designed. Structural analysis, on the other hand, studies the mathematical structure underlying the successive values produced by the generator, most often over its entire period length. For example, vectors of t successive output values of a LCG can be viewed as points in the t-dimensional unit hypercube $[0, 1]^t$. It turns out that all these points, over the entire period of the generator, form a regular *lattice* structure. As a result, all the points lie in a limited number of equidistant parallel hyperplanes, in each dimension t. Computing certain numerical figures of merit for these lattices (e.g., computing the distances between neighboring hyperplanes) is an example of structural analysis. Statistical testing and structural analysis are discussed more extensively in forthcoming sections. We emphasize that all these methods are in a sense heuristic: None ever proves that a particular generator is perfectly random or fully reliable for simulation. The best they can do is improve our confidence in the generator.

4.1.5 Overview of What Follows

We now give an overview of the remainder of this chapter. In the next section we portray our ideal random number generator. The desired properties include uniformity, independence, long period, rapid jump-ahead capability, ease of implementation, and efficiency in terms of speed and space (memory size used). In certain situations, unpredictability is also an issue. We discuss the scope and significance of structural analysis as a guide to select families of generators and choose specific parameters. Section 4.3 covers generators based on linear recurrences. This includes the linear congruential, multiple recursive, multiply-with-carry, Tausworthe, generalized feedback shift register generators, all of which have several variants, and also different types of combinations of these. We study their structural properties at length. Section 4.4 is devoted to methods based on nonlinear recurrences, such as inversive and quadratic congruential generators, as well as other types of methods originating from the field of cryptology. Section 4.5 summarizes the ideas of statistical testing. In Section 4.6 we outline the specifications of a modern uniform random number package and refer to available implementations. We also discuss parallel generators briefly.

4.2 DESIRED PROPERTIES

4.2.1 Unpredictability and "True" Randomness

From the user's perspective, an ideal random number generator should be like a black box producing a sequence that cannot be distinguished from a truly random one. In other words, the goal is that given the output sequence $(u_0, u_1, \ldots)$ and an infinite sequence of IID $U(0, 1)$ random variables, no statistical test (or computer program) could tell which is which with probability larger than $1/2$. An equivalent requirement is that after observing any finite number of output values, one cannot guess any given bit of any given unobserved number better than by flipping a fair coin. But this is an impossible dream. The pseudorandom sequence can always be determined by observing it sufficiently, since it is periodic. Similarly, for any periodic sequence, if enough computing time is allowed, it is always possible to construct a statistical test that the sequence will fail spectacularly.

To dilute the goal we may limit the time of observation of the sequence and the computing time for the test. This leads to the introduction of *computational complexity* into the picture. More specifically, we now consider a *family* of generators, $\{\mathcal{G}_k, k = 1, 2, \ldots\}$, indexed by an integral parameter k equal to the number of bits required to represent the state of the generator. We assume that the time required to compute the functions T and G is (at worst) polynomial in k. We also restrict our attention to the class of statistical tests whose running time is polynomial in k. Since the period length typically increases as 2^k, this precludes the tests that exhaust the period. A test is also allowed to toss coins at random, so its outcome is really a random variable. We say that the family $\{\mathcal{G}_k\}$ is *polynomial-time perfect* if, for any polynomial-time statistical test trying to distinguish the output sequence of the generator from an infinite sequence of IID $U(0, 1)$ random variables, the probability that the test makes the right guess does not exceed $1/2 + e^{-k\epsilon}$, where ϵ is a positive constant. An equivalent requirement is that no polynomial-time algorithm can predict any given bit of u_n with probability of success larger than $1/2 + e^{-k\epsilon}$, after observing $u_0, \ldots, u_{n-1}$, for some $\epsilon > 0$. This setup is based on the idea that what cannot be computed in polynomial time is practically impossible to compute if k is reasonably large. It was introduced in cryptology, where unpredictability is a key issue (see [4, 6, 59, 78] and other references given there).

Are efficient polynomial-time perfect families of generators available? Actually, nobody knows for sure whether or not such a family *exists*. But some generator families are *conjectured* to be polynomial-time perfect. The one with apparently the best behavior so far is the BBS, introduced by Blum, Blum, and Shub [4], explained in the next example.

Example 2 The BBS generator of size k is defined as follows. The state space S_k is the set of triplets (p, q, x) such that p and q are $(k/2)$-bit prime integers, $p + 1$ and $q + 1$ are both divisible by 4, and x is a quadratic residue modulo $m = pq$, relatively prime to m (i.e., x can be expressed as $x = y^2 \bmod m$ for some integer y that is not divisible by p or q). The initial state (seed) is chosen randomly from S_k, with the uniform distribution. The state then evolves as follows: p and q remain unchanged and the successive values of x follow the recurrence

$$x_n = x_{n-1}^2 \bmod m.$$

At each step, the generator outputs the ν_k least significant bits of x_n (i.e., $u_n = x_n \bmod 2^{\nu_k}$), where $\nu_k \leq K \log k$ for some constant K. The relevant conjecture here is that with probability at least $1 - e^{-k\epsilon}$ for some $\epsilon > 0$, factoring m (i.e., finding p or q, given m) cannot be done in polynomial time (in k). Under this conjecture, the BBS generator has been proved polynomial-time perfect [4, 124]. Now, a down-to-earth question is: How large should be k to be safe in practice? Also, how small should be K? Perhaps no one really knows. A k larger than a few thousands is probably pretty safe but makes the generator too slow for general simulation use.

Most of the generators discussed in the remainder of this chapter are known *not* to be polynomial-time perfect. However, they seem to have good enough statistical properties for most reasonable simulation applications.

4.2.2 What Is a Random Sequence?

The idea of a truly random sequence makes sense only in the (abstract) framework of probability theory. Several authors (see, e.g., [57]) give definitions of a random sequence, but these definitions require nonperiodic infinite-length sequences. Whenever one selects a generator with a fixed seed, as in Definition 1, one always obtains a deterministic sequence of finite length (the length of the period) which repeats itself indefinitely. Choosing such a random number generator then amounts to selecting a finite-length sequence. But among all sequences of length ρ of symbols from the set U, for given ρ and finite U, which ones are better than others? Let $|U|$ be the cardinality of the set U. If all the symbols are chosen uniformly and independently from U, each of the $|U|^\rho$ possible sequences of symbols from U has the same probability of occurring, namely $|U|^{-\rho}$. So it appears that no particular sequence (i.e., no generator) is better than any other. A pretty disconcerting conclusion! To get out of this dead end, one must take a different point of view.

Suppose that a starting index n is randomly selected, uniformly from the set $\{1, 2, \ldots, \rho\}$, and consider the output vector (or subsequence) $\mathbf{u}_n = (u_n, \ldots, u_{n+t-1})$, where

$t \ll \rho$. Now, $\mathbf{u}_n$ is a (truly) random vector. We would like $\mathbf{u}_n$ to be uniformly distributed (or almost) over the set U^t of all vectors of length t. This requires $\rho \geq |U|^t$, since there are at most ρ different values of $\mathbf{u}_n$ in the sequence. For $\rho < |U|^t$, the set $\Psi = \{\mathbf{u}_n, 1 \leq n \leq \rho\}$ can cover only part of the set U^t. Then one may ask Ψ to be uniformly spread over U^t. For example, if U is a discretization of the unit interval [0,1], such as $U = \{0, 1/m, 2/m, \ldots, (m-1)/m\}$ for some large integer m, and if the points of Ψ are evenly distributed over U^t, they are also (pretty much) evenly distributed over the unit hypercube $[0, 1]^t$.

Example 3 Suppose that $U = \{0, 1/100, 2/100, \ldots, 99/100\}$ and that the period of the generator is $\rho = 10^4$. Here we have $|U| = 100$ and $\rho = |U|^2$. In dimension 2, the pairs $\mathbf{u}_n = (u_n, u_{n+1})$ can be uniformly distributed over U^2, and this happens if and only if each pair of successive values of the form $(i/100, j/100)$, for $0 \leq i, j < 100$ occurs exactly once over the period. In dimension $t > 2$, we have $|U|^t = 10^{2t}$ points to cover but can cover only 10^4 of those because of the limited period length of our generator. In dimension 3, for instance, we can cover only 10^4 points out of 10^6. We would like those 10^4 points that are covered to be very uniformly distributed over the unit cube $[0, 1]^3$.

An even distribution of Ψ over U^t, in all dimensions t, will be our basis for discriminating among generators. The rationale is that under these requirements, subsequences of any t successive output values produced by the generator, from a random seed, should behave much like random points in the unit hypercube. This captures both *uniformity* and *independence*: If $\mathbf{u}_n = (u_n, \ldots, u_{n+t-1})$ is generated according to the uniform distribution over $[0, 1]^t$, the components of $\mathbf{u}_n$ are independent and uniformly distributed over [0,1]. This idea of looking at what happens when the seed is random, for a given finite sequence, is very similar to the *scanning ensemble* idea of Compagner [11, 12], except that we use the framework of probability theory instead.

The reader may have already noticed that under these requirements, Ψ will not look at all like a random set of points, because its distribution over U^t is too even (or *superuniform*, as some authors say [116]). But what the foregoing model assumes is that only a few points are selected at random from the set Ψ. In this case, the best one can do for these points to be distributed approximately as IID uniforms is to take Ψ superuniformly distributed over U^t. For this to make some sense, ρ must be several orders of magnitude larger than the number of output values actually used by the simulation.

To assess this even distribution of the points over the entire period, some (theoretical) understanding of their structural properties is necessary. Generators whose structural properties are well understood and precisely described may look less random, but those that are more complicated and less understood are not necessarily better. They may hide strong correlations or other important defects. One should avoid generators without convincing theoretical support. As a basic requirement, the period length must be known and huge. But this is not enough. Analyzing the equidistribution of the points as just discussed, which is sometimes achieved by studying the lattice structure, usually gives good insight on how the generator behaves. Empirical tests can be applied thereafter, just to improve one's confidence.

4.2.3 Discrepancy

A well-established class of measures of uniformity for finite sequences of numbers are based on the notion of *discrepancy*. This notion and most related results are well covered by Niederreiter [102]. We only recall the most basic ideas here.

Consider the N points $\mathbf{u}_n = (u_n, \ldots, u_{n+t-1})$, for $n = 0, \ldots, N-1$, in dimension t, formed by (overlapping) vectors of t successive output values of the generator. For any hyper-rectangular box aligned with the axes, of the form $R = \prod_{j=1}^{t} [\alpha_j, \beta_j)$, with $0 \le \alpha_j < \beta_j \le 1$, let $I(R)$ be the number of points $\mathbf{u}_n$ falling into R, and $V(R) = \prod_{j=1}^{t} (\beta_j - \alpha_j)$ be the volume of R. Let $\mathcal{R}$ be the set of all such regions R, and

$$D_N^{(t)} = \max_{R \in \mathcal{R}} \left| V(R) - \frac{I(R)}{N} \right| .$$

This quantity is called the t-dimensional *(extreme) discrepancy* of the set of points $\{\mathbf{u}_0, \ldots, \mathbf{u}_{N-1}\}$. If we impose $\alpha_j = 0$ for all j, that is, we restrict $\mathcal{R}$ to those boxes which have one corner at the origin, then the corresponding quantity is called the *star discrepancy*, denoted by $D_N^{*(t)}$. Other variants also exist, with richer $\mathcal{R}$.

A low discrepancy value means that the points are very evenly distributed in the unit hypercube. To get superuniformity of the sequence over its entire period, one might want to *minimize* the discrepancy $D_\rho^{(t)}$ or $D_\rho^{*(t)}$ for $t = 1, 2, \ldots$. A major practical difficulty with discrepancy is that it can be computed only for very special cases. For LCGs, for example, it can be computed efficiently in dimension $t = 2$, but for larger t, the computing cost then increases as $O(N^t)$. In most cases, only (upper and lower) bounds on the discrepancy are available. Often, these bounds are expressed as orders of magnitude as a function of N, are defined for $N = \rho$, and/or are averages over a large (specific) class of generators (e.g., over all full-period MLCGs with a given prime modulus). Discrepancy also depends on the rectangular orientation of the axes, in contrast to other measures of uniformity, such as the distances between hyperplanes for LCGs (see Section 4.3.4). On the other hand, it applies to all types of generators, not only those based on linear recurrences.

We previously argued for superuniformity over the entire period, which means seeking the lowest possible discrepancy. When a subsequence of length N is used (for $N \ll \rho$), starting, say, at a random point along the entire sequence, the discrepancy of that subsequence should behave (viewed as a random variable) as the discrepancy of a sequence of IID $U(0, 1)$ random variables. The latter is (roughly) of order $O(N^{-1/2})$ for both the star and extreme discrepancies.

Niederreiter [102] shows that the discrepancy of full-period MLCGs *over their entire period* (of length $\rho = m-1$), on the average over multipliers a, is of order $O(m^{-1}(\log m)^t \log\log(m+1))$. This order is much smaller (for large m) than $O(m^{-1/2})$, meaning superuniformity. Over small fractions of the period length, the available bounds on the discrepancy are more in accordance with the law of the iterated logarithm [100]. This is yet another important justification for never using more than a negligible fraction of the period.

Suppose now that numbers are generated in [0, 1] with L fractional binary digits. This gives *resolution* 2^{-L}, which means that all u_n's are multiples of 2^{-L}. It then follows [102] that $D_N^{*(t)} \le 2^{-L}$ for all $t \ge 1$ and $N \ge 1$. Therefore, as a *necessary* condition for the discrepancy to be of the right order of magnitude, the resolution 2^{-L} must be small enough for the number of points N that we plan to generate: 2^{-L} should be much smaller than $N^{-1/2}$. A too coarse discretization implies a too large discrepancy.

4.2.4 Quasi-Random Sequences

The interest in discrepancy stems largely from the fact that deterministic error bounds for (Monte Carlo) numerical integration of a function are available in terms of $D_N^{(t)}$ and of a certain measure of variability of the function. In that context, the smaller the discrepancy, the better [because the aim is to minimize the numerical error, not really to imitate IID $U(0, 1)$ random variables]. Sequences for which the discrepancy of the first N values is small for all N are called *low-discrepancy* or *quasi-random* sequences [102]. Numerical integration using such sequences is called *quasi-Monte Carlo integration*. To estimate the integral using N points, one simply evaluates the function (say, a function of t variables) at the first N points of the sequence, takes the average, multiplies by the volume of the domain of integration, and uses the result as an approximation of the integral. Specific low-discrepancy sequences have been constructed by Sobol', Faure, and Niederreiter, among others (see ref. 102). Owen [106] gives a recent survey of their use. In this chapter we concentrate on *pseudorandom* sequences and will not discuss *quasi-random* sequences further.

4.2.5 Long Period

Let us now return to the desired properties of pseudorandom sequences, starting with the length of the period. What is long enough? Suppose that a simulation experiment takes N random numbers from a sequence of length ρ. Several reasons justify the need to take $\rho \gg N$ (see, e.g., refs. 21, 64, 86, 102, 112). Based on geometric arguments, Ripley [112] suggests that $\rho \gg N^2$ for linear congruential generators. The papers [75, 79] provide strong experimental support for this, based on extensive empirical tests. Our previous discussion also supports the view that ρ must be huge in general.

Period lengths of 2^{32} or smaller, which are typical for the default generators of many operating systems and software packages, are unacceptably too small. Such period lengths can be exhausted in a matter of minutes on today's workstations. Even $\rho = 2^{64}$ is a relatively small period length. Generators with period lengths over 2^{200} are now available.

4.2.6 Efficiency

Some say that the speed of a random number generator (the number of values that it can generate per second, say) is not very important for simulation, since generating the numbers typically takes only a tiny fraction of the simulation time. But there are several counterexamples, such as for certain large simulations in particle physics [26] or when using intensive Monte Carlo simulation to estimate with precision the distribution of a statistic that is fast to compute but requires many random numbers. Moreover, even if a fast generator takes only, say, 5% of the simulation time, changing to another one that is 20 times slower will approximately double the total simulation time. Since simulations often consume several hours of CPU time, this is significant.

The memory size used by a generator might also be important in general, especially since simulations often use several generators in parallel, for instance to maintain synchronization for variance reduction purposes (see Section 4.6 and refs. 7 and 60 for more details).

4.2.7 Repeatability, Splitting Facilities, and Ease of Implementation

The ability to replicate exactly the same sequence of random numbers, called *repeatability*, is important for program verification and to facilitate the implementation of certain variance reduction techniques [7, 55, 60, 113]. Repeatability is a major advantage of pseudorandom sequences over sequences generated by physical devices. The latter can of course be stored on disks or other memory devices, and then reread as needed, but this is less convenient than a good pseudorandom number generator that fits in a few lines of code in a high-level language.

A code is said to be *portable* if it works without change and produces exactly the same sequence (at least up to machine accuracy) across all "standard" compilers and computers. A portable code in a high-level language is clearly much more convenient than a machine-dependent assembly-language implementation, for which repeatability is likely to be more difficult to achieve.

Ease of implementation also means the ease of *splitting* the sequence into (long) disjoint substreams and jumping quickly from one substream to the next. In Section 4.6 we show why this is important. For this, there should be an efficiency way to compute the state $s_{n+\nu}$ for any large ν, given s_n. For most linear-type generators, we know how to do that. But for certain types of nonlinear generators and for some methods of combination (such as *shuffling*), good jump-ahead techniques are unknown. Implementing a random number package as described in Section 4.6 requires efficient jump-ahead techniques.

4.2.8 Historical Accounts

There is an enormous amount of scientific literature on random number generation. Law and Kelton [60] present a short (but interesting) historical overview. Further surveys and historical accounts of the old days are provided in refs. 47, 53, and 119.

Early attempts to construct pseudorandom number generators have given rise to all sorts of bad designs, sometimes leading to disatrous results. An illustrative example is the *middle-square* method, which works as follows (see, e.g., ref. 57, 60). Take a b-digit number x_{i-1} (say, in base 10, with b even), square it to obtain a $2b$-digit number (perhaps with zeros on the left), and extract the b middle digits to define the next number x_i. To obtain an output value u_i in [0,1), divide x_i by 10^b. The period length of this generator depends on the initial value and is typically very short, sometimes of length 1 (such as when the sequence reaches the absorbing state $x_i = 0$). Hopefully, it is no longer used. Another example of a bad generator is RANDU (see G4 in Table 1).

4.3 LINEAR METHODS

4.3.1 Multiple-Recursive Generator

Consider the linear recurrence

$$x_n = (a_1 x_{n-1} + \cdots + a_k x_{n-k}) \bmod m, \qquad (2)$$

where the *order k* and the *modulus m* are positive integers, while the *coefficients* $a_1, \ldots, a_k$ are integers in the range $\{-(m-1), \ldots, m-1\}$. Define $\mathbb{Z}_m$ as the set $\{0, 1, \ldots, m-1\}$ on which operations are performed modulo m. The state at step n of the *multi-*

ple recursive generator (MRG) [57, 62, 102] is the vector $s_n = (x_n, \ldots, x_{n+k-1}) \in \mathbb{Z}_m^k$. The output function can be defined simply by $u_n = G(s_n) = x_n/m$, which gives a value in [0,1], or by a more refined transformation if a better resolution than $1/m$ is required. The special case where $k = 1$ is the MLCG mentioned previously.

The characteristic polynomial P of (2) is defined by

$$P(z) = z^k - a_1 z^{k-1} - \cdots - a_k \tag{3}$$

The maximal period length of (2) is $\rho = m^k - 1$, reached if and only if m is prime and P is a primitive polynomial over $\mathbb{Z}_m$, identified here as the finite field with m elements. Suppose that m is prime and let $r = (m^k - 1)/(m - 1)$. The polynomial P is primitive over $\mathbb{Z}_m$ if and only if it satisfies the following conditions, where everything is assumed to be modulo m (see ref. 57)

(a) $[(-1)^{k+1}a_k]^{(m-1)/q} \neq 1$ for each prime factor q of $m - 1$
(b) $z^r \bmod P(z) = (-1)^{k+1}a_k$
(c) $z^{r/q} \bmod P(z)$ has degree > 0 for each prime factor q of r, $1 < q < r$

For $k = 1$ and $a = a_1$ (the MLCG case), these conditions simplify to $a \neq 0 \pmod m$ and $a^{(m-1)/q} \neq 1 \pmod m$ for each prime factor q of $m - 1$. For large r, finding the factors q to check condition (c) can be too difficult, since it requires the factorization of r. In this case, the trick is to choose m and k so that r is prime (this can be done only for prime k). Testing primality of large numbers (usually probabilistic algorithms, for example, as in [73, 111]) is much easier than factoring. Given m, k, and the factorizations of $m - 1$ and r, primitive polynomials are generally easy to find, simply by random search.

If m is not prime, the period length of (2) has an upper bound typically much lower than $m^k - 1$. For $k = 1$ and $m = 2^e$, $e \geq 4$, the maximum period length is 2^{e-2}, which is reached if $a_1 = 3$ or 5 (mod 8) and x_0 is odd [57, p. 20]. Otherwise, if $m = p^e$ for p prime and $e \geq 1$, and $k > 1$ or $p > 2$, the upper bound is $(p^k - 1)p^{e-1}$ [36]. Clearly, $p = 2$ is very convenient from the implementation point of view, because the modulo operation then amounts to chopping-off the higher-order bits. So to compute $ax \bmod m$ in that case, for example with $e = 32$ on a 32-bit computer, just make sure that the overflow-checking option or the compiler is turned off, and compute the product ax using unsigned integers while ignoring the overflow.

However, taking $m = 2^e$ imposes a big sacrifice on the period length, especially for $k > 1$. For example, if $k = 7$ and $m = 2^{31} - 1$ (a prime), the maximal period length is $(2^{31} - 1)^7 - 1 \approx 2^{217}$. But for $m = 2^{31}$ and the same value of k, the upper bound becomes $\rho \leq (2^7 - 1)2^{31-1} < 2^{37}$, which is more than 2^{180} times shorter. For $k = 1$ and $p = 2$, an upper bound on the period length of the ith least significant bit of x_n is $\max(1, 2^{i-2})$ [7], and if a full cycle is split into 2^d equal segments, all segments are identical except for their d most significant bits [20, 26]. For $k > 1$ and $p = 2$, the upper bound on the period length of the ith least significant bit is $(2^k - 1)2^{i-1}$. So the low-order bits are typically much too regular when $p = 2$. For $k = 7$ and $m = 2^{31}$, for example, the least significant bit has period length at most $2^7 - 1 = 127$, the second least significant bit has period length at most $2(2^7 - 1) = 254$, and so on.

Example 4 Consider the recurrence $x_n = 10{,}205x_{n-1} \bmod 2^{15}$, with $x_0 = 12{,}345$. The first eight values of x_n, in base 10 and in base 2, are

$$
\begin{aligned}
x_0 &= 12,345 = 011000000111001_2\\
x_1 &= 20,533 = 101000000110101_2\\
x_2 &= 20,673 = 101000011000001_2\\
x_3 &= 7,581 = 001110110011101_2\\
x_4 &= 31,625 = 111101110001001_2\\
x_5 &= 1,093 = 000010001000101_2\\
x_6 &= 12,945 = 011001010010001_2\\
x_7 &= 15,917 = 011111000101101_2
\end{aligned}
$$

The last two bits are always the same. The third least significant bit has a period length of 2, the fourth least significant bit has a period length of 4, and so on.

Adding a constant c as in (1) can slightly increase the period length. The LCG with recurrence (1) has period length m if and only if the following conditions are satisfied ([57, p. 16])

1. c is relatively prime to m.
2. $a - 1$ is a multiple of p for every prime factor p of m (including m itself if m is prime).
3. If m is a multiple of 4, then $a - 1$ is also a multiple of 4.

For $m = 2^e \geq 4$, these conditions simplify to c is odd and $a \bmod 4 = 1$. But the low-order bits are again too regular: The period length of the ith least significant bit of x_n is at most 2^i.

A constant c can also be added to the right side of the recurrence (2). One can show (see ref. [62]) that a linear recurrence of order k with such a constant term is equivalent to some linear recurrence of order $k + 1$ with no constant term. As a result, an upper bound on the period length of such a recurrence with $m = p^e$ is $(p^{k+1} - 1)p^{e-1}$, which is much smaller than m^k for large e and k.

All of this argues against the use of power-of-2 moduli in general, despite their advantage in terms of implementation. It favors prime moduli instead. Later, when discussing combined generators, we will also be interested in moduli that are the products of a few large primes.

4.3.2 Implementation for Prime *m*

For $k > 1$ and prime m, for the characteristic polynomial P to be primitive, it is necessary that a_k and at least another coefficient a_j be nonzero. From the implementation point of view, it is best to have only two nonzero coefficients; that is, a recurrence of the form

$$x_n = (a_r x_{n-r} + a_k x_{n-k}) \bmod m \tag{4}$$

with characteristic trinomial P defined by $P(z) = z^k - a_r z^{k-r} - a_k$. Note that replacing r by $k - r$ generates the same sequence in reverse order.

When m is not a power of 2, computing and adding the products modulo m in (2) or (4) is not necessarily straightforward, using ordinary integer arithmetic, because of the possibility of overflow: The products can exceed the largest integer representable on the computer. For example, if $m = 2^{31} - 1$ and $a_1 = 16{,}807$, then x_{n-1} can be as large as $2^{31} - 2$, so the product $a_1 x_{n-1}$ can easily exceed 2^{31}. L'Ecuyer and Côté [76] study and compare different techniques for computing a product modulo a large integer m, using only integer arithmetic, so that no intermediate result ever exceeds m. Among the *general* methods, working for all representable integers and easily implementable in a high-level language, *decomposition* was the fastest in their experiments. Roughly, this method simply decomposes each of the two integers that are to be multiplied in two blocks of bits (e.g., the 15 least significant bits and the 16 most significant ones, for a 31-bit integer) and then cross-multiplies the blocks and adds (modulo m) just as one does when multiplying large numbers by hand.

There is a faster way to compute $ax \bmod m$ for $0 < a, x < m$, called *approximate factoring*, which works under the condition that

$$a(m \bmod a) < m. \tag{5}$$

This condition is satisfied if and only if $a = i$ or $a = \lfloor m/i \rfloor$ for $i < \sqrt{m}$ (here $\lfloor x \rfloor$ denotes the largest integer smaller or equal to x, so $\lfloor m/i \rfloor$ is the integer division of m by i). To implement the approximate factoring method, one initially precomputes (once for all) the constants $q = \lfloor m/a \rfloor$ and $r = m \bmod a$. Then, for any positive integer $x < m$, the following instructions have the same effect as the assignment $x \leftarrow ax \bmod m$, but with all intermediate (integer) results remaining strictly between $-m$ and m [7, 61, 107]:

```
y ← ⌊x/q⌋;
x ← a(x − yq) − yr;
IF x < 0 THEN x ← x + m END.
```

As an illustration, if $m = 2^{31} - 1$ and $a = 16{,}807$, the generator satisfies the condition, since $16{,}807 < \sqrt{m}$. In this case, one has $q = 127{,}773$ and $r = 2836$. Hörmann and Derflinger [51] give a different method, which is about as fast, for the case where $m = 2^{31} - 1$. Fishman [41, p. 604] also uses a different method to implement the LCG with $m = 2^{31} - 1$ and $a = 95{,}070{,}637$, which does not satisfy (5).

Another approach is to represent all the numbers and perform all the arithmetic modulo m in double-precision floating point. This works provided that the multipliers a_i are small enough so that the integers $a_i x_{n-i}$ and their sum are always represented *exactly* by the floating-point values. A sufficient condition is that the floating-point numbers are represented with at least

$$\lceil \log_2((m-1)(a_1 + \cdots + a_k)) \rceil$$

bits of precision in their mantissa, where $\lceil x \rceil$ denotes the smallest integer larger or equal to x. On computers with good 64-bit floating-point hardware (most computers nowadays), this approach usually gives by far the fastest implementation (see, e.g., [68] for examples and timings).

4.3.3 Jumping Ahead

To jump ahead from x_n to $x_{n+\nu}$ with an MLCG, just use the relation

$$x_{n+\nu} = a^{\nu} x_n \bmod m = (a^{\nu} \bmod m) x_n \bmod m$$

If many jumps are to be performed with the same ν, the constant $a^{\nu} \bmod m$ can be precomputed once and used for all subsequent computations.

Example 5 Again, let $m = 2{,}147{,}483{,}647$, $a = 16{,}807$, and $x_0 = 12{,}345$. Suppose that we want to compute x_3 directly from x_0, so $\nu = 3$. One easily finds that $16,807^3 \bmod m = 1{,}622{,}650{,}073$ and $x_3 = 1{,}622{,}650{,}073 x_0 \bmod m = 2{,}035{,}175{,}616$, which agrees with the value given in Example 1. Of course, we are usually interested in much larger values of ν, but the method works the same way.

For the LCG, with $c \neq 0$, one has

$$x_{n+\nu} = \left[a^{\nu} x_n + \frac{c(a^{\nu} - 1)}{a - 1} \right] \bmod m$$

To jump ahead with the MRG, one way is to use the fact that it can be represented as a matrix MLCG: $X_n = AX_{n-1} \bmod m$, where X_n is s_n represented as a column vector and A is a $k \times k$ square matrix. Jumping ahead is then achieved in the same way as for the MLCG:

$$X_{n+\nu} = A^{\nu} X_n \bmod m = (A^{\nu} \bmod m) X_n \bmod m$$

Another way is to transform the MRG into its polynomial representation [64], in which jumping ahead is easier, and then apply the inverse transformation to recover the original representation.

4.3.4 Lattice Structure of LCGs and MRGs

A lattice of dimension t, in the t-dimensional real space $\mathbb{R}^t$, is a set of the form

$$L = \left\{ V = \sum_{j=1}^{t} z_j V_j \mid \text{each } z_j \in \mathbb{Z} \right\} \tag{6}$$

where $\mathbb{Z}$ is the set of all integers and $\{V_1, \ldots, V_t\}$ is a basis of $\mathbb{R}^t$. The lattice L is thus the set of all *integer* linear combinations of the vectors $V_1, \ldots, V_t$, and these vectors are called a *lattice basis* of L. The basis $\{W_1, \ldots, W_t\}$ of $\mathbb{R}^t$ which satisfies $V_i' W_j = \delta_{ij}$ for all $1 \leq i, j \leq t$ (where the prime means "transpose" and where $\delta_{ij} = 1$ if $i = j$, 0 otherwise) is called the *dual* of the basis $\{V_1, \ldots, V_t\}$ and the lattice generated by this dual basis is called the dual lattice to L.

Consider the set

$$T_t = \{\mathbf{u}_n = (u_n, \ldots, u_{n+t-1}) | n \geq 0, s_0 = (x_0, \ldots, x_{k-1}) \in \mathbb{Z}_m^k\} \tag{7}$$

of all overlapping t-tuples of successive values produced by (2), with $u_n = x_n/m$, from all possible initial seeds. Then this set T_t is the intersection of a lattice L_t with the t-dimensional unit hypercube $I^t = [0, 1)^t$. For more detailed studies and to see how to construct a basis for this lattice L_t and its dual, see refs. 23, 57, 73, 77. For $t \leq k$ it is clear from the definition of T_t that each vector $(x_0, \ldots, x_{t-1})$ in $\mathbb{Z}_m^t$ can be taken as s_0, so $T_t = \mathbb{Z}_m^t/m = (\mathbb{Z}^t/m) \cap I^t$; that is, L_t is the set of all t-dimensional vectors whose coordinates are multiples of $1/m$, and T_t is the set of m^t points in L_t whose coordinates belong to $\{0, 1/\mathrm{m}, \ldots, (m-1)/m\}$. For a full-period MRG, this also holds if we fix s_0 in the definition of T_t to any nonzero vector of $\mathbb{Z}_m^k$, and then add the zero vector to T_t. In dimension $t > k$, the set T_t contains only m^k points, while $\mathbb{Z}_m^t/m$ contains m^t points. Therefore, for large t, T_t contains only a small fraction of the t-dimensional vectors whose coordinates are multiples of $1/m$.

For full-period MRGs, the generator covers all of T_t except the zero state in one cycle. In other cases, such as for MRGs with nonprime moduli or MLCGs with power-of-2 moduli, each cycle covers only a smaller subset of T_t, and the lattice generated by that subset is often equal to L_t, but may in some cases be a strict *sublattice* or *subgrid* (i.e., a *shifted lattice* of the form $V_0 + L$ where $V_0 \in \mathbb{R}^t$ and L is a lattice). In the latter case, to analyze the structural properties of the generator, one should examine the appropriate sublattice or subgrid instead of L_t. Consider, for example, an MLCG for which m is a power of 2, $a \bmod 8 = 5$, and x_0 is odd. The t-dimensional points constructed from successive values produced by this generator form a subgrid of L_t containing one-fourth of the points [3, 50]. For a LCG with m a power of 2 and $c \neq 0$, with full period length $\rho = m$, the points all lie in a grid that is a shift of the lattice L_t associated with the corresponding MLCG (with the same a and m). The value of c determines only the shifting and has no other effect on the lattice structure.

Example 6 Figures 4.1 to 4.3 illustrate the lattice structure of a small, but instructional, LCGs with (prime) modulus $m = 101$ and full period length $\rho = 100$, in dimension $t = 2$.

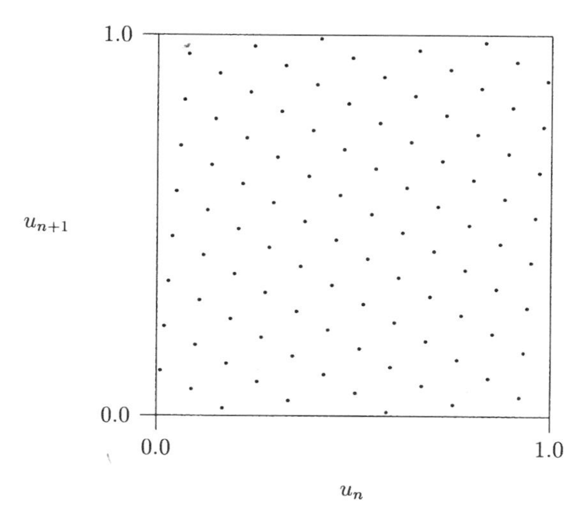

Figure 4.1 All pairs (u_n, u_{n+1}) for the LCG with $m = 101$ and $a = 12$.

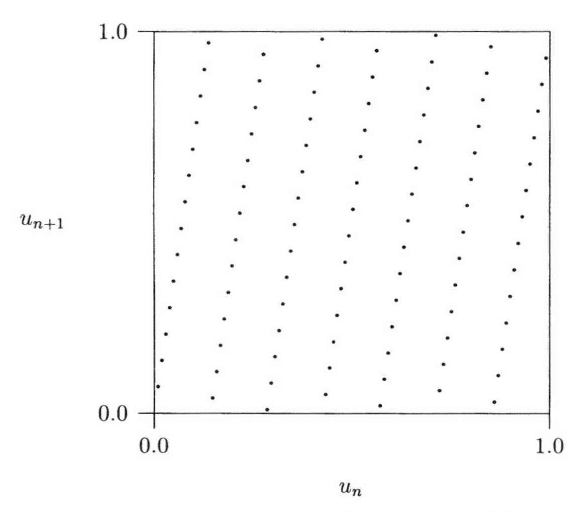

Figure 4.2 All pairs (u_n, u_{n+1}) for the LCG with $m = 101$ and $a = 7$.

They show all 100 pairs of successive values (u_n, u_{n+1}) produced by these generators, for the multipliers $a = 12$, $a = 7$, and $a = 51$, respectively. In each case, one clearly sees the lattice structure of the points. Any pair of vectors forming a basis determine a parallelogram of area $1/101$. This holds more generally: In dimension t, the vectors of any basis of L_t determine a parallelepiped of volume $1/m^k$. Conversely, any set of t vectors that determine such a parallelepiped form a lattice basis.

The points are much more evenly distributed in the square for $a = 12$ than for $a =$

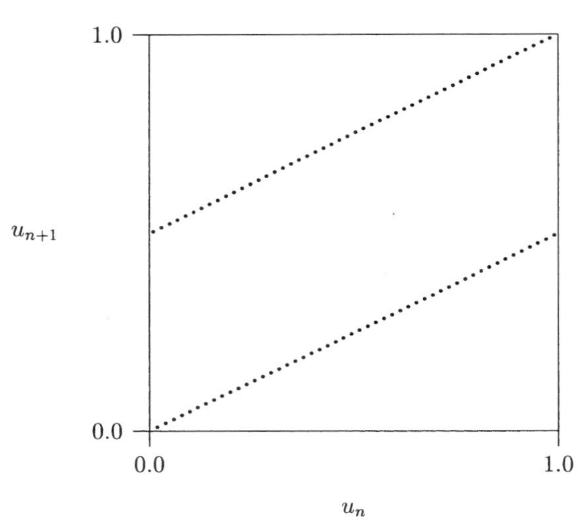

Figure 4.3 All pairs (u_n, u_{n+1}) for the LCG with $m = 101$ and $a = 51$.

51, and slightly more evenly distributed for $a = 12$ than for $a = 7$. The points of L_t are generally more evenly distributed when there exists a basis comprised of vectors of similar lengths. One also sees from the figures that all the points lie in a relative small number of equidistant parallel lines. In Figure 3, only two lines contain all the points and this leaves large empty spaces between the lines, which is bad.

In general, the lattice structure implies that all the points of T_t lie on a family of equidistant parallel hyperplanes. Among all such families of parallel hyperplanes that cover all the points, take the one for which the successive hyperplanes are farthest apart. The distant d_t between these successive hyperplanes is equal to $1/l_t$, where l_t is the length of a shortest nonzero vector in the *dual* lattice to L_t. Computing a shortest nonzero vector in a lattice L means finding the combination of values of z_j in (6) giving the shortest V. This is a quadratic optimization problem with integer variables and can be solved by a branch-and-bound algorithm, as in [15, 40]. In these papers the authors use an ellipsoid method to compute the bounds on the z_j for the branch-and-bound. This appears to be the best (general) approach known to date and is certainly much faster than the algorithm given in [23] and [57]. This idea of analyzing d_t was introduced by Coveyou and MacPherson [18] through the viewpoint of spectral analysis. For this historical reason, computing d_t is often called the *spectral test*.

The shorter the distance d_t, the better, because a large d_t means thick empty slices of space between the hyperplanes. One has the theoretical lower bound

$$d_t \geq d_t^* = \frac{1}{\gamma_t m^{k/t}} \tag{8}$$

where γ_t is a constant which depends only on t and whose exact value is currently known only for $t \leq 8$ [57]. So, for $t \leq 8$ and $T \leq 8$, one can define the figures of merit $S_t = d_t^*/d_t$ and $M_T = \min_{k \leq t \leq T} S_t$, which lie between 0 and 1. Values close to 1 are desired. Another lower bound on d_t, for $t > k$, is (see [67])

$$d_t \geq \left(1 + \sum_{j=1}^{k} a_j^2\right)^{-1/2} \tag{9}$$

This means that an MRG whose coefficients a_j are small is guaranteed to have a large (bad) d_t.

Other figures of merit have been introduced to measure the quality of random number generators in terms of their lattice structure. For example, one can count the minimal number of hyperplanes that contain all the points or compute the ratio of lengths of the shortest and longest vectors in a Minkowski-reduced basis of the lattice. For more details on the latter, which is typically much more costly to compute than d_t, the reader can consult [77] and the references given there. These alternative figures of merit do not tell us much important information in addition to d_t.

Tables 4.1 and 4.2 give the values of d_t and S_t for certain LCGs and MRGs. All these generators have full period length. The LCGs of the first tables are well known and most are (or have been) heavily used. For $m = 2^{31} - 1$, the multiplier $a = 742{,}938{,}285$ was found by Fishman and Moore [42] in an exhaustive search for the MLCGs with the best value of M_6 for this value of m. It is used in the GPSS/H simulation environment. The second multiplier, $a = 16807$, was originally proposed in [83], is suggested

TABLE 4.1 Distances Between Hyperplanes for Some LCGs

	G1	G2	G3	G4	G5	G6	G7
m	$2^{31}-1$	$2^{31}-1$	$2^{31}-1$	2^{31}	2^{32}	2^{48}	$10^{12}-11$
k	1	1	1	1	1	1	1
a	742,938,285	16,807	630,360,016	65,539	69,069	25,214,903,917	427,419,669,081
c	0	0	0	0	0	11	0
ρ	$2^{31}-2$	$2^{31}-2$	$2^{31}-2$	2^{29}	2^{32}	2^{48}	$10^{12}-12$
S_2	0.8673	0.3375	0.8212	0.9307	0.6541	0.5110	0.7513
S_3	0.8607	0.4412	0.4317	0.0119	0.4971	0.8030	0.7366
S_4	0.8627	0.5752	0.7833	0.0595	0.6223	0.4493	0.6491
S_5	0.8319	0.7361	0.8021	0.1570	0.6583	0.5847	0.7307
S_6	0.8341	0.6454	0.5700	0.2927	0.3356	0.6607	0.6312
S_7	0.6239	0.5711	0.6761	0.4530	0.4499	0.8025	0.5598
S_8	0.7067	0.6096	0.7213	0.6173	0.6284	0.5999	0.5558
$1/m$	4.65E-10	4.65E-10	4.65E-10	4.65E-10	2.33E-10	3.55E0-15	1.00E-12
d_2	2.315E-5	5.950E-5	2.445E-5	4.315E-5	3.070E-5	1.085E-7	1.239E-6
d_3	8.023E-4	1.565E-3	1.599E-3	0.0921	1.389E-3	1.693E-5	1.209E-4
d_4	4.528E-3	6.791E-3	4.987E-3	0.0928	6.277E-3	4.570E-4	1.295E-3
d_5	0.0133	0.0150	0.0138	0.0928	0.0168	1.790E-3	4.425E-3
d_6	0.0259	0.0334	0.0379	0.0928	0.0643	4.581E-3	0.0123
d_7	0.0553	0.0604	0.0510	0.0928	0.0767	7.986E-3	0.0256
d_8	0.0682	0.0791	0.0668	0.0928	0.0767	0.0184	0.0402
d_9	0.1060	0.1125	0.0917	0.0928	0.1000	0.0314	0.0677

d_{10}	0.1085	0.1250	0.1155	0.1543	0.1387	0.0374	0.0702
d_{11}	0.1690	0.1429	0.1270	0.1543	0.1443	0.0541	0.0778
d_{12}	0.2425	0.1961	0.2132	0.1622	0.1581	0.0600	0.1005
d_{13}	0.2425	0.1961	0.2132	0.1961	0.1826	0.0693	0.1336
d_{14}	0.2425	0.2000	0.2132	0.2132	0.1961	0.0928	0.1336
d_{15}	0.2425	0.2000	0.2182	0.2132	0.2041	0.0953	0.1361
d_{16}	0.2425	0.2085	0.2294	0.2357	0.2236	0.1000	0.1414
d_{17}	0.2425	0.2425	0.2357	0.2673	0.2236	0.1291	0.1690
d_{18}	0.2500	0.2500	0.2500	0.2673	0.2236	0.1291	0.1690
d_{19}	0.2673	0.2500	0.2500	0.2673	0.2500	0.1471	0.1961
d_{20}	0.2673	0.2887	0.2673	0.2887	0.2500	0.1508	0.2041
d_{21}	0.2673	0.2887	0.2673	0.2887	0.3162	0.1667	0.2294
d_{22}	0.2887	0.2887	0.2774	0.2887	0.3162	0.1768	0.2294
d_{23}	0.2887	0.2887	0.2774	0.3162	0.3162	0.1890	0.2294
d_{24}	0.3015	0.2887	0.3015	0.3162	0.3162	0.1961	0.2294
d_{25}	0.3015	0.2887	0.3015	0.3162	0.3162	0.1961	0.2425
d_{26}	0.3015	0.2887	0.3015	0.3162	0.3162	0.1961	0.2425
d_{27}	0.3015	0.3015	0.3015	0.3162	0.3162	0.1961	0.2500
d_{28}	0.3015	0.3015	0.3333	0.3162	0.3162	0.2132	0.2673
d_{29}	0.3162	0.3015	0.3333	0.3162	0.3162	0.2236	0.2673
d_{30}	0.3162	0.3162	0.3333	0.3536	0.3162	0.2236	0.2673

TABLE 4.2 Distances Between Hyperplanes for Some MRGs

	G8	G9	G10	G11
m	$2^{31}-19$	$2^{31}-19$	$(2^{31}-1)(2^{31}-2000169)$	$(2^{31}-85)(2^{31}-249)$
k	7	7	3	1
a_1	1,975,938,786	1071064	2,620,007,610,006,878,699	1,968,402,271,571,654,650
a_2	875,540,239	0	4,374,377,652,968,432,818	
a_3	433,188,390	0	667,476,516,358,487,852	
a_4	451,413,575	0		
a_5	1,658,907,683	0		
a_6	1,513,645,334	0		
a_7	1,428,037,821	2113664		
S_2				0.66650
S_3				0.76439
S_4			0.75901	0.39148
S_5			0.77967	0.74850
S_6			0.75861	0.67560
S_7			0.76042	0.61124
S_8	0.73486	0.00696	0.74215	0.56812
$1/m$	4.6E-10	4.6E-10	4.6E-10	4.6E-10
d_2				6.5E-10
d_3				7.00E-7
d_4			1.1E-14	4.63E-5
d_5			6.6E-12	2.00E-4
d_6			4.7E-10	8.89E-4
d_7			9.80E-9	2.62E-3
d_8	6.57E-9	6.94E-7	9.55E-8	5.78E-3
d_9	5.91E-8	4.58E-6	6.00E-7	9.57E-3
d_{10}	2.87E-7	8.38E-6	2.24E-6	1.73E-2
d_{11}	1.08E-6	1.10E-5	8.41E-6	2.36E-2
d_{12}	3.85E-6	1.10E-5	2.66E-5	3.07E-2
d_{13}	9.29E-6	1.26E-5	4.68E-5	3.47E-2
d_{14}	1.99E-5	2.17E-5	1.05E-4	3.96E-2
d_{15}	4.17E-5	4.66E-5	1.60E-4	5.98E-2
d_{16}	7.63E-5	8.36E-5	2.68E-4	6.07E-2
d_{17}	1.33E-4	1.31E-4	4.26E-4	6.51E-2
d_{18}	2.77E-4	2.04E-4	7.05E-4	7.43E-2
d_{19}	2.95E-4	3.50E-4	1.03E-3	8.19E-2
d_{20}	4.62E-4	4.17E-4	1.32E-3	8.77E-2

in many simulation books and papers (e.g., [7, 107, 114]) and appears in several software systems such as the AWESIM and ARENA simulation systems, MATLAB [94], the IMSL statistical library [54], and in operating systems for the IBM and Macintosh computers. It satisfies condition (5). The IMSL library also has available the two multipliers 397,204,094 and 950,706,376, with the same modulus, as well as the possibility of adding a shuffle to the LCG. The multiplier $a = 630{,}360{,}016$ was proposed in [108], is recommended in [60, 92] among others, and is used in software such as the SIMSCRIPT II.5 and INSIGHT simulation programming languages. Generator G4, with modulus $m = 2^{31}$ and multiplier $a = 65{,}539$, is the infamous RANDU generator, used for a long time in the IBM/360 operating system. Its lattice structure is particularly bad in dimension 3, where all the points lie in only 15 parallel planes. Law and Kelton [60] give a graphical illustration. Generator G5, with $m = 2^{32}$, $a = 69{,}069$, and $c = 1$, is used in the VAX/VMS operating system. The LCG G6, with modulus $m = 2^{48}$, multiplier $a = 25{,}214{,}903{,}917$, and constant $c = 11$, is the generator implemented in the procedure `drand48` of the SUN Unix system's library [117]. G7, whose period length is slightly less than 2^{40}, is used in the *Maple* mathematical software. We actually recommend *none* of the generators G1 to G7. Their period lengths are too short and they fail many statistical tests (see Section 4.5).

In Table 4.2, G8 and G9 are two MRGs of order 7 found by a random search for multipliers with a "good" lattice structure in all dimensions $t \leq 20$, among those giving a full period with $m = 2^{31} - 19$. For G9 there are the additional restrictions that a_1 and a_7 satisfy condition (5) and $a_i = 0$ for $2 \leq i \leq 6$. This m is the largest prime under 2^{31} such that $(m^7 - 1)/(m - 1)$ is also prime. The latter property facilitates the verification of condition (c) in the full-period conditions for an MRG. These two generators are taken from [73], where one can also find more details on the search and a precise definition of the selection criterion. It turns out that G9 has a very bad figure of merit S_8, and larger values of d_t than G8 for t slightly larger than 7. This is due to the restrictions $a_i = 0$ for $2 \leq i \leq 6$, under which the lower bound (9) is always much larger than d_t^* for $t = 8$. The distances between the hyperplanes for G9 are nevertheless much smaller than the corresponding values of any LCG of Table 4.1, so this generator is a clear improvement over those. G8 is better in terms of lattice structure, but also much more costly to run, because there are seven products modulo m to compute instead of two at each iteration of the recurrence. The other generators in this table are discussed later.

4.3.5 Lacunary Indices

Instead of constructing vectors of successive values as in (7), one can (more generally) construct vectors with values that are a fixed distance apart in the sequence, using *lacunary indices*. More specifically, let $I = \{i_1, i_2, \ldots, i_t\}$ be a given set of integers and define, for an MRG,

$$T_t(I) = \{(u_{i_1+n}, \ldots, u_{i_t+n}) | n \geq 0, s_0 = (x_0, \ldots, x_{k-1}) \in \mathbb{Z}_m^k\}$$

Consider the lattice $L_t(I)$ spanned by $T_t(I)$ and $\mathbb{Z}^t$, and let $d_t(I)$ be the distance between the hyperplanes in this lattice. L'Ecuyer and Couture [77] show how to construct bases for such lattices, how to compute $d_t(I)$, and so on. The following provides "quick-and-dirty" lower bounds on $d_t(I)$ [13, 67]:

1. If I contains all the indices i such that $a_{k-i+1} \neq 0$, then

$$d_t(I) \geq \left(1 + \sum_{j=1}^{k} a_i^2\right)^{-1/2} \tag{10}$$

In particular, if $x_n = (a_r x_{n-r} + a_k x_{n-k}) \bmod m$ and $I = \{0, k-r, k\}$, then $d_3(I) \geq (1 + a_r^2 + a_k^2)^{-1/2}$.

2. If m can be written as $m = \sum_{j=1}^{t} c_{i_j} a^{i_j}$ for some integers c_{i_j}, then

$$d_t(I) \geq \left(\sum_{j=1}^{t} c_{ij}^2 \right)^{-1/2} \tag{11}$$

As a special case of (10), consider the *lagged-Fibonacci generator*, based on a recurrence whose only two nonzero coefficients satisfy $a_r = \pm 1$ and $a_k = \pm 1$. In this case, for $I = \{0, k-r, k\}$, $d_3(I) \geq 1/\sqrt{3} \approx 0.577$. The set of all vectors $(u_n, u_{n+k-r}, u_{n+k})$ produced by such a generator lie in successive parallel planes that are at distance $1/\sqrt{3}$ to each other, and orthogonal to the vector (1,1,1). Therefore, apart from the vector (0,0,0), all other vectors of this form are contained in only two planes! Specific instances of this generator are the one proposed by Mitchell and Moore and recommended by Knuth [57], based on the recurrence $x_n = (x_{n-24} + x_{n-55}) \bmod 2^e$ for e equal to the computer's word length, as well as the `addrans` function in the SUN Unix library [117], based on $x_n = (x_{n-5} + x_{n-17}) \bmod 2^{24}$. These generators should not be used, at least not in their original form.

4.3.6 Combined LCGs and MRGs

Several authors advocated the idea of combining in some way different generators (e.g., two or three different LCGs), hoping that the composite generator will behave better than any of its components alone. See refs. 10, 57, 60, 62, and 87, and dozens of other references given there. Combination can provably increase the period length. Empirical tests show that it typically improves the statistical behavior as well. Some authors (e.g., refs. 8, 46, and 87) have also given theoretical results which (on the surface) appear to "prove" that the output of a combined generator is "more random" than (or at least "as random" as) the output of each of its components. However, these theoretical results make sense only for random variables defined in a probability space setup. For (deterministic) pseudorandom sequences, they prove nothing and can be used only as heuristic arguments to support the idea of combination. To assess the quality of a specific combined generator, one should make a structural analysis of the combined generator itself, not only analyze the individual components and assume that combination will make things more random. This implies that the structural effect of the combination method must be well understood. Law and Kelton [60, Prob. 7.6] give an example where combination makes things worse.

The two most widely known combination methods are:

1. Shuffling one sequence with another or with itself.
2. Adding two or more integer sequences modulo some integer m_0, or adding sequences of real numbers in [0,1] modulo 1, or adding binary fractions bitwise modulo 2.

Shuffling one LCG with another can be accomplished as follows. Fill up a table of size d with the first d output values from the first LCG (suggested values of d go from 2 up to 128 or more). Then each time a random number is needed, generate an index $I \in \{1, \ldots, d\}$ using the $\log_2(d)$ most significant bits of the next output value from the *second* LCG, return (as output of the combined generator) the value stored in the table at position I, then replace this value by the next output value from the *first* LCG. Roughly, the first LCG produces the numbers and the second LCG changes the order of their occurrence. There are several variants of this shuffling scheme. In some of them, the same LCG that produces the numbers to fill up the table is also used to generate the values of I. A large number of empirical investigations performed over the past 30 years strongly support shuffling and many generators available in software libraries use it (e.g., [54, 110, 117]). However, it has two important drawbacks: (1) the effect of shuffling is not well-enough understood from the theoretical viewpoint, and (2) one does not know how to jump ahead quickly to an arbitrary point in the sequence of the combined generator.

The second class of combination method, by modular addition, is generally better understood theoretically. Moreover, jumping ahead in the composite sequence amounts to jumping ahead with each of the individual components, which we know how to do if the components are LCGs or MRGS.

Consider J MRGs evolving in parallel. The jth MRG is based on the recurrence:

$$x_{j,n} = (a_{j,1}x_{j,n-1} + \cdots + a_{j,k}x_{j,n-k}k_j) \bmod m_j$$

for $j = 1, \ldots, J$. We assume that the moduli m_j are pairwise relatively prime and that each recurrence is purely periodic (has zero transient) with period length ρ_j. Let $\delta_1, \ldots, \delta_J$ be arbitrary integers such that for each j, δ_j and m_j have no common factor. Define the two combinations

$$z_n = \left(\sum_{j=1}^{J} \delta_j x_{j,n}\right) \bmod m_1 \qquad u_n = \frac{z_n}{m_1} \tag{12}$$

and

$$w_n = \left(\sum_{j=1}^{J} \delta_j \frac{x_{j,n}}{m_j}\right) \bmod 1 \tag{13}$$

Let $k = \max(k_1, \ldots, k_J)$ and $m = \prod_{j=1}^{J} m_j$. The following results were proved in ref. 80 for the case of MLCG components ($k = 1$) and in ref. 65 for the more general case:

1. The sequences $\{u_n\}$ and $\{w_n\}$ both have period length $\rho = \mathrm{LCM}(\rho_1, \ldots, \rho_J)$ (the least common multiple of the period lengths of the components).
2. The w_n obey the recurrence

$$x_n = (a_1x_{n-1} + \cdots + a_kx_{n-k}) \bmod m \qquad w_n = \frac{x_n}{m} \tag{14}$$

   where the a_i can be computed by a formula given in [65] and do not depend on the δ_j.
3. One has $u_n = w_n + \epsilon_n$, with $\Delta^- \leq \epsilon_n \leq \Delta^+$, where Δ^- and Δ^+ can be computed as explained in ref. 65 and are generally extremely small when the m_j are close to each other.

The combinations (12) and (13) can then be viewed as efficient ways to implement an MRG with very large modulus m. A structural analysis of the combination can be done by analyzing this MRG (e.g., its lattice structure). The MRG components can be chosen with only two nonzero coefficients a_{ij}, both satisfying condition (5), for ease of implementation, and the recurrence of the combination (14) can still have all of its coefficients nonzero and large. If each m_j is an odd prime and each MRG has maximal period length $\rho_j = m_j^{k_j} - 1$, each ρ_j is even, so $\rho \le (m_1^{k_1} - 1)\cdots(m_J^{k_J} - 1)/2^{J-1}$ and this upper bound is attained if the $(m_j^{k_j} - 1)/2$ are pairwise relatively prime [65]. The combination (13) generalizes an idea of Wichmann and Hill [126], while (12) is a generalization of the combination method proposed by L'Ecuyer [61]. The latter combination somewhat scrambles the lattice structure because of the added "noise" ϵ_n.

Example 7 L'Ecuyer [65] proposes the following parameters and gives a computer code in the C language that implements (12). Take $J = 2$ components, $\delta_1 = -\delta_2 = 1$, $m_1 = 2^{31} - 1$, $m_2 = 2^{31} - 2{,}000{,}169$, $k_1 = k_2 = 3$, $(a_{1,1}, a_{1,2}, a_{1,3}) = (0, 63{,}308, -183{,}326)$, and $(a_{2,1}, a_{2,2}, a_{2,3}) = (86{,}098, 0, -539{,}608)$. Each component has period length $\rho_j = m_j^3 - 1$, and the combination has period length $\rho = \rho_1\rho_2/2 \approx 2^{185}$. The MRG (14) that corresponds to the combination is called G10 in Table 2, where distances between hyperplanes for the associated lattice are given. Generator G10 requires four modular products at each step of the recurrence, so it is slower than G9 but faster than G8. The combined MLCG originally proposed by L'Ecuyer [61] also has an approximating LCG called G11 in the table. Note that this combined generator was originally constructed on the basis of the lattice structure of the components only, *without* examining the lattice structure of the combination. Slightly better combinations of the same size have been constructed since this original proposal [80, 77]. Other combinations of different sizes are given in [68].

4.3.7 Matrix LCGs and MRGs

A natural way to generalize LCGs and MRGs is to consider linear recurrences for vectors, with matrix coefficients

$$X_n = (A_1 X_{n-1} + \cdots + A_k X_{n-k}) \bmod m \tag{15}$$

where $A_1, \ldots, A_k$ are $L \times L$ matrices and each X_n is an L-dimensional vector of elements of $\mathbb{Z}_m$, which we denote by

$$X_n = \begin{pmatrix} x_{n,1} \\ \vdots \\ x_{n,L} \end{pmatrix}$$

At each step, one can use each component of X_n to produce a uniform variate: $u_{nL+j-1} = x_{n,j}/m$. Niederreiter [105] introduced this generalization and calls it the *multiple recursive matrix method* for the generation of vectors. The recurrence (15) can also be written as a *matrix LCG* of the form $\mathbf{X}_n = \mathbf{A}\mathbf{X}_{n-1} \bmod m$, where

$$\mathbf{A} = \begin{pmatrix} 0 & I & \cdots & 0 \\ \vdots & \vdots & \ddots & \vdots \\ 0 & 0 & \cdots & I \\ A_k & A_{k-1} & \cdots & A_1 \end{pmatrix} \quad \text{and} \quad \mathbf{X}_n = \begin{pmatrix} X_n \\ X_{n+1} \\ \vdots \\ X_{n+k-1} \end{pmatrix} \tag{16}$$

are a matrix of dimension $kL \times kL$ and a vector of dimension kL, respectively (here I is the $L \times L$ identity matrix). This matrix notation applies to the MRG as well, with $L = 1$.

Is the matrix LCG more general than the MRG? Not much. If a k-dimensional vector X_n follows the recurrence $X_n = AX_{n-1} \bmod m$, where the $k \times k$ matrix A has a primitive characteristic polynomial $P(z) = z^k - a_1 z^{k-1} - \cdots - a_k$, then X_n also follows the recurrence [48, 62, 101]

$$X_n = (a_1 X_{n-1} + \cdots + a_k X_{n-k}) \bmod m \tag{17}$$

So each component of the vector X_n evolves according to (2). In other words, one simply has k copies of the same MRG sequence in parallel, usually with some shifting between those copies. This also applies to the matrix MRG (15), since it can be written as a matrix LCG of dimension kL and therefore corresponds to kL copies of the same MRG of order kL (and maximal period length $m^{kL} - 1$). The difference with the single MRG (2) is that instead of taking successive values from a single sequence, one takes values from different copies of the same sequence, in a round-robin fashion. Observe also that when using (17), the dimension of X_n in this recurrence (i.e., the number of parallel copies) does not need to be equal to k.

4.3.8 Linear Recurrences with Carry

Consider a generator based on the following recurrence:

$$x_n = (a_1 x_{n-1} + \cdots + a_k x_{n-k} + c_{n-1}) \bmod b \tag{18}$$

$$c_n = (a_1 x_{n-1} + \cdots + a_k x_{n-k} + c_{n-1}) \text{ div } b \tag{19}$$

$$u_n = \frac{x_n}{b}.$$

where "div" denotes the integer division. For each n, $x_n \in \mathbb{Z}_b$, $c_n \in \mathbb{Z}$, and the state at step n is $s_n = (x_n, \ldots, x_{n+k-1}, c_n)$. As in [14, 16, 88], we call this a *multiply-with-carry* (MWC) generator. The idea was suggested in [58, 91]. The recurrence looks like that of an MRG, except that a *carry* c_n is propagated between the steps. What is the effect of this carry?

Assume that b is a power of 2, which is very nice form the implementation viewpoint. Define $a_0 = -1$,

$$m = \sum_{l=0}^{k} a_l b^l$$

and let a be such that $ab \bmod m = 1$ (a is the inverse of b in arithmetic modulo m). Note that m could be either positive or negative, but for simplicity we now assume that $m > 0$. Consider the LCG:

$$z_n = az_{n-1} \bmod m \qquad w_n = \frac{z_n}{m} \tag{20}$$

There is a close correspondence between the LCG (20) and the MWC generator, assuming that their initial states agree [16]. More specifically, if

$$w_n = \sum_{i=1}^{\infty} x_{n+i-1} b^{-i} \tag{21}$$

holds for $n = 0$, then it holds for all n. As a consequence, $|u_n - w_n| \leq 1/b$ for all n. For example, if $b = 2^{32}$, then u_n and w_n are the same up to 32 bits of precision! The MWC generator can thus be viewed as just another way to implement (approximately) a LCG with huge modulus and period length. It also inherits from this LCG an approximate lattice structure, which can be analyzed as usual.

The LCG (20) is purely periodic, so each state z_n is *recurrent* (none is transient). On the other hand, the MWC has an infinite number of states (since we imposed no bound on c_n) and most of them turn out to be transient. How can one characterize the recurrent states? They are (essentially) the states s_0 that correspond to a given z_0 via (20)–(21). Couture and L'Ecuyer [16] give necessary and sufficient conditions for a state s_0 to be recurrent. In particular, if $a_l \geq 0$ for $l \geq 1$, all the recurrent states satisfy $0 \leq c_n < a_1 + \cdots + a_k$. In view of this inequality, we want the a_l to be small, for their sum to fit into a computer word. More specifically, one can impose $a_1 + \cdots + a_k \leq b$. Now b is a nice upper bound on the c_n as well as on the x_n.

Since b is a power of 2, a is a quadratic residue and so cannot be primitive mod m. Therefore, the period length cannot reach $m - 1$ even if m is prime. But if $(m - 1)/2$ is odd and 2 is primitive mod m [e.g., if $(m - 1)/2$ is prime], then (20) has period length $\rho = (m - 1)/2$.

Couture and L'Ecuyer [16] show that the lattice structure of the LCG (20) satisfies the following: In dimensions $t \leq k$, the distances d_t do not depend on the parameters $a_1, \ldots, a_k$, but only on b, while in dimension $t = k + 1$, the shortest vector in the dual lattice to L_t is $(a_0, \ldots, a_k)$, so that

$$d_t = (1 + a_1^2 + \cdots + a_k^2)^{-1/2} \tag{22}$$

The distance d_{k+1} is then minimized if we put all the weight on one coefficient a_l. It is also better to put more weight on a_k, to get a larger m. So one should choose a_k close to b, with $a_0 + \cdots + a_k \leq b$. Marsaglia [88] proposed two specific parameter sets. They are analyzed in [16], where a better set of parameters in terms of the lattice structure of the LCG is also given.

Special cases of the MWC include the add-with-carry (AWC) and subtract-with-borrow (SWB) generators, originally proposed by Marsaglia and Zaman [91] and subsequently analyzed in [13, 122]. For the AWC, put $a_r = a_k = -a_0 = 1$ for $0 < r < k$ and all other a_l equal to zero. This gives the simple recurrence

$$x_n = (x_{n-r} + x_{n-k} + c_{n-1}) \bmod b$$
$$c_n = I[x_{n-r} + x_{n-k} + c_{n-1} \geq b]$$

where I denotes the indicator function, equal to 1 if the bracketted inequality is true and to 0 otherwise. The SWB is similar, except that either a_r or a_k is -1 and the carry c_n is 0 or -1. The correspondence between AWC/SWB generators and LCGs was established in ref. 122.

Equation (22) tells us very clearly that all AWC/SWB generators have a bad lattice structure in dimension $k + 1$. A little more can be said when looking at the lacunary indices: For $I = \{0, r, k\}$, one has $d_3(I) = 1/\sqrt{3}$ and all vectors of the form (w_n, w_{n+r}, w_{n+k}) produced by the LCG (20) lie in only two planes in the three-dimensional unit cube, exactly as for the lagged-Fibonacci generators discussed in Section 4.3.5. Obviously, this is bad.

Perhaps one way to get around this problem is to take only k successive output values, then skip (say) ν values, take another k successive ones, skip another ν, and so on. Lüscher [85] has proposed such an approach, with specific values of ν for a specific SWB generator, with theoretical justification based on chaos theory. James [56] gives a Fortran implementation of Lüscher's generator. The system *Mathematica* uses a SWB generator ([127, p. 1019]), but the documentation does not specify if it skips values.

4.3.9 Digital Method: LFSR, GFSR, TGFSR, etc., and Their Combination

The MRG (2), matrix MRG (15), combined MRG (12), and MWC (18–19) have resolution $1/m$, $1/m$, $1/m_1$, and $1/b$, respectively. (The *resolution* is the largest number x such that all output values are multiples of x.) This could be seen as a limitation. To improve the resolution, one can simply take several successive x_n to construct each output value u_n. Consider the MRG. Choose two positive integers s and $L \leq k$, and redefine

$$u_n = \sum_{j=1}^{L} x_{ns+j-1} m^{-j} \tag{23}$$

Call s the *step size* and L the *number of digits* in the m-adic expansion. The state at step n is now $s_n = (x_{ns}, \ldots, x_{ns+k-1})$. The output values u_n are multiples of m^{-L} instead of m^{-1}. This output sequence, usually with $L = s$, is called a *digital multistep sequence* [64, 102]. Taking $s > L$ means that $s - L$ values of the sequence $\{x_n\}$ are skipped at each step of (23). If the MRG sequence has period ρ and if s has no common factor with ρ, the sequence $\{u_n\}$ also has period ρ.

Now, it is no longer necessary for m to be large. A small m with large s and L can do as well. In particular, one can take $m = 2$. Then $\{x_n\}$ becomes a sequence of bits (zeros and ones) and the u_n are constructed by juxtaposing L successive bits from this sequence. This is called a *linear feedback shift register* (LFSR) or *Tausworthe generator* [41, 64, 102, 118], although the bits of each u_n are often filled in reverse order than in (23). An efficient computer code that implements the sequence (23), for the case where the recurrence has the form $x_n = (x_{n-r} + x_{n-k}) \bmod 2$ with $s \leq r$ and $2r > k$, can be found in refs. 66, 120, and 121. For specialized jump-ahead algorithms, see [22, 66]. Unfortunately, such simple recurrences lead to LFSR generators with bad structural properties (see refs. 11, 66, 97, and 120 and other references therein), but combining several recurrences of this type can give good generators.

Consider J LFSR generators, where the jth one is based on a recurrence $\{x_{j,n}\}$ with primitive characteristic polynomial $P_j(z)$ of degree k_j (with binary coefficients), an m-adic expansion to L digits, and a step size s_j such that s_j and the period length $\rho_j = 2^{k_j} - 1$ have no common factor. Let $\{u_{j,n}\}$ be the output sequence of the jth generator and define u_n as the bitwise exclusive-or (i.e., bitwise addition modulo 2) of $u_{1,n}, \ldots, u_{J,n}$. If the polynomials $P_1(z), \ldots, P_J(z)$ are pairwise relatively prime (no pair of polynomials has a common factor), the period length ρ of the combined sequence $\{u_n\}$ is equal to the least common multiple of the individual periods $\rho_1, \ldots, \rho_J$. These ρ_j can be relatively prime, so it is possible here to have $\rho = \prod_{j=1}^{J} \rho_j$. The resulting combined generator is also exactly equivalent to a LFSR generator based on a recurrence with characteristic polynomial $P(z) = P_1(z) \cdots P_J(z)$. All of this is shown in [121], where specific combinations with two components are also suggested. For good combinations with more components, see ref. 66. Wang and Compagner [125] also suggested similar combinations, with much longer periods. They recommended constructing the combination so that the polynomial $P(z)$ has approximately half of its coefficients equal to 1. In a sense, the main justification for combined LFSR generators is the efficient implementation of a generator based on a (reducible) polynomial $P(z)$ with many nonzero coefficients.

The digital method can be applied to the matrix MRG (15) or to the parallel MRG (17) by making a digital expansion of the components of X_n (assumed to have dimension L):

$$u_n = \sum_{j=1}^{L} x_{n,j} m^{-j} \tag{24}$$

The combination of (15) with (24) gives the *multiple recursive matrix method* of Niederreiter [103]. For the matrix LCG, L'Ecuyer [64] shows that if the shifts between the successive L copies of the sequence are all equal to some integer d having no common factor with the period length $\rho = m^k - 1$, the sequence (24) is exactly the same as the digital multistep sequence (23) with s equal to the inverse of d modulo m. The converse also holds. In other words, (23) and (24), with these conditions on the shifts, are basically two different implementations of the same generator. So one can be analyzed by analyzing the other, and vice versa. If one uses the implementation (24), one must be careful with the initialization of $X_0, \ldots, X_{k-1}$ in (17) to maintain the correspondence: The shift between the states $(x_{0,j}, \ldots, x_{k-1,j})$ and $(x_{0,j+1}, \ldots, x_{k-1,j+1})$ in the MRG sequence must be equal to the proper value d for all j.

The implementation (24) requires more memory than (23), but may give a faster generator. An important instance of this is the *generalized feedback shift register* (GFSR) *generator* [43, 84, 123], which we now describe. Take $m = 2$ and L equal to the computer's word length. The recurrence (17) can then be computed by a bitwise exclusive-or of the X_{n-j} for which $a_j = 1$. In particular, if the MRG recurrence has only two nonzero coefficients, say a_k and a_r, we obtain

$$X_n = X_{n-r} \oplus X_{n-k}$$

where $\oplus$ denotes the bitwise exclusive-or. The output is then constructed via the binary fractional expansion (24). This GFSR can be viewed as a different way to implement a LFSR generator, provided that it is initialized accordingly, and the structural properties of the GFSR can then be analyzed by analyzing those of the corresponding LFSR generator [44, 64].

For the recurrence (17), we need to memorize kL integers in $\mathbb{Z}_m$. With this memory size, one should expect a period length close to m^{kL}, but the actual period length cannot exceed $m^k - 1$. A big waste! Observe that (17) is a special case of (15), with $A_i = a_i I$. An interesting idea is to "twist" the recurrence (17) slightly so that each $a_i I$ is replaced by a matrix A_i such that the corresponding recurrence (15) has full period length $m^{kL} - 1$ while its implementation remains essentially as fast as (17). Matsumoto and Kurita [95, 96] proposed a specific way to do this for GFSR generators and called the resulting generators *twisted* GFSR (TGFSR). Their second paper and ref. 98 and 120 point out some defects in the generators proposed in their first paper, proposes better specific generators, and give nice computer codes in C. Investigations are currently made to find other twists with good properties. The multiple recursive matrix method of ref. 103 is a generalization of these ideas.

4.3.10 Equidistribution Properties for the Digital Method

Suppose that we partition the unit hypercube $[0, 1)^t$ into m^{tl} cubic cells of equal size. This is called a *(t, l)-equidissection in base m*. A set of points is said to be (t, l)-equidistributed if each cell contains the same number of points from that set. If the set contains m^k points, the (t, l)-equidistribution is possible only for $l \le \lfloor k/t \rfloor$. For a given digital multistep sequence, let

$$T_t\{\mathbf{u}_0 = (u_0, \ldots, u_{t-1}) | (x_0, \ldots, x_{k-1}) \in \mathbb{Z}_m^k\} \tag{25}$$

(where repeated points are counted as many times as they appear in T_t) and $l_t = \min(L, \lfloor k/t \rfloor)$. If the set T_t is (t, l_t)-equidistributed for all $t \le k$, we call it a *maximally equidistributed* (ME) set and say that the generator is ME. If it has the additional property that for all t, for $l_t < l \le L$, no cell of the (t, l)-equidissection contains more than one point, we also call it collision-free (CF). ME-CF generators have their sets of points T_t very evenly distributed in the unit hypercube, in all dimensions t.

Full-period LFSR generators are all $(\lfloor k/s \rfloor, s)$-equidistributed. Full-period GFSR generators are all $(k, 1)$-equidistributed, but their (k, l)-equidistribution for $l > 1$ depends on the initial state (i.e., on the shifts between the different copies of the MRG). Fushimi and Tezuka [45] give a necessary and sufficient condition on this initial state for (t, L)-equidistribution, for $t = \lfloor k/L \rfloor$. The condition says that the tL bits $(x_{0,1}, \ldots, x_{0,L}, \ldots, x_{t-1,1}, \ldots, x_{t-1,L})$ must be independent, in the sense that the $tL \times k$ matrix which expresses them as a linear transformation of $(x_{0,1}, \ldots, x_{k-1,1})$ has (full) rank tL. Fushimi [44] gives an initialization procedure satisfying this condition.

Couture et al. [17] show how the (t, l)-equidistribution of simple and combined LFSR generators can be analyzed via the lattice structure of an equivalent LCG in a space of formal series. A different (simpler) approach is taken in ref. 66: Check if the matrix that expresses the first l bits of $\mathbf{u}_n$ as a linear transformation of $(x_0, \ldots, x_{k-1})$ has full rank. This is a necessary and sufficient condition for (t, l)-equidistribution.

An ME LFSR generator based on the recurrence $x_n = (x_{n-607} + x_{n-273}) \bmod 2$, with $s = 512$ and $L = 23$, is given in ref. 23. But as stated previously, only two nonzero coefficients for the recurrence is much too few. L'Ecuyer [66, 70] gives the results of computer searches for ME and ME-CF combined LFSR generators with $J = 2, 3, 4, 5$ components, as described in subSection 4.3.9. Each search was made within a class with each component j based on a characteristic trinomial $P_j(z) = z^{k_j} - z^{r_j} - 1$, with $L = 32$ or $L = 64$, and step size s_j such that $s_j \leq r_j$ and $2r_j > k_j$. The period length is $\rho = (2^{k_1} - 1) \cdots (2^{k_J} - 1)$ in most cases, sometimes slightly smaller. The searches were for good parameters r_j and s_j. We summarize here a few examples of search results. For more details, as well as specific implementations in the C language, see refs. 66 and 70.

Example 8

(a) For $J = 2$, $k_1 = 31$, and $k_2 = 29$, there are 2565 parameter sets that satisfy the conditions above. None of these combinations is ME. Specific combinations which are nearly ME, within this same class, can be found in ref. 21.

(b) Let $J = 3$, $k_1 = 31$, $k_2 = 29$, and $k_3 = 28$. In an exhaustive search among 82,080 possibilities satisfying our conditions within this class, 19 ME combinations were found, and 3 of them are also CF.

(c) Let $J = 4$, $k_1 = 31$, $k_2 = 29$, $k_3 = 28$, and $k_4 = 25$. Here, in an exhaustive search among 3,283,200 possibilities, we found 26,195 ME combinations, and 4,744 of them also CF.

These results illustrate the fact that ME combinations are much easier to find as J increases. This appears to be due to more possibilities to "fill up" the coefficients of $P(z)$ when it is the product of more trinomials. Since GFSR generators can be viewed as a way to implement fast LFSR generators, these search methods and results can be used as well to find good combined GFSRs, where the combination is defined by a bitwise exclusive-or as in the LFSR case.

One may strengthen the notion of (t, l)-equidistribution as follows: Instead of looking only at equidissections comprised of cubic volume elements of identical sizes, look at more general partitions. Such a stronger notion is that of a *(q, k, t)-net in base m*, where there should be the same number of points in each box for *any* partition of the unit hypercube into rectangular boxes of identical shape and equal volume m^{q-k}, with the length of each side of the box equal to a multiple of $1/m$. Niederreiter [102] defines a figure of merit $r^{(t)}$ such that for all $t > \lfloor k/L \rfloor$, the m^k points of T_t for (23) form a (q, k, t)-net in base m with $q = k - r^{(t)}$. A problem with $r^{(t)}$ is the difficulty to compute it for medium and large t (say, $t > 8$).

4.4 NONLINEAR METHODS

An obvious way to remove the linear (and perhaps too regular) structure is to use a *nonlinear* transformation. There are basically two classes of approaches:

1. Keep the transition function T linear, but use a nonlinear transformation G to produce the output.
2. Use a nonlinear transition function T.

Several types of nonlinear generators have been proposed over the last decade or so, and an impressive volume of theoretical results have been obtained for them. See, for example, refs. 31, 34, 59, 78, 102, and 104 and other references given there. Here, we give a brief overview of this rapidly developing area.

Nonlinear generators avoid lattice structures. Typically, no t-dimensional hyperplane contains more than t overlapping t-tuples of successive values. More important, their output behaves much like "truly" random numbers, even over the entire period, with respect to discrepancy. Roughly, there are lower and upper bounds on their discrepancy (or in some cases on the average discrepancy over a certain set of parameters) whose asymptotic order (as the period length increases to infinity) is the same as that of an IID $U(0, 1)$ sequence of random variables. They have also succeeded quite well in empirical tests performed so far [49]. Fast implementations with specific well-tested parameters are still under development, although several generic implementations are already available [49, 71].

4.4.1 Inversive Congruential Generators

To construct a nonlinear generator with long period, a first idea is simply to add a nonlinear twist to the output of a known generator. For example, take a full-period MRG with prime modulus m and replace the output function $u_n = x_n/m$ by

$$z_n = (\tilde{x}_{n+1}\tilde{x}_n^{-1}) \bmod m \quad \text{and} \quad u_n = \frac{z_n}{m} \tag{26}$$

where $\tilde{x}_i$ denotes the ith nonzero value in the sequence $\{x_n\}$ and $\tilde{x}_n^{-1}$ is the inverse of $\tilde{x}_n$ modulo m. (The zero values are skipped because they have no inverse.) For $x_n \neq 0$, its inverse x_n^{-1} can be computed by the formula $x_n^{-1} = x_n^{m-2} \bmod m$, with $O(\log m)$ multiplications modulo m. The sequence $\{z_n\}$ has period m^{k-1}, under conditions given in refs. 31 and 102. This class of generators was introduced and first studied in refs. 28, 27, and 30. For $k = 2$, (26) is equivalent to the recurrence

$$z_n = \begin{cases} (a_1 + a_2 z_{n-1}^{-1}) \bmod m & \text{if } z_{n-1} \neq 0 \\ a_1 & \text{if } z_{n-1} = 0 \end{cases} \tag{27}$$

where a_1 and a_2 are the MRG coefficients.

A more direct approach is the *explicit inversive congruential* method of ref. 32, defined as follows. Let $x_n = an + c$ for $n \geq 0$, where $a \neq 0$ and c are in $\mathbb{Z}_m$ and m is prime. Then, define

$$z_n = x_n^{-1} = (an + c)^{m-2} \bmod m \quad \text{and} \quad u_n = \frac{z_n}{m} \tag{28}$$

This sequence has period $\rho = m$. According to ref. 34, this family of generators seems to enjoy the most favorable properties among the currently proposed inversive and quadratic families. As a simple illustrative example, take $m = 2^{31} - 1$ and $a = c = 1$. (However, at the moment, we are not in a position to recommend these particular parameters nor any other specific ones.)

Inversive congruential generators with power-of-2 moduli have also been studied [30, 31, 35]. However, they have have more regular structures than those based on prime moduli [31, 34]. Their low-order bits have the same short period lengths as for the LCGs. The idea of combined generators, discussed earlier for the linear case, also applies to nonlinear generators and offers some computational advantages. Huber [52] and Eichenauer-Herrmann [33] introduced and analyzed the following method. Take J inversive generators as in (27), with distinct prime moduli $m_1, \ldots, m_J$, all larger than 4, and full period length $\rho_j = m_j$. For each generator j, let $z_{j,n}$ be the state at step n and let $u_{j,n} = z_{j,n}/m_j$. The output at step n is defined by the following combination:

$$u_n = (u_{1,n} + \cdots + u_{J,n}) \bmod 1$$

The sequence $\{u_n\}$ turns out to be equivalent to the output of an inversive generator (27) with modulus $m = m_1 \cdots m_J$ and period length $\rho = m$. Conceptually, this is pretty similar to the combined LCGs and MRGs discussed previously, and provides a convenient way to implement an inversive generator with large modulus m. Eichenauer-Herrmann [33] shows that this type of generator has favorable asymptotic discrepancy properties, much like (26–28).

4.4.2 Quadratic Congruential Generators

Suppose that the transformation T is *quadratic* instead of linear. Consider the recurrence

$$x_n = (ax_{n-1}^2 + bx_{n-1} + c) \bmod m$$

where $a, b, c \in \mathbb{Z}_m$ and $x_n \in \mathbb{Z}_m$ for each n. This is studied in refs. 29, 37, 57, and 102. If m is a power of 2, this generator has full period ($\rho = m$) if and only if a is even, $(b - a) \bmod 4 = 1$, and c is odd. Its t-dimensional points turn out to lie on a union of grids. Also, the discrepancy tends to be too large. Our usual caveat against power-of-2 moduli applies again.

4.4.3 BBS and Other Cryptographic Generators

The BBS generator, explained in Section 4.2, is conjectured to be polynomial-time perfect. This means that for a large enough size k, a BBS generator with properly (randomly) chosen parameters is practically certain to behave very well from the statistical point of view. However, it is not clear how large k must be and how K can be chosen in practice for the generator to be really safe. The speed of the generator slows down with k, since at each step we must square a $2k$-bit integer modulo another $2k$-bit integer. An implementation based on fast modular multiplication is proposed by Moreau [99].

Other classes of generators, conjectured to be polynomial-time perfect, have been proposed. From empirical experiments, they have appeared no better than the BBS. See refs. 5, 59, and 78 for overviews and discussions. An interesting idea, pursued for instance in ref. 1, is to combine a slow but cryptographically strong generator (e.g., a polynomial-time perfect one) with a fast (but unsecure) one. The slow generator is used sparingly, mostly in a preprocessing step. The result is an interesting compromise between speed, size, and security. In ref. 1, it is also suggested to use a block cipher encryption algorithm for the slow generator. These authors actually use triple-DES (three passes over the well-known data encryption standard algorithm, with three different keys), combined with a linear hashing function defined by a matrix. The keys and the hashing matrix must be (truly) random. Their fast generator is implemented with a six-regular expander graph (see their paper for more details).

4.5 EMPIRICAL STATISTICAL TESTING

Statistical testing of random number generators is indeed a very empirical and heuristic activity. The main idea is to seek situations where the behavior of some function of the generator's output is significantly different than the normal or expected behavior of the same function applied to a sequence of IID uniform random variables.

Example 9 As a simple illustration, suppose that one generates n random numbers from a generator whose output is supposed to imitate IID $U(0,1)$ random variables. Let T be the number of values that turn out to be below $\frac{1}{2}$, among those n. For large n, T should normally be not too far from $n/2$. In fact, one should expect T to behave like a binomial random variable with parameters $(n, \frac{1}{2})$. So if one repeats this experiment several times (e.g., generating N values of T), the distribution of the values of T obtained should resemble that of the binomial distribution (and the normal distribution with mean $n/2$ and standard deviation $\sqrt{n}/2$ for large n). If $N = 100$ and $n = 10000$, the mean and standard deviation are 5000 and 50, respectively. With these parameters, if one observes, for instance, that 12 values of T are less than 4800, or that 98 values of T out of 100 are less than 5000, one would readily conclude that something is wrong with the generator. On the other hand, if the values of T behave as expected, one may conclude that the generator seems to reproduce the correct behavior *for this particular statistic* T (and for this particular sample size). But nothing prevents *other* statistics than this T to behave wrongly.

4.5.1 General Setup

Define the null hypothesis H_0 as: "The generator's output is a sequence of IID $U(0,1)$ random variables." Formally, this hypothesis is false, since the sequence is periodic and usually deterministic (except perhaps for the seed). But if this cannot be detected by reasonable statistical tests, one may assume that H_0 holds anyway. In fact, what really counts in the end is that the statistics of interest in a given simulation have (sample) distributions close enough to their theoretical ones.

A statistical test for H_0 can be defined by any function T of a finite number of $U(0,1)$ random variables, for which the distribution under H_0 is known or can be approximated well enough. The random variable T is called the *test statistic*. The statistical test tries to find empirical evidence against H_0.

When applying a statistical test to a random number generator, a *single-level* procedure computes the value of T, say t_1, then computes the p-value

$$\delta_1 = P[T > t_1 | H_0]$$

and, in the case of a two-sided test, rejects H_0 if δ_1 is too close to either 0 or 1. A single-sided test will reject only if δ_1 is too close to 0, or only if it is too close to 1.

The choice of rejection area depends on what the test aims to detect. Under H_0, δ_1 is a $U(0, 1)$ random variable.

A *two-level* test obtains (say) N "independent" copies of T, denoted $T_1, \ldots, T_N$, and computes their empirical distribution $\hat{F}_N$. This empirical distribution is then compared to the theoretical distribution of T under H_0, say F, via a standard goodness-of-fit test, such as the Kolmogorov–Smirnov (KS) or Anderson-Darling tests [25, 115]. One version of the KS goodness-of-fit test uses the statistic

$$D_n = \sup_{-\infty < x < \infty} |\hat{F}_N(x) - F(x)|$$

for which an approximation of the distribution under H_0 is available, assuming that the distribution F is continuous [25]. Once the value d_N of the statistic D_N is known, one computes the p-value of the test, defined as

$$\delta_2 = P[D_N > d_N | H_0]$$

which is again a $U(0, 1)$ random variable under H_0. Here one would reject H_0 if δ_2 is too close to 0.

Choosing $N = 1$ yields a single-level test. For a given test and a fixed computing budget, the question arises of what is best: To choose a small N (e.g., $N = 1$) and base the test statistic T on a large sample size, or the opposite? There is no universal winner. It depends on the test and on the alternative hypothesis. The rationale for two-level testing is to test the sequence not only globally, but also locally, by looking at the distribution of values of T over shorter subsequences [57]. In most cases, when testing random number generators, $N = 1$ turns out to be the best choice because the same regularities or defects of the generators tend to repeat themselves over all long-enough subsequences. But it also happens for certain tests that the cost of computing T increases faster than linearly with the sample size, and this gives another argument for choosing $N > 1$.

In statistical analyses where a limited amount of data is available, it is common practice to fix some significance level α in advance and reject H_0 when and only when the p-value is below α. Popular values of α are 0.05 and 0.01 (mainly for historical reasons). When testing random number generators, one can always produce an arbitrary amount of data to make the test more powerful and come up with a clean-cut decision when suspicious p-value occur. We would thus recommend the following strategy. If the outcome is clear, for example if the p-value is less than 10^{-10}, reject H_0. Otherwise, if the p-value is suspicious (0.005, for example), then increase the sample size or repeat the test with other segments of the sequence. In most cases, either suspicion will disappear or clear evidence against H_0 will show up rapidly.

When H_0 is not rejected, this somewhat improves confidence in the generator but never proves that it will always behave correctly. It may well be that the next test T to be designed will be the one that catches the generator. Generally speaking, the more extensive and varied is the set of tests that a given generator has passed, the more faith we have in the generator. For still better confidence, it is always a good idea to run important simulations twice (or more), using random number generators of totally different types.

4.5.2 Available Batteries of Tests

The statistical tests described by Knuth [57] have long been considered the "standard" tests for random number generators. A Fortran implementation of (roughly) this set of tests is given in the package TESTRAND [24]. A newer battery of tests is DIEHARD, designed by Marsaglia [87, 89]. It contains more stringent tests than those in ref. 57, in the sense that more generators tend to fail some of the tests. An extensive testing package called `TestU01` [71], that implements most of the tests proposed so far, as well as several classes of generators implemented in generic form, is under development. References to other statistical tests applied to random number generators can be found in refs. 63, 64, 71, 75, 74, 69, 79, and 116.

Simply testing uniformity, or pair correlations, is far from enough. Good tests are designed to catch higher-order correlation properties or geometric patterns of the successive numbers. Such patterns can easily show up in certain classes of applications [39, 49, 75]. Which are the best tests? No one can really answer this question. If the generator is to be used to estimate the expectation of some random variable T by generating replicates of T, the best test would be the one based on T as a statistic. But this is impractical, since if one knew the distribution of T, one would not use simulation to estimate its mean. Ideally, a good test for this kind of application should be based on a statistic T' whose distribution is known and resembles that of T. But such a test is rarely easily available. Moreover, only the user can apply it. When designing a general purpose generator, one has no idea of what kind of random variable interests the user. So, the best the designer can do (after the generator has been properly designed) is to apply a wide variety of tests that tend to detect defects of different natures.

4.5.3 Two Examples of Empirical Tests

For a short illustration, we now apply two statistical tests to some of the random number generators discussed previously. The first test is a variant of the well-known *serial test* and the second one is a *close-pairs* test. More details about these tests, as well as refined variants, can be found in refs. 57, 74, 75, 79.

Both tests generate n nonoverlapping vectors in the t-dimensional unit cube $[0, 1)^t$. That is, they produce the point set:

$$P_t = \{\mathbf{U}_i = (U_{t(i-1)}, \ldots, U_{ti-1}), i = 1, \ldots, n\}$$

where $U_0, U_1, \ldots$ is the generator's output. Under H_0, P_t contains n IID random vectors uniformly distributed over the unit hypercube.

For the serial test, we construct a (t, l)-equidissection in base 2 of the hypercube (see Section 4.3.10), and compute how many points fall in each of the $k = 2^{tl}$ cells. More specifically, let X_j be the number of points $\mathbf{U}_i$ falling in cell j, for $j = 1, \ldots, k$, and define the chi-square statistic

$$X^2 = \sum_{j=1}^{k} \frac{(X_j - n/k)^2}{n/k} \tag{29}$$

Under H_0, the exact mean and variance of X^2 are $\mu = E[X^2] = k - 1$ and $\sigma^2 = \text{Var}[X^2] = 2(k-1)(n-1)/n$, respectively. Moreover, if $n \to \infty$ for fixed k, X^2 converges in distribution to a chi-square random variable with $k-1$ degrees of freedom, whereas if $n \to \infty$ and $k \to \infty$ simultaneously so that $n/k \to \gamma$ for some constant γ, $(X^2 - \mu)/\sigma$ converges in distribution to a $N(0, 1)$ (a standard normal) random variable. Most authors use a chi-square approximation to the distribution of X^2, with $n/k \geq 5$ (say) and very large n. But one can also take $k \gg n$ and use the normal approximation, as in the forthcoming numerical illustration.

For the close-pairs test, let $D_{n,i,j}$ be the Euclidean distance between the points $\mathbf{U}_j$ and $\mathbf{U}_i$ in the unit *torus*, i.e., where the opposite faces of the hypercube are identified so that points facing each other on opposite sides become close to each other. For $s \geq 0$ let $Y_n(s)$ be the number of distinct pairs of points $i < j$ such $D^t_{n,i,j} V_t n(n-1)/2 \leq s$, where V_t is the volume of a ball of radius 1 in the t-dimensional real space. Under H_0, for any constant $s_1 > 0$, as $n \to \infty$, the process $\{Y_n(s), 0 \leq s \leq s_1\}$ converges weakly to a Poisson process with unit rate. Let $0 = T_{n,0} \leq T_{n,1} \leq T_{n,2} \leq \cdots$ be the jump times of the process Y_n, and let $W_{n,i} = 1 - \exp[-(T_{n,i} - T_{n,i-1})]$. For a fixed integer $m > 0$ and large enough n, the random variables $W_{n,1}, \ldots, W_{n,m}$ are approximately IID $U(0, 1)$ under H_0. To compare their empirical distribution to the uniform, one can compute, for example, the Anderson-Darling statistic

$$A_m^2 = -m - \frac{1}{m} \sum_{i=1}^{m} \{(2i-1)\ln(W_{(n,i)}) + (2m+1-2i)\ln(1 - W_{(n,i)}\}$$

and reject H_0 if the p-value is too small (i.e., if A_m^2 is too large).

These tests have been applied to the generators G1 to G11 in Tables 4.1 and 4.2. We took $N = 1$ and dimension $t = 3$. We applied two instances of the serial test, one named ST1, with $n = 2^{20}$ and $l = 9$, which gives $k = 2^{27}$ and $n/k = 1/128$, and the second one named ST2, with $n = 2^{22}$ and $l = 10$, so $k = 2^{30}$ and $n/k = 1/256$. For the close-pairs (CP) test, we took $n = 2^{18}$ and $m = 32$. In each case, $3n$ random numbers were used, and this value is much smaller than the period length of the generators tested. For all generators, at the beginning of the first test, we used the initial seed 12345 when a single integer was needed and the vector (12,345, ... , 12,345) when a vector was needed. The seed was not reset between the tests. Table 4.3 gives the p-values of these tests for G1 to G5. For G6 to G11, all p-values remained inside the interval (0.01, 0.99).

For the serial test, the p-values that are too close to 1 (e.g., ST1 and ST2 for G1) indicate that the n points are too evenly distributed among the k cells compared to what one would expect from random points (X^2 is too small). On the other hand, the very small p-values indicate that the points tend to go significantly more often in certain cells

TABLE 4.3 The p-Values of Two Empirical Tests Applied to Generators G1 to G11

Generator	ST1	ST2	CP
G1	$1 - 9.97 \times 10^{-6}$	$>1 - 10^{-15}$	$<10^{-15}$
G2	0.365	$<10^{-15}$	$<10^{-15}$
G3	$1 - 2.19 \times 10^{-4}$	$<10^{-15}$	$<10^{-15}$
G4	$< 10^{-15}$	$<10^{-15}$	$<10^{-15}$
G5	0.950	$>1 - 10^{-15}$	$<10^{-15}$

than in others (X^2 is too large). The p-values less than 10^{-15} for the CP test stem from the fact that the jumps of the process Y_n tend to be clustered (and often superposed), because there are often equalities (or almost) among the small $D_{n,i,j}$'s, due to the lattice structure of the generator [75, 112]. This implies that several $W_{n,i}$ are very close to zero, and the Anderson-Darling statistic is especially sensitive for detecting this type of problem. As a general rule of thumb, *all* LCGs and MRGs, whatever be the quality of their lattice structure, fail spectacularly this close-pairs test with $N = 1$ and $m = 32$ when n exceeds the square root of the period length [75].

G6 and G7 pass these tests, but will soon fail both tests if we increase the sample size. For G8 to G11, on the other hand, the sample size required for clear failure is so large that the test becomes too long to run in reasonable time. This is especially true for G8 and G10.

One could raise the issue of whether these tests are really relevant. As mentioned in the previous subsection, the relevant test statistics are those that behave similarly as the random variable of interest to the user. So, relevance depends on the application. For simulations that deal with random points in space, the close-pairs test could be relevant. Such simulations are performed, for example, to estimate the (unknown) distribution of certain random variables in spatial statistics [19]. As an illustration, suppose one wishes to estimate the distribution of $\min_{i,j} D_{n,i,j}$ for some fixed n, by Monte Carlo simulation. For this purpose, generators G1 to G5 are not to be trusted. The effect of failing the serial or close-pairs test in general is unclear. In many cases, if not so many random numbers are used and if the application does not interact constructively with the structure of the point set produced by the generator, no bad effect will show up. On the other hand, simulations using more than, say, 2^{32} random numbers are becoming increasingly common. Clearly, G1 to G5 and all other generators of that size are unsuitable for such simulations.

4.5.4 Empirical Testing: Summary

Experience from years of empirical testing with different kinds of tests and different generator families provides certain guidelines [49, 63, 69, 74, 75, 81, 89]. Some of these guidelines are summarized in the following remarks.

1. Generators with period length less than 2^{32} (say) can now be considered as "baby toys" and should not be used in general software packages. In particular, all LCGs of that size fail spectacularly several tests that run in a reasonably short time and use much less random numbers than the period length.
2. LCGs with power-of-2 moduli are easier to crack than those with prime moduli, especially if we look at lower-order bits.
3. LFSRs and GFSRs based on primitive trinomials, or lagged-Fibonacci and AWC/SWB generators, whose structure is too simple in moderately large dimension, also fail several simple tests.
4. Combined generators with long periods and good structural properties do well in the tests. When a large fraction of the period length is used, nonlinear inversive generators with prime modulus do better than the linear ones.
5. In general, generators with good theoretical figures of merit (e.g., good lattice structure or good equidistribution over the entire period, when only a small fraction of the period is used) behave better in the tests. As a crude general rule,

generators based on more complicated recurrences (e.g., combined generators) and good theoretical properties perform better.

4.6 PRACTICAL RANDOM NUMBER PACKAGES

4.6.1 Recommended Implementations

As stated previously, no random number generator can be guaranteed against all possible defects. However, there are generators with fairly good theoretical support, that have been extensively tested, and for which computer codes are available. We now give references to such implementations. Some of them are already mentioned earlier. We do not reproduce the computer codes here, but the user can easily find them from the references. More references and pointers can be found from the pages `http://www.iro.umontreal.ca/~lecuyer` and `http://random.mat.sbg.ac.at` on the World Wide Web.

Computer implementations that this author can suggest for the moment include those of the MRGs given in [73], the combined MRGs given in [65, 68], the combined Tausworthe generators given in [66, 70], the twisted GFSRs given in [96, 98], and perhaps the RANLUX code of [56].

4.6.2 Multigenerator Packages with Jump-Ahead Facilities

Good simulation languages usually offer many (virtual) random number generators, often numbered 1, 2, 3, ... In most cases this is the same generator but starting with different seeds, widely spaced in the sequence. L'Ecuyer and Côté [76] have constructed a package with 32 generators (which can be easily extended to 1024). Each generator is in fact based on the same recurrence (a combined LCG of period length near 2^{61}), with seeds spaced 2^{50} values apart. Moreover, each subsequence of 2^{50} values is split further into 2^{20} segments of length 2^{30}. A simple procedure call permits one to have any of the generators jump ahead to the beginning of its next segment, or its current segment, or to the beginning of its first segment. The user can also set the initial seed of the first generator to any admissible value (a pair of positive integers) and all other initial seeds are automatically recalculated so that they remain 2^{50} values apart. This is implemented with efficient jump-ahead tools. A boolean switch can also make any generator produce antithetic variates if desired.

To illustrate the utility of such a package, suppose that simulation is used to compare two similar systems using common random numbers, with n simulation runs for each system. To ensure proper synchronization, one would typically assign different generators to different streams of random numbers required by the simulation (e.g., in a queueing network, one stream for the interarrival times, one stream for the service times at each node, one stream for routing decisions, etc.), and make sure that for each run, each generator starts at the same seed and produces the same sequence of numbers for the two systems. Without appropriate tools, this may require tricky programming, because the two systems do not necessarily use the same number of random numbers in a given run. But with the package in ref. 76, one can simply assign each run to a segment number. With the first system, use the initial seed for the first run, and before each new run, advance each generator to the beginning of the next segment. After the nth run, reset the generators to their initial seeds and do the same for the second system.

The number and length of segments in the package of ref. 76 are now deemed too small for current and future needs. A similar package based on a combined LCG with period length near 2^{121} in given in ref. 72, and other systems of this type, based on generators with much larger periods, are under development. In some of those packages, generators can be seen as objects that can be created by the user as needed, in practically unlimited number.

When a generator's sequence is cut into subsequences spaced, say, ν values apart as we just described, to provide for multiple generators running in parallel, one must analyze and test the vectors of nonsuccessive output values (with lacunary indices; see Section 4.3.5) spaced ν values apart. For LCGs and MRGs, for example, the lattice structure can be analyzed with such lacunary indices. See refs. 38 and 77 for more details and numerical examples.

4.6.3 Generators for Parallel Computers

Another situation where multiple random number generators are needed is for simulation on parallel processors. The same approach can be taken: Partition the sequence of a single random number generator with very long period into disjoint subsequences and use a different subsequence on each processor. So the same packages that provide multiple generators for sequential computers can be used to provide generators for parallel processors. Other approaches, such as using completely different generators on the different processors or using the same type of generator with different parameters (e.g., changing the additive term or the multiplier in a LCG), have been proposed but appear much less convenient and sometimes dangerous [62, 64]. For different ideas and surveys on parallel generators, the reader can consult refs. 2, 9, 22, 93, and 109.

ACKNOWLEDGMENTS

This work has been supported by NSERC-Canada Grant ODGP0110050 and SMF0169893, and FCAR-Québec Grant 93ER1654. Thanks to Christos Alexopoulos, Jerry Banks, Raymond Couture, Hannes Leeb, Thierry Moreau, and Richard Simard for their helpful comments.

REFERENCES

1. Aiello, W., S. Rajagopalan, and R. Venkatesan (1996). Design of practical and provably good random number generators. Manuscript (contact venkie@bellcore.com).
2. Anderson, S. L. (1990). Random number generators on vector supercomputers and other advanced architecture. *SIAM Review*, Vol. 32, pp. 221–251.
3. Atkinson, A. C. (1980). Tests of pseudo-random numbers. *Applied Statistics*, Vol. 29, pp. 164–171.
4. Blum, L., M. Blum, and M. Schub (1986). A simple unpredictable pseudo-random number generator. *SIAM Journal on Computing*, Vol. 15, No. 2, pp. 364–383.
5. Boucher, M. (1994). La génération pseudo-aléatoire cryptographiquement sécuritaire et ses considérations pratiques. Master's thesis, Département d'I.R.O., Université de Montréal.
6. Brassard, G. (1988). *Modern Cryptology—A Tutorial*, Vol. 325 of *Lecture Notes in Computer Science*. Springer-Verlag, New York.

7. Bratley, P., B. L. Fox, and L. E. Schrage (1987). *A Guide to Simulation*, 2nd ed., Springer-Verlag, New York.
8. Brown, M. and H. Solomon (1979). On combining pseudorandom number generators. *Annals of Statistics*, Vol. 1, pp. 691–695.
9. Chen, J. and P. Whitlock (1995). Implementation of a distributed pseudorandom number generator. In H. Niederreiter and P. J.-S. Shiue, editors, *Monte Carlo and Quasi-Monte Carlo Methods in Scientific Computing*, number 106 in Lecture Notes in Statistics, pp. 168–185. Springer-Verlag, New York.
10. Collings, B. J. (1987). Compound random number generators. *Journal of the American Statistical Association*, Vol. 82, No. 398, pp. 525–527.
11. Compagner, A. (1991). The hierarchy of correlations in random binary sequences. *Journal of Statistical Physics*, Vol. 63, pp. 883–896.
12. Compagner, A. (1995). Operational conditions for random number generation. *Physical Review E*, Vol. 52, No. 5-B, pp. 5634–5645.
13. Couture, R. and P. L'Ecuyer (1994). On the lattice structure of certain linear congruential sequences related to AWC/SWB generators. *Mathematics of Computation*, Vol. 62, No. 206, pp. 798–808.
14. Couture, R. and P. L'Ecuyer (1995). Linear recurrences with carry as random number generators. In *Proceedings of the 1995 Winter Simulation Conference*, pp. 263–267.
15. Couture, R. and P. L'Ecuyer (1996). Computation of a shortest vector and Minkowski-reduced bases in a lattice. In preparation.
16. Couture, R. and P. L'Ecuyer (1997). Distribution properties of multiply-with-carry random number generators. *Mathematics of Computation*, Vol. 66, No. 218, pp. 591–607.
17. Couture, R., P. L'Ecuyer and S. Tezuka (1993). On the distribution of k-dimensional vectors for simple and combined Tausworthe sequences. *Mathematics of Computation*, Vol. 60, No. 202, pp. 749–761, S11–S16.
18. Coveyou, R. R. and R. D. MacPherson (1967). Fourier analysis of uniform random number generators. *Journal of the ACM*, Vol. 14, pp. 100–119.
19. Cressie, N. (1993). *Statistics for Spatial Data*. Wiley, New York.
20. De Matteis, A. and S. Pagnutti (1988). Parallelization of random number generators and long-range correlations. *Numerische Mathematik*, Vol. 53, pp. 595–608.
21. De Matteis, A. and S. Pagnutti (1990). A class of parallel random number generators. *Parallel Computing*, Vol. 13, pp. 193–198.
22. Deák, I. (1990). Uniform random number generators for parallel computers. *Parallel Computing*, Vol. 15, pp. 155–164.
23. Dieter, U. (1975). How to calculate shortest vectors in a lattice. *Mathematics of Computation*, Vol. 29, No. 131, pp. 827–833.
24. Dudewicz, E. J. and T. G. Ralley (1981). *The Handbook of Random Number Generation and Testing with TESTRAND Computer Code*. American Sciences Press, Columbus, Ohio.
25. Durbin, J. (1973). *Distribution Theory for Tests Based on the Sample Distribution Function.* SIAM CBMS-NSF Regional Conference Series in Applied Mathematics. SIAM, Philadelphia.
26. Durst, M. J. (1989). Using linear congruential generators for parallel random number generation. In *Proceedings of the 1989 Winter Simulation Conference*, pp. 462–466. IEEE Press.
27. Eichenauer, J., H. Grothe, J. Lehn, and A. Topuzŏglu (1987). A multiple recursive nonlinear congruential pseudorandom number generator. *Manuscripta of Computation*, Vol. 60, pp. 375–384.
33. Eichenauer-Herrmann, J. (1994). On generalized inversive congruential pseudorandom numbers. *Mathematics of Computation*, Vol. 63, pp. 293–299.

34. Eichenauer-Herrmann, J. (1995). Pseudorandom number generation by nonlinear methods. *International Statistical Reviews*, Vol. 63, pp. 247–255.

35. Eichenauer-Herrmann, J. and H. Grothe (1992). A new inversive congruential pseudorandom number generator with power of two modulus. *ACM Transactions on Modeling and Computer Simulation*, Vol. 2, No. 1, pp. 1–11.

36. Eichenauer-Herrmann, J., H. Grothe, and J. Lehn (1989). On the period length of pseudorandom vector sequences generated by matrix generators. *Mathematics of Computation*, Vol. 52, No. 185, pp. 145–148.

37. Eichenauer-Herrmann, J. and H. Niederreiter (1995). An improved upper bound for the discrepancy of quadratic congruential pseudorandom numbers. *Acta Arithmetica*, Vol. LXIX.2, pp. 193–198.

38. Entacher, K. (1998). Bad subsequences of well-known linear congruential pseudorandom number generators. *ACM Transactions on Modeling and Computer Simulation*, Vol. 8, No. 1, pp. 61–70.

39. Ferrenberg, A. M., D. P. Landau, and Y. J. Wong (1992). Monte Carlo simulations: Hidden errors from "good" random number generators. *Physical Review Letters*, Vol. 69, No. 23, pp. 3382–3384.

40. Fincke, U. and M. Pohst (1985). Improved methods for calculating vectors of short length in a lattice, including a complexity analysis. *Mathematics of Computation*, Vol. 44, pp. 463–471.

41. Fishman, G. S. (1996). *Monte Carlo: Concepts, Algorithms, and Applications*. Springer Series in Operations Research. Springer-Verlag, New York.

42. Fishman, G. S. and L. S. Moore III (1986). An exhaustive analysis of multiplicative congruential random number generators with modulus $2^{31} - 1$. *SIAM Journal on Scientific and Statistical Computing*, Vol. 7, No. 1, pp. 24–45.

43. Fushimi, M. (1983). Increasing the orders of equidistribution of the leading bits of the Tausworthe sequence. *Information Processing Letters*, Vol. 16, pp. 189–192.

44. Fushimi, M. (1989). An equivalence relation between Tausworthe and GFSR sequences and applications. *Applied Mathematics Letters*, Vol. 2, No. 2, pp. 135–137.

45. Fushimi, M. and S. Tezuka (1983). The k-distribution of generalized feedback shift register pseudorandom numbers. *Communications of the ACM*, Vol. 26, No. 7, pp. 516–523.

46. Good, I. J. (1950). *Probability and the Weighting of Evidence*. Griffin, London.

47. Good, I. J. (1969). How random are random numbers? *The American Statistician*, Vol. , pp. 42–45.

48. Grothe, H. (1987). Matrix generators for pseudo-random vectors generation. *Statistische Hefte*, Vol. 28, pp. 233–238.

49. Hellekalek, P. (1995). Inversive pseudorandom number generators: Concepts, results, and links. In C. Alexopoulos, K. Kang, W. R. Lilegdon, and D. Goldsman, editors, *Proceedings of the 1995 Winter Simulation Conference*, pp. 255–262. IEEE Press.

50. Hoaglin, D. C. and M. L. King (1978). A remark on algorithm AS 98: The spectral test for the evaluation of congruential pseudo-random generators. *Applied Statistics*, Vol. 27, pp. 375–377.

51. Hörmann, W. and G. Derflinger (1993). A portable random number generator well suited for the rejection method. *ACM Transactions on Mathematical Software*, Vol. 19, No. 4, pp. 489–495.

52. Huber, K. (1994). On the period length of generalized inversive pseudorandom number generators. *Applied Algebra in Engineering, Communications, and Computing*, Vol. 5, pp. 255–260.

53. Hull, T. E. (1962). Random number generators. *SIAM Review*, Vol. 4, pp. 230–254.

54. IMSL (1987). *IMSL Library User's Manual, Vol. 3.* IMSL, Houston, Texas.
55. James, F. (1990). A review of pseudorandom number generators. *Computer Physics Communications*, Vol. 60, pp. 329–344.
56. James, F. (1994). RANLUX: A Fortran implementation of the high-quality pseudorandom number generator of Lüscher. *Computer Physics Communications*, Vol. 79, pp. 111–114.
57. Knuth, D. E. (1981). *The Art of Computer Programming, Volume 2: Seminumerical Algorithms*, second edition. Addison-Wesley, Reading, Mass.
58. Koç, C. (1995). Recurring-with-carry sequences. *Journal of Applied Probability*, Vol. 32, pp. 966–971.
59. Lagarias, J. C. (1993). Pseudorandom numbers. *Statistical Science*, Vol. 8, No. 1, pp. 31–39.
60. Law, A. M. and W. D. Kelton (1991). *Simulation Modeling and Analysis*, second edition. McGraw-Hill, New York.
61. L'Ecuyer, P. (1988). Efficient and portable combined random number generators. *Communications of the ACM*, Vol. 31, No. 6, pp. 742–749 and 774. See also the correspondence in the same journal, Vol. 32, No. 8 (1989), pp. 1019–1024.
62. L'Ecuyer, P. (1990). Random numbers for simulation. *Communications of the ACM*, Vol. 33, No. 10, pp. 85–97.
63. L'Ecuyer, P. (1992). Testing random number generators. In *Proceedings of the 1992 Winter Simulation Conference*, pp. 305–313. IEEE Press.
64. L'Ecuyer, P. (1994). Uniform random number generation. *Annals of Operations Research*, Vol. 53, pp. 77–120.
65. L'Ecuyer, P. (1996). Combined multiple recursive random number generators. *Operations Research*, Vol. 44, No. 5, pp. 816–822.
66. L'Ecuyer, P. (1996). Maximally equidistributed combined Tausworthe generators. *Mathematics of Computation*, Vol. 65, No. 213, pp. 203–213.
67. L'Ecuyer, P. (1997). Bad lattice structures for vectors of non-successive values produced by some linear recurrences. *INFORMS Journal on Computing*, Vol. 9, No. 1, pp. 57–60.
68. L'Ecuyer, P. (1997). Good parameters and implementations for combined multiple recursive random number generators. Manuscript.
69. L'Ecuyer, P. (1997). Tests based on sum-functions of spacings for uniform random numbers. *Journal of Statistical Computation and Simulation*, Vol. 59, pp. 251–269.
70. L'Ecuyer, P. (1998). Tables of maximally equidistributed combined LFSR generators. *Mathematics of Computation*, To appear.
71. L'Ecuyer, P. (Circa 2000). TestU01: Un logiciel pour appliquer des tests statistiques à des générateurs de valeurs aléatoires. In preparation.
72. L'Ecuyer, P. and T. H. Andres (1997). A random number generator based on the combination of four LCGs. *Mathematics and Computers in Simulation*, Vol. 44, pp. 99–107.
73. L'Ecuyer, P., F. Blouin, and R. Couture (1993). A search for good multiple recursive random number generators. *ACM Transactions on Modeling and Computer Simulation*, Vol. 3, No. 2, pp. 87–98.
74. L'Ecuyer, P., A. Compagner, and J.-F. Cordeau (1997). Entropy tests for random number generators. Manuscript.
75. L'Ecuyer, P., J.-F. Cordeau, and R. Simard (1997). Close-point spatial tests and their application to random number generators. Submitted.
76. L'Ecuyer, P. and S. Côté (1991). Implementing a random number package with splitting facilities. *ACM Transactions on Mathematical Software*, Vol. 17, No. 1, pp. 98–111.
77. L'Ecuyer, P. and R. Couture (1997). An implementation of the lattice and spectral tests for

multiple recursive linear random number generators. *INFORMS Journal on Computing*, Vol. 9, No. 2, pp. 206–217.

78. L'Ecuyer, P. and R. Proulx (1989). About polynomial-time "unpredictable" generators. In *Proceedings of the 1989 Winter Simulation Conference*, pp. 467–476. IEEE Press.
79. L'Ecuyer, P., R. Simard, and S. Wegenkittl (1998). Sparse serial tests of randomness. In preparation.
80. L'Ecuyer, P. and S. Tezuka (1991). Structural properties for two classes of combined random number generators. *Mathematics of Computation*, Vol. 57, No. 196, pp. 735–746.
81. Leeb, H. and S. Wegenkittl (1997). Inversive and linear congruential pseudorandom number generators in empirical tests. *ACM Transactions on Modeling and Computer Simulation*, Vol. 7, No. 2, pp. 272–286.
82. Lehmer, D. H. (1951). Mathematical methods in large scale computing units. *Annals Comp. Laboratory Harvard University*, Vol. 26, pp. 141–146.
83. Lewis, P. A. W., A. S. Goodman, and J. M. Miller (1969). A pseudo-random number generator for the system/360. *IBM System's Journal*, Vol. 8, pp. 136–143.
84. Lewis, T. G. and W. H. Payne (1973). Generalized feedback shift register pseudorandom number algorithm. *Journal of the ACM*, Vol. 20, No. 3, pp. 456–468.
85. Lüscher, M. (1994). A portable high-quality random number generator for lattice field theory simulations. *Computer Physics Communications*, Vol. 79, pp. 100–110.
86. MacLaren, N. M. (1992). A limit on the usable length of a pseudorandom sequence. *Journal of Statistical Computing and Simulation*, Vol. 42, pp. 47–54.
87. Marsaglia, G. (1985). A current view of random number generators. In *Computer Science and Statistics, Sixteenth Symposium on the Interface*, pp. 3–10, North-Holland, Amsterdam. Elsevier Science Publishers.
88. Marsaglia, G. (1994). Yet another rug. Posted to the electronic billboard `sci.stat.math`, August 1.
89. Marsaglia, G. (1996). DIEHARD: a battery of tests of randomness. See `http://stat.fsu.edu/~geo/diehard.html`.
90. Marsaglia, G. (1996). The Marsaglia random number CDROM. See `http://stat.fsu.edu/~geo/`.
91. Marsaglia, G. and A. Zaman (1991). A new class of random number generators. *The Annals of Applied Probability*, Vol. 1, pp. 462–480.
92. Marse, K. and S. D. Roberts (1983). Implementing a portable FORTRAN uniform (0,1) generator. *Simulation*, Vol. 41, No. 4, pp. 135–139.
93. Mascagni, M., M. L. Robinson, D. V. Pryor, and S. A. Cuccaro (1995). Parallel pseudorandom number generation using additive lagged-fibonacci recursions. In H. Niederreiter and P. J.-S. Shiue, editors, *Monte Carlo and Quasi-Monte Carlo Methods in Scientific Computing*, number 106 in Lecture Notes in Statistics, pp. 263–277. Springer-Verlag.
94. MATLAB (1992). *MATLAB Reference Manual*. The MathWorks Inc., Natick, Mass.
95. Matsumoto, M. and Y. Kurita (1992). Twisted GFSR generators. *ACM Trasactions on Modeling and Computer Simulation*, Vol. 2, No. 3, pp. 179–194.
96. Matsumoto, M. and Y. Kurita (1994). Twisted GFSR generators II. *ACM Transactions on Modeling and Computer Simulation*, Vol. 4, No. 3, pp. 254–266.
97. Matsumoto, M. and Y. Kurita (1996). Strong deviations from randomness in m-sequences based on trinomials. *ACM Transactions on Modeling and Computer Simulation*, Vol. 6, No. 2, pp. 99–106.
98. Matsumoto, M. and T. Nishimura (1998). Mersenne twister: A 623-dimensionally equidistributed uniform pseudorandom number generator. *ACM Transactions on Modeling and Computer Simulation*, Vol. 8, No. 1, pp. 31–42.

99. Moreau, T. (1996). A practical "perfect" pseudo-random number generator. Manuscript.
100. Niederreiter, H. (1985). The serial test for pseudorandom numbers generated by the linear congruential method. *Numerische Mathematik*, Vol. 46, pp. 51–68.
101. Niederreiter, H. (1986). A pseudorandom vector generator based on finite field arithmetic. *Mathematica Japonica*, Vol. 31, pp. 759–774.
102. Niederreiter, H. (1992). *Random Number Generation and Quasi-Monte Carlo Methods*, volume 63 of *SIAM CBMS-NSF Regional Conference Series in Applied Mathematics*. SIAM, Philadelphia.
103. Niederreiter, H. (1995). The multiple-recursive matrix method for pseudorandom number generation. *Finite Fields and their Applications*, Vol. 1, pp. 3–30.
104. Niederreiter, H. (1995). New developments in uniform pseudorandom number and vector generation. In H. Niederreiter and P. J.-S. Shiue, editors, *Monte Carlo and Quasi-Monte Carlo Methods in Scientific Computing*, number 106 in Lecture Notes in Statistics, pp. 87–120. Springer-Verlag.
105. Niederreiter, H. (1995). Pseudorandom vector generation by the multiple-recursive matrix method. *Mathematics of Computation*, Vol. 64, No. 209, pp. 279–294.
106. Owen, A. B. (1998). Latin supercube sampling for very high dimensional simulations. *ACM Transactions of Modeling and Computer Simulation*, Vol. 8, No. 1, pp. 71–102.
107. Park, S. K. and K. W. Miller (1988). Random number generators: Good ones are hard to find. *Communications of the ACM*, Vol. 31, No. 10, pp. 1192–1201.
108. Payne, W. H., J. R. Rabung, and T. P. Bogyo (1969). Coding the Lehmer pseudorandom number generator. *Communications of the ACM*, Vol. 12, pp. 85–86.
109. Percus, D. E. and M. Kalos (1989). Random number generators for MIMD parallel processors. *Journal of Parallel and Distributed Computation*, Vol. 6, pp. 477–497.
110. Press, W. H. and S. A. Teukolsky (1992). Portable random number generators. *Computers in Physics*, Vol. 6, No. 5, pp. 522–524.
111. Rabin, M. O. (1980). Probabilistic algorithms for primality testing. *J. Number Theory*, Vol. 12, pp. 128–138.
112. Ripley, B. D. (1987). *Stochastic Simulation*. Wiley, New York.
113. Ripley, B. D. (1990). Thoughts on pseudorandom number generators. *Journal of Computational and Applied Mathematics*, Vol. 31, pp. 153–163.
114. Schrage, L. (1979). A more portable fortran random number generator. *ACM Transactions on Mathematical Software*, Vol. 5, pp. 132–138.
115. Stephens, M. S. (1986). Tests based on EDF statistics. In R. B. D'Agostino and M. S. Stephens, editors, *Goodness-of-Fit Techniques*. Marcel Dekker, New York and Basel.
116. Stephens, M. S. (1986). Tests for the uniform distribution. In R. B. D'Agostino and M. S. Stephens, editors, *Goodness-of-Fit Techniques*, pp. 331–366. Marcel Dekker, New York and Basel.
117. Sun Microsystems (1991). *Numerical Computations Guide*. Document number 800-5277-10.
118. Tausworthe, R. C. (1965). Random numbers generated by linear recurrence modulo two. *Mathematics of Computation*, Vol. 19, pp. 201–209.
119. Teichroew, D. (1965). A history of distribution sampling prior to the era of computer and its relevance to simulation. *Journal of the American Statistical Association*, Vol. 60, pp. 27–49.
120. Tezuka, S. (1995). *Uniform Random Numbers: Theory and Practice*. Kluwer Academic Publishers, Norwell, Mass.
121. Tezuka, S. and P. L'Ecuyer (1991). Efficient and portable combined Tausworthe random number generators. *ACM Transactions on Modeling and Computer Simulation*, Vol. 1, No. 2, pp. 99–112.
122. Tezuka, S., P. L'Ecuyer, and R. Couture (1994). On the add-with-carry and subtract-with-bor-

row random number generators. *ACM Transactions of Modeling and Computer Simulation*, Vol. 3, No. 4, pp. 315–331.

123. Tootill, J. P. R., W. D. Robinson, and D. J. Eagle (1973). An asymptotically random Tausworthe sequence. *Journal of the ACM*, Vol. 20, pp. 469–481.
124. Vazirani, U. and V. Vazirani (1984). Efficient and secure pseudo-random number generation. In *Proceedings of the 25th IEEE Symposium on Foundations of Computer Science*, pp. 458–463.
125. Wang, D. and A. Compagner (1993). On the use of reducible polynomials as random number generators. *Mathematics of Computation*, Vol. 60, pp. 363–374.
126. Wichmann, B. A. and I. D. Hill (1982). An efficient and portable pseudo-random number generator. *Applied Statistics*, Vol. 31, pp. 188–190. See also corrections and remarks in the same journal by Wichmann and Hill, Vol. 33 (1984) p. 123; McLeod Vol. 34 (1985) pp. 198–200; Zeisel Vol. 35 (1986) p. 89.
127. Wolfram, S. (1996). *The Mathematica Book*, third edition. Wolfram Media/Cambridge University Press, Champaign, USA.

CHAPTER 5

Random Variate Generation

RUSSELL C. H. CHENG

University of Kent at Canterbury

5.1 INTRODUCTION

In Chapter 4 the generation of (pseudo)*random numbers* was discussed. In this chapter, by *random number* is always meant a uniform random variable, denoted by RN(0,1), whose distribution function is

$$F_U(u) = \begin{cases} 0 & u \leq 0 \\ u & 0 < u < 1 \\ 1 & u \geq 1 \end{cases} \tag{1}$$

Thus random numbers are always uniformly distributed on the unit interval (0,1). Random number generation is an important topic in its own right (see Chapter 4). In contrast, random variate generation always refers to the generation of variates whose probability distribution is different from that of the uniform on the interval (0,1). The basic problem is therefore to generate a random variable, X, whose distribution function

$$F(x) = \Pr(X \leq x) \qquad -\infty < x < \infty \tag{2}$$

is assumed to be completely known, and which is different from that of (1).

Most mainframe computing systems contain libraries with implementations of generators for the more commonly occurring distributions. Generators are also available in many of the existing statistical and simulation packages for personal computers. Choice of a generator often has as much to do with properties of the distribution to be used, as with the properties of the generator itself. In the description of specific generators given below, emphasis is therefore given to the underlying characteristics of the distributions being considered and in the relations between different distributions. This information should be useful in aiding the understanding of properties of generators. The intention is to allow an informed choice to be made, and the algorithms will then not need to be

Handbook of Simulation, Edited by Jerry Banks.
ISBN 0-471-13403-1

treated so much as black box implementations. Additionally, general principles and general methods of random variate generation are described which will enable the reader to construct generators for nonstandard distributions.

Random variable generators invariably use as their starting point a random number generator, which yields RN(0,1), with distribution given by (1). The technique is to transform or manipulate one or more such uniform random variables in some efficient way to obtain a variable with the desired distribution (2). In this chapter it will be assumed that a source of random numbers is available. In addition to being uniform, the random numbers must also be mutually independent. A significant problem, discussed in Chapter 4, is to construct generators which produce random numbers that can safely be treated as being independent. It will be assumed in what follows that such a random number generator is available. In what follows it is assumed that a sequence of numbers produced by such a generator: $\{U_1, U_2, \ldots\}$ behaves indistinguishably from a sequence of independently distributed RN(0,1) variates.

As with all numerical techniques, there will usually be more than one method available to generate variates from some prescribed distribution. Four factors should be considered in selecting an appropriate generator:

1. *Exactness.* This refers to the distribution of the variates produced by the generator. The generator is said to be *exact* if the distribution of variates generated has the exact form desired. In certain applications where the precise distribution is not critical, a method may be acceptable which may be good as far as the other factors are concerned, but which produces variates whose distribution only approximates that of the desired one.

2. *Speed.* This refers to the computing time required to generate a variate. There are two contributions to the overall time. Usually, an initial *setup time* is needed to construct constants or tables used in calculating the variate. Calculation of the variate itself then incurs a further *variable generation time*. The relative importance of these two contributions depends on the application. There are two cases. The more usual one is where a sequence of random variates is needed, all with exactly the same distribution. Then the constants or tables need be setup only once, as the same values can be used in generating each variate of the sequence. In this case the setup time is negligible and can be ignored. However, if each variate has a different distribution, this setup time will be just as important, since in this case it has to be added to every marginal generation time.

3. *Space.* This is simply the computer memory requirement of the generator. Most algorithms for random variate generation are short. However, some make use of extensive tables, and this can become significant if different tables need to be held simultaneously in memory.

4. *Simplicity.* This refers to both the algorithmic simplicity and the implementational simplicity. Its importance depends on the context of the application. If, for example, the generator is needed in a single study, a simple generator that is easy to understand and to code will usually be more attractive than a complicated alternative even if the latter has better characteristics as far as other factors are concerned. If variates from a particular distribution are needed often and there is not already a library implementation, it will be worthwhile selecting a generator that is exact and fast and to write an efficiently coded version for permanent use.

In general, random variate generation will be only an incidental part of a computer simulation program. The users who need to make their own generator will usually therefore not wish to spend an inordinate time writing and testing such generators. Preference has thus been given below to specific methods which are simple and reasonably fast without necessarily being the fastest.

Section 5.2 covers general principles commonly used in variate generation. Section 5.3 then discusses how these ideas are specialized for continuous random variables, together with specific algorithms for the more commonly occurring continuous distributions. Section 5.4 repeats this discussion for the case of discrete random variables. Section 5.5 covers extensions to multivariate distributions. In Section 5.6 we discuss the generation of stochastic processes.

Good references to the subject are refs. 1 to 3. A good elementary introduction to the subject is given in ref. 4, Chapter 9. A more advanced discussion is given in ref. 5, Chapter 3.

5.2 GENERAL PRINCIPLES

As mentioned in Section 5.1, from now on it is assumed that a (pseudo) random number generator is available that produces a sequence of independent RN(0,1) variates, with the understanding that each time this function is called, a new RN(0,1) variate is returned that is independent of all variates generated by previous calls to this function.

5.2.1 Inverse Transform Method

When X is a continuously distributed random variable, the distribution can be defined by its *density function* $f(x)$. The density function allows the probability that X lies between two given values a and $b(> a)$ to be evaluated as

$$\Pr\{a \le X \le b\} = \int_a^b f(x)\,dx$$

From its definition, (2), it will be seen that the distribution function in this case can be evaluated from the density function as the integral

$$F(x) = \int_{-\infty}^{x} f(z)\,dz$$

The distribution function is thus strictly increasing and continuous (Figure 5.1). In this case, for any $0 < u < 1$, the equation $F(x) = u$ can be solved to give a unique x. If (and this is the critical proviso) the inverse of a distribution function can be expressed in a simple form, this process can be denoted by rewriting the equation in the equivalent form $x = F^{-1}(u)$; so that x is expressed as an explicit function of u. This function F^{-1} is called the *inverse of* F. This yields the following convenient method for generating variables with the given distribution function F.

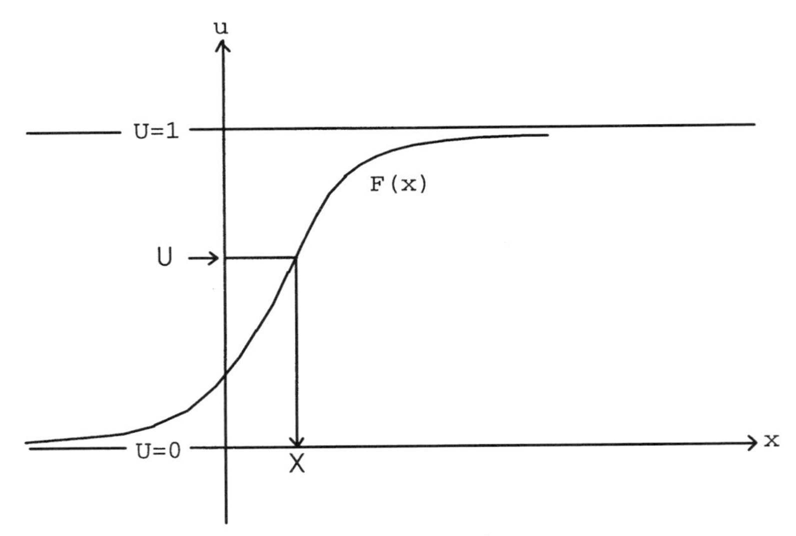

Figure 5.1 Inverse transform method $X = F^{-1}(U)$, continuous distribution.

Inverse Transform Method (Continuous Case)

```
Let U = RN(0,1)
Return X = F^-1(U)
```

To show that it works, all that is needed is to verify that the distribution function of X is indeed F [i.e., that $\Pr(X \le x) = F(x)$]. Now

$$\Pr(X \le x) = \Pr[F^{-1}(U) \le x] = \Pr[U \le F(x)]$$

But the right-hand probability is the distribution function of a uniform random variable evaluated at the value $F(x)$, and from (1) this is equal to $F(x)$, as required.

The best known and most useful example is when X is an exponentially distributed random variable with mean $a > 0$. Its distribution function is

$$F(x) = \begin{cases} 1 - \exp\left(\dfrac{-x}{a}\right) & x > 0 \\ 0 & \text{otherwise} \end{cases} \tag{3}$$

Solving $u = F(x)$ for x in this case yields

$$x = F^{-1}(u) = -a \ln(1 - u) \tag{4}$$

A random variate with distribution function (3) is therefore obtained by using this formula (4) to calculate X, with u generated as a $U(0, 1)$ variate.

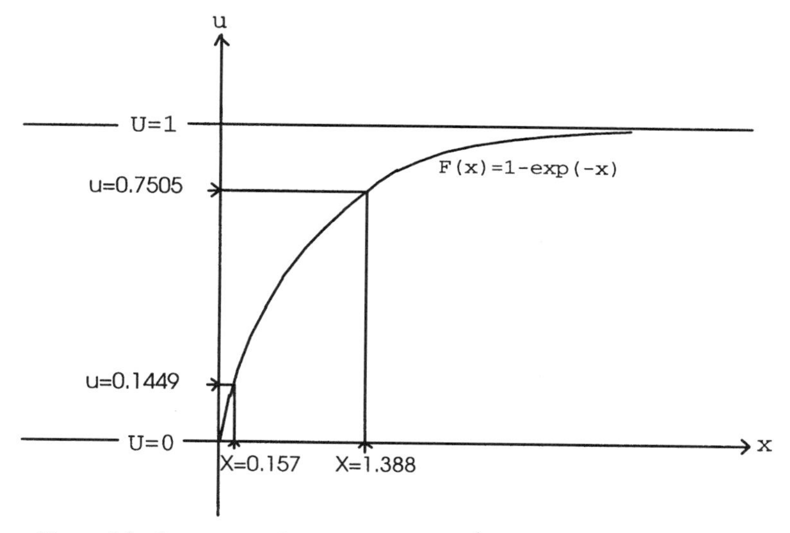

Figure 5.2 Inverse transform method $X = F^{-1}(U)$, exponential distribution.

Example 1 Figure 5.2 illustrates the use of the formula in the case when $a = 1$. The curve is the distribution function $F(x) = 1 - e^{-x}$. Two random variates have been generated using (4). For the first variate the random number is $u_1 = 0.7505$ and this is depicted on the vertical (y) scale. The horizontal distance at this vertical height from the *y-axis* to the graph of $F(x)$ is $x_1 = -\ln(1 - 0.7505) = 1.388$. This is the corresponding exponential variate that is generated. Similarly, the random number $u_2 = 0.1449$ generates the exponential variate value $x_2 = 0.1565$.

Random variates from many distributions can be generated in this way. The most commonly occurring are listed in Section 5.5.2. Johnson [6] lists the following additional cases.

1. Burr:

$$f(x) = \frac{ckx^{c-1}}{(1 + x^c)^{k+1}} \qquad x > 0$$

$$F^{-1}(u) = [(1 - u)^{-1/k} - 1]^{1/c}$$

2. Laplace:

$$f(x) = \tfrac{1}{2}\exp(-|x|) \qquad -\infty < x < \infty$$

$$F^{-1}(u) = \begin{cases} \ln(2u) & u \le \frac{1}{2} \\ -\ln[2(1 - u)] & u > \frac{1}{2} \end{cases}$$

3. Logistic:

$$f(x) = \frac{\exp(-x)}{[1+\exp(-x)]^2} \qquad -\infty < x < \infty$$

$$F^{-1}(u) = \ln \frac{1-u}{u}$$

4. Pareto:

$$f(x) = cx^{-c-1} \qquad x \geq 1$$
$$F^{-1}(u) = (1-u)^{-1/c}$$

In all these cases further flexibility is afforded by rescaling (a) and relocation (b) using the linear transformation

$$Y = aX + b$$

although some care may be needed in interpreting the meaning of a and b in particular contexts.

The inverse transform method extends to *discrete* random variables. Suppose that X takes only the values $x_1, x_2, \ldots, x_n$ with probabilities $p_i = \Pr(X = x_i)$ such that $\sum_{i=1}^{n} p_i = 1$. The distribution function for such a variable is

$$F(x) = \Pr(X \leq x) = \sum_{i: x_i \leq x} p_i$$

It can be verified (see, e.g., ref. 2) that if the inverse is defined by

$$F^{-1}(u) = \min\{x | u \leq F(x)\}$$

the inverse transform method will still work despite the discontinuities in $F(x)$ (see Figure 5.3). Moreover, we still have $\Pr(X \leq x) = F(x)$, so that X generated in this way has the required discrete distribution. The method reduces to:

Inverse Transform Method (Discrete Case)

```
Let U = RN(0,1)
Let i = 1
While (F(x_i) < U) {i = i + 1}
Return X = x_i
```

Because the method uses a linear search, it can be inefficient if n is large. More efficient methods are described later.

If a table of x_i values with the corresponding $F(x_i)$ values is stored, the method is also called the *table look-up method*. The method runs through the table comparing U with each $F(x_i)$, and returning, as X, the first x_i encountered for which $F(x_i) \geq U$.

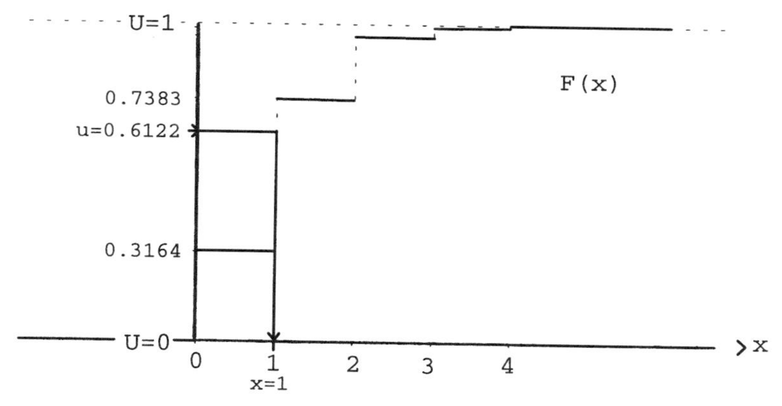

Figure 5.3 Inverse transform method $X = F^{-1}(U)$, Bin(4, 0.25) distribution.

Example 2 The well-known binomial distribution Bin(n, p), described later, is an example of a discrete distribution. The possible values that X can take in this case are 0, 1, ... n, with probabilities that depend on the parameters p and n (the general formula is given below). For the case $n = 4$ and $p = 0.25$, the possible values are $x_i = i$ for $i =$ 0, 1, 2, 3, 4; and the distribution function is given in Table 5.1.

The table look-up method works as follows. Suppose that $u = 0.6122$ is a given random number. Looking along the row of $F(x_i)$ values, it will be seen that $F(x_0) = 0.3164 < u = 0.6122 < F(x_1) = 0.7383$. Thus $x_1 = 1$ is the first x_i encountered for which $u \leq F(x_i)$. The variate generated in this case is therefore $X = 1$. The generation of this particular variate is depicted geometrically in Figure 5.3.

Sometimes the test of whether $F(x_i) < u$ takes a simple and explicit algebraic form. An example is the geometric distribution described below.

5.2.2 Acceptance–Rejection Method

The acceptance–rejection method is most easily explained for a continuous distribution. Suppose that we wish to sample from the distribution with density $f(x)$ but that it is difficult to do so by the inverse transform method. Suppose now that the following three assumptions hold (see also Figure 5.4):

1. There is another function $e(x)$ that dominates $f(x)$ in the sense that $e(x) \geq f(x)$ for all x.

TABLE 5.1 Distribution of Bin(4,0.25)

i	0	1	2	3	4
p_i	0.3164	0.4219	0.2109	0.0469	0.0039
$F(x_i)$	0.3164	0.7383	0.9492	0.9961	1.0000

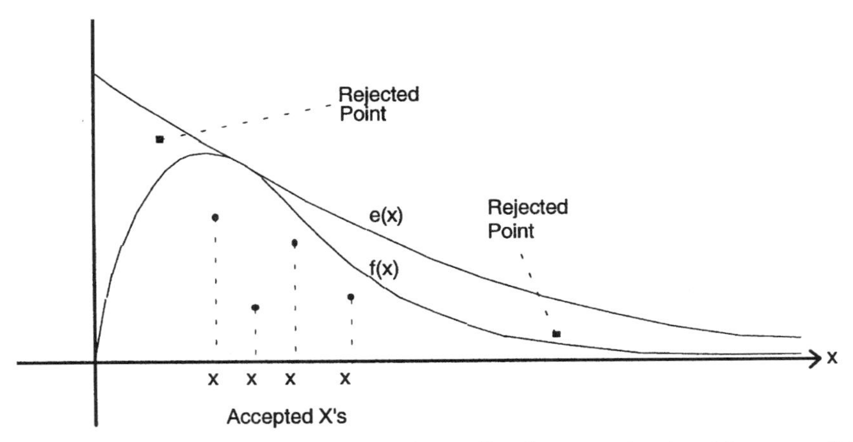

Figure 5.4 Acceptance–rejection method, points uniformly scattered under $e(x)$, gamma distribution with $b = 2$.

2. It is possible to generate points uniformly scattered under the graph of $e(x)$ (above the x-axis). Denote the coordinates of such a typical point by (X, Y).
3. If the graph of $f(x)$ is drawn on the same diagram, the point (X, Y) will be above or below it according as $Y > f(X)$ or $Y \leq f(X)$.

The acceptance–rejection method works by generating such points (X, Y) and returning the X coordinate as the generated X value, but only if the point lies under the graph of $f(x)$ [i.e., only if $Y \leq f(X)$]. Intuitively it is clear that the density of X is proportional to the height of f, so that X has the correct density. A formal proof can be found in ref. 3.

There are many ways of constructing $e(x)$. One requirement is that the area between the graphs of f and e be small, to keep the proportion of points rejected small, as such points represent wasted computing effort. A second requirement is that it should be easy to generate points uniformly distributed under $e(x)$. The average number of points (X, Y) needed to produce one acceptable X will be called the *trials ratio*. Clearly, the trials ratio is always greater than or equal to unity. The closer the trials ratio is to unity, the more efficient is the resulting generator.

A neat way of constructing a suitable $e(x)$ is to take $e(x) = Kg(x)$, where $g(x)$ is the density of a distribution for which an easy way of generating variates already exists. It can be shown that if X is a variate from this density, then points of the form $(X, Y) = (X, KUg(X))$, where U is a RN(0,1) variable that is generated independently of X, will be uniformly distributed under the graph of $e(x)$. For a proof of this result see Devroye [1]. Usually, K is taken just large enough to ensure that $e(x) \geq f(x)$ (see assumption 1 above).

The trials ratio is precisely K. The method is thus dependent on being able to find a density $g(x)$ for which K can be kept small.

A nice example of this last technique is the third method given below for generating from the gamma distribution GAM$(1, b)$, due to Fishman [7]. The method is valid for the case $b > 1$. The distribution has density $f(x) = x^{b-1}\exp(-x)/\Gamma(b)$, $x > 0$. There is no general closed-form expression for the distribution function, so the inverse transform method is not easy to implement for this distribution. Fishman's method takes an exponential envelope $e(x) = K\exp(-x/b)/b$. A little calculation shows that if we set

$$K = \frac{b^b \exp(1 - b)}{\Gamma(b)}$$

then $e(x) \geq f(x)$ for all $x \geq 0$. For values of b close to unity, K also remains close to unity and K does not increase all that fast as b increases ($K = 1$ when $b = 1$, $K = 1.83$ when $b = 3$, and $K = 4.18$ when $b = 15$). The method is thus convenient when gamma variates are needed for b not too large, which is often the case. Figure 5.4 illustrates the envelope and the gamma distribution for the case $b = 2$, showing how the envelope fits quite neatly over the gamma density function, even though its shape is somewhat different.

5.2.3 Composition Method

Suppose that a given density f can be written down as the weighted sum of r other densities:

$$f(x) = \sum_{i=1}^{r} p_i f_i(x)$$

where the weights, p_i, satisfy $p_i > 0$ and $\sum_{i=1}^{r} p_i = 1$. The density f is then said to be a *mixture* or a *compound* density. An example of where such a mixture of distributions occurs is a queueing simulation where customer arrivals are composed of a mixture of different customer types, each with its own arrival pattern. If methods exist for the generation of variates from the component densities f_i, the following method can be used to generate from the mixture density.

Composition Method

```
Setup: Let F_j = Σ_{i=1}^{j} p_i for j = 1, 2, . . . , r
Let U = RN(0,1)
Set i = 1
While (F_i < U) {i = i + 1}
Return X = X_i, a variate drawn from the density f_i
```

The method simply selects the ith component distribution with probability p_i and then returns a variate from this distribution.

One use of the method is to split up the range of X into different intervals in such a way that it is easy to sample from each interval. Another common use is where the first component is made easy to generate from and also has a high probability of being chosen. Then the generator will probably generate from this easy distribution and will only occasionally have to resort to generation from one of the other distributions. Several fast normal variate generators have been constructed in this way [1].

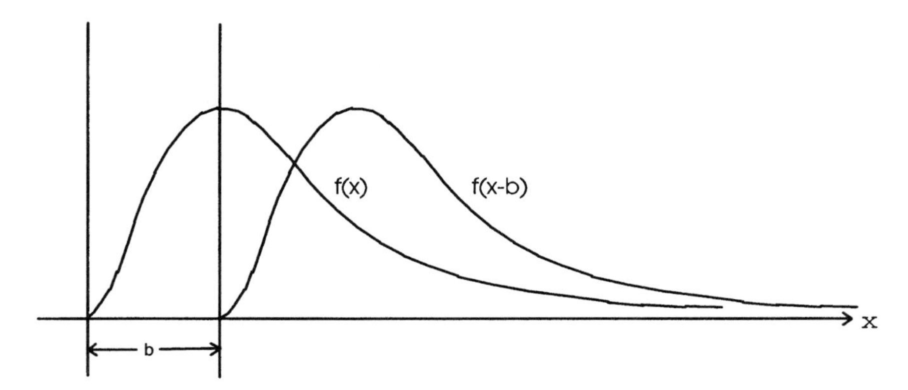

Figure 5.5 Translation of a distribution using a location parameter.

5.2.4 Translations and Other Simple Transforms

Although not strictly a method in its own right, often a random variable can be obtained by some elementary transformation of another. Many explicit generators fall into this category. For example, a *lognormal variable* is simply an exponentiated normal variable; a *chi-squared variable* with one degree of freedom (χ_1^2) is simply a standard normal variable that has been squared. Even more elementary than such transformations, but forming a broad class, are the *location-scale models*. If X is a continuously distributed random variable with density function $f(x)$, say, we can rescale and reposition the distribution by the linear transform

$$Y = aX + b$$

where a is a given *scale* constant (typically, $a > 0$) and b is a given *location* constant. The variable Y is thus a location-scale version of X and has density $g(y)$ given by

$$g(y) = a^{-1} f\left(\frac{y-b}{a}\right)$$

Note that if X has a restricted range, Y will be restricted, too. For example, if $X > 0$, then $Y > b$ (if $a > 0$). Figure 5.5 illustrates a random variate X that has been translated to a new location that is b units away from its original position.

Below we shall give distributions in their most commonly used form. Sometimes this form incorporates the location-scale formulation; an example is the normal distribution. Sometimes this form incorporates the scale parameter but not the location parameter; an example is the Weibull distribution. It is left to the reader to extend these forms to their full version if needed.

5.3 CONTINUOUS DISTRIBUTIONS

5.3.1 Inverse Transform by Numerical Solution of $F(X) = U$

In this section we list generators for a number of commonly occurring continuous distributions. If a generator is required for a distribution not listed and the inverse distribution is not expressible in simple terms, the inverse transform method can still be employed by solving $F(X) = U$ numerically for X. The following bisection method is one way of doing this.

Inverse Transform Method by Bisection

```
Setup: Let δ = 0.001, say
Let U = RN(0,1)
Let a = -1.0 While(F(a) > U) {a = 2a}
Let b = +1.0 While(F(b) < U) {b = 2b}
While(b - a > δ){
  X = (a + b)/ 2
  If(F(X) ≤ U) a = X
  Else b = X
  }
```

Notes:

1. The generator assumes that $F(x)$ can be calculated for all x.
2. The method first finds an interval (a, b) in which X lies and then checks the midpoint of the interval to see which side of this X lies on. This reduces the length of the interval of uncertainty for X by half. The process is repeated by considering the midpoint of this reduced interval, and so on, until the width of the interval of uncertainty for X is reduced to a prescribed, sufficiently small, value δ.
3. The initial interval (a, b) in which X lies is obtained by setting arbitrary initial values for a and b (i.e., $a = -1$ and $b = 1$). These values are then doubled repeatedly until $a \leq X \leq b$. This procedure can be replaced by a more efficient one if a convenient one exists. For example, if X is positive, the setup for a can become

```
Set a = 1.0 While(F(a) > U) {a = a/2}
```

For unimodal densities with known mode X_m, the following alternative is quicker. (See ref. 1 for details on how the method works.)

Inverse Transform Method by Newton–Raphson Iteration

```
Setup: δ = 0.001, Y_m = F(X_m)
Let U = RN(0,1) Set X = X_m, Y = Y_m, h = Y - U
While(|h| > δ){
  X = X - h/f(X)
  h = F(X) - U
  }
Return X
```

Notes:

1. Convergence is guaranteed for unimodal densities because $F(x)$ is convex for $x \in (-\infty, X_m)$, and concave for $x \in (X_m, \infty)$.
2. The tolerance criterion guarantees that the value of $F(X)$ for the X value returned will be within δ of U. However, it does not guarantee that X will be close to the exact solution of $F(X) = U$.

With both previous generators, if $F(x)$ is not known, it can be evaluated from the integral of the density $f(x)$:

$$F(x) = \int_{-\infty}^{x} f(u)\, du$$

using numerical quadrature. This is likely to be a slow procedure.

5.3.2 Specific Continuous Distributions

Uniform U(a, b), a < b

Density:

$$f(x) = \begin{cases} \dfrac{1}{b-a} & a < x < b \\ 0 & \text{otherwise} \end{cases}$$

Distribution Function:

$$F(x) = \begin{cases} 0 & x \leq a \\ \dfrac{x-a}{b-a} & a < x < b \\ 1 & x > b \end{cases}$$

Generator: Inverse transform method:

```
Let U = RN(0,1)
Return X = a + (b − a)U
```

Exponential EXP*(a), a > 0*

Density:

$$f(x) = \begin{cases} a^{-1} \exp\left(-\dfrac{x}{a}\right) & x > 0 \\ 0 & \text{otherwise} \end{cases}$$

Distribution Function:

$$F(x) = \begin{cases} 1 - \exp\left(-\dfrac{x}{a}\right) & x > 0 \\ 0 & \text{otherwise} \end{cases}$$

Generator: Inverse transform method:

```
Generate U = RN(0,1)
Return X = -a ln(1 - U)
```

Note: The parameter $a > 0$ is the mean of the distribution. The rate parameter is a^{-1}.

Weibull WEIB*(a, b), a, b* > *0*

Density:

$$f(x) = \begin{cases} ba^{-b}x^{b-1} \exp\left[-\left(\dfrac{x}{a}\right)^b\right] & x > 0 \\ 0 & \text{otherwise} \end{cases}$$

Distribution Function:

$$F(x) = \begin{cases} 1 - \exp\left[-\left(\dfrac{x}{a}\right)^b\right] & x > 0 \\ 0 & \text{otherwise} \end{cases}$$

Generator: Inverse transform method:

```
Generate U = RN(0,1)
Return X = a[-ln(1 - U)]^(1/b)
```

Notes:

1. Some references replace $1 - U$ with U in the final formula for X, as this also has the $U(0, 1)$ distribution. However, this is not usually recommended; see ref. 3.
2. The parameters $a, b > 0$ are scale and shape parameters, respectively.
3. The special case $b = 1$ is the exponential distribution.
4. If X is a Weibull variate, X^b is an exponential variate with mean a^b. Conversely, if E has exponential distribution with mean a^b, then $E^{1/b}$ is Weibull with scale and shape parameters a and b, respectively.

Extreme Value EXTREME(μ, σ), $\sigma > 0$

Density:

$$f(x) = \sigma^{-1} \exp\left(-\frac{x-\mu}{\sigma}\right) \exp\left[-\exp\left(-\frac{x-\mu}{\sigma}\right)\right] \qquad -\infty < x < \infty$$

Distribution Function:

$$F(x) = \exp\left\{-\exp\left[\left(-\frac{x-\mu}{\sigma}\right)\right]\right\} \qquad -\infty < x < \infty$$

Generator: Inverse transform method:

```
Let U = RN(0,1)
Return X = −σ ln[−ln(U)] + μ
```

Note: The extreme value distribution is sometimes given in the form where X is the negative of that given here.

Gamma GAM(*a,b*), *a,b* > 0

Density:

$$f(x) = \begin{cases} \dfrac{(x/a)^{b-1}}{a\Gamma(b)} \exp\left(-\dfrac{x}{a}\right) & x > 0 \\ 0 & \text{otherwise} \end{cases}$$

where $\Gamma(b)$ is the *gamma function*:

$$\Gamma(b) = \int_0^\infty u^{b-1} e^{-u}\, du$$

Distribution Function: No simple closed form.

Here $a > 0$ and $b > 0$ are scale and shape parameters, respectively. The key parameter is the shape parameter, and no single method of generation is satisfactory for all values of b. The following methods cover different ranges of b.

Generator 1. Acceptance–rejection given by Ahrens and Dieter [8]; requires that $0 < b < 1$:

```
Setup: β = (e + b)/e where e = 2.71828 ... is the base of the natural
logarithm.
While(True){
  Let U = RN(0,1), W = β U
  If (W < 1){
    Let Y = W^{1/b}, V = RN(0,1)
    If (V ≤ e^{-Y}) Return X = aY
```

```
    }
  Else{
    Let Y = −ln[(β− W)/ b], V = RN(0,1)
    If (V ≤ Y^(b−1)) Return X = aY
    }
}
```

Generator 2. Acceptance–rejection Cheng [9]; requires that $b > 1$:

```
Setup: α = (2b −1)^(−1/2), β = b − ln 4, γ = b + α^(−1), δ = 1 + ln 4.5.
While(True){
  Let U_1 = RN(0,1), U_2 = RN(0,1)
  Let V = αln[U_1/(1− U_1)], Y = be^V, Z = U_1^2 U_2, W = β + γ V − Y
  If (W + δ− 4.5Z ≥ 0){
    Return X = aY
    }
  Else{
    If (W ≥ ln Z) Return X = aY
    }
}
```

Note: The trials ratio improves from $4/e \simeq 1.47$ to $(4/\pi)^{1/2} \simeq 1.13$ as b increases from 1 to ∞.

Generator 3. Acceptance–rejection Fishman [7]; requires that $b > 1$. The method is simple and is efficient for values of $b < 5$, say:

```
While(True){
  Let U_1 = RN(0,1), U_2 = RN(0,1), V_1 = −ln U_1, V_2 = −ln U_2
  If (V_2 > (b − 1)(V_1 − ln V_1 − 1)){
    Return X = aV_1
    }
}
```

Note: The trials ratio degrades from unity at $b = 1$, to 2.38 at $b = 5$, to 2.84 at $b = 7$.

k-Erlang ERL(*m*, *k*), *m* > 0, *k* a Positive Integer

Density: Same as that of GAM$(m/k, k)$ [i.e., GAM(a, b) with $a = m/k$ and $b = k$ (an integer)]

Distribution Function: No closed form except for the case $k = 1$.

Generator 1. If X is a k-Erlang variate with mean m, it is the same as the sum of k independent exponential variates each with mean m/k:

```
Let U_1 = RN(0,1), U_2 = RN(0,1), . . . , U_k = RN(0,1)
Return X = −(m/k) ln[(1 − U_1)(1 − U_2) . . . (1 − U_k)]
```

Generator 2. Generate X as a gamma variate with $a = m/k$ and $b = k$:

```
Return X = GAM(m/k, k)
```

Note: The first method is more efficient for small values of k ($k < 10$, say). For larger values of k, the second method is faster and not to subject to finite arithmetic error through repeated multiplication of quantities all less than unity as might occur with the first method.

Normal N(μ, σ²), σ > 0

Density:

$$f(x) = (2\pi\sigma^2)^{-1/2} \exp\left[-\frac{(x-\mu)^2}{2\sigma^2}\right] \qquad -\infty < x < \infty$$

Here μ is the mean and σ^2 is the variance of the distribution.

Distribution Function: No closed-form expression.

Generator: Polar version of the Box–Muller [10]; special transform:

```
While(True){
  Generate U_1, U_2, RN(0,1) variates.
  Let V_1 = 2U_1 - 1, V_2 = 2U_2 - 1, W = V_1^2 + V_2^2
  If(W < 1){
    Let Y = [(-2 ln W)/ W]^{1/2}
    Return X_1 = μ + σ V_1 Y and X_2 = μ + σ V_2 Y
    }
}
```

Notes:

1. The method returns pairs of independent normal variates, each with mean μ and variance σ^2. If only one variate is needed each time, just X_1 can be returned; the method is then not as efficient, as X_2 is not used even though most of the work to get it is done.
2. There are many alternative methods. The one shown above is one of the simplest and is reasonably fast. The *original* Box–Muller method simply returns $X_1 = R\cos T$ and $X_2 = R\sin T$, where $R = [-2\ln(U_1)]$ and $T = 2\pi U_2$. This is elegantly simple but does require the relatively slow calculation of a sine and a cosine.
3. The standard normal distribution is where $\mu = 0$ and $\sigma^2 = 1$.

Chi-Squared with k Degrees of Freedom χ²(k), k a Positive Integer

Density and Distribution Function: These are the same as those of GAM$(2, k/2)$.

Generator 1: Use the relationship with the gamma distribution:

```
Return X = GAM(2, k/2)
```

Generator 2: When k is even, use the relationship with the k-Erlang distribution:

```
Return X = ERL(k, k/2)
```

Generator 3: When k is small and a fast normal variate generator is available, use the fact that $\chi^2(k)$ is the sum of k independent squared standard normal variates:

```
X = 0
For(i = 1 to k){X = X + [N(0,1)]^2}
Return X
```

t-Distribution $t(\nu)$, ν a Positive Integer

Density:

$$f(x) = \frac{\Gamma[(\nu+1)/2]}{\sqrt{\pi\nu}\,\Gamma(\nu/2)} \left(1 + \frac{x^2}{\nu}\right)^{-(\nu+1)/2} \qquad -\infty < x < \infty$$

This is called the *t-distribution with ν degrees of freedom.*

Generator: Use the fact that $X = t(\nu)$ has the same distribution as the ratio $Z/\sqrt{Y/\nu}$, where Z is a standard normal variate and Y is a $\chi^2(\nu)$ variate independent of Z:

```
Let Z = N(0,1), W = √(χ²(ν)/ν)
Return X = Z/W
```

Lognormal LN(μ, σ^2), $\sigma > 0$

Density:

$$f(x) = \frac{1}{x\sqrt{2\pi\sigma^2}} \exp\left[-\frac{(\ln x - \mu)^2}{2\sigma^2}\right] \qquad x > 0$$

Distribution Function: No simple closed form.

Generator: Use the property that if Y is $N(\mu, \sigma^2)$, then $X = e^Y$ is LN(μ, σ^2):

```
Let Y = N(μ,σ²)
Return X = e^Y
```

Note: If θ and τ^2 are the mean and variance of the lognormal, respectively, they are related to μ and σ^2 by the formulas

$$\mu = \ln \frac{\theta^2}{\sqrt{\theta^2 + \tau^2}} \quad \text{and} \quad \sigma^2 = \ln \frac{\theta^2 + \tau^2}{\theta^2}$$

So to generate lognormals with given mean and variance, these formulas should be used in computing the correct μ and σ^2 to use in the generator.

Beta BETA(*p*, *q*), *p*, *q* > 0

Density:

$$f(x) = \begin{cases} \dfrac{x^{p-1}(1-x)^{q-1}}{B(p,q)} & \text{if } 0 < x < 1 \\ 0 & \text{otherwise} \end{cases}$$

where $B(p, q)$ is the beta function:

$$B(p,q) = \int_0^1 z^{p-1}(1-z)^{q-1}\,dz$$

and $p, q > 0$ are shape parameters.

Distribution Function: No closed form in general.

Generator 1: If G_1 and G_2 are independent GAM(a, p) and GAM(a, q) gamma variates, then $X = G_1/(G_1 + G_2)$ is a BETA(p, q) variate:

```
Let G1 = GAM(a,p)
Let G2 = GAM(a,q)
Return X = G1/(G1 + G2)
```

Generator 2: For $p, q > 1$, Cheng [11] gives the following acceptance–rejection method with bounded trials ratio ($< 4/e \simeq 1.47$):

```
Setup: α = p + q, β = √((α − 2)/(2pq − α)), γ = p + β^(−1).
Do{
  Let U1 = RN(0,1), U2 = RN(0,1)
  V = β ln[U1/(1− U1)], W = pe^V
}While(α ln[α/(q + W)] + γV − ln 4 < ln[U1^2 U2])
Return X = W/(q + W)
```

Generator 3: For $p, q < 1$, Jöhnk [12] gives the following acceptance–rejection method:

```
Do{
  Let U = RN(0,1), V = RN(0,1)
  Y = U^(1/p), Z = V^(1/q)
}While(Y + Z > 1)
Return X = Y/(Y + Z)
```

Inverse Gaussian IG(μ, λ), λ > 0, μ > 0

Density:

$$f(x) = \sqrt{\frac{\lambda}{2\pi x^3}}\exp\left[-\frac{\lambda(x-\mu)^2}{2\mu^2 x}\right] \qquad x > 0$$

Distribution Function:

$$F(x) = \Phi\left[\sqrt{\frac{\lambda}{x}}\left(\frac{x}{\mu} - 1\right)\right] + e^{2\lambda/\mu}\Phi\left[-\sqrt{\frac{\lambda}{x}}\left(\frac{x}{\mu} + 1\right)\right]$$

where Φ is the distribution function of $N(0, 1)$, the standard normal distribution.

Generator: Use the many-to-one transformation of Michael et al. [13]:

```
Setup: Let φ = λ/μ
Let Z = N(0,1), Y = Z²
Let T = 1 + (Y − √(4φY + Y²))/(2φ)
Let U = RN(0,1)
If(U ≤ 1/(1 + T)) Return X = μT
Else Return X = μ/T
```

Notes:

1. The method is based on the fact that

$$Y = \frac{\lambda(X - \mu)^2}{\mu^2 X} \tag{5}$$

is a chi-squared variable with one degree of freedom when X is an $IG(\mu, \lambda)$ variable. If Y is generated as a chi-squared variate and equation (5) solved for X, this gives two possible values for X: $X = \mu T$ or $X = \mu/T$. The randomized choice ensures that the resulting overall distribution of X is $\text{IG}(\mu, \lambda)$.

2. The parameterization using $\phi = \lambda/\mu$ as shape parameter rather than λ is often preferable, as the shape, when measured by ϕ, is invariant as μ varies.

Pearson Type V PT5(a, b), a, b > 0

Density:

$$f(x) = \begin{cases} \dfrac{a^b x^{-b-1}}{\Gamma(b)} \exp\left(-\dfrac{a}{x}\right) & x > 0 \\ 0 & \text{otherwise} \end{cases}$$

where a is a scale parameter and b a shape parameter.

Generator: $X = \text{PT5}(a, b)$ is precisely the same as the reciprocal of a gamma variate with scale $1/a$ and shape b:

```
Return X = 1/GAM(1/a,b)
```

Note: The mean is $a/(b - 1)$, which exists only when $b > 1$. So b must be set greater than unity if the variates generated are to have finite mean.

Pearson Type VI PT*6(a, p, q), a, p, q* > *0*

Density:

$$f(x) = \begin{cases} \dfrac{(x/a)^{p-1}}{aB(p,q)[1 + (x/a)]^{p+q}} & x > 0 \\ 0 & \text{otherwise} \end{cases}$$

Scale parameter $a > 0$ and shape parameters $p, q > 0$.

Generator 1: Use the fact that $X = \mathrm{PT6}(1, p, q)$ has precisely the same distribution as $Y/(1 - Y)$, where $Y = \mathrm{BETA}(p, q)$. This is known as a *beta variate of the second kind* [14]:

```
Let Y = BETA (p, q)
Return X = aY/(1 - Y)
```

Generator 2: Use the fact that if $Y_1 = \mathrm{GAM}(a,p)$ and $y_2 = \mathrm{GAM}(a, q)$, with Y_1, Y_2 independent, then Y_1/Y_2 has the same distribution as $\mathrm{PT6}(a, p, q)$:

```
Let Y1 = GAM (a, p), Y2 = GAM (a, q)
Return X = Y1/Y2
```

Note: The mean is $ap/(q - 1)$, which exists only when $q > 1$.

F-Distribution F(ν_1, ν_2), ν_1, ν_2 Positive Integers

Density and Distribution Function: This is a distribution frequently used in statistical tests. It is known as the *F-distribution with ν_1 and ν_2 degrees of freedom.* It is a special case of PT6, being the same distribution as $\mathrm{PT6}(a, p, q)$ with $a = \nu_2/\nu_1$, $p = \nu_1/2$, $q = \nu_2/2$.

Generator: Use the fact that $X = F(\nu_1, \nu_2)$ has the same distribution as the ratio of two independent χ^2 variates each scaled by its own mean:

```
Let Y1 = χ²(ν1)/ν1, Y2 = χ²(ν2)/ν2
Return X = Y1/Y2
```

Triangular TRI*(a, b, c), a* < *b* < *c*

Density:

$$f(x) = \begin{cases} \dfrac{2(x - a)}{(b - a)(c - a)} & a \le x \le b \\ \dfrac{2(c - x)}{(c - a)(c - b)} & b \le x \le c \\ 0 & \text{otherwise} \end{cases}$$

The shape of the density gives the distribution its name (see Figure 5.6).

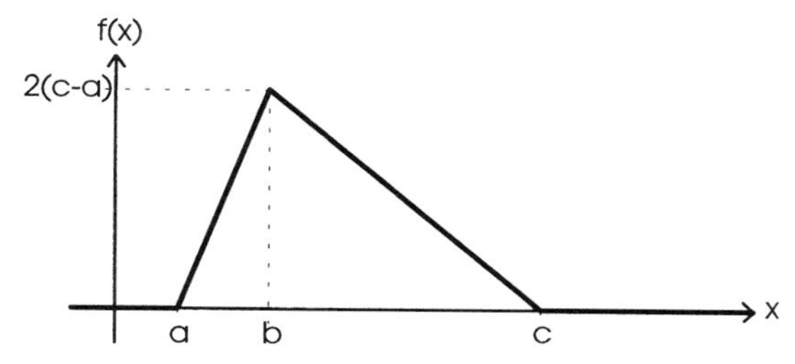

Figure 5.6 Density function for the triangular distribution.

Distribution Function:

$$F(x) = \begin{cases} 0 & x < a \\ \dfrac{(x-a)^2}{(b-a)(c-a)} & a \le x \le b \\ 1 - \dfrac{(c-x)^2}{(c-a)(c-b)} & b \le x \le c \\ 1 & x > c \end{cases}$$

Generator: $F(x)$ can be inverted. The inverse transform method gives

```
Setup: β = (b − a)/(c − a)
Let U = RN(0,1)
If(U < β) T = √(βU)
Else T = 1 − √((1 − β)(1 − U))
Return X = a + (c − a)T
```

Cauchy CAUCHY*(a, b), b > 0*

Density:

$$f(x) = \frac{b}{\pi[b^2 + (x-a)^2]} \qquad -\infty < x < \infty$$

Distribution Function:

$$F(x) = \frac{1}{2} + \frac{1}{\pi} \arctan \frac{x-a}{b} \qquad -\infty < x < \infty$$

Generator: The distribution function can be inverted. The inverse transform method gives

```
Let U = RN(0,1)
Return X = a + b tan πU
```

Notes:

1. The distribution does not have a finite mean, even though it is symmetric about a. The distribution has long tails.
2. The ratio of two independent standard normal variates has the CAUCHY(0,1) distribution.

von Mises VM(κ), $\kappa > 0$

Density:

$$f(x) = \frac{\exp(\kappa \cos x)}{2\pi I_0(\kappa)} \qquad -\pi < x < \pi$$

where I_0 is the modified Bessel function of the first kind of order zero (see Abramowitz and Stegun [15] for details of the Bessel function).

Generator: Use the Best and Fisher [16] rejection method:

```
Setup r = 1 + √(1 + 4κ²)  ρ = (r − √(2r))/(2κ)  s = (1 + ρ²)/(2ρ)
Do{
  U = U(-1, 1), V = U(-1, 1)
  Z = cos πU
  W = (1 + sZ)/(s + Z)
  Y = κ(s − W)
  }While (W(2 − W) < V AND ln(W/V) + 1 < W)
Return X = sgn U/ cos W
```

Note: X is the random angle of the direction on a circle.

Empirical Distribution EDF($x_1, x_2, \ldots, x_n$)

Distribution: Suppose that $x_1, x_2, \ldots, x_n$ is a random sample of size n. A piecewise linear distribution function can be constructed corresponding to this sample as follows. First sort the x's into ascending order: $x_{(1)} \leq x_{(2)} \leq \cdots \leq x_{(n)}$. The smoothed empirical distribution function is then

$$F(x) = \begin{cases} 0 & x < x_{(1)} \\ \dfrac{i-1}{n-1} + \dfrac{x - x_{(i)}}{(n-1)(x_{(i+1)} - x_{(i)})} & x_{(i)} \leq x \leq x_{(i+1)}, \quad i = 1, \ldots, n-1 \\ 1 & x > x_{(n)} \end{cases}$$

Generator: The smoothed distribution function can be inverted. Using the inverse transform method gives

```
Let U = RN(0,1), A = (n − 1)U, i = trunc (A) + 1
Return X = x(i) + (A − i + 1)(x(i+1) − x(i))
```

Here trunc(A) denotes integer part of A.

5.4 DISCRETE DISTRIBUTIONS

The general methods of Section 5.2 are in principle available for constructing discrete variate generators. However, the special characteristics of discrete variables imply that modifications of the general techniques are usually necessary. Two general methods are especially useful: the *look-up table method* and the *alias method*. In what follows trunc(x) and frac(x) mean the integer and fractional part of x.

5.4.1 Look-up Tables

The basic technique was described in Section 5.2.1. If the table is large, the look-up procedure can be slow, with the ith value requiring i steps to find. A simple alternative is to use a binary search to home in more rapidly on the value to be returned. We assume that the distribution has the form

$$p_i = \Pr(X = x_i) \qquad i = 1, 2, \ldots, n$$

$$P_i = \sum_{j=1}^{i} p_j = \Pr(X \leq x_i) \qquad i = 1, 2, \ldots, n \tag{6}$$

For distributions where the number of points extends to infinity, an appropriate cutoff for the distribution must be made. For example, set n so that

$$P_n > 1 - \delta = 0.99999, \text{say}$$

The precise value of the cutoff has to be selected carefully. For example, if the simulation involves tail probabilities that are important, δ must be chosen small enough to enable these probabilities to be estimated sufficiently accurately.

Look-up by Binary Search

```
Let U = RN(0,1), A = 0, B = n
While(A < B - 1){
  i = trunc[(A + B)/2]
  If(U > P_i) A = i
  Else B = i
  }
Return X = X_i
```

An alternative is to make a table of the starting points approximately every (n/m)th entry, in the same way that the letters of the alphabet form convenient starting points for search in a dictionary.

Look-up by Indexed Search. Set up the index table:

```
i = 0
For (j = 0 to m - 1){
  While(P_i < j/m){i = i + 1}
  Q_j = i
  }
```

```
Let U = RN(0,1), j = trunc(mU), i = Q_j
While(U ≥ P_i){i = i + 1}
Return X = X_i
```

5.4.2 Alias Method

The alias method, proposed by Walker [17], gives a way of returning a value from a table without searching. For a more recent reference, clear explanation is given in ref. 3. Assume that the distribution still has the form (6). We select a value of j in the range(1, 2, ..., n) all with equal probability. We then return one of two values: either the selected j which we return with probability q_j, or a precomputed alternative called the *alias* of j: $a(j)$, which is returned with probability $1 - q_j$. The $a(j)$ and q_j are selected to ensure that overall, the probability of returning each j is p_j, as required. A setup that does not need linked lists is given below. For a discussion of how it works and faster versions of the setup using linked lists, see ref. 2. (The definition and implementation of linked lists is discussed in ref. 3.)

Setup for Alias Method

```
For( j = 1 to n){
  Let q_j = 1, r_j = p_j, t_j = True
  }
For(k = 1 to n − 1){
  r_min = 2.0
  For(l = 1 to n){
    If(r_l < r_min AND t_l = True){r_min = r_l, i = l}
    }
  r_max = −1.0
  For(l = 1 to n){
    If(r_l > r_max AND t_l = True){r_max = r_l, j = l}
    }
If(r_i = r_j) Exit
t_i = False, a(i) = j, q_i = nr_i, r_j = r_j − (1 − q_i)/ n
}
```

Alias Method

```
Let U = RN(0,1), j = 1 + trunc(nU), p = frac(nU)
If(p ≤ q_j) Return X = x_j
Else Return X = x_a(j)
```

5.4.3 Empirical Distribution

Empirical distributions can be handled as a special case of a discrete distribution. Suppose that $x_1, x_2, \dots, x_n$ is a random sample of size n. Assume that each value has equal probability of occurring:

$$\Pr(X = x_i) = \frac{1}{n} \qquad i = 1, 2, \ldots, n$$

Then variates can be generated from this discrete distribution using

```
Let U = RN(0,1), i = trunc(nU) + 1
Return X = x_i
```

Sampling Without Replacement; Permutations

The following modification (see ref. 18) allows sampling of $m \leq n$ items from the random sample $x_1, x_2, \ldots, x_n$ of size n, *without replacement*. To avoid losing the original order, assume that the x's are held in the array a_i, $i = 1, 2, \ldots, n$:

```
For( j = 1 to m){
  Let U = RN(0,1), i = trunc[(n - j + 1)U] + j
  a = a_j, a_j = a_i, a_i = a
  }
Return a_1, a_2, . . . , a_m
```

The routine progressively swaps each entry with one drawn from the remaining list. At the end of the call the entries in the first m positions (i.e., $a_1, a_2, \ldots, a_m$) contain the elements sampled without replacement. The advantage of this algorithm is that repeated calls to it give further samples. The special case $m = n$ generates a random permutation of the initial sample.

5.4.4 Specific Discrete Distributions

The general look-up table or alias method is available for all the distributions listed below. So these two general methods will not be stated explicitly in individual cases unless there is a special interest in doing so.

Bernoulli BER(p), $0 < p < 1$

Probability Mass Function:

$$X = \begin{cases} 1 & \text{with probability } p \\ 0 & \text{with probability } 1 - p \end{cases}$$

This represents a trial with probability p of success and probability $(1 - p)$ of failure.

Generator: Elementary look-up table:

```
Let U = RN(0,1)
If (U ≤ p) Return X = 1
Else Return X = 0
```

Binomial BIN*(n, p), n a Positive Integer, 0 < p < 1*

Probability Mass Function:

$$p(x) = \begin{cases} \binom{n}{x} p^x (1-p)^{n-x} & x = 0, 1, \ldots, n \\ 0 & \text{otherwise} \end{cases}$$

where

$$\binom{n}{x} = \frac{n!}{x!\,(n-x)!}$$

is the *binomial coefficient.*

Generator: Use the special property that $X = \text{BIN}(n, p)$ if it is the sum of n independent $\text{BER}(p)$ variables:

```
X = 0
For(i = 1 to n){
  Let B = BER(p), X = X + B
  }
Return X
```

Note: The generation time increases linearly with n. One of the general methods will be preferable for n large (> 20, say).

Geometric GEOM*(p), 0 < p < 1*

Probability Mass Function:

$$p(x) = \begin{cases} p(1-p)^x & x = 0, 1, \ldots \\ 0 & \text{otherwise} \end{cases}$$

This gives the number of failures before the first success is encountered in a sequence of independent Bernoulli trials.

Generator: The distribution function is invertible:

```
Setup: a = 1/ln(1 - p)
Let U = RN(0,1)
Return X = trunc(a ln U)
```

Negative Binomial NEGBIN*(n, p), n an Integer, 0 < p < 1*

Probability Mass Function:

$$p(x) = \begin{cases} \binom{n+x-1}{x} p^n (1-p)^x & x = 0, 1, \ldots \\ 0 & \text{otherwise} \end{cases}$$

This gives the number of failures before the nth success is encountered in a sequence of independent Bernoulli trials.

Generator: By definition this variable is the sum of n independent GEOM(p) variables:

```
X = 0
For(i = 1 to n){
  Let Y = GEOM(p), X = X + Y
  }
Return X
```

Note: The generation time increases linearly with n. One of the general methods will be preferable for n large (> 10, say).

Hypergeometric HYP*(a, b), a, b Positive Integers*

Probability Mass Function:

$$p(x) = \begin{cases} \binom{a}{x}\binom{b}{n-x} \Big/ \binom{a+b}{n} & x = 0, 1, \ldots, n \\ 0 & \text{otherwise} \end{cases}$$

Consider a population where a individuals have a particular characteristic and b do not. If $n \leq a+b$ individuals are selected without replacement, the number X with the characteristic has the given hypergeometric distribution.

Generator: Fishman [19] gives the following inverse transform method:

```
Setup: α = p0 = [b!(a + b − n)!]/[(b − n)!(a + b)!]
Let A = α, B = α, X = 0
Let U = RN(0,1)
While(U > A){
  X = X + 1, B = B(a − X)(n − X)/[(X + 1)(b − n + X + 1), A = A + B
  }
Return X
```

Poisson POIS*(λ), λ > 0*

Probability Mass Function:

$$p(x) = \begin{cases} \dfrac{e^{-\lambda}\lambda^x}{x!} & x = 0, 1, \ldots \\ 0 & \text{otherwise} \end{cases}$$

If events occur randomly in time at rate r and X is the number of events that occurs in a time period t, then X = POIS(rt).

Generator 1: The direct method is to count the number of events in an appropriate time period, as indicated above:

```
Setup: a = e^-λ
Let p = 1, X = -1
While(p > a'){
  Let U = RN(0,1), p = pU, X = X + 1
  }
Return X
```

Generator 2: The time taken by generator 1 to return a variate increases approximately as λ. So for large λ (> 30, say) it is slow. The following acceptance–rejection technique given by Atkinson [20] is then preferable:

```
Setup: a = π√(λ/3), b = a/λ, c = 0.767 − 3.36/λ, d = ln c − ln b −λ
Do{
  Do{
    U = RN(0,1), Y = [a − ln((1 − U)/ U)]/ b
    } While(Y ≤ −1/2)
  Let X = trunc(Y + 1/2), V = RN(0,1)
  } While(a − bY + ln[V/(1 + e^(a − bY))^2] > d + X ln λ− ln X!)
Return X
```

Generator 3: An alternative to generator 2 uses the fact that for large λ, the distribution of $\lambda^{-1/2}(X - \lambda)$ tends to that of the standard normal. For large λ (> 20, say) we then have the following:

```
Setup: a = λ^(1/2)
Let Z = N(0,1)
Let X = max[0, trunc(0.5 + λ + aZ)]
Return X
```

5.5 MULTIVARIATE DISTRIBUTIONS

5.5.1 General Methods

The generation of multivariate distributions is not nearly as well developed as that of the univariate case. The key requirement in the generation of multivariate samples is the need to ensure an appropriate correlation structure among the components of the multivariate vector. Often the correlation arises because the variates are the output of a stochastic model with a certain structure; for example, the variates may describe the state of some stochastic process at given points in time. Correct sampling from the underlying form of the multivariate distribution can then be obtained simply by ensuring that the variates are generated according to the definition of the process. It is often not only more natural but considerably easier to generate multivariate samples in this way rather than to attempt to derive the distribution itself and then generate variates directly from it. There is one general approach that is sometimes useful, however [6].

Conditional Sampling

Let $\mathbf{X} = (X_1, X_2, \ldots, X_n)^T$ be a random vector with joint distribution function $F(x_1, x_2, \ldots, x_n)$. Suppose that the conditional distribution of X_j given that $X_i = x_i$, for $i =$

1, 2, ..., $j-1$, is known, for each j. Then the vector $\mathbf{X}$ can be built up one component at a time, with each component obtained by sampling from a *univariate* distribution:

```
Generate x_1 from the distribution F_1(x)
Generate x_2 from the distribution F_2(x|X_1 = x_1)
Generate x_3 from the distribution F_3(x|X_1 = x_1, X_2 = x_2)
...
Generate x_n from the distribution F_n(x|X_1 = x_1, X_2 = x_2, . . . , X_{n-1} =
x_{n-1})
Return X = (x_1, x_2, . . . , x_n)^T
```

The usefulness of this method is dependent on the availability of the conditional distributions, and on the ease of sampling from them.

5.5.2 Special Distributions

Multivariate Normal MVN($\boldsymbol{\mu}, \boldsymbol{\Sigma}$)

Density:

$$f(x) = (2\pi|\boldsymbol{\Sigma}|)^{-n/2} \exp[-\tfrac{1}{2}(\mathbf{x}-\boldsymbol{\mu})^{\mathrm{T}}\boldsymbol{\Sigma}^{-1}(\mathbf{x}-\boldsymbol{\mu})]$$
$$-\infty < x_i < \infty, \quad i = 1, 2, \ldots, n$$

where $\boldsymbol{\mu}$ is a $n \times 1$ vector and $\boldsymbol{\Sigma}$ is a $n \times n$ positive-definite symmetric matrix.

Generator: The simplest method uses a linear transformation of a set of independent $N(0,1)$ variates: $\mathbf{Z} = (Z_1, Z_2, \ldots, Z_n)^{\mathrm{T}}$ is transformed into $\mathbf{X} = \mathbf{LZ} + \boldsymbol{\mu}$, where $\mathbf{L}$ is a lower triangular matrix satisfying $\boldsymbol{\Sigma} = \mathbf{LL}^{\mathrm{T}}$. $\mathbf{L}$ can be obtained by the standard Choleski decomposition [19].

```
Setup: L (see below)
Let a = sqrt(σ_11)
For(i = 1 to n){L_i1 = σ_i1/ a}
Let i = 2
While(True){
  S = 0
  For( j = 1 to i - 1) S = S + L_ij^2
  L_ii = (σ_ii - S)^1/2
  If(i = n) Return L
  i = i + 1
  For( j = 2 to i - 1){
    S = 0
    For(k = 1 to j - 1) S = S + L_ik L_jk
    L_ij = (σ_ij - S)/ L_jj
    }
}
```

MVN Generator:

```
For(i = 1 to n) Z_i = N(0,1)
For(i = 1 to n){
  X_i = μ_i
  For(j = 1 to i) X_i = X_i + L_ij Z_j
  }
Return X = (X_1, X_2, . . . , X_n)
```

Uniform Distribution on the n-Dimensional Sphere. If the components of MVN($\mathbf{0}, \mathbf{I}$) are treated as a direction vector in n-dimensional Euclidean space, all directions are equally likely. Rescaling the vector to unit length therefore gives a point uniformly distributed on the unit n-dimensional sphere.

Generator:

```
S = 0
For(i = 1 to n){
  Z_i = N(0,1), S = S + Z_i^2
  }
S = √S
For(i = 1 to n){
  X_i = Z_i/S
  }
Return X = (X_1, X_2, . . . , X_n)
```

Order Statistics. The order statistics of a random sample $X_1, X_2, \ldots, X_n$ of size n are just the individual variates arranged by value in ascending order:

$$X_{(1)} \leq X_{(2)} \leq \cdots \leq X_{(n)}$$

This can be done by generating the sample and then reordering. The fastest sorting routines are $O(n \ln n)$, and the sorting dominates once n is large.

If the X's can be generated by the inverse transform $X = F^{-1}(U)$, the sample can be generated in order from the order statistics of a uniform sample

$$U_{(1)} \leq U_{(2)} \leq \cdots \leq U_{(n)} \tag{7}$$

This uses the fact that (1) the largest uniform order statistic $U_{(n)}$ has an invertible distribution function, and (2) $U_{(1)}, U_{(2)}, \ldots, U_{(i)}$ are the order statistics of a sample of size i drawn from the uniform distribution $U(0, U_{(i+1)})$.

```
Let U = RN(0,1), U_(n) = U^(1/n)
For(i = n - 1 down to 1){
  Let U = RN(0,1)
  U_(i) = U_(i+1) U^(1/i)
  }
```

An alternative way of generating (7) is given in (2):

```
Let E_1 = EXP(1), S_1 = E_1
For(i = 2 to n + 1){
  Let E_i = EXP(1), S_i = S_{i-1} + E_i
  }
For(i = 1 to n) Let U_(i) = S_i/S_{n+1}
```

Dirichlet

Density:

$$f(x) = Cx_1^{p_1 - 1}x_2^{p_2 - 1} \cdots x_k^{p_k - 1}(1 - x_1 - x_2 - \cdots - x_k)^{p_{k+1} - 1}$$

$$\text{all } p_i > 0, \quad x_1, x_2, \ldots, x_k > 0, \quad \sum_{i=1}^{k} x_i \leq 1$$

where

$$\frac{1}{C} = \frac{\Gamma(p_1)\Gamma(p_2)\cdots\Gamma(p_k)}{\Gamma(p_1 + p_2 + \cdots + p_k)} \int_0^1 u^{p_1 + p_2 + \cdots + p_k - 1}(1 - u)^{p_{k+1} - 1}\, du$$

Generator: Use the fact that a set of gamma variates scaled by their sum, so that they sum to unity, has the Dirichlet distribution:

```
Y = 0
For(i = 1 to k + 1){
Let G_i = GAM(1, p_i), Y = Y + G_i
}
For(i = 1 to k + 1){
Let X_i = G_i/Y
}
Return (X_1, X_2, . . . , X_{k+1})
```

Note: The $X_1, X_2, \ldots, X_k$ have the distribution given above. Including X_{k+1} gives the sum $\Sigma_{i=1}^{k+1} X_i = 1$. The distribution is therefore useful in representing random proportions that have to sum to unity.

5.6 STOCHASTIC PROCESSES

5.6.1 Point Processes

A sequence of points $t_0 = 0, t_1, t_2, \ldots$ in time is known as a *point process*. The times between occurrences $x_i = t_i - t_{i-1}$ are usually random. Examples are where the t_i are arrival times of customers and the x_i are interarrival times, or where the t_i are moments of breakdowns and the x_i are lifetimes.

Poisson Process. When the x_i are independent EXP($1/\lambda$) variables, the t_i sequence is known as a *Poisson process with rate* λ. To generate the next time point, assuming that t_{i-1} has already been generated:

```
Let U = RN(0,1)
Return t_i = t_{i-1} -λ^{-1} ln U
```

Nonstationary Poisson Process. Suppose that the Poisson process has $\lambda = \lambda(t)$; that is, the rate varies with time. One way to generate such a nonstationary process is via an analog of the inverse transform technique. Define the cumulative rate

$$\Lambda(t) = \int_0^t \lambda(u)\, du$$

and suppose this is invertible with inverse $\Lambda^{-1}(\cdot)$.

To generate the next time point, assume that s_{i-1}, the previous point of a unit rate Poisson process, has already been generated. Then the next point of the nonstationary process is given by

```
Let U = RN(0,1), s_i = s_{i-1}- ln U
Return t_i = Λ^{-1}(s_i)
```

An alternative is to use an analog of the acceptance–rejection method proposed by Lewis and Shedler [21], called *thinning*. Suppose that $\lambda_M = \max_t \lambda(t)$. Then, assuming that t_{i-1} has already been generated, the next point of the nonstationary process is given by

```
Let t = t_{i-1}
Do{
  U = RN(0,1) t = t -λ_M^{-1} ln U  V = RN(0,1)
  }While(V>λ(t)/λ_M)
Return t_i = t
```

Markov Process. The simplest Markov process is the *discrete-time* Markov chain. Here time is advanced one unit at a time: $t = 0, 1, 2, \ldots$. At each time point the system is assumed to be in one of n states: $X = 1, 2, \ldots, n$, say. Given that $X_t = i$, the next state X_{t+1} is selected according to the discrete probability distribution

$$\Pr(X_{t+1} = j | X_t = i) = p_{ij} \qquad j = 1, 2, \ldots, n$$

Continuous-time Markov chains are best simulated slightly differently. Assume that the system has just entered state i at time t_k. Then the next change of state occurs at $t_{k+1} = t_k + \text{EXP}(1/\lambda_i)$. The state entered is j with probability p_{ij} ($j = 1, 2, \ldots, n$).

5.6.2 Time-Series Models and Gaussian Processes

A stochastic process $X(t)$ all of whose joint distributions are multivariate normal (i.e., $X_{t_1}, X_{t_2}, \ldots, X_{t_r}$ is multivariate normal for any given set of times $t_1, t_2, \ldots, t_r$) is said to be a *Gaussian process*.

Many time-series models use normal perturbations (usually called *innovations*) and are Gaussian.

Moving Average. A moving-average process X_t is defined by

$$X_t = Z_t + \beta_1 Z_{t-1} + \cdots + \beta_q Z_{t-q} \qquad t = 1, 2, 3, \ldots$$

where the Z's are all independent $N(0, \sigma^2)$ normal variates and the β's are user-prescribed coefficients. The X's can be generated directly from this definition.

Autoregressive Process. An autoregressive process X_t is defined by

$$X_t = \alpha_1 X_{t-1} + \cdots + \alpha_p X_{t-p} + Z_t \qquad t = 1, 2, 3, \ldots$$

where the Z's are all independent $N(0, \sigma^2)$ normal variates and the α's are user prescribed coefficients. The X's can again be generated directly from this definition, but in this case the initial values $X_0, X_{-1}, \ldots, X_{1-p}$ need to be obtained. Now

$$(X_0, X_{-1}, \ldots, X_{1-p}) = \text{MVN}(\mathbf{0}, \mathbf{\Sigma})$$

where $\mathbf{\Sigma}$ satisfies

$$\mathbf{\Sigma} = \mathbf{A\Sigma A}^{\mathrm{T}} + \mathbf{B} \tag{8}$$

with

$$\mathbf{A} = \begin{bmatrix} \alpha_1 & \alpha_2 \cdots \alpha_{p-1} & \alpha_p \\ 1 & 0 \cdots 0 & 0 \\ 0 & 1 \cdots 0 & 0 \\ \cdot & \cdots & \cdot \\ 0 & 0 \cdots 1 & 0 \end{bmatrix} \qquad \mathbf{B} = \begin{bmatrix} \sigma^2 & 0 \cdots 0 \\ 0 & 0 \cdots 0 \\ 0 & \cdots 0 \\ \cdots & \cdots \\ 0 & 0000 \end{bmatrix}$$

Once $\mathbf{\Sigma}$ has been found from (8), $(X_0, X_{-1}, \ldots, X_{1-p})$ can be generated using $\text{MVN}(0, \mathbf{\Sigma})$; see ref. 22. An alternative is to set $X_0 = X_{-1} = \cdots = X_{1-p} = 0$ and run the sequence for a settling in period before collecting the results. For more complicated Gaussian models, see ref. 2.

REFERENCES

1. Devroye, L. (1986). *Non-uniform Random Variate Generation*, Springer-Verlag, New York.
2. Ripley, B. D. (1987). *Stochastic Simulation*, Wiley, New York.
3. Law, A. M., and W. D. Kelton (1991). *Simulation Modeling and Analysis*, 2nd ed., McGraw-Hill, New York.
4. Banks, J., J. S. Carson II, and B. L. Nelson (1996). *Discrete-Event Simulation*, 2nd ed., Prentice Hall, Upper Saddle River, N.J.
5. Fishman, G. S. (1996). *Monte Carlo: Concepts, Algorithms, and Applications*, Springer-Verlag, New York.
6. Johnson, M. E. (1987). *Multivariate Statistical Simulation*, Wiley, New York.
7. Fishman, G. S. (1976). Sampling from the gamma distribution on a computer, *Communications of the ACM*, Vol. 19, pp. 407–409.
8. Ahrens, J. H., and U. Dieter (1974). Computer methods for sampling from gamma, beta, Poisson and binomial distributions, *Computing*, Vol. 12, pp. 223–246.
9. Cheng, R. C. H. (1977). The generation of gamma variables with non-integral shape parameter, *Applied Statistics*, Vol. 26, pp. 71–75.

10. Box, G. E. P., and M. E. Muller (1958). A note on the generation of random normal deviates, *Annals of Mathematical Statistics*, Vol. 29, pp. 610–611.
11. Cheng, R. C. H. (1978). Generating beta variables with nonintegral shape parameters, *Communications of the ACM*, Vol. 21, pp. 317–322.
12. Jöhnk, M. D. (1964). Generation of beta distribution and gamma distribution random variates (in German), *Metrika*, Vol. 8, pp. 5–15.
13. Michael, J. R., W. R. Schucany, and R. W. Haas (1976). Generating random variates using transformations with multiple roots, *The American Statistician*, Vol. 30, pp. 88–90.
14. Stuart, A., and J. K. Ord (1987). *Kendall's Advanced Theory of Statistics*, 5th ed., Vol. 1, Griffin, London.
15. Abramowitz, M., and I. A. Stegun (1965). *Handbook of Mathematical Functions*, Dover, New York.
16. Best, D. J., and N. I. Fisher (1979). Efficient simulation of the von Mises distribution, *Applied Statistics*, Vol. 28, pp. 152–157.
17. Walker, A. J. (1977). An efficient method for generating discrete random variables with general distributions, *ACM Transactions on Mathematical Software*, Vol. 3, pp. 253–256.
18. Moses, L. E., and R. V. Oakford (1963). *Tables of Random Permutations*, Stanford University Press, Stanford, Calif.
19. Fishman, G. S. (1978). *Principles of Discrete Event Simulation*, Wiley, New York.
20. Atkinson, A. C. (1979). The computer generation of Poisson random variables, *Applied Statistics*, Vol. 28, pp. 29–35.
21. Lewis, P. A. W., and G. S. Shedler (1979). Simulation of non-homogeneous Poisson processes by thinning, *Naval Research Logistics Quarterly*, Vol. 26, pp. 403–413.
22. Gardner, G., A. C. Harvey, and G. D. A. Phillips (1979). Algorithm AS-154: an algorithm for the exact maximum likelihood estimation of autoregressive–moving average models by means of Kalman filtering, *Applied Statistics*, Vol. 29, pp. 311–322.

CHAPTER 6

Experimental Design for Sensitivity Analysis, Optimization, and Validation of Simulation Models

JACK P. C. KLEIJNEN
Tilburg University

6.1 INTRODUCTION

In this chapter we survey *design of experiments* (DOE), which includes designs such as 2^{k-p} designs, applied to simulation. The related term *experimental design* may suggest that this subdiscipline is still experimental, but it is not; see early publications such as that of Plackett and Burman (1946). DOE is a subdiscipline within mathematical statistics. This chapter is a tutorial that discusses not only methodology but also applications. These applications come from the author's experience as a consultant and from publications by others in the United States and Europe. The reader is assumed to have a basic knowledge of mathematical statistics and simulation.

In this section we address the questions of *what* DOE is and *why* DOE is needed. These questions are illustrated using two case studies. The first case concerns an ecological study that uses a deterministic simulation model (consisting of a set of nonlinear difference equations) with 281 parameters. The ecological experts are interested in the effects of these parameters on the response: namely, future carbon dioxide (CO_2) concentration, CO_2 being the major cause of the greenhouse effect. The pilot phase of this study aims at *screening*: which factors among the many potentially important factors are really important. Details are given in Section 6.2.

The second case study concerns a decision support system (DDS) for production planning in a Dutch steel tube factory. The DSS and the factory are modeled through a stochastic, discrete-event simulation (stochastic or random simulations have a special variable, the (pseudo)random number seed). The DSS to be optimized has 14 input or decision variables and two response variables, productive hours and lead time. Simulation of one combination of these 14 inputs takes 6 hours on the computer available

Handbook of Simulation, Edited by Jerry Banks.
ISBN 0-471-13403-1

at that time, so searching for the optimal combination must be performed with care. Details are given in Section 6.5.

Decision making is an important use of simulation. In particular, factor screening and optimization concern that topic. Note that closely related to optimization is *goal seeking*: Given a target value for the response variable, find the corresponding input values; the regression metamodel of this chapter can be used to find these desired input values. Much of this chapter is about the use of simulation to improve decision making. This will be demonstrated by several case studies; more case studies, especially in manufacturing, are reviewed by Yu and Popplewell (1994).

It is convenient now to introduce some DOE *terminology*, defined in a simulation context. A *factor* is a parameter, an input variable, or a module of a simulation model (or simulation computer program). Examples of parameters and input variables were in the preceding discussion of two case studies. Other examples are provided by classic queueing simulations: A parameter may be a customer arrival rate or a service rate, an input variable may be the number of parallel servers, and a module may be the submodel for the priority rules [first in, first out (FIFO), shortest processing time (SPT), etc.].

By definition, factors are changed during an experiment; they are not kept constant during the entire experiment. Hence a factor takes at least two *levels* or values during the experiment. The factor may be *qualitative*, as the priority rules exemplified. A detailed discussion of qualitative factors and various measurement scales is given in Kleijnen (1987, pp. 138–142).

The central problem in DOE is the astronomically great *number of combinations of factor levels*. For example, in the ecological case study at least 2^{281} ($>10^{84}$) combinations may be distinguished; a queueing network also has many factors. DOE can be defined as selecting the combinations of factor levels that will actually be simulated in an experiment with the simulation model.

After this selection of factor combinations, the simulation program is executed or run for these combinations. Next, DOE analyzes the resulting input–output (I/O) data of the experiment, to derive conclusions about the importance of the factors. In simulation this is also known as *what-if analysis*: What happens if the analysts change parameters, input variables, or modules of the simulation model? This question is closely related to sensitivity analysis, optimization, and validation/verification, as we show in detail in this chapter.

Unfortunately, the vast literature on simulation does not provide a standard definition of *sensitivity analysis*. In this chapter, sensitivity analysis is interpreted as the systematic investigation of the reaction of the simulation responses to *extreme* values of the model's input or to *drastic* changes in the model's structure. For example, what happens to the customers' mean waiting time when their arrival rate doubles; what happens if the priority rule changes from FIFO to SPT? So in this chapter we do not focus on marginal changes or perturbations in the input values.

For this what-if analysis, DOE uses *regression analysis*, also known as analysis of variance (ANOVA). This analysis is based on a *metamodel*, which is defined as a model of the underlying simulation model (Friedman, 1996; Kleijnen, 1975b). In other words, a metamodel is an approximation of the simulation program's I/O transformation; it is also called a response surface. Typically, this regression metamodel belongs to one of the following three classes: (1) a first-order polynomial, which consists of main effects only, besides an overall or grand mean; (2) a first-order polynomial augmented with interactions between pairs of factors (two-factor interactions); and (3) a second-order polynomial, which also includes purely quadratic effects [see also equation (1) in Section 6.3.2].

Most simulation models have *multiple outputs*, also called responses or criteria. For example, outputs may be both customer's queueing time and server's idle time, or both mean and 90% quantile of the waiting time. In practice, multiple outputs are handled through the application of the techniques of this chapter *per* output type. [Ideally, however, such simulations should be studied through multivariate regression analysis, and the design should also account for the presence of multiple outputs (Khuri, 1996; Kleijnen, 1987).] Simultaneous inference may be taken care of through Bonferroni's inequality (see Appendix 6.1). Optimization in the presence of multiple responses is discussed in Section 6.5. Optimization accounting for both the mean and the variance of the response is the focus of Taguchi's methods (Ramberg et al., 1991). Note that the term *multiple regression analysis* refers not to the number of outputs but to the presence of multiple inputs, or better, multiple independent variables [see also the definition of z following equation (4)].

A metamodel treats the simulation model as a *black box*: the simulation model's inputs and outputs are observed, and the factor effects in the metamodel are estimated. This approach has the following *advantages* and *disadvantages*. An advantage is that DOE can be applied to all simulation models, either deterministic or stochastic. Further, DOE gives better estimates of the factor effects than does the intuitive approach often followed in practice, namely the one-factor-at-a-time approach, which is discussed in Section 6.4.2.

A drawback is that DOE cannot take advantage of the specific structure of a given simulation model, so it takes more simulation runs than do *perturbation analysis* and modern *importance sampling*, also known as *likelihood ratio* or *score function*. These alternative methods usually require a single run (by definition, a simulation run is a single time path with fixed values for all its inputs and parameters). Such a run, however, may be much longer than a run in DOE. Moreover, these alternatives require more mathematical sophistication, and they must satisfy more mathematical assumptions. Importance sampling as a variance reduction technique (not a what-if technique) is discussed in Section 6.3.3 and Appendix 6.2. There is much literature on these alternative methods (see Chapter 9; Glynn and Iglehart, 1989; Ho and Cao, 1991; Kleijnen and Rubinstein, 1996; Rubinstein and Shapiro, 1993).

DOE may be used not only for sensitivity analysis and optimization, but also for *validation*. In this chapter we address only part of the validation and verification problem (see Section 6.6). A detailed discussion of validation and verification is provided in Chapter 10.

In summary, in this chapter we discuss DOE as a method for answering what-if questions in simulation. It is not surprising that DOE is important in simulation: By definition, simulation means that a model is used not for mathematical analysis or numerical methods but for experimentation. But experimentation requires good design and good analysis!

DOE with its concomitant regression analysis is a standard topic in statistics. However, the standard statistical techniques must be adapted such that they account for the *peculiarities of simulation*:

1. There are a great many factors in many practical simulation models. Indeed, the ecological case study (mentioned above) has 281 factors, whereas standard DOE assumes only up to (say) 15 factors.
2. Stochastic simulation models use (pseudo)random numbers, which means that the analysts have much more control over the noise in their experiments than the

investigators have in standard statistical applications. For example, common and antithetic seeds may be used (see Section 6.3.5 and Appendix 6.2).

3. Randomization is of major concern in DOE outside simulation: Assign the experimental units (e.g., patients) to the treatments (say, types of medication) in a random, nonsystematic way so as to avoid bias (healthy patients receive medication of type 1 only). In simulation, however, this randomization problem disappears: Pseudorandom number streams take over.
4. Outside simulation the application of blocking is an important technique to reduce systematic differences among experimental units; for example, tire wear differs among the four positions on the car: left front, ... , right rear. In simulation, however, complete control over the experiment eliminates the need for blocking. Yet the blocking concept may be used to assign common and antithetic (pseudo)random numbers, applying the Schruben–Margolin approach (see Section 6.3.5).

The *main conclusions* of this chapter will be:

1. Screening may use a novel technique, called *sequential bifurcation*, that is simple, efficient, and effective.
2. Regression metamodeling generalizes the results of a simulation experiment with a small number of factors, since a regression metamodel estimates the I/O transformation specified by the underlying simulation model.
3. Statistical designs give good estimators of main (first-order) effects, interactions between factors, and quadratic effects; these designs require fewer simulation runs than intuitive designs do.
4. Optimization may use RSM, which combines regression analysis and statistical designs with steepest ascent (see conclusions 2 and 3).
5. Validation may use regression analysis and statistical designs.
6. These statistical techniques have already been applied many times in practical simulation studies in many domains; these techniques make simulation studies give more general results in less time.

The remainder of this chapter is *organized* as follows. Section 6.2 covers the screening phase of a simulation study. After an introduction (Section 6.2.1), in Section (6.2.2) we discuss a special screening technique, sequential bifurcation.

In Section 6.3 we discuss how to approximate the I/O transformation of simulation models by regression analysis. First we discuss a graphical method, scatter plots (Section 6.3.1). Next, in Section 6.3.2 we present regression analysis, which formalizes the graphical approach, including generalized least squares (GLS). Then in Section 6.3.3 we show how to estimate the variances of individual simulation responses, including means, proportions, and quantiles, in either steady-state or transient-state simulations. These estimates lead to estimated GLS (Section 6.3.4). Variance reduction techniques (VRTs), such as common random numbers, complicate the regression analysis; VRTs are discussed in Section 6.3.5. The estimated regression model may inadequately approximate the underlying simulation model: Section 6.3.6 covers several lack-of-fit tests. The section closes with a numerical example in Section 6.3.7 and a case study in Section 6.3.8.

In Section 6.4 we discuss statistical designs. After an introduction (Section 6.4.1), the focus is first on designs that assume only main effects (Section 6.4.2). Then follow designs that give unbiased estimators for the main effects even if there are interactions between factors (Section 6.4.3). Further, in this section we discuss designs that allow estimation of interactions between pairs of factors (Section 6.4.4), interactions among any subset of factors (Section 6.4.5), and quadratic effects (Section 6.4.6). All these designs are based on certain assumptions; how to satisfy these assumptions is discussed in Section 6.4.7. Optimal designs are discussed in Section 6.4.8. This section ends with a case study in Section 6.4.9.

Section 6.5 covers optimization of simulated systems. RSM is discussed in Section 6.5.1. Two case studies are summarized in Sections 6.5.2 and 6.5.3. Section 6.6 proceeds with the role of sensitivity analysis in validation, emphasizing the effects of data availability. In Section 6.7 we provide a summary and conclusions.

Three appendixes cover tactical issues in addition to the strategic issues addressed by DOE. Appendix 6.1 summarizes confidence intervals for expected values, proportions, and quantiles in terminating and steady-state simulations (see also Chapter 7). This appendix makes the chapter self-sufficient. Appendix 6.2 gives more details on VRTs, because VRTs are important when designing simulation experiments. This appendix covers four VRTs: common (pseudo)random numbers, antithetic numbers, control variates or regression sampling, and importance sampling. Appendix 6.3 covers jackknifing, which is a general method that may reduce bias of estimated simulation responses and may give robust confidence intervals.

Nearly 100 references conclude the chapter. To reduce the number of references, only the most recent references for a topic are given unless a specific older reference is of great historical value.

6.2 SCREENING

6.2.1 Introduction

In the pilot phase of a simulation study there are usually a great many potentially important factors; for example, the ecological case study has 281 parameters. It is the mission of science to come up with a *short list* of the most important factors; this is sometimes called the *principle of parsimony* or *Occam's razor*.

Obviously, a full factorial design requires an astronomically great number of factor combinations: in the case study, at least 2^{281} ($>10^{84}$). Even a design with only as many combinations as there are factors (see Section 6.4.2) may require too much computer time. Therefore, many practitioners often restrict their study to a few factors, usually no more than 15. Those factors are selected through intuition, prior knowledge, and the like. The factors that are ignored (kept constant), are (explicitly or implicitly) assumed to be unimportant. For example, in queueing networks the analysts may assume equal service rates for different servers. Of course, such an assumption severely restricts the generality of the conclusions from the simulation study!

The statistics literature does include screening designs: random designs, supersaturated designs, group screening designs, and so on (Kleijnen, 1987). Unfortunately, too little attention is paid to screening designs in the statistics literature. The reason for this neglect is that outside simulation, it is virtually impossible to control hundreds of factors; (say) 15 is difficult enough.

In simulation, however, models may have hundreds of parameters, yet their control is simple: Specify which combinations of parameter values to simulate. Nevertheless, screening applications in simulation are still scarce, because most analysts are not familiar with these designs. Recently, screening designs have been improved and new variations have been developed; details are given in Bettonvil and Kleijnen (1997) and Saltelli et al. (1995). In the next section we describe a promising screening technique, sequential bifurcation.

6.2.2 Sequential Bifurcation

Sequential bifurcation is a group-screening technique; that is, it uses *aggregation* (which is often applied in science when studying complicated systems). Hence at the start of the simulation experiment, sequential bifurcation groups the individual factors into clusters. A specific group is said to be at its high level (denoted as + or *on*) if each of its components or individual factors is at a level that gives a higher or equal response (not a lower response). Analogously, a group is at its low level (– or *off*) if each component gives a lower or equal response. To know with certainty that individual factor effects within a group do not cancel out, sequential bifurcation must assume that analysts know whether a specific individual factor has a positive or negative effect on the simulation response; that is, the factor effects have *known signs*. In practice this assumption may not be too restrictive. For example, in specific queueing simulations it may be known that increasing the service rates while keeping all other factors constant (*ceteris paribus* assumption) decreases waiting time (but it is unknown how big this decrease is; therefore, the analysts use a simulation model). In the ecological case study the experts could indeed specify in which direction a specific parameter affects the response (CO_2 concentration). Moreover, if a few individual factors have unknown signs, these factors can be investigated separately, outside sequential bifurcation!

Sequentialization means by definition that factor combinations to be simulated are selected as the experimental results become available. So as simulation runs are executed, insight into factor effects is accumulated and used to select the next run. It is well known that in general, sequentialization requires fewer observations; the price is more cumbersome analysis and data handling. Sequential bifurcation eliminates groups of factors as the experiment proceeds because the procedure concludes that these clusters contain no important factors.

Also, as the experiment proceeds, the groups become smaller. More specifically, each group that seems to include one or more important factors is split into two subgroups of the same size: *bifurcation*. At the end of bifurcation, individual factor effects are estimated. As a numerical example, consider a simple academic exercise with 128 factors, of which only three factors are important, factors 68, 113, and 120. Let the symbol $y_{(h)}$ denote the simulation output when the factors $1, \ldots, h$ are switched on and the remaining factors $(h + 1, \ldots, k)$ are off. Consequently, the sequence $\{y_{(h)}\}$ is nondecreasing in h. The main effect of factor h is denoted as β_h [see also (1) in Section 6.3.2]. Let the symbol $\beta_{h-h'}$ denote the sum of individual effects β_h through $\beta_{h'}$ with $h' > h$; for example, β_{1-128} denotes the sum of β_1 through β_{128} (Figure 6.1, line 1).

At the start (stage 0) of the procedure, sequential bifurcation always observes the two "extreme" factor combinations: $y_{(0)}$ (no factor high) and $y_{(k)}$ (all factors high). The presence of (three) important factors gives $y_{(0)} < y_{(128)}$. Hence sequential bifurcation concludes that the sum of all individual main effects is important: $\beta_{1-128} > 0$. Sequential bifurcation works such that any important sum of effects leads to a new observation that splits that sum into two subsums (see the symbol $\downarrow$ in Figure 6.1). Because stage 0 gives $\beta_{1-128} > 0$, sequential bifurcation proceeds to the next stage.

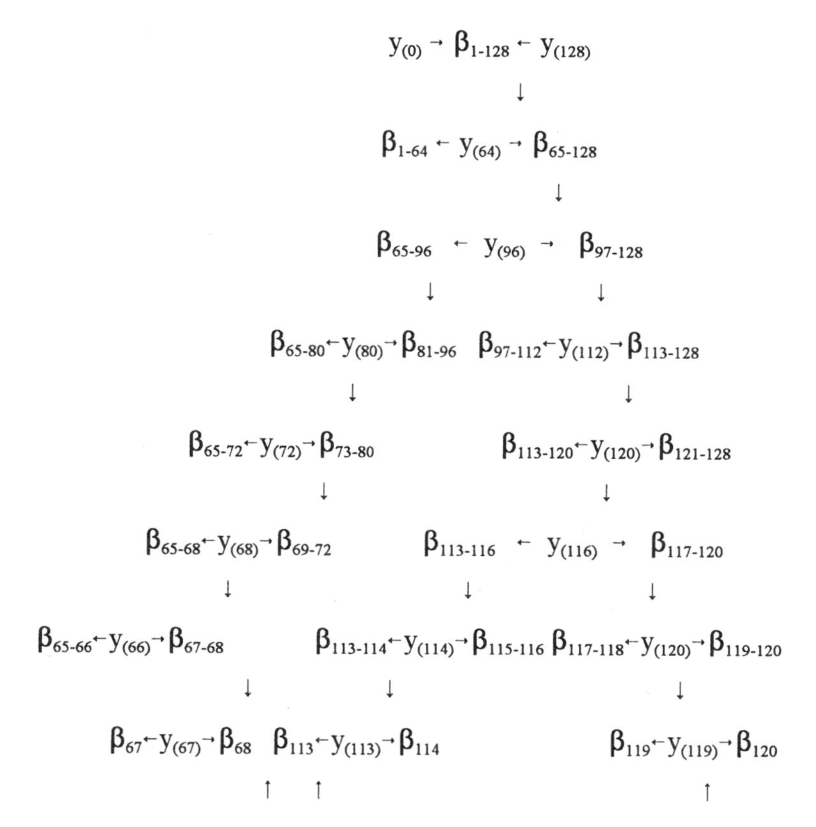

Figure 6.1 Finding $k = 3$ important factors among $K = 128$ factors in Jacoby and Harrison's (1962) example.

Stage 1 gives $y_{(64)}$. The analysis first compares $y_{(64)}$ with $y_{(0)}$ and notices that these two outputs are equal (remember that only factors 68, 113, and 120 are important). Hence the procedure concludes that the first 64 individual factors are unimportant! So after only three simulation runs and based on the comparison of two runs, sequential bifurcation eliminates all factors in the first half of the total group of 128 factors. Next, sequential bifurcation compares $y_{(64)}$ with $y_{(128)}$ and notices that these two outputs are not equal. Hence the procedure concludes that the second subgroup of 64 factors is important; that is, there is at least one important factor in the second half of the group of 128 factors.

In stage 2 sequential bifurcation concentrates on the remaining factors (65 through 128). That subgroup is again bifurcated, and so on. At the end, sequential bifurcation finds the three important factors (68, 113, and 120). In total, sequential bifurcation requires only 16 observations. The procedure also determines the individual main effects of the important factors (see the symbol $\uparrow$ in the last line of Figure 6.1).

It can be proven that if analysts assume that there are interactions between factors, the number of runs required by sequential bifurcation is double the number required in the main-effects-only case (see also the foldover principle in Section 6.4.3). In general, it is wise to accept this doubling to obtain estimators of main effects that are not biased by interactions between factors.

The ecological case study also uses the sequential bifurcation algorithm that was indicated in Figure 6.1. It took 154 simulation runs to identify and estimate the 15 most important factors among the original 281 factors. Some of these 15 factors surprised the ecological experts, so sequential bifurcation turns out to be a powerful statistical (black box) technique. Notice that on hindsight it turned out that there are no important interactions between factors, so only $154/2 = 77$ runs would have sufficed.

Another case study is the building thermal deterministic simulation in De Wit (1997). In his simulation, sequential bifurcation gave the 16 most important inputs among the 82 factors after only 50 runs. He checked these results by applying a different screening technique, the randomized one-factor-at-a time designs of Morris (1991), which took 328 runs.

These two case studies concern deterministic simulation models. In *stochastic simulation* the signal/noise ratio can be controlled by selecting an appropriate run length. Then sequential bifurcation may be applied as in deterministic simulation. Bettonvil (1990, pp. 49–142) has investigated further the use of sequential bifurcation in random simulation. More research is needed to find out whether in practice this method performs well in random simulations (Cheng, 1997).

6.3 REGRESSION METAMODELS

6.3.1 Introduction: Graphical Methods

Suppose that the number of factors to be investigated is small, for example, 15 (this small number may be reached after a screening phase; see Section 6.2). Suppose further that the simulation has been run for several combinations of factor levels. How should these I/O data be analyzed?

Practitioners often make a *scatter plot*, which has on the x-axis the values of one factor (e.g., traffic rate) and on the y-axis the simulation response (e.g., average waiting time). This graph indicates the I/O transformation of the simulation model treated as a black box! The plot shows whether this factor has a positive or negative effect on the response and whether that effect remains constant over the domain (experimental area) of the factor.

The practitioners may further analyze this scatter plot: They may fit a curve to these (x,y) data: for example, a straight line (say) $y = \beta_0 + \beta_1 x$. Of course, they may fit other curves (such as a quadratic curve: second-degree polynomial), or they may use paper with one or both scales logarithmic.

To study interactions between factors, the practitioners may combine several of these scatter plots (each drawn per factor). For example, the scatter plot for different traffic rates was drawn, given a certain priority rule (say, FIFO). Plots for different priority rules can now be superimposed. Intuitively, the average waiting time curve for SPT lies below the curve for FIFO (if not, either this intuition or the simulation model is wrong; see the discussion on validation in Section 6.6). If the response curves are not parallel, then by definition there is interaction between priority rule and traffic rate.

However, superimposing many plots is cumbersome. Moreover, their interpretation is subjective: are the response curves really parallel and straight lines? These shortcomings are removed by regression analysis. (Kleijnen and Van Groenendaal, 1992; Kleijnen and Sargent, 1997).

6.3.2 GLS, OLS, and WLS

A regression metamodel may be used to approximate the I/O transformation of the simulation model that generates the data to which the regression analysis is applied. [A different use of regression analysis is to obtain control variates, which are a specific type of VRT; see (15).] Draper (1994) provides a bibliography on applied regression analysis, outside simulation. Consider the *second-degree polynomial*

$$Y_{i,j} = \beta_0 + \sum_{h=1}^{k} \beta_h x_{i,h} + \sum_{h=1}^{k} \sum_{h'=h}^{k} \beta_{h,h'} x_{i,h} x_{i,h'} + E_{i,j}$$

$$(i = 1, \ldots, n; j = 1, \ldots, m_i) \tag{1}$$

where stochastic variables are shown as uppercase letters; the specific symbols are

$Y_{i,j}$ = simulation response of factor combination i, replication j
k = number of factors in the simulation experiment
β_0 = overall mean response or regression intercept
β_h = main effect or first-order effect of factor h
x_{ih} = value of standardized factor h in combination i [see (2) below]
$\beta_{h,h'}$ = interaction between factors h and h' with $h < h'$
$\beta_{h,h}$ = quadratic effect of factor h
$E_{i,j}$ = fitting error of the regression model in combination i, replication j
n = number of simulated factor combinations
m_i = number of replications for combination i

To interpret this equation, it is convenient first to ignore interactions and quadratic effects. Then the *relative importance* of a factor is obtained by sorting the absolute values of the main effects β_h, provided that the factors are *standardized*. Let the original (nonstandardized) factor h be denoted by w_h. In the simulation experiment w_h ranges between a lowest value l_h and an upper value u_h; that is, the simulation model is not valid outside that range (see the discussion on validation in Section 6.6) or in practice that factor can range over that domain only (e.g., because of space limitations the number of servers can vary only between 1 and 5). Measure the variation or spread of that factor by the half-range $a_h = (u_h - l_h)/2$, and its location by the mean $b_h = (u_h + l_h)/2$. Now the following standardization is appropriate:

$$x_{ih} = \frac{w_{ih} - b_h}{a_h} \tag{2}$$

Notice that importance and significance are related but different concepts. Significance is a statistical concept. An important factor may be declared nonsignificant if the variance of the estimated effect is high (because the simulation response has high variance σ^2 and the total sample size $N = \sum_{i=1}^{n} m_i$ is small): this is called *type II* or β *error*. Importance depends on the practical problem that is to be solved by simulation. An unimportant factor may be declared significant if the variance of the estimated factor effect is small (in simulation, large sample sizes do occur).

It is convenient to use the following notation: Denote the *general linear regression model* by

$$Y_{i,j} = \sum_{g=1}^{q} \gamma_g z_{ig} + E_{i,j} \tag{3}$$

Comparison of (1) and (3) gives the following identities: $\gamma_1 = \beta_0, \gamma_2 = \beta_1, \ldots, \gamma_q = \beta_{k,k}$, and $z_{i,1} = 1, z_{i,2} = x_{i,1}, \ldots, z_{i,q} = x_{i,k}^2$. It is also convenient to use matrix notation for equation (3):

$$\mathbf{Y} = \boldsymbol{\gamma}\,\mathbf{z} + \mathbf{E} \tag{4}$$

where bold symbols denote matrices including vectors (vectors are matrices with a single column; in linear algebra it is customary to denote matrices by uppercase letters and vectors by lowercase letters, but in this chapter uppercase letters denote random variables); the specific symbols are

N = total number of simulation responses, $\sum_{i=1}^{n} m_i$

$\mathbf{Y}$ = vector with N components the first m_1 elements are the m_1 replicated simulation responses for input combination 1, ..., the last m_n elements are the m_n replications for input combination n

$\boldsymbol{\gamma}$ = vector of q regression, $(\gamma_1, \gamma_2, \ldots, \gamma_g, \ldots, \gamma_q)'$

$\mathbf{z}$ = $N \times q$ matrix of independent variables; the first m_1 rows are the same and denote factor combination 1, ... ; the last m_n rows are the same and denote factor combination n

$\mathbf{E}$ = vector with N fitting errors, $(E_{1,1}, \ldots, E_{n,m_n})$

An alternative notation is

$$\overline{\mathbf{Y}} = \boldsymbol{\gamma}\,\overline{\mathbf{z}} + \overline{\mathbf{E}} \tag{5}$$

where a bar denotes the average per factor combination; each matrix (including vector) now has only n (instead of N) rows. For example, the first element of the vector $\overline{\mathbf{Y}}$ is $\overline{Y}_1 = \sum_{j=1}^{m_1} Y_{1,j}/m_1$. *Note:* Nonlinear regression models and the concomitant DOE are discussed in Ermakov and Melas (1995, pp. 167–187).

The *generalized least squares* (GLS) estimator of the parameter vector $\boldsymbol{\gamma}$, denoted by $\mathbf{C}_{\text{GLS}} = (C_1, C_2, \ldots, C_g, \ldots, C_q)'$, is

$$\mathbf{C}_{\text{GLS}} = (\mathbf{z}'\boldsymbol{\sigma}_y^{-1}\mathbf{z})^{-1}\mathbf{z}'\boldsymbol{\sigma}_y^{-1}\mathbf{Y} \tag{6}$$

where $\boldsymbol{\sigma}_y$ denotes the $N \times N$ covariance matrix of $\mathbf{Y}$, which is assumed to be nonsingular, so that its inverse $\boldsymbol{\sigma}_y^{-1}$ exists (in Section 6.3.5 this assumption is revisited for the estimate $\hat{\boldsymbol{\sigma}}_y^{-1}$); it is further assumed that $\mathbf{z}$ is such that $\mathbf{z}'\boldsymbol{\sigma}_y^{-1}\mathbf{z}$ is regular (further discussion of $\mathbf{z}$ follows in Section 6.4); to simplify the notation, we sometimes denote $\mathbf{C}_{\text{GLS}}$ by $\mathbf{C}$. It is not too difficult to derive an alternative expression for the GLS estimator that uses the alternative notation in (5) (Kleijnen, 1987, p. 195; 1992).

GLS gives the *best linear unbiased estimator* (BLUE), where "best" means minimum variance. The individual variances can be found on the main diagonal of the covariance matrix for $\mathbf{C}$:

$$\boldsymbol{\sigma}_c = (\mathbf{z}'\boldsymbol{\sigma}_y^{-1}\mathbf{z})^{-1} \tag{7}$$

A $1-a$ confidence interval per parameter follows from the well-known Student statistic. The general formula for this statistic (used repeatedly below) is

$$T_\nu = \frac{\overline{Y} - E(Y)}{S_{\overline{y}}} \tag{8}$$

where ν denotes the number of degrees of freedom. Notice that a statistic is called *Studentized* if the numerator is divided by its standard error or standard deviation. For the GLS estimator, $\overline{Y}$ in (8) is replaced by C_g ($g = 1, \ldots, q$) and $S_{\overline{y}}$ by the estimator of the standard deviation of C_g; this estimator will be discussed in Section 6.3.3. Software for GLS estimation is abundant (Swain, 1996).

A special, classic case of GLS is *ordinary least squares* (OLS). OLS remains BLUE if the simulation responses have *white noise*; that is, they are independent and have constant variances; that is, $\boldsymbol{\sigma}_y = \sigma^2\mathbf{1}_{N\times N}$, where $\mathbf{1}_{N\times N}$ denotes the $N \times N$ identity matrix (usually denoted by I in linear algebra). Then the GLS estimator in (6) reduces to

$$\mathbf{C}_{\text{OLS}} = (\mathbf{z}'\mathbf{z})^{-1}\mathbf{z}'\mathbf{Y} \tag{9}$$

Even if the OLS assumptions do not hold, an alternative to GLS is the OLS point estimator in (9), but with the correct covariance matrix

$$\boldsymbol{\sigma}_{c\text{OLS}} = (\mathbf{z}'\mathbf{z})^{-1}\mathbf{z}'\boldsymbol{\sigma}_y\mathbf{z}(\mathbf{z}'\mathbf{z})^{-1} \tag{10}$$

If the OLS assumptions ($\boldsymbol{\sigma}_y = \sigma^2\mathbf{1}_{N\times N}$) do hold, equation (10) simplifies to $\boldsymbol{\sigma}_{c\text{OLS}} = (\mathbf{z}'\mathbf{z})^{-1}\boldsymbol{\sigma}^2$. The white noise assumption may be used in deterministic simulation (see Section 6.3.8).

In random simulation, however, it is realistic to assume that the response variances vary with the input combinations: $\text{Var}(Y_{i,j}) = \sigma_i^2$. (So $Y_{i,j}$ has a mean and a variance that both depend on the input.) In other words, $\boldsymbol{\sigma}_y$ (the covariance matrix for the simulation responses) becomes a diagonal matrix with the first m_1 elements equal to $\sigma_1^2, \ldots,$ the last m_n elements equal to σ_n^2. Then a special case of GLS applies, namely *weighted least squares* (WLS): Substitute this diagonal matrix $\boldsymbol{\sigma}_y$ into GLS formulas (6) and (7). The interpretation of the resulting formula is that WLS uses the standard deviations σ_i

as weights. In case of variance heterogeneity, the WLS estimator is BLUE; both OLS and WLS still give unbiased estimators.

6.3.3 Estimation of Response Variances

The preceding formulas feature $\boldsymbol{\sigma}_y$ (the covariance matrix of the simulation responses), which unfortunately is unknown [as are the means $E(Y_i) = \mu_i$]. The estimation of the standard deviations σ_i is a classic *tactical* problem in simulation. To solve this problem, the analysts should distinguish terminating and steady-state simulations (see Chapter 7), and expected values, proportions, and quantiles. In this section we consider four cases: (1) mean response of terminating simulation, (2) mean response of steady-state simulation, (3) proportions, and (4) quantiles.

Mean Response of Terminating Simulation. By definition, a terminating simulation has an event that stops the simulation run; an example is a queueing simulation of a bank that opens at 9 A.M. and closes at 4 P.M. In such a situation, the simulation runs for one specific combination (say) i may give independently and identically distributed (IID) responses Y_i, for example, average waiting time per day. Then the estimator of the variance of Y_i, given a sample size of m_i replications (e.g., m_i days) with integer $m_i \geq 2$, is

$$S_i^2 = \sum_{j=1}^{m_i} \frac{(Y_{i,j} - \overline{Y}_i)^2}{m_i - 1} \tag{11}$$

where the average of the m_i replications is $\overline{Y}_i = \sum_{j=1}^{m_i} Y_{i,j}/m_i$; each replication uses a nonoverlapping sequence of (pseudo)random numbers.

Note that in the simulation literature on tactical issues it is classic to focus on $1 - \alpha$ confidence intervals. If a Gaussian distribution is assumed ($N(\mu_i, \sigma_i^2)$), the confidence interval should use the Student statistic in (8), now with $\nu = m_i - 1$ degrees of freedom. If, however, Y_i has an asymmetric distribution, then Johnson's modified Student statistic is a good alternative; this statistic includes an estimator for the skewness of the distribution of Y_i; see Appendix 6.1. One more alternative is a distribution-free or nonparametric confidence interval such as the sign test or the signed rank test. Conover (1971) gives an excellent discussion of distribution-free statistics. Kleijnen (1987) discusses the application in simulation. Other alternatives are jackknifing and bootstrapping (see Section 6.3.4). Appendix 6.1 gives more details on $1-\alpha$ confidence intervals for the response of an individual factor combination. (This appendix covers "per comparison" intervals, not "familywise" or "experimentwise" intervals; see Bonferroni's inequality at the end of Appendix 6.1, and Chapter 8.) In DOE, however, confidence intervals are desired, not for the individual simulation responses, but for the individual factor effects. For that purpose, variances of the simulation responses are needed.

Mean Response of Steady-State Simulation. Some practical problems require steady-state simulations; examples are strategic decisions on production-facility layout, assuming static environments and long-term reward systems. Suppose that the simulationists execute a single long run (not several replicated runs) per factor combination. Assume that a simulation run yields a time series that (possibly, after elimination of the startup phase) gives a stationary process in the wide sense. Then there are several methods for the estimation of the variance σ_i^2: batching (or subruns), renewal (or regenerative) analysis, spectral analysis, standardized time series (see Chapter 7). Appendix 6.1 gives formulas for renewal analysis only.

Proportions. Let p_a denote the probability of the response exceeding a given value a; for example, the probability of waiting time exceeding 5 minutes in a simulation run. This leads to a binomially distributed variable, and the estimation of its mean p_a and variance $p_a\ (1 - p_a)/m$. Actually, the subscript i must be added to the parameters p_a and m [see (11)].

When estimating an extremely small probability, importance sampling is needed: The probability distribution of some input variable is changed such that the probability of the event of interest increases; for example, the mean (say) $1/\lambda$ of the Poisson service process is increased so that the probability of buffer overflow increases (see Appendix 6.2; also, Heidelberger, 1995; Heidelberger et al., 1996).

Quantiles. A response closely related to a proportion is a quantile: What is the value not exceeded by (say) 80% of the waiting times? Quantile estimation requires sorting the m observations y_j, which yields the order statistics $y_{(j)}$; that is, $y_{(1)}$ is the smallest observation, ... , $y_{(m)}$ is the largest observation. Appendix 6.1 gives formulas. Notice that the median is a good alternative for the mean when quantifying the location of a random variable; the regression metamodel may explain how the various factors affect the median simulation response.

6.3.4 Estimated GLS and WLS

Cases 1 through 4 in Section 6.3.3 show that there are different types of responses, each requiring its own procedures for the estimation of the variances σ_i^2. But whenever the analysts use estimated response variances, WLS becomes *estimated WLS* or EWLS. This gives a *nonlinear* estimator for the factor effects γ [see (6) with σ_y replaced by a diagonal matrix $\mathbf{S}_y$ with main-diagonal elements S_i^2 defined in (11)]. But the properties of nonlinear estimators are not well known: Is the estimator still unbiased; does it have minimum variance; what is its variance?

Jackknifing is a general computer-intensive technique that the analysts should consider whenever they presume the estimator under consideration to be biased or whenever they do not know how to construct a valid confidence intervals. In the EWLS case, *jackknifed EWLS* (JEWLS) may be defined as follows. Suppose that there are m replications of Y_i. This yields (say) C_{EWLS}, the EWLS estimator of the regression parameter $\boldsymbol{\gamma}$. Next, for each factor combination eliminate one replication, say, replication j (with $j = 1, \dots, m$). Calculate the EWLS estimator from the remaining $m - 1$ replications; this gives m estimators $C_{\text{EWLS},-j}$. The pseudovalue (say) P_j is defined as the following linear combination of the original and the jth estimator:

$$P_j = mC_{\text{EWLS}} - (m - 1)C_{\text{EWLS},-j} \qquad (j = 1, \dots, m) \tag{12}$$

The jackknifed estimator is defined as the average pseudovalue:

$$\overline{P} = \sum_{j=1}^{m} \frac{P_j}{m} \tag{13}$$

If the original estimator is biased, the jackknifed estimator may have less bias, albeit at the expense of a higher variance. Moreover, jackknifing gives the following robust confidence interval. Treat the m pseudovalues P_j as m IID variables [see the Student statistic in (8) with Y replaced by P and $\nu = m - 1$].

Efron (1982) and Miller (1974) give classic reviews of jackknifing. Jackknifing of EWLS is discussed further in Kleijnen et al. (1987). Jackknifed renewal analysis is discussed in Kleijnen and Van Groenendaal (1992, pp. 202–203) (see also Appendix 6.3). Jackknifing is related to *bootstrapping*, which samples from the set of m observations (Cheng, 1995; Efron, 1982; Efron and Tibshirani, 1993).

6.3.5 Variance Reduction Techniques

VRTs are supposed to decrease the variances of estimators such as the estimated mean simulation response, the estimated differences among mean simulation responses, and the estimated factor effects [which are linear combinations of simulation responses; see the GLS estimator in (6)]. This section covers the following three VRTs that have relations with GLS: (1) common random numbers, (2) antithetics, and (3) control variates.

Common Random Numbers. Practitioners often use common (pseudo)random numbers to simulate different factor combinations (see Appendix 6.2 for details). When responses are statistically dependent, GLS gives BLUE. But in practice the covariances between simulation responses of different factor combinations are unknown, so these covariances must be estimated. Assume m independent replications per factor combination, so that Y_{ij} and $Y_{i'j}$ are dependent but Y_{ij} and $Y_{ij'}$ are not, when $j \neq j'$ and $j' = 1, \ldots, m$. Then the classic covariance estimator is

$$S_{i,i'} = \sum_{j=1}^{m} \frac{(Y_{i,j} - \overline{Y}_i)(Y_{i',j} - \overline{Y}_{i'})}{m-1} \tag{14}$$

Notice that this equation reduces to the classic variance estimator in (11) if $i = i'$ and S_{ii} is defined as S_i^2. Dykstra (1970) proves that the estimated covariance matrix $\mathbf{S}_y$ is singular if $m \leq n$. *Estimated GLS* (EGLS) gives good results (Kleijnen, 1992). A detailed example is given in Chapter 8.

The simulation literature has ignored the estimation of covariances in *steady-state* simulations. If renewal analysis is used, the renewal cycles get out of step; for example, when a factor combination uses a lower traffic rate, its cycles get shorter. However, if subruns are used, the estimators are strictly analogous to (14).

Antithetic Random Numbers. Closely related to common (pseudo)random numbers are antithetic (pseudo)random numbers: To realize negative correlation between pairs of replications, use the (pseudo)random numbers r for one replication and the complements or *antithetics* $1 - r$ for the other replication. Then m replications give $m/2$ independent pairs $(\overline{Y}_1, \ldots, \overline{Y}_{m/2})$, where $\overline{Y}_i = (Y_{2i-1} + Y_{2i})/2$. Hence to estimate the variance, use (11) but replace m_i by $m/2$ and Y_j by $\overline{Y}_r$ with $r = 1, \ldots, m/2$.

Notice that these antithetic (pseudo)random numbers do create negative correlation between two variables sampled from the same distribution if this sampling uses the *inverse transformation* technique: $X = f(R)$, where f denotes the inverse (cumulative) distribution function (which is monotonically increasing). An example is provided by the exponential distribution: $X = -\ln(R)/\lambda_x$ with $1/\lambda_x = \mathrm{E}(X)$. A counterexample is provided by the sampling of two independent standard normal variables (say) V_1 and V_2 through the well-known Box–Muller transformation: $V_1 = \cos(2\pi R_1)[-2\ln(R_2)]^{1/2}$ and $V_2 = \sin(2\pi R_1)[-2\ln(R_2)]^{1/2}$. But then the basic idea can still be applied: $X_1 = \mathrm{E}(X) + \sigma_x V_1$ and $X_2 = 2\mathrm{E}(X) - X_1$ (V_2 can be used for the next sample of the pair, X_1 and X_2). Appendix 6.2 gives some details on antithetic variates.

Since common and antithetic (pseudo)random numbers are so closely related, Kleijnen (1975c) investigated their combination. Later Schruben and Margolin (1978) examined this combination in the case of a first-order polynomial metamodel. Their rule is: Treat the selection of common and antithetic seeds as a separate factor -say, factor k- in a two-level design. Associate one level of that factor with common seeds; that is, use common seeds for the $n/2$ combinations that have factor k at its plus level. Associate the other level with antithetic seeds; that is, use the same antithetic seeds for the remaining $n/2$ combinations that have factor k at its minus level. Later, the Schruben–Margolin strategy was investigated further for second-order metamodels (Donohue, 1995).

Control Variates. Regression models can be used not only as metamodels for what-if questions, but also as a VRT: control variates or regression sampling. Whereas antithetics makes the companion replication compensate "overshoot" [i.e., $y > \mathrm{E}(Y)$], control variates corrects the response of a given replication as follows.

A random variable (say) X can serve as a control variate if its mean $\mathrm{E}(X)$ is known and it is strongly correlated with the response Y: $|\rho_{x,y}| \gg 0$, where $\rho_{x,y}$ denotes the linear correlation coefficient between X and Y. As an example, consider the simulation of an M/M/1 queueing system. On first reading the subscript i may be ignored in the following definitions and formulas; actually, a control variate estimator may be computed for each factor combination i. Denote the average input 1 (say, average service time) per replication in combination i by $X_{i,1}$. Obviously, this input and the output Y_i (say, either the average or the 90% quantile of waiting time in combination i) are positively correlated: $\rho(X_{i,1}, Y_i) > 0$. Hence in case of an overshoot, y should be corrected downward:

$$\overline{Y}_{i,c} = \overline{Y}_i + C_{i,1,\mathrm{OLS}}[E(\overline{X}_{i,1}) - \overline{X}_{i,1}] \tag{15}$$

where $C_{i,1,\mathrm{OLS}}$ denotes the OLS estimator when output Y_i is regressed on $X_{i,1}$, which denotes input 1 in combination i [this OLS estimator is computed from m_i replications of the pair $(Y_i, X_{i,1})$]; the input averaged over m_i replicates is $\overline{X}_{i,1} = \sum_{j=1}^{m_i} X_{i,1,j}$. Therefore, the technique of control variates is also called *regression sampling*. Notice that in this equation the input is a random variable (uppercase X), whereas in the regression metamodel this input is fixed at its extreme values (lowercase x) [see (1); see also Appendix 6.2].

The single control variate estimator in (15) can be extended to *multiple* control variates; for example, service time X_1 and arrival time (say) X_2. This requires multiple regression analysis. Actually, a better control variate may be traffic load, $\mathrm{E}(X_1)/\mathrm{E}(X_2)$. In general, the explanatory variables in this regression model may be selected such that the well-known multiple correlation coefficient R^2 is maximized. $R^2 = 1$ means perfect fit. However, because R^2 increases as the number of independent variables increases, the adjusted R^2 is preferred as a criterion for the selection of control variates.

A complication is that the estimator $C_{i,1,\text{OLS}}$ in (15) leads to a nonlinear estimator (see the product $C_{i,1,\text{OLS}}\overline{X}_{i,1}$), which in general is biased. Moreover, the construction of a confidence interval for $E(\overline{Y}_{i,c})$ becomes problematic. These problems can be solved either assuming multivariate normality for $(Y_i, X_{i,1}, X_{i,2}, \ldots)$ or using the robust technique of jackknifing. Details of jackknifing the control variates are shown in Appendix 6.3.

Joint application of common, antithetic, and control variates is examined in Tew and Wilson (1994).

6.3.6 Lack of Fit of the Regression Metamodel

Once having specified a regression metamodel and having estimated its parameters (with or without applying VRTs), it becomes necessary to check possible lack of fit of the regression metamodel: Is the estimated regression model an adequate approximation of the I/O transformation of the specific simulation model, given a specific experimental domain? In other words, the simulation model is supposed to be valid only within a certain area of its parameters and input variables (see also Section 6.6). Similarly, the metamodel is supposed to be valid only for those simulation I/O data that lead to its estimated parameter values. Consequently, the metamodel is more reliable when used for interpolation; it may be dangerous when used to extrapolate the simulated behavior far outside the domain simulated in the DOE.

Notice that in practice, analysts often try to interpret individual effects before they check whether the regression model as a whole makes sense. However, first the analysts should check if the estimated regression model is a *valid* approximation of the simulation model's I/O transformation. If the metamodel seems valid, its individual effects are to be examined.

To test the adequacy of the metamodel, this model might be used to predict the outcomes for *new* factor combinations of the simulation model. For example, in (1) replace β by its estimate, and substitute new combinations of x (remember that there are n old combinations). Compare the predictions $\hat{y}$ with the simulation response y.

A refinement is *cross-validation*. The idea is as follows:

1. Eliminate one combination (say) i with $i = 1, \ldots, n$, instead of adding new combinations, which require more computer time.
2. Reestimate the regression model from the remaining $n - 1$ combinations
3. Repeat this elimination for all values of i.

Notice that cross-validation resembles jackknifing.

The formulas are as follows. The predictor for the simulation response given $\mathbf{C}$ (the estimator of the regression parameters or factor effects) and $\mathbf{z}_i$ (the vector of independent variables) is

$$\hat{Y}_i = \mathbf{z}_i'\mathbf{C} \tag{16}$$

The variance of this predictor is

$$\text{Var}(\hat{Y}_i) = \mathbf{z}_i' \boldsymbol{\sigma}_C \mathbf{z}_i \tag{17}$$

The estimator of this variance follows from substitution of the estimator for $\boldsymbol{\sigma}_C$. Hence, for OLS the estimator of (17) follows immediately from (10), (11), and (14). For GLS the estimator of (17) might use (7), (11), and (14); this gives an asymptotically valid estimator.

After elimination of I/O combination i, the new vector of estimated effects is based on the GLS formula in (6):

$$C_{-i} = (\mathbf{z}_{-i}' \boldsymbol{\sigma}_{y_{-i}}^{-1} \mathbf{z}_{-i})^{-1} \mathbf{z}_{-i}' \sigma_{y_{-i}}^{-1} Y_{-i} \tag{18}$$

Substitution into (16) gives the predictor $\hat{Y}_{-i} = y(\mathbf{C}_{-i}, \mathbf{z}_i)$. The Studentized cross-validation statistic [see (8)] is

$$\tilde{T}_\nu = \frac{Y_i - \hat{Y}_{-i}}{[\widehat{\text{Var}}(Y_i) + \widehat{\text{Var}}(\hat{Y}_{-i})]^{1/2}} \tag{19}$$

where $\widehat{\text{Var}}(Y_i) = S_i^2$ and $\text{Var}(\hat{Y}_{-i})$ follows from (17), replacing $\mathbf{C}$ by $\mathbf{C}_{-i}$, and so on. The degrees of freedom ν in (19) are unknown; Kleijnen (1992) uses $\nu = m - 1$. When this Studentized prediction error is significant, the analysts should revise the regression metamodel they specified originally. When judging this significance, they may apply Bonferroni's inequality, since there are multiple runs, namely n (see Appendix 6.1). In their revision they may use transformations of the original inputs, such as logarithmic transformations and cross-products or interactions.

An alternative to cross-validation is *Rao's lack-of-fit test.* To understand this test, it is convenient first to consider the classic OLS case: normally distributed simulation responses Y_i with white noise. Then there are the following two estimators of the common response variance σ^2. The first estimator is based on replication: See the classic variance estimator S_i^2 defined in (11). Because the true variance is constant, these estimators are averaged or pooled: $\sum_{i=1}^n S_i^2/n$; if m_i is not constant, a weighted average is used, with the degrees of freedom $m_i - 1$ as weights. Next consider the n estimated residuals, $\hat{E}_i = \overline{Y}_i - \hat{Y}_i$ with $i = 1, \ldots, n$. These residuals give the second variance estimator, $\sum_{i=1}^n \hat{E}_i^2 m/(n - q)$. The latter estimator is unbiased if and only if (iff) the regression model is specified correctly; otherwise, this estimator overestimates the true variance. Hence the two estimators are compared statistically through the well-known F-statistic, namely $F_{n-q,n(m-1)}$.

Rao (1959) extends this test from OLS to GLS:

$$F_{n-q,m-n-q} = \frac{m-n+q}{(n-q)(m-1)} (\overline{\mathbf{Y}} - \overline{\mathbf{z}}\mathbf{C})' \mathbf{S}_{\overline{y}}^{-1} (\overline{\mathbf{Y}} - \overline{\mathbf{z}}\mathbf{C}) \tag{20}$$

where the n estimated GLS residuals are collected in the vector $\overline{\mathbf{Y}} - \overline{\mathbf{z}}\mathbf{C}$ and where not only response variances are estimated but also response covariances, collected in the estimated covariance matrix $\mathbf{S}_{\overline{y}} = \mathbf{S}_y/m$ [see also (14)].

Kleijnen (1992) shows that Rao's test is better than cross-validation if the simulation responses are distributed symmetrically, for example, normally or uniformly distributed. Lognormally distributed responses, however, are better analyzed through cross-validation.

In *deterministic* simulation, studentizing the prediction errors as in (19) gives misleading conclusions. In such simulations the constant error σ^2 is estimated from the residuals. Hence the worse the metamodel is, the bigger this estimate becomes. But then the denominator in (19) increases [$\widehat{\text{Var}}(Y_i) = 0$], so the probability of rejecting this false model decreases! Therefore, relative prediction errors $\hat{y}_i/y_i$ are of more interest to practitioners. Examples are presented in the coal transport and FMS case studies in Sections 6.3.8 and 6.4.9.

It is also interesting to observe how the *estimated individual input effects* change, as combinations are deleted: see $c_{h,-i}$. Obviously, if the specified regression model is a good approximation, the estimates remain stable. Examples are given for the same coal transport and FMS case studies. Notice that the $q \times n$ matrix $\mathbf{C}_{-i} = (c_{h,-i})$ concerns the use of the simulation model and its concomitant metamodel for explanation, whereas the vector with the n elements $\hat{Y}_i/Y_i$ concerns the use for prediction purposes. Other diagnostic statistics are PRESS, DEFITS, DFBETAS, and Cook's D (see the general literature on regression analysis; also, Kleijnen and Van Groenendaal, 1992, p. 157).

6.3.7 Numerical Example: Multiple Server System

Kleijnen and Van Groenendaal (1992, pp. 150–151, 159–162) consider the well-known class of server systems with (say) s servers in parallel, and Markovian arrival and service processes: exponential interarrival times with rate λ and exponential service times with rate μ, which are IID. These systems are denoted as M/M/s.

Suppose that the response of interest is the steady-state mean queue length. Estimate this mean by the run average (say) $\overline{\nu}$. Start each run in the empty state. Stop each run after 2000 customers. (Better solutions for these tactical problems are discussed in Appendix 6.1.) Simulate six intuitively selected combinations of λ, μ, and s. Replicate each combination 20 times. This gives Table 6.1. (In Section 6.4 we show that much better designs are possible.)

Specify a regression metamodel for the M/M/s simulation model, with both the response and the inputs logarithmically transformed [see (1) with $Y = \ln(\overline{\nu})$, $x_1 = \ln(\lambda)$, $x_2 = \ln(\mu)$, and $x_3 = \ln(s)$]. Use of the SAS package for the regression analysis of this problem gives Table 6.2, which shows $N = 6 \times 20 = 120$ and $q = 1 + 3 = 4$, so 116 degrees of freedom remain to estimate the common response variance under the classical OLS assumptions. Corrected least squares (CLS) denotes the OLS point estimates with correct standard errors [see (10)]. The classical OLS computations give wrong standard errors for the estimated factor effects that are slightly smaller than the standard errors that use the unbiased estimated response variances.

The point estimates for OLS and EWLS do not differ much in this example: The standard errors are small, and both estimators have the same expectation. EWLS gives only slightly smaller standard errors. The explanation is that the logarithmic transformation reduces variance heterogeneity: Estimated response variances range only between 0.17 and 0.26 (in Section 6.4 we return to transformations).

All values for the multiple correlation coefficient R^2, adjusted or not, are very high. This numerical example does not formally test the goodness of fit. The individual input effects have the same absolute values, roughly speaking. Their signs are as expected intuitively. Kleijnen and Van Groenendaal (1992) also examine a simpler regression metamodel that uses a single input, namely the traffic rate $\lambda/\mu s$.

TABLE 6.1 Average Queue Length $\bar{v}$ of 2000 Customers in M/M/s Simulation Started in Empty State

				Replication									
				1	2	3	4	5	6	7	8	9	10
Combination	λ	μ	s	11	12	13	14	15	16	17	18	19	20
1	1	0.9	2	0.56	0.55	0.47	0.52	0.42	0.55	0.44	0.46	0.31	0.52
				0.38	0.56	0.66	0.64	0.56	0.45	0.51	0.49	0.38	0.47
2	1	0.38	4	1.14	0.72	0.78	0.74	1.14	0.67	0.51	1.04	0.86	0.97
				0.82	0.71	0.63	0.81	0.60	0.67	0.49	0.47	0.66	0.62
3	2	1.6	2	1.18	0.78	0.80	1.60	0.87	0.81	0.93	0.71	0.62	0.87
				0.69	1.15	0.98	1.00	0.72	0.84	0.67	0.99	0.90	0.69
4	2	1	4	0.11	0.19	0.16	0.14	0.20	0.14	0.13	0.11	0.20	0.16
				0.17	0.18	0.18	0.23	0.12	0.14	0.16	0.13	0.18	0.23
5	4	2.9	2	1.74	1.10	1.24	1.23	1.31	1.27	1.07	1.64	1.70	1.57
				1.07	1.27	0.90	1.12	1.29	1.13	1.08	1.41	1.19	1.17
6	4	1.69	4	0.41	0.39	0.60	0.31	0.37	0.36	0.42	0.26	0.42	0.40
				0.43	0.57	0.27	0.36	0.38	0.35	0.29	0.29	0.32	0.41

6.3.8 Case Study: Coal Transport

In this section we summarize Kleijnen (1995d), which concerns the following real-life system. A certain British coal mine has three coalfaces, each linked to its own bunker. These bunkers have specific capacities. Each bunker receives coal from a single coalface and discharges this coal onto a conveyor belt that serves all three bunkers. The belt transports the coal to the surface of the mine. Whenever a bunker is full, the corresponding coalface must stop; obviously, this congestion decreases the efficiency.

Wolstenholme (1990, p. 115) considers three inputs: total belt capacity, maximum discharge rate per bunker, and bunker capacities (which are assumed to be equal). He further presents three control rules for managing the discharge rate of the bunkers. For example, under policy I the discharge rate of each bunker can only be either zero or maximal (no intermediate values). The maximum is used as long as there is coal in the bunker and room on the conveyor belt. For this chapter it suffices to understand that policies II and III are more sophisticated than policy I. Policy is a qualitative factor with three levels.

Vital questions are: What are the efficiency effects of changing inputs; are there interactions among inputs? So this case study is representative of many problems that arise in real life, especially in physical distribution and production planning.

Wolstenholme (1990) develops a *system dynamics model*, which is a particular type of simulation, i.e., deterministic nonlinear difference equations with feedback relations. As software he uses STELLA, whereas Kleijnen (1995d) uses POWERSIM 1.1. Kleijnen (1995d) runs eight combinations of the three quantitative inputs (w_1, w_2, w_3); the output is the efficiency y (see Table 6.3; the selection of input combinations is discussed in Section 6.4).

The problem is how to find a *pattern* in the I/O behavior of the simulation model. To solve this problem, Wolstenholme (1990, pp. 116–121) uses intuition and common sense, studying run after run. In this section, however, we use regression metamodeling. Because qualitative factors are slightly more difficult to represent in a regression model,

Table 6.2 Regression Analysis of M/M/s Example Using Standard SAS Software

Linear regression: OLS
R-square 0.9078
ADJ R-SQ 0.9054

Parameter Estimates

Variable	D.F.	Parameter Estimate	Standard Error	t for HO: Parameter = 0	Prob > |t|
INTERCEPT	1	2.82523873	0.10951557	25.798	0.0001
LOGLABDA	1	5.10760536	0.19460843	26.246	0.0001
LOGMU	1	−5.21545	0.19938765	−26.157	0.0001
LOGSERVERS	1	−5.90304	0.18832235	−31.345	0.0001

Covariance of Estimates

COVB	INTERCEPT	LOGLABDA	LOGMU	LOGSERVERS
INTERCEPT	0.01199366	0.01576176	−0.0172908	−0.0189349
LOGLABDA	0.01576176	0.03787244	−0.0381753	−0.034371
LOGMU	−0.0172908	−0.0381753	0.03975544	0.035793
LOGSERVERS	−0.0189349	−0.034371	0.0357937	0.03546531

Linear regression: CLS
R-square 0.9491

Parameter	Standard Errors	t Statistic
2.82524	0.11376	24.83509
5.10761	0.20000	25.53805
−5.21545	0.21361	−24.41576
−5.90304	0.19607	−30.10680

COVB

0.012942	0.0180258	−0.020285	−0.02091
0.018026	0.0399989	−0.042134	−0.03712
−0.020285	−0.042134	0.045630	0.040085
−0.020909	−0.03712	0.040085	0.038444

Linear regression: EWLS
R-square 0.9477
ADJ R-SQ 0.9463

Parameter	Standard Errors	t Statistic
2.80905	0.11300	24.85885
5.08647	0.19873	25.59488
−5.19890	0.21285	−24.42518
−5.87790	0.19492	−30.15545

COVB

0.0127697	0.0178555	−0.020153	−0.020676
0.0178555	0.0394948	−0.04173	−0.036675
−0.020153	−0.04173	0.0453059	0.0397319
−0.020676	−0.036675	0.0397319	0.0379927

TABLE 6.3 Input–Output per Policy for Wolstenholme's (1990) Coal Transport Model

Run	w_1	w_2	w_3	y		
				Policy I	Policy II	Policy III
1	2000	700	150	55.78	56.87	65.87
2	3500	700	150	66.34	57.65	66.34
3	2000	1000	150	56.48	62.09	66.42
4	3500	1000	150	72.63	65.48	72.63
5	2000	700	1200	62.62	74.61	86.16
6	3500	700	1200	87.94	74.76	87.94
7	2000	1000	1200	78.47	80.73	85.60
8	3500	1000	1200	100	93.18	100

the effects of the three quantitative factors are first examined per policy. Next, policy is incorporated as a factor.

Assume a regression metamodel with main effects and two-factor interactions [see (1) with $k = 3$ and no quadratic effects]. Hence $q = 7$ effects need to be estimated. OLS is used because the simulation model is deterministic. To check the *validity* of this metamodel, R^2 and cross-validation are used. The first measure is computed by all standard statistical software, whereas the second measure is supported by modern software only. In cross-validation the regression model is estimated using only seven of the eight combinations in Table 6.3. First combination 1 is deleted and the regression parameters are estimated from the remaining seven combinations. This new estimator (say) $\hat{\boldsymbol{\beta}}_{-1}$ is used to predict the simulation response through $\hat{y}_1$ [see (18) and (16), respectively]. The actual simulation response is already known: $y_1 = 55.78$ for policy I (see Table 6.3). Hence prediction errors can be computed. In this case the eight relative prediction errors $\hat{y}_i/y_i$ vary between 0.77 and 1.19 for policy I.

As combinations are deleted, estimated individual input effects change. But the estimates remain stable if the regression model specified is a good approximation. Table 6.4 gives results for the metamodel with main effects only, still using the I/O data of Table 6.3. The three estimated main effects have the correct positive signs: increasing input capacities increase efficiency. Moreover, an intuitive analysis of the I/O data in Table 6.3 suggests that input 3 has more effect than input 1, which in turn exceeds the effect of input 2; the formal analysis agrees with this intuitive analysis.

The I/O data in Table 6.3 are also analyzed through other regression metamodels. Searching for a good regression model requires intuition, common sense, and knowledge of the underlying system that generated the I/O data (the system dynamics model and the real system). This search receives more attention in econometrics than in DOE [see also the case study in Van Groenendaal and Kleijnen (1996)]. To save space, these details are skipped. The final conclusion is that in this case study a regression model with the three main effects seems best: Interactions turn out to be insignificant, whereas deleting the main effect of input 2 (which is the smallest estimated main effect) increases the relative prediction error.

Next *individual* estimated input effects are tested statistically, assuming white noise. Then classic OLS yields the standard errors $s(\hat{\beta}_h)$. To test if β_h is zero (unimportant main effect), OLS uses Student's t-statistic. The critical value of this statistic is determined by the significance level α. A usual value is 0.10, but to reduce the probability of falsely

TABLE 6.4 Estimates of Main Effects β_h, upon Deleting a Combination, in Policy I[a]

Combination Deleted	β_0	β_1	β_2	β_3	R^2	Adj. R^2
1	70.900	10.823	5.995	11.358	0.9772	0.9543
2	72.858	9.520	4.038*	9.400	0.9316	0.8632
3	72.906	8.821	4.736*	9.351	0.9202	0.8404
4	73.466	10.129	5.300	8.791	0.9478	0.8955
5	74.053	7.675	2.843*	11.245	0.9730	0.9461
6	72.320	8.983	4.575*	9.613	0.9194	0.8387
7	72.271	9.456	4.101	9.463	0.9310	0.8620
8	71.486	8.149		8.679	0.9026	0.8052
None	72.535	9.195	4.363	9.725	0.9314	0.8799

[a] A blank denotes a nonsignificant effect; *denotes an estimated effect significant at $\alpha = 0.20$; all other estimated effects are significant at $\alpha = 0.10$.

eliminating important inputs, the value 0.20 is also used in this study. A higher α means a smaller critical value. This yields the blanks and the symbol * in Table 6.4.

Next consider policy II (for which no table is displayed). The best metamodel turns out to have main effects only for inputs 2 and 3: Deleting the nonsignificant main effect of input 1 decreases the maximum relative prediction error; interactions are not significant.

For policy III a model with the three main effects gives a good approximation. Note that such simple metamodels do not always hold: in the case study on a flexible manufacturing system (FMS), only a regression model *with* interactions, in addition to main effects, gives valid predictions and sound explanations (see Section 6.4.9).

Finally, consider regression metamodels with policy as a *qualitative factor*. So now there are three quantitative inputs, each simulated for only two values, and there is one qualitative factor with three levels, denoted I, II, and III. Technically, regression analysis handles this qualitative factor through two binary (0,1) variables (Kleijnen, 1987). Now the best regression model includes the main effects of all four factors. Policy III is the best policy; policy II is worse than policy I, even though policy I is the simplest policy. These regression results agree with an intuitive analysis of the I/O data in Table 6.3: Calculate the efficiency per policy, averaged over all eight combinations of the three other factors. These averages are 72.50, 70.75, 78.75.

6.4 DESIGN OF EXPERIMENTS

6.4.1 Introduction

In Section 6.3 we assumed that the matrix of independent variables $\mathbf{z}$ is such that the corresponding inverse matrices are not singular when computing GLS, WLS, and OLS point estimates and their estimated covariance matrices. An obvious condition seems to be that the number of observations is not smaller than the number of regression parameters. But does this mean that $N \geq q$ or $n \geq q$ (with $N = \sum_{i=1}^{n} m_i$)?

Consider a simple example, a second-order polynomial with a single factor: $Y = \beta_0 + \beta_1 X_1 + \beta_{1,1} X_1^2$ [see (1) with $k = 1$]. Obviously, simulating only two values of X_1—corresponding with $n = 2$—does not give a unique estimate of this regression model, whatever the number of replicates m is. Hence the condition is $n \geq q$, not $N \geq q$. See also the alternative notation of the regression model in (5).

Which n combinations to simulate (provided that $n \geq q$) can be determined such that the variances of the estimated factor effects are minimized. This is one of the main goals of the statistical theory on DOE; other goals follow in Section 6.4.8.

In this section we first cover *classical DOE*, which assumes white noise or $\boldsymbol{\sigma}_y = \sigma^2 \mathbf{1}_{N \times N}$ and a correctly specified regression model or $\mathrm{E}(E_{i,j}) = 0$ in (3). This gives well-known designs such as 2^{k-p} and central composite design (Sections 6.4.2 to 6.4.6). Next, we present ways to design simulation experiments such that these assumptions do hold (Section 6.4.7). *Optimal DOE* does account for heterogeneous variances and correlated simulation responses (Section 6.4.8). Finally we present an FMS case study in Section 6.4.9.

In this chapter we do not cover all types of designs. For example, we do not discuss "mixture" designs, which imply that the factor values are fractions that sum up to the value 1: $\sum_{h=1}^{k} x_{i,h} = 1$ (see Myers et al., 1989, p. 142). We do not cover Taguchi's DOE (see Donohue, 1994). We assume further that the experimental area is a k-dimensional hypercube in the standardized factors. In practice, however, certain corners of this area may represent unfeasible combinations (Kleijnen, 1987, p. 319; Nachtsheim, 1987; Nachtsheim et al., 1996).

Classical designs are tabulated in many publications. Two authoritative textbooks are Box and Draper (1987) and Box et al. (1978). Two textbooks by Kleijnen (1975a, 1987) focus on DOE in simulation. The analysts may also learn how to construct those designs; see the preceding references. In the next section we show which types of designs are available, but we only indicate how to construct these designs (see the generators in Section 6.4.2).

Recently, software, including artificial intelligence and expert systems, has been developed to help analysts specify these designs. Expert systems for DOE, however, are still in the prototype phase (Nachtsheim et al., 1996). "Nonintelligent" software for DOE outside the simulation domain is supported by several commercial software products, such as CADEMO (ProGAMMA, 1997), ECHIP (Nachtsheim, 1987), and RS/1 (BBN, 1989). The need for DOE software in simulation was articulated in a panel discussion at the 1994 Winter Simulation Conference (Sanchez et al., 1994). DOE software for simulation is investigated in Ören (1993), Ozdemirel et al. (1996), and Tao and Nelson (1997). Simulation applications of classical designs are referenced in Donohue (1994), Kleijnen (1987, 1995c), and Kleijnen and Van Groenendaal (1992). A FMS case study is presented in Section 6.4.9.

6.4.2 Main Effects Only: Resolution 3 Designs

According to Box and Hunter's (1961) definition, resolution 3 designs give unbiased estimators of the parameters of a first-order polynomial regression model. These parameters are the k main effects, plus the overall mean; see the first two terms on the right-hand side of (1). Sometimes these designs are called screening designs (Nachtsheim, 1987, p. 133). This chapter, however, reserves the term *screening* for designs with fewer runs than factors: $n < k$ (see Section 6.2).

In practice, analysts often simulate the base situation first and then change *one factor at a time*. This approach implies that $n = 1 + k$. However, consider *orthogonal* designs, that is, designs that satisfy

TABLE 6.5 Plackett–Burman Design with 12 Combinations[a]

1	2	3	4	5	6	7	8	9	10	11
+	−	+	−	−	−	+	+	+	−	+
+	+	−	+	−	−	−	+	+	+	−
−	+	+	−	+	−	−	−	+	+	+
+	+	+	−	+	+	−	−	−	+	+
+	−	+	+	−	−	−	−	−	−	+
+	+	−	+	+	+	+	+	−	−	−
−	+	+	+	−	+	+	−	+	−	−
−	−	+	+	+	−	+	+	−	+	−
−	−	−	+	+	+	−	+	+	−	+
+	−	−	−	+	+	+	−	+	+	−
−	+	−	−	−	+	+	+	−	+	+
−	−	−	−	−	−	−	−	−	−	−

[a] + denotes +1; − denotes −1.

$$\mathbf{d}'\mathbf{d} = n\mathbf{1}_{n \times n} \tag{21}$$

with design matrix $\mathbf{d} = (d_{ij})$ and $i = 1, \ldots, n$, $j = 1, \ldots, k$, and $n > k$. Notice that the matrix of independent variables $\mathbf{z}$ becomes $(\mathbf{1}_{n \times 1}, \mathbf{d})$, where $\mathbf{1}_{n \times 1}$ corresponds with the dummy factor $x_{i0} = 1$, which has effect β_0. Box (1952) proves that an orthogonal design minimizes the variances of the estimated regression parameters. Moreover, the effect estimators become independent: Use (21) in the covariance matrix in (10) with $\boldsymbol{\sigma}_y = \sigma^2 \mathbf{1}_{N \times N}$. Obviously, both orthogonal and one-factor-at-a-time designs give unbiased estimators.

How do we obtain the desired orthogonal matrices? Plackett and Burman (1946) derive orthogonal designs for k up to 99 and n equal to $k + 1$ rounded upward to a multiple of 4. For example, k equal to 8, 9, 10 or 11 requires that $n = 12$. This design is displayed in Table 6.5.

Another example is k equal to 4, 5, 6 or 7, which requires that $n = 8$. Writing $n = 2^{7-4}$ symbolizes that a fraction 2^4 is not simulated; that is, only 2^3 combinations are simulated. This design is displayed in Table 6.6, where the symbol $\mathbf{4} = \mathbf{1} \cdot \mathbf{2}$ means that $d_{i,4} = d_{i,1}d_{i,2}, \ldots$; the symbol $\mathbf{7} = \mathbf{1} \cdot \mathbf{2} \cdot \mathbf{3}$ means that $d_{i,7} = d_{i,1}d_{i,2}d_{i,3}$ with $i = 1, \ldots, n$. These symbols are called the *generators* of the design.

In general, 2^{k-p} designs with nonnegative integer p smaller than k are a subclass of Plackett–Burman designs. These designs have p generators. These generators determine how effects are confounded with each other; that is, they fix the bias pattern among factor effects. For example, $\mathbf{4} = \mathbf{1} \cdot \mathbf{2}$ implies that the estimator of the main effect of factor 4 is biased by the interaction between factors 1 and 2. Of course, if the analysts assume that only main effects are important, this bias is unimportant (Kleijnen, 1987).

The examples of Tables 6.5 and 6.6 give *saturated* designs when k is 11 or 7. Smaller k values (i.e., 8, 9, 10 and 4, 5, 6, respectively) enable cross-validation, to check if the first-order regression metamodel is adequate (see Section 6.3.6).

So resolution 3 designs are useful when a first-order polynomial seems an adequate regression model a priori. This will be the case in the first stages of RSM used in opti-

TABLE 6.6 2^{7-4} Design with Generators $4 = 1 \cdot 2$, $5 = 1 \cdot 3$, $6 = 2 \cdot 3$, $7 = 1 \cdot 2 \cdot 3$[a]

1	2	3	$4 = 1 \cdot 2$	$5 = 1 \cdot 3$	$6 = 2 \cdot 3$	$7 = 1 \cdot 2 \cdot 3$
−	−	−	+	+	+	−
+	−	−	−	−	+	+
−	+	−	−	+	−	+
+	+	−	+	−	−	−
−	−	+	+	−	−	+
+	−	+	−	+	−	−
−	+	+	−	−	+	−
+	+	+	+	+	+	+

[a]+ denotes +1; − denotes −1.

mization (see Section 6.5). Moreover, these resolution 3 designs are useful as building blocks for the next type of design, namely resolution 4 designs. An example of this building block approach is given in the FMS case study of Section 6.4.9.

6.4.3 Main Effects Against Two-Factor Interactions: Resolution 4 Designs

According to Box and Hunter (1961), resolution 4 designs give unbiased estimators of all k main effects, even if there are interactions between pairs of factors. Box and Wilson (1951, p. 35) prove that this design property can be achieved through the *foldover* principle: To the original resolution 3 design with design matrix (say) $\mathbf{d}_3$, now add the "mirror" or "negative" image of $\mathbf{d}_3$, namely $-\mathbf{d}_3$. Obviously, the foldover principle implies doubling the number of simulated factor combinations; for example, $k = 7$ requires that $n = 2 \times 8 = 16$. A subclass of resolution 4 designs are 2^{k-p} designs with the proper choice of p. An example is the 2^{8-4} design in Table 6.7. Notice that this design implies that the estimator of the main effect of factor 4 is biased by the interaction among the factors 1, 2, and 8, but not by any interactions between pairs of factors. [The estimators of two-factor interactions are biased by each other; for example, the estimated interaction between the factors 1 and 2 is biased by the interaction between the factors 4 and 8 (Kleijnen, 1987).]

Another example of a resolution 4 design for a great many factors is given in Kleijnen et al. (1992, p. 416): $k = 62$ factors requires $p = 55$ generators, which are specified in that reference. Fürbringer and Roulet (1995) give a simulation application with $k = 24$ and $p = 16$.

A different subclass are the *nonorthogonal* designs derived by Webb (1968). These designs are specified only for k is 3, 5, 6, or 7 with n equal to $2k$. Details are given in Kleijnen (1987, pp. 303–309) and in the other references.

Obviously, resolution 4 designs leave degrees of freedom over since $n > 1 + k$. Hence these designs can give an indication of the importance of two-factor interactions. Actually, these designs give estimators of certain sums of two-factor interactions; for example, $\beta_{1,2} + \beta_{4,8} + \beta_{3,7} + \beta_{5,6}$ (Kleijnen, 1987, pp. 304–305).

TABLE 6.7 2^{8-4} Foldover Design with Generators 4 = 1 · 2 · 8, 5 = 1 · 3 · 8, 6 = 2 · 3 · 8, 7 = 1 · 2 · 3

Run	1	2	3	4	5	6	7	8
1	−	−	−	+	+	+	−	+
2	+	−	−	−	−	+	+	+
3	−	+	−	−	+	−	+	+
4	+	+	−	+	−	−	−	+
5	−	−	+	+	−	−	+	+
6	+	−	+	−	+	−	−	+
7	−	+	+	−	−	+	−	+
8	+	+	+	+	+	+	+	+
9	+	+	+	−	−	−	+	−
10	−	+	+	+	+	−	−	−
11	+	−	+	+	−	+	−	−
12	−	−	+	−	+	+	+	−
13	+	+	−	−	+	+	−	−
14	−	+	−	+	−	+	+	−
15	+	−	−	+	+	−	+	−
16	−	−	−	−	−	−	−	−

[a]+ denotes +1; − denotes −1.

6.4.4 Individual Two-Factor Interactions: Resolution 5 Designs

Resolution 5 designs give estimators of main effects and two-factor interactions that are not biased by each other; they may be biased by interactions among three or more factors. Obviously, there are $k(k-1)/2$ such interactions [see $\beta_{h,h'}$ in (1)].

One subclass is again 2^{k-p} designs with a proper choice of the p generators. An example is a 2^{8-2} design with the generators **7** = **1** · **2** · **3** · **4** and **8** = **1** · **2** · **5** · **6**. In this design no two-factor interaction estimator is biased by another two-factor interaction or main effect estimator; the estimator of the interaction between the factors 1 and 2, however, is biased by the interaction among the factors 3, 4, and 7, and among the factors 5, 6, and 8 (Kleijnen, 1987).

Rechtschaffner (1967) gives *saturated* resolution 5 designs: $n = 1 + k + k(k-1)/2$ (Kleijnen, 1987, pp. 309–311). In general, resolution 5 designs require many factor combinations. Therefore, in practice, these designs are used only for small values of k (see Section 6.4.5).

6.4.5 High-Order Interactions: Full Factorial Designs

If k is very small (say, $k = 3$), all 2^k combinations can be simulated. Then all interactions (not only two-factor interactions) can be estimated. In practice, these full factorial designs are indeed sometimes used.

Although high-order interactions can be defined mathematically, they are hard to interpret. Therefore, a better metamodel may be specified, using *transformations*. For example, replace Y or x in a first-order polynomial by log Y or log x, so elasticity coefficients and decreasing marginal responses may be represented. We presented a numerical example in Section 6.3.7, concerning an M/M/s simulation model. A recent application is a simulation of Japanese production control systems known as *Kanban sys-*

TABLE 6.8 Central Composite Designs for Two Factors[a]

Run	1	2	3	4	5	6	7	8	9
1′	+	–	+	–	c	$-c$	0	0	0
2′	–	+	–	+	0	0	c	$-c$	0

[a]+ denotes +1; – denotes –1.

tems (Aytug et al., 1996). Transformations are further discussed in Cheng and Kleijnen (accepted); (see also Section 6.4.7).

6.4.6 Quadratic Effects: Central Composite Designs

Obviously, if quadratic effects are to be estimated, at least k extra runs are needed [see $\beta_{h,h}$ in (1) with $h = 1, \ldots, k$]. Moreover, each factor must be simulated for more than two values. Designs that are popular in both statistics and simulation are *central composite designs*. These designs combine resolution 5 designs (see Section 6.4.4) with one-factor-at-a-time star designs; that is, each factor is simulated for two more values while the other $k - 1$ factors are kept at their base values. In symbols: Taking standardized values for all factor values, each factor is simulated for the values c and $-c$ with c not equal to 1 or zero, so these values are symmetrical relative to zero. The selection of an appropriate value for c is surveyed in Myers et al. (1989). Besides the resolution 5 and the star design, simulate the central point $x_h = 0$ with $h = 1, 2, \ldots, k$. In other words, each factor is simulated for five values: the standardized values -1, $+1$, $-c$, $+c$, 0. These designs require relatively many runs: $n \gg q$. An example is displayed in Table 6.8: $k = 2$ gives $n = 9$, whereas $q = 6$. These designs imply nonorthogonal columns for the $k + 1$ independent variables that correspond with the quadratic and the overall effects. Other designs for second-order polynomials, including saturated and fractional 3^k designs, are summarized in Kleijnen (1987, pp. 314–316) and Myers et al. (1989, p. 140). A case study that uses the design of Table 6.8 will follow in Section 6.6.5.

6.4.7 Satisfying the Classical DOE Assumptions

Classical DOE assumes white noise and a correctly specified regression model: $\boldsymbol{\sigma}_y = \sigma^2 \mathbf{1}_{N \times N}$ and $\mathrm{E}(E_{i,j}) = 0$. In simulation, responses are *independent* if the (pseudo)random number streams do not overlap and if the (pseudo)random number generator is adequate (see Chapter 4). Practitioners, however, often use *common* (pseudo)random numbers. In that case the resulting correlations should be estimated and incorporated through either EGLS (see Section 6.3.5) or OLS with corrected covariance matrix [see (10)]. Orthogonal designs, however, no longer give independent estimators of the regression parameters [see (7) and (10)]. Whether other "optimality" characteristics still hold requires more research (see Section 6.4.8).

The classical DOE literature tries to realize *constant variances* through variance stabilizing transformations (Box and Cox, 1964). A major problem is that such a transformation may result in a regression metamodel that has lack of fit or that is hard to interpret (Dagenais and Dufour, 1994). A counterexample is the successful M/M/s study in Section 6.3.7.

In simulation, however, there is another way to realize constant variances, as simulation proceeds sequentially (apart from simulation on multiprocessors). So simulate so many replications that the average response per factor combination has a constant variance (say) c_0:

$$\sigma^2_{\bar{y}_i} = \frac{\sigma^2_{y_i}}{m_i} = c_0 \tag{22}$$

The variances in this equation can be estimated either sequentially or in a pilot phase; Kleijnen and Van Groenendaal (1995) report good results for this heuristic. In steady-state simulations this equation needs slight adjustment.

Whether the regression metamodel is specified correctly can be verified through *lack-of-fit* tests, which were discussed at length in Section 6.3.6. These tests assume that degrees of freedom are left over after estimation of the factor effects ($n > q$). Saturated designs violate this assumption. Many designs, however, are not saturated, as the preceding subsections demonstrated. If, however, a design is saturated, one or more extra factor combinations may be simulated (Kleijnen and Sargent, 1997).

If the regression model shows significant lack of fit, a polynomial regression model may be augmented with higher-order terms: See the high-order interactions in Section 6.4.5. Fortunately, many classical designs can be simulated *stagewise*. For example, a resolution 3 design can easily be augmented to a resolution 4 design: See the foldover principle in Section 6.4.3; and a resolution 5 design can easily be augmented to a central composite design (see Section 6.4.6). An example is provided by the case study on FMS in Section 6.4.9 (see also Kleijnen, 1987, pp. 329–333); Kleijnen and Sargent, 1997). Lack of fit may also be reduced through transformations of the dependent or independent regression variables (see Section 6.3.7).

6.4.8 Optimal DOE

There are several optimality criteria in DOE: See St. John and Draper (1975); many more references are given in Kleijnen (1987, p. 357, footnote 65). Here only the following three related criteria are discussed.

1. *A-Optimality or Trace of* $\boldsymbol{\sigma}_C$. An optimal design may minimize the average variance of the estimated regression parameters given the number of parameters q and the number of factor combinations n. In other words, the trace of the covariance matrix $\boldsymbol{\sigma}_C$ may be minimized.

2. *D-Optimality or Determinant of* $\boldsymbol{\sigma}_C$. An error in the estimate for one parameter may affect the error for another parameter, so the off-diagonal elements of the covariance matrix $\boldsymbol{\sigma}_C$ may also be relevant. The determinant of $\boldsymbol{\sigma}_C$ gives a scalar criterion, which may be minimized.

3. *G-Optimality or Maximum Mean-Squared Error (MSE).* The regression metamodel specified may be wrong. Therefore, the design may minimize the maximum value of the MSE between the true regression model and the fitted model (Box and Draper, 1987; Myers et al., 1989).

Kiefer and Wolfowitz (1959, p. 272) prove that under certain conditions a saturated design ($n = q$) gives "optimal" results. They further started the investigation of *which values* of the independent variables to observe and *how many replicates* to take (Fedorov, 1972). Recently, Cheng and Kleijnen (accepted) extended Kiefer and Wolfowitz's results to "nearly saturated" queueing simulations (traffic rates close to one), which have variance heterogeneity. Their main results are that the highest traffic rate actually simulated should be much lower than the highest traffic rate that is of interest; and the highest traffic rate simulated should be simulated for more customers than the lower traffic rates. The latter result implies that the simulation budget should not be allocated equally to the various traffic rates simulated.

Sacks et al. (1989) assume that fitting errors are *positively correlated*: The closer two combinations are in k-dimensional space, the more the fitting errors are correlated. This assumption is realistic if the response function is smooth. Technically speaking, they assume a covariance stationary process, as an alternative to white noise (Welch et al., 1992).

Optimal DOE gives designs that do not have the standard geometric patterns of classical designs. Optimal designs cannot be looked up in a table; they are generated by means of a computer (Nachtsheim, 1987, pp. 146, 151). Additional research is needed to facilitate more frequent application of these designs in simulation practice. An overview of optimal DOE theory, including many references to the older literature, is Myers et al. (1989, pp. 140–141). More recent literature is cited in Atkinson and Cook (1995) and Ermakov and Melas (1995).

6.4.9 Case Study: FMS

A machine-mix problem for a particular FMS is studied in Kleijnen and Standridge (1988) and summarized in Kleijnen and Van Groenendaal (1992, pp. 162–164). Analysts wish to determine the number of machines, per type of machine, such that a given production volume is realized. There are four machine types, w_1 through w_4. Only type 4 is a flexible machine; the other types are dedicated to a particular operation. A more complicated simulation model would allow for random service and arrival times, and random machine breakdowns. Yet the deterministic simulation model actually used demonstrates many main issues to be solved in DOE. Issues typical for random simulation are run-length determination and estimation of the response covariance matrix $\boldsymbol{\sigma}_y$. These tactical issues are not addressed in this case study. (Yet this study was initiated at a major supplier of discrete-event simulation software, previously Pritsker Corporation now Symix.)

Because these inputs must be integers, there are only 24 feasible combinations in the experimental domain. The actual values are given in Kleijnen and Standridge (1988), but they are not relevant here. Initially, eight intuitively selected combinations are simulated (Table 6.9). Then DOE is applied, initially assuming a first-order polynomial regression metamodel for these four factors. A possible resolution 3 design is a 2^{4-1} design with $\mathbf{3} = \mathbf{1} \cdot \mathbf{2} \cdot \mathbf{4}$ (see Section 6.4.2). This design is not saturated, so lack of fit can be examined. Because the simulation model is deterministic, OLS is used to estimate the factor effects. Table 6.10 shows that the orthogonal design indeed gives more accurate estimates.

In the remainder of this section we concentrate on the results of the more accurate 2^{4-1} design. Cross-validation gives Table 6.11, which displays the estimated factor effects that remain significantly different from zero when $\alpha = 0.30$. Besides the stability of individual effects in cross-validation, the relative prediction errors $\hat{y}_i/y_i$ are of interest. These errors turn out to range between -38 and $+33\%$. Altogether, these

TABLE 6.9 Intuitive Design for the Four Factors in the FMS Simulation in Kleijnen and Standridge (1988)[a]

Run i	w_1	w_2	w_3	w_4
1	−	−	−	−
2	+	−	+	0
3	−	+	+	0
4	+	+	−	−
5	−	−	+	+
6	+	−	−	0
7	−	+	−	0
8	+	+	+	+

[a]+ denotes +1; − denotes −1.

cross-validation results lead to rejection of this first-order polynomial regression metamodel.

Table 6.11 does suggest that factors 1 and 3 are *not* important. Therefore, a metamodel for the remaining factors 2 and 4 is formulated, but now including their interaction. Using the old I/O data, which resulted from the 2^{4-1} design, gives Table 6.12. This table shows that in this new metamodel all estimated effects remain significant in cross-validation! Further, the relative prediction errors $\hat{y}_i/y_i$ become much smaller: They now vary between −16 and +14%. So the conclusions of this case study are:

1. Machines of types 2 and 4 are bottlenecks, but not types 1 and 3: $\hat{\beta}_1$ and $\hat{\beta}_3$ are not significant in Table 6.11.
2. There is a trade-off between machine of types 2 and 4, as there is a significant negative interaction (see $\hat{\beta}_{2,4}$ in Table 6.12).
3. The regression metamodel helps to understand how the FMS works.

6.5 OPTIMIZATION OF SIMULATED SYSTEMS: RSM

Whereas the previous designs were meant to gain understanding of the simulation model through what-if analysis, DOE may also aim at optimizing the simulated system. There

TABLE 6.10 Estimated Variances of Estimated Effects in First-Order Polynomial Regression Metamodel for FMS Simulation

Effect	Intuitive Design	Formal Design
$\hat{\beta}_1$	0.5	0.5
$\hat{\beta}_2$	0.5	0.5
$\hat{\beta}_3$	1.0	0.5
$\hat{\beta}_4$	0.5	0.13
$\hat{\beta}_0$	20.6	19.6

TABLE 6.11 Cross-Validation and Estimated Factor Effects Significantly Different from Zero When $\alpha = 0.30$; First-Order Polynomial Metamodel for Four Factors in FMS Simulation

Run Deleted	$\hat{\beta}_1$	$\hat{\beta}_2$	$\hat{\beta}_3$	$\hat{\beta}_4$	$\hat{\beta}_0$
1		557		577	
2		712		500	2032
3		640		700	
4		629		694	
5				658	1962
6				736	
7				536	
None				541	3288
				541	3288

are many mathematical techniques for optimizing the decision variables of nonlinear implicit functions; simulation models are indeed examples of such functions. These functions may include stochastic noise, as is the case in random simulation. Examples of such optimization techniques are sequential simplex search (Nachtsheim, 1987), genetic algorithms, simulated annealing, and tabu search (Nash, 1995). Software is given in Chapter 25, for example, ProModel's SimRunner, Witness' Optimizer, and Micro-Saint's OptQuest. There is virtually no software for optimization in the presence of multiple responses (Khuri, 1996, pp. 240, 242). Software for the optimization of system dynamics includes DYSMOD's pattern search (Dangerfield and Roberts, 1996). System dynamics is a special kind of simulation, which may inspire developers of discrete-event simulation software. We, however, concentrate on RSM.

Note that some authors outside the discrete-event simulation area speak of RSM but mean what we call the what-if regression-metamodeling approach, not the sequential optimization approach (see, e.g., Olivi, 1980). Further, RSM assumes that the decision variables are quantitative. Systems that differ qualitatively are discussed in Chapter 8.

Classical RSM assumes a *single* type of response. However, the two case studies in

TABLE 6.12 Cross-Validation and Stability of $\hat{\beta}_2$, $\hat{\beta}_4$, $\hat{\beta}_{2,4}$, and $\hat{\beta}_0$ in Metamodel with Interaction for FMS

Run Deleted	$\hat{\beta}_2$	$\hat{\beta}_4$	$\hat{\beta}_{2,4}$	$\hat{\beta}_0$
1	952	1364	−492	776
2	952	1300	−460	776
3	952	1324	−468	776
4	952	1340	−484	776
5	1152	1432	−576	576
6	752	1232	−376	976
7	952	1332	−476	776
8	952	1332	−476	776
None	952	1332	−476	776

Sections 6.5.2 and 6.5.3 have two response types. In these case studies one response is to be maximized, whereas the other response must meet a side condition. In the case study in Keijzer et al. (1981), however, three response types are considered: namely, waiting time for each of three priority classes. A single overall criterion function is formulated by quantifying trade-offs among the waiting times per job class.

In Section 6.5.1 we first give general characteristics of RSM, then some details on RSM. Sections 6.5.2 and 6.5.3 give case studies.

6.5.1 Response Surface Methodology

RSM has the following four general characteristics:

1. RSM relies on first- and second-order polynomial regression metamodels or response surfaces; the responses are assumed to have white noise (see Section 6.3).
2. RSM uses classical designs (see Section 6.4).
3. RSM adds to regression models and DOE the mathematical (not statistical) technique of steepest ascent; that is, the estimated gradient determines in which direction the decision variables are changed.
4. RSM uses the mathematical technique of canonical analysis to analyze the shape of the optimal region: Does that region have a unique maximum, a saddle point, or a ridge (stationary points)?

Now we consider some details. Suppose that the goal of the simulation project is to maximize the response; minimization is strictly analogous.

RSM begins by selecting a *starting point*. Because RSM is a heuristic, success is not guaranteed. Therefore, several starting points may be tried later, if time permits.

RSM explores the *neighborhood* of the starting point. The response surface is approximated locally by a first-order polynomial in the decision variables, as the Taylor series expansion suggests. This gives the regression metamodel in (1) but with all cross-products eliminated. Hence k main effects β_h with $h = 1, \ldots, k$ are to be estimated. For that purpose a resolution-3 design should be used; this gives $n \approx k + 1$ (see Section 6.4.2).

If (say) the estimated effects are such that $\hat{\beta}_1 \gg \hat{\beta}_2 > 0$, obviously the increase of w_1 (decision variable 1) should be larger than that of w_2; the symbol w refers to the original, nonstandardized variables [see (2)]. The *steepest ascent path* means $\Delta w_1/\Delta w_2 = \hat{\beta}_1/\hat{\beta}_2$; in other words, steepest ascent uses the local gradient.

Unfortunately, the steepest ascent technique does not quantify the *step size* along this path. Therefore, the analysts may try a specific value for the step size. If that value yields a lower response, the step size should be reduced. Otherwise, one more step is taken. Note that there are more sophisticated mathematical procedures for selecting step sizes (Safizadeh and Signorile, 1994). Notice that special designs have been developed to estimate the slope accurately; that is, these designs are alternatives to the classical resolution 3 designs mentioned above (Myers et al., 1989, pp. 142–143).

Ultimately, the simulation response must decrease, since the first-order polynomial is only a local approximation to the real I/O transformation of the simulation model. In that case the procedure is repeated. So around the best point so far, the next $n \approx k + 1$ combinations of w_1 through w_k are simulated. Note that the same resolution 3 design may be used; only the locations and spreads of the original variables are adjusted [see

(2)]. Next, the factor effects in the new local first-order polynomial are estimated; and so RSM proceeds.

A first-order polynomial or hyperplane, however, cannot adequately represent a hilltop. So in the neighborhood of the optimum, a first-order polynomial may show serious lack of fit. To detect such specification error, the analysts might use cross-validation. In RSM, however, simple diagnostic measures are more popular: $R^2 \ll 1$, $\widehat{\text{Var}}(\hat{\beta}_h) \gg \hat{\beta}_h$ (see also Section 6.3.6).

So next a second-order polynomial is fitted (see Sections 6.3 and 6.4). Finally, the *optimal* values of the decision variables w_h are found by straightforward differentiation of this polynomial. A more sophisticated evaluation uses canonical analysis (Myers et al., 1989, p. 146).

Software for RSM is available. Much software referenced in the section on DOE (Section 6.4), also handles RSM. Nachtsheim (1987) discusses special RSM software: SIMPLEX-V and ULTRAMAX. This software provides designs such as central composite designs (see Section 6.4.6), and contour plots of the fitted surface, even for multiple responses (Myers et al., 1989, pp. 145–146, 148).

RSM is discussed further in Fu (1994), Ho et al. (1992), Khuri and Cornell (1996), and Sheriff and Boice (1994). Applications of RSM to simulated systems can be found in Hood and Welch (1993), Kleijnen (1987), and Kleijnen and Van Groenendaal (1992). Numerous applications outside the simulation field are surveyed in Myers et al. (1989, pp. 147–151). In Section 6.5.2 we summarize a case study illustrating how RSM climbs a hill; Section 6.5.3 gives a different case study illustrating how a hilltop can be explored further when there are two response types.

6.5.2 Case Study: Production Planning DSS

Kleijnen (1993) studies a DSS for production planning in a specific Dutch steel tube factory that has already been mentioned in Section 6.1. Both the DSS and the factory are simulated; the DSS is to be optimized. This DSS has 14 decision variables; for example, w_1 denotes "penalty for producing class 2 products on the next-best machine". There are two response variables, the total number of productive hours and the 90% quantile of lead time. Simulation of one input combination takes 6 hours on the computer available at that time. Consequently, searching for the optimal combination must be performed with care.

The original team of operations researchers planned to fit a local first-order polynomial with 14 inputs. They decided to simulate the base case first. Next they planned to change one factor at a time, once making each factor 20% higher than its base value, and once, 20% lower. Obviously, this design implies n, number of simulated factor combinations, equal to $1 + 2 \times 14 = 29$. Kleijnen (1993), however, uses a 2^{14-10} design, so $n = 16 > q = 15$. The specific $n = 16$ combinations are specified by writing all 2^4 combinations of the first four factors and then using the generators $\mathbf{5} = \mathbf{1} \cdot \mathbf{2}$, $\mathbf{6} = \mathbf{1} \cdot \mathbf{3}$, $\mathbf{7} = \mathbf{1} \cdot \mathbf{4}$, $\mathbf{8} = \mathbf{2} \cdot \mathbf{3}$, $\mathbf{9} = \mathbf{2} \cdot \mathbf{4}$, $\mathbf{10} = \mathbf{3} \cdot \mathbf{4}$, $\mathbf{11} = \mathbf{1} \cdot \mathbf{2} \cdot \mathbf{3}$, $\mathbf{12} = \mathbf{1} \cdot \mathbf{2} \cdot \mathbf{4}$, $\mathbf{13} = \mathbf{1} \cdot \mathbf{3} \cdot \mathbf{4}$, and $\mathbf{14} = \mathbf{2} \cdot \mathbf{3} \cdot \mathbf{4}$.

The resulting OLS estimates are *not* tested for significance, as a small parameter value may become a big value in a next local area. Table 6.13 gives the local effects on productive hours and on lead time, respectively. This table demonstrates that some decision variables have a favorable local effect on both response types. For example, raising w_1 by 1 unit increases production by $\hat{\beta}_1 = 0.52$; at the same time, this raise improves lead time by $\hat{\gamma}_1 = -0.054$. Lowering w_4 by 1 unit changes production by $\hat{\beta}_4$

TABLE 6.13 Estimated Local Effects on Productive Hours and Lead Time in Simulated Production Planning DSS; Unit Effects $\hat{\beta}_h$ and $\hat{\gamma}_h$; Base Values $w_{0,h}$

	Effect on Productive Hours		Effect on Lead Time	
h	$\hat{\beta}_h$	$\hat{\beta}_h w_{0,h}$	$\hat{\gamma}_h$	$\hat{\gamma}_h w_{0,h}$
1	0.52	62.40	−0.054	−6.48
2	−39.30	−117.90	−1.504	−4.51
3	0.65	78.0	0.072	8.64
4	−18.07	−0.90	150.583	7.53
5	−128.96	−64.48	−16.519	−8.26
6	0.00	0.00	−0.102	−29.38
7	−0.22	−132.00	−0.006	−3.60
8	13.88	20.82	2.963	4.44
9	−1.53	−38.25	1.311	32.78
10	1.39	139.00	0.072	7.20
11	0.03	9.00	0.037	11.10
12	527.23	158.17	8.485	2.55
13	−9.27	−46.35	−6.351	−31.76
14	−0.46	−55.20	−0.145	−17.40

$= -18.07$ while decreasing lead time by $\hat{\gamma}_4 = 150.583$. Because the decision variables have different scales and ranges, this table gives not only the unit effects but also the unit effects multiplied by the base values (say) $w_{0,1}$ through $w_{0,14}$.

The step size in RSM is determined heuristically. In this case the step size is selected such that at least one of the decision variables is doubled. Table 6.14 shows that w_{12} changes from 0.33 to 0.5936; the other 13 variables do not change much. These changes result from a step size of 0.0005; that is, the base values $w_{0,h}$ become $w_{0,h}$ + 0.0005 $\hat{\beta}_h$. Further results are given in Kleijnen (1993).

6.5.3 Case Study: Coal Transport

In this section we return to the coal transport model already presented in Section 6.3.8, but now we focus on optimization instead of sensitivity analysis. Wolstenholme (1990, pp. 125–127) restricts the optimization to the best control rule, policy III. Obviously, efficiency denoted by (say) $y^{(1)}$, cannot exceed 100%. Therefore, the goal is to minimize total costs, denoted by $y^{(2)}$, under the condition that the efficiency remains at its maximum of 100%. Wolstenholme assumes that one input is fixed: The maximum discharge rate is fixed at an average of 1000 tons/hr. He wishes to optimize the two remaining inputs, w_1, total belt capacity, and w_2, capacity per bunker. (The two symbols w_2 and w_3 have different meanings in Section 6.3.8 and this section. The cost parameters are £1000 per ton per hour for w_1 and £2000 per ton for w_2. The total costs are a linear function of the decision variables: $y^{(2)} = 1000w_1 + 2000w_2$. Efficiency, however, is a complicated nonlinear function specified by the system dynamics model (see Section 6.3.8).

Kleijnen (1995d) develops the following heuristic, inspired by RSM. Classic RSM, however, maximizes a single criterion, ignoring restrictions such as the one on efficiency: namely, $y^{(1)} = 1$.

TABLE 6.14 Values of Decision Variables in Base Run and After a Step of 0.005 on Steepest Ascent Path

h	Unit Effect, $\hat{\beta}_h$	Value of Variable: Base Run	Value of Variable: Steepest Ascent
1	0.52	132	132.0003
2	−39.30	3.3	3.28035
3	0.65	132	132.0003
4	−18.07	0.075	0.065965
5	−128.96	0.55	0.48552
6	0.00	316.8	316.8
7	−0.22	660	659.9999
8	13.88	1.65	1.65694
9	1.53	27.5	27.44992
10	1.39	110	110.0007
11	0.03	330	330
12	527.23	0.33	0.5936
13	−9.27	5.5	5.4954
14	−0.46	132	131.9998

Step 1. Find an initial combination $\mathbf{w} = (w_1, w_2)'$ that yields a simulated efficiency of 100%. Such a combination is already available; see the element in the last row and column of Table 6.3, discussed in Section 6.3.8.

Step 2. Reduce each input by (say) 10%. Simulate the system dynamics model with this input. Obtain the corresponding output.

Step 3. If the output of step 2 is still 100%, return to step 2; else, proceed to the next step.

Step 4. Find the most recent input combination that satisfies the efficiency restriction $y^{(1)} = 1$ (see steps 1 and 2). Reduce the step size to (say) 5%. Simulate the model with this new input combination, and obtain the corresponding output.

Step 5. Further explore the most recent local area that includes a combination with $y = 1$. In other words, simulate the model for the 4 input combinations that are specified by the 2^2 design. In Figure 6.2 these 4 combinations form the rectangle with lower left corner (2693, 923) and upper right corner (2835, 972).

Since this heuristic does not result in further progress, the optimum seems to be close. Now a second-order polynomial is specified as an approximation to the production function $y^{(1)}(w_1, w_2)$. To estimate the 6 parameters in this polynomial, the 2^2 design is expanded to the central composite design of Table 6.8, discussed in Section 6.4.6. Figure 6.2 shows the central point (2764, 948). The standardized axial value c is set at 0.75; if $c > 1$, then $(0, c)$ and $(c, 0)$ give too-high costs, and $(0, -c)$ and $(-c, 0)$ give too-low efficiencies. Table 6.15 shows the standardized and original values of the two decision variables, and the resulting costs and efficiencies. From the I/O data in the table the second-order polynomial can be estimated. Because the simulation model is deterministic, OLS is used. This gives

$$\hat{y}^{(1)} = 601.138514 - 0.1239693w_1 - 0.7161188w_2 - 0.0000334w_1w_2 + 0.0000291w_1^2 + 0.0004282w_2^2 \tag{23}$$

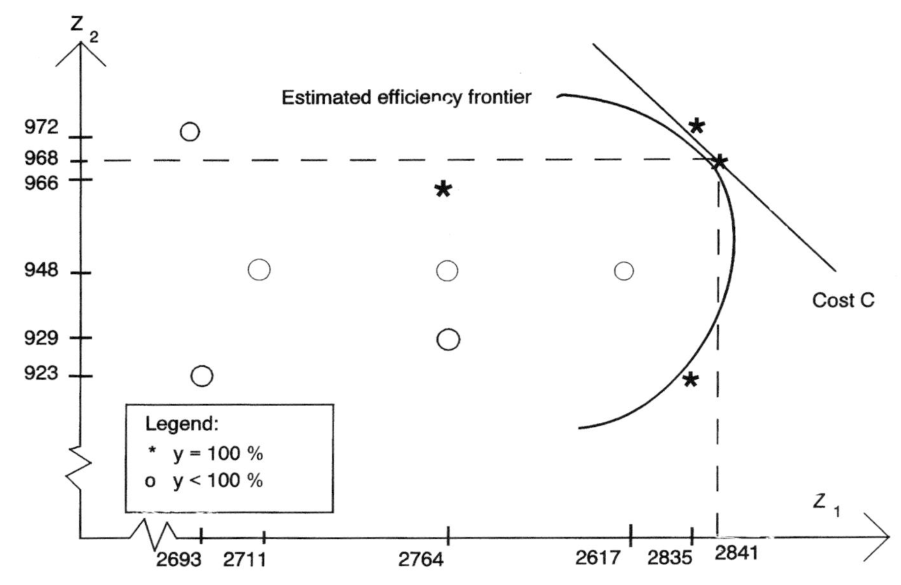

Figure 6.2 Central composite design, estimated efficiency frontier, and optimal isocost line; * means $y^{(1)} = 1$; O means $y^{(1)} < 1$.

Although the quadratic and interaction coefficients look small, they should not be ignored, since they are multiplied by w_1^2, w_2^2, and $w_1 w_2$, which are large.

Step 6. Combine this second-order polynomial approximation with the cost restriction $y^{(2)} = 1000w_1 + 2000w_2$; that is, replace $\hat{y}^{(1)}$ on the left-hand side of this polynomial by the value 1. Mathematically, this results in an ellipsoid: Figure 6.2 shows only a small part of that ellipsoid. In economic jargon, the ellipsoid gives the *efficiency*

TABLE 6.15 Input Combinations in the Central Composite Design with Corresponding Costs and Efficiencies for the Coal Transport System

Input Combination Standard Variables, (x_1, x_2)	Input Combination Original Variables, (w_1, w_2)	Cost, $y^{(2)}$ (£)	Efficiency $y^{(1)}$ (%)
(1.00, 1.00)	(2835.00, 972.00)	4.7790	100.00
(−1.00, 1.00)	(2693.25, 972.00)	4.63725	99.59
(1.00, −1.00)	(2835.00, 923.40)	4.68180	100.00
(−1.00, −1.00)	(2693.25, 923.40)	4.54005	99.36
(0.00, 0.00)	(2764.13, 947.70)	4.65953	99.35
(0.00, 0.75)	(2764.13, 965.97)	4.69607	99.59
(0.75, 0.00)	(2817.28, 947.70)	4.71268	100.00
(0.00, −0.75)	(2764.13, 929.48)	4.62309	99.36
(−0.75, 0.00)	(2710.97, 947.70)	4.60637	98.83

frontier: Outside the ellipsoid, inputs are wasted; inside that ellipsoid the efficiency is too low.

Minimizing the total cost $y^{(2)} = 1000w_1 + 2000w_2$ under the efficiency restriction $y^{(1)} = 1$ can be done through a Lagrangean multiplier. This yields the estimated optimal input combination $(\hat{w}_1^*, \hat{w}_2^*) = (2841.35, 968.17)$. Graphically, this is the point in which the efficiency frontier is touched by an *iso-cost line* with the angle $-1000/2000 = -\frac{1}{2}$.

This optimum is based on an approximation: The estimated second-order polynomial has a multiple correlation coefficient R^2 of only 0.80. Therefore, this solution is checked: Simulate the system dynamics model with the estimated optimal input combination. This simulation indeed gives 100% efficiency. However, its cost is £4.78 million, which exceeds the lowest cost in Table 6.15 that corresponds with a combination that also gives 100% efficiency. In that table the combination (2835.00, 923.40) gives a cost of £4.68 million, which is 2% lower than the estimated optimal cost. Compared with Wolstenholme (1990, pp. 125–127), this final solution gives a substantial cost reduction, namely £0.72 million, or 15%.

6.6 VALIDATION AND VERIFICATION

6.6.1 Overview

We limit our discussion of validation and verification (V&V) to the role of regression analysis and DOE; V&V is discussed further in Chapter 10 (see also Kleijnen, 1995a). Obviously, V&V is one of the first questions that must be answered in a simulation study. For didactic reasons, however, we discuss V&V at the end of this chapter.

True validation requires that *data on the real system* be available. In practice, the amount of such data varies greatly: Data on failures of nuclear installations are rare, whereas electronically captured data on computer performance and on supermarket sales are abundant.

If data are available, many statistical techniques can be applied. Assume that the simulation is fed with real-life input data: This is known as *trace-driven simulation*. Then simulated and real responses (say) Y and X, respectively, might be compared through the Student t-statistic for paired observations, assuming (say) m normally and independently distributed (NID) observations on X and Y [see (8) with $\overline{Y}$ replaced by $\overline{Y} - \overline{X}$ and $\nu = m - 1$].

A better test, however, uses *regression analysis*, as follows. Regress $Y - X$ on $Y + X$; that is, in the regression model (1), replace Y by $Y - X$ and x by $Y + X$. Use a first-order polynomial; that is, delete cross-products in (1). Next test the null hypothesis H_0: $\beta_0 = 0$ and $\beta_1 = 0$. This hypothesis implies equal means and equal variances of X and Y, as is easily derived assuming bivariate normality for (X, Y). This hypothesis is tested using the familiar F statistic. Whenever the simulation responses are nonnormal, a normalizing transformation should be applied. Kleijnen et al. (1998) give details, including numerical examples for single-server queueing simulations. This reference also demonstrates that a valid simulation model is rejected too often when simply regressing Y on X and testing for zero intercept and unit slope.

If *no data* are available, DOE can be used in the following way. The simulationists and their clients do have *qualitative* knowledge about certain parts of the simulated and the corresponding real system; that is, they do know in which direction certain factors affect the response of the corresponding module in the simulation model; see also the discussion on known signs in sequential bifurcation in Section 6.2.2. If the regression metamodel discussed in Section 6.3 gives an estimated factor effect with the wrong sign, this is a strong indication of a wrong simulation model or a wrong computer program!

To obtain a valid simulation model, some inputs may need to be restricted to a certain domain of factor combinations. This domain corresponds with the *experimental frame* in Zeigler (1976), a seminal book on modeling and simulation.

The regression metamodel shows which factors are most important; that is, which factors have highly significant regression estimates in the metamodel. If possible, information on these factors should be collected, for validation purposes. The importance of sensitivity analysis in V&V is also emphasized by Fossett et al. (1991, p. 719). They invesigate three military case studies, but do not present any details.

One application of DOE and regression analysis in V&V is the ecological study in Kleijnen et al. (1992, p. 415), which concerns the same greenhouse problem examined in Bettonvil and Kleijnen (1997). The regression metamodel helped to detect a serious error in the simulation model: One of the original modules should be split into two modules. This application further shows that some factors are more important than the ecological experts originally expected. This "surprise" gives more insight into the simulation model. Another application is given in Section 6.6.2.

6.6.2 Case Study: Mine Hunting at Sea

Kleijnen (1995b) considers a simulation model for the study of the search for explosive mines on the sea bottom by means of sonar. This model was developed for the Dutch navy, by TNO-FEL (Applied Scientific Research–Physics and Electronics Laboratory); this is a major military research institute in the Netherlands. The model is called HUNTOP (mine HUNTing OPeration). Other countries have similar simulation models for naval mine hunting; the corresponding literature is classified.

In this case study, V&V proceeds in two stages: in stage 1 individual modules are validated; in stage 2 the entire simulation model is treated as one black box and is validated. The latter stage, however, is not discussed here, but in Kleijnen (1995b).

Some modules within the model give *intermediate* output that is hard to observe in practice, and hence hard to validate. Sensitivity analysis is applied to such modules to check if certain factor effects have signs or directions that agree with experts' prior qualitative knowledge. For example, deeper water gives a wider sonar window (see β_2 in the sonar window module below). Because of time constraints, only the following two modules are examined in the HUNTOP case study.

Sonar Window Module. The sonar rays hit the bottom under the grazing angle. This angle is determined deterministically by three factors: w_1, the sound velocity profile (SVP), which maps sound velocity as a function of depth; w_2, the average water depth; and w_3, the tilt angle. SVP is treated as a qualitative factor.

The sonar window module has as response variables $y^{(1)}$, the minimum distance of the area on the sea bottom that is insonified by the sonar beam, and $y^{(2)}$, the maximum distance of that same area. Consider a second-degree polynomial in the two quantitative factors w_2 and w_3, namely one such polynomial for each SVP type. Note that a first-degree polynomial misses interactions and has constant marginal effects; a third-order polynomial is more difficult to interpret and needs many more simulation runs. So a second-order polynomial seems a good compromise.

To estimate the $q = 6$ regression parameters of this polynomial, use the classical central composite design with $n = 9$ input combinations, already displayed in Table 6.8. The fitted polynomial turns out to give an acceptable approximation: The multiple correlation coefficient R^2 ranges between 0.96 and 0.98 for the four SVPs simulated.

Expert knowledge suggests that certain factor effects have specific signs: $\beta_2 > 0$, $\beta_3 < 0$, and $\beta_{2,3} < 0$. Fortunately, the corresponding estimates turn out to have the correct signs. So this module has the correct I/O transformation, and the validity of this module need not be questioned. The quadratic effects are not significantly different from zero. So on hindsight, simulation runs could have been saved, since a resolution 5 design instead of a central composite design would have sufficed.

For $y^{(2)}$, maximum distance, similar results hold. The exception is one SVP that results in an R^2 of only 0.68 and a nonsignificant β_2.

Visibility Module. An object (a mine, a garbage can, etc.) is visible if it is within the sonar window and it is not concealed by the bottom profile. HUNTOP represents the bottom profile through a simple geometric pattern: hills of fixed heights with constant upward slopes and constant downward slopes. A fixed profile is used within a single simulation run. Intuitively, the orientation of the hills relative to the ship's course and to the direction of the sonar beam is important: Does the sonar look down a valley or is its view blocked by a hill? The response variable of this module is the time that the object is visible, expressed as a percentage of the time it would have been visible were the bottom flat; obviously, a flat bottom does not conceal an object. Six inputs are varied: water depth, tilt angle, hill height, upward hill slope, and object's position on the hill slope (top, bottom, or in between).

A second-order polynomial regression metamodel is also used for this module. To estimate its 28 regression parameters, a central composite design is used. This design has 77 input combinations. The simulation's I/O gives an R^2 of 0.86. Furthermore, the upward hill slope has no significant effects at all: no main effect, no interactions with the other factors, and no quadratic effect. These effects agree with the experts' qualitative knowledge. So the validity of this module is not questioned either.

6.7 CONCLUSIONS

In Section 6.1 we raised the following questions:

1. *What If:* What happens if analysts change parameters, input variables, or modules of a simulation model? This question is closely related to sensitivity analysis and optimization.
2. *Validation:* Is the simulation model an adequate representation of the corresponding system in the real world?

These questions are answered in the remainder of this chapter.

In the initial phase of a simulation it is often necessary to perform *screening*: Which factors among the multitude of potential factors are really important? The goal of screening is to reduce the number of factors to be explored further in the next phase. The technique of sequential bifurcation is simple and efficient. This technique also seems to be effective.

Once the important factors are identified, further study requires fewer assumptions;

no known signs are assumed. This study may use *regression analysis*. It generalizes the results of the simulation experiment since it characterizes the I/O transformation of the simulation model.

Design of experiments (DOE) gives better estimators of the main effects, interactions, and quadratic effects in this regression metamodel. So DOE improves the effectiveness of simulation experimentation. DOE requires relatively few simulation runs, which means improved efficiency. Once the factor effects are quantified through the corresponding regression estimates, they can be used in *V&V*, especially if there are no data on the I/O of the simulation model or its modules, and in *optimization* through RSM, which augments regression analysis and DOE with the steepest-ascent hill-climbing technique.

The goal of the statistical techniques of this chapter is to make simulation studies give more general results in less time. These techniques have already been applied many times in practical simulation studies in many domains, as the various case studies demonstrated. We hope that this chapter will stimulate more analysts to apply these techniques. In the meantime, research on statistical techniques adapted to simulation should continue. We did mention several items that require further research.

APPENDIX 6.1: CONFIDENCE INTERVALS FOR INDIVIDUAL RESPONSES

The following formulas are taken from Kleijnen (1996).

A6.1.1 Mean of Terminating Simulation

A $1 - \alpha$ one-sided confidence interval for $\mathrm{E}(Y)$ is

$$P\left[\mathrm{E}(Y) > \overline{Y} - t_{\alpha;m-1}\frac{S_y}{\sqrt{m}}\right] = 1 - \alpha \tag{24}$$

where $t_{\alpha,m-1}$ denotes the $1 - \alpha$ quantile of the Student statistic T_{m-1}. This interval assumes that the simulation response Y is normally, independently distributed (NID). The Student statistic is known to be not very sensitive to nonnormality; the average $\overline{Y}$ is asymptotically normally distributed (central limit theorem).

Johnson (1978) modifies the Student statistic in case Y has a distribution with *asymmetry coefficient* μ_3 (Kleijnen, 1987, pp. 22–23):

$$\tilde{T}_{m-1} = [\overline{Y} - \mathrm{E}(Y)] + \frac{S_3}{6S^2m} + \frac{S_3}{3(S^2)^2}[\overline{Y} - \mathrm{E}(Y)]^2\left(\frac{S^2}{m}\right)^{-1} \tag{25}$$

where S_3 denotes the asymmetry estimator:

$$S_3 = \frac{m \sum_{j=1}^{m} (Y_{i,j} - \overline{Y}_i)^3}{(m-1)(m-2)} \tag{26}$$

Kleijnen et al. (1986) discuss this statistic in detail.

A.6.1.2 Mean of Steady-State Simulation

For steady-state simulations the analysts may apply *renewal analysis* (see Chapter 7). Denote the length of the renewal cycle by L and the *total* cycle response (e.g., total waiting time over the entire cycle) by W. Then the steady-state mean response is

$$\mathrm{E}(Y) = \frac{\mathrm{E}(W)}{\mathrm{E}(L)} \tag{27}$$

This analysis uses *ratio estimators*; see $\overline{W}/\overline{L}$ in the next equation. Crane and Lemoine (1977, pp. 39–46) derive the following asymptotic $1 - \alpha$ confidence interval for the mean:

$$P\left[\mathrm{E}(Y) > \frac{\overline{W}}{\overline{L}} - \frac{z_\alpha S}{\sqrt{m}/\overline{L}}\right] = 1 - \alpha \tag{28}$$

where z_α denotes the $(1 - \alpha)$ quantile of the standard normal variate $N(0, 1)$, m is now the number of cycles (in terminating simulations m was the number of replications), and S^2 is a shorthand notation:

$$S^2 = \left[S_w^2 + \left(\frac{\overline{W}}{\overline{L}}\right)^2 S_L^2 - 2\left(\frac{\overline{W}}{\overline{L}}\right) S_{W,L}\right]^{1/2} \tag{29}$$

where the (co)variances are estimated analogous to (14).

In a Markov system, *any* state can be selected as the renewal state. A practical problem, however, is that it may take a long time before the renewal state selected occurs again; for example, if the traffic rate is high, it takes a long time for all servers to be idle again. Also, if there are very many states (as may be the case in network systems), it may take a long time before a specific state occurs again. In those cases *nearly renewal* states may be used; for example, define "many servers busy" as the set of (say) two states, "all servers busy" or "all minus one servers busy." This approximate renewal state implies that the cycles are not exactly IID. However, for practical purposes they may be nearly IID, which may be tested through the von Neumann statistic for successive differences:

$$\sum_{j=2}^{m} \frac{(W_j - W_{j-1})^2}{[(m-1)S_w^2]} \tag{30}$$

This statistic is approximately normally distributed with mean 2 and variance $4(m - 2)/(m^2 - 1)$, provided that the W_j are NID. Since W is a sum, normality may apply. However, when testing the mutual independence of the cycle lengths L_j and L_{j-1}, a complication is that L probably has an asymmetric distribution. Then the rank version of the von Neumann statistic may be applied (Bartels, 1982). To ensure that the von Neumann test has a reasonable chance of detecting dependence, at least 100 cycles are needed: For $m < 100$ the test has low power (Kleijnen, 1987, p. 68).

The proposed approximate renewal analysis of simulation requires more research (Gunther, 1975; Gunther and Wolff, 1980; Pinedo and Wolff, 1982). A different method for accelerating the renewal process is proposed in Andradóttir et al. (1995).

A6.1.3 Proportions

A proportion (say) p_a is estimated by comparing the simulation outputs y_j ($j = 1, \ldots, m$) with the prespecified constant a, which leads to the binomially distributed variable (say) $B = \sum_{j=1}^{m} D_j$ with $D_j = 0$ if $Y_j < a$; else $D_j = 1$. This binomial variable has variance $p_a(1 - p_a)m$. Obviously, B/m is an estimator of p_a with variance $p_a(1 - p_a)/m$.

A6.1.4 Quantiles

The following notation ignores the fact that pm, l, and u are not necessarily integers; actually, these three real variables must be replaced by their integer parts. The pth quantile z_p (with $0 < p < 1$) may be estimated by the order statistic $y_{(pm)}$. The $1 - \alpha$ confidence interval is

$$P(z_{(l)} < z_p < z_{(u)}) = 1 - \alpha \tag{31}$$

where the lower limit is the lth-order statistic with

$$l = pm - z_{\alpha/2}\sqrt{p(1-p)m} \tag{32}$$

and the upper limit is the uth-order statistic that follows from equation (32) replacing the minus sign in front of $z_{\alpha/2}$ by a plus sign. Proportions and quantiles in terminating and steady-state simulations are discussed further in Kleijnen (1987, pp. 36–40) and Kleijnen and Van Groenendaal (1992, pp. 195–197).

A6.1.5 Multiple Responses

In the presence of multiple responses, each individual $1 - \alpha$ confidence interval has a coverage probability of $1 - \alpha$, but the simultaneously or joint coverage probability is lower. If the intervals were independent and there were (say) two responses, this probability would be $(1 - \alpha)^2$. Bonferroni's inequality implies that if the individual confidence intervals use α, the probability that both intervals hold simultaneously is at least $1 - 2\alpha$. In general, this conservative procedure implies that the simultaneous type I error rate (say) α_E is divided by the number of confidence intervals, in order to guarantee a joint probability of α_E.

APPENDIX 6.2: VARIANCE REDUCTION TECHNIQUES

VRTs are discussed in detail in Ermakov and Melas (1995), Fishman (1989), Kleijnen (1974, pp. 105–285), Kleijnen and Van Groenendaal (1992, pp. 197–201), Tew and Wilson (1994), and in the references mentioned below.

A6.2.1 Common Random Numbers

In the what-if approach there is more interest in the differences than in the absolute magnitudes of the simulation outputs. Intuitively, it seems appropriate to examine simulated systems under equal conditions, that is, in the same environments. This implies the use of the same stream of (pseudo)random numbers for two different factor combinations. Then the two simulation responses (say) Y_1 and Y_2 become statistically dependent. A general relationship is

$$\sigma^2_{y_1 - y_2} = \sigma^2_{y_1} + \sigma^2_{y_2} - 2\rho_{y_1,y_2}\sigma_{y_1}\sigma_{y_2} \tag{33}$$

So if the use of the same (pseudo)random numbers does result in positive correlation, the variance of the difference decreases.

In complicated models, however, it may be difficult to realize a strong positive correlation. Therefore, separate sequences of (pseudo)random numbers are used per process; for example, in a queueing network a separate seed is used per server. One seed may be sampled through the computer's internal clock. However, sampling the other seeds in this way may cause overlap among the various streams, which makes times at different servers statistically dependent. For certain generators, there are tables with seeds 100,000 apart. For other generators such seeds may be generated in a separate computer run; see Kleijnen and Van Groenendaal (1992, pp. 29–30), Chapter 4.

The advantage of a smaller variance comes at a price: The analysis of the simulation results becomes more complicated, since the responses are not independent anymore. Now the analysts should use either GLS or OLS with adjusted standard errors for the estimated regression parameters; this implies that the analysts should estimate the covariances between simulation responses. In practice this complication is often overlooked.

A6.2.2 Antithetic Random Numbers

The intuitive idea behind antithetic (pseudo)random numbers (briefly, antithetics) is as follows. When replication 1 samples many long service times, the average waiting time y is higher than expected. So it is nice if replication 2 compensates this overshoot. Statistically, this compensation means negative correlation between the responses of replications 1 and 2. The variance of their average (say) $\overline{Y}$, taking into account that both replications have the same variance, follows from (33):

$$\sigma^2_{\overline{y}} = \frac{\sigma^2_y(1 + \rho_{y_1,y_2})}{2} \tag{34}$$

So the variance of the average $\overline{Y}$ decreases as the correlation ρ_{y_1,y_2} becomes more negative.

To realize a strong negative correlation, use the (pseudo)random numbers r for replication 1 and the antithetics $1 - r$ for replication 2. Actually, the computer does not need to calculate the complements $1 - r$ if it uses a multiplicative congruential generator. Then it suffices to replace the seed r_0 by its complement $e - r_0$, where e stands for the generator's modulo; that is, $r_t = f r_{t-1} \bmod e$, where f denotes the generator's multiplier (Kleijnen, 1974, pp. 254–256).

A6.2.3 Control Variates or Regression Sampling

Consider the following linear correction:

$$Y_c = Y + \gamma_1[\mathrm{E}(X_1) - X_1] \tag{35}$$

where Y_c is called the linear control variate estimator. Obviously, this new estimator remains unbiased. It is easy to derive that this control variate estimator has minimum variance if the correction factor γ_1 equals

$$\gamma_1^* = \frac{\rho_{y,x_1}\sigma_y}{\sigma_{x_1}} \tag{36}$$

In practice, however, the correlation ρ_{y,x_1} is unknown, so it is estimated. Actually, replacing the three factors on the right-hand side of equation (36) by their classic estimators results in the OLS estimator (say) $C_{1,\mathrm{OLS}}$ of the regression parameter γ_1 in the regression model

$$Y_{x_1} = \gamma_0 + \gamma_1 x_1 + U \tag{37}$$

where U denotes the fitting error of this regression model, analogous to E in (1). Obviously, these two regression parameters (γ_0, γ_1) are estimated from the m replications that give m IID pairs $(X_{1,h}, Y_h)$ with $h = 1, \ldots, m$.

The OLS estimator of γ_1^* defined in (36) gives a new control variate estimator. Let $\overline{Y}$ denote the average over m replications of the responses Y_j, $\overline{X}_1$ the average over m replications of X_1 (average service time per run), and $C_{1,\mathrm{OLS}}$ the OLS estimator of γ_1 in (37) or γ_1^* in (36) based on the m pairs (Y, X_1). Then the new control variate estimator is given in (15). This estimator is easy to interpret, noticing that the estimated regression line goes through the point of gravity, $(\overline{X}_1, \overline{Y})$.

A6.2.4 Importance Sampling

The preceding VRTs relied on the correlation between (1) the responses of systems with comparable simulated environments realized through common seeds, (2) the responses of antithetic runs, or (3) the output and inputs or control variates. In other words, the simulation model itself was not affected; the computer program might be adapted slightly to increase the positive and negative correlations (seeds were changed or inputs were monitored). Importance sampling, however, drastically changes the sampling process of the simulation model. This technique is more sophisticated, but it is necessary when simulating *rare events*; for example, buffer overflow may occur with a probability of (say) 1 in a million replicated months, so 1 million replicated months must be simulated to expect to see a single breakdown of the system!

The basic idea of importance sampling can be explained simply in the case of static, nondynamic simulation, also known as *Monte Carlo sampling*. Consider the example integral

$$\xi = \int_{\nu}^{\infty} \frac{1}{x} \lambda e^{-\lambda x} \, dx \qquad \text{with} \quad \lambda > 0, \quad \nu > 0 \tag{38}$$

This ξ can be estimated through crude Monte Carlo as follows:

1. Sample x from the negative exponential distribution with parameter λ; that is, $x \sim \text{Ne}(\lambda)$.
2. Substitute the sampled value x into the response $Y = g(X)$ with

$$g(x) = \begin{cases} \dfrac{1}{x} & \text{if } x > \nu \\ 0 & \text{otherwise} \end{cases} \tag{39}$$

Obviously, $g(X)$ is an unbiased estimator of ξ. Notice that the event "$g(x) > \nu$" becomes a rare event as $\nu \uparrow \infty$.

Importance sampling does not sample x from the original distribution $f(x)$ [in the example, $\text{Ne}(\lambda)$], but from a different distribution (say) $h(x)$. The resulting x is substituted into the response function $g(x)$. However, $g(x)$ is corrected by the *likelihood ratio* $f(x)/h(x)$. This gives the corrected response

$$g^*(x) = g(x) \frac{f(x)}{h(x)} \tag{40}$$

This estimator is an unbiased estimator of ξ:

$$\text{E}[g^*(X)] = \int_{\nu}^{\infty} g(x) \frac{f(x)}{h(x)} h(x) \, dx = \int_{\nu}^{\infty} g(x) f(x) \, dx = \xi \tag{41}$$

It is quite easy to derive the optimal form of $h(x)$, which results in minimum variance.

For *dynamic systems* (such as queueing systems) a sequence of inputs must be sampled; for example, successive service times $X_{1,t}$ with $t = 1, 2, \ldots$. If these inputs are assumed to be IID and $\text{Ne}(\lambda)$), their joint density function is given by

$$f(x_{1,1}, x_{1,2}, \ldots) = (\lambda e^{-\lambda x_{1,1}})(\lambda e^{-\lambda x_{1,2}}) \cdots \tag{42}$$

Suppose that crude Monte Carlo and importance sampling use the same type of input distribution, namely, negative exponential but with different parameters λ and λ_0, respectively. Then the likelihood ratio becomes

$$\frac{f(x_{1,1}, x_{1,2}, \ldots)}{h(x_{1,1}, x_{1,2}, \ldots)} = \frac{(\lambda e^{-\lambda x_{1,1}})(\lambda e^{-\lambda x_{1,2}}) \cdots}{(\lambda_0 e^{-\lambda_0 x_{1,1}})(\lambda_0 e^{-\lambda_0 x_{1,2}}) \cdots} \tag{43}$$

Obviously, this expression can be reformulated to make the computations more efficient.

In the simulation of dynamic systems it is much harder to obtain the optimal new density. Yet in some applications, distributions were derived that did give drastic variance reductions [see Heidelberger (1995), Heidelberger et al. (1996), Rubinstein and Shapiro (1993), and the literature mentioned at the beginning of this appendix].

APPENDIX 6.3: JACKKNIFING CONTROL VARIATES

Control variates are based on m IID pairs, say, $(Y_j, X_{1,j})$ with $j = 1, \ldots, m$; see Appendix 6.2. Now eliminate pair j and calculate the following control variate estimator [see (35)]:

$$\overline{Y}_{-j,c} = \overline{Y}_{-j} + C_{-j,\text{OLS}}[\text{E}(\overline{X}_{-j,1}) - \overline{X}_{-j,1}] \tag{44}$$

where $\overline{Y}_{-j}$ denotes the sample average of the responses after elimination of Y_j: further, $\overline{X}_{-j,1}$ denotes the average service time after eliminating replication j, and $C_{-j,\text{OLS}}$ denotes the OLS estimator based on the remaining $m - 1$ pairs. Note that $\text{E}(\overline{X}_{-j,1}) = \text{E}(X_1) = 1/\lambda$. This $\overline{Y}_{-j,c}$ gives the pseudovalue

$$P_j = m\overline{Y}_c - (m-1)\overline{Y}_{-j,c} \tag{45}$$

where $\overline{Y}_c$ is the control variate estimator based on all m pairs.

ACKNOWLEDGMENT

Jerry Banks (Georgia Tech, Atlanta), Bert Bettonvil and Gül Gürkan (both at Tilburg University, Tilburg), Barry Nelson (Northwestern, Evanston), Moshe Pollatschek (Technion, Haifa), and one anonymous referee gave useful comments on preliminary versions of this chapter.

REFERENCES

Andradóttir, S., J. M. Calvin, and P. W. Glynn (1995). Accelerated regeneration for Markov chain simulations, *Probability in the Engineering and Information Sciences*, Vol. 9, pp. 497–523.

Atkinson, A. C., and R. D. Cook (1995). D-optimum designs for heteroscedastic linear models, *Journal of the American Statistical Association*, Vol. 90, No. 429, pp. 204–212.

Aytug, H., C. A. Dogan, and G. Bezmez (1996). Determining the number of kanbans: a simulation metamodeling approach, *Simulation*, Vol. 67, No. 1, pp. 23–32.

Bartels, R. (1982). The rank version of von Neumann's ratio test for randomness, *Journal of the American Statistical Association*, Vol. 77, No. 377, pp. 40–46.

BBN (1989). *RS/1 Software: Data Analysis and Graphics Software*, BBN, Cambridge, Mass.

Bettonvil, B. (1990). *Detection of Important Factors by Sequential Bifurcation*, Tilburg University Press, Tilburg, The Netherlands.

Bettonvil, B., and J. P. C. Kleijnen (1997). Searching for important factors in simulation models with many factors: sequential bifurcation, *European Journal of Operational Research*, Vol. 96, No. 1, pp. 180–194.

Box, G. E. P. (1952). Multi-factor designs of first order. *Biometrika*, Vol. 39, No. 1, pp. 49–57.

Box, G. E. P., and D. R. Cox (1964). An analysis of transformations, *Journal of the Royal Statistical Society, Series B*, Vol. 26, pp. 211–252.

Box, G. E. P., and N. R. Draper (1987). *Empirical Model-Building with Response Surfaces*, Wiley, New York.

Box, G. E. P., and J. S. Hunter (1961). The 2^{k-p} fractional factorial designs, Part 1, *Technometrics*, Vol. 3, pp. 311–351.

Box, G. E. P., and K. B. Wilson (1951). On the experimental attainment of optimum conditions, *Journal of the Royal Statistical Society, Series B*, Vol. 13, No. 1, pp. 1–38.

Box, G. E. P., W. G. Hunter, and J. S. Hunter (1978). *Statistics for Experimenters: An Introduction to Design, Data Analysis and Model Building*, Wiley, New York.

Cheng, R. C. H. (1995). Bootstrap methods in computer simulation experiments, *Proceedings of the 1995 Winter Simulation Conference*, C. Alexopoulos, K. Kang, W. R. Lilegdon, and D. Goldsman, eds., IEEE, Piscataway, N.J., pp. 171–177.

Cheng, R. C. H. (1997). *Searching for Important Factors: Sequential Bifurcation Under Uncertainty*, Institute of Mathematics and Statistics, The University of Kent at Canterbury, Canterbury, Kent, England.

Cheng, R. C. H., and J. P. C. Kleijnen (accepted). Improved designs of queueing simulation experiments with highly heteroscedastic responses, *Operations Research*.

Conover, W. J. (1971). *Practical Non-parametric Statistics*, Wiley, New York.

Crane, M. A., and A. J. Lemoine (1977). *An Introduction to the Regenerative Method for Simulation Analysis*, Springer-Verlag, Berlin.

Dagenais, M. G., and J. M. Dufour (1994). Pitfalls of rescaling regression models with Box–Cox transformations, *Review of Economics and Statistics*, Vol. 76, No. 3, pp. 571–575.

Dangerfield, B., and C. Roberts (1996). An overview of strategy and tactics in system dynamics optimization, *Journal of the Operational Research Society*, Vol. 47, pp. 405–423.

De Wit, M. S. (1997). Uncertainty analysis in building thermal modelling, *Journal of Statistical Computation and Simulation*, Nos. 1–4, pp. 305–320.

Donohue, J. M. (1994). Experimental designs for simulation, in *Proceedings of the 1994 Winter Simulation Conference*, J. D. Tew, S. Manivannan, D. A. Sadowski, and A. F. Seila, eds., IEEE, Piscataway, N.J.

Donohue, J. M. (1995). The use of variance reduction techniques in the estimation of simulation metamodels, in *Proceedings of the 1995 Winter Simulation Conference*, C. Alexopoulos, K. Kang, W. R. Lilegdon, and D. Goldsman, eds., IEEE, Piscataway, N.J., pp. 195–199.

Draper, N. R. (1994). Applied regression analysis; bibliography update 1992–93, *Communications in Statistics, Theory and Methods*, Vol. 23, No. 9, pp. 2701–2731.

Dykstra, R. L. (1970). Establishing the positive definiteness of the sample covariance matrix, *Annals of Mathematical Statistics*, Vol. 41, No. 6, pp. 2153–2154.

Efron, B. (1982). *The Jackknife, the Bootstrap, and Other Resampling Plans*, CBMS-NSF Series, SIAM, Philadelphia, Pa.

Efron, B., and R. J. Tibshirani (1993). *Introduction to the Bootstrap*, Chapman & Hall, London.

Ermakov, S. M., and V. B. Melas (1995). *Design and Analysis of Simulation Experiments*. Kluwer, Dordrecht, The Netherlands.

Fedorov, V. V. (1972). *Optimal Experimental Design*, Wiley, New York.

Fishman, G. S. (1989). Focussed issue on variance reduction methods in simulation: introduction, *Management Science*, Vol. 35, p. 1277.

Fossett, C. A., Harrison, D., Weintrob, H., and Gass, S. I. (1991). An assessment procedure for simulation models: a case study, *Operations Research*, Vol. 39, pp. 710–723.

Friedman, L. W. (1996). *The Simulation Metamodel*, Kluwer, Dordrecht, The Netherlands.

Fu, M. C. (1994). Optimization via simulation: a review, *Annals of Operations Research*, Vol. 53, pp. 199–247.

Fürbringer, J.-M., and C. A. Roulet (1995). Comparison and combination of factorial and Monte-Carlo design in sensitivity analysis, *Building and Environment*, Vol. 30, pp. 505–519.

Glynn, P. W., and D. L. Iglehart (1989). Importance sampling for stochastic simulation, *Management Science*, Vol. 35, pp. 1367–1392.

Gunther, F. L. (1975). The almost regenerative method for stochastic system simulations, *Research Report 75-21*, Operations Research Center, University of California, Berkeley, Calif.

Gunther, F. L., and R. W. Wolff (1980). The almost regenerative method for stochastic system simulations, *Operations Research*, Vol. 28, No. 2, pp. 375–386.

Heidelberger, P. (1995). Fast simulation of rare events in queueing and reliability models, *ACM Transactions on Modeling and Computer Simulation*, Vol. 5, No. 1, pp. 43–85.

Heidelberger, P., P. Shahabuddin, and V. Nicola (1996). Bounded relative error in estimating transient measures of highly dependable non-Markovian systems, in *Reliability and Maintenance of Complex Systems*, NATO ASI Series, Springer-Verlag, Berlin.

Ho, Y., and X. Cao (1991). *Perturbation Analysis of Discrete Event Systems*, Kluwer, Dordrecht, The Netherlands.

Ho, Y. C., L. Shi, L. Dai, and W. Gong (1992). Optimizing discrete event dynamic systems via the gradient surface method, *Discrete Event Dynamic Systems: Theory and Applications*, Vol. 2, pp. 99–120.

Hood, S. J., and P. D. Welch (1993). Response surface methodology and its application in simulation, in *Proceedings of the 1993 Winter Simulation Conference*, G. W. Evans, M. Mollaghasemi, E. C. Russell, and W. E. Biles, eds., IEEE, Piscataway, N.J.

Jacoby, J. E. and Harrison, S. (1962), "Multi-variable experimentation and simulation models," *Naval Research Logistic Quarterly*, vol. 9, pp. 121–136.

Johnson, N. J. (1978). Modified *t* tests and confidence intervals for asymmetric populations, *Journal of the American Statistical Association*, Vol. 73, pp. 536–544.

Keijzer, F., J. Kleijnen, E. Mullenders, and A. van Reeken (1981). Optimization of priority class queues, with a computer center case study, *American Journal of Mathematical and Management Sciences*, Vol. 1, No. 4, pp. 341–358. Reprinted in E. J. Dudewicz and Z. A. Karian, *Modern Design and Analysis of Discrete-Event Computer Simulations.* IEEE Computer Society Press, Washington, D.C., 1985, pp. 298–310.

Khuri, A. I. (1996). Analysis of multiresponse experiments: a review, in *Statistical Design and Analysis of Industrial Experiments*, S. Ghosh, ed., Marcel Dekker, New York, pp. 231–246.

Khuri, A. I., and J. A. Cornell (1996). *Response Surfaces: Designs and Analyses, 2nd ed.*, Marcel Dekker, New York.

Kiefer, J., and J. Wolfowitz (1959). Optimum designs in regression problems, *Annals of Mathematical Statistics*, Vol. 30, pp. 271–294.

Kleijnen, J. P. C. (1974). *Statistical Techniques in Simulation*, Part I, Marcel Dekker, New York.

Kleijnen, J. P. C. (1975a). *Statistical Techniques in Simulation*, Part II, Marcel Dekker, New York.

Kleijnen, J. P. C. (1975b). A comment on Blanning's metamodel for sensitivity analysis: the regression metamodel in simulation, *Interfaces*, Vol. 5, No. 3, pp. 21–23.

Kleijnen, J. P. C. (1975c). Antithetic variates, common random numbers and optimum computer time allocation, *Management Science, Application Series*, Vol. 21, No. 10, pp. 1176–1185.

Kleijnen, J. P. C. (1987). *Statistical Tools for Simulation Practitioners*, Marcel Dekker, New York.

Kleijnen, J. P. C. (1992). Regression metamodels for simulation with common random numbers: comparison of validation tests and confidence intervals, *Management Science*, Vol. 38, No. 8, pp. 1164–1185.

Kleijnen, J. P. C. (1993). Simulation and optimization in production planning: a case study, *Decision Support Systems*, Vol. 9, pp. 269–280.

Kleijnen, J. P. C. (1995a). Verification and validation of simulation models, *European Journal of Operational Research*, Vol. 82, pp. 145–162.

Kleijnen, J. P. C. (1995b). Case study: statistical validation of simulation models, *European Journal of Operational Research*, Vol. 87, No. 1, pp. 21–34.

Kleijnen, J. P. C. (1995c). Sensitivity analysis and optimization in simulation: design of experiments and case studies, *Proceedings of the 1995 Winter Simulation Conference*, C. Alexopoulos, K. Kang, W. R. Lilegdon, and D. Goldsman, eds., IEEE, Piscataway, N.J., pp. 133–140.

Kleijnen, J. P. C. (1995d). Sensitivity analysis and optimization of system dynamics models: regression analysis and statistical design of experiments, *System Dynamics Review*, Vol. 11, No. 4, pp. 275–288.

Kleijnen, J. P. C. (1996). Simulation: runlength selection and variance reduction techniques, in *Reliability and Maintenance of Complex Systems*, S. Ozekici et al., eds., Springer-Verlag, Berlin, pp. 411–428.

Kleijnen, J. P. C., and R. Y. Rubinstein (1996). Optimization and sensitivity analysis of computer simulation models by the score function method, *European Journal of Operational Research*, Vol. 88, pp. 413–427.

Kleijnen, J. P. C., and R. G. Sargent (1997). *A Methodology for the Fitting and Validation of Metamodels*, Tilburg University, Tilburg, The Netherlands.

Kleijnen, J. P. C., and C. R. Standridge (1988). Experimental design and regression analysis in simulation: an FMS case study, *European Journal of Operational Research*, Vol. 33, pp. 257–261.

Kleijnen, J. P. C., and W. Van Groenendaal (1992). *Simulation: A Statistical Perspective*, Wiley, Chichester, West Sussex, England.

Kleijnen, J. P. C., and W. Van Groenendaal (1995). Two-stage versus sequential sample-size determination in regression analysis of simulation experiments, *American Journal of Mathematical and Management Sciences*, Vol. 15, No. 1–2, pp. 83–114.

Kleijnen, J. P. C., G. L. J. Kloppenburg, and F. L. Meeuwsen (1986). Testing the mean of an asymmetric population: Johnson's modified t test revisited, *Communications in Statistics, Simulation and Computation*, Vol. 15, No. 3, pp. 715–732.

Kleijnen, J. P. C., P. C. A. Karremans, W. K. Oortwijn, and W. J. H. Van Groenendaal (1987). Jackknifing estimated weighted least squares: JEWLS, *Communications in Statistics, Theory and Methods*, Vol. 16, No. 3, pp. 747–764.

Kleijnen, J. P. C., G. Van Ham, and J. Rotmans (1992). Techniques for sensitivity analysis of simulation models: a case study of the CO_2 greenhouse effect, *Simulation*, Vol. 58, pp. 410–417.

Kleijnen, J. P. C., B. Bettonvil, and W. Van Groenendaal (1998). Validation of trace-driven simulation models: a novel regression test, *Management Science*, Vol. 44, No. 6.

Miller, R. G. (1974). The jackknife: a review, *Biometrika*, Vol. 61, pp. 1–15.

Morris, M. D. (1991). Factorial plans for preliminary computational experiments, *Technometrics*, Vol. 33, No. 2, pp. 161–174.

Myers, R. H., A. I. Khuri, and W. H. Carter (1989). Response surface methodology: 1966–1988. *Technometrics*, 31, No. 2, pp. 137–157.

Nachtsheim, C. J. (1987). Tools for computer-aided design of experiments, *Journal of Quality Technology*, Vol. 19, No. 3, pp. 132–160.

Nachtsheim, C. J., P. E. Johnson, K. D. Kotnour, R. K. Meyer, and I. A. Zualkernan (1996). Expert systems for the design of experiments, in *Statistical Design and Analysis of Industrial Experiments*, S. Ghosh, ed., Marcel Dekker, New York, pp. 109–131.

Nash, S. G. (1995). Software survey NLP, *OR/MS Today*, Vol. 22, pp. 60–71.

Olivi, L. (1980). Response surface methodology in risk analysis, in *Synthesis and Analysis Methods for Safety and Reliability Studies*, G. Apostolakis, S. Garibba, and G. Volta, eds., Plenum Press, New York.

Ören, T. I. (1993). Three simulation experimentation environments: SIMAD, SIMGEST, and E/SLAM, in *Proceedings of the 1993 European Simulation Symposium*, Society for Computer Simulation, La Jolla, Calif.

Ozdemirel, N. E., G. Y. Yurttas, and G. Koksal (1996). Computer aided planning and design of manufacturing simulation experiments, *Simulation*, Vol. 67, No. 3, pp. 171–191.

Pinedo, M., and R. W. Wolff (1982). A comparison between tandem queues with dependent and interdependent service times, *Operations Research*, Vol. 30, No. 3, pp. 464–479.

Plackett, R. L., and J. P. Burman (1946). The design of optimum multifactorial experiments, *Biometrika*, Vol. 33, pp. 305–325.

ProGAMMA (1997). *CADEMO: Computer Aided Design of Experiments and Modeling*, ProGAMMA, Groningen, The Netherlands.

Ramberg, J. S., S. M. Sanchez, P. J. Sanchez, and L. J. Hollick (1991). Designing simulation experiments: Taguchi methods and response surface metamodels, *Proceedings of the 1991 Winter Simulation Conference*, B. L. Nelson, W. D. Kelton, and G. M. Clark, eds., IEEE, Piscataway, N.J., pp. 167–176.

Rao, C. R. (1959). Some problems involving linear hypothesis in multivariate analysis, *Biometrika*, Vol. 46, pp. 49–58.

Rechtschaffner, R. L. (1967). Saturated fractions of 2^n and 3^n factorial designs, *Technometrics*, Vol. 9, pp. 569–575.

Rubinstein, R. Y., and A. Shapiro (1993). *Discrete Event Systems: Sensitivity Analysis and Stochastic Optimization via the Score Function Method*, Wiley, New York.

Sacks, J., W. J. Welch, T. J. Mitchell, and H. P. Wynn (1989). Design and analysis of computer experiments (includes comments and rejoinder), *Statistical Science*, Vol. 4, No. 4, pp. 409–435.

Safizadeh, M. H., and R. Signorile (1994). Optimization of simulation via quasi-Newton methods, *ORSA Journal on Computing*, Vol. 6, No. 4, pp. 398–408.

Saltelli, A., T. H. Andres, and T. Homma (1995). Sensitivity analysis of model output: performance of the iterated fractional factorial design method, *Computational Statistics and Data Analysis*, Vol. 20, pp. 387–407.

Sanchez, P. J., F. Chance, K. Healy, J. Henriksen, W. D. Kelton, and S. Vincent (1994). Simulation statistical software: an introspective appraisal, *Proceedings of the 1994 Winter Simulation Conference*, J. D. Tew, S. Manivannan, D. A. Sadowski, and A. F. Seila, eds., IEEE, Piscataway, N.J., pp. 1311–1315.

Schruben, L. W., and B. H. Margolin (1978). Pseudorandom number assignment in statistically designed simulation and distribution sampling experiments, *Journal of the American Statistical Association*, Vol. 73, No. 363, pp. 504–525.

Sheriff, Y. S., and B. A. Boice (1994). Optimization by pattern search, *European Journal of Operational Research*, Vol. 78, No. 3, pp. 277–303.

St. John, R. C., and N. R. Draper (1975). *D*-optimality for regression designs: a review, *Technometrics*, Vol. 17, No. 1, pp. 15–23.

Swain, J. J. (1996). Number crunching: 1996 statistics survey, *OR/MS Today*, Vol. 23, No. 1, pp. 42–55.

Tao, Y.-H., and B. L. Nelson (1997). Computer-assisted simulation analysis, *IEE Transactions*, Vol. 29, No. 3, pp. 221–231.

Tew, J. D., and J. R. Wilson (1994). Estimating simulation metamodels using combined correlation-based variance reduction techniques, *IIE Transactions*, Vol. 26, No. 3, pp. 2–16.

Van Groenendaal, W., and J. P. C. Kleijnen (1996). Regression metamodels and design of experiments, *Proceedings of the 1996 Winter Simulation Conference*, J. M. Charnes, D. J. Morrice, D. T. Brunner, and J. J. Swain, eds., IEEE, Piscataway, N.J., pp. 1433–1439.

Webb, S. (1968). Non-orthogonal designs of even resolution, *Technometrics*, Vol. 10, pp. 291–299.

Welch, W. J., R. J. Buck, J. Sacks, H. P. Wynn, et al. (1992). Screening, predicting, and computer experiments, *Technometrics*, Vol. 34, No. 1, pp. 15–25.

Wolstenholme, E. F. (1990). *System Enquiry: A System Dynamics Approach*, Wiley, Chichester, West Sussex, England.

Yu, B., and K. Popplewell (1994). Metamodel in manufacturing: a review, *International Journal of Production Research*, Vol. 32, No. 4, pp. 787–796.

Zeigler, B. (1976). *Theory of Modelling and Simulation*, Wiley-Interscience, New York.

CHAPTER 7

Output Data Analysis

CHRISTOS ALEXOPOULOS
Georgia Institute of Technology

ANDREW F. SEILA
University of Georgia

7.1 INTRODUCTION

The primary purpose of most simulation studies is the approximation of prescribed system parameters with the objective of identifying parameter values that optimize some system performance measures. If some of the input processes driving a simulation are random, the output data are also random and runs of the simulation result in *estimates* of performance measures. Unfortunately, a simulation run does not usually produce independent, identically distributed observations; therefore, "classical" statistical techniques are not directly applicable to the analysis of simulation output.

A simulation study consists of several steps, such as data collection, coding and verification, model validation, experimental design, output data analysis, and implementation. This chapter focuses on statistical methods for computing confidence intervals for system performance measures from output data. Several aspects of output analysis, such as comparison of systems, design of simulation experiments, and variance reduction methods, are not discussed. These subjects are treated in other chapters of this book and in several texts, including Bratley et al. (1987), Fishman (1978b, 1996), Kleijnen (1974, 1975), and Law and Kelton (1991).

The reader is assumed to be comfortable with probability theory and statistics at the level of Hogg and Craig (1978), and stochastic processes at the level of Ross (1993). A reader who is only interested in computational methods can skip the technical parts of this chapter. In Sections 7.1.1 and 7.1.2 we review definitions and results that are essential for the study of this chapter. In Section 7.2 we discuss methods for analyzing output from finite-horizon simulations. Techniques for point and interval estimation of steady-state parameters are presented in Sections 7.3 and 7.4.

Handbook of Simulation, Edited by Jerry Banks.
ISBN 0-471-13403-1

7.1.1 Limit Theorems and Their Statistical Implications

In this section we review the tools needed to establish asymptotic (as the sample size increases) properties of estimators and to obtain confidence intervals. Consider the following three forms of convergence for sequences of random variables on the same probability space: The first form is the strongest while the last form is the weakest (and easiest to establish). For additional forms of convergence as well as the relationships between the forms, see Karr (1993, Chap. 5).

Almost Sure Convergence The sequence $X_1, X_2, \ldots$ converges to the random variable X *almost surely* (or *with probability* 1) (we write $X_n \xrightarrow{\text{a.s.}} X$) if $P(X_n \rightarrow X$ as $n \rightarrow \infty) = 1$.

Convergence in Probability The sequence $X_1, X_2, \ldots$ converges to X *in probability* (we write $X_n \xrightarrow{\mathcal{P}} X$) if for every $\epsilon > 0$),

$$P(|X_n - X| \leq \epsilon) \rightarrow 1 \qquad \text{as } n \rightarrow \infty$$

Convergence in Distribution The sequence $X_1, X_2, \ldots$ converges to X *in distribution* (we write $X_n \xrightarrow{\mathcal{D}} X$) if

$$P(X_n \leq x) \rightarrow P(X \leq x) \qquad \text{as } n \rightarrow \infty$$

at all points x where the cumulative distribution function $P(X \leq x)$ is continuous.

Now suppose that the random variables $X_1, X_2, \ldots$ are from some distribution with an unknown parameter θ and the objective is to estimate a function $g(\theta)$. For fixed n, let $\delta_n = \delta_n(X_1, \ldots, X_n)$ be an estimator of $g(\theta)$. If $\text{E}(\delta_n) = g(\theta)$, δ_n is called an *unbiased* estimator. Furthermore, δ_n is said to be a *consistent* (respectively, *strongly consistent*) estimator of $g(\theta)$ if $\delta_n \xrightarrow{\mathcal{P}} g(\theta)$ [respectively, $\delta_n \xrightarrow{\text{a.s.}} g(\theta)$]. If δ_n is unbiased for each n and $\text{Var}(\delta_n) \rightarrow 0$ as $n \rightarrow \infty$, then δ_n is also consistent (Lehmann, 1991, pp. 331–333).

In the remainder of this section we illustrate the foregoing concepts with a few classical results. Suppose that $X_1, X_2, \ldots, X_n$ are independent, identically distributed (IID) random variables with finite mean μ and variance σ^2. Let

$$\overline{X}_n = \frac{1}{n} \sum_{i=1}^{n} X_i$$

be the sample mean of the X_i's. Since $\text{E}(\overline{X}_n) = \mu$, $\overline{X}_n$ is an unbiased estimator of μ. $\overline{X}_n$ is also a strongly consistent estimator of μ by the *strong law of large numbers*:

$$\overline{X}_n \xrightarrow{\text{a.s.}} \mu \qquad \text{as } n \rightarrow \infty$$

(Karr, 1993, pp. 188–189).

If $0 < \sigma^2 < \infty$, the *central limit theorem* (Karr, 1993, p. 174) states that

$$\frac{\overline{X}_n - \mu}{\sigma/\sqrt{n}} \xrightarrow{\mathcal{D}} N(0,1) \qquad \text{as } n \to \infty$$

where $N(0,1)$ denotes a normal random variable with mean 0 and variance 1. In other words,

$$P\left(\frac{\overline{X}_n - \mu}{\sigma/\sqrt{n}} \le z\right) \to \Phi(z) \qquad \text{as } n \to \infty \tag{1}$$

where Φ is the distribution function of the standard normal random variable.

The central limit theorem remains valid when the potentially unknown parameter σ^2 is replaced by its unbiased and consistent estimator

$$S_n^2(X) = \frac{1}{n-1} \sum_{i=1}^{n} (X_i - \overline{X}_n)^2$$

Therefore, for sufficiently large n,

$$P\left(\frac{|\overline{X}_n - \mu|}{S_n(X)/\sqrt{n}} \le z_{1-\alpha/2}\right) \approx 1 - \alpha \tag{2}$$

where $z_{1-\alpha/2}$ denotes the $1 - \alpha/2$ quantile of $N(0,1)$.

Now suppose that the mean μ is unknown. Solving the inequality on the left-hand side of (2) for μ, one has the well-known approximate (two-sided) $1 - \alpha$ confidence interval

$$\overline{X}_n \pm z_{1-\alpha/2} \frac{S_n(X)}{\sqrt{n}} \tag{3}$$

The left-hand side of (2) is the probability that the confidence interval (3) contains the true mean μ. Denote this probability by $p_{n,\alpha}$ and call it the *coverage probability* of (3). One interprets this confidence interval as follows: Suppose that a large number of independent trials is performed; in each trail, n observations are collected and a confidence interval for μ is computed using (3). As the number of trials grows, the proportion of confidence intervals that contain μ approaches $1 - \alpha$.

The number of observations n required for $p_{n,\alpha} \approx 1 - \alpha$ depends on the symmetry of the distribution of X_i. The more skewed (asymmetric) the density/probability function of X_i, the larger n required. To reduce undercoverage problems ($p_{n,\alpha} < 1 - \alpha$) for small n, one may replace the normal quantile $z_{1-\alpha/2}$ by the larger quantile $t_{n-1,1-\alpha/2}$ of the t distribution with $n - 1$ degrees of freedom. This choice for degrees of freedom is due to the fact that for IID normally distributed X_i,

$$\frac{\overline{X}_n - \mu}{S_n(X)/\sqrt{n}} \sim t_{n-1}$$

where the notation $X \sim Y$ is used to indicate that the random variables X and Y have the same distribution.

7.1.2 Stochastic Processes

Simulation output data are realizations (or *sample paths*) of *stochastic processes*. A stochastic process is a probabilistic model of a system that evolves randomly. More formally, a stochastic process is a collection $X = \{X(u), u \in T\}$ of random variables indexed by a parameter u taking values in the set T. The random variables $X(u)$ take values in a set S, called the *state space* of the process X. Throughout this chapter, u will represent time, and we encounter the following two cases:

1. $T = \{0, 1, 2, \ldots\}$, for which the notation $X = \{X_i, i \geq 0\}$ will be used. For example, X_i may represent the price of a stock at the end of day i or the time in queue of the ith customer at a post office.
2. $T = [0, \infty)$. In this case, the notation $X = \{X(t), t \geq 0\}$ will be used. Some examples of $X(t)$ would be the number of failed machines in a shop at time t, the throughput of a shop at time t, or the price of a stock at time t.

One way to describe a stochastic process is to specify the joint distribution of $X(t_1)$, $X(t_2)$, ... , $X(t_n)$ for each set of times $t_1 < t_2 < \cdots < t_n$ and each n. This approach is typically too complicated to be attempted in practice. An alternative, and simpler approach, is to specify the first and second *moment* functions of the process. These functions are the *mean* function $\mu(t) = \text{E}[X(t)]$, the *variance* function $\sigma^2(t) = \text{Var}[X(t)]$, and the *autocovariance* function

$$C(t_1, t_2) = \text{Cov}[X(t_1), X(t_2)]$$

Notice that $C(t_1, t_2) = C(t_2, t_1)$ and $C(t, t) = \sigma^2(t)$. (For a discrete-time process, the notation μ_t, σ_t^2, and C_{t_1, t_2} will be used.)

To analyze a simulation output process, one must make some structural assumptions. The following are the two most frequently used assumptions.

Strict Stationarity. The process X is called (*strictly*) *stationary* if the joint distribution of $X(t_1)$, $X(t_2)$, ... , $X(t_n)$ is the same as the joint distribution of $X(t_1 + s)$, $X(t_2 + s)$, ... , $X(t_n + s)$ for all t_1, t_2, ... , t_n, and s. In simpler terms, shifting the time origin from zero to any other value s, has no effect on the joint distributions. An immediate result is that the joint distribution of $X(t_1)$, $X(t_2)$, ... , $X(t_n)$ depends only on the intervals between t_1, t_2, ... , t_n.

Example 1 (The M/M/1 queue) Consider an M/M/1 queueing system with IID interarrival times A_i, $i \geq 1$, from the exponential distribution with rate τ and IID service times S_i, $i \geq 1$, from the exponential distribution with rate $\omega(\tau < \omega)$. The ratio $\nu = \tau/\omega$ is called the traffic intensity or (long-run) server utilization. Suppose that the service discipline is first come, first served. Let D_i be the delay time in queue of the ith customer and assume that the system starts empty. The first of Lindley's recursive equations (Lindley, 1952)

$$\begin{aligned} D_1 &= 0 \\ D_{i+1} &= \max\{D_i + S_i - A_{i+1}, 0\} \qquad i \geq 1 \end{aligned} \tag{4}$$

implies that $\mathrm{E}(D_1) = 0$, whereas $P(D_2 > 0) = P(S_1 > A_2) = \tau/(\tau + \omega) > 0$ implies that $\mathrm{E}(D_2) > 0$. Therefore, the delay process $\{D_i, i \geq 1\}$ is not stationary. Using queueing theory (Ross, 1993, Chap. 8) one has

$$\lim_{i \to \infty} P(D_i \leq x) = 1 - \nu + \nu(1 - e^{-(\omega - \tau)x}) \qquad x \geq 0$$

$$\mu = \lim_{i \to \infty} \mathrm{E}(D_i) = \frac{\nu}{(1-\nu)\omega} \quad \text{and} \quad \sigma^2 = \lim_{i \to \infty} \mathrm{Var}(D_i) = \frac{\nu(2-\nu)}{\omega^2(1-\nu)^2} \tag{5}$$

Equation (5) suggests that the delay process becomes asymptotically stationary. Indeed, if D_1 has the distribution on the right-hand side of (5), equations (4) imply (after some work) that all D_i have the same distribution and the delay process is stationary.

Weak Stationarity. In practice it is often necessary to consider a less restricted form of stationarity. The process X is said to be *weakly stationary* if its mean and variance functions are constant (equal to μ and σ^2, respectively) and its autocovariance function satisfies

$$\mathrm{Cov}[X(t), X(t+s)] = C(s) \qquad t \geq 0, \quad s \geq 0$$

that is, it depends only on the *lag* s. In this case, the *autocorrelation* function is defined by

$$\rho(s) = \mathrm{Corr}[X(t), X(t+s)] = \frac{C(s)}{\sigma^2} \qquad s \geq 0$$

Example 2 (A Stationary M/M/1 Queue) The autocorrelation function of the delay process $\{D_i\}$ in a stationary M/M/1 queueing system is given by (Blomqvist, 1967)

$$\rho_j = \frac{(1-\nu)^3(1+\nu)}{(2-\nu)\nu^3} \sum_{k=j+3}^{\infty} \left[\frac{\nu}{(\nu+2)^2}\right]^k \times \frac{(2k-3)!}{k!\,(k-2)!}(k-j-1)(k-j-2) \qquad j = 0, 1, \ldots$$

This function is monotone decreasing with a very long tail that increases as the server utilization ν increases (e.g., $\rho_{200} \approx 0.30$ when $\nu = 0.9$). This makes the M/M/1 system a good test bed for evaluating simulation methodologies.

Example 3 (Moving-Average Process) A well-studied example of a discrete-time weakly stationary process is the moving-average process of order q [often abbreviated to MA(q)]:

$$X_i = \beta_0 Z_i + \beta_1 Z_{i-1} + \cdots + \beta_q Z_{i-q} \qquad i \geq 0$$

where the coefficients β_i are constants and $\{Z_i, i = 0, \pm 1, \pm 2, \ldots\}$ are IID random variables with mean zero and finite variance a^2. MA processes have applications in several areas, particularly econometrics (Chatfield, 1989).

Clearly,

$$\mathrm{E}(X_i) = 0 \qquad \mathrm{Var}(X_i) = a^2 \sum_{i=0}^{q} \beta_i^2$$

while some algebra yields the autocovariance function

$$C_j = \begin{cases} a^2 \sum_{i=0}^{q-j} \beta_i \beta_{i+j} & j = 0, 1, \ldots, q \\ 0 & j > q \end{cases}$$

which "cuts off" at lag q. If, in addition, the Z_i's are normally distributed, the MA(q) process is stationary.

Now suppose that one observes the portion $X_1, \ldots, X_n$ of a discrete-time weakly stationary process for the purpose of estimating the mean μ. Clearly, $\overline{X}_n$ is an unbiased estimator of μ, while some algebra yields

$$\mathrm{Var}(\overline{X}_n) = \frac{\sigma^2}{n} \left[1 + 2 \sum_{j=1}^{n-1} \left(1 - \frac{j}{n} \right) \rho_j \right] \equiv \frac{\sigma^2}{n} (1 + \gamma_n) \tag{6}$$

In order for $\overline{X}_n$ to be a consistent estimator of μ, we require that $\lim_{n \to \infty} \mathrm{Var}(\overline{X}_n) = 0$. The last condition holds if $\lim_{n \to \infty} n \mathrm{Var}(\overline{X}_n) < \infty$ or, equivalently,

$$\lim_{n \to \infty} \gamma_n < \infty \tag{7}$$

For (7) to hold, $\lim_{j \to \infty} C_j = 0$ is necessary but not sufficient. A necessary and sufficient condition is

$$\lim_{n \to \infty} \sum_{j=-(n-1)}^{n-1} C_j = \sum_{j=-\infty}^{\infty} C_j < \infty \tag{8}$$

In simple terms, the covariance between X_i and X_{i+j} must dissipate sufficiently fast so that the summation in (8) remains bounded.

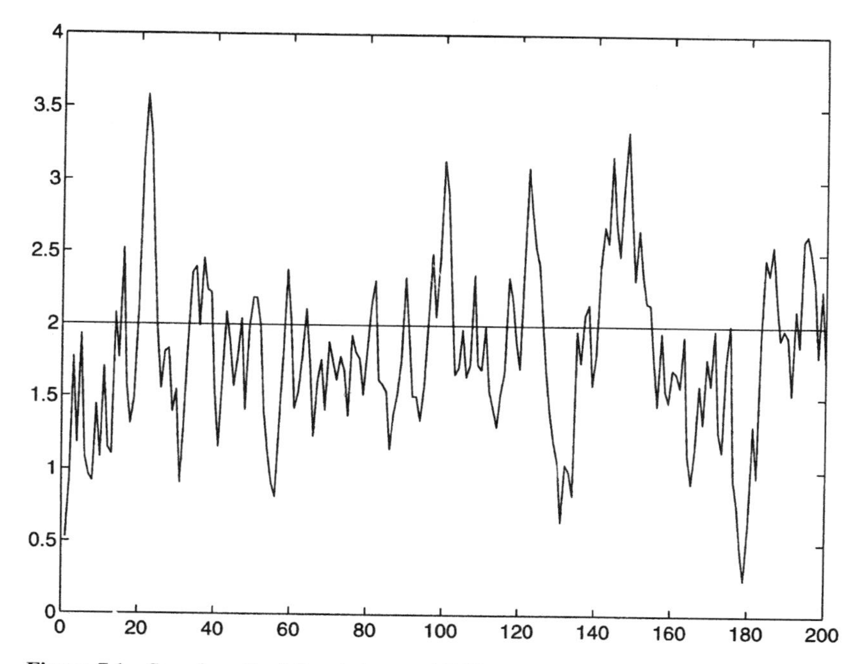

Figure 7.1 Sample path of the stationary AR(1) process $X_i = 2 + 0.8(X_{i-1} - 2) + Z_i$.

Example 4 (First-Order Autoregressive Process) Another well-known stationary process is the autoregressive process of order 1, denoted by AR(1), and often called the Markov process in the time-series literature,

$$X_i = \mu + \rho(X_{i-1} - \mu) + Z_i \qquad i \geq 1$$

where $|\rho| < 1$, $X_0 \sim N(\mu, 1)$, and the Z_i's are IID $N(0, 1 - \rho^2)$ (see Figure 7.1).

The autocorrelation function of this process

$$\rho_j = \rho^j \qquad j \geq 0$$

is monotone decreasing if $\rho > 0$ with a tail that becomes longer as ρ increases, and exhibits damped harmonic behavior around the zero axis if $\rho < 0$.

Applying equation (6), one has

$$n\text{Var}(\overline{X}_n) = 1 + 2\sum_{j=1}^{n-1}\left(\frac{1-j}{n}\right)\rho^j \rightarrow \frac{1+\rho}{1-\rho} \qquad \text{as } n \rightarrow \infty$$

Hence $\overline{X}_n$ is a consistent estimator of the mean $\mu = E(X_i)$. The limit $(1 + \rho)/(1 - \rho)$ is often called the *time-average process variance*.

Brownian Motion and Brownian Bridge. A continuous-time stochastic process with frequent use in simulation output analysis (see Section 7.3.4) is the standard Brownian motion $\{W(t), t \geq 0\}$. This process has the following properties:

1. $W(0) = 0$.
2. W has independent *increments*, that is, for $0 \leq t_0 \leq t_1 \leq \cdots \leq t_n$,

$$P[W(t_j) - W(t_{j-1}) \leq w_j, 1 \leq j \leq n] = \prod_{j=1}^{n} P[W(t_j) - W(t_{j-1}) \leq w_j]$$

3. For $0 \leq s < t$, the increment $W(t) - W(s)$ has the $N(0, t - s)$ distribution.

A well-known function of the Brownian motion is the (standard) Brownian bridge process defined by

$$B(t) = W(t) - tW(1) \qquad 0 \leq t \leq 1$$

Figures 7.2 and 7.3 depict sample paths of $W(t)$ and $B(t)$ in the interval [0,1]. Notice that $B(0) = B(1) = 0$.

7.1.3 Types of Simulations

There are two types of simulations with regard to output analysis:

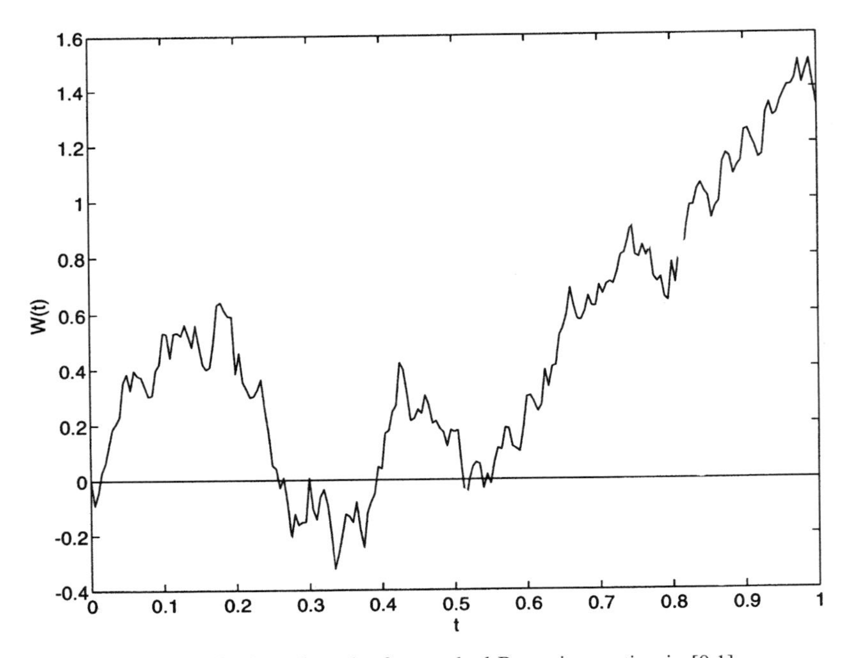

Figure 7.2 Sample path of a standard Brownian motion in [0,1].

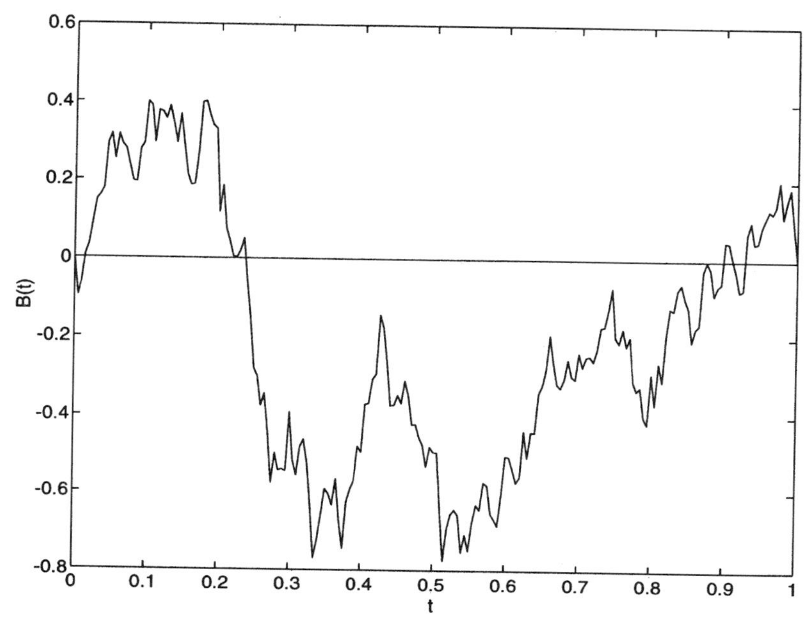

Figure 7.3 Respective Brownian bridge for the Brownian motion in Figure 7.2.

1. *Finite-Horizon Simulations.* In this case the simulation starts in a specific state, such as the empty and idle state, and is run until some terminating event occurs. The output process is not expected to achieve any steady-state behavior and any parameter estimated from the output data will be transient in the sense that its value will depend on the initial conditions. An example is the simulation of a computer network, starting empty, until n jobs are completed. One might wish to estimate the mean time to complete n jobs, or the mean of the average waiting time for the n jobs.

2. *Steady-State Simulations.* The purpose of a steady-state simulation is the study of the long-run behavior of the system of interest. A performance measure of a system is called a *steady-state parameter* if it is a characteristic of the equilibrium distribution of an output stochastic process (Law and Kelton, 1991). An example is the simulation of a continuously operating communication system where the objective is the computation of the mean delay of a data packet.

7.2 FINITE-HORIZON SIMULATIONS

Suppose that one starts in a specific state and simulates a system until n output data $X_1, X_2, \ldots, X_n$ are collected with the objective of estimating $f(X_1, \ldots, X_n)$, where f is a "nice"* function of the data. For example, X_i may be the transit time of unit i

*Formally, f must be a measurable function. In practice, all functions encountered in simulation output analysis are measurable.

through a network of queues or the total time station i is busy during the ith hour and $f(X_1, \ldots, X_n) = \overline{X}_n = (1/n) \sum_{i=1}^{n} X_i$ is the average transit time for the n jobs.

7.2.1 Estimation of the Mean via Independent Replications

In this section we focus on the estimation of $\mu = \mathrm{E}(\overline{X}_n)$. By definition, $\overline{X}_n$ is an unbiased estimator for μ. Unfortunately, the X_i's are generally dependent random variables, which makes the estimation of the variance $\mathrm{Var}(\overline{X}_n)$ a nontrivial problem. In many queueing systems the X_i's are positively correlated. When this is the case, the familiar estimator

$$\frac{S_n^2(X)}{n} = \frac{1}{n(n-1)} \sum_{i=1}^{n} (X_i - \overline{X}_n)^2$$

is a highly biased estimator of $\mathrm{Var}(\overline{X}_n)$.

Example 5 Consider a stationary M/M/1 queueing system (see Examples 1 and 2) with service rate $\omega = 1$ and server utilization $\nu = 0.9$. Using the formulas for ρ_j, one can show that

$$\mathrm{E}\left[\frac{S_{10}^2(D)}{10}\right] = 0.033\sigma^2$$

where $\sigma^2 = \mathrm{Var}(D_i) = 99$. As a result, the $1 - \alpha$ confidence interval $\overline{D}_{10} \pm t_{9,1-\alpha/2} S_{10}(D)/\sqrt{10}$ for the mean delay $\mu = 0.9/(1-0.9) = 9$ computed from 10 consecutive delay observations from a single replication will probably be unacceptably narrow.

To overcome this problem, one can run k independent replications of the system simulation. Each replication starts in the same state and uses a portion of the random number stream that is different from the portions used to run the other replications. Assume that replication i produces the output data $X_{i1}, X_{i2}, \ldots, X_{in}$. Then the sample means

$$Y_i = \frac{1}{n} \sum_{j=1}^{n} X_{ij} \quad i = 1, \ldots, k$$

are IID random variables,

$$\overline{Y}_k = \frac{1}{k} \sum_{i=1}^{k} Y_i$$

is also an unbiased estimator of μ, and the sample variance of the Y_i's

$$S_k^2(Y) = \frac{1}{k-1} \sum_{i=1}^{k} (Y_i - \overline{Y}_k)^2$$

is an unbiased estimator of $\mathrm{Var}(\overline{X}_n)$. If, in addition, n and k are sufficiently large, an approximate $1 - \alpha$ confidence interval for μ is

$$\overline{Y}_k \pm t_{k-1,1-\alpha/2} \frac{S_k(Y)}{\sqrt{k}} \tag{9}$$

Denote the half-width of the interval (9) by $\delta(k, \alpha) = t_{k-1,1-\alpha/2} S_k(Y)/\sqrt{k}$.

7.2.2 Sequential Estimation

A fundamental problem is the estimation of μ within a tolerance $\pm d$, where d is user specified. More formally, one would like to make k runs, so that

$$P(\overline{Y}_k - d \leq \mu \leq \overline{Y}_k + d) \geq 1 - \alpha \tag{10}$$

where $\alpha \in (0,1)$. The sequential procedure of Chow and Robbins (1965) (see also Nadas, 1969) is to run one replication at a time and stop at run k^* such that

$$k^* = \min\left[k : k \geq 2, \delta(k, \alpha) \leq \sqrt{\frac{k}{k-1} d^2 - \frac{t^2_{k-1,1-\alpha/2}}{k(k-1)}} \right] \tag{11}$$

The *stopping rule* (11) is based on the limiting result

$$\lim_{d \to 0} P(\overline{Y}_{k*} - d \leq \mu \leq \overline{Y}_{k*} + d) = 1 - \alpha \tag{12}$$

Equation (12) indicates that as the tolerance d decreases, the probability that the interval $\overline{Y}_{k*} \pm d$ contains μ converges to $1 - \alpha$. Notice that as k increases, the right-hand side of the last inequality in (11) approaches d.

Now suppose that $Y_1, \ldots, Y_k$ are normally distributed. Starr (1966) showed that the stopping rule

$$k^* = \min[k : k \geq 3, k \text{ odd}, \delta(k, \alpha) \leq d]$$

yields

$$P(\overline{Y}_{k*} - d \leq \mu \leq \overline{Y}_{k*} + d) \geq \begin{cases} 0.928 & \text{if } \alpha = 0.05 \\ 0.985 & \text{if } \alpha = 0.01 \end{cases}$$

The last inequalities indicate little loss in the confidence level for arbitrary d. Based on Starr's result and (12), Fishman (1978b) recommended the simpler and more intuitive stopping rule

$$k^* = \min[k : k \geq 2, \delta(k, \alpha) \leq d]$$

An alternative *two-stage* approach for computing a confidence interval for μ with half-width at most d works as follows: The first stage uses k_0 replications to compute a variance estimate $S^2_{k_0}(Y)$ and a confidence interval with half-width $\delta(k_0, \alpha)$. Assume that the estimate $S^2_{k_0}(Y)$ does not change significantly as k_0 increases. If $\delta(k_0, \alpha) \leq d$, the procedure terminates. Otherwise, an estimate of the total number of replications required to obtain a half-width of at most d is computed from

$$\hat{k} = \min\left[k : k \geq k_0, t_{k-1,1-\alpha/2} \frac{S_{k_0}(Y)}{\sqrt{k}} \leq d\right]$$

The efficacy of this method depends on the closeness of $S^2_{k_0}(Y)$ to the unknown $\mathrm{Var}(Y_i)$. If $S^2_{k_0}(Y)$ underestimates $\mathrm{Var}(Y_i)$, then $\hat{k}$ will be smaller than actually needed. Conversely, if $S^2_{k_0}(Y)$ overestimates $\mathrm{Var}(Y_i)$, unnecessary replications will have to be made.

Example 6 Table 7.1 summarizes the results of experiments that were run to estimate the mean number of customers that complete service during the first hour in an M/M/1 queueing system with arrival rate 0.9 per hour, service rate 1, and empty initial state. The sequential procedure was implemented with the stopping rule

$$k^* = \min[k : k \geq k_0, \delta(k, \alpha) \leq d]$$

and initial sample sizes $k_0 = 2, 3, 4, 5$. The two-stage procedure used initial samples of size 4, 5, and 10. For each experiment, 100 independent replications were run.

Based on Table 7.1, the sequential procedure with an initial sample of at least five replications appears to outperform the two-stage procedure. The advantages of the sequential procedure are (1) the resulting confidence interval half-width is always less than or equal to the target value; and (2) the variation in the final sample sizes and confidence interval half-widths is substantially smaller.

An alternative problem is the computation of an estimate for μ with relative error $|\overline{Y}_k - \mu|/|\mu| \leq c$, where c is a positive constant. Formally, one requests that

TABLE 7.1 Comparisons Between Sequential and Two-Stage Confidence Interval Procedures

		Final Sample Size		Interval Half-width	
Procedure	Initial Sample Size	Mean	Standard Deviation	Mean	Standard Deviation
Sequential	2	91.7	35.0	0.945	0.198
	3	94.7	26.5	0.981	0.077
	4	99.0	16.8	0.996	0.005
	5	97.2	19.6	0.995	0.006
Two-stage	4	88.0	83.9	1.362	0.685
	5	92.1	57.7	1.200	0.425
	10	101.9	48.5	1.060	0.226

$$P\left(\frac{|\overline{Y}_k - \mu|}{|\mu|} \leq c\right) \geq 1 - \alpha$$

Using some algebra, one can show that

$$P\left(\frac{|\overline{Y}_k - \mu|}{|\mu|} \leq c'\right) \geq P\left(\frac{|\overline{Y}_k - \mu|}{|\mu|} \leq \frac{\delta(k, \alpha)}{|\overline{Y}_k|}\right)$$

where $c' = c/(1 + c)$. Based on these observations, one can use the following stopping rule:

$$k^* = \min\left[k : k \geq k_0, \frac{\delta(k, \alpha)}{|\overline{Y}_k|} \leq c'\right] \qquad (13)$$

Law et al. (1981) showed that when c is close to 0, the coverage of the confidence interval $\overline{Y}_k \pm \delta(k, \alpha)$ can be arbitrarily close to $1 - \alpha$. They recommend that (13) be used with $c \leq 0.15$ and $k_0 \geq 10$.

7.2.3 Quantile Estimation

The method of replications can also be used to implement nonparametric methods for estimating performance measures other than means. For example, suppose that we want to estimate the p-quantile ($0 < p < 1$), say ξ_p, of the maximum queue size Y in a single-server queueing system during a fixed time window. Let $F(y) = P(Y \leq y)$ be the cumulative distribution function of Y. Then ξ_p is defined as

$$\xi_p = \inf[\, y : F(\, y) \geq p]$$

If the distribution F of Y is monotone increasing, ξ_p is the unique solution to the equation $P(Y \leq y) = p$. Let $Y_1, \ldots, Y_k$ be a random sample from F obtained by performing k independent replications, and let $Y_{(1)} < Y_{(2)} < \cdots < Y_{(k)}$ be the order statistics corresponding to the Y_i's. Then a point estimator for ξ_p is

$$\hat{\xi}_p = \begin{cases} Y_{(kp)} & \text{if } kp \text{ is integer} \\ Y_{(\lfloor kp+1 \rfloor)} & \text{otherwise} \end{cases}$$

where $\lfloor x \rfloor$ is the greatest integer that is less than or equal to x.

Now the event $Y_{(i)} < \xi_p < Y_{(j)}$ has the binomial probability

$$P(Y_{(i)} < \xi_p < Y_{(j)}) = \sum_{l=i}^{j-1} \binom{k}{l} p^l (1-p)^{k-l}$$

$$\approx \Phi\left(\frac{j-1-kp}{\sqrt{kp(1-p)}}\right) - \Phi\left(\frac{i-1-kp}{\sqrt{kp(1-p)}}\right)$$

where the normal approximation is recommended for $kp \geq 5$ (see Hogg and Craig, 1978, pp. 196–198). To compute a $1-\alpha$ confidence interval for ξ_p, one identifies indices $i < j$ such that $P(Y_{(i)} < \xi_p < Y_{(j)}) \geq 1 - \alpha$. Then $(Y_{(i)}, Y_{(j)})$ is the required interval. Notice that several index pairs can satisfy the last inequality. Normally, one would choose a symmetric range of indices. In this case, the indices would be

$$i = \left\lfloor kp + 1 - \Phi^{-1}\left(1 - \frac{\alpha}{2}\right) \sqrt{kp(1-p)} \right\rfloor$$

and

$$j = \left\lfloor kp + 1 + \Phi^{-1}\left(1 - \frac{\alpha}{2}\right) \sqrt{kp(1-p)} \right\rfloor$$

It should be noted that quantile estimation is more difficult than estimation of the mean because point estimates for quantiles are biased, and significantly larger sample sizes are required to obtain reasonably tight confidence intervals. These problems are much more severe for more extreme quantiles (i.e., for p closer to 1). An introduction to nonparametric interval estimation methods is given in Hogg and Craig (1978, pp. 304–311).

7.3 STEADY-STATE ANALYSIS

Several methods have been developed for the estimation of steady-state system parameters. In this section we review these methods and provide the interested reader with an extensive list of references. We consider primarily the estimation of the steady-state mean μ of a discrete-time output process $\{X_i, i \geq 1\}$. Analogous methods for analyzing continuous-time output data are described in a variety of texts (Bratley et al., 1987; Fishman, 1978b; Law and Kelton, 1991).

7.3.1 Removal of Initialization Bias

One of the hardest problems in steady-state simulations is the removal of the *initialization bias*. Let I be the set of initial conditions for the simulation model and assume that as $n \to \infty$, $P(X_n \leq x | I) \to P(X \leq x)$, where X is the corresponding steady-state random variable. The steady-state mean of the process $\{X_i\}$ is $\mu = \lim_{n \to \infty} \mathrm{E}(X_n | I)$. The problem with the use of the estimator $\overline{X}_n$ for a finite n is that $\mathrm{E}(X_n | I) \neq \mu$ [and thus $\mathrm{E}(\overline{X}_n | I) \neq \mu$].

The most commonly used method for reducing the bias of $\overline{X}_n$ involves identifying an index l, $1 \leq l \leq n-1$, and *truncating* the observations $X_1, \ldots, X_l$. Then the estimator

$$\overline{X}_{n,l} = \frac{1}{n-l} \sum_{i=l+1}^{n} X_i$$

is generally less biased than $\overline{X}_n$ because the initial conditions primarily affect data at the beginning of a run. Several procedures have been proposed for the detection of a cutoff index l (see Fishman, 1972; Gafarian et al., 1978; Kelton and Law, 1983; Schruben, 1982; Schruben et al., 1983; Wilson and Pritsker, 1978a,b). The procedure of Kelton (1989) uses a *pilot* run to estimate the steady-state distribution and starts a production run by sampling from the estimated distribution. More sophisticated truncation rules and initialization bias tests have recently been proposed by Chance and Schruben (1992), Goldsman et al. (1994), and Ockerman (1995).

The graphical procedure of Welch (1981, 1983) is popular due to its generality and ease of implementation. Another graphical method has been proposed by Fishman (1978a,b) in conjunction with the batch means method (see Remark 1 in Section 7.3.4). Welch's method uses k independent replications with the ith replication producing observations $X_{i1}, X_{i2}, \ldots, X_{in}$ and computes the averages:

$$\overline{X}_j = \frac{1}{k} \sum_{i=1}^{k} X_{ij} \qquad j = 1, \ldots, n \tag{14}$$

Then for a given *time window* w, the procedure plots the moving averages

$$\overline{X}_j(w) = \begin{cases} \dfrac{1}{2w+1} \displaystyle\sum_{m=-w}^{w} \overline{X}_{j+m} & w+1 \leq j \leq n-w \\[2ex] \dfrac{1}{2j-1} \displaystyle\sum_{m=-j+1}^{j-1} \overline{X}_{j+m} & 1 \leq j \leq w \end{cases}$$

against j. For example, when $w = 2$,

$$\begin{aligned} \overline{X}_1(2) &= \overline{X}_1 \\ \overline{X}_2(2) &= \tfrac{1}{3}(\overline{X}_1 + \overline{X}_2 + \overline{X}_3) \\ \overline{X}_3(2) &= \tfrac{1}{5}(\overline{X}_1 + \overline{X}_2 + \overline{X}_3 + \overline{X}_4 + \overline{X}_5) \\ &\vdots \\ \overline{X}_{n-2}(2) &= \tfrac{1}{5}(\overline{X}_{n-4} + \overline{X}_{n-3} + \overline{X}_{n-2} + \overline{X}_{n-1} + \overline{X}_n) \end{aligned}$$

If the plot is reasonably smooth, l is chosen to be the value of j beyond which the sequence of moving averages converges. Otherwise, a different time window is chosen and a new plot is drawn. The choice of w is similar to the choice of an interval width for a histogram. Since the truncation index is selected visually, the user will generally have to try several window sizes.

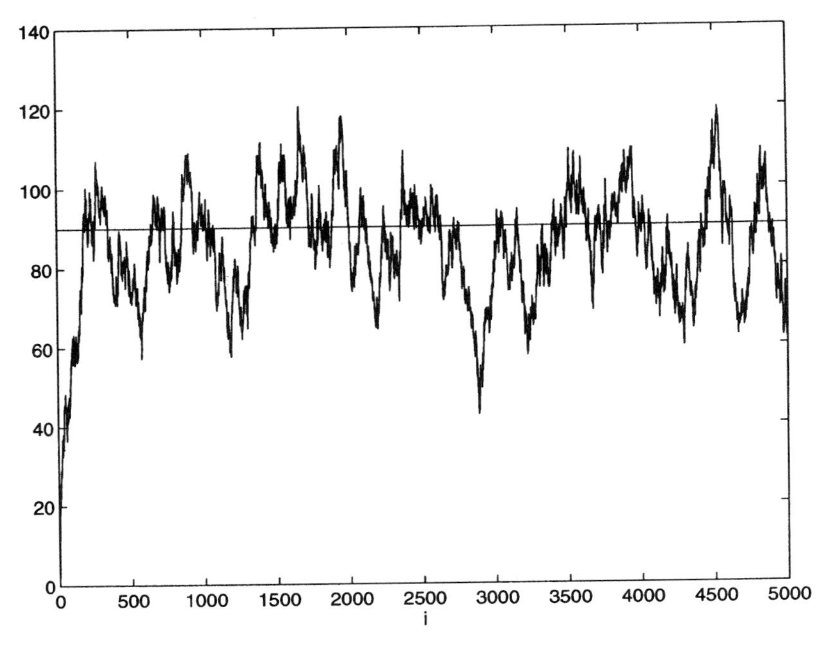

Figure 7.4 Average delay times $\overline{D}_j$ for the first 5000 customers in an M/M/1 queue from 50 independent replications.

Example 7 (The M/M/1 Queue Revisited) Consider an M/M/1 queueing system with interarrival rate $\tau = 0.09$ and service rate $\omega = 0.1$. The limiting mean customer delay is $\mu = 90$. Assume that the system starts empty. 50 independent replications of the first 5000 delays were run by using equation (4). Figure 7.4 depicts the plot of the averages $\overline{D}_j$, $1 \leq j \leq 5000$, computed as in (14).

Figures 7.5 and 7.6 show the plots of the moving averages $\overline{D}_j(w)$, $1 \leq w \leq 5000 - w$, for window sizes $w = 100$ and 500. The transient period is long as the plots of $\overline{D}_j(w)$ first exceed μ for $j \approx 250$. Notice that a large window is required to get a reasonably smooth moving-average plot for this system. In the absence of the horizontal line $\mu = 90$, one would hesitate to choose a truncation index $l \geq 2000$ as all $\overline{D}_j(500)$, $j \geq 2500$, are smaller than $\overline{D}_{2000}(500)$, giving the impression that the actual mean is less than 90. Similarly, the plot of $\overline{D}_j(100)$ is asymmetric with respect to μ with more observations smaller than 90. It should be noted that the method of Welch may be difficult to apply in congested systems with output time series having autocorrelation functions with very long tails.

7.3.2 Replication–Deletion Approach

This approach runs k independent replications, each of length n observations, and uses the method of Welch (1981, 1983) or some other method to discard the first l observations from each run. One then uses the IID sample means

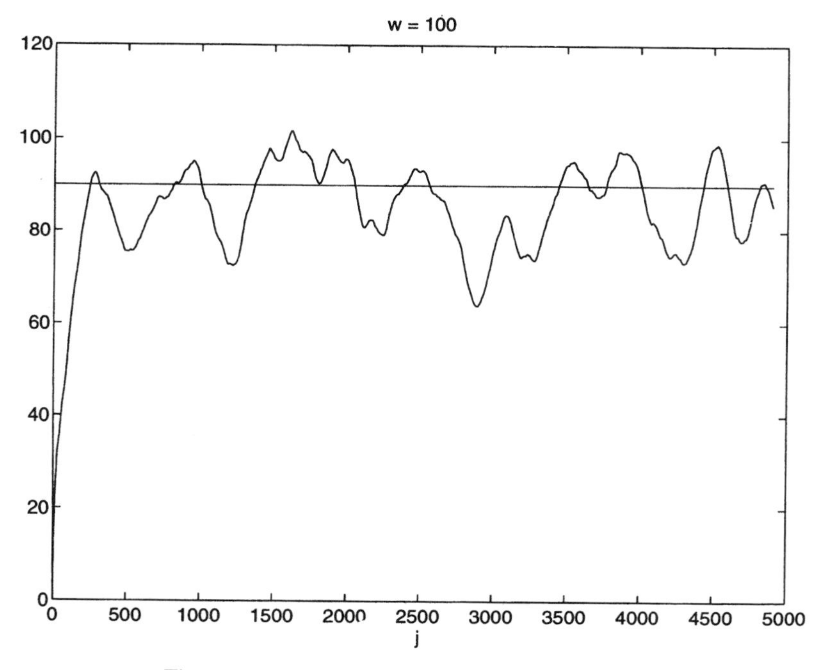

Figure 7.5 Moving averages with window $w = 100$.

$$Y_i = \frac{1}{n-l} \sum_{j=l+1}^{n} X_{ij}$$

to compute point and interval estimators for the steady-state mean μ (see Section 7.2). The method is characterized by its simplicity and generality. The following list contains important observations about l, n, and k.

1. As l increases for fixed n, the "systematic" error in each Y_i due to the initial conditions decreases. However, the sampling error increases because of the smaller number of observations [the variance of Y_i is proportional to $1/(n-l)$].
2. As n increases for fixed l, the systematic and sampling errors in Y_i decrease.
3. The systematic error in the sample means Y_i cannot be reduced by increasing the number of replications k.

Overall, one must be aware that the replication–deletion approach can require a substantial amount of effort to find a "good" truncation index l (as evidenced by Example 7) as well as a large sample size n and a large number of replications to obtain a confidence interval with the required coverage. This approach is also potentially wasteful of data as the truncated portion is removed from each replication. The regenerative method (Section 7.3.3) and the batch means method (Section 7.3.4) seek to overcome these dis-

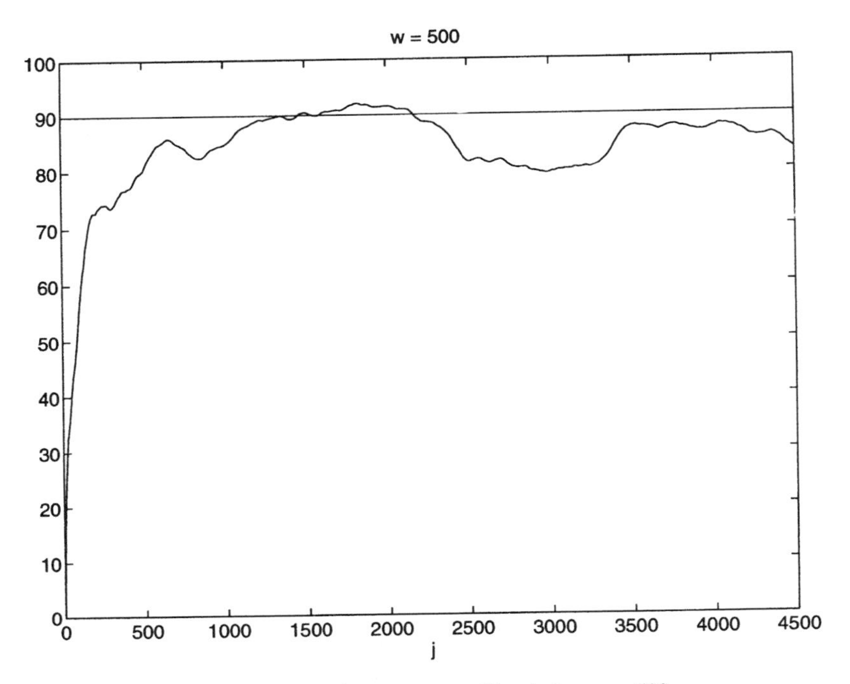

Figure 7.6 Moving averages with window $w = 500$.

advantages. The graph of the batch means (see Remark 1) provides an easy means to assess the effect of the initial conditions at a small incremental cost.

7.3.3 Regenerative Method

This method assumes the identification of time indices at which the process $\{X_i\}$ probabilistically *starts over* and uses these regeneration epochs for obtaining IID random variables that can be used to compute point and interval estimates for the mean μ. As a result, it eliminates the need to detect the length of the initial transient period. The method was proposed by Crane and Iglehart (1974a,b, 1975, 1978) and Fishman (1973, 1974). [For a complete treatment, see Crane and Lemoine (1977).]

More precisely, assume that there are (random) time indices $1 \leq T_1 < T_2 < \cdots$ such that the portion $\{X_{T_i+j}, j \geq 0\}$ has the same distribution for each i and is independent of the portion prior to time T_i. The portion of the process between two successive regeneration epochs is called a *cycle*. Let $Y_i = \sum_{j=T_i}^{T_{i+1}-1} X_j$ and $Z_i = T_{i+1} - T_i$ for $i = 1, 2, \ldots$ and assume that $\text{E}(Z_i) < \infty$. Then the mean μ is given by

$$\mu = \frac{\text{E}(Y_1)}{\text{E}(Z_1)}$$

In addition, the long-run fraction of time the process spends in a set of states E is equal to

$$\lim_{n \to \infty} P(X_n \in E) = \frac{\text{E(total time the process } X \text{ spends in } E \text{ during a cycle)}}{\text{E}(Z_1)}$$

Example 8 [An (s, S) Inventory System] The demand for an item on day i at a store is a nonnegative integer random variable L_i with positive mean. The store uses the following inventory management policy: If the inventory at the end of the day (after the demand is met) is at least s ($s \leq 0$), the store takes no action. If, however, the inventory is less than s, the store orders enough to bring the inventory at the beginning of the next day to level S ($S > s$).

Assume that the inventory is replenished instantaneously (overnight) and let X_i be the level of inventory at the start of day i (before demands occur but after inventory replenishment). Then $\{X_i, i \geq 1\}$ is a regenerative process with return state S. If T_i is the day of the ith return to state S ($T_1 = 1$), the steady-state probability of stockout can be computed by

$$\lim_{n \to \infty} P(X_n < 0) = \frac{\text{E(total time stockout occurs during a cycle)}}{\text{E}(Z_1)}$$

$$= \frac{P(X_{T_2 - 1} < L_{T_2 - 1})}{\text{E}(Z_1)}$$

Estimation of the Mean. Suppose that one simulates the process $\{X_i\}$ over k cycles and collects the observations $Y_1, \ldots, Y_k$ and $Z_1, \ldots, Z_k$. Then

$$\hat{\mu} = \frac{\overline{Y}_k}{\overline{Z}_k}$$

is a strongly consistent, although typically biased for finite k, estimator of μ.

Confidence intervals for μ can be constructed by using the random variables $V_i = Y_i - \mu Z_i$, $i = 1, \ldots, k$, and the central limit theorem. Indeed, $\text{E}(V_i) = 0$ and

$$\sigma^2 = \text{Var}(V_i) = \text{Var}(Y_i) - 2\mu\text{Cov}(Y_i, Z_i) + \mu^2\text{Var}(Z_i)$$

By the central limit theorem $\overline{V}_k/\sqrt{\text{Var}(\overline{V}_k)}$ asymptotically has the standard normal distribution and for large k

$$P\left(\frac{\sqrt{k}|\overline{V}_k|}{\sigma} \leq z_{1-\alpha/2}\right) \approx 1 - \alpha$$

The classical, and most commonly used, approach estimates σ^2 by

$$S_k^2(V) = S_k^2(Y) - 2\hat{\mu}S_k(Y, Z) + \hat{\mu}^2 S_k^2(Z)$$

where

$$S_k(Y, Z) = \frac{1}{k-1} \sum_{i=1}^{k} (Y_i - \overline{Y}_k)(Z_i - \overline{Z}_k)$$

is the sample covariance of Y_i and Z_i, to produce the approximate $1 - \alpha$ confidence interval

$$\hat{\mu} \pm z_{1-\alpha/2} \frac{S_k(V)}{\overline{Z}_k \sqrt{k}}$$

For a small sample size k, Iglehart (1975) showed that the approximate confidence interval

$$\hat{\mu}_J \pm z_{1-\alpha/2} \frac{S_J}{\sqrt{k}}$$

where

$$\hat{\mu}_J = \frac{1}{k} \sum_{i=1}^{k} \theta_i$$

$$\theta_i = k \frac{\overline{Y}_k}{\overline{Z}_k} - (k-1) \frac{\sum_{j \neq i} Y_j}{\sum_{j \neq i} Z_j}$$

is a *jackknife* estimator of μ (with smaller bias than $\hat{\mu}$) and

$$S_J^2 = \frac{1}{k-1} \sum_{i=1}^{k} (\theta_i - \hat{\mu}_J)^2$$

often provides better coverage than the classical regenerative confidence interval. However, its evaluation requires substantial bookkeeping in addition to $O(n^2)$ operations, making its use costly for large sample sizes. The jackknife method also generally increases the width of the confidence interval. A comprehensive review of the jackknife method is given by Efron (1982).

For small sample sizes and bounded Y_i and Z_i, one can also compute the confidence interval in Alexopoulos (1993), which provides superior coverage over confidence intervals based on the central limit theorem at the expense of increased width.

The regenerative method is difficult to apply in practice in simulations that have either no regenerative points or very long cycle lengths. Two classes of systems the regenerative method has been applied to successfully are inventory systems and highly reliable communications systems with repairs.

Quantile Estimation. Iglehart (1976), Moore (1980), and Seila (1982a,b) have proposed methods for computing confidence intervals for quantiles when the output process is regenerative. Seila's method will be presented because it is somewhat simpler to apply than the other methods and has been shown to produce intervals that are as reliable. The method is illustrated by describing the estimation of the p-quantile, say ξ, of the steady-state customer delay distribution in an M/M/1 queueing system that starts empty. In this case, the regeneration epochs are the times the system returns to the empty state.

The method begins by simulating a total of $r \cdot m$ cycles and then partitions the data into r contiguous "batches" with m cycles per batch. For example, the ith batch contains the delay times from the cycles $(i-1)m+1, \ldots, im$. Denote the delay times in this cycle by $D_{i1}, \ldots, D_{iM_i}$ and notice that M_i is generally a random variable. Now

$$\hat{\xi}_i = \begin{cases} D_{i,(M_i p)} & \text{if } M_i p \text{ is integer} \\ D_{i,(\lfloor M_i p+1 \rfloor)} & \text{otherwise} \end{cases}$$

is the quantile estimator for this batch and the overall estimator for q is

$$\bar{\xi} = \frac{1}{r} \sum_{i=1}^{r} \hat{\xi}_i$$

To reduce the bias of the estimators $\hat{\xi}_i$, one can use the jackknife method by forming the estimators

$$\hat{\xi}_{J,i} = 2\hat{\xi}_i - \frac{\hat{\xi}_i^{(1)} + \hat{\xi}_i^{(2)}}{2}$$

where $\hat{\xi}_i^{(1)}$ (respectively, $\hat{\xi}_i^{(2)}$) is the sample quantile computed from the first (respectively, second) half of the m cycles in the ith batch. Then the overall jackknife estimator for ξ is

$$\bar{\xi}_J = \frac{1}{r} \sum_{i=1}^{r} \hat{\xi}_{J,i}$$

and for large m and r,

$$\frac{\bar{\xi}_J - \xi}{S_r(\hat{\xi}_J)/\sqrt{r}} \approx t_{r-1}$$

where $S_r^2(\hat{\xi}_J)$ is the sample variance of the sample quantiles $\hat{\xi}_{J,i}$. The resulting approximate $1-\alpha$ confidence interval for ξ is

$$\bar{\xi}_J \pm t_{r-1,1-\alpha/2} \frac{S_r(\hat{\xi}_J)}{\sqrt{r}} \tag{15}$$

Experiments in Seila (1982a,b) indicate that the confidence interval (15) has better coverage than the confidence interval resulting from the estimator $\bar{\xi}$.

The selection of m and r is an open research problem. Based on practical experience, m should be large enough so that $\mathrm{E}(M_i) \geq 100$ and that $r \geq 10$. The mean $\mathrm{E}(M_i)$ can be estimated by the sample mean $\overline{M}_r$ of the M_i's.

7.3.4 Batch Means Method

The method of batch means is frequently used to estimate the steady-state mean μ or the $\mathrm{Var}(\overline{X}_n)$ and owes its popularity to its simplicity and effectiveness. Original accounts on the method were given by Conway (1963), Fishman (1978a,b), and Law and Carson (1979).

The classical approach divides the output $X_1, \ldots, X_n$ of a long simulation run into a number of contiguous *batches* and uses the sample means of these batches (or *batch means*) to produce point and interval estimators.

To motivate the method, assume temporarily that the process $\{X_i\}$ is weakly stationary with $\lim_{n \to \infty} n\mathrm{Var}(\overline{X}_n) < \infty$ and split the data into k batches, each consisting of b observations. (Assume that $n = kb$.) The ith batch consists of the observations

$$X_{(i-1)b+1}, X_{(i-1)b+2}, \ldots, X_{ib}$$

for $i = 1, 2, \ldots, k$ and the ith batch mean is given by

$$Y_i(b) = \frac{1}{b} \sum_{j=1}^{b} X_{(i-1)b+j}$$

For fixed m, let $\sigma_m^2 = \mathrm{Var}(\overline{X}_m)$. Since the batch means process $\{Y_i(b)\}$ is also weakly stationary, some algebra yields

$$\begin{aligned}\sigma_n^2 = \mathrm{Var}(\overline{X}_n) &= \frac{\sigma_b^2}{k} + \frac{1}{k^2} \sum_{i \neq j} \mathrm{Cov}[Y_i(b), Y_j(b)] \\ &= \frac{\sigma_b^2}{k} \left(1 + \frac{n\sigma_n^2 - b\sigma_b^2}{b\sigma_b^2} \right)\end{aligned}$$

Since $n \geq b$, $(n\sigma_n^2 - b\sigma_b^2)/n\sigma_b^2 \to 0$ as first $n \to \infty$ and then $b \to \infty$. As a result, σ_b^2/k approximates σ_n^2 with error that diminishes as b and n approach infinity. Equivalently, the correlation among the batch means diminishes as b and n approach infinity.

To use the last limiting property, one forms the grand batch mean

$$\overline{Y}_k = \overline{X}_n = \frac{1}{k} \sum_{i=1}^{k} Y_i(b)$$

estimates the σ_b^2 by

$$\hat{V}_B(n,k) = \frac{1}{k-1} \sum_{i=1}^{k} (Y_i(b) - \overline{Y}_k)^2 \tag{16}$$

and computes the approximate $1 - \alpha$ confidence interval for μ:

$$\overline{Y}_k \pm t_{k-1,1-\alpha/2} \sqrt{\frac{\hat{V}_B(n,k)}{k}} \tag{17}$$

The main problem with the application of the batch means method in practice is the choice of the batch size b. If b is small, the batch means $Y_i(b)$ can be highly correlated and the resulting confidence interval will frequently have coverage below the user-specified nominal coverage $1-\alpha$. Alternatively, a large batch size can result in very few batches and potential problems with the application of the central limit theorem to obtain (17).

The method of Fishman (1978a,b) selects the smallest batch size from the set $\{1, 2, 4, \ldots, n/8\}$ that passes the test of independence based on von Neumann's statistic (see "Test for Correlation," below). A variant of this method was proposed by Schriber and Andrews (1979). Mechanic and McKay (1966) choose a batch size from the set $\{16b_1, 64b_1, 256b_1, \ldots, n/25\}$ (usually $b_1 = 1$) and select the batch size that passes an alternative test for independence. The procedure of Law and Carson (1979) starts with 400 batches of size 2. Then it considers sample sizes that double every two iterations until an estimate for lag-1 correlation among 400 batch means becomes smaller than 0.4 and larger than the estimated lag-1 correlation among 200 batch means. The procedure stops when the confidence interval (17) computed with 40 batches satisfies a relative width criterion. Schmeiser (1982) reviews the foregoing procedures and concludes that selecting between 10 and 30 batches should suffice for most simulation experiments. The major drawback of these methods is their inability to yield a consistent variance estimator.

Remark 1 For fixed sample size, a plot of the batch means is a very useful tool for checking the effects of initial conditions, nonnormality of batch means, and existence of correlation between hatch means. For example, consider the M/M/1 queueing system in Example 7. A sample of 100,000 customer delays was generated by means of (4), starting with an empty system. Figure 7.7 shows the plot of the batch means $Y_1(2000)$, $\ldots$, $Y_{50}(2000)$ for batch size $b = 2000$. The first batch mean is small but not the smallest, relaxing one's worries about the effect of the initial transient period. This also hints that $l = 2000$ is a reasonable truncation index for Welch's method. Had the first batch mean been smaller than the other batch means, one can assess the effect of the initial conditions by removing the first batch and comparing the new grand batch mean with the old. Although the plot does not indicate the presence of serious autocorrelation among

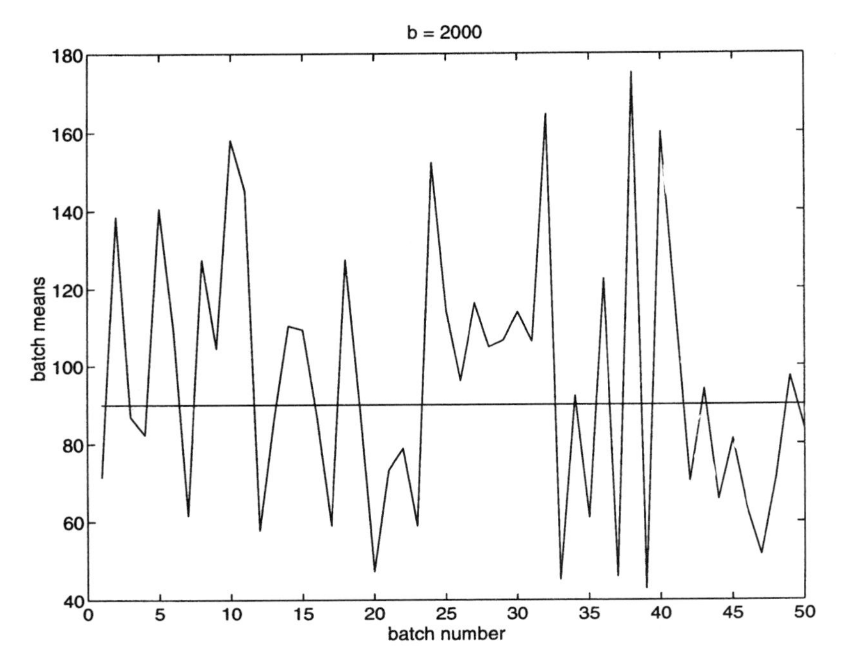

Figure 7.7 Batch means for delay times in an M/M/1 queue.

the batch means, the asymmetric dispersion of the means about the actual mean should make the experimenter concerned about the coverage of the confidence interval (17).

Example 9 shows how an asymptotically optimal batch size can be obtained in special cases.

Example 9 For the AR(1) process in Example 4, Carlstein (1986) showed that

$$\mathrm{Bias}[\hat{V}_B(n,k)] = -\frac{2\rho}{(1-\rho)^3(1+\rho)}\,\frac{1}{b} + o\left(\frac{1}{b}\right) \tag{18}$$

and

$$\mathrm{Var}[\hat{V}_B(n,k)] = \frac{2}{(1-\rho)^4}\,\frac{b}{n} + o\left(\frac{b}{n}\right)$$

where $o(h)$ is a function such that $\lim_{h \to 0} o(h)/h = 0$. Then the batch size that minimizes the asymptotic (as $n \to \infty$ and $k \to \infty$) mean-squared error $\mathrm{MSE}[\hat{V}_B(n,k)] = \mathrm{Bias}^2[\hat{V}_B(n,k)] + \mathrm{Var}[\hat{V}_B(n,k)]$ is

$$b_0 = \left(\frac{2|\rho|}{1-\rho^2}\right)^{2/3} n^{1/3} \tag{19}$$

Clearly, the optimal batch size increases with the absolute value of the correlation ρ between successive observations.

The reader should keep in mind that the optimal batch size may differ substantially from (19) for a finite sample size (e.g., Song and Schmeiser, 1995), and the model generally does not apply to the analysis of data from queueing systems. Furthermore, it is not evident that this strategy for batch size selection allows the space and time complexities achievable by the LBATCH and ABATCH strategies described below for generating an assessment of the stability of the variance of the sample mean.

Consistent Estimation Batch Means Methods. *Consistent estimation* batch means methods assume the existence of a parameter σ_∞^2 (the time-average variance of the process $\{X_i\}$) such that a central limit theorem holds:

$$\sqrt{n}(\overline{X}_n - \mu) \xrightarrow{\mathcal{D}} \sigma_\infty N(0,1) \qquad \text{as } n \to \infty \tag{20}$$

and aim at constructing a consistent estimator for σ_∞^2 and an asymptotically valid confidence interval for μ. [Notice that the X_i's in (20) need not be IID]. Consistent estimation methods are often preferable to methods that "cancel" σ_∞^2 (see Glynn and Iglehart, 1990) because (1) the expectation and variance of the half-width of the confidence interval resulting from (20) is asymptotically smaller for consistent estimation methods, and (2) under reasonable assumptions $n\text{Var}(\overline{X}_n) \to \sigma_\infty^2$ as $n \to \infty$.

Example 10 The delay process $\{D_i\}$ of a stationary M/M/1 system has

$$\sigma_\infty^2 = \frac{\nu}{\omega^2(1-\nu)^4}(\nu^3 - 4\nu^2 + 5\nu + 2)$$

(Blomqvist, 1967) whereas a stationary AR(1) process has $\sigma_\infty^2 = (1+\rho)/(1-\rho)$.

Chien et al. (1997) considered stationary processes and, under quite general moment and sample path conditions, showed that as both $b \to \infty$ and $k \to \infty$, $\text{E}[b\hat{V}_B(n,k)] \to \sigma_\infty^2$, $k\text{Var}[\hat{V}_B(n,k)] \to 2\sigma_\infty^2$, and $\text{MSE}[b\hat{V}_B(n,k)] \to 0$. Notice that the last limiting property of $b\hat{V}_B(n,k)$ differs from consistency.

The limiting result (20) is implied under the following two assumptions.

Assumption of Weak Approximation (AWA) There exist finite constants μ and $\sigma_\infty > 0$ such that

$$\frac{n(\overline{X}_n - \mu)}{\sigma_\infty} \xrightarrow{\mathcal{D}} W(n) \qquad \text{as } n \to \infty$$

Assumption of Strong Approximation (ASA) There exist finite constants μ, $\sigma_\infty > 0$, $\lambda \in (0, \frac{1}{2}]$, and a finite random variable C such that, with probability 1,

$$|n(\overline{X}_n - \mu) - \sigma_\infty W(n)| \le Cn^{1/2-\lambda} \qquad \text{as } n \to \infty$$

Both AWA and ASA state that the process $\{n(\overline{X}_n - \mu)/\sigma_\infty\}$ is close to a standard Brownian motion. However, the stronger ASA addresses the convergence rate of (20).

The ASA is not restrictive, as it holds under relatively weak assumptions for a variety of stochastic processes, including Markov chains, regenerative processes, and certain queueing systems (see Damerdji, 1994a, for details). The constant λ is closer to $\frac{1}{2}$ for processes having little autocorrelation, while it is closer to zero for processes with high autocorrelation. In the former case the "distance" between the processes $\{n(\overline{X}_n - \mu)/\sigma_\infty\}$ and $\{W(n)\}$ "does not grow" as n increases.

Batching Rules. Fishman (1996, Chap. 6) and Fishman and Yarberry (1997) present a thorough discussion of batching rules. Our account in this section and below in "Implementing the ABATCH Strategy" parallels their development. Both references contain detailed instructions for obtaining FORTRAN, C, and SIMSCRIPT II.5 implementations for various platforms via anonymous ftp from the site `ftp.or.unc.edu`.

The discussion prior to the derivation of (17) suggests that fixing the number of batches and letting the batch size grow as $n \to \infty$ ensures that $\sigma_b^2/k \to \sigma_n^2$. This motivates the following rule.

Fixed Number of Batches (FNB) Rule Fix the number of batches at k. For sample size n, use batch size $b_n = \lfloor n/k \rfloor$.

The FNB rule along with AWA lead to the following result.

Theorem 1 (Glynn and Iglehart (1990) If $\{X_i\}$ satisfies AWA, then as $n \to \infty$, $\overline{X}_n \xrightarrow{\mathcal{P}} \mu$ and (20) holds. Furthermore, if k is constant and $\{b_n, n \geq 1\}$ is a sequence of batch sizes such that $b_n \to \infty$ as $n \to \infty$, then

$$\frac{\overline{X}_n - \mu}{\sqrt{\hat{V}_B(n,k)/k}} \xrightarrow{\mathcal{D}} t_{k-1} \qquad \text{as } n \to \infty$$

The primary implication of Theorem 1 is that (17) is an asymptotically valid confidence interval for μ. Unfortunately, the FNB rule has two major limitations: (1) $b_n \hat{V}_B(n,k)$ is not a consistent estimator of σ_∞^2 [therefore, the confidence interval (17) tends to be wider than the interval a consistent estimation method would produce], and (2) after some algebra (see Fishman, 1996, Chap. 6) one has

$$\lim_{n \to \infty} \frac{\operatorname{Var}\left[\sqrt{\hat{V}_B(n,k)/k}\right]}{\operatorname{Var}(\overline{X}_n)} = \frac{1}{2(k-1)} - \frac{1}{8(k-1)^2} - \frac{1}{16(k-1)^3} - \cdots$$

so that statistical fluctuations in the half-width of the confidence interval (17) do not diminish relative to statistical fluctuation in the sample mean.

The following theorem proposes batching assumptions which along with ASA yield a strongly consistent estimator for σ_∞^2.

Theorem 2 (Damerdji, 1994a) If $\{X_i\}$ satisfies ASA, then $\overline{X}_n \xrightarrow{\text{a.s.}} \mu$ as $n \to \infty$. Furthermore, suppose that $\{(b_n, k_n), n \geq 1\}$ is a batching sequence satisfying

(A.1) $b_n \to \infty$ and $k_n \to \infty$ monotonically as $n \to \infty$.

(A.2) $b_n^{-1} n^{1-2\lambda} \ln n \to 0$ as $n \to \infty$.

(A.3) There exists a finite positive integer m such that

$$\sum_{n=1}^{\infty} \left(\frac{b_n}{n} \right)^m < \infty$$

Then, as $n \to \infty$,

$$b_n \hat{V}_B(n, k_n) \xrightarrow{\text{a.s.}} \sigma_\infty^2 \tag{21}$$

and

$$Z_{k_n} = \frac{\overline{X}_n - \mu}{\sqrt{\hat{V}_B(n, k_n)/k_n}} \xrightarrow{\mathcal{D}} N(0,1) \tag{22}$$

Equation (22) implies that

$$\overline{X}_n \pm t_{k_n - 1, 1 - \alpha/2} \sqrt{\frac{\hat{V}_B(n, k_n)}{k_n}}$$

is an asymptotically valid $1 - \alpha$ confidence interval for μ.

Theorem 2 motivates the consideration of batch sizes of the form $b_n = \lfloor n^\theta \rfloor$, $0 < \theta < 1$. In this case one can show that the conditions (A.1) to (A.3) are met if $\theta \in (1 - 2\lambda, 1)$. In particular, the assignment $\theta = \frac{1}{2}$ and the SQRT rule below are valid if $\frac{1}{4} < \lambda < \frac{1}{2}$. Notice that the last inequality is violated by processes having high autocorrelation ($\lambda \approx 0$).

Square Root (SQRT) Rule For sample size n, use batch size $b_n = \lfloor \sqrt{n} \rfloor$ and number of batches $k_n = \lfloor \sqrt{n} \rfloor$.

Under some additional moment conditions, Chien (1989) showed that the convergence of Z_{k_n} to the $N(0,1)$ distribution is fastest if b_n and k_n grow proportionally to $\sqrt{n}$. Unfortunately, in practice the SQRT rule tends to seriously underestimate the $\text{Var}(\overline{X}_n)$ for fixed n.

Example 11 (The M/M/1 Queue Revisited) Consider an M/M/1 queueing system with interarrival rate $\tau = 0.9$ and service rate $\omega = 1$, and assume that the system starts empty. Table 7.2 contains performance statistics for 0.95 confidence intervals for the steady-state mean customer delay $\mu = 0.9/[1 \times (1 - 0.9)] = 9$. The confidence intervals resulted from 500 independent replications. Within each replication, the delays were generated by means of (4). The FNB rule used 16 batches and batch sizes 2^m, $m \geq 0$. The SQRT rule started with batch size $b_1 = 1$ and number of batches $k_1 = 8$, and computed confidence intervals with batch sizes

TABLE 7.2 Performance Statistics for the FNB and SQRT Rules on 0.95 Confidence Intervals for the Mean Customer Delay in an M/M/1 Queue with Utilization $\nu = 0.9$

	FNB Rule		SQRT Rule	
$\log_2 n$	Coverage	Average Half-width	Coverage	Average Half-width
10	0.544	3.244	0.326	1.694
11	0.640	3.506	0.366	1.665
12	0.746	3.304	0.414	1.437
13	0.798	2.963	0.466	1.271
14	0.838	2.435	0.498	1.063
15	0.880	1.901	0.604	0.904
16	0.912	1.437	0.664	0.738
17	0.944	1.053	0.778	0.599
18	0.934	0.756	0.810	0.471
19	0.950	0.541	0.854	0.369
20	0.940	0.385	0.858	0.283

$$b_l = 2^{(l-1)/2} \times \begin{cases} b_1 & \text{if } l \text{ is odd} \\ \dfrac{3}{(2\sqrt{2})} & \text{otherwise} \end{cases}$$

and numbers of batches

$$k_l = 2^{(l-1)/2} \times \begin{cases} k_1 & \text{if } l \text{ is odd} \\ \dfrac{11}{\sqrt{2}} & \text{otherwise} \end{cases}$$

The resulting sample sizes $n_l = k_l b_l$ are roughly powers of 2 (see the section below, "Implementing the ABATCH Strategy" for details).

The second and fourth columns contain the estimated coverage probabilities of the confidence intervals produced by the FNB rule and the SQRT rule, respectively. The third and fifth columns display the respective average interval half-widths. Specifically, for sample size $n \approx 2^{17} = 131{,}072$, roughly 94% of the confidence intervals resulting from the FNB rule contained μ, whereas only 78% of the confidence intervals resulting from the SQRT rule contained μ. However, the latter intervals were 43% narrower. Experiments by Fishman and Yarberry showed that the disparity in coverage between the two rules grows with increasing traffic intensity $\nu = \tau/\omega$.

With the contrasts between the FNB and SQRT rules in mind, Fishman and Yarberry proposed two strategies that dynamically shift between the two rules. Both strategies perform "interim reviews" and compute confidence intervals at times $n_l \approx n_1 2^{l-1}$, $l = 1, 2, \ldots$.

LBATCH Strategy At time n_l, if an hypothesis test detects autocorrelation between the batch means, the batching for the next review is determined by the FNB rule. If the test fails to detect correlation, all future reviews omit the test and employ the SQRT rule.

ABATCH Strategy If at time n_l the hypothesis test detects correlation between the batch means, the next review employs the FNB rule. If the test fails to detect correlation, the next review employs the SQRT rule.

Both strategies LBATCH and ABATCH yield random sequences of batch sizes. Under relatively mild assumptions, these sequences imply convergence results analogous to (21) and (22) (see Fishman, 1996; Fishman and Yarberry, 1997).

Test for Correlation. We review a test for the hypothesis

$$H_0 : \text{the batch means } Y_1(b), \ldots, Y_k(b) \text{ are uncorrelated}$$

The test is due to von Neumann (1941) and is very effective when the number of batches k is small.

Assume that the process $\{X_i\}$ is weakly stationary and let

$$\rho_l(b) = \mathrm{Corr}[Y_i(b), Y_{i+l}(b)] \qquad l = 0, 1, \ldots$$

be the autocorrelation function of the batch means process. The von Neumann test statistic for H_0 is

$$\Gamma_k = \sqrt{\frac{k^2 - 1}{k - 2}} \left[\hat{\rho}_1(b) + \frac{[Y_1(b) - \overline{X}_n]^2 + [Y_k(b) - \overline{X}_n]^2}{2 \sum_{i=1}^{k} [Y_i(b) - \overline{X}_n]^2} \right] \tag{23}$$

where

$$\hat{\rho}_1(b) = \frac{\sum_{i=1}^{k-1} [Y_i(b) - \overline{X}_n][Y_{i+1}(b) - \overline{X}_n]}{\sum_{i=1}^{k} [Y_i(b) - \overline{X}_n]^2}$$

is an estimator for the lag-1 autocorrelation $\rho_1(b)$. The rightmost ratio in equation (23) carries more weight when k is small, but it approaches zero as $k \to \infty$.

Suppose that H_0 is true. If the batch means are IID normal, the distribution of Γ_k is very close to $N(0, 1)$ for as few as $k = 8$ batches (von Neumann, 1941; Young, 1941, Table 1). On the other hand, if the batch means are IID but nonnormal, the first four cumulants of Γ_k converge to the respective cumulants of the $N(0, 1)$ distribution as $k \to \infty$. This discussion suggests the approximation

$$\Gamma_k \approx N(0, 1)$$

for large b (the batch means become approximately normal) or large k (by the central limit theorem).

If $\{X_i\}$ has a monotone-decreasing autocorrelation function (e.g., the delay process for an M/M/1 queueing system), the batch means process also has a monotone-decreasing autocorrelation function. As a result, one rejects H_0 at level β if

$$\Gamma_k > z_{1-\beta}$$

Alternatively, if $\{X_i\}$ has an autocorrelation function with damped harmonic behavior around the zero axis [e.g., the AR(1) process with $\rho < 0$], the test can lead to erroneous conclusions. In this case, repeated testing under the ABATCH strategy reduces this possibility.

The p-value of the test, $1 - \Phi(\Gamma_k)$, is the largest value of the type I error $\beta = P$(reject $H_0 | H_0$ is true) for which H_0 is rejected given the observed value of Γ_k. Equivalently, H_0 is accepted if the p-value is larger than β. Hence, a p-value close to zero implies low credibility for H_0. The plot of the p-value versus the batch size is a useful graphical device.

Implementing the ABATCH Strategy. Next we present a pseudocode for implementing the ABATCH strategy. Implementation of the LBATCH strategy is discussed in short after the pseudocode.

To understand the role of the hypothesis test in the LBATCH and ABATCH algorithms, define the random variables

$$R_l = \begin{cases} 1 & \text{if } H_0 \text{ is rejected on review } l \\ 0 & \text{otherwise} \end{cases}$$

and

$$\overline{R}_l = \frac{R_1 + \cdots + R_l}{l} = \text{fraction of rejected tests for } H_0 \text{ on reviews } 1, \ldots, l$$

A sufficient condition for strong consistency [equation (21)] and asymptotic normality [equation (22)] is $\beta_0 > 1 - 4\lambda$ [or $\lambda > (1 - \beta_0)/4$], where $\beta_0 = \lim_{l \to \infty} \overline{R}_l$ is the long-run fraction of rejections. In practice, β_0 differs from but is expected to be close to the type I error β. Clearly, $\lambda > \frac{1}{4}$ guarantees (21) and (22) regardless of β_0. However, β_0 plays a role when $\lambda \leq \frac{1}{4}$. Specifically, for β_0 equal to 0.05 or 0.10, the lower bound $(1 - \beta_0)/4$ becomes 0.2375 or 0.2225, respectively, a small reduction from $\frac{1}{4}$.

On review l, the ABATCH strategy induces batch size

$$b_l = 2^{(l-1)(1+\overline{R}_{l-1})/2} \times \begin{cases} b_1 & \text{if } (l-1)(1+\overline{R}_{l-1}) \text{ is even} \\ \dfrac{\tilde{b}_1}{\sqrt{2}} & \text{otherwise} \end{cases}$$

where

$$\tilde{b}_1 = \begin{cases} \dfrac{3}{2} & \text{if } b_1 = 1 \\ \lfloor \sqrt{2} b_1 + 0.5 \rfloor & \text{if } b_1 > 1 \end{cases}$$

and number of batches

$$k_l = 2^{(l-1)(1-\overline{R}_{l-1})/2} \times \begin{cases} k_1 & \text{if } (l-1)(1-\overline{R}_{l-1}) \text{ is even} \\ \dfrac{\tilde{k}_1}{\sqrt{2}} & \text{otherwise} \end{cases}$$

where $\tilde{k}_1 = \lfloor \sqrt{2} k_1 + 0.5 \rfloor$.

The resulting sample sizes are

$$n_l = k_l b_l = \begin{cases} 2^{l-1} k_1 b_1 & \text{if } (l-1)(1+\overline{R}_{l-1}) \text{ is even} \\ 2^{l-2} \tilde{k}_1 \tilde{b}_1 & \text{otherwise} \end{cases}$$

and the definitions for $\tilde{b}_1$ and $\tilde{k}_1$ guarantee that if H_0 is never rejected, then both b_l and k_l grow approximately as $\sqrt{2}$ with l (i.e., they follow the SQRT rule).

Suppose that $L + 1$ reviews are performed. The final implementation issue for the ABATCH strategy is the relative difference between the potential terminal sample sizes

$$\Delta(b_1, k_1) = \frac{|2^L k_1 b_1 - 2^{L-1} \tilde{k}_1 \tilde{b}_1|}{2^L k_1 b_1} = \frac{|2k_1 b_1 - \tilde{k}_1 \tilde{b}_1|}{2k_1 b_1}.$$

This quantity is minimized (i.e., the final sample size is deterministic) when $2k_1 b_1 = \tilde{k}_1 \tilde{b}_1$. Pairs (b_1, k_1), with small b_1, satisfying the last equality are (1, 3), (1, 6), (2, 3), and (2, 6). Unfortunately, the condition $2k_1 b_1 = \tilde{k}_1 \tilde{b}_1$ excludes several practical choices for b_1 and k_1, such as $b_1 = 1$ (to test the original sample for independence) and $8 \le k_1 \le 10^5$. Overall, $\Delta(b_1, k_1)$ remains small for numerous choices of b_1 and k_1. For example, $b_1 = 1$ and $8 \le k_1 \le 32$ ensure that $\Delta(b_1, k_1) \le 0.078$.

Algorithm ABATCH

Source: Fishman (1996, Chap. 6) and Fishman and Yarberry (1997). Minor notational changes have been made.

Input: Minimal number of batches k_1, minimal batch size b_1, desired sample size $n = 2^L k_1 b_1$ (L is a positive integer), and confidence level $1 - \alpha$.

Output: Sequences of point estimates and confidence intervals for sample sizes $N \le n$.

Method:

1. $b \leftarrow b_1$ and $k \leftarrow k_1$.
2. If $b_1 = 1$, $\tilde{b}_1 \leftarrow \frac{3}{2}$; otherwise, $\tilde{b}_1 \leftarrow \lfloor \sqrt{2} b_1 + 0.5 \rfloor$.
3. $\tilde{k}_1 \leftarrow \lfloor \sqrt{2} k_1 + 0.5 \rfloor$.

4. $g \leftarrow \tilde{b}_1/b_1$ and $f \leftarrow \tilde{k}_1/k_1$.
5. $i \leftarrow 0$.
6. $\tilde{n} \leftarrow 2^{L-1}\tilde{k}_1\tilde{b}_1$.

Until $N = n$ or $N = \tilde{n}$:

7. $N \leftarrow kb$.
8. Randomly generate $X_{i+1}, \ldots, X_N$.

Compute:

9. The batch means $Y_1(b), \ldots, Y_k(b)$.
10. $\overline{X}_N$ as a point estimate of μ.
11. The sample variance $\hat{V}_B$ of the batch means.
12. The half-width $\delta = t_{k-1, 1-\alpha/2}\sqrt{\hat{V}_B/k}$ of the confidence interval (17).
13. Print $N, k, b, \overline{X}_N, \overline{X}_N - \delta, \overline{X}_N + \delta, \hat{V}_B$.
14. $i \leftarrow N$.
15. Test $H_0 : Y_1(b), \ldots, Y_k(b)$ are uncorrelated. Print the p-value of this test.
16. If H_0 is rejected, $b \leftarrow 2b$. (FNB rule)

If H_0 is accepted:

17. If $b = 1$, $b \leftarrow 2$. (FNB rule)

Otherwise: (SQRT rule)

18. $b \leftarrow bg$ and $k \leftarrow kf$.
19. If $g = \tilde{b}_1/b_1$, $g \leftarrow 2b_1/\tilde{b}_1$ and $f \leftarrow 2k_1/\tilde{k}_1$; otherwise, $g \leftarrow \tilde{b}_1/b_1$ and $f \leftarrow \tilde{k}_1/k_1$.

Remark 2 Algorithm ABATCH requires $O(n)$ time and $O(\log_2 n)$ space. For details, see Yarberry (1993, Chap. 5).

Remark 3 The implementation of strategy LBATCH is simpler. Once H_0 is accepted in step 15, steps 17 to 19 are ignored for the remainder of the execution.

Tests for the Batching Rules. The experiments in Examples 12 to 14 compare the FNB rule and the LBATCH and ABATCH strategies by means of three queueing systems with traffic intensity $\nu = 0.9$. Each system starts empty and has a first come, first served discipline. Each experiment computed 0.95 confidence intervals for the long-run mean customer delay from 500 independent replications. The FNB rule relied on 16 batches, whereas the LBATCH and ABATCH strategies started with $k_1 = 8$ batches of size $b_1 = 1$ and used type I error $\beta = 0.1$ for H_0.

Example 12 (Example 11 Continued) The entries of Tables 7.2 and 7.3 indicate that the ABATCH strategy comes closer to the FNB rule's superior coverage with shorter confidence intervals.

TABLE 7.3 Performance Statistics for the LBATCH and ABATCH Strategies on 0.95 Confidence Intervals for the Mean Customer Delay in an M/M/1 Queue with Utilization $\nu = 0.9$

	LBATCH Strategy		ABATCH Strategy		
$\log_2 n$	Coverage	Average Half-width	Rejection Proportion	Coverage	Average Half-width
10	0.398	2.085	0.622	0.562	3.384
11	0.420	1.992	0.552	0.632	3.450
12	0.464	1.693	0.458	0.712	3.100
13	0.518	1.477	0.394	0.760	2.686
14	0.562	1.227	0.340	0.816	2.168
15	0.652	1.029	0.266	0.850	1.708
16	0.714	0.834	0.206	0.902	1.296
17	0.808	0.663	0.200	0.932	0.955
18	0.852	0.513	0.176	0.938	0.688
19	0.866	0.395	0.154	0.930	0.493
20	0.876	0.298	0.156	0.936	0.353

Example 13 (An M/G/1 Queue) Consider an M/G/1 queueing system with IID interarrival times from the exponential distribution with parameter $\tau = 0.9$ and IID service times S_i from the hyperexponential distribution with density function

$$f(x) = 0.9\left(\frac{1}{0.5}\, e^{-x/0.5}\right) + 0.1\left(\frac{1}{5.5}\, e^{-x/5.5}\right) \qquad x \geq 0$$

This distribution applies when customers are classified into two types, 1 and 2, with respective probabilities 0.9 and 0.1; type 1 customers have exponential service times with mean 0.5, and type 2 customers have exponential service times with mean 5.5. The service times have mean $\mathrm{E}(S) = 0.9(0.5) + 0.1(5.5) = 1$, second moment $\mathrm{E}(S^2) = 0.9 \times 2(0.5^2) + 0.1 \times 2(5.5^2) = 6.5$, and coefficient of variation

$$\frac{\sqrt{\mathrm{Var}(S)}}{\mathrm{E}(S)} = 2.739$$

which is larger than 1, the coefficient of variation of the exponential distribution. Then the traffic intensity is $\nu = \tau\mathrm{E}(S) = 0.9$.

The long-run mean delay time in queue is given by the Pollaczek–Khintchine formula (Ross, 1993, Chap. 8)

$$\mu = \lim_{i \to \infty} \mathrm{E}(D_i) = \frac{\tau \mathrm{E}(S^2)}{2(1-\nu)} = 29.25 \qquad (24)$$

Notice that the M/M/1 system in Example 11 with the same arrival rate and traffic intensity has a much smaller long-run mean delay time.

Table 7.4 displays the results of this experiment. As n increases, the conservative

TABLE 7.4 Performance Statistics for the FNB, LBATCH, and ABATCH Algorithms on 0.95 Confidence Intervals for the Mean Customer Delay in an M/G/1 Queue with Hyperexponential Service Times and Utilization $\nu = 0.9$

	FNB Rule		LBATCH Strategy		ABATCH Strategy		
$\log_2 n$	Coverage	Average Half-width	Coverage	Average Half-width	Rejection Proportion	Coverage	Average Half-width
10	0.324	9.251	0.204	5.865	0.742	0.356	10.305
11	0.420	10.756	0.254	6.962	0.724	0.436	11.426
12	0.560	11.482	0.294	5.552	0.614	0.566	11.635
13	0.688	11.711	0.354	5.083	0.584	0.652	11.166
14	0.774	11.021	0.392	4.418	0.502	0.746	10.147
15	0.832	9.715	0.452	3.863	0.386	0.794	8.658
16	0.886	8.053	0.540	3.215	0.344	0.856	7.057
17	0.908	6.208	0.620	2.678	0.300	0.898	5.483
18	0.900	4.542	0.632	2.178	0.202	0.896	4.090
19	0.914	3.262	0.694	1.761	0.148	0.900	2.997
20	0.934	2.339	0.748	1.387	0.134	0.924	2.145
21	0.942	1.669	0.806	1.083	0.118	0.926	1.525

ABATCH strategy produces 0.95 confidence intervals for μ that are roughly 50 to 100% wider than the respective confidence intervals produced by the LBATCH strategy but have coverage rates that are acceptably close to 0.95 for substantially smaller sample sizes (as small as $2^{17} = 131{,}072$).

Example 14 (An M/D/1 Queue) Consider an M/G/1 queueing system with IID interarrival times from the exponential distribution with parameter $\tau = 0.9$ and fixed unit service times. Then, by (24), the long-run mean delay time in queue is $\mu = 4.5$.

The results of this experiment are contained in Table 7.5. As in Examples 12 and 13, the performance of the ABATCH strategy makes it an attractive compromise between the "extreme" FNB and SQRT rules.

TABLE 7.5 Performance Statistics for the FNB, LBATCH, and ABATCH Algorithms on 0.95 Confidence Intervals for the Mean Customer Delay in an M/D/1 Queue with Unit Service Times and Utilization $\nu = 0.9$

	FNB Rule		LBATCH Strategy		ABATCH Strategy		
$\log_2 n$	Coverage	Average Half-width	Coverage	Average Half-width	Rejection Proportion	Coverage	Average Half-width
10	0.618	1.626	0.460	1.062	0.552	0.616	1.631
11	0.740	1.620	0.548	0.962	0.504	0.720	1.538
12	0.826	1.529	0.598	0.842	0.396	0.788	1.391
13	0.858	1.222	0.648	0.686	0.334	0.842	1.101
14	0.878	0.946	0.696	0.556	0.274	0.858	0.852
15	0.908	0.707	0.794	0.445	0.220	0.884	0.632
16	0.920	0.517	0.808	0.351	0.222	0.924	0.472
17	0.946	0.375	0.862	0.271	0.140	0.942	0.343

TABLE 7.6 Performance Statistics for the FNB, LBATCH, and ABATCH Algorithms on 0.95 Confidence Intervals for the Mean Customer $\mu = 0$ of the Stationary AR(1) Process $X_i = -0.9X_{i-1} + Z_i$

	FNB Rule		LBATCH Strategy		ABATCH Strategy		
$\log_2 n$	Coverage	Average Half-width	Coverage	Average Half-width	Rejection Proportion	Coverage	Average Half-width
5	0.954	0.0515	1.000	0.1217	0.020	1.000	0.1153
6	0.978	0.0364	0.978	0.0364	0.102	0.980	0.0367
7	0.980	0.0248	0.980	0.0244	0.060	0.980	0.0246
8	0.974	0.0164	0.982	0.0166	0.038	0.980	0.0167
9	0.960	0.0108	0.972	0.0111	0.018	0.966	0.0111
10	0.966	0.0071	0.984	0.0076	0.022	0.980	0.0075
11	0.944	0.0048	0.976	0.0051	0.012	0.978	0.0050
12	0.962	0.0034	0.982	0.0035	0.014	0.984	0.0035
13	0.938	0.0023	0.964	0.0024	0.012	0.962	0.0023
14	0.942	0.0017	0.960	0.0016	0.020	0.962	0.0016

Example 15 tests the FNB, LBATCH, and ABATCH methods by means of an AR(1) process.

Example 15 [The AR(1) Process Revisited] Consider the stationary AR(1) process $X_i = -0.9X_{i-1} + Z_i$ with mean 0 (see Example 4). The autocorrelation function $\rho_j = (-0.9)^j$, $j \geq 0$, of this process oscillates around the zero axis and the time-average process variance is $\sigma_\infty^2 = (1 - 0.9)/(1 + 0.9) = 0.053$.

The entries of Table 7.6 were obtained from 500 independent replications. The FNB rule used 16 batches and the type I error for H_0 was $\beta = 0.1$. The 0.95 confidence intervals for μ produced by the three methods have roughly equal half-widths and coverages. In fact, almost all coverage estimates are greater than the nominal coverage 0.95. This behavior is due to the fact that $b\text{Var}(\hat{V}_B(n,k))$ tends to overestimate σ_∞^2 [the coefficient of $1/b$ in equation (18) is $2.624 > 0$].

From equation (19), the batch size that minimizes $\text{MSE}(\hat{V}_B(n,k))$ is $b_0 = 113.71$. Five hundred independent replications with 144 batches of size 114 (sample size 16,416) produced 0.95 confidence intervals with estimated coverage 0.958 and average half-width 0.0016—not a substantial improvement over the statistics in the last row of Table 7.6 (for sample size roughly equal to $2^{14} = 16{,}384$).

Based on Examples 12 to 14 (Tables 7.3 to 7.5), the ABATCH strategy appears to provide approximately 10% reduction in confidence interval width over the FNB rule for sample sizes large enough to achieve the nominal coverage probability.

Overlapping Batch Means. An interesting variation of the traditional batch means method is the method of *overlapping* batch means (OBM) proposed by Meketon and Schmeiser (1984). For given batch size b, this method uses all $n - b + 1$ overlapping batches to estimate μ and $\text{Var}(\overline{X}_n)$. The first batch consists of observations $X_1, \ldots, X_b$, the second batch consists of $X_2, \ldots, X_{b+1}$, and so on. The OBM estimator of μ is

$$\overline{Y}_O = \frac{1}{n-b+1} \sum_{i=1}^{n-b+1} Y_i(b)$$

where

$$Y_i(b) = \frac{1}{b} \sum_{j=i}^{i+b-1} X_j \qquad i = 1, \ldots, n-b+1$$

are the respective batch means, and has sample variance

$$\hat{V}_O = \frac{1}{n-b} \sum_{i=1}^{n-b+1} [Y_i(b) - \overline{Y}_O]^2$$

The following list contains properties of the estimators $\overline{Y}_O$ and $\hat{V}_O$:

1. The OBM estimator is a weighted average of nonoverlapping batch means estimators.
2. Asymptotically (as n, $b \to \infty$ and $b/n \to 0$), the OBM variance estimator $\hat{V}_O$ and the nonoverlapping batch means variance estimator $\hat{V}_B \equiv \hat{V}_B(n, k)$ have the same expectation. Furthermore,

$$\frac{\text{Var}(\hat{V}_O)}{\text{Var}(\hat{V}_B)} \to \frac{2}{3}$$

   In words, the asymptotic ratio of the mean-squared error of $\text{Var}(\hat{V}_O)$ to the mean-squared error of $\text{Var}(\hat{V}_B)$ is equal to $\frac{2}{3}$ (Meketon and Schmeiser, 1984).
3. The behavior of $\text{Var}(\hat{V}_O)$ appears to be less sensitive to the choice of the batch size than the behavior of $\text{Var}(\hat{V}_B)$ (Song and Schmeiser, 1995, Table 1).
4. If $\{X_i\}$ satisfies ASA and $\{(b_n, k_n), n \geq 1\}$ is a sequence that satisfies the assumptions (A.1)–(A.3) in Theorem 2 and

$$\frac{b_n^2}{n} \to 0 \qquad \text{as } n \to \infty$$

   then (Damerdji, 1994a)

$$b_n \hat{V}_O \xrightarrow{\text{a.s.}} \sigma_\infty^2$$

Song and Schmeiser (1995) considered weakly stationary processes with $\gamma_m = \sum_{j=-\infty}^{\infty} j^m C_j < \infty$ for $m = 0$, 1 and studied batch means variance estimators with

$$\text{Bias}(\hat{V}) = -c_b \gamma_1 \frac{1}{b} + o\left(\frac{1}{b}\right)$$

and

$$\text{Var}(\hat{V}) = c_v \gamma_0^2 \frac{b}{n} + o\left(\frac{b}{n}\right)$$

The constants c_b and c_v depend on the amount of overlapping between the batches. In particular, the estimator $\hat{V}_B$ has $c_b = 1$ and $c_v = 2$, while $\hat{V}_O$ has $c_b = 1$ and $c_v = \frac{4}{3}$. Then the asymptotic batch size that minimizes $\text{MSE}(\hat{V}) = \text{Bias}^2(\hat{V}) + \text{Var}(\hat{V})$ is

$$b^* = \left(\frac{2c_b^2 \gamma_1^2}{c_v \gamma_0^2}\right)^{1/3} n^{1/3} \tag{25}$$

Pedrosa and Schmeiser (1994) and Song (1996) developed methods for estimating the ratio $(\gamma_1/\gamma_0)^2$ for a variety of processes, including moving average processes and autoregressive processes. Then one can obtain an estimator for b^* by plugging the ratio estimator into equation (25). Sherman (1995) proposed a method that does not rely on the estimation of $(\gamma_1/\gamma_0)^2$.

Welch (1987) noted that both traditional batch means and overlapping batch means are special cases of spectral estimation (see Section 7.3.6) at frequency 0 and, more important, suggested that overlapping batch means yield near-optimal variance reduction when one forms subbatches within each batch and applies the method to the subbatches. For example, a batch of size 64 is split into four sub-batches, and the first (overlapping) batch consists of observations $X_1, \ldots, X_{64}$, the second consists of observations $X_{17}, \ldots, X_{80}$, and so on.

7.3.5 Standardized Time Series Method

This method was proposed by Schruben (1983). The standardized time series is defined by

$$T_n(t) = \frac{\lfloor nt \rfloor (\overline{X}_n - \overline{X}_{\lfloor nt \rfloor})}{\sigma_\infty \sqrt{n}} \qquad 0 \le t \le 1$$

and under some mild assumptions (e.g., stationarity and ϕ-mixing),

$$(\sqrt{n}(\overline{X}_n - \mu), \sigma_\infty T_n) \xrightarrow{\mathcal{D}} (\sigma_\infty W(1), \sigma_\infty B)$$

where $\{B(t) : t \le 0\}$ is the Brownian bridge process (see Billingsley, 1968). Informally, $\{X_i\}$ is ϕ-mixing if X_i and X_{i+j} are approximately independent for large j. Figure 7.8 shows the standardized time series for the AR(1) sample path in Figure 7.1.

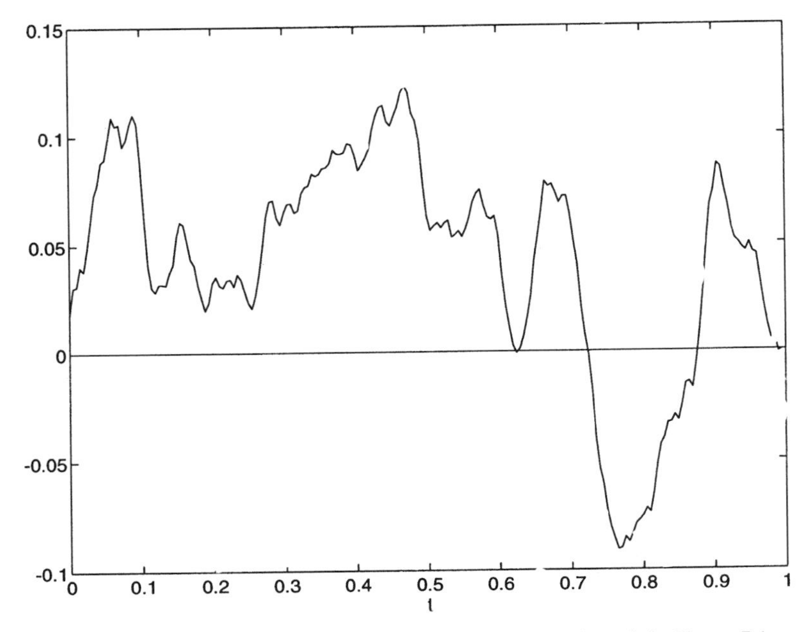

Figure 7.8 Standardized time series for the AR(1) sample path in Figure 7.1.

If $A = \int_0^1 \sigma_\infty B(t)\, dt$ is the area under B, the identity

$$\mathrm{E}(A^2) = \frac{\sigma_\infty^2}{12}$$

implies that σ_∞^2 can be estimated by multiplying an estimator of $\mathrm{E}(A^2)$ by 12. Suppose that the data $X_1, \ldots, X_n$ are divided into k (contiguous) batches, each of size b. Then for sufficiently large n the random variables

$$A_i = \sum_{j=1}^{b} \left(\frac{n+1}{2} - j \right) X_{(i-1)b+j} \qquad i = 1, \ldots, k$$

become approximately IID normal and an estimator of $\mathrm{E}(A^2)$ is

$$\hat{\mathrm{E}}(A^2) = \frac{1}{(b^3 - b)k} \sum_{i=1}^{k} A_i^2$$

Hence an (approximate) $1 - \alpha$ confidence interval for μ is

$$\overline{Y}_k \pm t_{k,1-\alpha/2}\sqrt{\frac{\hat{V}_T}{n}}$$

where

$$\hat{V}_T = 12\hat{\mathrm{E}}(A^2)$$

The standardized time-series method is easy to implement and has asymptotic advantages over the batch means method (see Goldsman and Schruben, 1984). However, in practice it can require prohibitively long runs as noted by Sargent et al. (1992). Some useful theoretical foundations of the method are given in Glynn and Iglehart (1990). Additional developments on the method, as well as other standardized time-series estimators, are contained in Goldsman et al. (1990) and Goldsman and Schruben (1984, 1990). Finally, Damerdji (1994a,b) shows that under the assumption of strong approximation in Section 7.3.4, batching sequences satisfying assumptions (A.1)–(A.3) yield consistent estimators for the process variance σ_∞^2.

7.3.6 Spectral Estimation Method

This method also assumes that the process $\{X_i\}$ is weakly stationary. Under this assumption, the variance of $\overline{X}_n$ is given by (6). The name of the method is due to the fact that if $\sum_{j=-\infty}^{\infty} |C_j| < \infty$, then $n\mathrm{Var}(\overline{X}_n) \to 2\pi g(0)$ as $n \to \infty$, where $g(\lambda)$ is the *spectrum* of the process at frequency λ and is defined by

$$g(\lambda) = \frac{1}{2\pi} \sum_{j=-\infty}^{\infty} C_j e^{-i\lambda j} \qquad |\lambda| \le \pi$$

where $i = \sqrt{-1}$. Therefore, for large n the estimation of $\mathrm{Var}(\overline{X}_n)$ can be viewed as that of estimating $g(0)$. Estimators of this variance have the form

$$\hat{V}_S = \frac{1}{n}\left(\hat{C}_0 + 2\sum_{j=1}^{p-1} w_j \hat{C}_j\right)$$

where p and the *weights* w_j are chosen to improve the properties of the variance estimator $\hat{V}_S$. The selection of these parameters is discussed in Fishman (1978b) and Law and Kelton (1984). Further discussions of spectral methods are given in Heidelberger and Welch (1981a,b, 1983) and Damerdji (1991).

7.3.7 Quantile Estimation from Stationary Output Data

Heidelberger and Lewis (1984) proposed three methods for computing confidence intervals for quantiles when the output process is stationary but not regenerative. Only the average group quantile method will be presented because it is simpler to implement than the competitors and has performed as well or better than the others in terms of the width of confidence intervals and the coverage probabilities. This method makes use of the *maximum transform*.

Maximum Transform. The purpose of the maximum transform is to convert the problem of computing an extreme quantile to one of computing a quantile close to the median. The transform works as follows: Let $X_1, X_2, \ldots, X_v$ be IID random variables with p-quantile equal to ξ, and let $Y = \max\{X_1, X_2, \ldots, X_v\}$. Then

$$P(Y \leq \xi) = F_Y(\xi) = P(X_1 \leq \xi, \ldots, X_v \leq \xi) = [F_X(\xi)]^v = p^v \equiv p'$$

Thus the p'-quantile of Y is the p^v-quantile of X. The idea is to choose v such that $p^v \approx 0.5$, since estimators for the median will generally have smaller bias and variance than estimators for more extreme quantiles. For example, if $p = 0.99$, then $v \approx \ln(.5)/\ln(0.99) = 6.58$. So choosing $v = 7$ gives $p' = 0.99^7 = 0.48$. Notice that by choosing groups of seven observations and applying the maximum transform, the amount of data that must be processed is reduced by a factor of 7.

Applying the maximum transform generally results in inflation of the variance by a factor of approximately 1.4 (see Heidelberger and Lewis, 1984). It is also possible to use other schemes, such as the next to maximum. The maximum transform is clearly applicable to quantile estimation by means of independent replications (see "Quantile Estimation" in Section 7.3.3). For processes that are stationary but not IID, Heidelberger and Lewis apply the maximum transform to observations at least m positions apart, where $m = n/v$, n is the sample size of the output, and v is an integer such that $p^v \approx 0.5$.

Average Group Quantile Method. This method works as follows: First, determine a v so that $p^v \approx 0.5$; that is, $v \approx \lfloor \ln(0.5)/\ln(p) \rfloor$. Then form k contiguous batches of $m \cdot v$ observations each. Within each batch, form m sub-batches of v observations. The first sub-batch consists of observations $1, m + 1, 2m + 1, \ldots, (v - 1)m + 1$, the second of observations $2, m + 2, 2m + 2, \ldots, (v - 1)m + 2$, and so on. All of the observations in each sub-batch are m positions apart. The maximum transform is applied within each sub-batch, producing m maximum transformed observations within the sub-batch. The p'-quantile is then computed from these observations, producing a quantile from each batch. Denote these batch quantiles by $\hat{\xi}_1, \hat{\xi}_2, \ldots, \hat{\xi}_k$. Then the overall quantile estimate is the sample mean of these batch quantiles, and an approximate $1 - \alpha$ confidence interval for ξ is computed by treating $\hat{\xi}_1, \hat{\xi}_2, \ldots, \hat{\xi}_k$ as a set of IID observations and applying the usual confidence interval estimator for the mean

$$\bar{\xi} \pm t_{k-1,\alpha/2} \frac{S_k(\hat{\xi})}{\sqrt{k}}$$

Heidelberger and Lewis compared this estimator to two competitors, one based on estimation of the spectral density of a binary process and another based on nested group quantiles. The average group quantile method performed well relative to the other methods, was not dominated by any of the other methods, and has the advantage that it is the easiest method to implement. The performance of the method depends on the choice of the quantities m and k. While Heidelberger and Lewis do not provide a specific method or specific guidelines for choosing these parameters, they do recommend making m as large as possible to assure that the observations used in the maximum transform have a maximum distance between them and make the spaced observations approximately independent.

7.4 MULTIVARIATE ESTIMATION

Frequently, the output from a single simulation run is used to estimate several system parameters. The estimators of these parameters are typically correlated. As an example, one might wish to estimate simultaneously the average delays for customers at three stations in a queueing network.

Let $\boldsymbol{\theta} = (\theta_1, \theta_2, \ldots, \theta_h)$ be a vector of h parameters that will be estimated using simulation output data. Two types of multivariate interval estimation procedures are generally used: *simultaneous confidence intervals* and *confidence regions*. The set of intervals $\{I_i = (\hat{\theta}_{il}, \hat{\theta}_{iu}, i = 1, 2, \ldots, h\}$ is said to be a set of $1-\alpha$ simultaneous confidence intervals for $\boldsymbol{\theta}$ if

$$P\left(\bigcap_{i=1}^{h} \theta_i \in I_i\right) = 1 - \alpha$$

A region $\boldsymbol{\Theta} \subset \mathbb{R}^h$ is said to be a $1 - \alpha$ confidence region for $\boldsymbol{\theta}$ if $P(\boldsymbol{\theta} \in \boldsymbol{\Theta}) = 1 - \alpha$. Note that simultaneous confidence intervals form a rectangular region in $\mathbb{R}^h$. In general, a confidence region will not be rectangular but will have smaller volume.

There are many articles in the literature concerning multivariate estimation in general and multivariate methods for simulation in particular. For a general introduction to multivariate statistical methods, see Anderson (1984). Charnes (1989, 1995) and Seila (1984) survey multivariate methods for simulation, primarily methods for estimating the mean, and provide extensive lists of references.

7.4.1 Bonferroni Intervals

Bonferroni's inequality provides a means for computing a lower bound on the simultaneous confidence coefficient for any set of confidence intervals. Let $E_1, E_2, \ldots, E_h$ be any set of events. Bonferroni's inequality states that

$$P(E_1 \cap E_2 \cap \cdots \cap E_h) \geq 1 - \sum_{j=1}^{h} [1 - P(E_j)]$$

To apply this inequality to a set of confidence intervals, let I_j be a $1 - \alpha_j$ confidence interval for θ_j, $j = 1, 2, \ldots, h$, and let E_j represent the event $\theta_j \in I_j$. Then $P(E_j) = 1 - \alpha_j$. By Bonferroni's inequality, the simultaneous confidence coefficient is

$$1 - \alpha = P(E_1 \cap E_2 \cap \cdots \cap E_h) \geq 1 - \sum_{j=1}^{h} \alpha_j \tag{26}$$

Bonferroni's inequality applies in very general circumstances. No conditions or restrictions are placed on the population, the parameters, or the methods of computing the intervals $I_1, I_2, \ldots, I_h$. Normally, to apply this approach one would compute a $1 - \alpha/h$ confidence interval for each parameter θ_i. Then by (26) the simultaneous confidence coefficient is at least $1 - \alpha$. The correctness of this simultaneous confidence coefficient depends on the correctness of the individual confidence coefficients for the individual intervals, however. See the last paragraph in Section 7.4.2.

7.4.2 Multivariate Inference for the Mean Using Independent Replications

Suppose that the simulation run consists of n identical, independent replications, and that replication i produces output data vector $\mathbf{X}_i = (X_{i1}, X_{i2}, \ldots, X_{ih})$, where X_{ij} is an observation that will be used to estimate θ_j. Thus the output of the entire simulation experiment consists of n IID vectors of observations $\mathbf{X}_1, \mathbf{X}_2, \ldots, \mathbf{X}_n$. Let $\boldsymbol{\mu} = \mathrm{E}(\mathbf{X}_i)$ be the vector of population means, and $\Sigma = \mathrm{E}[(\mathbf{X}_i - \boldsymbol{\mu})(\mathbf{X}_i - \boldsymbol{\mu})']$ be the variance–covariance matrix for each $\mathbf{X}_i$ with components $\sum_{jk} = \mathrm{Cov}(X_{ij}, X_{ik})$. The point estimator for $\boldsymbol{\mu}$ is the multivariate sample mean

$$\hat{\boldsymbol{\mu}} = \overline{\mathbf{X}}_n = \frac{1}{n} \sum_{i=1}^{n} \mathbf{X}_i$$

with components

$$\hat{\mu}_j = \overline{X}_{nj} = \frac{1}{n} \sum_{i=1}^{n} X_{ij} \qquad j = 1, 2, \ldots, h$$

and the estimator of Σ is

$$\mathbf{S} = \frac{1}{n-1} \sum_{i=1}^{n} (\mathbf{X}_i - \overline{\mathbf{X}}_n)(\mathbf{X}_i - \overline{\mathbf{X}}_n)'$$

Here $\overline{\mathbf{X}}_n$ and Σ are the basic sample statistics that are used for multivariate inference for the mean $\boldsymbol{\mu}$. If $\mathbf{X}_1, \mathbf{X}_2, \ldots, \mathbf{X}_n$ have a multivariate normal distribution, then a $1 - \alpha$ confidence region for $\boldsymbol{\mu}$ is given by all vectors $\mathbf{x}$ for which

$$n(\overline{\mathbf{X}}_n - \mathbf{x})'\mathbf{S}^{-1}(\overline{\mathbf{X}}_n - \mathbf{x}) \leq \frac{(n-1)h}{n-h} F_{h, n-h, 1-\alpha}$$

where $F_{h,n-h,1-\alpha}$ is the $1 - \alpha$ quantile of the F distribution with h and $n - h$ degrees of freedom in the numerator and denominator, respectively. More generally, if $\boldsymbol{\pi}_1, \ldots, \boldsymbol{\pi}_d$ are h-dimensional nonnull vectors of constants, then a $1 - \alpha$ confidence region for $(\phi_1, \ldots, \phi_d)$ with

$$\phi_l = \boldsymbol{\pi}_l' \boldsymbol{\mu} = \sum_{j=1}^{l} \pi_{lj} \mu_j \qquad l = 1, 2, \ldots, d$$

is given by all vectors $\mathbf{x} \in \mathbb{R}^d$ such that

$$n(\hat{\phi} - \mathbf{x})'[\boldsymbol{\pi}'\mathbf{S}\boldsymbol{\pi}]^{-1}(\hat{\phi} - \mathbf{x}) \leq \frac{(n-1)r}{n-r} F_{r,n-r,1-\alpha}$$

where $\hat{\phi}_l = \boldsymbol{\pi}_l'\hat{\boldsymbol{\mu}}$ is the estimator for ϕ_l. If the data $\mathbf{X}_1, \mathbf{X}_2, \ldots, \mathbf{X}_n$ are not multivariate normal but are approximately multivariate normal, the regions above may be used as approximate $1 - \alpha$ confidence regions for $\boldsymbol{\mu}$. This would be the case, for example, if $\mathbf{X}_i$ were the sample mean for a sequence of observations and conditions for a central limit theorem were met.

Two methods are available for computing simultaneous confidence intervals for the mean. The approach using Bonferroni's inequality has already been discussed. A second method, originally proposed by Roy and Bose (1953) and Scheffé (1953), computes the intervals

$$\overline{X}_{nj} \pm T_{h,n-h,1-\alpha/2} \frac{S_{jj}}{\sqrt{n}} \qquad j = 1, 2, \ldots, h$$

where $T_{h,n-h,1-\alpha/2}$ is the $1 - \alpha/2$ quantile of Hotelling's T^2 distribution with h and $n-h$ degrees of freedom in the numerator and denominator, respectively (see Anderson, 1984, Chap. 5, or Hotelling, 1931).

Bonferroni's inequality is rather tight; Scheffé intervals are very conservative. Therefore, Bonferroni intervals will normally be shorter than Scheffé intervals. However, if the true confidence coefficients of the individual intervals are less than the nominal values, Scheffé intervals may be preferred to protect against an unacceptably small simultaneous confidence coefficient. For example, suppose that $h = 5$ parameters are to be estimated using simultaneous confidence intervals with simultaneous confidence coefficient $1-\alpha = 0.95$. To use Bonferroni's inequality, one would compute a 0.99 confidence interval for each parameter. If, in fact, the parameter estimators were independent, the true simultaneous confidence coefficient would be $0.99^5 = 0.951$. However, if the true coverage probability for each confidence interval were actually 0.98 instead of 0.99, the simultaneous confidence coefficient would be only $0.98^5 = 0.904$, considerably below the desired value of 0.95.

7.4.3 Multivariate Inference for the Mean Using Stationary Data

Methods for computing simultaneous confidence intervals and confidence regions for the mean have been developed when output data are stationary. As in the case with univariate inference, the initial transient portion must be identified and removed, leaving observations that are approximately stationary. One option is to run a series of n independent replications, compute the mean from each replication and use the methods of Section 7.4.2 to compute either approximate simultaneous confidence intervals or an approximate confidence region for μ. Like the replication–deletion approach for univariate inference described in Section 7.3.2, this approch is wasteful of data and will result in a biased point estimator if the initial transient portion is judged too short. Bias in the point estimator of the mean or variance will reduce the coverage probability of the interval or regions.

Multivariate Batch Means. An alternative to independent replications is to apply a generalization of the univariate batch means method. As in the univariate batch means procedure, this method divides a long run into batches of multivariate observations.

These vectors could be produced because the output process is naturally in the form of a vector. Such a process would result, for example, in a queueing network if an observation is produced each time a customer leaves the system, and X_{ij} is the time required by customer i to travel path j in the network. Vector observations could also be produced by sampling continuous-time processes. More generally, if only means for continuous-time processes are to be estimated, batches could be formed using continuous data accumulated every t time units.

It should be noted that for certain combinations of parameters, one can encounter synchronization problems. Suppose, for example, that a queueing system has two classes of customers, A and B, and suppose that 90% of the customers are of class A while the remaining 10% are of class B. The objective is to estimate the mean waiting times simultaneously for each class, say μ_A and μ_B. Then, if the batch size is set to 100, for example, the amount of simulation time required to collect a batch of 100 observations for class B customers will be approximately nine times that for class A customers. One can easily see that the batches for class A customers will be completed long before those for class B customers, and the relationship between batches for classes A and B customers will change over time. In the following, the observation processes are assumed to be synchronous, in the sense that for any batch size the statistical relationship among batch means does not change.

The multivariate batch means method is applied analogously to the univariate batch means procedure. Suppose that a stationary multivariate output process is divided into k batches of b vectors each, and let $\overline{\mathbf{Y}}_1, \ldots, \overline{\mathbf{Y}}_k$ be the sequence of batch means. If

$$\sum_{l=-\infty}^{\infty} \text{Cov}(X_{ij}, X_{i+l,j}) < \infty \qquad \text{for all } j = 0, 1, \ldots, h$$

then the vectors $\overline{\mathbf{Y}}_1, \ldots, \overline{\mathbf{Y}}_k$ are asymptotically uncorrelated and their sample mean vector is a consistent estimator of the steady-state mean vector $\boldsymbol{\mu}$. The multivariate batch means method then treats $\overline{\mathbf{Y}}_1, \ldots, \overline{\mathbf{Y}}_k$ as a sequence of IID random vectors and applies the methods of Section 7.4.2 to compute a confidence region or simultaneous confidence intervals. One is left with the problem of determining the appropriate batch size and number of batches. This problem is complicated by the fact since the batch means are vectors, the autocorrelation function will be a sequence of correlation matrices. Chen and Seila (1987) developed a procedure that is based on fitting a first-order autoregressive process to the sequence of batch means to test for autocorrelation and determine the batch size. This procedure has been shown to work well in a variety of systems.

ACKNOWLEDGMENTS

The authors would like to thank George Fishman and David Goldsman for their many fruitful discussions. The research of the first author was partially supported by the Air Force Office of Scientific Research under Contract 93-0043.

REFERENCES

Alexopoulos, C. (1993). Distribution-free confidence intervals for conditional probabilities and ratios of expectations, *Management Science*, Vol. 40, No. 12, pp. 1748–1763.

Anderson, T. W. (1984). *An Introduction to Multivariate Statistical Analysis*, Wiley, New York.

Billingsley, P. (1968). *Convergence of Probability Measures*, Wiley, New York.

Blomqvist, N. (1967). The covariance function of the M/G/1 queueing system, *Skandinavisk Aktuarietidskrift*, Vol. 50, pp. 157–174.

Bratley, P., B. L. Fox, and L. E. Schrage (1987). *A Guide to Simulation*, 2nd ed., Springer-Verlag, New York.

Carlstein, E. (1986). The use of subseries for estimating the variance of a general statistic from a stationary sequence, *Annals of Mathematical Statistics*, Vol. 14, pp. 1171–1179.

Chance, F., and L. W. Schruben (1992). Establishing a truncation point in simulation output, *Technical Report*, School of Operations Research and Industrial Engineering, Cornell University, Ithaca, N.Y.

Charnes, J. M. (1989). Statistical analysis of multivariate discrete-event simulation output, Ph.D. thesis, Department of Operations and Management Science, University of Minnesota, Minneapolis, Minn.

Charnes, J. M. (1995). Analyzing multivariate output, in *Proceedings of the 1995 Winter Simulation Conference*, C. Alexopoulos, K. Kang, W. R. Lilegdon, and D. Goldsman, eds., IEEE, Piscataway, N.J., pp. 201–208.

Chatfield, C. (1989). *The Analysis of Time Series: An Introduction*, 4th ed., Chapman & Hall, New York.

Chen, R. D., and A. F. Seila (1987). Multivariate inference in stationary simulation using batch means, in *Proceedings of the 1987 Winter Simulation Conference*, A. Thesen, H. Grant, and W. D. Kelton, eds., IEEE, Piscataway, N.J., pp. 302–304.

Chien, C.-H. (1989). Small sample theory for steady state confidence intervals, *Technical Report 37*, Department of Operations Research, Stanford University, Palo Alto, Calif.

Chien, C., D. Goldsman, and B. Melamed (1997). Large-sample results for batch means, *Management Science*, Vol. 43, pp. 1288-1295.

Chow, Y. S., and H. Robbins (1965). On the asymptotic theory of fixed-width sequential confidence intervals for the mean, *Annals of Mathematical Statistics*, Vol. 36, pp. 457–462.

Conway, R. W. (1963). Some tactical problems in digital simulation, *Management Science*, Vol. 10, pp. 47–61.

Crane, M. A., and D. L. Iglehart (1974a). Simulating stable stochastic systems, I: General multiserver queues, *Journal of the ACM*, Vol. 21, pp. 103–113.

Crane, M. A., and D. L. Iglehart (1974b). Simulating stable stochastic systems, II: Markov chains, *Journal of the ACM*, Vol. 21, pp. 114–123.

Crane, M. A., and D. L. Iglehart (1975). Simulating stable stochastic systems, III: Regenerative processes and discrete-event simulations, *Operations Research*, Vol. 23, pp. 33–45.

Crane, M. A., and A. J. Lemoine (1977). *An Introduction to the Regenerative Method for Simulation Analysis*, Springer-Verlag, New York.

Damerdji, H. (1991). Strong consistency and other properties of the spectral variance estimator, *Management Science*, Vol. 37, pp. 1424–1440.

Damerdji, H. (1994a). Strong consistency of the variance estimator in steady-state simulation output analysis, *Mathematics of Operations Research*, Vol. 19, pp. 494–512.

Damerdji, H. (1994b). On the batch means and area variance estimators, in *Proceedings of the 1994 Winter Simulation Conference*, S. Manivannan, J. D. Tew, D. A. Sadowski, and A. F. Seila, eds., IEEE, Piscataway, N.J., pp. 340–344.

Efron, B. (1982). *The Jackknife, the Bootstrap and Other Resampling Plans*, SIAM, Philadelphia, Pa.

Fishman, G. S. (1972). Bias considerations in simulation experiments, *Operations Research*, Vol. 20, pp. 785–790.

Fishman, G. S. (1973). Statistical analysis for queueing simulations, *Management Science*, Vol. 20, pp. 363–369.

Fishman, G. S. (1974). Estimation of multiserver queueing simulations, *Operations Research*, Vol. 22, pp. 72–78.

Fishman, G. S. (1978a). Grouping observations in digital simulation, *Management Science*, Vol. 24, pp. 510–521.

Fishman, G. S. (1978b). *Principles of Discrete Event Simulation*, Wiley, New York.

Fishman, G. S. (1996). *Monte Carlo: Concepts, Algorithms and Applications*, Chapman & Hall, New York.

Fishman, G. S., and L. S. Yarberry, (1997). An implementation of the batch means method, *INFORMS Journal on Computing*, Vol. 9, pp. 296–310.

Gafarian, A. V., C. J. Ancker, and F. Morisaku (1978). Evaluation of commonly used rules for detecting steady-state in computer simulation, *Naval Research Logistics Quarterly*, Vol. 25, pp. 511–529.

Glynn, P. W., and D. L. Iglehart (1990). Simulation analysis using standardized time series, *Mathematics of Operations Research*, Vol. 15, pp. 1–16.

Goldsman, D., and L. W. Schruben (1984). Asymptotic properties of some confidence interval estimators for simulation output, *Management Science*, Vol. 30, pp. 1217–1225.

Goldman, D., and L. W. Schruben (1990). New confidence interval estimators using standardized time series, *Management Science*, Vol. 36, pp. 393–397.

Goldsman, D., M. Meketon, and L. W. Schruben (1990). Properties of standardized time series weighted area variance estimators, *Management Science*, Vol. 36, pp. 602–612.

Goldsman, D., L. W. Schruben, and J. J. Swain (1994). Tests for transient means in simulated time series, *Naval Research Logistics Quarterly*, Vol. 41, pp. 171–187.

Heidelberger, P., and P. A. W. Lewis (1984). Quantile estimation in dependent sequences, *Operations Research*, Vol. 32, pp. 185–209.

Heidelberger, P., and P. D. Welch (1981a). A spectral method for confidence interval generation and run length control in simulations, *Communications of the ACM*, Vol. 24, pp. 233–245.

Heidelberger, P., and P. D. Welch (1981b). Adaptive spectral methods for simulation output analysis, *IBM Journal of Research and Development*, Vol. 25, pp. 860–876.

Heidelberger, P., and P. D. Welch (1983). Simulation run length control in the presence of an initial transient, *Operations Research*, Vol. 31, pp. 1109–1144.

Hogg, R. V., and A. T. Craig (1978). *Introduction to Mathematical Statistics*, 4th ed., Macmillan, New York.

Hotelling, H. (1931). The generalization of Student's ratio, *Annals of Mathematical Statistics*, Vol. 2, pp. 360–378.

Iglehart, D. L. (1975). Simulating stable stochastic systems, V: Comparison of ratio estimators, *Naval Research Logistics Quarterly*, Vol. 22, pp. 553–565.

Iglehart, D. L. (1976). Simulating stable stochastic systems, VI: Quantile estimation, *Journal of the ACM*, Vol. 23, pp. 347–360.

Iglehart, D. L. (1978). The regenerative method for simulation analysis, in *Current Trends in Programming Methodology, Vol. III*, K. M. Chandy and K. M. Yeh, eds., Prentice-Hall, Upper Saddle River, N.J., pp. 52–71.

Karr, A. F. (1993). *Probability*, Springer-Verlag, New York.

Kelton, W. D. (1989). Random initialization methods in simulation, *IIE Transactions*, Vol. 21, pp. 355–367.

Kelton, W. D., and A. M. Law (1983). A new approach for dealing with the startup problem in discrete event simulation, *Naval Research Logistics Quarterly*, Vol. 30, pp. 641–658.

Kleijnen, J. P. C. (1974). *Statistical Techniques in Simulation, Part I*, Marcel Dekker, New York.

Kleijnen, J. P. C. (1975). *Statistical Techniques in Simulation, Part II*, Marcel Dekker, New York.

Law, A. M., and J. S. Carson (1979). A sequential procedure for determining the length of a steady-state simulation, *Operations Research*, Vol. 27, pp. 1011–1025.

Law, A. M., and W. D. Kelton (1984). Confidence intervals of steady-state simulations, I: A survey of fixed sample size procedures, *Operations Research*, Vol. 32, pp. 1221–1239.

Law, A. M., and W. D. Kelton (1991). *Simulation Modeling and Analysis*, 2nd ed., McGraw-Hill, New York.

Law, A. M., W. D. Kelton, and L. W. Koenig (1981). Relative width sequential confidence intervals for the mean, *Communications in Statistics B*, Vol. 10, pp. 29–39.

Lehmann, E. L. (1991). *Theory of Point Estimation*, 2nd ed., Wadsworth, Belmont, Calif.

Lindley, D. V. (1952). The theory of queues with a single server, *Proceedings of the Cambridge Philosophical Society*, Vol. 48, pp. 277–289.

Mechanic, H., and W. McKay (1966). Confidence intervals for averages of dependent data in simulations II, *Technical Report ASDD 17-202*, IBM Corporation, Yorktown Heights, N.Y.

Meketon, M. S., and B. W. Schmeiser (1984). Overlapping batch means: something for nothing? in *Proceedings of the 1984 Winter Simulation Conference*, S. Sheppard, U. W. Pooch, and C. D. Pegden, eds., IEEE, Piscataway, N.J., pp. 227–230.

Moore, L. W. (1980). Quantile estimation in regenerative processes, Ph.D. thesis, Curriculum in Operations Research and Systems Analysis, University of North Carolina, Chapel Hill, N.C.

Nadas, A. (1969). An extension of the theorem of Chow and Robbins on sequential confidence intervals for the mean, *Annals of Mathematical Statistics*, Vol. 40, pp. 667–671.

Ockerman, D. H. (1995). Initialization bias tests for stationary stochastic processes based upon standardized time series techniques, Ph.D. thesis, School of Industrial and Systems Engineering, Georgia Institute of Technology, Atlanta, Ga.

Pedrosa, A. C., and B. W. Schmeiser (1994). Estimating the variance of the sample mean: Optimal batch size estimation and 1–2–1 overlapping batch means, *Technical Report SMS94–3*, School of Industrial Engineering, Purdue University, West Lafayette, In.

Ross, S. M. (1993). *Introduction to Probability Models*, 5th ed., Academic Press, San Diego, Calif.

Roy, S. N., and R. C. Bose (1953). Simultaneous confidence interval estimation, *Annals of Mathematical Statistics*, Vol. 24, pp. 513–536.

Sargent, R. G., K. Kang, and D. Goldsman (1992). An investigation of finite-sample behavior of confidence interval estimators, *Operations Research*, Vol. 40, pp. 898–913.

Scheffé, H. (1953). A method of judging all contrasts in analysis of variance, *Biometrica*, Vol. 40, pp. 87–104.

Schmeiser, B. W. (1982). Batch size effects in the analysis of simulation output, *Operations Research*, Vol. 30, pp. 556–568.

Schriber, T. J., and R. W. Andrews (1979). Interactive analysis of simulation output by the method of batch means, in *Proceedings of the 1979 Winter Simulation Conference*, M. G. Spiegel, N. R. Nielsen, and H. J. Highland, eds., IEEE, Piscataway, N.J., pp. 513–525.

Schruben, L. W. (1982). Detecting initialization bias in simulation output, *Operations Research*, Vol. 30, pp. 569–590.

Schruben, L. W. (1983). Confidence interval estimation using standardized time series, *Operations Research*, Vol. 31, pp. 1090–1108.

Schruben, L. W., H. Singh, and L. Tierney. (1983). Optimal tests for initialization bias in simulation output, *Operations Research*, Vol. 31, pp. 1167–1178.

Seila, A. F. (1982a). A batching approach to quantile estimation in regenerative simulations, *Management Science*, Vol. 28, pp. 573–581.

Seila, A. F. (1982b). Percentile estimation in discrete event simulation, *Simulation*, Vol. 39, pp. 193–200.

Seila, A. F. (1984). Multivariate simulation output analysis, *American Journal of Mathematical and Management Sciences*, Vol. 4, pp. 313–334.

Sherman, M. (1995). On batch means in the simulation and statistical communities, in *Proceedings of the 1995 Winter Simulation Conference*, C. Alexopoulos, K. Kang, W. R. Lilegdon, and D. Goldsman, eds., IEEE, Piscataway, N.J., pp. 297–302.

Song, W.-M. T. (1996). On the estimation of optimal batch sizes in the analysis of simulation output analysis, *European Journal of Operations Research*, Vol. 88, pp. 304–309.

Song, W.-M. T., and B. W. Schmeiser (1995). Optimal mean-squared-error batch sizes, *Management Science*, Vol. 41, pp. 110–123.

Starr, N. (1966). The performance of a statistical procedure for the fixed-width interval estimation for the mean, *Annals of Mathematical Statistics*, Vol. 37, No. 1, pp. 36–50.

von Neumann, J. (1941). Distribution of the ratio of the mean square successive difference and the variance, *Annals of Mathematical Statistics*, Vol. 12, pp. 367–395.

Welch, P. D. (1981). On the problem of the initial transient in steady state simulations, *Technical Report*, IBM Watson Research Center, Yorktown Heights, N.Y.

Welch, P. D. (1983). The statistical analysis of simulation results, in *The Computer Performance Modeling Handbook*, S. Lavenberg, ed., Academic Press, San Diego, Calif., pp. 268–328.

Welch, P. D. (1987). On the relationship between batch means, overlapping batch means and spectral estimation, in *Proceedings of the 1987 Winter Simulation Conference*, A. Thesen, H. Grant, and W. D. Kelton, IEEE, Piscataway, N.J., pp. 320–323.

Wilson, J. R., and A. A. B. Pritsker (1978a). A survey of research on the simulation startup problem, *Simulation*, Vol. 31, pp. 55–58.

Wilson, J. R., and A. A. B. Pritsker (1978b). Evaluation of startup policies in simulation experiments, *Simulation*, Vol. 31, pp. 79–89.

Yarberry, L. S. (1993). Incorporating a dynamic batch size selection mechanism in a fixed-sample-size batch means procedure, Ph.D. thesis, Department of Operations Research, University of North Carolina, Chapel Hill, N.C.

Young, L. C. (1941). Randomness in ordered sequences, *Annals of Mathematical Statistics*, Vol. 12, pp. 293–300.

CHAPTER 8

Comparing Systems via Simulation

DAVID GOLDSMAN
Georgia Institute of Technology

BARRY L. NELSON
Northwestern University

8.1 INTRODUCTION

Simulation experiments are typically performed to compare, in some fashion, two or more system designs. The statistical methods of *ranking and selection* and *multiple comparisons* are applicable when comparisons among a finite and typically small number of systems (say, 2 to 20) are required. The particular method that is appropriate depends on the type of comparison desired and properties of the simulation output data. In this chapter we describe methods for five classes of problems: screening a substantial number of system designs, selecting the best system, comparing all systems to a standard, comparing alternatives to a default and comparing systems that are functionally related. For optimization with respect to a (conceptually) infinite number of systems, see Chapter 9.

Ranking and selection procedures (R&S) are statistical methods specifically developed to select the best system, or a subset of systems that includes the best system, from among a collection of competing alternatives. Provided that certain assumptions are met, these methods guarantee that the probability of a correct selection will be at least some user-specified value. Multiple-comparison procedures (MCPs) treat the comparison problem as an inference problem on the performance measures of interest. MCPs account for the error that arises when simultaneously estimating the differences in performance among several systems. Both types of procedures are relevant in the context of computer simulation because the assumptions behind the procedures can frequently be (approximately) satisfied: (1) the assumption of normally distributed data can often be secured by batching large numbers of outputs (see Section 8.2.4), (2) independence can be obtained by controlling random-number assignments (see Section 8.2.3), and (3) multiple-stage sampling—which is required by some methods—is feasible in computer simulation because a subsequent stage can be initialized simply by retaining the final random number seeds from the pre-

Handbook of Simulation, Edited by Jerry Banks.
ISBN 0-471-13403-1

ceding stage, or by regenerating the entire sample. The procedures presented in this chapter include R&S, MCPs, and combinations of the two.

Most readers should begin by scanning Section 8.2 for unfamiliar material, since knowledge of these basic topics is essential for the remainder of the chapter. The core comparison methods are contained in Section 8.3, which is organized according to five types of comparison problems. The material on each problem type is self-contained so that the reader can proceed directly from Section 8.2 to the material that is relevant for the problem at hand.

- *Section 8.3.1—Screening Problems:* relevant when the goal is to compare a substantial number of system designs in order to group those with similar performance and eliminate clearly inferior performers.
- *Section 8.3.2—Selecting the Best:* relevant when the goal is to find the system with the largest or smallest performance measure.
- *Section 8.3.3—Comparisons with a Standard:* relevant when the goal is to find the best system, provided that its performance exceeds a known, fixed performance standard.
- *Section 8.3.4—Comparisons with a Default:* relevant when the goal is to compare alternative systems to the current system (not necessarily a fixed, numerical performance standard as in Section 8.3.3).
- *Section 8.3.5—Estimating Functional Relationships:* relevant when the goal is to represent the difference between systems in terms of the parameters of a linear model.

Each of these subsections contains procedures, tables of key critical constants, and numerical examples. A detailed case study is included in Section 8.4. This case illustrates selecting the best system, but the approach applies to all types of comparison problems and should be useful to all readers.

8.2 BACKGROUND

In this section we introduce notation and provide background material on simulation output processes, random number assignment, batching, and comparing two systems.

8.2.1 Notation

The goal is to compare k different systems via simulation. For compatibility with statistics textbooks, sometimes the k simulated systems are called *design points*.

Let Y be a random variable that represents the output (sample performance) of a simulation generically. For example, Y might be the cost to operate an inventory system for a month, the average flow time of parts in a manufacturing system, or the time a customer has to wait to be seated in a restaurant. Let Y_{ij} represent the jth simulation output from system design i, for $i = 1, 2, \ldots, k$ alternatives and $j = 1, 2, \ldots$. For fixed i it will always be assumed that the outputs from system i, $Y_{i1}, Y_{i2}, \ldots$, are identically distributed (have the same probability distribution, and therefore the same mean, variance, etc.). This assumption is plausible if $Y_{i1}, Y_{i2}, \ldots$ are outputs across independent replications of system i, or if they are outputs from within a single replication of a steady-state simulation after accounting for initialization effects (see Chapter 7 and Section 8.2.2).

Let $\mu_i = \mathrm{E}[Y_{ij}]$ denote the expected value of an output from the ith system, and let

$$p_i = \Pr\{Y_{ij} > \max_{l \neq i} Y_{lj}\}$$

be the probability that Y_{ij} is the largest of the jth outputs across all systems. The procedures described in this chapter provide comparisons based on either μ_i, the long-run or expected performance of system i, or p_i, the probability that system i will actually be the best performer (if bigger is better).

In some contexts it is useful to think of the k systems as functionally related. Let

$$\mathbf{Y}_j = \begin{bmatrix} Y_{1j} \\ Y_{2j} \\ \vdots \\ Y_{kj} \end{bmatrix}$$

be a $k \times 1$ vector of outputs across all k design points on the jth replication. The relationship among the k design points can sometimes be approximated by a general linear model:

$$\mathbf{Y} = \mathbf{X}\boldsymbol{\beta} + \boldsymbol{\epsilon}$$

where $\mathbf{X}$ is a $k \times p$ fixed *design matrix*, $\boldsymbol{\beta}$ is a $p \times 1$ vector of unknown constants, and $\boldsymbol{\epsilon}$ is a $k \times 1$ vector of random errors with expectation $\mathbf{0}$. The differences between systems (design points) is captured in the parameter $\boldsymbol{\beta}$. In this chapter we discuss estimation of, and inference about, $\boldsymbol{\beta}$. For that portion of classical experiment design that addresses the specification of $\mathbf{X}$, see Chapter 6.

8.2.2 Simulation Output Processes

There are two broad (not necessarily exhaustive) classes of system performance parameters of interest: those defined with respect to prespecified initial and final conditions for the system of interest, and those defined over a (conceptually) infinite time horizon. Simulation experiments that estimate the former are called *terminating* simulation experiments, while the latter are called *steady-state* simulation experiments. Examples of terminating simulations include simulation of a store that is open each day from 9 A.M. to 9 P.M., and simulation of an inventory policy that begins with the current inventory position and has a planning horizon of 1 month. Examples of steady-state simulations include the simulation of a computer system under sustained peak load, and simulation of an inventory policy over an infinite planning horizon.

The experiment design for terminating simulations always calls for multiple replications, and the length of each replication is determined by the prespecified initial and final conditions. Therefore, for a terminating simulation of system i, the $Y_{i1}, Y_{i2}, \ldots$ represent observed system performance *across* different replications of the system and are thus independent.

The experiment design for steady-state simulation may call for one or more replications, and the length of each replication is a design decision. Therefore, for a steady-state simulation of system i, the $Y_{i1}, Y_{i2}, \ldots$ may represent outputs from *within* a single replication—outputs that are typically dependent—or they may represent summary outputs from *across* multiple replications. The decision between single or multiple replications typically depends on the severity of the initial-condition bias (see Chapter 7).

For the purposes of this chapter, *experiment design* includes the following:

- Specifying the number of replications for each system or design point
- Specifying the length of each replication for each system or design point when it is a steady-state simulation
- Assigning the pseudorandom numbers to each system or design point

8.2.3 Controlling Randomness

Uncertainty (*randomness*) in a simulation experiment is derived from the pseudorandom numbers, typically numbers in the interval (0,1) that are difficult to distinguish from independent and identically distributed (IID) uniform random numbers. A useful way to think about the random numbers is as a large, ordered table, where the number of entries in the table is often around $2^{31} \approx 2 \times 10^9$. Given a starting point in the table, a simulation uses the pseudorandom numbers in order until the experiment is completed. If the end of the table is encountered, numbers starting from the beginning of the table are used. Simulation languages do not actually store the pseudorandom numbers in a table (they are generated by a recursive function as needed), but the table of random numbers is a good conceptual representation of how the simulation language works.

Although the (conceptual) table of pseudorandom numbers is ordered, the order does not matter. As long as the numbers are used without replacement, they can be taken in any manner or starting from any position in the table and still appear to be a sample of IID random numbers. An important feature of most simulation languages is that they permit control of the pseudorandom numbers through *seeds* or *streams*. The seeds or streams are nothing more than different starting points in the table, typically spaced far apart. For example, stream 2 might correspond to entering the table at the 131,072nd random number.

The assignment of random number seeds or streams is part of the design of a simulation experiment. All subsequences within the (conceptual) table appear to be IID random numbers, so assigning a different seed or stream to different systems guarantees that the outputs from different systems will be statistically independent. Similarly, assigning the same seed or stream to different systems induces dependence among the corresponding outputs, since they all have the same source of randomness. Controlling the dependence between systems is the primary reason for the existence of seeds or streams.

CAUTION. Assigning different seeds or streams to different systems does not guarantee independence if, say, stream 1 is assigned to system 1, but the simulation of system 1 uses so many random numbers that values from stream 2—which is assigned to system 2—are also used. Remember that the streams are just starting points in an ordered table. If independence is critical, it is worthwhile to know the spacing between seeds or streams in a simulation language and to make a rough estimate of the number of pseudorandom numbers needed at each design point.

On the other hand, to obtain independent replications it is typically not necessary to assign different seeds or streams to different replications for a single system. Simulation languages generally begin subsequent replications using random numbers from where the previous replication finished, implying that different replications use different random numbers and are therefore independent.

Although many statistical procedures call for independence across systems or design points, in comparison problems it is often useful to assign the same random number seeds or streams to all of the design points; this technique is called *common random numbers* (CRNs). The discussion that follows is in terms of a multiple-replication experiment, but the ideas apply to a single-replication experiment when batch means (see Section 8.2.4) play the role of replications.

The intuition behind CRN is that a fairer comparison among systems is achieved if all of the systems are subjected to the same experimental conditions, specifically the same source of randomness. CRN can ensure this.

The mathematical justification for CRN is as follows: Suppose that the sample mean, $\overline{Y}_i = \sum_{j=1}^{n} Y_{ij}/n$, is used to estimate the unknown expected performance, μ_i, from system i, where n is the number of independent replications taken across system i. Then for systems i and l, the (unknown) variance of the estimated difference in performance is

$$\text{Var}[\overline{Y}_i - \overline{Y}_l] = \text{Var}[\overline{Y}_i] + \text{Var}[\overline{Y}_l] - 2\,\text{Cov}[\overline{Y}_i, \overline{Y}_l]$$

where Var denotes variance and Cov denotes covariance. If different seeds or streams are assigned to systems i and l, then $\text{Cov}[\overline{Y}_i, \overline{Y}_l] = 0$; if common seeds or streams are assigned, then frequently $\text{Cov}[\overline{Y}_i, \overline{Y}_l] > 0$, reducing the variance of the difference, which leads to a more precise comparison.

The effect of CRN can be enhanced by *synchronizing* the random numbers, which means forcing the random numbers to be used for the same purpose in each system. The primary technique for achieving synchronization is to assign a different seed or stream to each random input process and then to use the same collection of seeds or streams across all systems. For example, in a simulation of a queue, this means assigning a stream to the arrival process and a different stream to the service process. When common streams are used across systems, the same random numbers will generate arrivals and service times for each one.

Sometimes synchronization is facilitated by generating entity–transaction attributes at the time the entity–transaction is created. For instance, the entire sequence of processing times for a job at several stations can be generated when the job arrives at the shop rather than generating each processing time as the job arrives at each station.

One must also take care to synchronize the random numbers across replications. To be specific, replication 2 of systems i and l should both begin with the same random numbers. This may not happen automatically, since replication 1 of system i may require a different quantity of random numbers than replication 1 of system l. The problem of ensuring that all replications across all systems begin with the same random numbers is simulation-language dependent. If a large number of seeds or streams can be created, one approach is to assign different seeds or streams to each replication.

An exploratory experiment can be used to verify that CRN is having the desired effect by estimating the covariance between outputs from different systems or design points. The sample covariance between systems i and l is

$$C_{il} = \frac{1}{n-1} \sum_{j=1}^{n} (Y_{ij} - \overline{Y}_i)(Y_{lj} - \overline{Y}_l) = \frac{1}{n-1} \left(\sum_{j=1}^{n} Y_{ij} Y_{lj} - n \overline{Y}_i \overline{Y}_l \right) \tag{1}$$

The covariance terms should be positive; if they are negative, CRN may inflate variance and should not be used.

8.2.4 Batching

For system or design point i, suppose that n outputs $Y_{i1}, Y_{i2}, \ldots, Y_{in}$ are available, either representing the results from across n replications, or n outputs from within a single replication of a steady-state simulation. The following aggregation of the data is sometimes useful: Set

$$\overline{Y}_{ih} = \frac{1}{m} \sum_{j=1}^{m} Y_{(h-1)m+j}$$

for $h = 1, 2, \ldots, b$, so that $n = bm$. Simply stated, $\overline{Y}_{ih}$ is the sample mean of the "batch" of outputs $Y_{i,(h-1)m+1}, \ldots, Y_{i,hm}$. There are several reasons to consider basing statistical analysis on the *batch means* $\overline{Y}_{i1}, \ldots, \overline{Y}_{ib}$ rather than the original outputs $Y_{i1}, \ldots, Y_{in}$ (see also the discussion in Chapter 7):

1. The batch means tend to be more nearly normally distributed than the original outputs, since they are averages. This property is useful if the statistical analysis is based on normal distribution theory.

2. When the data are from within a single replication of a steady-state simulation and are therefore dependent, the batch means tend to be more nearly independent than the original outputs. This property is useful if only a single replication is obtained from each system and the statistical analysis is based on having IID data.

3. By using different batch sizes, m, for different systems, the variances of the batch means across systems can be made more nearly equal than the variances of the original outputs; equal variances is a standard assumption behind many statistical procedures used for comparisons. To be specific, suppose that $S_i^2(n)$ and $S_l^2(n)$ are the sample variances of systems i and l for the original outputs. If $S_i^2(n) > S_l^2(n)$, a batch size of $m \approx S_i^2(n)/S_l^2(n)$ for system i and a batch size $m = 1$ for system l will cause the batch means from i and l to have approximately equal variance. Of course, the same will be true for batch sizes cm and c, respectively, for any positive integer c.

4. Saving all of the batch means may be possible when it is impossible or inconvenient to save all of the original, raw output data. Frequently, the output data are aggregated into summary statistics, such as the sample mean and variance. But it is often useful to be somewhere in between having all the data and having only the summary statistics, especially if it is necessary to test the data for normality, equality of variances, and so on. Batch means provide the middle ground. Surprisingly, it is not necessary to maintain a large number of batch means, b, even if the number of original observations, n, is large. One way to see this is to look at $t_{1-\alpha/2,\nu}$, the $1 - \alpha/2$ quantile of the t-distribution with ν degrees of freedom (available in standard tables): $t_{0.975,30} = 2.04$,

while $t_{0.975,\infty} = 1.96$, a very small difference. Therefore, batching a large number of replications into more than $b = 30$ batch means will not have much of an effect on, say, confidence intervals for the steady-state mean of a system, which are typically based on the t-distribution with $b - 1$ degrees of freedom.

8.2.5 Comparing Two Systems

The simplest comparison problem is to estimate the difference in expected performance of $k = 2$ systems. The appropriate procedure depends on whether the systems are simulated independently or with common random numbers; both procedures are reviewed in this section. The presentation assumes that n replications are made of each system, but it is equally valid if the outputs are batch means from a single replication of each system. The procedures to be described below also make use of the sample mean and variance of n outputs $X_1, X_2, \ldots, X_n$:

$$\overline{X} = \frac{1}{n} \sum_{j=1}^{n} X_j$$

$$S^2 = \frac{1}{n-1} \sum_{j=1}^{n} (X_j - \overline{X})^2 = \frac{1}{n-1} \left(\sum_{j=1}^{n} X_j^2 - n\overline{X}^2 \right)$$

When two systems are simulated, the output data can be organized as follows, where $D_j = Y_{1j} - Y_{2j}$, the difference in performance between systems 1 and 2 on replication j:

	Replication						
System	1	2	...	n	Sample Mean	Sample Variance	Expected Performance
1	Y_{11}	Y_{12}	...	Y_{1n}	$\overline{Y}_1$	S_1^2	μ_1
2	Y_{21}	Y_{22}	...	Y_{2n}	$\overline{Y}_2$	S_2^2	μ_2
1 − 2	D_1	D_2	...	D_n	$\overline{D}$	S_D^2	$\mu_1 - \mu_2$

The estimator $\overline{D} = \overline{Y}_1 - \overline{Y}_2$ can be used to estimate the expected difference in performance, $\mu_1 - \mu_2$. The estimate is almost certainly wrong, so a measure of error, typically a confidence interval, is needed to bound the error with high probability.

If the systems are simulated independently, a $(1 - \alpha)100\%$ confidence interval for $\mu_1 - \mu_2$ is

$$\overline{Y}_1 - \overline{Y}_2 \pm t_{1-\alpha/2, 2n-2} \sqrt{\frac{S_1^2 + S_2^2}{n}} \tag{2}$$

where $t_{1-\alpha/2,\nu}$ is the $1 - \alpha/2$ quantile of the t-distribution with ν degrees of freedom. If, on the other hand, CRN is used, the appropriate confidence interval is

$$\overline{Y}_1 - \overline{Y}_2 \pm t_{1-\alpha/2, n-1} \sqrt{\frac{S_D^2}{n}} \tag{3}$$

This confidence interval accounts for the positive covariance due to CRN.

Both confidence intervals are valid with approximate coverage probability $1 - \alpha$ if the output data are nearly normally distributed, or the sample size is large. Provided that n is not too small, say $n > 10$, the loss of degrees of freedom in going from (2) to (3) is more than compensated by even a small positive covariance due to CRN. In other words, CRN can lead to a shorter confidence interval and therefore a tighter bound on the error of the estimate. The shorter the confidence interval, the easier it is to detect differences in system performance.

When there are $k > 2$ systems, difficulties arise in extending the analysis above to all $d = k(k - 1)/2$ differences $\mu_i - \mu_l$, for all $i \neq l$. A standard approach is to form each confidence interval at level $1 - \alpha/d$, rather than $1 - \alpha$, which guarantees that the overall confidence level for all d intervals is at least $1 - \alpha$ by the Bonferroni inequality. Unfortunately, this procedure is so conservative when k is large that the confidence intervals may be too wide to detect differences in expected performance. Therefore, other procedures, described in Section 8.3, are required when $k > 2$ systems are compared.

8.3 PROBLEMS AND SOLUTIONS

In this section we address a variety of important problem formulations: screening of systems, selection of the best competitor, comparisons with a standard, comparisons with a default, and comparison of systems that are functionally related. Each problem formulation is accompanied by a number of solution methods along with numerical examples.

8.3.1 Screening Problems

Example 1 A brainstorming session produces 15 potential designs for the architecture of a new computer system. Response time is the performance measure of interest, but there are so many designs that a careful simulation study will be deferred until a pilot simulation study determines which designs are worth further scrutiny. A shorter response time is preferred.

If the expected response time is the performance measure of interest, the goal of the pilot study is to determine which designs are the better performers, which have similar performance, and which can be eliminated as clearly inferior.

Multiple Comparisons Approach. Let μ_i denote the expected response time for architecture i. Multiple comparisons approaches the screening problem by forming simultaneous confidence intervals on the parameters $\mu_i - \mu_l$ for all $i \neq l$. These $k(k-1)/2$ confidence intervals indicate the magnitude and direction of the difference between each pair of alternatives. The most widely used method for forming the intervals is Tukey's procedure, which is implemented in many statistical software packages. General references include Hochberg and Tamhane (1987) and Miller (1981).

Suppose that the systems are simulated independently to obtain IID normal outputs $Y_{i1}, Y_{i2}, \ldots, Y_{in_i}$ from system i. Let $\overline{Y}_i = \sum_{j=1}^{n_i} Y_{ij}/n_i$ be the sample mean from system i, and let

$$S^2 = \frac{1}{k} \sum_{i=1}^{k} \frac{1}{n_i - 1} \sum_{j=1}^{n_i} (Y_{ij} - \overline{Y}_i)^2 = \frac{1}{k} \sum_{i=1}^{k} \frac{1}{n_i - 1} \left(\sum_{j=1}^{n_i} Y_{ij}^2 - n_i \overline{Y}_i^2 \right)$$

be the pooled sample variance. Tukey's simultaneous confidence intervals for $\mu_i - \mu_l$ are

$$\overline{Y}_i - \overline{Y}_l \pm \frac{Q_{k,\nu}^{(\alpha)}}{\sqrt{2}} S \sqrt{\frac{1}{n_i} + \frac{1}{n_l}}$$

for all $i \neq l$, where $Q_{k,\nu}^{(\alpha)}$ is the $1 - \alpha$ quantile of the Studentized range distribution with parameter k and $\nu = \sum_{i=1}^{k} (n_i - 1)$ degrees of freedom [see Table 8.1 for critical values when $\alpha = 0.05$; for more complete entries, see Hochberg and Tamhane (1987, App. 3, Table 8)].

When the Y_{ij} are normally distributed with common (unknown) variance, and $n_1 = n_2 = \cdots = n_k$, these intervals achieve simultaneous coverage probability $1 - \alpha$. Hayter (1984) showed that the coverage probability is strictly greater than $1 - \alpha$ when the sample sizes are not equal.

Numerical Example. Suppose that there are only $k = 4$ computer architectures. For each of them, $n = 6$ replications (each simulating several hours of computer use) are obtained, giving the following summary data on response time in milliseconds:

$$\overline{Y}_1 = 72, \quad \overline{Y}_2 = 85, \quad \overline{Y}_3 = 76, \quad \overline{Y}_4 = 62, \quad S^2 = 100.9$$

Suppose that the objective is to determine, with confidence 0.95, bounds on the difference between the expected response times of each alternative. Then Tukey's procedure forms confidence intervals with half-widths

$$\frac{3.96}{\sqrt{2}} \sqrt{100.9 \left(\frac{1}{6} + \frac{1}{6} \right)} = 16$$

where $Q_{4,20}^{(0.05)} = 3.96$ is from Table 8.1. For instance, a confidence interval for $\mu_2 - \mu_4$, the difference in the expected response times of architectures 2 and 4, is $85 - 62 \pm 16$, or 23 ± 16 ms. Since this confidence interval does not contain 0, and since shorter response time is better, we can informally screen out architecture 2 from further consideration.

Subset Selection Approach. The subset selection approach is a screening device that attempts to select a (random-size) *subset* of the $k = 15$ competing designs of Example 1 that contains the design with the smallest expected response time. Gupta (1956,

TABLE 8.1 95% Critical Values $Q^{(0.05)}_{k,\nu}$ of the Studentized Range Distribution

	k												
ν	3	4	5	6	7	8	9	10	11	12	13	14	15
1	27.0	32.8	37.1	40.4	43.1	45.4	47.4	49.1	50.6	52.0	53.2	54.3	55.4
2	8.33	9.80	10.9	11.7	12.4	13.0	13.5	14.0	14.4	14.8	15.1	15.4	15.7
3	5.91	6.83	7.50	8.04	8.48	8.85	9.18	9.46	9.72	9.95	10.2	10.4	10.5
4	5.04	5.76	6.29	6.71	7.05	7.35	7.60	7.83	8.03	8.21	8.37	8.53	8.66
5	4.60	5.22	5.67	6.03	6.33	6.58	6.80	7.00	7.17	7.32	7.47	7.60	7.72
6	4.34	4.90	5.31	5.63	5.90	6.12	6.32	6.49	6.65	6.79	6.92	7.03	7.14
7	4.17	4.68	5.06	5.36	5.61	5.82	6.00	6.16	6.30	6.43	6.55	6.66	6.76
8	4.04	4.53	4.89	5.17	5.40	5.60	5.77	5.92	6.05	6.18	6.29	6.39	6.48
9	3.95	4.42	4.76	5.02	5.24	5.43	5.60	5.74	5.87	5.98	6.09	6.19	6.28
10	3.88	4.33	4.65	4.91	5.12	5.31	5.46	5.60	5.72	5.83	5.94	6.03	6.11
11	3.82	4.26	4.57	4.82	5.03	5.20	5.35	5.49	5.61	5.71	5.81	5.90	5.98
12	3.77	4.20	4.51	4.75	4.95	5.12	5.27	5.40	5.51	5.62	5.71	5.80	5.88
13	3.74	4.15	4.45	4.69	4.89	5.05	5.19	5.32	5.43	5.53	5.63	5.71	5.79
14	3.70	4.11	4.41	4.64	4.83	4.99	5.13	5.25	5.36	5.46	5.55	5.64	5.71
15	3.67	4.08	4.37	4.60	4.78	4.94	5.08	5.20	5.31	5.40	5.49	5.57	5.65
16	3.65	4.05	4.33	4.56	4.74	4.90	5.03	5.15	5.26	5.35	5.44	5.52	5.59
17	3.63	4.02	4.30	4.52	4.71	4.86	4.99	5.11	5.21	5.31	5.39	5.47	5.54
18	3.61	4.00	4.28	4.50	4.67	4.82	4.96	5.07	5.17	5.27	5.35	5.43	5.50
19	3.59	3.98	4.25	4.47	4.65	4.79	4.92	5.04	5.14	5.23	5.32	5.39	5.46
20	3.58	3.96	4.23	4.45	4.62	4.77	4.90	5.01	5.11	5.20	5.28	5.36	5.43
24	3.53	3.90	4.17	4.37	4.54	4.68	4.81	4.92	5.01	5.10	5.18	5.25	5.32
30	3.49	3.85	4.10	4.30	4.46	4.60	4.72	4.82	4.92	5.00	5.08	5.15	5.21
40	3.44	3.79	4.04	4.23	4.39	4.52	4.64	4.74	4.82	4.90	4.98	5.04	5.11
60	3.40	3.74	3.98	4.16	4.31	4.44	4.55	4.65	4.73	4.81	4.88	4.94	5.00
120	3.36	3.69	3.92	4.10	4.24	4.36	4.47	4.56	4.64	4.71	4.78	4.84	4.90
∞	3.31	3.63	3.86	4.03	4.17	4.29	4.39	4.47	4.55	4.62	4.69	4.74	4.80

1965) proposed a single-stage procedure for this problem that is applicable in cases when the data from the competing designs are independent, balanced (i.e., $n_1 = \cdots = n_k = n$) and are normally distributed with common (unknown) variance σ^2.

First specify the desired probability $1 - \alpha$ of actually including the best design in the selected subset. Then simulate the systems independently to obtain IID normal outputs $Y_{i1}, Y_{i2}, \ldots, Y_{in}$ for $i = 1, 2, \ldots, k$. Let $\overline{Y}_i = \sum_{j=1}^{n} Y_{ij}/n$ be the sample mean from system i, and let

$$S^2 = \frac{1}{k} \sum_{i=1}^{k} \frac{1}{n-1} \sum_{j=1}^{n} (Y_{ij} - \overline{Y}_i)^2 = \frac{1}{k} \sum_{i=1}^{k} \frac{1}{n-1} \left(\sum_{j=1}^{n} Y_{ij}^2 - n\overline{Y}_i^2 \right)$$

be the pooled sample variance, an unbiased estimator of σ^2.

Include the ith design in the selected subset if

$$\overline{Y}_i \leq \min_{1 \leq j \leq k} \overline{Y}_j + gS\sqrt{\frac{2}{n}}$$

where $g = T_{k-1, k(n-1)}^{(\alpha)}$ is a critical value from a multivariate t-distribution. Table 8.2 gives values of this constant for $\alpha = 0.05$; more complete tables can be found in Hochberg and Tamhane (1987, App. 3, Table 4); Bechhofer, Santner, and Goldsman (BSG) (1995); or by using the Fortran program `AS251` of Dunnett (1989). Gupta and Huang (1976) proposed a similar procedure (requiring more obscure tables) for the unbalanced case.

Notice that if a larger value of the system performance measure is better, the ith design is selected if $\overline{Y}_i \geq \max_{1 \leq j \leq k} \overline{Y}_j - gS\sqrt{2/n}$.

Numerical Example. Continuing the example above, suppose that the objective is to determine, with confidence 0.95, a subset of architectures that contains the one with the shortest response time. Then the Gupta procedure selects those architectures for which

$$\overline{Y}_i \leq 62 + 2.19\sqrt{100.9\left(\frac{2}{6}\right)} = 74.7$$

where $T_{3,20}^{(0.05)} = 2.19$ is from Table 8.2. Therefore, the procedure selects architectures 1 and 4.

8.3.2 Selecting the Best

Example 2 For the purpose of evaluation prior to purchase, simulation models of four different airline-reservation systems have been developed. The single measure of system performance is the time to failure (TTF), so that larger TTF is better. A reservation system works if either of two computers works. The four systems arise from variations in parameters affecting the TTF and time-to-repair distributions. Differences of less than about 2 days are considered practically equivalent.

TABLE 8.2 95% Critical Values $T_{p,\nu}^{(0.05)}$ of the Multivariate t-Distribution with Common Correlation 1/2

	p								
ν	1	2	3	4	5	6	7	8	9
1	6.31	9.51	11.58	13.10	14.27	15.23	16.04	16.73	17.34
2	2.92	3.80	4.34	4.71	5.00	5.24	5.44	5.60	5.75
3	2.35	2.94	3.28	3.52	3.70	3.85	3.97	4.08	4.17
4	2.13	2.61	2.88	3.08	3.22	3.34	3.44	3.52	3.59
5	2.02	2.44	2.68	2.85	2.98	3.08	3.16	3.24	3.30
6	1.94	2.34	2.56	2.71	2.83	2.92	3.00	3.06	3.12
7	1.89	2.27	2.48	2.62	2.73	2.82	2.89	2.95	3.00
8	1.86	2.22	2.42	2.55	2.66	2.74	2.81	2.87	2.92
9	1.83	2.18	2.37	2.50	2.60	2.68	2.75	2.81	2.86
10	1.81	2.15	2.34	2.47	2.56	2.64	2.70	2.76	2.81
11	1.80	2.13	2.31	2.43	2.53	2.60	2.67	2.72	2.77
12	1.78	2.11	2.29	2.41	2.50	2.58	2.64	2.69	2.74
13	1.77	2.09	2.27	2.39	2.48	2.55	2.61	2.66	2.71
14	1.76	2.08	2.25	2.37	2.46	2.53	2.59	2.64	2.69
15	1.75	2.07	2.24	2.36	2.44	2.52	2.57	2.62	2.67
16	1.75	2.06	2.23	2.34	2.43	2.50	2.56	2.61	2.65
17	1.74	2.05	2.22	2.33	2.42	2.49	2.54	2.59	2.64
18	1.73	2.04	2.21	2.32	2.41	2.48	2.53	2.58	2.62
19	1.73	2.03	2.20	2.31	2.40	2.47	2.52	2.57	2.61
20	1.72	2.03	2.19	2.30	2.39	2.46	2.51	2.56	2.60
25	1.71	2.00	2.17	2.27	2.36	2.42	2.48	2.52	2.56
30	1.70	1.99	2.15	2.25	2.34	2.40	2.45	2.50	2.54
35	1.69	1.98	2.13	2.24	2.32	2.38	2.44	2.48	2.52
40	1.68	1.97	2.13	2.23	2.31	2.37	2.42	2.47	2.51
45	1.68	1.96	2.12	2.22	2.30	2.36	2.41	2.46	2.50
50	1.68	1.96	2.11	2.22	2.29	2.36	2.41	2.45	2.49
55	1.67	1.96	2.11	2.21	2.29	2.35	2.40	2.44	2.48
60	1.67	1.95	2.10	2.21	2.28	2.35	2.40	2.44	2.48
120	1.66	1.93	2.08	2.18	2.26	2.32	2.37	2.41	2.45
∞	1.65	1.92	2.06	2.16	2.23	2.29	2.34	2.38	2.42

Indifference-Zone Selection Approach. If expected TTF is taken as the performance measure of interest, the goal in this example is to select the system with the largest expected TTF. In a stochastic simulation such a "correct selection" can never be guaranteed with certainty. A compromise solution offered by *indifference-zone selection* is to guarantee to select the best system with high probability, say $1 - \alpha$, whenever it is at least a user-specified amount better than the others; this practically significant difference is called the indifference zone. In the example the indifference zone is $\delta = 2$ days. What happens if, unknown to the user, some system happens to be within δ of the best (i.e., within the indifference zone)? Then it can (usually) be shown that the probability of selecting a *good* system (i.e., one of the systems within the indifference zone) is *at least* $1 - \alpha$.

Law and Kelton (1991) present indifference-zone procedures that have proven useful in simulation, while Bechhofer et al. (1995) provide a comprehensive review of R&S

procedures. In this section we present three procedures, one due to Rinott (1978) that is applicable when the output data are normally distributed and all systems are simulated independently of each other, and two others, due to Nelson and Matejcik (1995), that work in conjunction with common random numbers. See Section 8.4 for a numerical example employing the first procedure.

Multiple Comparisons Approach. Multiple comparisons addresses the problem of determining the best system by forming simultaneous confidence intervals on the parameters $\mu_i - \max_{l \neq i} \mu_l$ for $i = 1, 2, \ldots, k$, where μ_i denotes the expected TTF for the ith reservation system. These confidence intervals are known as *multiple comparisons with the best* (MCB), and they bound the difference between the expected performance of each system and the best of the others, with probability $1 - \alpha$. The first MCB procedures were developed by Hsu (1984); a thorough review is provided in Hochberg and Tamhane (1987).

Matejcik and Nelson (1995) and Nelson and Matejcik (1995) established a fundamental connection between indifference-zone selection and MCB by showing that *most indifference-zone procedures can simultaneously provide MCB confidence intervals with the width of the intervals corresponding to the indifference zone.* The procedures displayed below are combined indifference-zone selection and MCB procedures. The advantage of a combined procedure is that it not only selects a system as best but also provides information about how close each of the inferior systems is to being the best. This information is useful if secondary criteria that are not reflected in the performance measure (such as ease of installation, cost to maintain, etc.) may tempt one to choose an inferior system if it is not deficient by much, say less than δ.

The combined procedures below use the convention that a "." subscript indicates averaging with respect to that subscript. For example, $\overline{Y}_{i\cdot}$ is the sample average of Y_{i1}, $Y_{i2}, \ldots$.

The first procedure, Rinott + MCB, takes observations in two stages. The first stage uses $n_0 \geq 2$ observations from each system to estimate marginal variances; an initial sample size of at least $n_0 = 10$ is recommended. These estimates establish the number of observations to be taken in the second stage of sampling in order to meet the indifference-zone probability requirement.

Procedure Rinott + MCB (Independent Sampling)

1. Specify δ, α, and the first-stage sample size n_0.
2. Take an IID sample $Y_{i1}, Y_{i2}, \ldots, Y_{in_0}$ from each of the k systems *simulated independently*.
3. Compute the marginal sample variances

$$S_i^2 = \frac{\sum_{j=1}^{n_0} (Y_{ij} - \overline{Y}_{i\cdot})^2}{n_0 - 1} = \frac{\sum_{j=1}^{n_0} Y_{ij}^2 - n_0 \overline{Y}_{i\cdot}^2}{n_0 - 1}$$

   for $i = 1, 2, \ldots, k$.
4. Find h from Table 8.3 if $\alpha = 0.1$ or 0.05 [otherwise, use the more detailed tables in Wilcox (1984) or Bechhofer et al. (1995)]. Compute the final sample sizes

TABLE 8.3 Values of h Required by the Procedure Rinott + MCB

		k								
α	n_0	2	3	4	5	6	7	8	9	10
0.10	5	2.291	3.058	3.511	3.837	4.093	4.305	4.486	4.644	4.786
	6	2.177	2.871	3.270	3.552	3.771	3.951	4.103	4.235	4.352
	7	2.107	2.758	3.126	3.384	3.582	3.744	3.881	3.999	4.103
	8	2.059	2.682	3.031	3.273	3.459	3.609	3.736	3.845	3.941
	9	2.025	2.628	2.963	3.195	3.372	3.515	3.635	3.738	3.829
	10	1.999	2.587	2.913	3.137	3.307	3.445	3.560	3.659	3.746
	11	1.978	2.556	2.874	3.092	3.258	3.391	3.503	3.598	3.682
	12	1.962	2.531	2.843	3.056	3.218	3.349	3.457	3.551	3.632
	13	1.948	2.510	2.817	3.027	3.186	3.314	3.420	3.512	3.592
	14	1.937	2.493	2.796	3.003	3.160	3.285	3.390	3.480	3.558
	15	1.928	2.479	2.779	2.983	3.138	3.261	3.364	3.453	3.530
	16	1.919	2.467	2.764	2.966	3.119	3.241	3.343	3.430	3.506
	17	1.912	2.456	2.751	2.951	3.102	3.223	3.324	3.410	3.485
	18	1.906	2.447	2.739	2.938	3.088	3.208	3.308	3.393	3.467
	19	1.901	2.438	2.729	2.926	3.075	3.194	3.293	3.378	3.451
	20	1.896	2.431	2.720	2.916	3.064	3.182	3.280	3.364	3.437
	30	1.866	2.387	2.666	2.855	2.997	3.110	3.204	3.284	3.354
	40	1.852	2.366	2.641	2.827	2.966	3.077	3.169	3.247	3.315
	50	1.844	2.354	2.627	2.810	2.948	3.057	3.148	3.225	3.292
0.05	5	3.107	3.905	4.390	4.744	5.025	5.259	5.461	5.638	5.797
	6	2.910	3.602	4.010	4.303	4.533	4.722	4.884	5.025	5.150
	7	2.791	3.424	3.791	4.051	4.253	4.419	4.559	4.681	4.789
	8	2.712	3.308	3.649	3.889	4.074	4.225	4.353	4.463	4.561
	9	2.656	3.226	3.550	3.776	3.950	4.091	4.210	4.313	4.404
	10	2.614	3.166	3.476	3.693	3.859	3.993	4.106	4.204	4.290
	11	2.582	3.119	3.420	3.629	3.789	3.918	4.027	4.121	4.203
	12	2.556	3.082	3.376	3.579	3.734	3.860	3.965	4.055	4.135
	13	2.534	3.052	3.340	3.539	3.690	3.812	3.915	4.003	4.080
	14	2.517	3.027	3.310	3.505	3.654	3.773	3.874	3.960	4.035
	15	2.502	3.006	3.285	3.477	3.623	3.741	3.839	3.924	3.998
	16	2.489	2.988	3.264	3.453	3.597	3.713	3.810	3.893	3.966
	17	2.478	2.973	3.246	3.433	3.575	3.689	3.785	3.867	3.938
	18	2.468	2.959	3.230	3.415	3.556	3.669	3.763	3.844	3.914
	19	2.460	2.948	3.216	3.399	3.539	3.650	3.744	3.824	3.894
	20	2.452	2.937	3.203	3.385	3.523	3.634	3.727	3.806	3.875
	30	2.407	2.874	3.129	3.303	3.434	3.539	3.626	3.701	3.766
	40	2.386	2.845	3.094	3.264	3.392	3.495	3.580	3.652	3.716
	50	2.373	2.828	3.074	3.242	3.368	3.469	3.553	3.624	3.687

$$N_i = \max\left\{n_0, \left\lceil\left(\frac{hS_i}{\delta}\right)^2\right\rceil\right\}$$

for $i = 1, 2, \ldots, k$, where $\lceil\cdot\rceil$ means to round up.

5. Take $N_i - n_0$ additional IID observations from system i, independently of the first-stage sample and the other systems, for $i = 1, 2, \ldots, k$.

6. Compute the overall sample means

$$\overline{\overline{Y}}_{i\cdot} = \frac{1}{N_i} \sum_{j=1}^{N_i} Y_{ij}$$

for $i = 1, 2, \ldots, k$.

7. Select the system with the largest $\overline{\overline{Y}}_{i\cdot}$ as best.
8. Simultaneously form the MCB confidence intervals for $\mu_i - \max_{l \neq i} \mu_l$ as

$$\min(0, \overline{\overline{Y}}_{i\cdot} - \max_{l \neq i} \overline{\overline{Y}}_{l\cdot} - \delta), \quad \max(0, \overline{\overline{Y}}_{i\cdot} - \max_{l \neq i} \overline{\overline{Y}}_{l\cdot} + \delta) \tag{4}$$

for $i = 1, 2, \ldots, k$.

Notice that in step 5 it is equally valid to generate the N_i observations by restarting the simulation of system i from the beginning, thereby regenerating the initial n_0 observations. In some simulation languages this approach is easier than restarting the simulation from the end of the first stage.

If a smaller performance measure is better, Rinott + MCB (and the two procedures that follow) change only in the final two steps, which become

7. Select the system with the smallest $\overline{\overline{Y}}_{i\cdot}$ as best.
8. Simultaneously form the MCB confidence intervals for $\mu_i - \min_{l \neq i} \mu_l$ as

$$\min(0, \overline{\overline{Y}}_{i\cdot} - \min_{l \neq i} \overline{\overline{Y}}_{l\cdot} - \delta), \quad \max(0, \overline{\overline{Y}}_{i\cdot} - \min_{l \neq i} \overline{\overline{Y}}_{l\cdot} + \delta)$$

for $i = 1, 2, \ldots, k$.

Rinott's procedure, and the accompanying MCB intervals, simultaneously guarantee a probability of correct selection and confidence-interval coverage greater than or equal to $1 - \alpha$ under the stated assumptions. A fundamental assumption of the Rinott + MCB procedure is that the k systems are simulated independently (see step 2 above). In practice this means that different streams of (pseudo)random numbers are assigned to the simulation of each system. However, under fairly general conditions, assigning common random numbers (CRN) to the simulation of each system decreases the variances of estimates of the pairwise differences in performance (see Section 8.2.3). The following procedure provides the same guarantees as Rinott + MCB under a more complex set of conditions, but it has been shown to be quite robust to departures from those conditions. Unlike Rinott + MCB, it is designed to exploit the use of CRN to reduce the total number of observations required to make a correct selection.

Procedure NM + MCB (Common Random Numbers)

1. Specify δ, α, and the first-stage sample size n_0.
2. Take an IID sample $Y_{i1}, Y_{i2}, \ldots, Y_{in_0}$ from each of the k systems *using CRN across systems*.

3. Compute the approximate sample variance of the difference of the sample means

$$S^2 = \frac{2\sum_{i=1}^{k}\sum_{j=1}^{n_0}(Y_{ij} - \overline{Y}_{i\cdot} - \overline{Y}_{\cdot j} + \overline{Y}_{\cdot\cdot})^2}{(k-1)(n_0-1)}$$

$$= \frac{2}{(k-1)(n_0-1)}\left[\sum_{i=1}^{k}\left(\sum_{j=1}^{n_0} Y_{ij}^2 - n_0\overline{Y}_{i\cdot}^2\right) - k\left(\sum_{j=1}^{n_0}\overline{Y}_{\cdot j}^2 - n_0\overline{Y}_{\cdot\cdot}^2\right)\right]$$

4. Let $g = T^{(\alpha)}_{k-1,(k-1)(n_0-1)}$ [see Table 8.2; or Hochberg and Tamhane (1987, App. 3, Table 4); or Bechhofer et al. (1995)]. Compute the final sample size

$$N = \max\left\{n_0, \left\lceil\left(\frac{gS}{\delta}\right)^2\right\rceil\right\}$$

5. Take $N - n_0$ additional IID observations from each system, using CRN across systems.
6. Compute the overall sample means

$$\overline{\overline{Y}}_{i\cdot} = \frac{1}{N}\sum_{j=1}^{N} Y_{ij}$$

   for $i = 1, 2, \ldots, k$.
7. Select the system with the largest $\overline{\overline{Y}}_{i\cdot}$ as best.
8. Simultaneously form the MCB confidence intervals as in Rinott + MCB.

The following procedure provides the same guarantees as Rinott + MCB under more general conditions than NM + MCB; in fact, it requires only normally distributed data. However, because it is based on the Bonferroni inequality, it tends to require more observations than NM + MCB, especially when k is large.

Procedure Bonferroni + MCB (Common Random Numbers)

1. Specify δ, α and n_0. Let $t = t_{1-\alpha/(k-1), n_0-1}$.
2. Take an IID sample $Y_{i1}, Y_{i2}, \ldots, Y_{in_0}$ from each of the k systems *using CRN across systems*.
3. Compute the sample variances of the differences

$$S_{il}^2 = \frac{1}{n_0 - 1}\sum_{j=1}^{n_0}[Y_{ij} - Y_{lj} - (\overline{Y}_{i\cdot} - \overline{Y}_{l\cdot})]^2$$

   for all $i \neq l$.
4. Compute the final sample size

$$N = \max\left\{n_0, \left\lceil \max_{l \neq i}\left(\frac{tS_{il}}{\delta}\right)^2 \right\rceil\right\}$$

5. Take $N - n_0$ additional IID observations from each system, using CRN across systems.
6. Compute the overall sample means

$$\overline{\overline{Y}}_{i\cdot} = \frac{1}{N}\sum_{j=1}^{N} Y_{ij}$$

for $i = 1, 2, \ldots, k$.
7. Select the system with the largest $\overline{\overline{Y}}_{i\cdot}$ as best.
8. Simultaneously form the MCB confidence intervals as in Rinott + MCB.

For a numerical example, see Section 8.4.

Multinomial Selection Approach. Another approach to the airline-reservation problem is to select the system that is most likely to have the largest *actual* TTF. To this end, one can define p_i as the probability that design i will produce the largest TTF from a given observation from each system, $(Y_{1j}, Y_{2j}, \ldots, Y_{kj})$ [i.e., $p_i = \Pr\{Y_{ij} > \max_{l \neq i} Y_{lj}\}$. The goal now is to select the design associated with the largest p_i-value.

More specifically, suppose that the goal is to select the best system with probability $1 - \alpha$ whenever the ratio of the largest to second-largest p_i is greater than some user-specified constant, say $\theta > 1$. The indifference constant θ can be regarded as the smallest ratio "worth detecting."

The following *single-stage* procedure was proposed by Bechhofer, Elmaghraby, and Morse (BEM) (1959) to guarantee the foregoing probability requirement.

Procedure BEM

1. For the given k and specified α and θ, find n from Table 8.4 for $\alpha = 0.25, 0.10,$ or 0.05 [or from the tables in Bechhofer et al. (1959), Gibbons et al. (1977), or Bechhofer et al. (1995)].
2. Take a random sample of n multinomial observations $\mathbf{X}_j = (X_{1j}, X_{2j}, \ldots, X_{kj})$, for $j = 1, 2, \ldots, n$ in a *single* stage, where

$$X_{ij} = \begin{cases} 1 & \text{if } Y_{ij} > \max_{l \neq i}\{Y_{lj}\} \\ 0 & \text{otherwise} \end{cases}$$

In other words, if on the jth replication system i is best, set $X_{ij} = 1$ and $X_{lj} = 0$ for all $l \neq i$ (all of the nonwinners).
3. Let $W_i = \sum_{j=1}^{n} X_{ij}$ for $i = 1, 2, \ldots, k$, the number of times system i is the best. Select the design that yielded the largest W_i as the one associated with the largest p_i (in the case of a tie, pick any of the systems with the largest W_i).

TABLE 8.4 Sample Size n for Multinomial Procedure BEM, and Truncation Numbers n_T for Procedure BG

		$k = 2$		$k = 3$		$k = 4$		$k = 5$	
α	θ	n	n_T	n	n_T	n	n_T	n	n_T
0.25	3.0	1	1	5	5	8	9	11	12
	2.0	5	5	12	13	20	24	29	34
	1.8	5	7	17	18	29	35	41	50
	1.6	9	9	26	32	46	57	68	86
	1.4	17	19	52	71	92	124	137	184
	1.2	55	67	181	285	326	495	486	730
0.10	3.0	7	10	11	12	16	19	21	24
	2.0	15	15	29	34	43	53	58	71
	1.8	19	27	40	50	61	75	83	104
	1.6	31	41	64	83	98	126	134	172
	1.4	59	79	126	170	196	274	271	374
	1.2	199	267	437	670	692	1050	964	1460
0.05	3.0	9	11	17	20	23	26	29	34
	2.0	23	27	42	52	61	74	81	98
	1.8	33	35	59	71	87	106	115	142
	1.6	49	59	94	125	139	180	185	240
	1.4	97	151	186	266	278	380	374	510
	1.2	327	455	645	960	979	1500	1331	2000

Numerical Example. To select the airline reservation system that is most likely to have the largest actual TTF, suppose that management dictates a correct selection be made with probability at least 0.95 whenever the ratio of the largest to second-largest true (but unknown) probabilities of having the largest TTF is at least 1.2. From Table 8.4 with $k = 4$, $\alpha = 0.05$, and the ratio $\theta = 1.2$, one finds that $n = 979$ TTFs must be simulated for each system.

A more efficient but more complex procedure, due to Bechhofer and Goldsman (1986), uses *closed, sequential* sampling; the procedure stops when one design is "sufficiently ahead" of the others.

Procedure BG

1. For the k given, and the α and θ specified, find the *truncation number* (i.e., an upper bound on the number of vector observations) n_T from Table 8.4 for $\alpha =$ 0.25, 0.10, or 0.05 [or from the tables in Bechhofer and Goldsman (1986) or Bechhofer et al. (1995)].
2. At the mth stage of experimentation ($m \geq 1$), take the random multinomial observation $\mathbf{X}_m = (X_{1m}, X_{2m}, \ldots, X_{km})$ (defined in procedure BEM) and calculate the *ordered* category totals $W_{[1]m} \leq W_{[2]m} \leq \cdots \leq W_{[k]m}$, where $W_{im} = \sum_{j=1}^{m} X_{ij}$ is the number of times system i was the best in the first m stages, and $[i]$ indicates the index of the ith smallest total. Also calculate

$$Z_m = \sum_{i=1}^{k-1} \left(\frac{1}{\theta}\right)^{W_{[k]m} - W_{[i]m}}$$

3. Stop sampling at the first stage when *either*

$$Z_m \leq \frac{\alpha}{1-\alpha} \quad \text{or} \quad m = n_T \quad \text{or} \quad W_{[k]m} - W_{[k-1]m} \geq n_T - m$$

whichever occurs first.

4. Let N (a random variable) denote the stage at which the procedure terminates. Select the design that yielded the largest W_{iN} as the one associated with the largest p_i (in the case of a tie, pick any of the systems with the largest W_{iN}).

Numerical Example. Suppose that $k = 3$, $\alpha = 0.25$, and $\theta = 3.0$. Table 8.4 gives a truncation number of $n_T = 5$ observations. Consider the data

m	X_{1m}	X_{2m}	X_{3m}	W_{1m}	W_{2m}	W_{3m}
1	0	1	0	0	1	0
2	0	1	0	0	2	0

Since $Z_2 = (\frac{1}{3})^2 + (\frac{1}{3})^2 = \frac{2}{9} \leq \alpha/(1-\alpha) = \frac{1}{3}$, the first termination criterion for procedure BG dictates that the procedure stops sampling and selects system 2.

Numerical Example. Under the same setup as the preceding example, consider the following data.

m	X_{1m}	X_{2m}	X_{3m}	W_{1m}	W_{2m}	W_{3m}
1	0	1	0	0	1	0
2	1	0	0	1	1	0
3	0	1	0	1	2	0
4	1	0	0	2	2	0
5	1	0	0	3	2	0

Since $m = n_T = 5$ observations, sampling stops by the second criterion, and the procedure selects system 1.

Numerical Example. Again under the same setup, consider the following data.

m	X_{1m}	X_{2m}	X_{3m}	W_{1m}	W_{2m}	W_{3m}
1	0	1	0	0	1	0
2	1	0	0	1	1	0
3	0	1	0	1	2	0
4	1	0	0	2	2	0
5	0	0	1	2	2	1

The procedure stops according to the second criterion because $m = n_T = 5$. However, since there is a tie between W_{15} and W_{25}, the experimenter selects either system 1 or 2.

Numerical Example. Again under the same setup, consider the following data.

m	X_{1m}	X_{2m}	X_{3m}	W_{1m}	W_{2m}	W_{3m}
1	0	1	0	0	1	0
2	1	0	0	1	1	0
3	0	1	0	1	2	0
4	0	0	1	1	2	1

Because systems 1 and 3 can do no better than tie system 2 (even if the potential remaining $n_T - m = 5 - 4 = 1$ observation were taken), the third criterion instructs the user to select system 2.

8.3.3 Comparisons with a Standard

Example 3 Several different investment strategies will be simulated to evaluate their expected rate of return—the higher the better. None of the strategies will be chosen unless its expected return is larger than a zero-coupon bond that offers a known, fixed return. Since factors such as risk could also be considered, the strategy ultimately chosen may not be the one with the largest expected return.

Here the goal is to select the best investment strategy *only if it is better than the standard*; if no strategy is better than the standard, continue with the standard. More precisely, the following probability requirement should be satisfied: Denote the standard by μ_0 and the ordered means of the competitors by $\mu_{[1]} \leq \mu_{[2]} \leq \cdots \leq \mu_{[k]}$. For constants $(\delta, \alpha_0, \alpha_1)$ with $0 < \delta < \infty$, $2^{-k} < 1 - \alpha_0 < 1$ and $(1 - 2^{-k})/k < 1 - \alpha_1 < 1$, specified prior to the start of experimentation, the probability requirement must guarantee that

$$P\{\text{select the standard}\} \geq 1 - \alpha_0 \qquad \text{whenever} \quad \mu_{[k]} \leq \mu_0 \tag{5}$$

and

$$P\{\text{select the best strategy}\} \geq 1 - \alpha_1 \qquad \text{whenever} \quad \mu_{[k]} \geq \max\{\mu_0, \mu_{[k-1]}\} + \delta \tag{6}$$

Equation (5) requires that the standard be selected as best with probability at least $1 - \alpha_0$ whenever μ_0 exceeds all of the means of the other strategies; (6) requires that the strategy with the largest mean be selected with probability at least $1 - \alpha_1$ whenever its mean, $\mu_{[k]}$, exceeds both the standard and the $k - 1$ other means by at least δ.

Presented below is a procedure due to Bechhofer and Turnbull (1978) for selecting the best system relative to a given standard when the responses are normal with common unknown variance σ^2 and the systems are simulated independently. It requires that an initial sample of $n_0 \geq 2$ observations be taken from each system to estimate σ^2 in the first stage; an initial sample size of at least $n_0 = 10$ is recommended.

Procedure BT

1. For the given (k, μ_0) and specified $(\delta, \alpha_0, \alpha_1)$, fix a number of observations $n_0 \geq 2$ to be taken in stage 1.
2. Choose constants (g, h) from Table 8.5 for $1 - \alpha_0 = 1 - \alpha_1 = 1 - \alpha = 0.90$ or 0.95 (or Bechhofer and Turnbull, 1978) corresponding to the k, n_0, $1 - \alpha_0$, and $1 - \alpha_1$ of interest.
3. In stage 1, take a random sample of n_0 observations Y_{ij} $(j = 1, 2, \ldots, n_0)$ from the k strategies. Calculate the first-stage sample means,

$$\overline{Y}_i = \frac{1}{n_0} \sum_{j=1}^{n_0} Y_{ij}$$

   for $i = 1, 2, \ldots, k$, and the unbiased pooled estimate of σ^2,

$$S^2 = \frac{\sum_{i=1}^{k} \sum_{j=1}^{n_0} (Y_{ij} - \overline{Y}_i)^2}{k(n_0 - 1)} = \frac{\sum_{i=1}^{k} \left(\sum_{j=1}^{n_0} Y_{ij}^2 - n_0 \overline{Y}_i^2 \right)}{k(n_0 - 1)}.$$

4. In stage 2, take a random sample of $N - n_0$ *additional* observations from each of the strategies, where

$$N = \max\{n_0, \lceil (gS/\delta)^2 \rceil\}$$

   and $\lceil \cdot \rceil$ means to round up.
5. Calculate the overall sample means

$$\overline{\overline{Y}}_i = \frac{1}{N} \sum_{j=1}^{N} Y_{ij}$$

   for $i = 1, 2, \ldots, k$.
6. If the largest sample mean $\overline{\overline{Y}}_{[k]} > \mu_0 + h\delta/g$, select the strategy that yielded it as the one associated with $\mu_{[k]}$; otherwise, select no strategy and retain the standard as best.

Notice that in step 4 it is equally valid to generate the N observations by restarting the simulation of each system from the beginning, thereby regenerating the initial n_0 observations. In some simulation languages this approach is easier than restarting the simulation from the end of the first stage.

If a smaller performance measure is better, BT changes only in the final step, which becomes

6. If the smallest sample mean $\overline{\overline{Y}}_{[1]} < \mu_0 - h\delta/g$, select the strategy that yielded it as the one associated with $\mu_{[1]}$; otherwise, select no strategy and retain the standard as best.

TABLE 8.5 Values of g (Top Entries) and h (Bottom Entries) for Procedure BT for Comparing Normal Treatments with a Standard with $1 - \alpha \equiv 1 - \alpha_0 = 1 - \alpha_1$

		$k = 2$		$k = 3$		$k = 4$		$k = 5$	
n_0	$1 - \alpha$:	0.90	0.95	0.90	0.95	0.90	0.95	0.90	0.95
2		4.775	7.178	4.416	6.041	4.269	5.582	4.197	5.342
		2.743	4.075	2.648	3.551	2.620	3.340	2.616	3.234
3		3.679	4.920	3.688	4.701	3.714	4.619	3.743	4.584
		2.072	2.722	2.172	2.696	2.243	2.704	2.300	2.721
4		3.401	4.406	3.487	4.357	3.554	4.356	3.608	4.369
		1.905	2.417	2.042	2.479	2.136	2.531	2.207	2.576
5		3.276	4.181	3.394	4.201	3.478	4.239	3.544	4.267
		1.829	2.285	1.982	2.381	2.085	2.451	2.162	2.507
6		3.204	4.055	3.340	4.112	3.434	4.163	3.506	4.208
		1.787	2.211	1.947	2.325	2.055	2.405	2.136	2.468
8		3.126	3.920	3.281	4.014	3.384	4.085	3.463	4.142
		1.740	2.132	1.909	2.263	2.022	2.354	2.107	2.424
10		3.084	3.848	3.248	3.961	3.358	4.043	3.440	4.107
		1.715	2.090	1.888	2.230	2.004	2.327	2.091	2.400
12		3.058	3.804	3.227	3.928	3.341	4.016	3.426	4.084
		1.699	2.064	1.875	2.210	1.993	2.309	2.081	2.385
∞		2.945	3.615	3.139	3.786	3.266	3.900	3.361	3.986
		1.632	1.955	1.818	2.121	1.943	2.234	2.037	2.319

Numerical Example. Suppose that there are $k = 3$ investment strategies to compare to a zero-coupon bond that has a fixed return of $\mu_0 = \$25{,}000$. The investor requires that to merit the riskier investment, the best of the alternatives have at least a $\delta = \$500$ greater expected return than the bond. The investor wants to make the correct selection with probability at least 0.95.

An initial simulation of $n_0 = 10$ replications is executed for each investment strategy (excluding the standard, whose expected return is known), giving a pooled sample variance estimate of $S^2 = (827)^2$. From Table 8.5 with $1 - \alpha_0 = 1 - \alpha_1 = 0.95$, the critical values are $g = 3.961$ and $h = 2.230$. Therefore, a total of

$$N = \max\left\{10, \left\lceil \frac{(3.961)^2(827)^2}{500^2} \right\rceil\right\} = \max\{10, \lceil 42.9 \rceil\} = 43$$

replications is needed for each of the $k = 3$ investment strategies, which can be obtained by restarting the simulations from the beginning, or by simulating $43 - 10 = 33$ additional replications of each alternative. After obtaining these data, the investment with the largest sample mean return, $\overline{\overline{Y}}_{[3]}$, will be selected if

$$\overline{\overline{Y}}_{[3]} > \$25{,}000 + \frac{(2.230)(\$500)}{3.961} = \$25{,}281$$

If this inequality is not satisfied, the investor will stay with the zero-coupon bond.

8.3.4 Comparisons with a Default

Example 4 A manufacturing company will replace an existing storage-and-retrieval system if one can be found that is superior to the system currently in place. Five vendors have proposed hardware–software systems, and simulation models have been developed for each. Systems will be evaluated in terms of their retrieval times, but the system ultimately chosen might not be the one with the smallest retrieval time because of differences in cost, ease of installation, and so on.

If the expected retrieval time is the performance measure of interest, the goal of the simulation study is to determine which designs are better than the system currently in place, which is called the *default* (or *control*). Data on the performance of the default system may be obtained either from the system itself or from a simulation model of the system.

Notice that this problem of comparison with a default differs somewhat from that of comparison with a standard. In comparison with a default, the contending systems must compete against themselves as well as the (stochastic) system currently being used; in comparison with a standard, the contending systems compete against themselves and must also better a *fixed* hurdle.

Multiple Comparisons Approach. Let μ_i denote the expected retrieval time for system i, where $i = 0$ corresponds to the default system, and $i = 1, 2, \ldots, k-1$ correspond to $k-1$ competing systems (for a total of k stochastic systems). A fundamental principle of multiple comparisons is to make only the comparisons that are necessary for the decision at hand, because the fewer confidence statements that are required to be true simultaneously, the sharper the inference. Therefore, when comparing to a default, it is better to find simultaneous confidence intervals for $\mu_i - \mu_0$ for $i = 1, 2, \ldots, k-1$, rather than $\mu_i - \mu_l$ for all $i \neq l$. Such comparisons are called *multiple comparisons with a control* (MCC). Further, if differences in a *specified direction* are of interest—in the example only systems with smaller expected retrieval time than the default are of interest—one-sided confidence intervals should be formed.

Suppose that the data are acquired independently from each system (a necessity if data are collected from the default system itself) to obtain IID outputs $Y_{i1}, Y_{i2}, \ldots, Y_{in_i}$ from system i. Let

$$S_i^2 = \frac{1}{n_i - 1} \sum_{j=1}^{n_i} (Y_{ij} - \overline{Y}_i)^2 = \frac{1}{n_i - 1} \left(\sum_{j=1}^{n_i} Y_{ij}^2 - n_i \overline{Y}_i^2 \right)$$

be the sample variance from system i. MCC procedures are well known for the case when the variances across systems are equal and the data are normal (see, e.g., Hochberg and Tamhane, 1987; Miller 1981). Presented here is a simple procedure due to Tamhane (1977) that is valid when variances may not be equal.

The simultaneous, upper one-sided confidence intervals are

$$\mu_i - \mu_0 \leq \overline{Y}_i - \overline{Y}_0 + \sqrt{\frac{t_{1-\beta, n_i - 1}^2 S_i^2}{n_i} + \frac{t_{1-\beta, n_0 - 1}^2 S_0^2}{n_0}}$$

for all $i \neq 0$, where $t_{1-\beta,\nu}$ is the $1-\beta = (1-\alpha)^{1/(k-1)}$ quantile of the t-distribution with ν degrees of freedom. When the output data are normally distributed, these intervals are guaranteed to have coverage at least $1-\alpha$ regardless of the system variances. If the upper bound for $\mu_i - \mu_0$ is less than or equal to 0, one can conclude that system i has lower expected retrieval time than the default.

Numerical Example. Suppose that the daily average storage-and-retrieval time is obtained from 10 simulated days (replications) of each of the five proposed systems, each simulated independently. The corresponding data for the system currently in place are available for 40 days. Summary statistics are as follows, with storage-and-retrieval times in minutes. System 0 corresponds to the default.

	i					
	0	1	2	3	4	5
$\overline{Y}_i$	8.8	7.2	7.8	6.9	8.6	7.7
S_i^2	3.1	2.8	3.1	2.8	4.1	3.0
n_i	40	10	10	10	10	10

When $\alpha = 0.10$ and $k = 6$, $1-\beta = (1-0.1)^{1/5} = 0.98$, giving $t_{0.98,9} = 2.398$ and $t_{0.98,39} = 2.125$. Based on these results,

$$\mu_3 - \mu_0 \leq (6.9 - 8.8) + \sqrt{\frac{(2.398)^2(2.8)}{10} + \frac{(2.125)^2(3.1)}{40}} = -0.5$$

establishing that proposed system 3 has an expected storage-and-retrieval time that is at least 0.5 minute shorter than that of the system currently in place. Similarly,

$$\mu_4 - \mu_0 \leq (8.6 - 8.8) + \sqrt{\frac{(2.398)^2(4.1)}{10} + \frac{(2.125)^2(3.1)}{40}} = 1.4$$

which shows that proposed system 4 may not be better than the default. Both of these statements, together with analogous confidence intervals for $\mu_1 - \mu_0$, $\mu_2 - \mu_0$ and $\mu_5 - \mu_0$, are correct with overall confidence level $1 - \alpha = 0.9$.

Selection Procedure. Here the goal is to select the best storage-and-retrieval system *only if it is better than the current system*; if no alternative system is better than the default, continue with the default. (Recall that in this example, smaller expected retrieval time is better.) More precisely, the following probability requirement should be satisfied: Denote the expected storage-and-retrieval time of the default by μ_0 and the ordered means of the competitors by $\mu_{[1]} \leq \mu_{[2]} \leq \cdots \leq \mu_{[k-1]}$. For constants $(\delta, \alpha_0, \alpha_1)$ with $0 < \delta < \infty$, $2^{-k} < 1 - \alpha_0 < 1$, and $(1 - 2^{-k})/k < 1 - \alpha_1 < 1$, specified prior to the start of experimentation, the probability requirement must guarantee that

$$P\{\text{select the default}\} \geq 1 - \alpha_0 \qquad \text{whenever} \quad \mu_{[1]} \geq \mu_0 \tag{7}$$

and

TABLE 8.6 Sample Sizes n to Implement Procedure P for $\alpha_0 = \alpha_1 = 0.05$

	k								
δ/σ:	2	3	4	5	6	7	8	9	10
1.0	22	26	28	29	31	31	32	33	34
0.9	27	32	34	36	38	39	40	41	41
0.8	34	40	43	46	47	49	50	51	52
0.7	45	52	57	60	62	64	65	67	68
0.6	61	71	77	81	84	87	89	91	92
0.5	87	102	110	116	121	124	128	130	133
0.4	136	159	172	182	188	194	199	203	207
0.3	242	283	306	322	334	345	353	361	368
0.2	543	636	687	725	751	775	795	811	827
0.1	2172	2543	2747	2897	3004	3098	3177	3241	3306

$$P\{\text{select best alternative}\} \geq 1 - \alpha_1 \qquad \text{whenever} \quad \mu_{[1]} \leq \min\{\mu_0, \mu_{[2]}\} - \delta \tag{8}$$

Equation (7) requires that the default be selected as best with probability at least $1 - \alpha_0$ whenever μ_0 is smaller than all of the means of the alternatives; (8) requires that the alternative system with the smallest mean be selected with probability at least $1 - \alpha_1$ whenever its mean, $\mu_{[1]}$, is less than both the default and the $k - 2$ other means by at least δ.

The following single-stage procedure was proposed by Paulson (1952) for the case in which σ^2 is *known*. In practice this means that a pilot experiment must be run to estimate σ^2, after which it is treated as known.

Procedure P

1. For the k given and $(\delta, \alpha_0, \alpha_1)$ specified, let $h = T_{k-1,\infty}^{(\alpha_0)}$ be determined from the ∞ row of Table 8.2 for $\alpha_0 = 0.05$ [for other values of α_0, see Table A.1 of Bechhofer et al. (1995)]. Further, let n be the sample size from Table 8.6 for $\alpha_0 = \alpha_1 = 0.05$.
2. Take a random sample of n observations $Y_{i1}, Y_{i2}, \ldots, Y_{in}$ in a single stage from each system, including the default, 0.
3. Calculate the k sample means

$$\overline{Y}_i = \frac{1}{n} \sum_{j=1}^{n} Y_{ij}$$

for $i = 0, 1, 2, \ldots, k - 1$. Let $\overline{Y}_{[1]} = \min\{\overline{Y}_1, \ldots, \overline{Y}_{k-1}\}$ denote the smallest (nondefault) sample mean.

4. If $\overline{Y}_{[1]} < \overline{Y}_0 - h\sigma\sqrt{2/n}$, select the system associated with $\overline{Y}_{[1]}$; otherwise, select the default.

Notice that if larger performance is better, the only change in procedure P occurs in step 4, which becomes

4. Let $\overline{Y}_{[k-1]} = \max\{\overline{Y}_1, \ldots, \overline{Y}_{k-1}\}$ denote the largest (nondefault) sample mean. If $\overline{Y}_{[k-1]} > \overline{Y}_0 + h\sigma\sqrt{2/n}$, select the system associated with $\overline{Y}_{[k-1]}$; otherwise, select the default.

Numerical Example. Continuing the example above, suppose that the data from the previous experiment are used to derive a pooled estimate of $\sigma^2 = 3.15$, which will be treated as the known variance from here on. If a difference of less than 1 minute in storage-and-retrieval time means that it is not worth replacing the current system, $\delta = 1$ and the ratio $\delta/\sigma = 1/\sqrt{3.15} = 0.56$. If a confidence level at least $1 - \alpha_0 = 1 - \alpha_1 = 0.95$ is required for the decision, $h = T^{(0.05)}_{6-1,\infty} = 2.23$ (from Table 8.2) and a sample of size $n = 98$ must be taken from each system, according to the $k = 6$ column of Table 8.6 (after interpolation). After obtaining n replications of each system, including the default, the alternative with the smallest average storage-and-retrieval time is selected over the current system only if

$$\overline{Y}_{[1]} < \overline{Y}_0 - h\sigma\sqrt{\frac{2}{n}} = \overline{Y}_0 - 2.23\sqrt{3.15}\sqrt{\frac{2}{98}} = \overline{Y}_0 - 0.57$$

8.3.5 Estimating Functional Relationships

Example 5 A number of factors may affect the performance of the storage-and-retrieval system described in Section 8.3.4, including the number of retrieval devices, the configuration of the storage area, and the logic that determines which items are stored or retrieved next. A simulation study is performed to determine the impact of these and other factors on performance as a first step toward choosing a configuration.

In this example the "systems" are defined by different specifications for a single physical system. Questions that naturally arise include: Which factors have a significant impact and are therefore worthy of further study? Which combination of the factor settings gives the best performance, and how much better is it than other settings? And how sensitive is system performance to a particular factor? Rather than treating the factor settings as defining distinct, unrelated systems, it is statistically more efficient to exploit a functional relationship among them by fitting a model.

The relationship that is most often assumed is the general linear model (GLM)

$$\mathbf{Y} = \mathbf{X}\boldsymbol{\beta} + \boldsymbol{\epsilon} \tag{9}$$

where $\mathbf{Y}$ is a $k \times 1$ vector of outputs across k design points (systems), $\mathbf{X}$ is a $k \times p$ known design matrix of independent variables that defines the alternative systems, $\boldsymbol{\beta}$ is a $p \times 1$ vector of unknown parameters that captures the relationships among the systems, and $\boldsymbol{\epsilon}$ is a $k \times 1$ vector of mean-zero random errors. A model such as (9) is called a *metamodel* of the simulation. Fitting the model means estimating the value of $\boldsymbol{\beta}$ based on simulation output data. After fitting the model, certain comparisons among the elements of $\boldsymbol{\beta}$ may be of interest, as in Example 5, where there may be interest in comparing the impact of various factors, or comparing mean performance at different settings of the factors.

Suppose that n replications are made at each design point. Let $\overline{Y}_i = \sum_{j=1}^{n} Y_{ij}/n$ be the sample mean at design point i, and let

$$\overline{\mathbf{Y}} = \begin{bmatrix} \overline{Y}_1 \\ \overline{Y}_2 \\ \vdots \\ \overline{Y}_k \end{bmatrix}$$

be the vector of sample means across all k design points. Then as long as the same number of replications is made at each design point, an estimator of $\boldsymbol{\beta}$ is the ordinary-least-squares (OLS) estimator

$$\hat{\boldsymbol{\beta}} = (\mathbf{X}'\mathbf{X})^{-1}\mathbf{X}'\overline{\mathbf{Y}}$$

where $\mathbf{X}'$ indicates the transpose of a matrix $\mathbf{X}$.

The OLS point estimator is appropriate whether or not CRN is employed, but the associated statistical analysis is affected by CRN. Many procedures exist for analysis of a GLM under CRN, but they are complex and depend on a host of conditions. A straightforward procedure proposed by Kleijnen (1988) is presented here. The primary assumption behind Kleijnen's method is that the number of replications at each design point, n, is reasonably large, say $n \geq 25$. This section assumes that the reader is familiar with standard analysis techniques for the GLM when the design points are simulated independently (see Chapter 6).

Let $\hat{\boldsymbol{\Sigma}}_Y$ be the $k \times k$ matrix with i,lth element S_i^2, for $i = l$, and C_{il}, for $i \neq l$, where

$$S_i^2 = \frac{1}{n-1} \sum_{j=1}^{n} (Y_{ij} - \overline{Y}_i)^2 = \frac{1}{n-1} \left(\sum_{j=1}^{n} Y_{ij}^2 - n\overline{Y}_i^2 \right)$$

is the sample variance of the outputs at design point i, and C_{il} is the sample covariance between the outputs from design points i and l [see equation (1)]. In other words, $\hat{\boldsymbol{\Sigma}}_Y$ is an estimator of the variance–covariance matrix of $\mathbf{Y}$ based on n replications. Kleijnen proposes estimating the variance–covariance matrix of $\hat{\boldsymbol{\beta}}$ by

$$\hat{\boldsymbol{\Sigma}}_{\hat{\beta}} = \frac{1}{n} (\mathbf{X}'\mathbf{X})^{-1}\mathbf{X}'(\hat{\boldsymbol{\Sigma}}_Y)\mathbf{X}(\mathbf{X}'\mathbf{X})^{-1} \tag{10}$$

Inference is based on the elements of $\hat{\boldsymbol{\Sigma}}_{\hat{\beta}}$. In particular, an approximate $(1 - \alpha)100\%$ confidence interval for the mth element of $\boldsymbol{\beta}$, β_m, is

$$\hat{\beta}_m \pm t_{1-\alpha/2, n-1}\hat{\sigma}_m$$

where $\hat{\sigma}_m$ is the square root of the mth diagonal element of $\hat{\boldsymbol{\Sigma}}_{\hat{\beta}}$.

Numerical Example. The model $Y = \beta_0 + \beta_1 x + \epsilon$ is used to represent the relationship between the average time in system Y and release rate x in the simulation of a manufacturing facility. At each of $k = 3$ release rates, $x = 2, 5, 8$, the experimenter makes $n = 3$ replications, using CRN across systems on corresponding replications. With $\boldsymbol{\beta} = (\beta_0, \beta_1)'$, the design matrix is

$$\mathbf{X} = \begin{bmatrix} 1 & 2 \\ 1 & 5 \\ 1 & 8 \end{bmatrix}$$

which implies that

$$(\mathbf{X}'\mathbf{X})^{-1} = \begin{bmatrix} 3 & 15 \\ 15 & 93 \end{bmatrix}^{-1} = \begin{bmatrix} 1.722 & -0.278 \\ -0.278 & 0.056 \end{bmatrix}$$

The output data (Y_{ij}) obtained from the simulation are

	Replication, j		
Design Point, i	1	2	3
1	10	7	9
2	12	8	11
3	15	10	12

implying that $\overline{Y} = (8.667, 10.333, 12.333)'$. Therefore, the estimated coefficients are

$$\hat{\boldsymbol{\beta}} = (\mathbf{X}'\mathbf{X})^{-1}\mathbf{X}'\overline{\mathbf{Y}} = \begin{bmatrix} 7.389 \\ 0.611 \end{bmatrix}$$

Inference on these coefficients requires the estimated variance–covariance matrix of $\mathbf{Y}$, which is

$$\hat{\boldsymbol{\Sigma}}_Y = \begin{bmatrix} 2.333 & 3.167 & 3.667 \\ 3.167 & 4.333 & 4.833 \\ 3.667 & 4.833 & 6.333 \end{bmatrix}$$

For example, 2.333 is the sample variance of $\{10, 7, 9\}$, the simulation outputs when $x = 2$; and 3.667 is the covariance between the simulation outputs when $x = 2$ and $x = 8$. Notice that all of the estimated covariances are positive, which is the desired effect of CRN.

Applying (10) yields

$$\hat{\boldsymbol{\Sigma}}_{\hat{\beta}} = \begin{bmatrix} 0.605 & 0.043 \\ 0.043 & 0.012 \end{bmatrix}$$

Thus an approximate 95% confidence interval for β_1, the change in expected average system time per unit change in release rate, is $0.611 \pm (4.303)(0.012)$, or 0.611 ± 0.052. If CRN had been ignored in the analysis and a standard least-squares computation for independent simulations had been used, the standard error $\hat{\sigma}_1 = 0.012$ would become $\hat{\sigma}_1 = 0.263$, which is 20 times larger and would make it impossible even to be sure if $\hat{\beta}_1$ is positive.

CAUTION: This example is for illustrative purposes. Kleijnen recommends $n \geq 25$ replications at each design point for validity of inference based on $\hat{\mathbf{\Sigma}}_{\hat{\beta}}$.

8.4 CASE STUDY: AIRLINE-RESERVATION SYSTEM

The purpose of this section is to illustrate the implementation issues involved in a realistic comparison problem and to serve as an example of good practice. This case study was first described in Goldsman et al. (1991). The goal is to evaluate $k = 4$ different airline-reservation systems. The single measure of performance is the expected time-to-failure, E[TTF]—the larger the better. The system works if either of two computers works. Computer failures are rare, repair times are fast, and the resulting E[TTF] is large. The four systems arise from variations in parameters affecting the time-to-failure and time-to-repair distributions. From experience it is known that the E[TTF] values are roughly 100,000 minutes (about 70 days) for all four systems. The end users are indifferent to expected differences of less than 3000 minutes (about 2 days).

The large E[TTF] values, the highly variable nature of rare failures, the similarity of the systems, and (as it turns out) the relatively small *indifference zone* of 3000 minutes yield a problem with reasonably large computational costs. Although the similarity of the systems suggests the use of common random numbers, for the purpose of this case study the systems are simulated independently.

8.4.1 Method

Recall that the general goal behind R&S methods is to select the best system from among k competitors, $k = 4$ airline-reservation systems in this case. *Best* means the system having the largest underlying E[TTF]. Denote the E[TTF] arising from system i by μ_i, $i = 1, 2, \ldots, k$, and the associated ordered μ_i's by $\mu_{[1]} \leq \mu_{[2]} \leq \cdots \leq \mu_{[k]}$. The μ_i's, $\mu_{[i]}$'s, and their pairings are completely unknown. Since larger E[TTF] is preferred, the mean difference between the two best systems in the airline-reservation example is $\mu_{[k]} - \mu_{[k-1]}$. The smaller this difference is, the greater the amount of sampling that will be required to differentiate between the two best systems. Of course, if $\mu_{[k]} - \mu_{[k-1]}$ is very small, say less than δ, for all practical purposes it would not matter which of the two associated systems is chosen as best. In other words, δ is the smallest difference "worth detecting." In the airline-reservation example, $\delta = 3000$ minutes.

A R&S procedure assures that the probability of making a *correct selection* (CS) of the best system is at least a certain high value, $1 - \alpha$. The greater the value of $1 - \alpha$, the greater the number of observations that will be required. A value of $1 - \alpha = 0.90$ is used in this study.

The two-stage procedure Rinott + MCB in Section 8.3.2 is appropriate for the airline-reservation problem. The procedure assumes that system i produces independent and identically distributed (IID) normal (μ_i, σ_i^2) output, where μ_i and σ_i^2 are unknown, $i = 1, 2, \ldots, k$, and where the k systems are independent. If $\mu_{[k]} - \mu_{[k-1]} > \delta$, the procedure guarantees that $P\{CS\} \geq 1 - \alpha$.

The procedure runs as follows. In the first stage of sampling, take a random sample of n_0 observations from each of the k systems, using different random number streams for each system to ensure independence. To aid the assumption of normality, and to aggregate what turns out to be a very large number of replications, the observations from system i that are used in the analysis are batch means, $\overline{Y}_{i1}, \overline{Y}_{i2}, \ldots$, as defined in

Section 8.2.4. In other words, each of the basic data points $\overline{Y}_{ij}$ is actually an average of the TTF values observed across a batch of replications.

Calculate the first-stage sample means,

$$\overline{Y}_{i\cdot} = \frac{1}{n_0} \sum_{j=1}^{n_0} \overline{Y}_{ij}$$

and sample variances

$$S_i^2 = \frac{1}{n_0 - 1} \sum_{j=1}^{n_0} (\overline{Y}_{ij} - \overline{Y}_{i\cdot})^2$$

for $i = 1, 2, \ldots, k$. The sample variances are used to determine the number of observations that must be taken in the second stage of sampling; the larger a sample variance, the more observations must be taken in the second stage from the associated system. Now set $N_i = \max\{n_0, \lceil (hS_i/\delta)^2 \rceil\}$, where $\lceil \cdot \rceil$ means to round the number up, and h is a constant from Table 8.3. During the second stage of sampling, take $N_i - n_0$ *additional* observations from the ith system, $i = 1, 2, \ldots, k$, or regenerate all N_i observations starting from the beginning.

Finally, calculate the grand means $\overline{\overline{Y}}_{i\cdot} = \sum_{j=1}^{N_i} \overline{Y}_{ij}/N_i$, $i = 1, 2, \ldots, k$, and select the system having the largest $\overline{\overline{Y}}_{i\cdot}$ (i.e., the largest observed average TTF) as best. Also, bound the difference between the E[TTF] of each system and the largest E[TTF] of the others by forming MCB confidence intervals for $\mu_i - \max_{l \neq i} \mu_l$, $i = 1, 2, \ldots, k$, as given in equation (4).

8.4.2 Assumptions

The Rinott + MCB procedure requires that the observations taken within a particular system be IID and normally distributed. These assumptions are evaluated in this subsection.

- The outputs, $\overline{Y}_{i1}, \overline{Y}_{i2}, \ldots, \overline{Y}_{iN_i}$, from the ith system are assumed to be IID with expectation μ_i. This is true since the replications are independent of each other.
- The outputs from across all systems are assumed to be independent (i.e., if $i \neq i'$, all $\overline{Y}_{ij}$'s are independent of all $\overline{Y}_{i'j}$'s, $j = 1, 2, \ldots$). This requirement is also satisfied since different random number streams are chosen for each system's simulation.
- The estimators, $\overline{Y}_{ij}$, for $i = 1, 2, \ldots, k$ and $j = 1, 2, \ldots, N_i$, are assumed to be normally distributed. If the number of replications in a batch is large enough, say at least 20, the central limit theorem yields approximate normality for the batch means estimators.
- There are no assumptions concerning the variances of the outputs.

8.4.3 Experiment

The goal is to find the system having the largest E[TTF]. To achieve the goal, the following sequence of experiments was performed:

TABLE 8.7 Pilot Experiment (n_0 = 20 Batch Means)

	i			
	1	2	3	4
$\overline{Y}_{i\cdot}$	108,286	107,686	96,167.7	89.747.9
S_i	29,157.3	24,289.9	25,319.5	20.810.8
N_i	699	485	527	356

1. A debugging experiment to check the computer code and assess execution speed.
2. A pilot experiment to study characteristics of the data and aid in planning the production run.
3. A production run to produce the final results.

All experiments and analyses were performed on various SPARCstations.

Debugging Experiment. Five batches, each consisting of five replications of system 1, produced a sample mean TTF of 129,182 minutes and a sample standard deviation of 69,417.2 minutes. Each replication took about 24 seconds of real time on a (nondedicated) SPARCstation 1. Since the sample variance was so large, a somewhat larger pilot study was conducted; this would also serve as the first stage of the Rinott + MCB procedure. The pilot study would take 20 batch means, each consisting of 20 replications, for each of the $k = 4$ systems. It was anticipated that the pilot study would use at most 10 hours of real time.

Pilot Experiment. By dividing the pilot study among various SPARCstations it was completed in less than 3 hours. The results are given in Table 8.7. To check the normality assumption, data from the pilot study were used to conduct Shapiro–Wilk tests for normality on the 20 batch means from each system (see, e.g., Bratley et al., 1987, pp. 303–305); the tests passed at all reasonable levels.

For the case $k = 4$ and $1 - \alpha = 0.90$, the critical constant from Table 8.3 is $h = 2.720$. This enabled calculation of the N_i-values for the second stage of sampling. Since the pilot study was also intended to be used as the first stage of Rinott + MCB sampling, the resulting N_i-values are displayed in the table. For example, for system 2, $N_2 - n_0 = 465$ additional batch means are needed in stage 2, each consisting of 20 replications. The total number of *individual* replications over all four systems was about 40,000. A worst-case scenario of 24 seconds of real time per replication (as in the debugging experiment) implied that the production run might take 250 hours.

Final Results. By dividing the production runs among the various SPARCstations, they were completed in less than 2 days. The results are given in Table 8.8. These

TABLE 8.8 R&S Production Run

i	1	2	3	4
$\overline{\overline{Y}}_{i\cdot}$	110,816.5	106,411.8	99,093.1	86,568.9

TABLE 8.9 Two-Stage Rinott + MCB Intervals for the Airline-Reservation Problem

i	$\overline{\overline{Y}}_{i\cdot}$	MCB Lower Limit	$\overline{\overline{Y}}_{i\cdot} - \max_{l \neq i} \overline{\overline{Y}}_{l\cdot}$	MCB Upper Limit
1	110,816.5	0	4,405	7,405
2	106,411.8	−7,405	−4,405	0
3	99,093.1	−14,724	−11,723	0
4	86,568.9	−27,248	−24,248	0

results clearly establish system 1 as the winner. The formal statement is that with 90% confidence the correct selection has been made (with the proviso that the true difference between the best and second best E[TTF] values is at least $\delta = 3000$ minutes).

The associated MCB confidence intervals, shown in Table 8.9, bound the difference between each system and the best of the others. With a 90% confidence level these intervals exclude systems 2 to 4 from being the best and clearly identify system 1 as being the best. This statement is independent of the true difference between the best and second best. In addition, the upper bound on the MCB interval for $\mu_1 - \max_{l \neq 1} \mu_l$ indicates that the E[TTF] value for system 1 may be as much as 7405 minutes (approximately 5 days) longer than the second-best reservation system, while the lower bound for $\mu_4 - \max_{l \neq 4} \mu_l$ indicates that the E[TTF] value for system 4 may be as much as 27,248 minutes (approximately 19 days) shorter than the E[TTF] value of the best reservation system.

8.4.4 Discussion

There are a number of reasons to use R&S procedures when seeking the best of a number of competing systems. Procedures such as Rinott's guarantee the user of a correct selection with high probability when the true difference between the best and second-best system is at least δ; even when the true difference is less than δ, Rinott's procedure ensures selection with high probability of a "good" system (i.e., one that is within δ of the best). The associated MCB intervals bound the difference. These guarantees compare favorably to the simple "yes" or "no" answer that a classical hypothesis test is likely to provide. R&S procedures are also straightforward, as our case study demonstrated; little more than one tabled constant look-up and a sample-mean calculation is required.

One drawback to the Rinott + MCB procedure is that it tends to be conservative; that is, it sometimes takes more observations than necessary in the presence of "favorable" system mean configurations (i.e., configurations in which the largest mean and the others differ by more than δ). This drawback arises from the fact that Rinott + MCB guarantees $P\{CS\} \geq 1 - \alpha$ for *all* configurations of the system means for which the best is at least δ better than the second best.

ACKNOWLEDGMENTS

Portions of this chapter were published previously in Goldsman et al. (1991), Nelson (1992), and Goldsman and Nelson (1994). Our work was supported by National Science Foundation Grants DMI-9622065 and DMI-9622269.

REFERENCES

Bechhofer, R. E., and D. Goldsman (1986). Truncation of the Bechhofer–Kiefer–Sobel sequential procedure for selecting the multinomial event which has the largest probability, II: extended tables and an improved procedure, *Communications in Statistics—Simulation and Computation B*, Vol. 15, pp. 829–851.

Bechhofer, R. E., and B. W. Turnbull (1978). Two $(k+1)$-decision selection procedures for comparing k normal means with a specified standard, *Journal of the American Statistical Association*, Vol. 73, pp. 385–392.

Bechhofer, R. E., S. Elmaghraby, and N. Morse (1959). A single-sample multiple decision procedure for selecting the multinomial event which has the highest probability, *Annals of Mathematical Statistics*, Vol. 30, pp. 102–119.

Bechhofer, R. E., T. J. Santner, and D. Goldsman (1995). *Design and Analysis for Statistical Selection, Screening and Multiple Comparisons*, Wiley, New York.

Bratley, P., B. L. Fox, and L. E. Schrage (1987). *A Guide to Simulation*, 2nd ed., Springer-Verlag, New York.

Dunnett, C. W. (1989). Multivariate normal probability integrals with product correlation structure, *Applied Statistics*, Vol. 38, pp. 564–579. Correction: Vol. 42, p. 709.

Gibbons, J. D., I. Olkin, and M. Sobel (1977). *Selecting and Ordering Populations: A New Statistical Methodology*, Wiley, New York.

Goldsman, D., and B. L. Nelson (1994). Ranking, selection and multiple comparisons in computer simulation, in *Proceedings of the 1994 Winter Simulation Conference*, J. D. Tew, S. Manivannan, D. A. Sadowski, and A. F. Seila, eds., IEEE, Piscataway, N.J., pp. 192–199.

Goldsman, D., B. L. Nelson, and B. Schmeiser (1991). Methods for selecting the best system, in *Proceedings of the 1991 Winter Simulation Conference*, B. L. Nelson, W. D. Kelton, and G. M. Clark, eds., IEEE, Piscataway, N.J., pp. 177–186.

Gupta, S. S. (1956). On a decision rule for a problem in ranking means, Ph.D. dissertation (Mimeo. Ser. 150), Institute of Statistics, University of North Carolina, Chapel Hill, N.C.

Gupta, S. S. (1965). On some multiple decision (selection and ranking) rules, *Technometrics*, Vol. 7, pp. 225–245.

Gupta, S. S., and D.-Y. Huang (1976). Subset selection procedures for the means and variances of normal populations: unequal sample sizes case, *Sankhyā B*, Vol. 38, pp. 112–128.

Hayter, A. J. (1984). A proof of the conjecture that the Tukey–Kramer multiple comparisons procedure is conservative, *Annals of Statistics*, Vol. 12, pp. 61–75.

Hochberg, Y., and A. C. Tamhane (1987). *Multiple Comparison Procedures*, Wiley, New York.

Hsu, J. C. (1984). Constrained simultaneous confidence intervals for multiple comparisons with the best, *Annals of Statistics*, Vol. 12, pp. 1136–1144.

Kleijnen, J. P. C. (1988). Analyzing simulation experiments with common random numbers, *Management Science*, Vol. 34, pp. 65–74.

Law, A. M., and W. D. Kelton (1991). *Simulation Modeling and Analysis*, 2nd ed., McGraw-Hill, New York.

Matejcik, F. J., and B. L. Nelson (1995). Two-stage multiple comparisons with the best for computer simulation, *Operations Research*, Vol. 43, pp. 633–640.

Miller, R. G. (1981). *Simultaneous Statistical Inference*, 2nd ed., Springer-Verlag, New York.

Nelson, B. L. (1992). Designing efficient simulation experiments, in *Proceedings of the 1992 Winter Simulation Conference*, J. J. Swain, D. Goldsman, R. C. Crain, and J. R. Wilson, eds., IEEE, Piscataway, N.J., pp. 126–132.

Nelson, B. L., and F. J. Matejcik (1995). Using common random numbers for indifference-zone selection and multiple comparisons in simulation, *Management Science*, Vol. 41, pp. 1935–1945.

Paulson, E. (1952). On the comparison of several experimental categories with a control, *Annals of Mathematical Statistics*, Vol. 23, pp. 239–246.

Rinott, Y. (1978). On two-stage selection procedures and related probability-inequalities, *Communications in Statistics—Theory and Methods A*, Vol. 7, pp. 799–811.

Tamhane, A. C. (1977). Multiple comparisons in model I: One-way ANOVA with unequal variances, *Communications in Statistics—Theory and Methods A*, Vol. 6, pp. 15–32.

Wilcox, R. R. (1984). A table for Rinott's selection procedure, *Journal of Quality Technology*, Vol. 16, pp. 97–100.

CHAPTER 9

Simulation Optimization

SIGRÚN ANDRADÓTTIR
Georgia Institute of Technology

9.1 INTRODUCTION

In this chapter we consider how simulation can be used to design a system to yield optimal expected performance. More specifically, we assume that the performance of the system of interest depends on the values of the (input) parameters chosen for the system, and that we want to determine the optimal values of these parameters (possibly subject to some constraints). We are interested in the situation when the underlying system is complex enough that it is necessary to use simulation to evaluate its performance for each set of input parameter values. The technique used to optimize the expected system performance therefore needs to be robust enough to converge (i.e., locate an optimal solution) despite the noise in the performance evaluations. In addition, it would be desirable for the technique used to be reasonably efficient. However, this is in general difficult to accomplish in our setting because using simulation to evaluate the system performance for one set of parameter values with reasonable precision is often quite computer intensive, and to locate the optimal solution we obviously need to evaluate system performance for several different sets of parameter values.

We present an introduction to simulation optimization techniques and results. We consider both the case when the set of feasible input parameter values is continuous, as well as the case when this set is discrete. In the former case we focus on gradient-based techniques, whereas in the latter case we focus on random search techniques and other recent developments. The aim is to provide an introduction to major developments in the field of simulation optimization rather than a comprehensive survey of the current status and history of the field. The emphasis is on techniques that are relatively easy to apply and do not require extensive understanding of the structure of the stochastic system being optimized. Also, we do not provide a detailed discussion of when the methods presented are guaranteed to work, so a reader intending to apply the methods discussed should check the references for this information. Additional material on simulation optimization can be found in Chapters 6 and 8 of this book. Chapter 6

Handbook of Simulation, Edited by Jerry Banks.
ISBN 0-471-13403-1

covers experimental design, including response surface methodology, that can be used for continuous parameter simulation optimization. Chapter 8 is concerned with ranking and selection and multiple comparison procedures that can be used for discrete simulation optimization when the number of feasible alternatives is small (say, ≤ 20). For additional material on simulation optimization, the reader is referred to several recent review papers, including Glynn (1986a, 1989), Meketon (1987), Jacobson and Schruben (1989), Safizadeh (1990), Azadivar (1992), Gaivoronski (1992), Fu (1994), and Kleijnen (1995), and references therein.

The chapter is organized as follows: In Section 9.2, gradient-based techniques for solving simulation optimization problems with continuous decision parameters are discussed. First several different methods for using simulation to estimate the gradient of the expected system performance with respect to the input parameter values are presented. Then two classes of optimization techniques (stochastic approximation and sample path optimization) are discussed. In Section 9.3, recent developments aimed at solving simulation optimization problems with discrete decision parameters (including random search methods) are presented. Finally, some concluding remarks are presented in Section 9.4, and a large number of references on simulation optimization are provided.

9.2 CONTINUOUS DECISION PARAMETERS

In this section we discuss gradient-based techniques for solving continuous parameter simulation optimization problems. More specifically, we discuss two classes of methods that take very different approaches to solving such optimization problems. The first class of methods, stochastic approximation, is designed to address the randomness arising from using simulation to evaluate the system performance and its derivatives by moving conservatively through the feasible region, so that even large errors should not put these algorithms too far off course, and they should eventually converge to the optimal solution despite the noise. On the other hand, the second class of methods, sample path optimization, involves immediately converting the underlying (stochastic) simulation optimization problem to a deterministic optimization problem whose solution can be expected to be close to the solution of the original stochastic optimization problem. Then a standard deterministic (mathematical programming) optimization technique is used to solve the approximating deterministic optimization problem. Stochastic approximation is discussed in Section 9.2.2, and sample path optimization is discussed in Section 9.2.3.

Consider the optimization problem

$$\min_{\theta \in \Theta} f(\theta) \tag{1}$$

where θ is the (possibly vector-valued) decision parameter consisting of the input parameters of the simulation, the (real-valued) objective function f indicates how the expected system performance depends on the input parameter θ (i.e., $f(\theta)$ is the expected system performance when the input parameter values are given by θ), the feasible region $\Theta \subset \mathbb{R}^d$ is the set of possible values of the input parameters, the positive integer d is the dimension of the feasible region Θ, and $\mathbb{R}^d$ is the set of d-dimensional real vectors. We assume that the feasible region Θ is continuous and that the objective function f cannot be evaluated analytically, but instead, its values are estimated using simulation.

Both stochastic approximation and sample path optimization methods require using simulation to estimate the gradient (and possibly also higher derivatives) of the expected performance of the underlying stochastic system. [The gradient $\nabla f(\theta)$ of the performance measure $f(\theta)$ is the vector consisting of the elements $\partial f(\theta)/\partial\theta_i$ for $i = 1, \ldots, d$, where $\theta = (\theta_1, \ldots, \theta_d)$ and $\partial f(\theta)/\partial\theta_i$ is the (partial) derivative of $f(\theta)$ with respect to θ_i for $i = 1, \ldots, d$. In the one-dimensional case (when $d = 1$), the gradient reduces to the derivative $f'(\theta)$.] Gradient estimation using simulation is a challenging problem that has attracted a fair amount of attention in recent years. The underlying difficulty is that if it is necessary to use simulation to evaluate the system performance, then an analytical expression for the system performance obviously does not exist and hence calculus cannot be used to obtain the required derivatives. An overview of techniques for using simulation to obtain derivative estimates is given in Section 9.2.1.

Throughout this section we use an example that has become quite standard in the literature on continuous parameter simulation optimization. This example and variants thereof have been studied by Suri and Leung (1989), L'Ecuyer et al. (1994), L'Ecuyer and Glynn (1994), Fu (1994), Healy and Xu (1994), and Andradóttir (1995a, 1996a). The example is described below.

Example 1 Consider the problem of determining the optimal level of service in a GI/G/1 queueing system; this is a single-server queue in which the interarrival times between successive customers are independent and identically distributed (with an arbitrary distribution) and the service times of the successive customers are also independent and identically distributed (with an arbitrary distribution). Assume that the costs arise from two sources: There is a cost associated with providing the service (this cost increases as the level of service becomes better) and a cost associated with the average sojourn time of a customer in the queue (this cost estimates customer satisfaction at the service level θ; it decreases as the level of service is improved). More specifically, if θ is the mean service time per customer in a GI/G/1 queueing system with mean interarrival time $1/\lambda$, and if $w(\theta)$ is the average steady-state system time (sojourn time) per customer in this queueing system, then we want to solve an optimization problem of the form given in equation (1) with

$$\Theta = [\underline{\theta}, \overline{\theta}] \quad \text{and} \quad f(\theta) = \frac{\alpha}{\theta} + \beta w(\theta) \qquad \text{for all } \theta \in \Theta \tag{2}$$

(so that Θ is the closed interval from $\underline{\theta}$ to $\overline{\theta}$). The mean interarrival time $1/\lambda$ and the scalars $\alpha, \beta > 0$ and $\underline{\theta} < \overline{\theta}$ are assumed to be given. Moreover, to ensure that the queue is stable for all $\theta \in \Theta$, we require that $0 < \underline{\theta} < \overline{\theta} < 1/\lambda$. Note that the term α/θ in the objective function f represents the cost of providing service at the level θ and that the term $\beta w(\theta)$ represents the cost associated with the customer satisfaction at the service level θ. Therefore, the values of α and β represent management assessment of the relative importance of keeping service and customer satisfaction costs down.

9.2.1 Gradient Estimation

In this section we review techniques for applying simulation to obtain estimates of the gradient of the expected system performance with respect to the (continuous) input parameters to the simulation. The emphasis is on using finite differences to obtain the gradient estimates because we feel that this is the easiest approach to use. A brief review of other, more advanced gradient estimation techniques is also given.

Finite Differences. The most obvious way of using simulation to estimate the derivative of the expected system performance involves approximating the derivative using finite differences. In the one-dimensional case when the feasible region Θ is a subset of $\mathbb{R}$ (so that $d = 1$), the derivative of f is defined as follows:

$$f'(\theta) = \lim_{c \to 0} \frac{f(\theta + c) - f(\theta)}{c}$$

for all $\theta \in \mathbb{R}$. Therefore, if c is small, it is reasonable to expect that

$$f'(\theta) \simeq \frac{f(\theta + c) - f(\theta)}{c} \tag{3}$$

Therefore, one can estimate the derivative $f'(\theta)$ by conducting one simulation with input parameter $\theta + c$ and obtain an estimate $\hat{f}(\theta + c)$ of $f(\theta + c)$, and a second simulation with input parameter θ and obtain an estimate $\hat{f}(\theta)$ of $f(\theta)$ and then use

$$\frac{\hat{f}(\theta + c) - \hat{f}(\theta)}{c}$$

as an estimate of $f'(\theta)$.

In the general case (when $d \geq 1$), suppose that $e_i = (0, \ldots, 0, 1, 0, \ldots, 0)$ is the ith coordinate vector (with one in the ith position and zeros everywhere else) for $i = 1, \ldots, d$. Then the value of the gradient $g(\theta) = \nabla f(\theta)$ can be estimated by $\hat{g}(\theta)$, where

$$\hat{g}(\theta) = (\hat{g}_1(\theta), \ldots, \hat{g}_d(\theta))$$

and for $i = 1, \ldots, d$,

$$\hat{g}_i(\theta) = \frac{\hat{f}(\theta + ce_i) - \hat{f}(\theta)}{c} \tag{4}$$

This is the finite-difference gradient estimator obtained using *forward* differences. If *central* differences are used, then, for $i = 1, \ldots, d$, $\hat{g}_i(\theta)$ is instead defined as follows:

$$\hat{g}_i(\theta) = \frac{\hat{f}(\theta + ce_i) - \hat{f}(\theta - ce_i)}{2c} \tag{5}$$

In both cases, the symbols $\hat{f}(\theta)$, $\hat{f}(\theta + ce_i)$, and $\hat{f}(\theta - ce_i)$ denote estimates of $f(\theta)$, $f(\theta + ce_i)$, and $f(\theta - ce_i)$ obtained by conducting simulations at the input parameter values θ, $\theta + ce_i$, and $\theta - ce_i$, respectively, for $i = 1, \ldots, d$.

From the previous discussion it is clear that to use finite differences to estimate the gradient of the expected system performance with respect to the input parameters, it is necessary to conduct several simulations at different sets of input parameter values. In particular, when forward differences are used, it is necessary to conduct simulations at $d + 1$ sets of parameter values, namely θ and $\theta + ce_i$, for $i = 1, \ldots, d$. When central differences are used, it is necessary to conduct simulations at $2d$ parameter values, namely

$\theta + ce_i$ and $\theta - ce_i$, for $i = 1, \ldots, d$. Thus, when $d > 1$, more computational effort is usually required to obtain gradient estimates using central differences than when forward differences are used. On the other hand, estimators obtained using central differences usually have superior statistical properties to those obtained using forward differences (more details on this are provided below).

Gradient estimates obtained using finite differences are in general biased, even if the simulation results used to obtain these estimates are unbiased. This is because of the error involved in approximating the gradient using finite differences [see, e.g., equation (3)]. Usually, the bias is smaller when central differences are used than when forward differences are used.

Another difficulty in using finite differences for gradient estimation is that to reduce the bias, it is necessary to let the scalar c be small, but when c is small, the estimators obtained usually have a large variance. Therefore, it is necessary to balance the desire for small bias against the desire for small variance. Several studies addressing this problem have appeared recently (Glynn, 1989; Zazanis and Suri, 1993; L'Ecuyer and Perron, 1994). These studies discuss how the scalar c should be selected to minimize the asymptotic mean-squared error (the variance plus the squared bias). As expected, these studies show that in general better results are obtained for central differences than for forward differences.

One way of addressing the variance problem discussed in the preceding paragraph is to use common random numbers (CRN) in the different simulations required to obtain an estimate $\hat{g}(\theta)$ of the value of the gradient $g(\theta) = \nabla f(\theta)$. Glasserman and Yao (1992) present results showing when the use of common random numbers is guaranteed to achieve variance reduction in this setting. More details on common random numbers are given in Chapter 8.

In Example 1, let $\theta \in \Theta$ and suppose that we want to estimate the derivative $f'(\theta)$ using finite differences. This can be accomplished as follows: Select a small value of c having the property that $\theta + c \in (0, 1/\lambda)$ (so that the queue is stable when the mean service time per customer is $\theta + c$). Given a sequence of interarrival times $\{T_j\}$ and service times $\{S_j(\theta)\}$ (so that T_j is the interarrival time between customers $j - 1$ and j and $S_j(\theta)$ is the service time of customer j when the mean service time is θ), let $W_0(\theta) = S_0(\theta)$ and for $j = 1, \ldots, N$, let

$$W_j(\theta) = \max\{0, W_{j-1}(\theta) - T_j\} + S_j(\theta) \tag{6}$$

[for all j, $W_j(\theta)$ is the system time of customer j when the mean service time per customer is θ]. We can now use $\hat{g}(\theta)$ to estimate the derivative $f'(\theta)$, where

$$\begin{aligned}\hat{g}(\theta) &= -\frac{\alpha}{\theta^2} + \frac{\beta}{cN} \sum_{j=1}^{N} [W_j(\theta + c) - W_j(\theta)] \\ &= -\frac{\alpha}{\theta^2} + \beta \frac{\overline{W}(\theta + c) - \overline{W}(\theta)}{c} \end{aligned} \tag{7}$$

where $\overline{W}(\theta + c) = \sum_{j=1}^{N} W_j(\theta + c)/N$ and $\overline{W}(\theta) = \sum_{j=1}^{N} W_j(\theta)/N$. This is a forward difference estimator for $f'(\theta)$; a central difference estimator can be obtained by replacing $W_j(\theta)$ by $W_j(\theta - c)$ in the numerator of the second term on the right-hand side in equation (7) [assuming that $\theta - c \in (0, 1/\lambda)$] and by replacing c by $2c$ in the denominator of the second term on the right-hand side in equation (7). The estimator $\hat{g}(\theta)$ could be improved through initial transient deletion; see Chapter 7 for details. As discussed previously, using common random numbers can be expected to improve the estimator $\hat{g}(\theta)$. In this example, using common random numbers would involve conducting the simulations at the different parameter values with the same sequence of interarrival times $\{T_j\}$, and using the same stream of uniform random numbers to generate the needed service time sequences . (If forward differences are used, the sequences $\{S_j(\theta)\}$ and $\{S_j(\theta + c)\}$ should be generated using the same stream of uniform random numbers; if central differences are used, the sequences $\{S_j(\theta - c)\}$ and $\{S_j(\theta + c)\}$ should be generated using the same stream of uniform random numbers.) For example, if the interarrival and service times are exponentially distributed (so that the system under study is an M/M/1 queue), then two independent sequences $\{U_j\}$ and $\{V_j\}$ of independent and $U[0,1]$-distributed (uniformly distributed on the interval [0,1]) random variables are needed. Then, for all j, we can let $T_j = -\ln(U_j)/\lambda$, $S_j(\theta) = -\ln(V_j) \times \theta$, $S_j(\theta + c) = -\ln(V_j) \times (\theta + c)$, and $S_j(\theta - c) = -\ln(V_j) \times (\theta - c)$. (In this example it is possible to implement the common random numbers approach in such a way that each random number is used for exactly the same purpose in all simulation runs. This is not always possible for more complicated systems.)

To illustrate the behavior of finite-difference derivative estimators, we now present simulation results for the derivative $w'(\theta)$ of the steady-state system time per customer in an M/M/1 queue [this involves using $\alpha = 0$ and $\beta = 1$ in equation (7)]. We let $\lambda = 1$, $\theta \in \{0.25, 0.5, 0.75\}$, $c = 0.01$, and $N = 10{,}000$. We compare the estimators obtained using forward and central differences with and without common random numbers. Note that in this case it is known theoretically that we have $w(\theta) = \theta/(1 - \theta)$ for all $\theta \in (0, 1)$, so that we can compare the simulation results obtained with the true value of the derivative $w'(\theta) = 1/(1 - \theta)^2$ for all $\theta \in (0, 1)$. The results obtained using 10 replications of each approach with the same seeds are given in Tables 9.1 and 9.2. As expected, the tables show that using common random numbers vastly improves the precision of the resulting derivative estimator, and that the derivative estimates obtained using central differences are less biased than the corresponding derivative estimates using finite differences.

Advanced Gradient Estimation Techniques. We reviewed above how finite differences can be used for gradient estimation in simulation. This approach involves conducting simulations at several different sets of input parameter values in order to estimate the gradient. We now review briefly gradient estimation techniques that require only a single simulation run to estimate the gradient. These techniques are perturbation analysis, the likelihood ratio method, and frequency-domain experimentation. Perturbation analysis and the likelihood ratio method both require only one simulation run at one set of parameter values to obtain gradient estimates. On the other hand, frequency-domain experimentation involves oscillating the values of the input parameters during a single simulation run. For additional material on gradient estimation, the reader is referred to Glynn (1989), L'Ecuyer (1991), Fu (1994), and Fu and Hu (1997), and references therein.

As mentioned previously, the gradient estimation problem involves finding ways of using simulation to estimate the value of the gradient $g(\theta) = \nabla f(\theta)$ of the expected system performance $f(\theta)$ with respect to the input parameter θ. To illustrate the difficulty of this problem, assume for simplicity that

TABLE 9.1 Simulation Estimates of the Derivative $w'(\theta)$, for $\theta \in \{0.25, 0.5, 0.75\}$, Obtained Using Forward Differences With or Without Common Random Numbers

	$\theta = 0.25$		$\theta = 0.5$		$\theta = 0.75$	
	$w'(\theta) \simeq 1.778$		$w'(\theta = 4.000$		$w'(\theta) = 16.000$	
	$[w(\theta + c) - w(\theta)]/c \simeq 1.802$		$[w(\theta + c) - w(\theta)]/c \simeq 4.082$		$[w(\theta + c) - w(\theta)]/c \simeq 16.667$	
Replication	Without CRN	With CRN	Without CRN	With CRN	Without CRN	With CRN
1	1.330	1.864	5.432	4.027	20.320	14.982
2	2.401	1.812	8.414	4.293	57.411	15.651
3	2.350	1.724	8.891	3.629	59.916	14.510
4	0.419	1.760	−1.247	3.825	−16.868	13.768
5	1.649	1.832	7.261	4.014	33.372	20.571
6	1.191	1.835	−0.600	4.280	−28.554	16.357
7	2.105	1.719	8.700	3.797	32.676	14.797
8	1.894	1.790	4.842	4.308	−0.494	19.055
9	1.931	1.835	2.987	4.223	−8.983	15.964
10	1.524	1.773	2.439	3.936	−4.404	15.857
Average	1.679	1.794	4.712	4.033	14.439	16.151
Variance	0.359	2.454×10^{-3}	13.997	5.694×10^{-2}	955.066	4.439

$$f(\theta) = \mathrm{E}\{h(\theta,X)\} \qquad \text{for all } \theta \tag{8}$$

where h is a known deterministic and differentiable function and X is a random variable. For example, in the context of Example 1, X could be the average system time of the first N arriving customers with the mean service time per customer being θ, and h could be defined as $h(\theta,X) = \alpha/\theta + \beta X$, yielding an objective function $f(\theta)$ that strongly resembles the one given in equation (2). Similarly, consider a highly reliable system of d components and let θ_i denote the (known) probability that component i survives for at least T units of time for $i = 1, \ldots, d$, where $T > 0$ is fixed. Moreover, let $\theta = (\theta_1, \ldots, \theta_d)$, let $\alpha_i(\theta_i)$ be the cost of component i when its reliability is θ_i for $i = 1, \ldots, d$, let X be the time until the system of d components fails for the first time, and let $h(\theta,X) = \sum_{i=1}^{d} \alpha_i(\theta_i) + \beta I_{\{X<T\}}$, where $I_{\{X<T\}}$ equals 1 if $X < T$ and zero otherwise, and $\beta > 0$. Then the objective function f is of the form $f(\theta) = \sum_{i=1}^{d} \alpha_i(\theta_i) + \beta P\{X < T\}$, corresponding to trying to keep the cost of the components down while at the same time keeping the probability that the system functions for the desired T units of time large.

Suppose first that the cumulative distribution function F of the random variable X does not depend on the parameter θ. Then standard probability theory yields that $g(\theta) = \nabla f(\theta) = \nabla E\{h(\theta,X)\} = E\{\nabla h(\theta,X)\}$ under general conditions. Therefore, if independent and identically distributed observations $X_1, \ldots, X_N$ from the distribution F can be generated, we can use $\sum_{i=1}^{N} \nabla h(\theta,X_i)/N$ to estimate the gradient $g(\theta)$. The problem is that in a simulation application, the distributions of the random variables involved may depend on the parameter θ (this is the case in the two examples discussed at the end

TABLE 9.2 Simulation Estimates of the Derivative $w'(\theta)$, for $\theta \in \{0.25, 0.5, 0.75\}$, Obtained Using Forward Differences With or Without Common Random Numbers

	$\theta = 0.25$		$\theta = 0.5$		$\theta = 0.75$	
	$w'(\theta) \simeq 1.778$		$w'(\theta = 4.000$		$w'(\theta) = 16.000$	
	$[w(\theta + c) - w(\theta - c)]/2c \simeq 1.778$		$[w(\theta + c) - w(\theta - c)]/2c \simeq 4.002$		$[w(\theta + c) - w(\theta - c)]/2c \simeq 16.026$	
Replication	Without CRN	With CRN	Without CRN	With CRN	Without CRN	With CRN
1	2.130	1.837	7.695	3.932	29.175	14.492
2	2.017	1.789	2.836	4.225	19.491	15.004
3	1.466	1.702	3.328	3.559	27.210	14.051
4	0.990	1.743	0.908	3.721	2.155	13.404
5	1.399	1.806	3.064	3.928	4.140	18.996
6	1.740	1.812	4.976	4.195	18.035	15.797
7	2.545	1.700	7.310	3.726	36.314	14.245
8	2.271	1.766	6.737	4.207	18.589	18.338
9	1.213	1.809	1.474	4.144	5.213	15.564
10	1.377	1.750	3.036	3.874	−13.363	15.495
Average	1.715	1.771	4.137	3.951	14.696	15.539
Variance	0.257	2.246×10^{-3}	5.824	5.566×10^{-2}	225.423	3.298

of the preceding paragraph). In the example considered here [see equation (8)], this corresponds to the random variable X having cumulative distribution function F_θ (in mathematical notation, $X \sim F_\theta$) that depends on the input parameter value θ. Therefore, the expectation in equation (8) is now with respect to the distribution F_θ, and hence depends on the decision parameter θ; that is, we have

$$f(\theta) = \mathrm{E}_\theta\{h(\theta, X)\} \qquad \text{for all } \theta \tag{9}$$

Suppose that the random variable $X \sim F_\theta$ has density f_θ for all θ. Then we have

$$f(\theta) = \mathrm{E}_\theta\{h(\theta, X)\} = \int h(\theta, x) f_\theta(x)\, dx \qquad \text{for all } \theta \tag{10}$$

In this case it is clearly not possible to interchange the order of the gradient and the expectation (as we did previously when the distribution of X did not depend on θ) since the expectation depends on the parameter θ. This explains why specialized techniques are required for obtaining gradient estimates using simulation.

Perturbation analysis refers to a class of related gradient estimation techniques. Infinitesimal perturbation analysis (IPA) is the best known variant. Perturbation analysis addresses the difficulty discussed previously by transforming the original problem in such a way that it can be reformulated as the problem of estimating the gradient of an expected value involving a random variable whose distribution does not depend on the parameter θ [as in equation (8)]. Then the standard approach discussed previously (to interchange the order of the expected value and gradient) is used to obtain the gradient estimates. Expressions for the gradient of the random variable inside the expectation are obtained by considering how small perturbations in the underlying random variables generated in the simulation program affect the sample path of the stochastic system of interest generated using these random variables. To illustrate, consider Example 1. Suppose that $S_j(\theta)$ is exponentially distributed with mean θ for all j and $\theta \in \Theta$. Then, as discussed near the end of the subsection "Finite Differences" in Section 9.2.1, we can let $S_j(\theta) = -\ln(V_j) \times \theta$ for all j and $\theta \in \Theta$, where $\{V_j\}$ is a sequence of independent $U[0,1]$-distributed random variables. Hence the sequences of underlying random variables are now $\{T_j\}$ and $\{V_j\}$, and their distributions do not depend on the parameter θ. Note that when the service time $S_j(\theta)$ is differentiated with respect to the parameter θ, we get $S_j'(\theta) = -\ln(V_j) = S_j(\theta)/\theta$ for all j and $\theta \in \Theta$. From equation (6) we then get that $W_0'(\theta) = S_0'(\theta)$ for all $\theta \in \Theta$. The derivative of $W_1(\theta)$ with respect to θ depends on whether customer 1 has to wait in queue before receiving service. Note that for all j, the jth customer has to wait in queue if and only if $W_{j-1}(\theta) - T_j > 0$. From equation (6), we have that if $W_0(\theta) - T_1 > 0$ (so customer 1 has to wait before receiving service), then $W_1'(\theta) = S_0'(\theta) + S_1'(\theta)$; otherwise, $W_1'(\theta) = S_1'(\theta)$. Proceeding in the same manner, we see that the perturbations $S_j'(\theta)$ accumulate within each busy cycle (i.e., periods when the server is busy serving successive customers without interruption). In particular, if customer j is the kth arriving customer in a busy cycle, then $W_j'(\theta) = \sum_{n=j-k+1}^{j} S_n'(\theta)$. The IPA estimate of the derivative $g(\theta) = f'(\theta)$ can now be computed as follows:

$$\hat{g}(\theta) = -\frac{\alpha}{\theta^2} + \frac{\beta}{N} \sum_{j=1}^{N} W_j'(\theta) \tag{11}$$

While IPA is an intuitive approach for obtaining gradient estimates using simulation, it is unfortunately easy to develop examples where IPA is not guaranteed to work (because the interchange of the gradient and expectation is not valid). Several different variants of perturbation analysis have been developed to address this problem. For more material on IPA and other variants of perturbation analysis, the reader is referred to the references given at the beginning of this section, as well as Suri (1989), Glasserman (1991), Ho and Cao (1991), and references therein.

Applying the likelihood ratio method to estimate the derivative of the function f given in equations (9) and (10) involves differentiating in equation (10) directly, taking into account that f_θ will also need be differentiated with respect to θ. Assuming that we can exchange the order of integration and expectation, we have that

$$g(\theta) = f'(\theta) = \int h'(\theta, x) f_\theta(x)\, dx + \int h(\theta, x) f_\theta'(x)\, dx$$

(note that all the derivatives in this expression are with respect to the parameter θ). We now want to express the derivative $f'(\theta)$ as an expectation that can be estimated via simulation. The first integral in this expression equals $E_\theta\{h'(\theta, X)\}$ and it is easy to estimate through simulation (see below). The second integral in the expression is a little trickier to estimate, as we first need to rewrite it as an expectation [note that $f_\theta'(x)$ is usually not a density function]. It would be desirable if this expectation could be with respect to the same probability distribution as the first integral so that only one simulation is needed to estimate $g(\theta) = f'(\theta)$. This can be accomplished as follows:

$$\int h(\theta,x)f'_\theta(x)\,dx = \int h(\theta,x)f'_\theta(x)\,\frac{f_\theta(x)}{f_\theta(x)}\,dx = \int \left(h(\theta,x)\,\frac{f'_\theta(x)}{f_\theta(x)}\right) f_\theta(x)\,dx$$

$$= \mathrm{E}_\theta \left\{h(\theta,X)\,\frac{f'_\theta(X)}{f_\theta(X)}\right\}$$

[Note that we adopt the convention $0/0 = 1$. It is easy to show that division by zero will not occur in the equation above; i.e., if $f_\theta(x) = 0$, then $f'_\theta(x) = 0$.] We have shown that under general conditions we have

$$g(\theta) = f'(\theta) = \mathrm{E}_\theta \left\{h'(\theta,X) + h(\theta,X)\,\frac{f'_\theta(X)}{f_\theta(X)}\right\} \tag{12}$$

Therefore, we can estimate $g(\theta)$ as follows: Generate $X_1, \ldots, X_N$ independent observations from the distribution F_θ and use $\hat{g}(\theta) = \sum_{i=1}^N [h'(\theta,X_i) + h(\theta,X_i)f'_\theta(X_i)/f_\theta(X_i)]/N$ as our estimate.

To illustrate, consider the following extremely simple example. Suppose that $f(\theta) = E_\theta\{X\}$, where X has an exponential distribution with rate $\theta > 0$ [so that $h(\theta,X) = X$]. We have that $f_\theta(x) = \theta e^{-\theta x}$ when $x \geq 0$ and $f_\theta(x) = 0$ when $x < 0$, so that $f(\theta) = \int_0^\infty x\theta e^{-\theta x}\,dx$ and

$$f'(\theta) = \int_0^\infty x\left(\frac{1}{\theta} - x\right)\theta e^{-\theta x}\,dx = \mathrm{E}_\theta\left\{X\left(\frac{1}{\theta} - X\right)\right\}$$

corresponding to equation (12) with $h(\theta,X) = X$, $h'(\theta,X) = 0$, and $f'_\theta(X)/f_\theta(X) = 1/\theta - X$. For more material on the likelihood ratio gradient estimation technique (also called the score function method), the reader is referred to Glynn (1990), Rubinstein and Shapiro (1993), and Andradóttir (1996b), and references therein. For a discussion of how likelihood ratio derivative estimates can be obtained for Example 1, see, for example, Fu (1994) and L'Ecuyer and Glynn (1994). [The likelihood ratio $L(\theta,\theta_0,X) = f_\theta(X)/f_{\theta_0}(X)$, where $\theta,\theta_0 \in \Theta$, measures how likely the outcome X is to occur under the density f_θ relative to how likely it is to occur under the density f_{θ_0}. When the likelihood ratio is differentiated with respect to θ, we obtain $L'(\theta,\theta_0,X) = f'_\theta(X)/f_{\theta_0}(X)$. Evaluating the derivative with $\theta_0 = \theta$ yields $L'(\theta,\theta,X) = f'_\theta(X)/f_\theta(X)$, the ratio found in equation (12). This explains why the gradient estimation technique discussed in this and the preceding paragraph is referred to as the likelihood ratio method.]

The frequency-domain approach for gradient estimation involves oscillating the value of the (possibly vector-valued) input parameter in a sinusoidal fashion during a single simulation run. This can be used to estimate the sensitivity of the performance measure to the input parameter in a single simulation run, leading to estimates of the gradient of the performance measure relative to the input parameter. In particular, if it is of interest to estimate $g(\theta) = \nabla f(\theta)$, where $\theta = (\theta_1, \ldots, \theta_d)$, then the value of the input parameter used at "time" (index) t is $\theta_i(t) = \theta_i + \alpha_i \sin(\omega_i t)$, for all i and t, where $\omega = (\omega_1, \ldots, \omega_d)$ is the oscillation frequency and $\alpha = (\alpha_1, \ldots, \alpha_d)$ is the oscillation amplitude. The global simulation clock can often be used as the index t (Mitra and Park, 1991). Then the performance measure f is approximated around θ using a polynomial (response) function of the input parameters. This approximation can be used to derive frequency-domain (harmonic) estimators for $g(\theta)$. For more details on frequency-domain experimentation, including expressions for harmonic gradient estimators and a discussion of how the index t, amplitude α, and frequency ω should be selected, the reader is referred to Jacobson (1994) and references therein. For a discussion of how a harmonic estimator of the derivative can be obtained for Example 1, see, for example, Fu (1994).

We now briefly compare the four gradient estimation techniques we have discussed. The easiest method to use is the finite-difference approach. This approach should be implemented using common random numbers (so that the variance will not be too large). The disadvantage of using this approach is that it requires several simulation runs, and it yields biased estimates of the gradient in general. Unlike finite-difference methods, both perturbation analysis and the likelihood ratio method require knowledge of the structure of the stochastic system being simulated and hence are more difficult to use than finite-difference methods. But they require only a single simulation run and often produce estimators having desirable statistical features such as unbiasedness and strong consistency (unlike finite-difference approaches). The likelihood ratio method is more generally applicable than IPA (other variants of perturbation analysis are frequently needed in order to obtain unbiasedness or strong consistency) but often yields estimates with a larger variance than the estimates obtained using IPA (assuming both techniques are applicable). The frequency-domain experimentation approach resembles the finite-difference approach more than it resembles the perturbation analysis and likelihood ratio approaches (except that it only requires a single simulation run). In general, it produces biased estimates of the gradient.

9.2.2 Stochastic Approximation

In this section we discuss how stochastic approximation methods can be used to solve continuous parameter simulation optimization problems of the form given in equation (1). We emphasize the case when the simulation optimization problem at hand is either unconstrained (so that $\Theta = \mathbb{R}^d$), or constrained with known (not noisy) constraints and a closed, convex feasible region Θ [see, e.g., Bazaraa and Shetty (1979, Defs. 2.1.1 and 2.2.1), for the definitions of closed and convex sets].

The first stochastic approximation algorithm was proposed by Robbins and Monro (1951). This algorithm was originally designed to find a root of a noisy function. When applied to solve optimization problems of the form given in equation (1), this algorithm seeks to locate a root of the function ∇f, the gradient of the objective function. A projected, multivariate version of the Robbins–Monro algorithm is given below.

Algorithm 1

Step 0: Select (a) an initial estimate of the solution $\theta_1 \in \Theta$, (b) a (gain) sequence $\{a_n\}$ of positive real numbers such that $\sum_{n=1}^{\infty} a_n = \infty$ and $\sum_{n=1}^{\infty} a_n^2 < \infty$, and (c) a suitable stopping criterion. Let $n = 1$.

Step 1: Given θ_n, generate an estimate $\hat{g}(\theta_n)$ of $\nabla f(\theta_n)$.

Step 2: Compute

$$\theta_{n+1} = \pi_\Theta(\theta_n - a_n \hat{g}(\theta_n))$$

where for all $\theta \in \mathbb{R}^d$, $\pi_\Theta(\theta)$ is the point in Θ that is closest to θ [so that $\pi_\Theta(\theta) = \theta$ when $\theta \in \Theta$].

Step 3: If the stopping criterion is satisfied, then stop and return θ_{n+1} as the estimate of the optimal solution. Otherwise, let $n = n + 1$ and go to step 1.

When finite differences (see Section 9.2.1) are used to obtain the gradient estimates in step 1 of the Robbins–Monro algorithm, the resulting procedure is referred to as the Kiefer–Wolfowitz algorithm [Kiefer and Wolfowitz (1952) proposed and analyzed this method in the one-dimensional case using central differences to estimate the derivative]. In this case, one would select a sequence $\{c_n\}$ of positive real numbers such that $c_n \to 0$ as $n \to \infty$, $\sum_{n=1}^{\infty} a_n c_n < \infty$, and $\sum_{n=1}^{\infty} a_n^2/c_n^2 < \infty$ in step 0 of Algorithm 1. Then if forward differences are used to estimate the gradient, each gradient estimate $\hat{g}(\theta_n)$ in step 1 is obtained using equation (4) with $\theta = \theta_n$ and $c = c_n$. Otherwise, if central differences are used, each gradient estimate $\hat{g}(\theta_n)$ is obtained using equation (5) with $\theta = \theta_n$ and $c = c_n$. As in the section "Finite Differences" in Section 9.2.1, when finite differences are used to estimate the gradient, it is in general desirable to use common random numbers among the different simulations [see L'Ecuyer et al. (1994) for empirical results documenting this statement].

Obviously, the behavior of the stochastic approximation algorithms discussed previously depends on the choice of the sequences $\{a_n\}$ and $\{c_n\}$. These sequences are usually selected to maximize the asymptotic rate of convergence of the algorithms. The gain sequence $\{a_n\}$ is usually selected to be of the form $a_n = a/n$ for all n, where a is a positive scalar. With this choice of gain sequence, the Robbins–Monro algorithm will, under certain conditions, converge at the fastest possible asymptotic rate (i.e., at the rate $n^{-1/2}$). In the forward difference case, the sequence $\{c_n\}$ is usually chosen to be of the form $c_n = c/n^{1/4}$ for all n, yielding an asymptotic convergence rate of $n^{-1/4}$ under certain conditions. On the other hand, in the central difference case, one usually lets $c_n = c/n^{1/6}$ for all n, yielding an asymptotic convergence rate of $n^{-1/3}$ under certain conditions. In both cases the parameter c is a positive scalar. Even if the sequences $\{a_n\}$ and $\{c_n\}$ are selected in this manner, the behavior of the Robbins–Monro and Kiefer–Wolfowitz algorithms depends on how well the scalars a and c are chosen.

In general, the Robbins–Monro and Kiefer–Wolfowitz algorithms can only be expected to converge to a local optimal solution of the underlying optimization problem (1). Note that each iteration of these methods involves moving in the direction of $\hat{g}(\theta_n)$, where θ_n is the current estimate of the solution and $\hat{g}(\theta_n)$ is an estimate of $\nabla f(\theta_n)$. This is the direction along which we expect the objective function f to decrease the fastest. Therefore, both the Robbins–Monro algorithm (when applied to solve optimization problems) and the Kiefer–Wolfowitz algorithm are versions of the steepest descent algorithm for solving continuous deterministic optimization problems. For more discussion of when these algorithms are guaranteed to converge, the reader is referred to Kushner and Clark (1978), Benveniste et al. (1990), and Ljung et al. (1992). The (theoretical) convergence rates of these algorithms (see the previous discussion) can be derived using Theorem 2.2 of Fabian (1968). See also Ruppert (1991) for a recent review of stochastic approximation procedures.

The Robbins–Monro and Kiefer–Wolfowitz algorithms do not always work well, and in recent years, a number of new stochastic approximation algorithms have been proposed to address this problem. We now discuss very briefly several of these variants.

One problem with the Robbins–Monro and Kiefer–Wolfowitz procedures is that the unconstrained versions of these algorithms (with $\Theta = \mathbb{R}^d$) are not guaranteed to converge when the objective function f grows faster than quadratically in the decision parameter θ. Traditionally, this problem has been addressed by projecting the sequence generated by these algorithms onto a compact (i.e., closed and bounded), convex set, say $\tilde{\Theta}$, even though the original optimization problem is unconstrained. However, this is not a very satisfactory approach because these algorithms then cannot converge if the optimal solution does not lie in the set $\tilde{\Theta}$, and if the set $\tilde{\Theta}$ is selected to be large (so that the solution probably lies within the set $\tilde{\Theta}$, then these algorithms can converge very slowly. Approaches for addressing this problem have been proposed and analyzed by Andradóttir (1996a) using scaling, and by Chen and Zhu (1986), Yin and Zhu (1989), and Andradóttir (1995a) using projections involving an increasing sequence of sets.

From the discussion in the section "Finite Differences" in Section 9.2.1, it is clear that when the Kiefer–Wolfowitz algorithm is applied using forward differences to estimate the gradient, it is necessary to conduct simulations at $d + 1$ different parameter values in each iteration of the algorithm, where d is the dimension of the underlying stochastic optimization problem [see equation (1)]. Moreover, when central differences are used to estimate the gradient, the number of simulations that are needed in each iteration of the algorithm increases to $2d$. This obviously means that when the number of dimensions d is large, the Kiefer–Wolfowitz algorithm requires a lot of computational effort per iteration and may therefore converge slowly. To address this problem, Spall (1992) has proposed and analyzed the application of a stochastic approximation algorithm of the form given in Algorithm 1 to solve stochastic optimization problems using simultaneous perturbations to estimate the gradient. This procedure requires only two simulations per iteration, regardless of the dimension d. Similar ideas have also been proposed and analyzed by Kushner and Clark (1978) and Ermoliev (1983). In addition Pflug (1990) has proposed a method for simulation optimization that requires only two simulations per iteration. However, this approach differs from the other approaches discussed previously in that Pflug uses weak derivatives (which are defined in his paper), whereas Spall, Kushner and Clark, and Ermoliev all use gradient estimates that resemble standard finite-difference gradient estimates.

As mentioned previously, when the gain sequence $\{a_n\}$ is of the form $a_n = a/n$ for all n, where a is a positive scalar, the performance of the Robbins–Monro and Kiefer–Wolfowitz algorithms depends on the choice of the scalar a. A number of researchers have developed and analyzed adaptive procedures where the value of the multiplier a is updated throughout the optimization process. In general, when these methods are applied to optimize a function f, the aim is for the sequence of multipliers to converge to the inverse of the Hessian of the objective function f evaluated at the solution as the number of iterations grows (so the sequence of multipliers is matrix valued in this case). For more details, see Venter (1967), Nevel'son and Has'minskii (1973), Lai and Robbins (1979), Ruppert (1985), and Wei (1987).

Another difficulty experienced by the Robbins–Monro and Kiefer–Wolfowitz algorithms is that the gain sequence $\{a_n\}$ often decreases too quickly, forcing these algorithms to take very small steps and hence converge very slowly. The empirical speed of convergence can often be increased by decreasing the value of the gain sequence only when there is reason to believe the algorithm has reached a neighborhood of the optimal solution. This idea was proposed and analyzed in the one-dimensional case by Kesten (1958) and extended to higher dimensions by Delyon and Juditsky (1993).

Several other approaches for addressing the stepsize problem discussed in the preceding paragraph have been developed. One of these approaches involves using constant gain (i.e., $a_n = a$ for all n, where a is a positive scalar). Dupuis and Simha (1991) prove the convergence of a sampling controlled version of Algorithm 1 with constant gain. Another approach is to use ideas from deterministic optimization to select the step sizes $\{a_n\}$. Wardi (1990) and Yan and Mukai (1993) propose and analyze methods that resemble the steepest descent method for deterministic optimization using Armijo step-sizes. Also, Shapiro and Wardi (1996a) discuss the use of line search to select the step sizes. Another approach for addressing the step-size problem that has received a fair amount of attention in recent years is to use averaging. More specifically, this approach involves letting the gain sequence $\{a_n\}$ decrease to zero at a slower rate than $1/n$. This means that the convergence rate of the sequence $\{\theta_n\}$ generated by Algorithm 1 will equal $\sqrt{a_n}$ (under certain conditions), so the algorithm will converge at a rate that is slower than the optimal rate of $1/\sqrt{n}$. However, by using averaging to estimate the solution [i.e., let the estimate of the solution after n iterations have been completed be $\sum_{i=1}^{n+1} \theta_i/(n+1)$] the resulting algorithm can be shown to converge at the fastest possible rate of $1/\sqrt{n}$. This idea has been studied by Polyak (1990), Yin (1991), Polyak and Juditsky (1992), and Kushner and Yang (1993). [A different type of averaging has been studied by Ruszczynski and Syski (1983).]

The stochastic approximation algorithms discussed above are all designed to locate local solutions to either unconstrained optimization problems or optimization problems with a known closed and convex feasible region. For stochastic approximation algorithms that are designed to solve global optimization problems with continuous decision variables, the reader is referred to Gelfand and Mitter (1991). Moreover, for a discussion of stochastic approximation algorithms for solving constrained problems with noisy constraints, the reader is referred to Kushner and Clark (1978).

A number of researchers have studied the application of stochastic approximation algorithms to solve simulation optimization problems of the form given in equation (1). Essentially, this involves using gradient estimates derived using one of the methods discussed in Section 9.2.1 in step 1 of Algorithm 1. For example, Glynn (1986b) proved the convergence of stochastic approximation algorithms when applied to optimize the steady-state behavior of both regenerative stochastic processes (with finite-difference estimates of the gradient) and regenerative finite-state-space Markov chains (with either finite-difference or likelihood ratio gradient estimates). Andradóttir (1996b) has extended this work to show how the Robbins–Monro algorithm with likelihood ratio gradient estimates converges when applied to optimize the steady-state behavior of general state-space Markov chains that are regenerative in either the standard sense or the Harris sense. She also shows how the Robbins–Monro algorithm can be used to optimize the transient (finite-horizon) behavior of general state-space Markov chains, again using likelihood ratio estimates of the gradient. Moreover, Fu (1990), Chong and Ramadge (1992, 1993), and L'Ecuyer and Glynn (1994) have studied the application of stochastic approximation algorithms to optimize the steady-state behavior of single-server queues. Fu (1990) and Chong and Ramadge (1992, 1993) use IPA to estimate the gradient, whereas L'Ecuyer and Glynn (1994) consider gradient estimates obtained using either finite differences, the likelihood ratio method, or IPA.

9.2.3 Sample Path Optimization

The optimization techniques discussed in the preceding section (see Algorithm 1) are stochastic in that when these algorithms are used to solve simulation optimization problems, the estimators $\hat{g}(\theta_n)$ required in the different iterations are usually obtained using independent simulations [i.e., given the value of θ_n, the simulation used to estimate $\hat{g}(\theta_n)$ in iteration n does not depend on simulations conducted in the previous iterations $1, \ldots, n-1$ to estimate $\hat{g}(\theta_1), \ldots, \hat{g}(\theta_{n-1})$]. Recently, a different class of methods has been developed for solving continuous parameter simulation optimization problems. These methods involve converting the original simulation optimization problem into an approximate deterministic optimization problem, and then using standard mathematical programming techniques to locate the optimal solution.

To illustrate, suppose that the objective function f is of the form given in equation (8). Then we could generate independent observations $X_1, \ldots, X_N$ from the distribution F, and approximate the objective function using

$$\hat{f}_N(\theta) = \frac{1}{N} \sum_{i=1}^{N} h(\theta, X_i) \qquad \text{for all } \theta \in \Theta \tag{13}$$

Once the observations $X_1, \ldots, X_N$ have been generated, the approximate objective function $\hat{f}_N(\theta)$ is now a deterministic function of the decision parameter θ, so that we can approximate the original simulation optimization problem (1) with the deterministic optimization problem

$$\min_{\theta \in \Theta} \hat{f}_N(\theta) \tag{14}$$

Now a standard mathematical programming algorithm can be applied to solve this approximate deterministic optimization problem, yielding an estimate of the optimal solution $\theta_N^*(\omega)$, where $\omega = (X_1, \ldots, X_N)$. The issue at hand is to determine under what conditions, the solution $\theta_N^*(\omega)$ of the approximate deterministic optimization problem (14) is "close" to the solution θ^* of the original optimization problem (1).

As was discussed in the section "Advanced Gradient Estimation Techniques" in Section 9.2.1, assuming that the objective function f is such that it can be estimated using simulations involving exclusively random variables whose distribution does not depend on the decision parameter θ [as in equation (8)] is often not realistic. So suppose now that the objective function is instead of the form given in equation (9). Then, again as in the same section, it is necessary to transform the objective function in such a way that the distribution of the underlying random variables no longer depends on the parameter θ. This can be achieved using either IPA or likelihood ratios. We now illustrate how this can be accomplished using likelihood ratios when the objective function f is given by equation (9). Later in this section we illustrate the use of IPA for this purpose in the context of Example 1.

As in equation (10), assume that for all $\theta \in \Theta$, the underlying random variable X has density function f_θ. Moreover, suppose that there exists $\theta_0 \in \Theta$ such that $f_{\theta_0}(x) = 0$ implies that $f_\theta(x) = 0$ for all possible observations x of the random variable X and all $\theta \in \Theta$. Then we have that

$$\begin{aligned} f(\theta) &= \int h(\theta, x) f_\theta(x)\, dx \\ &= \int h(\theta, x) \frac{f_{\theta_0}(x)}{f_{\theta_0}(x)} f_\theta(x)\, dx \\ &= \int h(\theta, x) \frac{f_\theta(x)}{f_{\theta_0}(x)} f_{\theta_0}(x)\, dx \end{aligned}$$

for all $\theta \in \Theta$ (as in Section 9.2.1, we adopt the notation that $0/0 = 1$). Recall that the term $f_\theta(x)/f_{\theta_0}(x)$ is called the likelihood ratio. We can now approximate the original objective function f as follows. Generate independent observations $X_1, \ldots, X_N$ from the distribution F_{θ_0} and let

$$\hat{f}_N(\theta) = \frac{1}{N} \sum_{i=1}^{N} h(\theta, X_i) \frac{f_\theta(X_i)}{f_{\theta_0}(X_i)} \qquad \text{for all } \theta \in \Theta \tag{15}$$

[Note that this transformation of the objective function effectively involves using importance sampling to estimate the objective function for all $\theta \neq \theta_0$; see, for example, Glynn and Iglehart (1989), Shahabuddin (1994), and Andradóttir et al. (1995) for discussions of importance sampling.] As before, we can now apply a standard mathematical programming algorithm to solve the approximate deterministic optimization problem (14) with the objective function $\hat{f}_N(\theta)$ now given in equation (15), yielding an estimate of the optimal solution $\theta_N^*(\omega)$, where $\omega = (X_1, \ldots, X_N)$. Again, it is of interest to determine under what conditions the solution $\theta_N^*(\omega)$ of this approximate deterministic optimization problem is "close" to the solution θ^* of the original optimization problem (1).

The optimization approach discussed above is often called sample path optimization, because it involves using simulation to generate only one sample path $\omega = (X_1, \ldots, X_N)$ of the underlying stochastic process, and the estimate $\theta_N^*(\omega)$ of the solution depends on the particular sample path ω that was used to define the approximate deterministic optimization problem (14). A number of researchers have recently studied this approach in a variety of contexts. Rubinstein and Shapiro (1993) have proposed and analyzed this approach using importance sampling to obtain the approximate deterministic objective function [a special example of this approach is described in the preceding paragraph; see equation (15)]. Their approach is called the *stochastic counterpart method*. Plambeck et al. (1996) have applied this method to optimize stochastic systems using IPA gradient estimates. Healy and Xu (1994) have also discussed and analyzed this method; they call the method *retrospective optimization* [see also the earlier work of Healy and Schruben (1991)]. Moreover, Chen and Schmeiser (1994) have studied the application of a related approach to solve root-finding problems. Additional theoretical work regarding the convergence of the sample path approach has appeared in Robinson (1996) and Shapiro and Wardi (1996b).

One advantage of the sample path approach relative to stochastic approximation is that it is easy to extend it to situations where not only the objective function, but also the constraints, cannot be evaluated analytically, but instead, have to be estimated using simulation. (Recall that most stochastic approximation algorithms discussed in Section 9.2.2 assume that the underlying optimization problem is either unconstrained, or with known constraints.) This is because we can approximate the constraint functions using a deterministic (sample path based) approximate functions in the same fashion as we approximate the objective function [see equations (13) and (15)]. Then a standard mathematical programming algorithm for constrained optimization can be used to solve the resulting approximate constrained deterministic optimization problem. See also Shapiro (1996) for a comparison of the asymptotical behavior of stochastic approximation and sample path methods.

We now return to Example 1 and discuss how sample path optimization with IPA gradient estimates can be used to solve the simulation optimization problem (1) when the feasible set Θ and the objective function f are given in equation (2). Assume that the service times $S_i(\theta)$ are independent and exponentially distributed with mean θ for all i. Then we can let $S_i(\theta) = -\ln(V_i) \times \theta$ for all i, where $\{V_i\}$ is an independent sequence of $U[0, 1]$-distributed random variables (this same idea was used in Section 9.2.1 to obtain IPA estimates for the derivative $\hat{g}$ of the objective function f). It is clear that for all integers N, the following function $\hat{f}_N$ is an approximation of the original objective function f:

$$\hat{f}_N(\theta) = \frac{\alpha}{\theta} + \frac{\beta}{N} \sum_{i=1}^{N} W_i(\theta) \qquad \text{for all } \theta \in \Theta \tag{16}$$

where the system times $W_1(\theta), \ldots, W_N(\theta)$ are computed using equation (6). Moreover, once the first N interarrival times $T_1, \ldots, T_N$ and uniform random variables $V_1, \ldots, V_N$ have been generated, the function $\hat{f}_N(\theta)$ is a deterministic function of θ that depends on the parameters $\alpha, \beta, T_1, \ldots, T_N$, and $V_1, \ldots, V_N$. Therefore, we can now use a standard mathematical programming algorithm to solve the approximate deterministic optimization problem (14) with the objective function $\hat{f}_N$ defined in equation (16). Note that the derivative of the approximate objective function $\hat{f}_N$ with respect to θ equals the IPA estimate $\hat{g}$ of the derivative $g = f'$ of the original objective function f derived in Section 9.2.1 [see equation (11)]. Therefore, any derivative-based mathematical programming algorithm that is applied to solve the deterministic optimization problem (14) with the objective function $\hat{f}_N$ defined in equation (16) will use IPA derivative estimates to locate the optimal solution.

9.3 DISCRETE DECISION PARAMETERS

In this section we review briefly methods that have been proposed recently for solving simulation optimization problems with discrete decision parameters. More specifically, we are concerned with methods for solving the optimization problem (1) when the feasible region Θ is discrete. For example, in designing a manufacturing facility, one may need to determine which of, say, five different choices of a particular piece of equipment satisfies one's needs the best in terms of cost, capacity, reliability, and so on. This case when the decision parameter only takes a discrete set of values (five in the preceding example) has so far received less attention by the research community than the continuous-parameter case discussed in Section 9.2. When the number of alternatives to choose from is small, say no larger than 20, the ranking and selection and multiple comparison methods discussed in Chapter 8 can be used to locate the optimal alternative. However, these methods become computationally burdensome as the size of the feasible region grows. Related approaches that we do not discuss in this chapter include methods for solving the multiarmed-bandit problem and learning automata procedures. Instead, we first discuss some recently proposed random search methods for solving discrete simulation optimization problems, and then discuss other recent developments addressing the same problem.

The reason why we do not discuss methods for solving the multiarmed-bandit problem and learning automata procedures here is that these methods are designed to achieve the goal of spending as much time as possible at the optimal solution. Therefore, although these methods can be used to solve discrete simulation optimization problems, they are not designed to satisfy our goal, which is to find the optimal solution as quickly as possible. To illustrate, these methods are suitable for determining which of several treatment options is most likely to save a patient's life. Here the effectiveness of a particular treatment is determined by using it to treat patients, and it is obviously desirable to be conservative and apply the treatment that appears to be best to treat most of the patients, because the cost of applying a poor treatment to treat a patient may be the death of that patient. However, in the simulation optimization context, the cost of visiting the different alternatives does not depend on the quality of the alternatives. Therefore, methods that move aggressively around the feasible region attempting to locate the optimal alternative as quickly as possible are more suitable for solving simulation optimization problems than conservative methods (such as methods for solving the multiarmed-bandit problem and learning automata procedures) that may tend to spend too much time at local solutions and hence converge slowly.

9.3.1 Random Search

The algorithms discussed in this section all involve moving successively from a feasible point to a neighboring feasible point in search of the optimal solution. Therefore, for each $\theta \in \Theta$, it is necessary to specify a set $N(\theta) \subset \Theta \backslash \{\theta\}$ consisting of all the neighbors of the feasible alternative θ. The algorithms discussed in this section all have the feature that they generate a sequence $\{\theta_n\}$ taking values in the state space Θ, with $\theta_{n+1} \in N(\theta_n) \cup \{\theta_n\}$ for all n. The flexibility in choosing the neighborhood structure $\{N(\theta): \theta \in \Theta\}$ depends on the algorithm used, but in all cases the neighborhood structure must be connected, in the sense that for all $\theta, \theta' \in \Theta$, $\theta \neq \theta'$, there exists an integer l and feasible alternatives $\theta_0, \ldots, \theta_l$, such that $\theta_0 = \theta$, $\theta_l = \theta'$, and $\theta_{i+1} \in N(\theta_i)$ for all $i = 0, \ldots, l-1$. (In other words, it is possible to go from any feasible alternative θ to any other feasible alternative θ' by moving successively from a feasible alternative to a neighbor of that alternative.) In addition to the difference in flexibility in the choice of the neighborhood structure, these algorithms differ in how a decision to move from a current alternative θ_n to the next alternative θ_{n+1} is made. They also differ in how estimates of the optimal solution are obtained.

Andradóttir (1995b, 1996c) has proposed and analyzed two random search methods for discrete stochastic optimization. In each iteration of these methods, simulations are conducted to estimate the value of the objective function at two neighboring feasible alternatives, the resulting estimates of the objective function values are then compared, and the alternative that has the better observed objective function value is passed on to the next iteration. Also, both algorithms use the feasible alternative that the generated sequence $\{\theta_n\}$ has visited most often to estimate the optimal solution. One of these two methods is guaranteed to converge to a local solution of the underlying optimization problem, whereas the other one is globally convergent (under certain conditions). The details of how one version of the globally convergent method can be applied to solve the discrete optimization problem (1) are given below. This procedure assumes that the feasible set Θ is finite. Let $|\Theta| < \infty$ denote the number of feasible alternatives. Note that after n iterations, θ_n is the current feasible alternative, for all $\theta \in \Theta$, $V_n(\theta)$ is the number of times the algorithm has visited alternative θ so far, and θ_n^*, the estimate of the optimal solution after n iterations, is the alternative that the algorithm has visited most often so far.

Algorithm 2

Step 0: Select a starting point $\theta_0 \in \Theta$ and a suitable stopping criterion. Let $V_0(\theta_0) = 1$ and $V_0(\theta) = 0$ for all $\theta \in \Theta$, with $\theta \neq \theta_0$. Let $n = 0$ and $\theta_n^* = \theta_0$. Go to step 1.

Step 1: Generate a uniform random variable θ_n' such that for all $\theta \in \Theta$, with $\theta \neq \theta_n$, we select $\theta_n' = \theta$ with probability $1/(|\Theta| - 1)$. Go to step 2.

Step 2: Use simulation to generate an estimate $\hat{f}(\theta_n)$ of $f(\theta_n)$ and an estimate $\hat{f}(\theta_n')$ of $f(\theta_n')$. If $\hat{f}(\theta_n) > \hat{f}(\theta_n')$, then let $\theta_{n+1} = \theta_n'$. Otherwise, let $\theta_{n+1} = \theta_n$. Go to step 3.

Step 3: Let $n = n + 1$, $V_n(\theta_n) = V_{n-1}(\theta_n) + 1$, and $V_n(\theta) = V_{n-1}(\theta)$ for all $\theta \in \Theta$, with $\theta \neq \theta_n$. If $V_n(\theta_n) > V_n(\theta_{n-1}^*)$, then let $\theta_n^* = \theta_n$. Otherwise, let $\theta_n^* = \theta_{n-1}^*$. If the stopping criterion is satisfied, then stop and return θ_n^* as the estimate of the optimal solution. Otherwise, go to step 1.

Yan and Mukai (1992) have proposed and analyzed a method for (global) discrete stochastic optimization called the *stochastic ruler algorithm*. Essentially, the method proposed by Yan and Mukai compares observations of the objective function values with observations of a predetermined uniform random variable U called the stochastic ruler. The range of the stochastic ruler should include all possible observations of the objective function. Yan and Mukai use the current element θ_n of the sequence generated by the algorithm to estimate the optimal solution. They assume that the feasible region Θ is finite and that both the neighborhood structure N and the transition matrix R (see step 1 of Algorithm 3 below) are symmetric, which means that if $\theta' \in N(\theta)$, then $\theta \in N(\theta')$ and $R(\theta, \theta') = R(\theta', \theta)$. Moreover, the transition matrix R must satisfy $R(\theta, \theta') > 0$ if and only if $\theta' \in N(\theta)$ and $\sum_{\theta' \in N(\theta)} R(\theta, \theta') = 1$. The details of Yan and Mukai's procedure are given below. Note that in step 2 of this procedure, $\{M_n\}$ is a sequence of positive integers such that $M_n \to \infty$ as $n \to \infty$.

Algorithm 3 (Stochastic Ruler Algorithm)

Step 0: Select a starting point $\theta_0 \in \Theta$ and a suitable stopping criterion. Let $n = 0$ and go to step 1.

Step 1: Generate a neighbor $\theta_n' \in N(\theta_n)$ of the current alternative θ_n such that for all $\theta \in N(\theta_n)$, we have that $P\{\theta_n' = \theta\} = R(\theta_n, \theta)$. Go to step 2.

Step 2: FOR $i = 1, \ldots, M_n$, DO:

Use simulation to generate an estimate $\hat{f}^{(i)}(\theta_n')$ of $f(\theta_n')$.
Generate an observation $U_n^{(i)}$ of the stochastic ruler U.
If $\hat{f}^{(i)}(\theta_n') > U_n^{(i)}$, then let $\theta_{n+1} = \theta_n$ and go to step 3.

END DO

Let $\theta_{n+1} = \theta_n'$ and go to step 3.

Step 3: If the stopping criterion is satisfied, then stop and return θ_{n+1} as the estimate of the optimal solution. Otherwise, let $n = n + 1$ and go to step 1.

Alrefaei and Andradóttir (1996, submitted a) have developed a variant of the stochastic ruler algorithm (Algorithm 3) that is guaranteed to converge to the set of global optimal solutions under assumptions that are slightly more general than those of Yan and Mukai. This variant involves letting $M_n = M$, a constant positive integer, for all n, whereas Yan and Mukai assume that $M_n \to \infty$ as $n \to \infty$. Therefore, the new variant requires less computational effort per iteration than the original stochastic ruler method. Moreover, the approach of Alrefaei and Andradóttir involves using the number of visits to the different states to estimate the optimal solution, much as in Algorithm 2. This modification of the stochastic ruler approach appears to perform better in practice than the original version.

Gong et al. (1992) have also proposed and analyzed a (stochastic comparison) method for discrete stochastic optimization. Their method is motivated by the stochastic ruler method of Yan and Mukai, and can be thought of as a mixture of Algorithms 2 and 3. In particular, Gong et al. use the same neighborhood structure as Algorithm 2 [i.e., $N(\theta) = \Theta \backslash \{\theta\}$, for all $\theta \in \Theta$], and their algorithm differs from Algorithm 3 in that in step 2 of Algorithm 3 they do not compare $\hat{f}^{(i)}(\theta'_n)$ with $U_n^{(i)}$, but instead compare $\hat{f}^{(i)}(\theta'_n)$ with $\hat{f}^{(i)}(\theta_n)$ (similar to Algorithm 2).

The methods discussed so far in this section all use either the alternative that has been visited most often by the algorithm to estimate the optimal solution [this is the case of the methods proposed by Andradóttir (1995b, 1996c) and Alrefaei and Andradóttir (1996, submitted a); see Algorithm 2] or they use the current alternative being considered by the algorithm to estimate the optimal solution [this is the case of the methods proposed by Yan and Mukai (1992) and Gong et al. (1992); see Algorithm 3]. Andradóttir (submitted) has proposed another approach for estimating the optimal solution, namely to average all the estimates of the objective function values at the different feasible alternatives obtained so far by the algorithm and use the alternative with the best (smallest if one is minimizing, largest if one is maximizing) average estimated objective function value as the estimate of the optimal solution (this approach for estimating the optimal solution is used in Algorithm 4 below). She discusses the advantages of using this approach for estimating the optimal solution relative to the other approaches discussed previously and shows that the use of this approach for estimating the optimal solution appears to significantly accelerate the convergence of Algorithm 2 when applied to solve a simple discrete simulation optimization problem. She also presents and analyzes a variant of the stochastic comparison method of Gong et al. (1992) that uses this approach for estimating the optimal solution. Similarly, Alrefaei and Andradóttir (1997, submitted b) present and analyze a variant of the stochastic ruler method (Algorithm 3) that uses this approach for estimating the optimal solution; through a numerical example they show that this variant appears to perform better in practice than other variants of the stochastic ruler method. Finally, Alrefaei and Andradóttir (submitted c) have developed a variant of the simulated annealing algorithm that uses this approach for estimating the optimal solution (Algorithm 4 below); again this approach for estimating the optimal solution seems to yield improved performance relative to other approaches.

The remainder of this section is concerned with the application of simulated annealing to solve discrete simulation optimization problems. The simulated annealing algorithm was originally developed to solve (global) discrete deterministic optimization problems. The basic idea behind this algorithm is to allow hill-climbing moves so that the algorithm can escape from local solutions. More specifically, if in a given iteration n we are comparing the current alternative θ_n with a candidate alternative θ'_n, and if $f(\theta'_n) \leq f(\theta_n)$ [so that θ'_n is a better alternative than θ_n since we are trying to solve a minimization problem of the form (1)], then the algorithm will always move over to θ'_n (i.e., we will always have $\theta_{n+1} = \theta'_n$). On the other hand, if $f(\theta'_n) > f(\theta_n)$ (so that θ_n is a better alternative than θ'_n), then the algorithm will stay at the better alternative θ_n with a certain probability [with probability $1 - \exp[(f(\theta_n) - f(\theta'_n))/T_n]$, where $T_n > 0$ for all n] and it will make the hill-climbing move over to the worse alternative θ'_n with the remaining (positive) probability.

In recent years, a number of researchers have studied the use of simulated annealing to solve discrete simulation optimization problems. Bulgak and Sanders (1988) and Haddock and Mittenthal (1992) proposed heuristic versions of the simulated annealing approach and applied these to optimize certain manufacturing systems. The first rigorous application of simulated annealing to solve stochastic optimization problems that we are aware of was proposed and analyzed by Gelfand and Mitter (1989). Since then, other researchers have proposed and analyzed variants of the simulated annealing approach for discrete stochastic optimization. This includes Gutjahr and Pflug (1996), Fox and Heine (1996), Lee (1995), and Alrefaei and Andradóttir (1995, submitted c). Except for the methods of Alrefaei and Andradóttir (1995, submitted c), all of these methods (and the simulated annealing algorithm for discrete deterministic optimization described previously) use a slowly decreasing "cooling schedule" $\{T_n\}$ for deciding whether to stay at the current alternative θ_n or move to the candidate alternative θ'_n in iteration n of the algorithm [i.e., $\{T_n\}$ is a sequence of positive scalars such that $T_n \to 0$ as $n \to \infty$ and $T_n \geq C/\log(n+1)$ for all n, where C is a given constant that does not depend on n]. Also, it is usually necessary to assume that the estimates of the objective function values $f(\theta_n)$ and $f(\theta'_n)$ needed in iteration n of these methods become more and more precise as n grows. This means that the computational (simulation) effort required per iteration grows as the number of iterations grows.

We conclude this section by presenting one version of the simulated annealing approach of Alrefaei and Andradóttir (submitted c) that uses the approach for estimating the optimal solution proposed by Andradóttir (submitted). This approach involves using a constant temperature $T_n = T > 0$ for all n. This means that the probability of making a hill-climbing move (see step 2 of Algorithm 4) does not decrease as the number of iterations grows, allowing the procedure to continue to search the feasible set agressively for global solutions as the number of iterations grows. Note that for all $\theta \in \Theta$, $A_n(\theta)$ is the sum of all estimates of $f(\theta)$ obtained in the first n iterations of the algorithm and $C_n(\theta)$ is the number of such estimates. Moreover, θ_n^* denotes the estimate of the optimal solution after n iterations have been completed. Finally, the neighborhood structure N and transition matrix R can be chosen as in Algorithm 3.

Algorithm 4 (Simulated Annealing Algorithm)

Step 0: Select a starting point $\theta_0 \in \Theta$ and a suitable stopping criterion. For all $\theta \in \Theta$, let $A_0(\theta) = C_0(\theta) = 0$. Let $n = 0$, $\theta_0^* = \theta_0$, and go to step 1.

Step 1: Generate a neighbor $\theta'_n \in N(\theta_n)$ of the current alternative θ_n such that for all $\theta \in N(\theta_n)$, we have that $P\{\theta'_n = \theta\} = R(\theta_n, \theta)$. Go to step 2.

Step 2: Use simulation to generate estimates $\hat{f}(\theta_n)$ and $\hat{f}(\theta'_n)$ of $f(\theta_n)$ and $f(\theta'_n)$, respectively, that are independent of any estimates generated in previous iterations. If $\hat{f}(\theta'_n) \leq \hat{f}(\theta_n)$, then let $\theta_{n+1} = \theta'_n$. Otherwise, generate a uniform random variable $U_n \sim U[0,1]$. If

$$U_n \leq \exp\left[\frac{\hat{f}(\theta_n) - \hat{f}(\theta'_n)}{T}\right],$$

then let $\theta_{n+1} = \theta'_n$. Otherwise, let $\theta_{n+1} = \theta_n$. Go to step 3.

Step 3: Let $n = n+1$, $A_n(\theta) = A_{n-1}(\theta) + \hat{f}(\theta)$ and $C_n(\theta) = C_{n-1}(\theta) + 1$, for $\theta = \theta_n, \theta'_n$, and

$A_n(\theta) = A_{n-1}(\theta)$ and $C_n(\theta) = C_{n-1}(\theta)$, for all $\theta \in \Theta$, with $\theta \neq \theta_n, \theta'_n$. Let $\theta_n^* \in \Theta$ be a solution to the (deterministic) optimization problem $\min_{\theta \in \Theta} A_n(\theta)/C_n(\theta)$. If the stopping criterion is satisfied, then stop and return θ_n^* as the estimate of the optimal solution. Otherwise, go to step 1.

9.3.2 Other Recent Developments

The branch-and-bound method is a well-known approach for solving deterministic discrete optimization problems. Recently, Norkin et al. (1994) have proposed a version of this approach that is designed to solve discrete stochastic optimization problems. Their approach involves partitioning the feasible region Θ into smaller subsets and estimating upper and lower bounds on the values the objective function f can take within the individual subsets. These bounds are used for determining the most promising subset (which is divided into smaller subsets) and for removing from consideration nonprospective subsets. Details of the procedure, including a discussion of how the required upper and lower bounds can be computed, are given by Norkin et al.

Another approach for solving discrete stochastic optimization problems involves using simulation to generate observations of the objective function at the different feasible alternatives, and using these data for constructing a confidence set $S \subset \Theta$ having the feature that for any global solution θ^* to the discrete optimization problem (1), we have that $P\{\theta^* \in S\} \geq 1 - \alpha$, where $0 < \alpha < 1$ is a predetermined constant (usually, $\alpha \leq 0.1$). Obviously, it is desirable to conduct the simulation in such a way that the resulting confidence set is as small as possible (so that we will have obtained as much information as possible about which alternatives are optimal). Pflug (1994) has proposed a sequential procedure for achieving this in the case when the observations of the objective function values at the different feasible alternatives are all normally distributed (he also reviews some earlier approaches). In each iteration of his procedure, the expected size of the confidence set obtained after generating l additional observations of $f(\theta)$ is computed for all $\theta \in \Theta$ (here l is a positive integer chosen by the user). Then l additional observations of $f(\theta')$ are generated for the alternative $\theta' \in \Theta$ that yielded the smallest expected confidence set size. See also Futschik and Pflug (1995) for a two-step procedure for obtaining valid confidence sets.

Finally, Ho et al. (1992) discuss some ideas that are relevant in this context. They point out that it is often true that less computer time is required for ranking alternatives (according to the value the objective function takes at these alternatives) than for estimating the objective function values precisely at the different alternatives. (This is the basic idea behind both their ordinal optimization approach and several of the methods discussed in Section 9.3.1.) They also suggest that in situations where the underlying feasible region Θ is very large (so that the discrete optimization procedures discussed above are very computer intensive), one could quickly conduct simulations at the different alternatives $\theta \in \Theta$ to obtain a rough ranking of the alternatives. Then one could discard all but the r top alternatives in this rough ranking, where $r \ll |\Theta|$ is a positive integer chosen by the user, and use one of the methods discussed in Section 9.3 to locate the best alternative among the r remaining ones. Obviously, discarding all but r of the feasible alternatives carries the risk that none of the remaining alternatives is anywhere close to being optimal (in which case this approach would perform very poorly). However, Ho et al. show that the probability that the set consisting of the r remaining alternatives contains a near-optimal solution to the discrete optimization problem (1) is often surprisingly large.

9.4 CONCLUSIONS

In this chapter we have provided an introduction to the field of simulation optimization. The focus has been on gradient-based techniques for simulation optimization with respect to continuous decision variables and on random search methods for simulation optimization with respect to discrete decision variables. Related subjects are discussed in Chapters 6 and 8.

Meketon (1987) states that "optimization for simulation, to date, remains an art, not a science." Although a substantial amount of progress has been made since this was written, optimization in simulation remains a challenging problem. Most of the currently existing methods for simulation optimization require a fair amount of sophistication on the part of the user; a good understanding of both the structure of the underlying stochastic system and the optimization technique being applied is often required. In addition, a large amount of computer time is often required for locating a near-optimal solution of the problem at hand. Hence, increasing the efficiency and ease of application of simulation optimization techniques is an important area of current and future research.

ACKNOWLEDGMENT

This work was partially supported by the National Science Foundation under Grants DDM-9210679 and DMI-9523111.

REFERENCES

Alrefaei, M. H., and S. Andradóttir (1995). A new search algorithm for discrete stochastic optimization, in *Proceedings of the 1995 Winter Simulation Conference*, C. Alexopoulos, K. Kang, W. R. Lilegdon, and D. Goldsman, eds., IEEE, Piscataway, N.J., pp. 236–241.

Alrefaei, M. H., and S. Andradóttir (1996). Discrete stochastic optimization via a modification of the stochastic ruler method, in *Proceedings of the 1996 Winter Simulation Conference*, J. M. Charnes, D. J. Morrice, D. T. Brunner, and J. J. Swain, eds., IEEE, Piscataway, N.J., pp. 406–411.

Alrefaei, M. H., and S. Andradóttir (1997). Accelerating the convergence of the stochastic ruler method for discrete stochastic optimization, *Proceedings of the 1997 Winter Simulation Conference*, in S. Andradóttir, K. J. Healy, D. H. Withers, and B. L. Nelson, eds., IEEE, Piscataway, N.J., pp. 352–357.

Alrefaei, M. H., and S. Andradóttir (submitted a). A modification of the stochastic ruler method for discrete stochastic optimization.

Alrefaei, M. H., and S. Andradóttir (submitted b). Discrete stochastic optimization using variants of the stochastic ruler method.

Alrefaei, M. H., and S. Andradóttir (submitted c). A simulated annealing algorithm with constant temperature for discrete stochastic optimization.

Andradóttir, S. (1995a). A stochastic approximation algorithm with varying bounds, *Operations Research*, Vol. 43, pp. 1037–1048.

Andradóttir, S. (1995b). A method for discrete stochastic optimization, *Management Science*, Vol. 41, pp. 1946–1961.

Andradóttir, S. (1996a). A scaled stochastic approximation algorithm, *Management Science*, Vol. 42, pp. 475–498.

Andradóttir, S. (1996b). Optimization of the transient and steady-state behavior of discrete event systems, *Management Science*, Vol. 42, pp. 717–737.

Andradóttir, S. (1996c). A global search method for discrete stochastic optimization, *SIAM Journal on Optimization*, Vol. 6, pp. 513–530.

Andradóttir, S. (submitted). Accelerating the convergence of random search methods for discrete stochastic optimization.

Andradóttir, S., D. P. Heyman, and T. J. Ott (1995). On the choice of alternative measures in importance sampling with Markov chains, *Operations Research*, Vol. 33, pp. 509–519.

Azadivar, F. (1992). A tutorial on simulation optimization, in *Proceedings of the 1992 Winter Simulation Conference*, J. J. Swain, D. Goldsman, R. C. Crain, and J. R. Wilson, eds., IEEE, Piscataway, N.J., pp. 198–204.

Bazaraa, M. S., and C. M. Shetty (1979). *Nonlinear Programming: Theory and Algorithms*, Wiley, New York.

Benveniste, A., M. Métivier, and P. Priouret (1990). *Adaptive Algorithms and Stochastic Approximations*, Springer-Verlag, Berlin.

Bulgak, A. A., and J. L. Sanders (1988). Integrating a modified simulated annealing algorithm with the simulation of a manufacturing system to optimize buffer sizes in automatic assembly systems, in *Proceedings of the 1988 Winter Simulation Conference*, M. Abrams, P. Haigh, and J. Comfort, eds., IEEE, Piscataway, N.J., pp. 684–690.

Chen, H., and B. W. Schmeiser (1994). Retrospective approximation algorithms for stochastic root finding, in *Proceedings of the 1994 Winter Simulation Conference*, J. D. Tew, S. Manivannan, D. A. Sadowski, and A. F. Seila, eds., IEEE, Piscataway, N.J., pp. 255–261.

Chen, H. F., and Y. M. Zhu (1986). Stochastic approximation procedures with randomly varying truncations, *Scientia Sinica Series A*, Vol. 29, pp. 914–926.

Chong, E. K. P., and P. J. Ramadge (1992). Convergence of recursive algorithms using IPA derivative estimates, *Discrete Event Dynamic Systems*, Vol. 2, pp. 339–372.

Chong, E. K. P., and P. J. Ramadge (1993). Optimization of queues using an infinitesimal perturbation analysis-based stochastic algorithm with general update times, *SIAM Journal on Control and Optimization*, Vol. 31, pp. 698–732.

Delyon, B., and A. Juditsky (1993). Accelerated stochastic approximation, *SIAM Journal on Optimization*, Vol. 3, pp. 868–881.

Dupuis, P., and R. Simha (1991). On sampling controlled stochastic approximation, *IEEE Transactions on Automatic Control*, Vol. 36, pp. 915–924.

Ermoliev, Y. (1983). Stochastic quasigradient methods and their application to system optimization, *Stochastics*, Vol. 9, pp. 1–36.

Fabian, V. (1968). On asymptotic normality in stochastic approximation, *Annals of Mathematical Statistics*, Vol. 39, pp. 1327–1332.

Fox, B. L., and G. W. Heine (1996). Probabilistic search with overrides, *Annals of Applied Probability*, Vol. 6, pp. 1087–1094.

Fu, M. C. (1990). Convergence of a stochastic approximation algorithm for the GI/G/1 queue using infinitesimal perturbation analysis, *Journal of Optimization Theory and Applications*, Vol. 65, pp. 149–160.

Fu, M. C. (1994). Optimization via simulation: a review, *Annals of Operations Research*, Vol. 53, pp. 199–247.

Fu, M. C., and J. Q. Hu (1997). *Conditional Monte Carlo: Gradient Estimation and Optimization Applications*, Kluwer, Norwell, Mass.

Futschik, A., and G. Ch. Pflug (1995). Confidence sets for discrete stochastic optimization, *Annals of Operations Research*, Vol. 59, pp. 95–108.

Gaivoronski, A. A. (1992). Optimization of stochastic discrete event dynamic systems: a survey of some recent results, in *Simulation and Optimization: Proceedings of the International Workshop on Computationally Intensive Methods in Simulation and Optimization*, G. Ch. Pflug and U. Dieter, eds., Springer-Verlag, Berlin, pp. 24–44.

Gelfand, S. B., and Mitter, S. K. (1989). Simulated annealing with noisy or imprecise energy measurements, *Journal of Optimization Theory and Applications*, Vol. 62, pp. 49–62.

Gelfand, S. B., and Mitter, S. K. (1991). Recursive stochastic algorithms for global optimization in $\mathbb{R}^d$, *SIAM Journal on Control and Optimization*, Vol. 29, pp. 999–1018.

Glasserman, P. (1991). *Gradient Estimation via Perturbation Analysis*, Kluwer, Norwell, Mass.

Glasserman, P., and D. D. Yao (1992). Some guidelines and guarantees for common random numbers, *Management Science*, Vol. 38, pp. 884–908.

Glynn, P. W. (1986a). Optimization of stochastic systems, in *Proceedings of the 1986 Winter Simulation Conference*, J. R. Wilson, J. O. Henriksen, and S. D. Roberts, eds., IEEE, Piscataway, N.J., pp. 52–59.

Glynn, P. W. (1986b). Stochastic approximation for Monte Carlo optimization, in *Proceedings of the 1986 Winter Simulation Conference*, J. R. Wilson, J. O. Henriksen, and S. D. Roberts, eds., IEEE, Piscataway, N.J., pp. 356–365.

Glynn, P. W. (1989). Optimization of stochastic systems via simulation, in *Proceedings of the 1989 Winter Simulation Conference*, E. A. MacNair, K. J. Musselman, and P. Heidelberger eds., IEEE, Piscataway, N.J., pp. 90–105.

Glynn, P. W. (1990). Likelihood ratio gradient estimation for stochastic systems, *Communications of the ACM*, Vol. 33, pp. 75–84.

Glynn, P. W., and D. L. Iglehart (1989). Importance sampling for stochastic simulations, *Management Science*, Vol. 35, pp. 1367–1392.

Gong, W.-B., Y.-C. Ho, and W. Zhai (1992). Stochastic comparison algorithm for discrete optimization with estimation, *Proceedings of the 31st Conference on Decision Control*, pp. 795–800.

Gutjahr, W. J., and G. Ch. Pflug (1996). Simulated annealing for noisy cost functions, *Journal of Global Optimization*, Vol. 8, pp. 1–13.

Haddock, J., and J. Mittenthal (1992). Simulation optimization using simulated annealing, *Computers in Industrial Engineering*, Vol. 22, pp. 387–395.

Healy, K. J., and L. W. Schruben (1991). Retrospective simulation response optimization, in *Proceedings of the 1991 Winter Simulation Conference*, B. L. Nelson, W. D. Kelton, and G. M. Clark, eds., IEEE, Piscataway, N.J., pp. 901–906.

Healy, K. J., and Y. Xu (1994). Simulation based retrospective approaches to stochastic system optimization, preprint.

Ho, Y.-C., and X.-R. Cao (1991). *Perturbation Analysis of Discrete Event Dynamical Systems*, Kluwer, Norwell, Mass.

Ho, Y.-C., R. S. Sreenivas, and P. Vakili (1992). Ordinal optimization of DEDS, *Discrete Event Dynamic Systems*, Vol. 2, pp. 61–88.

Jacobson, S. H. (1994). Convergence results for harmonic gradient estimators, *ORSA Journal of Computing*, Vol. 6, pp. 381–397.

Jacobson, S. H., and L. W. Schruben (1989). Techniques for simulation response optimization, *Operations Research Letters*, Vol. 8, pp. 1–9.

Kesten, H. (1958). Accelerated stochastic approximation, *Annals of Mathematical Statistics*, Vol. 29, pp. 41–59.

Kiefer, J., and J. Wolfowitz (1952). Stochastic estimation of the maximum of a regression function, *Annals of Mathematical Statistics*, Vol. 23, pp. 462–466.

Kleijnen, J. P. C. (1995). Sensitivity analysis and optimization in simulation: design of experiments and case studies, in *Proceedings of the 1995 Winter Simulation Conference*, C. Alexopoulos, K. Kang, W. R. Lilegdon, and D. Goldsman eds., IEEE, Piscataway, N.J., pp. 133–140.

Kushner, H. J., and D. S. Clark (1978). *Stochastic Approximation Methods for Constrained and Unconstrained Systems*, Springer-Verlag, New York.

Kushner, H. J., and J. Yang (1993). Stochastic approximation with averaging of the iterates: optimal asymptotic rate of convergence for general processes, *SIAM Journal on Control and Optimization*, Vol. 31, pp. 1045–1062.

Lai, T. L., and H. Robbins (1979). Adaptive design and stochastic approximation, *Annals of Statistics*, Vol. 7, pp. 1196–1221.

L'Ecuyer, P. (1991). An overview of derivative estimation, in *Proceedings of the 1991 Winter Simulation Conference*, B. L. Nelson, W. D. Kelton, and G. M. Clark, eds., IEEE, Piscataway, N.J., pp. 207–217.

L'Ecuyer, P., and P. W. Glynn (1994). Stochastic optimization by simulation: convergence proofs for the GI/G/1 queue in steady-state, *Management Science*, Vol. 40, pp. 1562–1578.

L'Ecuyer, P., and G. Perron (1994). On the convergence rates of IPA and FDC derivative estimators, *Operations Research*, Vol. 42, pp. 643–656.

L'Ecuyer, P., N. Giroux, and P. W. Glynn (1994). Stochastic optimization by simulation: numerical experiments with the M/M/1 queue in steady-state, *Management Science*, Vol. 40, pp. 1245–1261.

Lee, J.-Y. (1995). Faster simulated annealing techniques for stochastic optimization problems, with application to queueing network simulation, Ph.D. dissertation, Statistics and Operations Research, North Carolina State University, Raleigh, N.C.

Ljung, L., G. Ch. Pflug, and H. Walk (1992). *Stochastic Approximation and Optimization of Random Systems*, Birkhauser Verlag, Basel, Switzerland.

Meketon, M. S. (1987). Optimization in simulation: a survey of recent results, in *Proceedings of the 1987 Winter Simulation Conference*, A. Thesen, H. Grant, and W. D. Kelton, eds., IEEE, Piscataway, N.J., pp. 58–67.

Mitra, M., and S. K. Park (1991). Solution to the indexing problem of frequency domain simulation experiments, in *Proceedings of the 1991 Winter Simulation Conference*, B. L. Nelson, W. D. Kelton, and G. M. Clark, eds., IEEE, Piscataway, N.J., pp. 907–915.

Nevel'son, M. B., and R. Z. Has'minskii (1973). An adaptive Robbins–Monro procedure, *Automation and Remote Control*, Vol. 34, pp. 1594–1607.

Norkin, V., Y. Ermoliev, and A. Ruszczynski (1994). On optimal allocation of indivisibles under uncertainty, preprint.

Pflug, G. Ch. (1990). On-line optimization of simulated Markovian processes, *Mathematics of Operations Research*, Vol. 15, pp. 381–395.

Pflug, G. Ch. (1994). Adaptive designs in discrete stochastic optimization, preprint.

Plambeck, E. L., B.-R. Fu, S. M. Robinson, and R. Suri (1996). Sample-path optimization of convex stochastic performance functions, *Mathematical Programming*, Vol. 75, pp. 137–176.

Polyak, B. T. (1990). New method of stochastic approximation type, *Automation and Remote Control*, Vol. 51, pp. 937–946.

Polyak, B. T., and A. B. Juditsky (1992). Acceleration of stochastic approximation by averaging, *SIAM Journal on Control and Optimization*, Vol. 30, pp. 838–855.

Robbins, H., and S. Monro (1951). A stochastic approximation method, *Annals of Mathematical Statistics*, Vol. 22, pp. 400–407.

Robinson, S. M. (1996). Analysis of sample-path optimization, *Mathematics of Operations Research*, Vol. 21, pp. 513–528.

Rubinstein, R. Y., and A. Shapiro (1993). *Discrete Event Systems: Sensitivity Analysis and*

Stochastic Optimization by the Score Function Method, Wiley, Chichester, West Sussex, England.

Ruppert, D. (1985). A Newton–Raphson version of the multivariate Robbins–Monro procedure, *Annals of Statistics*, Vol. 13, pp. 236–245.

Ruppert, D. (1991). Stochastic approximation, in *Handbook of Sequential Analysis*, B. K. Ghosh and P. K. Sen, eds., Marcel Dekker, New York, pp. 503–529.

Ruszczynski, A., and W. Syski (1983). Stochastic approximation method with gradient averaging for unconstrained problems, *IEEE Transactions on Automatic Control*, Vol. 29, pp. 1097–1105.

Safizadeh, M. H. (1990). Optimization in simulation: current issues and the future outlook, *Naval Research Logistics*, Vol. 37, pp. 807–825.

Shahabuddin, P. (1994). Importance sampling for the simulation of highly reliable Markovian systems, *Management Science*, Vol. 40, pp. 333–352.

Shapiro, A. (1996). Simulation-based optimization: convergence analysis and statistical inference, *Stochastic Models*, Vol. 12, pp. 425–454.

Shapiro, A., and Y. Wardi (1996a). Convergence analysis of stochastic algorithms, *Mathematics of Operations Research*, Vol. 21, pp. 615–628.

Shapiro, A., and Y. Wardi (1996b). Convergence analysis of gradient descent stochastic algorithms, *Journal of Optimization Theory and Applications*, Vol. 91 pp. 439–454.

Spall, J. C. (1992). Multivariate stochastic approximation using a simultaneous perturbation gradient approximation, *IEEE Transactions on Automatic Control*, Vol. 37, pp. 332–341.

Suri, R. (1989). Perturbation analysis: the state of the art and research issues explained via the GI/G/1 queue, *Proceedings of the IEEE*, Vol. 77, pp. 114–137.

Suri, R., and Y. T. Leung (1989). Single run optimization of discrete event simulations: an empirical study using the M/M/1 queue, *IEEE Transactions*, Vol. 21, pp. 35–49.

Venter, J. H. (1967). An extension of the Robbins–Monro procedure, *Annals of Mathematical Statistics*, Vol. 38, pp. 181–190.

Wardi, Y. (1990). Stochastic algorithms with Armijo stepsizes for minimization of functions, *Journal of Optimization Theory and Applications*, Vol. 64, pp. 399–417.

Wei, C. Z. (1987). Multivariate adaptive stochastic approximation, *Annals of Statistics*, Vol. 15, pp. 1115–1130.

Yan, D., and H. Mukai (1992). Stochastic discrete optimization, *SIAM Journal on Control and Optimization*, Vol. 30, pp. 594–612.

Yan, D., and H. Mukai (1993). Optimization algorithm with probabilistic estimation, *Journal of Optimization Theory and Applications*, Vol. 79, pp. 345–371.

Yin, G. (1991). On extensions of Polyak's averaging approach to stochastic approximation, *Stochastics and Stochastics Reports*, Vol. 36, pp. 245–264.

Yin, G., and Y. M. Zhu (1989). Almost sure convergence of stochastic approximation algorithms with non-additive noise, *International Journal of Control*, Vol. 49, pp. 1361–1376.

Zazanis, M. A., and R. Suri (1993). Convergence rates of finite-difference sensitivity estimates for stochastic systems, *Operations Research*, Vol. 41, pp. 694–703.

CHAPTER 10

Verification, Validation, and Testing

OSMAN BALCI
Virginia Polytechnic Institute and State University

10.1 INTRODUCTION

A simulation study is conducted for a variety of purposes, including problem solving and training. Starting with problem formulation and culminating with presentation of simulation study results, it consists of complex processes of formulation, analysis, modeling, and experimentation (or exercise). A typical simulation study requires multifaceted knowledge in diverse disciplines such as operations research, computer science, statistics, and engineering. Due to the complex processes and multifaceted knowledge requirements, simulation practitioners and managers face significant technical challenges in conducting *successful* simulation studies. A successful simulation study is defined to be the one that produces a sufficiently credible solution that is accepted and used by the decision makers.

To increase significantly the probability of success in conducting a simulation study, an organization must have a department or group called *simulation quality assurance* (SQA). The SQA group is responsible for total quality management and works closely with the simulation project managers in planning, preparing, and administering quality assurance activities throughout the simulation study. The SQA is a managerial approach that is critically essential for the success of a simulation study. Ören [1–3] presents concepts, criteria, and paradigms that can be used in establishing an SQA program within an organization.

Assuring total quality involves the measurement and assessment of a variety of quality characteristics such as accuracy, execution efficiency, maintainability, portability, reusability, and usability (human–computer interface). Simulation study objectives dictate a priority ordering of these quality characteristics since all of them cannot be achieved at the same level.

The purpose of this chapter is to present principles and techniques for the assessment of accuracy throughout the life cycle of a simulation study. The accuracy quality characteristic is assessed by conducting verification, validation, and testing (VV&T).

Handbook of Simulation, Edited by Jerry Banks.
ISBN 0-471-13403-1

Model verification is substantiating that the model is transformed from one form into another, as intended, with sufficient accuracy. Model verification deals with building the model *right*. The accuracy of transforming a problem formulation into a model specification or the accuracy of converting a model representation in a micro flowchart into an executable computer program is evaluated in model verification.

Model validation is substantiating that within its domain of applicability, the model behaves with satisfactory accuracy consistent with the study objectives. Model validation deals with building the *right* model.

An activity of accuracy assessment can be labeled as verification or validation based on an answer to the following question: In assessing the accuracy, is the model behavior compared with respect to the corresponding system behavior through mental or computer execution? If the answer is "yes," model validation is conducted; otherwise, it implies that the transformational accuracy is judged implying model verification.

Model testing is ascertaining whether inaccuracies or errors exist in the model. In model testing, the model is subjected to test data or test cases to determine if it functions properly. "Test failed" implies the failure of the model, not the test. A test is devised and testing is conducted to perform either validation or verification or both. Some tests are devised to evaluate the behavioral accuracy (i.e., validity) of the model, and some tests are intended to judge the accuracy of model transformation from one form into another (verification). Therefore, the entire process is commonly called *model VV&T*.

Testing should not be interpreted just as functional testing which requires computer execution of the model. Administering reviews, inspections, and walkthroughs is similar to devising a test under which model accuracy is judged. In this case, panel members become part of the devised test and the testing is conducted by each member executing a set of tasks. Therefore, informal techniques described in Section 10.4 are also considered testing techniques.

10.2 LIFE CYCLE AND A CASE STUDY

The processes and phases of the life cycle of a simulation study and a simulation and modeling case study are presented in this section. The case study is used throughout the chapter to illustrate the life cycle and the VV&T principles and techniques. The life cycle of a simulation study is presented in Figure 10.1 [4, 5]. The phases are shown by shaded oval symbols. The dashed arrows describe the processes that relate the phases to each other. The solid arrows refer to the credibility assessment stages. Banks et al. [6] and Knepell and Arangno [7] review other modeling processes for developing simulations.

The life cycle should not be interpreted as strictly sequential. The sequential representation of the dashed arrows is intended to show the direction of development throughout the life cycle. The life cycle is iterative in nature and reverse transitions are expected. Every phase of the life cycle has an associated VV&T activity. Deficiencies identified by a VV&T activity may necessitate returning to an earlier process and starting all over again.

The 10 processes of the life cycle are shown by the dashed arrows in Figure 10.1. Although each process is executed in the order indicated by the dashed arrows, an error identified may necessitate returning to an earlier process and starting all over again. Some guidelines are provided below for each of the 10 processes.

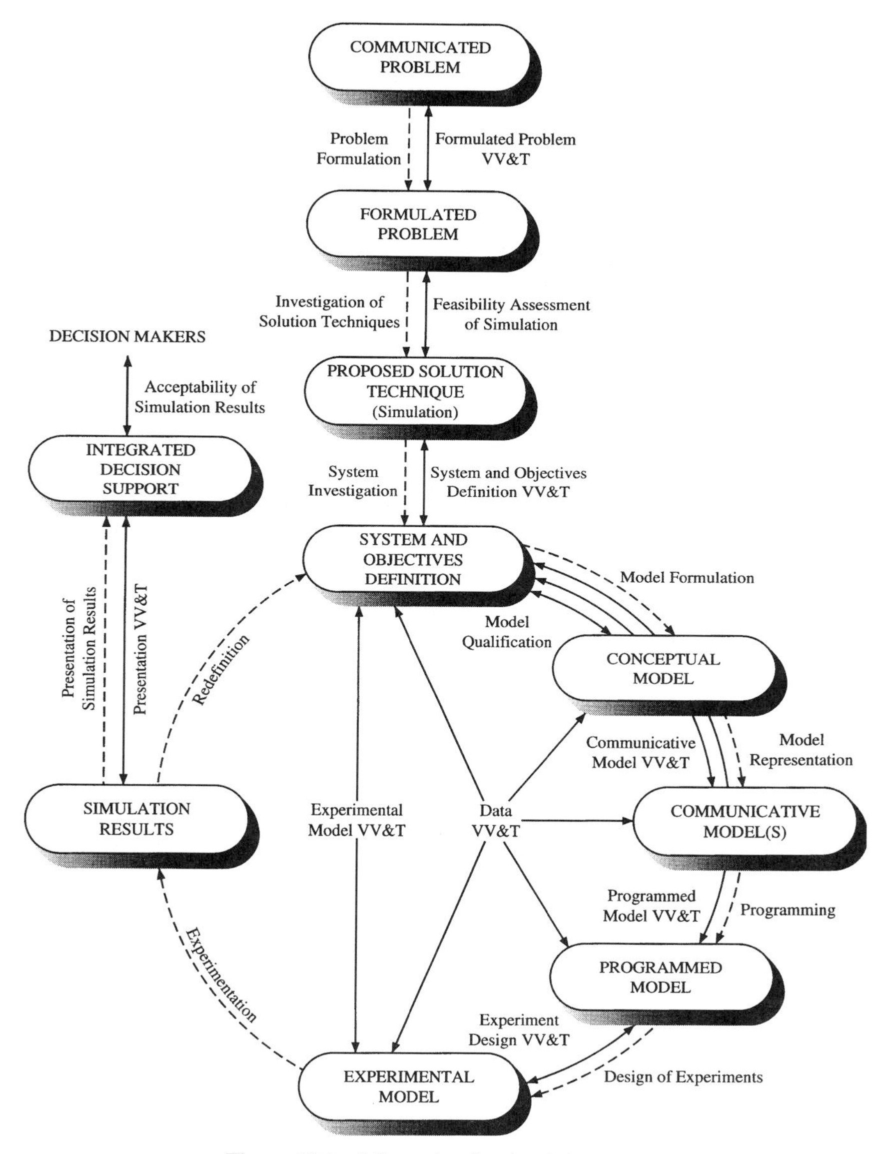

Figure 10.1 Life cycle of a simulation study.

10.2.1 Problem Formulation

When a problem is recognized, a decision maker (a client or sponsor group) initiates a study by communicating the problem to an analyst (a problem solver, contractor, or a consultant/research group). The problem communicated is rarely clear, specific, or organized. Hence an essential study to formulate the *actual* problem must follow. *Problem formulation* problem structuring or problem definition) is the process by which the initially communicated problem is translated into a formulated problem sufficiently well defined to enable specific research action [8].

Balci and Nance [9] present a high-level procedure that (1) guides the analyst during problem formulation, (2) structures the formulated problem VV&T, and (3) seeks to increase the likelihood that the study results are utilized by decision makers.

Case Study. The town of Blacksburg in Virginia (client) receives complaints from the drivers using the traffic intersection at Prices Fork Road and Toms Creek Road, shown in Figure 10.2, about too much waiting during rush-hour periods. The town hires a consulting company (the contractor) to conduct a study and propose a solution to the problem.

The contractor conducts the process of problem formulation and determines the study objective as follows:

> Identify which operating policy should be implemented at the traffic intersection so as to reduce the average waiting time of vehicles in each travel path to an acceptable level during rush-hour periods. Possible operating policies include different light timings, two-way stop signs, four-way stop signs, flashing red and yellow lights, and constructional changes such as adding new lanes.

10.2.2 Investigation of Solution Techniques

All alternative techniques that can be used to solve the formulated problem should be identified. A technique whose solution is estimated to be too costly or is judged not to be sufficiently beneficial with respect to the study objectives should be disregarded. Among the qualified ones, the technique with the highest expected benefits/cost ratio should be selected.

The statement "when all else fails, use simulation" is misleading if not invalid. The question is not to bring *a* solution to the problem, but to bring a sufficiently credible one that will be accepted and used by the decision maker(s). A technique other than simulation may provide a less costly solution, but it may not be as useful.

Sometimes, the problem communicated is formulated with the influence of a solution technique in mind. Occasionally, simulation is chosen without considering any other technique just because it is the only one the analyst(s) can handle. Skipping the investigation process may result in unnecessarily expensive solutions, sometimes to the wrong problems.

As a result of the investigation process, it is assumed here that simulation is chosen as the most appropriate solution technique. At this point, the simulation project team should be activated and be made responsible for the formulated problem VV&T and feasibility assessment of simulation before preceeding in the life cycle.

Case Study. The contractor investigates all possible solution techniques and selects discrete-event simulation as the one with the highest benefits/cost ratio.

10.2.3 System Investigation

Characteristics of the system that contains the problem formulated should be investigated for consideration in system definition and modeling. Shannon [10] identifies six

Figure 10.2 Traffic intersection at Prices Fork Road and Toms Creek Road.

major system characteristics: (1) change, (2) environment, (3) counterintuitive behavior, (4) drift to low performance, (5) interdependency, and (6) organization. Each characteristic should be examined with respect to the study objectives that are identified with the formulation of the problem.

In simulation, we deal primarily with stochastic and dynamic real systems that *change* over a period of time. How often and how much the system will change during the course of a simulation study should be estimated so that the model representation can be updated accordingly. Changes in the system may also change the study objectives.

A system's *environment* consists of all input variables that can affect its state significantly. The input variables are identified by assessing the significance of their influence on the system's state with regard to the study objectives. Underestimating the influence of an input variable may result in inaccurate environment definition.

Some complex systems may show *counterintuitive behavior*, which should be identified for consideration in defining the system. However, this is not an easy task, especially for those systems containing many subjective elements (e.g., social systems). Cause and effect are often not closely related in time or space. Symptoms may appear long after the primary causes [10]. To be able to identify counterintuitive behavior, it is essential that the simulation project employ people who have expert knowledge about the system under study.

A system may show a *drift to low performance* due to the deterioration of its components (e.g., machines in a manufacturing system) over a period of time. If this characteristic exists, it should be incorporated within the model representation especially if the model's intended use is forecasting.

The *interdependency* and *organization* characteristics of the system should be examined prior to the abstraction of the real system for the purpose of modeling. In a complex stochastic system, many activities or events take place simultaneously and influence each other. The system complexity can be overcome by way of decomposing the system into subsystems and subsystems into other subsystems. This decomposition can be carried out by examining how system elements or components are organized.

Once the system is decomposed into subsystems whose complexity is manageable and the system characteristics are documented, model formulation process can be started following the system and objectives definition VV&T.

Case Study. The contractor conducts the process of system investigation. It is determined that the traffic intersection will not change during the course of the study. The interarrival time of vehicles in lane L_j where $j = 1, 2, 3, \ldots, 11$ is identified as an input variable making up the environment, whereas pedestrians, emergency vehicles, and bicycles are excluded from the system definition due to their negligible effect on the system's state. No counterintuitive behavior can be identified. The system performance does not deteriorate over time.

10.2.4 Model Formulation

Model formulation is the process by which a conceptual model is envisioned to represent the system under study. The *conceptual model* is the model that is formulated in the mind of the modeler [5]. Model formulation and model representation constitute the process of model design.

Input data analysis and modeling [11, 12] is a subprocess of model formulation and is conducted with respect to the way the model is driven. Simulation models are classified as self-driven or trace-driven. A *self-driven* (distribution-driven or probabilistic) simulation model is the one that is driven by input values obtained via sampling from probability distributions using random numbers. A *trace-driven* (or retrospective) simulation model, on the other hand, is driven by input sequences derived from trace data obtained through measurement of the real system.

Under some study objectives (e.g., evaluation, comparison, determination of functional relations) and for model validation, input data model(s) are built to represent the system's input process. In a self-driven simulation (e.g., of a traffic intersection), we collect data on an input random variable (e.g., interarrival time of vehicles), identify the distribution, estimate its parameters, and conclude upon a probability distribution as the input data model to sample from in driving the simulation model [13]. In a trace-driven simulation, we trace the system (e.g., using hardware and software monitors) and

TABLE 10.1 Probabilistic Models of Vehicle Interarrival Times for Each Lane

Lane Number(s)	Probability Distribution	Location Parameter	Scale Parameter	Shape Parameter
1, 2	Gamma	0.06494	4.05843	1.1031
3	Weibull	1.99415	31.1737	0.79453
4	Weibull	0	8.71858	0.8773
5	Lognormal	0	1.42075	1.40056
6	Weibull	0	32.9441	1.14441
7	Weibull	0.99627	27.6663	0.70053
8	Weibull	0	33.1788	1.46385
9	Lognormal	0	1.93024	1.11273
10	Weibull	0	6.91658	0.78088
11	Weibull	0	5.57763	0.71616

utilize the refined trace data as the input data model to use in driving the simulation model.

Case Study. All assumptions made in abstracting the traffic intersection operation under the study objective are stated and listed explicitly. Data are collected on the interarrival times of vehicles, current traffic light timing, probabilities of turns, and travel times in each travel path. A single arrival process is observed for lanes 1 and 2 and divided probabilistically. ExpertFit software [13] is used to identify probabilistic models of the input process. The results of input data modeling are given in Table 10.1.

The estimated probabilities of right turns are presented in Table 10.2. The probability distributions identified characterize the rush-hour traffic conditions and are used to sample from in driving the self-driven simulation model built.

10.2.5 Model Representation

This is the process of translating the conceptual model into a communicative model. A *communicative model* is "a model representation which can be communicated to other humans, can be judged or compared against the system and the study objectives by more than one human" [5]. A communicative model (i.e., a simulation model design specification) may be represented in any of the following forms: (1) structured, computer-assisted graphs, (2) flowcharts, (3) structured English and pseudocode, (4) entity-cycle

TABLE 10.2 Estimated Probabilities of Right Turns

Location	Probability
Turning to lane 2 from the combined arrival process for lanes 1 and 2	0.634
Right turn in lane 2	0.346
Right turn in lane 5	0.160
Right turn in lane 11	0.140

(or activity-cycle) diagrams, (5) condition specification [14], and (6) more than a dozen diagramming techniques described in ref. 15.

Several communicative models may be developed: one in the form of structured English intended for nontechnical people, another in the form of a micro flowchart intended for a programmer. Different representation forms may also be integrated in a stratified manner. The representation forms should be selected based on (1) their applicability for describing the system under study, (2) the technical background of the people to whom the model is to be communicated, (3) how much they lend themselves to formal analysis and verification, (4) their support for model documentation, (5) their maintainability, and (6) their automated translatability into a programmed model.

Case Study. The Visual Simulation Environment (VSE) software product [16–18] is selected for simulation model development and experimentation. An aerial photograph of the traffic intersection obtained from the town of Blacksburg is scanned as shown in Figure 10.2. Vehicle images, tree branches, and light posts on the roads are removed from the scanned image using Adobe Photoshop software. The cleaned image is brought into the VSE Editor by clicking, dragging, and dropping. The photographic image is decomposed into components represented as circles as shown in Figure 10.3. The com-

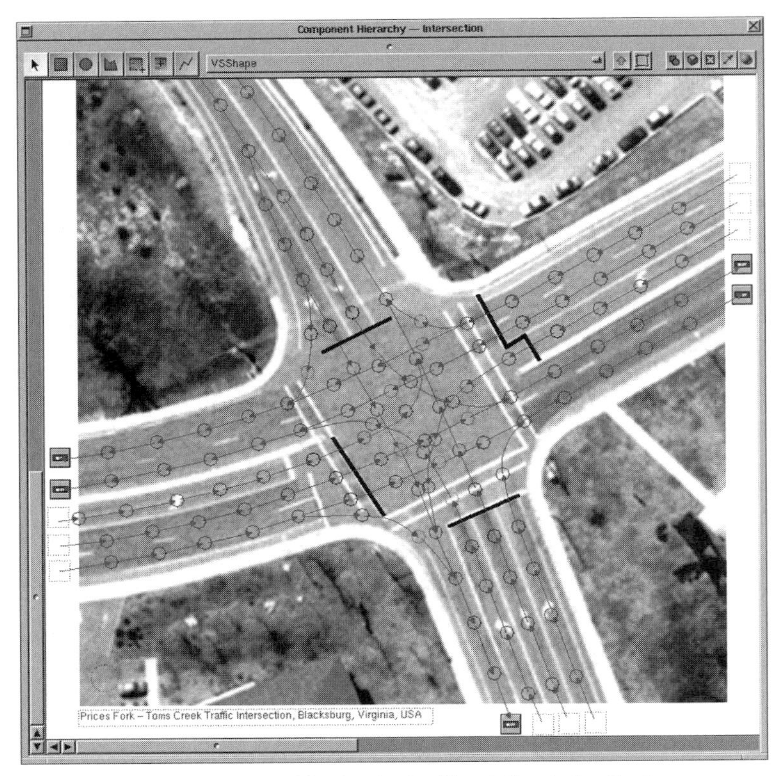

Figure 10.3 Model specification in the Visual Simulation Environment.

ponents are connected with each other using the path tool. The traffic light for each lane is depicted by a line which changes color during animation. New classes are created by inheriting from the built-in VSE class hierarchy. Methods in each class are specified by using VSE's very high-level object-oriented scripting language. Vehicles are modeled as dynamic objects and are instantiated at run time with respect to the interarrival times sampled from the probability distributions shown in Table 10.1. Each graphical object in the model representation is set to belong to a class to inherit the characteristics and behavior specified in that class.

10.2.6 Programming

Translation of the communicative model (model specification) into a programmed model (executable model) constitutes the process of programming. A *programmed model* is an executable simulation model representation which does not incorporate an experiment design. The process of programming can be performed by the modeler using a simulation software product [19], a simulation programming language [19], or a high-level programming language [20].

Case Study. The traffic intersection model specification is created by using the VSE Editor tool. Then, by selecting the "Prepare for Simulation" menu option, the model specification is automatically translated into an executable form. The VSE Simulator tool is used to execute and animate the model and conduct experiments.

10.2.7 Design of Experiments

This is the process of formulating a plan to gather the desired information at minimal cost and to enable the analyst to draw valid inferences [10]. An *experimental model* is the programmed model incorporating an executable description of operations presented in such a plan.

A variety of techniques are available for the design of experiments. *Response-surface methodologies* can be used to find the optimal combination of parameter values that maximize or minimize the value of a response variable [11]. *Factorial designs* can be employed to determine the effect of various input variables on a response variable (Chapter 6; [21]). *Variance reduction techniques* can be implemented to obtain greater statistical accuracy for the same amount of simulation [11]. *Ranking and selection techniques* can be utilized for comparing alternative systems (Chapter 8; [11, 19]). Several methods (e.g., replication, batch means, regenerative) can be used for statistical analysis of simulation output data (Chapter 7).

Case Study. The VSE Simulator's design of experiments panel is used to specify the method of replications for statistical analysis of simulation output data. Fourteen performance measures are defined for all travel paths:

W_{jL} = average waiting time of vehicles arriving and turning left in lane j,
$j = 1, 3, 6, 9$

W_{jS} = average waiting time of vehicles arriving and travelling straight in lane j,
$j = 2, 4, 5, 7, 10, 11$

W_{jR} = average waiting time of vehicles arriving and turning right in lane j,
$j = 2, 5, 8, 11$

The waiting time is the time spent by a vehicle in the entire traffic intersection, from arrival to the intersection to departure. The experimentation objective is to select the best traffic light timing out of three alternatives: the current light timing and two other alternatives suggested based on observation. The best light timing produces the lowest W_{jL}, W_{jS}, and W_{jR} for each lane j.

10.2.8 Experimentation

This is the process of experimenting with the simulation model for a specific purpose. Some purposes of experimentation are [10] (1) training, (2) comparison of different operating policies, (3) evaluation of system behavior, (4) sensitivity analysis, (5) forecasting, (6) optimization, and (7) determination of functional relations. The process of experimentation produces the simulation results.

Case Study. Using the VSE Editor, the model is instrumented to collect data on each performance measure. The model is warmed up for a total of 1000 vehicles passing through the intersection. Data are collected during the steady-state period of 10,000 vehicles. Identical experimental conditions are created by way of using the same random number generator seeds for each traffic light timing alternative. The model is replicated 30 times and each performance measure replication value is written to output file f, where $f = 1, 2, 3, \ldots, 14$. Then the VSE Output Analyzer tool is used to open the output files and construct confidence intervals and provide general statistics for each performance measure.

10.2.9 Redefinition

This is the process of (1) updating the experimental model so that it represents the current form of the system, (2) altering it for obtaining another set of results, (3) changing it for the purpose of maintenance, (4) modifying it for other use(s), or (5) redefining a new system to model for studying an alternative solution to the problem.

Case Study. Using the VSE Editor, the traffic light timing is modified and a new executable model is produced. The VSE Simulator is used to conduct experiments with the model under the new traffic light timing. The VSE Output Analyzer is used to construct confidence intervals and provide general statistics for each performance measure.

10.2.10 Presentation of Simulation Results

In this process, simulation results are interpreted and presented to the decision makers for their acceptance and implementation. Since all simulation models are descriptive, deciding on a solution to the problem requires rigorous analysis and interpretation of the results. The presentation should be made with respect to the intended use of the model. If the model is used in a "what if" environment, the results should be integrated to support the decision maker in the decision-making process. Complex simulation results may also necessitate such an integration. The report documenting the study and its results together with its presentation also constitutes a form of supporting the decision maker.

Case Study. The experimentation results under three traffic light timing alternatives are presented to the decision makers. There was not a single alternative that reduced the average waiting time in every travel path. However, alternative 1 was found to reduce the waiting times to acceptable levels in all travel paths and hence it is accepted and implemented by the decision makers.

10.3 VERIFICATION, VALIDATION, AND TESTING PRINCIPLES

According to Webster's Encyclopedic Unabridged Dictionary, a *principle* is defined as "1. an accepted or professed rule of action or conduct. 2. a fundamental, primary, or general law or truth from which others are derived. 3. a fundamental doctrine or tenet; a distinctive ruling opinion." All three definitions above apply to the way the term *principle* is used herein.

Principles are important to an understanding of the foundations of VV&T. The principles help researchers, practitioners, and managers better comprehend what VV&T is all about. They serve to provide the underpinnings for over 75 VV&T techniques, described in Section 10.4, that can be used throughout the life cycle. Understanding and applying these principles is crucially important for the success of a simulation study.

The 15 principles presented herein are established based on the experience described in the published literature and the author's experience during his VV&T research since 1978. The principles are listed below in no particular order.

Principle 1 VV&T must be conducted throughout the entire life cycle of a simulation study.

VV&T is not a phase or step in the life cycle but a continuous activity throughout the entire life cycle presented in Figure 10.1. Conducting VV&T for the first time in the life cycle when the experimental model is complete is analogous to a teacher who gives only a final examination [22]. No opportunity is provided throughout the semester to notify the student that he or she has serious deficiencies. Severe problems may go undetected until it is too late to do anything but fail the student. Frequent tests and homeworks throughout the semester are intended to inform students about their deficiencies so that they can study more to improve their knowledge as the course progresses.

The situation in VV&T is exactly analogous. The VV&T activities throughout the entire life cycle are intended to reveal any quality deficiencies that might be present as the simulation study progresses from problem definition to the presentation of simulation results. This allows us to identify and rectify quality deficiencies during the life-cycle phase in which they occur.

A simulation model goes through five levels of testing during its life cycle:

- *Level 1: Private Testing.* Performed by the modeler in private with no documentation. Although informal, private testing is strongly encouraged prior to formal submodel/module testing [23].
- *Level 2: Submodel (Module) Testing.* Planned, performed, and documented independently by the SQA group. Submodel testing treats each submodel as a stand-

alone unit, with its own input and output variables, that can be tested without other submodels.

- *Level 3: Integration Testing.* Planned, performed, and documented independently by the SQA group. Its objective is to substantiate that no inconsistencies in interfaces and communications between the submodels exist when the submodels are combined to form the model. It is assumed that each submodel has passed the submodel testing prior to integration testing.
- *Level 4: Model (Product) Testing.* Planned, performed, and documented independently by the SQA group. Its objective is to assess the validity of the overall model behavior.
- *Level 5: Acceptance Testing.* Planned, performed, and documented independently by the sponsor of the simulation study or the independent third party hired by the sponsor. Its objective is to establish the sufficient credibility of the simulation model so that its results can be accpeted and used by the sponsor.

Principle 2 The outcome of simulation model VV&T should not be considered as a binary variable where the model is absolutely correct or absolutely incorrect.

Since a model is an abstraction of a system, perfect representation is never expected. Shannon [10] indicates that "it is not at all certain that it is ever theoretically possible to establish if we have an absolutely valid model; even if we could, few managers would be willing to pay the price." The outcome of model VV&T should be considered as a degree of credibility on a scale from 0 to 100, where 0 represents absolutely incorrect and 100 represents absolutely correct. As depicted in Figure 10.4 [10, 24], as the degree of model credibility increases, so will the model development cost. At the same time, the model utility will also increase, but probably at a decreasing rate. The point of intersection of two curves changes from one model to another.

Principle 3 A simulation model is built with respect to the study objectives and its credibility is judged with respect to those objectives.

The objectives of a simulation study are identified in the formulated problem phase and explicitly and clearly specified in the system and objectives definition phase of the

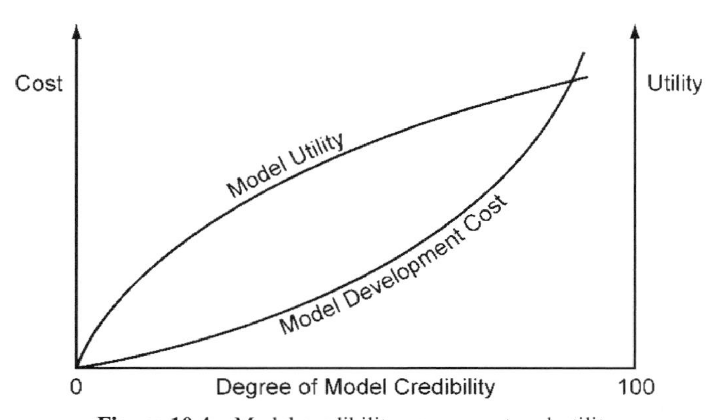

Figure 10.4 Model credibility versus cost and utility.

life cycle shown in Figure 10.1. Accurate specification of the study objectives is crucial for the success of a simulation study. The model is either developed from scratch or an existing model is modified for use or an available one is selected for use as is, *all* with respect to the study objectives.

The study objectives dictate how representative the model should be. Sometimes, 60% representation accuracy may be sufficient; sometimes, 95% accuracy may be required, depending on the importance of the decisions that will be made based on the simulation results. Therefore, model credibility must be judged with respect to the study objectives. The adjective *sufficient* must be used in front of the terms such as *model credibility*, *model validity*, or *model accuracy* to indicate that the judgment is made with respect to the study objectives. It is more appropriate to say that "the model is sufficiently valid" than to say that 'the model is valid." Here "sufficiently valid" implies that the validity is judged with respect to the study objectives and found to be sufficient.

Principle 4 Simulation model VV&T requires independence to prevent developer's bias.

Model testing is meaningful when conducted in an independent manner by an unbiased person. The model developer with the most knowledge of the model may be the least independent when it comes to testing. The developers are often biased because they fear that negative testing results can damage the credibility of the organization and may lead to the loss of future contracts.

The independence in model testing can be achieved in two ways: (1) establishing an SQA group within the organization conducting the simulation study, and (2) using an independent third party hired by the sponsor of the simulation study. The first one is required to achieve independence in level 3 and 4 testing within the organization as described under principle 1. The SQA group must be independent from the department in charge of conducting the simulation study and should report to the top management. The SQA group is responsible for planning, performing, and documenting all level 2, 3, and 4 tests in an unbiased manner. It should be made clear to the simulation project team that the main thrust of testing is to detect and document faults; it is *not* performance appraisal of the project team. This point must be communicated persuasively to everyone involved so that full cooperation is achieved in discovering and documenting errors.

The second one is required to achieve independence in level 5 (acceptance) testing as described under principle 1. The requirements for acceptance testing must be specified in the legal contract by the sponsor. The test cases to be used must be well documented. Although the sponsor can perform the acceptance testing, it is recommended that an independent third party contracted by the sponsor be responsible for the testing. In that way, the organization conducting the simulation study cannot claim that the sponsor is biased.

Principle 5 Simulation model VV&T is difficult and requires creativity and insight.

One must thoroughly understand the entire simulation model so as to design and implement effective tests and identify adequate test cases. Knowledge of the problem domain, expertise in the modeling methodology and prior modeling, and VV&T experience are required for successful testing. However, it is not possible for one person to fully understand all aspects of a large and complex model, especially if the model is a stochastic one containing hundreds of concurrent activities. The fundamental human limitation, called the *Hrair limit*, indicates that a human being cannot process more than 7 ± 2 entities simultaneously. Hence testing a complex simulation model is a very difficult task that requires creativity and insight.

A model's developers are usually the most qualified to show the creativity and insight required for successful testing since they are intimately knowledgeable about the internals of a model. However, they are usually biased when it comes to model testing and they cannot be fully utilized. Therefore, the inability to use model developers effectively for testing increases the difficulty of testing. False beliefs exist about testing, as indicated by Hetzel [22]: "Testing is easy; anyone can do testing; no training or prior experience is required." The difficulty of model VV&T must not be underestimated. The model testing must be well planned and administered by the SQA group.

Principle 6 Simulation model credibility can be claimed only for the conditions for which the model is tested.

The accuracy of the input–output transformation of a simulation model is affected by the characteristics of the input conditions. The transformation that works for one set of input conditions may produce absurd output when conducted under another set of input conditions. In the case study, for example, a stationary simulation model is built assuming constant arrival rate of vehicles during the evening rush hour, and its credibility may be judged sufficient with respect to the evening rush-hour input conditions. However, the simulation model will show invalid behavior when run under the input conditions of the same traffic intersection between 7:00 A.M. and 6:00 P.M. During this time period, the arrival rate of vehicles is not constant and a nonstationary simulation model is required. Hence establishing sufficient model credibility for the evening rush-hour conditions does not imply sufficient model credibility for input conditions during other times. The prescribed conditions for which the model credibility has been established is called the *domain of applicability* of the simulation model [25]. Model credibility can be claimed only for the domain of applicability of the model.

Principle 7 Complete simulation model testing is not possible.

Exhaustive (complete) testing requires that the model is tested under *all* possible input conditions. Combinations of feasible values of model input variables can generate millions of logical paths in the model execution. Due to time and budgetary constraints, it is impossible to test the accuracy of millions of logical paths. Therefore, in model testing, the purpose is to increase our confidence in model credibility as much as dictated by the study objectives rather than trying to test the model completely. How much to test or when to stop testing is dependent on the desired domain of applicability of the simulation model. The larger the domain, the more testing is required. The domain of applicability is determined with respect to the study objectives.

Hundreds of logical paths may need to be tested so as to substantiate model credibility under a set of prescribed conditions. Due to budgetary and time constraints, all logical paths may not be tested. Test data or test cases are prepared to test the logical paths in a random manner. Test data can be generated by using (1) random values, (2) deterministic values, (3) minimum values for all input variables, (4) maximum values for all input variables, (5) a mixture of minimum and maximum values for all input variables, (6) invalid values, and (7) simulated values.

When using test data, it must be noted that the law of large numbers simply does not apply. The question is not how much test data is used, but what percentage of the valid input domain is covered by the test data. The higher the percentage of coverage, the higher the confidence we can gain in model credibility.

Principle 8 Simulation model VV&T must be planned and documented.

Testing is not a phase or step in the model development life cycle; it is a continuous activity *throughout* the entire life cycle. The tests should be identified, test data or cases should be prepared, tests should be scheduled, and the entire testing process should be documented. Ad hoc or haphazard testing does not provide reasonable measurement of model accuracy. Hetzel [22] points out that "such testing may even be harmful in leading us to a false sense of security." Careful planning is required for successful testing.

Planning and documenting model testing involves at least three groups of people: (1) sponsor of the simulation study, (2) SQA group of the organization conducing the simulation study, and (3) simulation project management. The sponsor is responsible for documenting the tests and specifying the test cases or data with which the acceptance testing will be performed. It is recommended that a plan for acceptance testing be made part of the legal contract between sponsor and contractor of the simulation study. If the study is conducted internally within an organization, acceptance testing plan should be part of the requirements specification document. The SQA group is responsible for planning, performing, and documenting level 2 (module), level 3 (integration), and level 4 (model) testing.

A *test plan* is a document describing what is selected for testing, test database and code, test specifications, standards and conventions, test control, test configuration, test tools, and the results expected. An acceptance test plan is presented by Beizer [23].

Principle 9 Type I, II, and III errors must be prevented.

Three types of errors may be committed in conducting a simulation study as depicted in Figure 10.5 [9]. A *type I error* is committed when the simulation results are rejected when in fact they are sufficiently credible. A *type II error* occurs when invalid simulation results are accepted as if they are sufficiently valid. A *type III error* occurs if the wrong problem is solved and committed when the problem formulated does not completely contain the actual problem.

Committing a type I error unnecessarily increases the cost of model development. The consequences of type II and type III errors can be catastrophic, especially when critical decisions are made on the basis of simulation results. A type III error implies that the problem solution and the simulation study results will be irrelevant when it is committed.

The probability of committing a type I error is called *model builder's risk* and the probability of committing a type II error is called *model user's risk* [26]. VV&T activities must focus on minimizing these risks as much as possible. Balci and Sargent [26] show how to quantify these risks when using hypothesis testing for the validation of a simulation model with two or more output variables.

Figure 10.5 illustrates the occurrence of three types of errors, assuming that the simulation study results are certified by an independent organization. Whenever feasible, simulation results should be independently certified so as to remove the developer's bias and promote *independent* VV&T (see Principle 4).

Principle 10 Errors should be detected as early as possible in the life cycle of a simulation study.

A rush to model implementation is a common problem in simulation studies. Sometimes simulation models are built by direct implementation in a simulation system or (simulation) programming language with no or very little formal model specification. As a result of this harmful build-and-fix approach, experimental model VV&T becomes the only main credibility assessment stage.

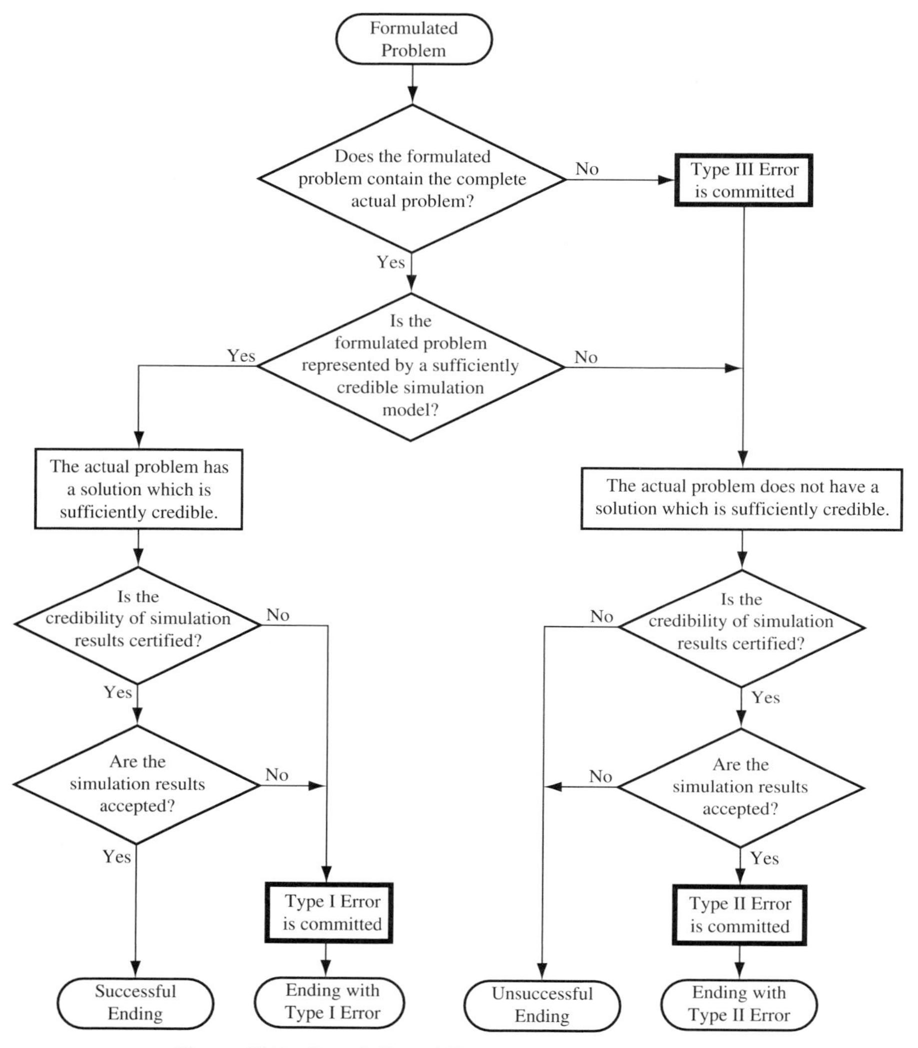

Figure 10.5 Type I, II, and III errors in a simulation study.

Detection and correction of errors as early as possible in the life cycle of a simulation study must be the primary objective. Sufficient time and energy must be expended for each VV&T activity shown in Figure 10.1. Nance [5] points out that detecting and correcting major modeling errors during the process of model implementation and in later phases is very time consuming, complex, and expensive. Some vital errors may not be detectable in later phases, resulting in the occurrence of a type II or III error.

Nance and Overstreet [27] advocate this principle and provide diagnostic testing techniques for models represented in the form of condition specification. A model analyzer software tool is included in the definition of a simulation model development environ-

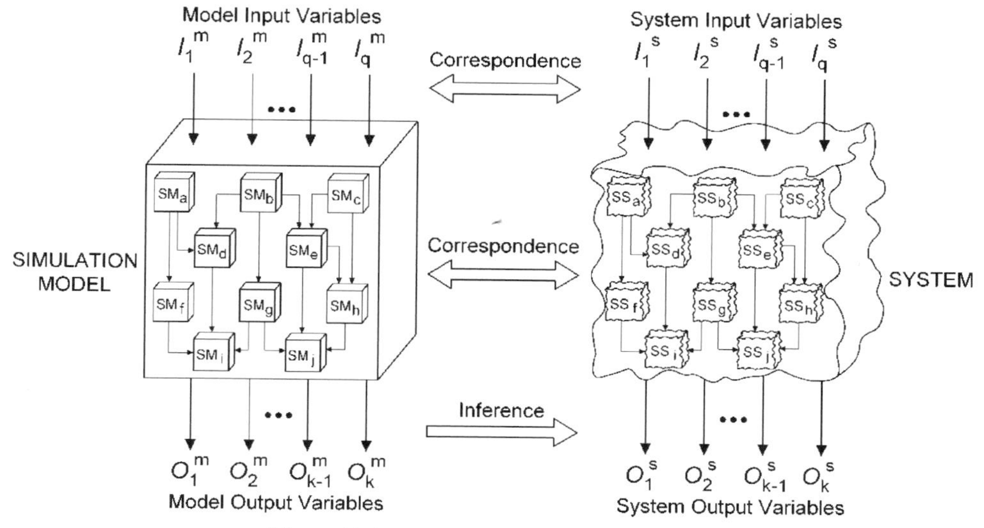

Figure 10.6 Model and system characteristics.

ment so as to provide effective early detection of model specification errors [16, 28, 29].

Principle 11 Multiple response problem must be recognized and resolved properly.

Figure 10.6 shows a simulation model with k output variables (responses or performance measures) and q input variables representing a system with corresponding k output variables and q input variables. Superscript m indicates model and s indicates system. SM stands for submodel and SS stands for subsystem.

Due to the multiple response problem described by Shannon [10], the validity of a simulation model with two or more output variables (responses) cannot be tested by comparing the corresponding model and system output variables one at a time (i.e., O_1^m versus O_1^s, O_2^m versus O_2^s, ..., O_k^m versus O_k^s, as shown in Figure 10.6) using a univariate statistical procedure. A multivariate statistical procedure must be used to incorporate the correlations among the output variables in the comparison. Two such multivariate statistical techniques are presented by Balci and Sargent [30] using Hotelling's T^2-statistic for constructing ellipsoidal joint confidence regions to assess model validity. The techniques are described below.

The first technique requires independence between the model and system output data and is intended for self-driven simulation models of observable systems. Assume that the model and system each has k output variables, as depicted in Figure 10.6, with n observations on each model output variable and N observations on each system output variable. Let $(\boldsymbol{\mu}^m)' = [\mu_1^m, \mu_2^m, \ldots, \mu_k^m]$ and $(\boldsymbol{\mu}^s)' = [\mu_1^s, \mu_2^s, \ldots, \mu_k^s]$ be the k-dimensional vectors of population means of model and system output variables, respectively. Let $(\overline{\mathbf{o}}^m)' = [\overline{\mathbf{o}}_1^m, \overline{\mathbf{o}}_2^m, \ldots, \overline{\mathbf{o}}_k^m]$ and $(\overline{\mathbf{o}}^s)' = [\overline{\mathbf{o}}_1^s, \overline{\mathbf{o}}_2^s, \ldots, \overline{\mathbf{o}}_k^s]$ be the k-dimensional vectors of sample means of observations on model and system output variables, respectively. Then, the $100(1-\gamma)\%$ joint confidence region is specified by those vectors $\boldsymbol{\delta} = \boldsymbol{\mu}^m - \boldsymbol{\mu}^s$ satisfying the inequality

$$(\overline{\mathbf{o}}^m - \overline{\mathbf{o}}^s - \boldsymbol{\delta})' S^{-1} (\overline{\mathbf{o}}^m - \overline{\mathbf{s}}^s - \boldsymbol{\delta}) \leq \frac{n+N}{nN} T^2_{\gamma; k, n+N-k-1} \qquad (1)$$

where S is the pooled variance–covariance matrix and $T^2_{\gamma; k, n+N-k-1}$ is the upper γ percentage point of Hotelling's T^2-distribution with degrees of freedom k and $n + N - k - 1$.

The second technique requires paired observations between the model and system output variables and is intended for trace-driven simulation models. Let $\overline{\mathbf{d}}' = [\overline{d}_1, \overline{d}_2, \ldots, \overline{d}_k]$ be the k-dimensional vector of sample means of differences between the paired observations on the model and system output variables with a sample size of N. The $100(1 - \gamma)\%$ joint confidence region consists of the vectors $\boldsymbol{\mu}^d$ satisfying the inequality

$$N(\overline{\mathbf{d}} - \boldsymbol{\mu}^d)' S_d^{-1} (\overline{\mathbf{d}} - \boldsymbol{\mu}^d) \leq T^2_{\gamma; k, N-k} \qquad (2)$$

where S_d is the variance–covariance matrix of the differences.

When $k = 3$, the joint confidence region can be presented visually as illustrated in Figure 10.7 and can be used to assess the model accuracy with an exact level of $100(1 - \gamma)\%$ confidence. We can conclude that we are $100(1 - \gamma)\%$ confident that the differences between the population means of corresponding model and system output variables are contained within the joint confidence region shown in Figure 10.7. Ideally, the joint confidence region contains zero at its center, and the smaller its size, the better it is. Any deviation from the idealistic case is an indication of the degree of invalidity. As k increases, interpretation of the joint confidence region becomes difficult but

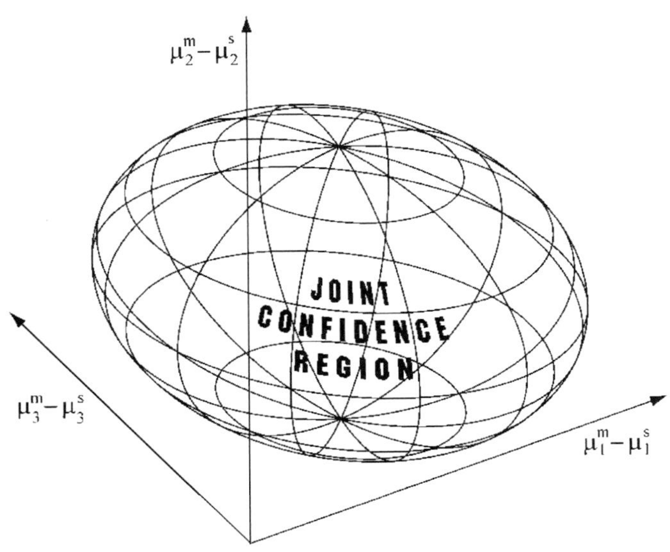

Figure 10.7 Joint confidence region representing the validity of a simulation model with three output variables.

not impossible. With the use of computer-aided assistance, the model validity can be examined.

Principle 12 Successfully testing each submodel (module) does not imply overall model credibility.

Suppose that a simulation model is composed of submodels (SM_x) representing subsystems (SS_x) respectively, as depicted in Figure 10.6. Submodel x can be tested individually by comparison to subsystem x, where $x = a, b, \ldots, j$, using many of the VV&T techniques described in Section 10.4. The credibility of each submodel is judged to be sufficient with some error that is acceptable with respect to the study objectives. We may find each submodel to be sufficiently credible, but this does not imply that the whole model is sufficiently credible. The allowable errors for the submodels may accumulate to be unacceptable for the entire model. Therefore, the entire model must be tested even if each submodel is found to be sufficiently credible.

Principle 13 Double validation problem must be recognized and resolved properly.

If data can be collected on both system input and output, model validation can be conducted by comparing model and system outputs obtained by running the model with the "same" input data that drives the system. Determination of the "same" is yet another validation problem within model validation. Therefore, this is called the *double validation problem*. This is an important problem that is often overlooked. It greatly affects the accuracy of model validation. If invalid input data models are used, we may still find the model and system outputs sufficiently matching each other and conclude incorrectly on the sufficient validity of the model.

The "same" is determined by validating the input data models. In the case study, the input data models are the probability distributions given in Table 10.1. We must substantiate that the input data models have sufficient accuracy in representing the system input process. Input data modeling deals with characterization of the system input data [13, 31]. Simulation models are categorized into two classes with respect to the way they are driven: trace-driven and self-driven. In *trace-driven simulation*, the system input is characterized by the trace data collected from the instrumented system. The trace data become the input data model which should be validated against the actual system input process.

In *self-driven simulations*, the simulation model is driven by randomly sampling from the probabilistic models developed to represent the data collected on the system input process. Usually, input data modeling is achieved by fitting standard probability distributions to observed data. The input data models should be constructed using a multivariate statistical approach if the input variables are correlated. Individually building a probabilistic model for each input variable does not incorporate the correlations among the input variables; therefore, a multivariate probabilistic approach should be used.

Principle 14 Simulation model validity does not guarantee the credibility and acceptability of simulation results.

Model validity is a necessary but not a sufficient condition for the credibility and acceptability of simulation results. We assess model validity with respect to the study objectives by comparing the model with the system as it is defined. If the study objectives are identified incorrectly and/or the system is defined improperly, the simulation results will be invalid; however, we may still find the model to be sufficiently valid by comparing it with the improperly defined system and with respect to the incorrectly identified objectives.

A distinct difference exists between the model credibility and the credibility of simulation results. Model credibility is judged with respect to the system (requirements) definition and the study objectives, whereas the credibility of simulation results is judged with respect to the actual problem definition and involves the assessment of system definition and identification of study objectives. Therefore, model credibility assessment is a subset of credibility assessment of simulation results.

Principle 15 Formulated problem accuracy greatly affects the acceptability and credibility of simulation results.

It has been said that a problem correctly formulated is half solved [32]. Albert Einstein once indicated that the correct formulation of a problem was even more crucial than its solution. The ultimate goal of a simulation study should not be just to produce a solution to a problem but to provide one that is sufficiently credible and accepted and implemented by the decision makers. We cannot claim that we conducted an excellent simulation study but the decision makers did not accept our results and we cannot do anything about it. Ultimately we are responsible for the acceptability and usability of our simulation solutions, although in some cases we cannot affect or control the acceptability.

Formulated problem accuracy assessed by conducting formulated problem VV&T greatly affects the credibility and acceptability of simulation results. Insufficient problem definition and inadequate sponsor involvement in defining the problem are identified as two important problems in the management of computer-based models. It must be recognized that if problem formulation is poorly conducted, resulting in an incorrect problem definition, no matter how fantastically we solve the problem, the simulation study results will be irrelevant. Balci and Nance [9] present an approach to problem formulation and 38 indicators for assessing the formulated problem accuracy.

10.4 VERIFICATION, VALIDATION, AND TESTING TECHNIQUES

More than 75 VV&T techniques are presented in this section. Most of these techniques come from the software engineering discipline and the remaining are specific to the modeling and simulation field. The software VV&T techniques selected which are applicable for simulation model VV&T are presented in a terminology understandable by a simulationist. Descriptions of some software VV&T techniques are changed so as to make them directly applicable and understandable for simulation model VV&T.

Figure 10.8 shows a taxonomy that classifies the VV&T techniques into four primary categories: informal, static, dynamic, and formal. A primary category is further divided into secondary categories, shown in italics. The use of mathematical and logic formalism by the techniques in each primary category increases from informal to formal from left to right. Similarly, the complexity also increases as the primary category becomes more formal.

It should be noted that some of the categories presented in Figure 10.8 possess similar characteristics and in fact have techniques that overlap from one category to another. However, a distinct difference between each classification exists, as it is evident in the discussion of each in this section. The categories and techniques in each category are described on page 355.

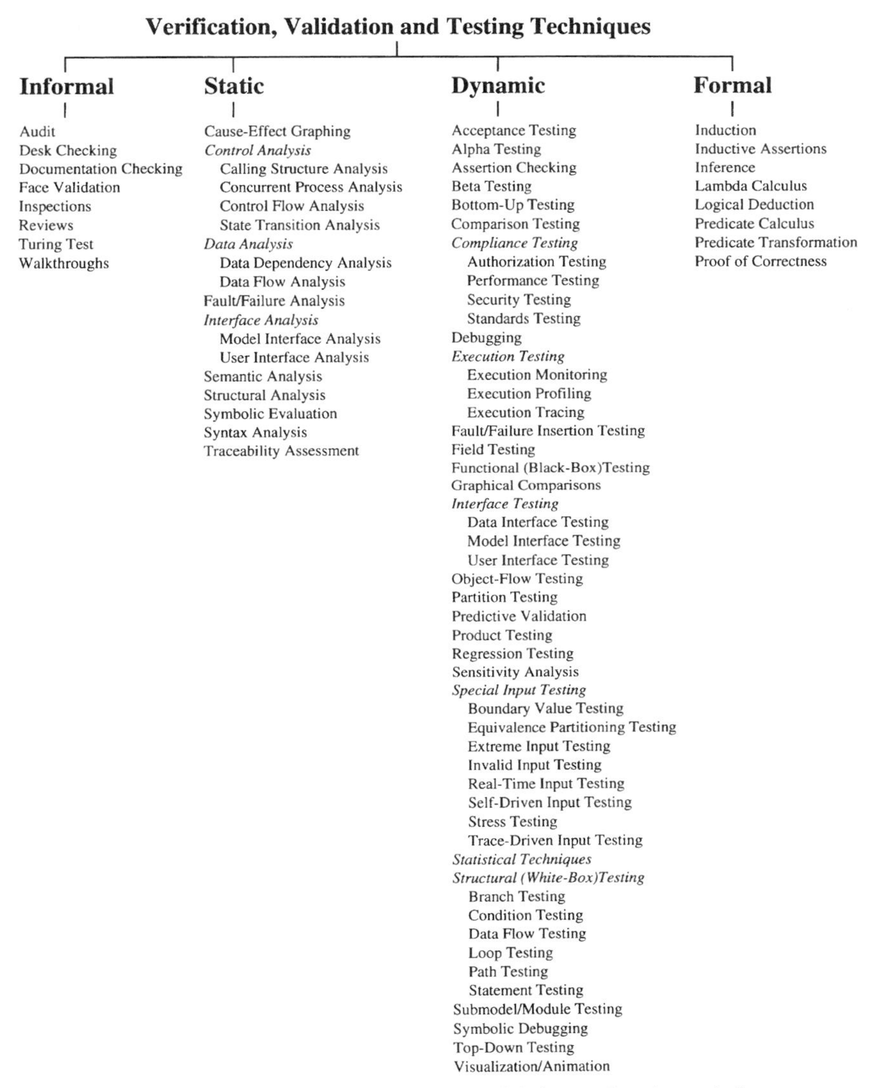

Figure 10.8 Taxonomy of verification, validation, and testing techniques.

10.4.1 Informal VV&T Techniques

Informal techniques are among the most commonly used. They are called informal because the tools and approaches used rely heavily on human reasoning and subjectivity without stringent mathematical formalism. The "informal" label does not imply a lack of structure or formal guidelines for the use of the techniques. In fact, these techniques are applied using well-structured approaches under formal guidelines and they can be very effective if employed properly.

Audit. An audit is undertaken to assess how adequately the simulation study is conducted with respect to established plans, policies, procedures, standards, and guidelines. The audit also seeks to establish traceability within the simulation study. When an error is identified, it should be traceable to its source via its audit trail. The process of documenting and retaining sufficient evidence about how the accuracy is substantiated is called an *audit trail* [33]. Auditing is carried out on a periodic basis through a mixture of meetings, observations, and examinations [34]. Audit is a staff function and serves as the "eyes and ears of management" [33].

Desk Checking. *Desk checking* (also known as *self-inspection*) is the process of thoroughly examining one's work to ensure correctness, completeness, consistency, and unambiguity. It is considered to be the very first step in VV&T and is particularly useful for the early stages of development. To be effective, desk checking should be conducted carefully and thoroughly, preferably by another person, since it is usually difficult to see one's own errors [35]. Syntax checking, cross-reference checking, convention violation checking, detailed comparison to specification, reading the code, the control flow graph analysis, and path sensitizing should all be conducted as part of desk checking [23].

Documentation Checking. Documentation checking is conducted to ensure correctness, completeness, consistency, and unambiguity of all model documentation and to justify that all documentation is up to date with respect to model logic specification. Often, a model component logic is modified but the component's documentation is not updated. Sometimes model logic is documented erroneously. In the case study, the documentation delivered to the decision makers must be an accurate and up-to-date description of model logic and its results.

Face Validation. The project team members, potential users of the model, people knowledgeable about the system under study, based on their estimates and intuition, subjectively compare model and system behaviors under identical input conditions and judge whether the model and its results are reasonable. Face validation is useful as a preliminary approach to validation [36]. In the case study, the confidence intervals for the 14 performance measures obtained under the currently used traffic light timing can be presented to experts. The experts can judge if average waiting times of vehicles are reasonable under the rush-hour traffic conditions observed.

Inspections. Inspections are conducted by a team of four to six members for any model development phase such as system and objectives definition, conceptual model design, or communicative model design. For example, in the case of communicative model design inspection, the team consists of:

1. *Moderator:* manages the inspection team and provides leadership.
2. *Reader:* narrates the model design (communicative model) and leads the team through it.
3. *Recorder:* produces a written report of detected faults.
4. *Designer:* representative of the team that created the model design
5. *Implementer:* translates the model design into code (programmed model).
6. *Tester:* SQA group representative.

An inspection goes through five distinct phases: overview, preparation, inspection, rework, and follow-up [37]. In phase I the designer gives an overview of the (sub)model design to be inspected. The (sub)model characteristics such as purpose, logic, and interfaces are introduced and related documentation is distributed to all participants to study. In phase II the team members prepare individually for the inspection by examining the documents in detail. The moderator arranges the inspection meeting with an established agenda and chairs it in phase III. The reader narrates the (sub)model design documentation and leads the team through it. The inspection team is aided by a checklist of queries during the fault-finding process. The objective is to find and document the faults, not to correct them. The recorder prepares a report of detected faults immediately after the meeting. Phase IV is for rework in which the designer resolves all faults and problems specified in the written report. In the final phase, the moderator ensures that all faults and problems have been resolved satisfactorily. All changes must be examined carefully to ensure that no new errors have been introduced as a result of a fix.

Major differences exist between inspections and walkthroughs. An inspection is a five-step process, but walkthroughs consist of only two steps. The inspection team uses the checklist approach for uncovering errors. The procedure used in each phase of the inspection technique is formalized. The inspection process takes much longer than a walkthrough; however, the extra time is justified because an inspection is a powerful and cost-effective way of detecting faults early in the model development life cycle [23, 33, 37–40].

Reviews. The review is conducted in a manner similar to that of the inspection and walkthrough except in the way the team members are selected. The review team also involves managers. The review is intended to give management and study sponsors evidence that the model development process is being conducted according to stated study objectives and to evaluate the model in light of development standards, guidelines, and specifications. As such, the review is a higher-level technique than the inspection and walkthrough.

Each review team member examines the model documentation prior to the review. The team then meets to evaluate the model relative to specifications and standards, recording defects and deficiencies. The review team may be given a set of indicators to measure, such as (1) appropriateness of the definition of system and study objectives, (2) adequacy of all underlying assumptions, (3) adherence to standards, (4) modeling methodology used, (5) model representation quality, (6) model structuredness, (7) model consistency, (8) model completeness, and (9) documentation. The result of the review is a document portraying the events of the meeting, deficiencies identified, and review team recommendations. Appropriate action may then be taken to correct any deficiencies.

In contrast with inspections and walkthroughs, which concentrate on correctness assessment, reviews seek to ascertain that tolerable levels of quality are being attained. The review team is more concerned with model design deficiencies and deviations from stated model development policy than it is with the intricate line-by-line details of the implementation. This does not imply that the review team is not concerned with discovering technical flaws in the model, only that the review process is oriented toward the early stages of the model development life cycle [33, 34, 41, 42].

Turing Test. The Turing test is based on the expert knowledge of people about the system under study. The experts are presented with two sets of output data obtained, one from the model and one from the system, under the same input conditions. Without identifying which one is which, the experts are asked to differentiate between the two. If they succeed, they are asked how they were able to do it. Their response provides valuable feedback for correcting model representation. If they cannot differentiate, our confidence in model validity is increased [44–45].

In the case study, two confidence intervals are first constructed for the average waiting time of vehicles arriving and turning left in lane 1: (1) confidence interval estimated via simulation, and (2) confidence interval constructed based on data collected at the traffic intersection. The two sets of confidence intervals are presented to an expert who has intimate knowledge of the traffic intersection operation. Without identifying which one is which, the expert is asked to differentiate between the two. If the expert cannot identify which one belongs to the real traffic intersection, the model is considered sufficiently valid with respect to that performance measure. This process is repeated for each of the 14 performance measures.

Walkthroughs. Walkthroughs are conducted by a team composed of a coordinator, model developer, and three to six other members. All members other than the model developer should not be directly involved in the development effort. A typical structured walkthrough team consists of:

1. *Coordinator:* most often the SQA group representative who organizes, moderates, and follows up the walkthrough activities
2. *Presenter:* most often the model developer
3. *Scribe:* documents the events of the walkthrough meetings
4. *Maintenance Oracle:* considers long-term implications
5. *Standards Bearer:* concerned with adherence to standards
6. *Client Representative:* reflects the needs and concerns of the client
7. *Other Reviewers:* such as simulation project manager and auditors

The main thrust of the walkthrough technique is to detect and document faults; it is *not* performance appraisal of the development team. This point must be made clear to everyone involved so that full cooperation is achieved in discovering errors. The coordinator schedules the walkthrough meeting, distributes the walkthrough material to all participants well in advance of the meeting to allow for careful preparation, and chairs the meeting. During the meeting, the presenter walks the other members through the walkthrough documents. The coordinator encourages questions and discussion so as to uncover any faults [35, 46–49].

10.4.2 Static VV&T Techniques

Static VV&T techniques are concerned with accuracy assessment on the basis of characteristics of the static model design and source code. Static techniques do not require machine execution of the model, but mental execution can be used. The techniques are very popular and widely used, with many automated tools available to assist in the VV&T process. The simulation language compiler is itself a static VV&T tool. Static VV&T techniques can obtain a variety of information about the structure of the model, modeling techniques and practices employed, data and control flow within the model, and syntactical accuracy [42].

Cause–Effect Graphing. Cause–effect graphing assists model correctness assessment by addressing the question of what causes what in the model representation. It is performed by first identifying causes and effects in the system being modeled and by examining if they are reflected accurately in the model specification. In the case study the following causes and effects may be identified: (1) the change of lane 1 light to red immediately causes the vehicles in lane 1 to stop; (2) an increase in the duration of lane 1 green light causes a decrease in the average waiting time of vehicles in lane 1; and (3) an increase in the arrival rate of lane 1 vehicles causes an increase in the average number of vehicles at the intersection.

As many causes and effects as possible are listed and the semantics are expressed in a cause–effect graph. The graph is annotated to describe special conditions or impossible situations. Once the cause–effect graph has been constructed, a decision table is created by tracing back through the graph to determine combinations of causes which result in each effect. The decision table is then converted into test cases with which the model is tested [42, 48, 50].

Control Analysis. The control analysis category consists of the following techniques that are used for the analysis of the control characteristics of the model:

1. *Calling Structure Analysis:* used to assess model accuracy by identifying who *calls* whom and who is *called* by whom. The "who" could be a module, procedure, subroutine, function, or a method in an object-oriented model [51]. In the case study, inaccuracies caused by message passing (e.g., sending a message to a nonexistent object) in the object-oriented traffic intersection VSE model can be revealed by analyzing which methods invoke a method and by which methods a method is invoked.

2. *Concurrent Process Analysis:* especially useful for parallel and distributed simulations presented in Chapter 13. Model accuracy is assessed by analyzing the overlap or concurrency of model components executed in parallel or as distributed. Such analysis can reveal synchronization problems such as deadlocks [52].

3. *Control Flow Analysis:* requires the development of a graph of the model where conditional branches and model junctions are represented by nodes and the model segments between such nodes are represented by links [23]. A node of the model graph usually represents a logical junction where the flow of control changes, while an edge represents toward which junction it changes. This technique examines sequences of control transfers and is useful for identifying incorrect or inefficient constructs within model representation.

Nance and Overstreet [27] propose several diagnostics based on analysis of graphs constructed from a particular form of model specification called *condition specification* [14, 53]. The diagnostic assistance is categorized into three parts:

1. *Analytical:* determination of the existence of a property
2. *Comparative:* measures of differences among multiple model representations
3. *Informative:* characteristics extracted or derived from model representations

Action cluster attribute graph, action cluster incidence graph, and run-time graph constitute the basis for the diagnosis.

The analytical diagnosis is conducted by measuring the following indicators: attribute utilization, attribute initialization, action cluster completeness, attribute consistency, connectedness, accessibility, out-complete, and revision consistency. The comparative diagnosis is done by measuring attribute cohesion, action cluster cohesion, and complexity. The following indicators are measured for the informative diagnosis: attribute classiciation, precedence structure, decomposition, and run-time graph [27].

4. *State Transition Analysis:* requires the identification of a finite number of states the model execution goes through. A state transition diagram is created showing how the model transitions from one state to another. Model accuracy is assessed by analyzing the conditions under which a state change occurs. This technique is especially effective for those simulation models created under the activity scanning, three-phase, and process interaction conceptual frameworks [20].

Data Analysis. The data analysis category consists of several techniques that are used to ensure that (1) proper operations are applied to data objects (e.g., data structures, event lists, linked lists), (2) the data used by the model are properly defined, and (3) the defined data are properly used [33]:

1. *Data Dependency Analysis:* involves the determination of what variables depend on what other variables [54]. For parallel and distributed simulations, the data dependency knowledge is critical for assessing the accuracy of process synchronization.

2. *Data Flow Analysis:* used to assess model accuracy with respect to the use of model variables. This assessment is classified according to the definition, referencing and unreferencing of variables [35] (i.e., when variable space is allocated, accessed, and deallocated). A data flow graph is constructed to aid in the data flow analysis. The nodes of the graph represent statements and corresponding variables. The edges represent control flow. Data flow analysis can be used to detect undefined or unreferenced variables (much as in static analysis) and, when aided by model instrumentation, can track minimum and maximum variable values, data dependencies, and data transformations during model execution. It is also useful in detecting inconsistencies in data structure declaration and improper linkages among submodels [42, 55].

Fault/Failure Analysis. Fault (incorrect model component)/failure (incorrect behavior of a model component) analysis uses model input–output transformation descriptions to identify how the model *might* logically fail. The model design specification is examined to determine if any failure-mode possibilities could logically occur and in what context and under what conditions. Such model examinations often lead to identification of model defects [51].

Interface Analysis. The interface analysis category consists of several techniques that are especially useful for verification and validation of interactive and distributed simulations:

1. *Model Interface Analysis:* conducted to examine the (sub)model-to-(sub)model interface and determine if the interface structure and behavior are sufficiently accurate.

2. *User Interface Analysis:* conducted to examine the user–model interface and determine if it is human engineered so as to prevent occurrences of errors during the user's interactions with the model. It is also used to assess how accurately the interface is integrated with the simulation model. This technique is particularly useful for accuracy assessment of interactive simulation models used for training purposes.

Semantic Analysis. Semantic analysis is conducted by the simulation system translator or a simulation programming language compiler and attempts to determine the modeler's intent in writing the code. The compiler informs the modeler about what is specified in the source code so that the modeler can verify that the true intent is accurately reflected.

The compiler generates a wealth of information to help the modeler determine if the true intent is accurately translated into the executable code [42]:

1. *Symbol Tables:* the elements or symbols that are manipulated in the model, function declarations, type and variable declarations, scoping relationships, interfaces, dependencies, and so on.
2. *Cross-Reference Tables:* describe called versus calling submodels (where each data element is declared, referenced, and altered), duplicate data declarations (how often and where occurring), and unreferenced source code.
3. *Subroutine Interface Tables:* describe the actual interfaces of the caller and the called.
4. *Maps:* relate the generated run-time code to the original source code.
5. *"Pretty Printers" or Source Code Formatters:* provide reformatted source listing on the basis of its syntax and semantics, clean pagination, highlighting of data elements, and marking of nested control structures.

Structural Analysis. Structural analysis is used to examine the model structure and to determine if it adheres to structured principles. It is conducted by constructing a control flow graph of the model structure and examining the graph for anomalies, such as multiple entry and exit points, excessive levels of nesting within a structure and questionable practices such as the use of unconditional branches (i.e., GOTOs). Yücesan and Jacobson [56, 57] apply the theory of computational complexity and show that the problem of verifying structural properties of simulation models is intractable. They illustrate that modeling issues such as accessibility of states, ordering of events, ambiguity of model specifications, and execution stalling are NP-complete decision problems.

Symbolic Evaluation. Symbolic evaluation is used to assess model accuracy by exercising the model using symbolic values rather than actual data values for input. It is performed by feeding symbolic inputs into the (sub)model and producing expressions for the output which are derived from the transformation of the symbolic data along model execution paths. Consider, for example, the following function:

```
function jobArrivalTime(arrivalRate,currentClock,randomNumber)
  lag = -10
  Y = lag * currentClock
  Z = 3 * Y
  if Z < 0 then
    arrivalTime = currentClock - log(randomNumber)/arrivalRate
  else
    arrivalTime = Z - log(randomNumber)/arrivalRate
  end if
  return arrivalTime
end jobArrivalTime
```

In symbolic execution, `lag` is substituted in `Y = -10*currentClock`. Substituting again, we find `Z = -30*currentClock`. Since `currentClock` is always zero or positive, an error is detected that `Z` will never be greater than zero.

When unresolved conditional branches are encountered, a decision must be made as to which path to traverse. Once a path is selected, execution continues down the new path. At some point in time, the execution evaluation will return to the branch point and the previously unselected branch will be traversed. Eventually, all paths are taken.

The result of the execution can be represented graphically as a symbolic execution tree [35, 58]. The branches of the tree correspond to the paths of the model. Each node of the tree represents a decision point in the model and is labeled with the symbolic values of data at that juncture. The leaves of the tree are complete paths through the model and depict the symbolic output produced.

Symbolic evaluation assists in showing path correctness for all computations regardless of test data and is also a great source of documentation. However, it has the following disadvantages: (1) the execution tree can explode in size and become too complex as the model grows; (2) loops cause difficulties, although inductive reasoning and constraint analysis may help; (3) loops make thorough execution impossible since all paths must be traversed; and (4) complex data structures may have to be excluded because of difficulties in symbolically representing particular data elements within the structure [58–60].

Syntax Analysis. Syntax analysis is carried on by the simulation software compiler or simulation programming language compiler to assure that the mechanics of the language are applied correctly [23]. In the case study, during model preparation, the VSE Editor lists all syntax errors in the preparation window as shown in Figure 10.9. Double-clicking a syntax error name displays the method where the error occurs, draws a rectangle around the statement, and highlights the token as a potential source of error.

Traceability Assessment. The traceability assessment is used to match, one to one, the elements of one form of the model to another. For example, the elements of the system and objectives definition (requirements specification) are matched one to one to the elements of the communicative model (design specification). Unmatched elements *may* reveal either unfulfilled requirements or unintended design functions [51].

10.4.3 Dynamic VV&T Techniques

Dynamic VV&T techniques require model execution and are intended for evaluating the model based on its execution behavior. Most dynamic VV&T techniques require model instrumentation. The insertion of additional code (probes or stubs) into the executable model for the purpose of collecting information about model behavior during execution is called *model instrumentation*. Probe locations are determined manually or automatically based on static analysis of model structure. Automated instrumentation is accomplished by a preprocessor which analyzes the model static structure (usually via graph-based analysis) and inserts probes at appropriate places.

Dynamic VV&T techniques are usually applied using the following three steps. In step 1 the executable model is instrumented, in step 2 the instrumented model is executed, and in step 3 the model output is analyzed and dynamic model behavior is evaluated. For example, consider the worldwide air traffic control and satellite communication object-oriented visual simulation model created by using the VSE [16–18] in

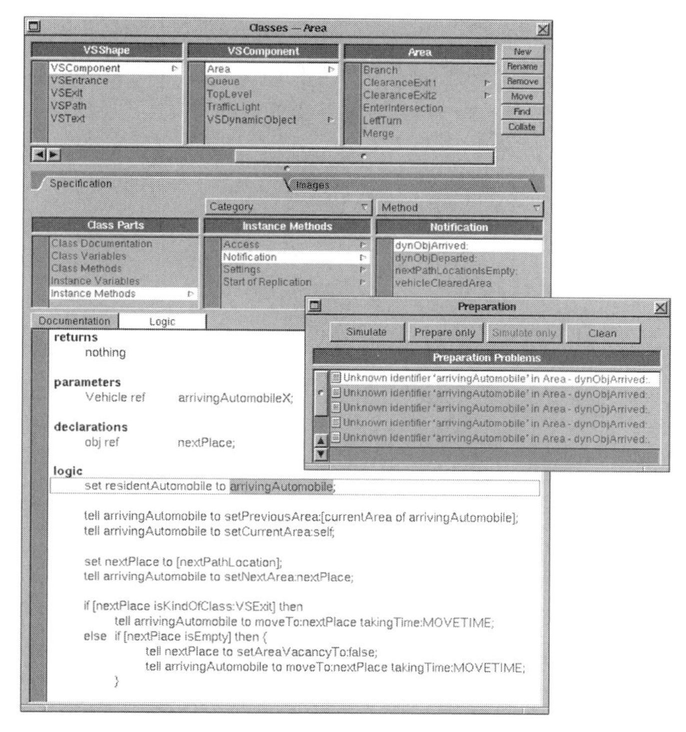

Figure 10.9 Identification of syntax errors during model preparation.

Figure 10.10. The model can be instrumented in step 1 to record the following information every time an aircraft enters the coverage area of a satellite: (1) aircraft tail number; (2) current simulation time; (3) aircraft's longitude, latitude, and altitude; and (4) satellite's position and identification number. In step 2 the model is executed and the information collected is written to an output file. In step the output file is examined to reveal discrepancies and inaccuracies in model representation.

Acceptance Testing. Acceptance testing is conducted either by the client organization, by the developer's SQA group in the presence of client representatives, or by an independent contractor hired by the client after the model is officially delivered and before the client officially accepts the delivery. The model is operationally tested by using the actual hardware and actual data to determine whether all requirements specified in the legal contract are satisfied [33, 37].

Alpha Testing. Alpha testing refers to the operational testing of the alpha version of the complete model at an inhouse site that is not involved with the model development [23].

Assertion Checking. An *assertion* is a statement that should hold true as the simulation model executes. Assertion checking is a verification technique used to check what

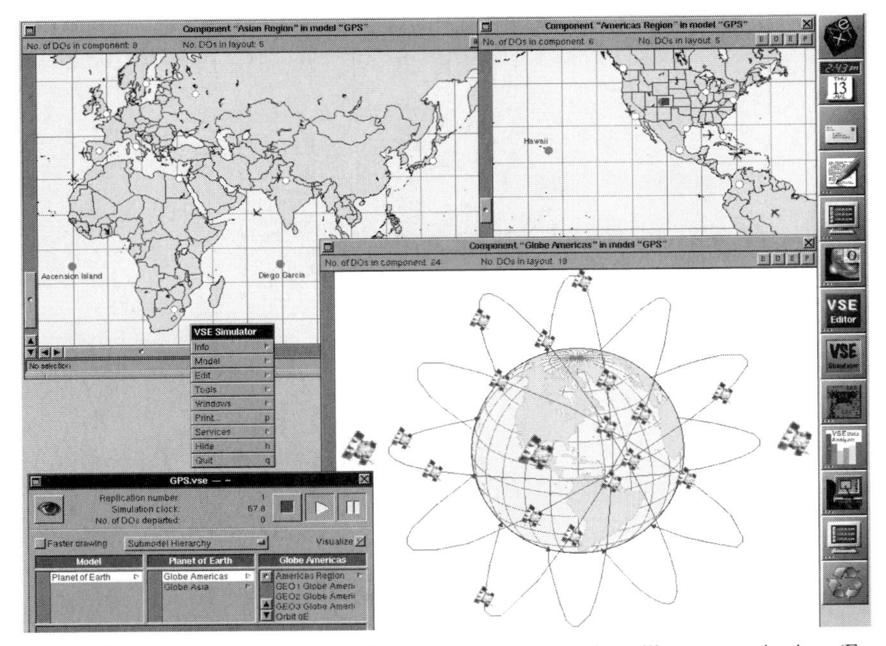

Figure 10.10 Visual simulation of global air traffic control and satellite communication. (From ref. 16.)

is happening against what the modeler *assumes* is happening so as to guard model execution against potential errors. The assertions are placed in various parts of the model to monitor model execution. They can be inserted to hold true *globally*—for the entire model; *regionally*—for some submodels; *locally*—within a submodel; or at *entry* and *exit* of a submodel. The assertions are similar in structure and the general format for a local assertion is [61]:

```
ASSERT LOCAL (extended-logical-expression) [optional-qualifiers]
[control-options]
```

The "optional-qualifiers" may be chosen such as `all, some, after jth aircraft, before time t`. The "control options" may have the following example syntax [61]:

$$\ldots \text{ LIMIT n [VIOLATIONS]} \left[\left\{ \begin{matrix} \underline{\text{HALT}} \\ \text{EXIT [VIA] procedure} - \text{name} \end{matrix} \right\} \right] \ldots$$

Consider, for example, the following pseudocode [42]:

```
Base := Hours * PayRate;
Gross := Base * (1 + BonusRate);
```

In just these two simple statements, several assumptions are being made. It is assumed that `Hours`, `PayRate`, `Base`, `BonusRate`, and `Gross` are all nonnegative. The following asserted code can be used to prevent execution errors due to incorrect values inputted by the user:

```
Assert Local (Hours ≥ 0 and PayRate ≥ 0 and BonusRate ≥ 0);
Base := Hours * PayRate;
Gross := Base * (1 + BonusRate);
```

Assertion checking is also used to prevent structural model inaccuracies. For example, the model in Figure 10.10 can contain assertions such as (1) a satellite communicates with the correct ground station, (2) an aircraft's tail number matches its type, and (3) an aircraft's flight path is consistent with the official airline guide.

Clearly, the assertion checking serves two important needs: (1) it verifies that the model is functioning within its acceptable domain, and (2) the assertion statement documents the intentions of the modeler. However, the assertion checking degrades the model execution performance forcing the modeler to make a trade-off between execution efficiency and accuracy. If the execution performance is critical, the assertions should be turned off but kept permanently to provide both documentation and means for maintenance testing [35].

Beta Testing. Beta testing refers to the operational testing of the beta version of the complete model as a "beta" user site under realistic field conditions [51].

Bottom-Up Testing. Bottom-up testing is used in conjunction with bottom-up model development strategy. In bottom-up development, model construction starts with the submodels at the base level (i.e., those that are not decomposed further) and culminates with the submodels at the highest level. As each submodel is completed, it is thoroughly tested. When submodels belonging to the same parent have been developed and tested, the submodels are integrated and integration testing is performed. This process is repeated in a bottom-up manner until the whole model has been integrated and tested. The integration of completed submodels need not wait for all "same level" submodels to be completed. Submodel integration and testing can be, and often is, performed incrementally [41].

Some of the advantages of bottom-up testing are (1) it encourages extensive testing at the submodel level; (2) since most well-structured models consist of a hierarchy of submodels, there is much to be gained by bottom-up testing; (3) the smaller the submodel and the more cohesion it has, the easier and more complete its testing will be; and (4) it is particularly attractive for testing distributed simulation models.

Major disadvantages of bottom-up testing include (1) individual submodel testing requires drivers, more commonly called *test harnesses*, which simulate the calling of the submodel and passing test data necessary to execute the submodel; (2) developing harnesses for every submodel can be quite complex and difficult; (3) the harnesses may themselves contain errors; and (4) faces the same cost and complexity problems as does top-down testing.

Comparison Testing. Comparison testing (also known as *back-to-back testing*) may be used when more than one version of a simulation model representing the same system is available for testing [41, 50]. For example, different simulation models may have been developed to simulate the same military combat aircraft by different organizations or different simulation models may have been developed to represent the U.S. economy by different economists. All versions of the simulation model built to represent exactly the same system are run with the same input data and the model outputs are compared. Differences in the outputs reveal problems with model accuracy.

Compliance Testing. The compliance type of testing is intended to test how accurately different levels of access authorization are provided, how closely and accurately dictated performance requirements are satisfied, how well the security requirements are met, and how properly the standards are followed. These techniques are particularly useful for testing the federation of distributed and interactive simulation models under the Defense Department's high-level architecture (HLA) and distributed interactive simulation (DIS) architecture [62].

1. *Authorization Testing:* used to test how accurately and properly different levels of access authorization are implemented in the simulation model and how properly they comply with the established rules and regulations. The test can be conducted by attempting to execute a classified model within a federation of distributed models or try to use classified input data for running a simulation model without proper authorization [33].

2. *Performance Testing:* used to test whether (1) all performance characteristics are measured and evaluated with sufficient accuracy, and (2) all established performance requirements are satisfied [33].

3. *Security Testing:* used to test whether all security procedures are implemented correctly and properly in conducting a simulation exercise. For example, the test can be conducted by attempting to penetrate the simulation exercise while it is ongoing and break into classified components such as secured databases. Security testing is applied to substantiate the accuracy and evaluate the adequacy of the protective procedures and countermeasures [33].

4. *Standards Testing:* used to substantiate that the simulation model is developed with respect to the required standards, procedures, and guidelines.

Debugging. Debugging is an iterative process whose purpose it is to uncover errors or misconceptions that cause the model's failure and to define and carry out the model changes that correct the errors. This iterative process consists of four steps. In step 1 the model is tested, revealing the existence of errors (bugs). Given the detected errors, the cause of each error is determined in step 2. In step 3 the model changes believed to be required for correcting the detected errors are identified. The identified model changes are carried out in step 4. Step 1 is reexecuted right after step 4 to ensure successful modification because a change correcting an error may create another one. This iterative process continues until no errors are identified in step 1 after sufficient testing [63].

Execution Testing. The execution testing category consists of several techniques that are used to collect and analyze execution behavior data for the purpose of revealing model representation errors:

1. *Execution Monitoring:* used to reveal errors by examining low-level information about activities and events that take place during model execution. It requires the instrumentation of a simulation model for the purpose of gathering data to provide activity- or

event-oriented information about the model's dynamic behavior. For example, the model in Figure 10.10 can be instrumented to monitor the arrivals and departures of aircrafts within a particular city and the results can be compared with respect to the official airline guide to judge model validity. The model can also be instrumented to provide other low-level information, such as number of late arrivals, average passenger waiting time at the airport, and average flight time between two locations.

2. *Execution Profiling:* used to reveal errors by examining high-level information (profiles) about activities and events that take place during model execution. It requires the instrumentation of an executable model for the purpose of gathering data to present profiles about the model's dynamic behavior. For example, the model in Figure 10.10 can be instrumented to produce the following profiles to assist in model VV&T: (1) a histogram of aircraft interdeparture times, (2) a histogram of arrival times, and (3) a histogram of passenger checkout times at an airport.

3. *Execution Tracing:* used to reveal errors by "watching" the line-by-line execution of a simulation model. It requires the instrumentation of an executable model for the purpose of tracing the model's line-by-line dynamic behavior. For example, the model in Figure 10.10 can be instrumented to record all aircraft arrival times at a particular airport. Then the trace data can be compared against the official airline guide to assess model validity. The major disadvantage of the tracing technique is that execution of the instrumented model may produce a large volume of trace data that may be too complex to analyze. To overcome this problem, the trace data can be stored in a database and the modeler can analyze it using a query language [64, 65].

Fault/Failure Insertion Testing. This technique is used to insert a kind of fault (incorrect model component) or a kind of failure (incorrect behavior of a model component) into the model and observe whether the model produces the invalid behavior as expected. Unexplained behavior may reveal errors in model representation.

Field Testing. Field testing places the model in an operational situation for the purpose of collecting as much information as possible for model validation. It is especially useful for validating models of military combat systems. Although it is usually difficult, expensive, and sometimes impossible to devise meaningful field tests for complex systems, their use wherever possible helps both the project team and decision makers to develop confidence in the model [10, 45].

Functional Testing. Functional testing (also known as *black-box testing*) is used to assess the accuracy of model input–output transformation. It is applied by beeding inputs (test data) to the model and evaluating the corresponding outputs. The concern is how accurately the model transforms a given set of input data into a set of output data.

It is virtually impossible to test all input–output transformation paths for a reasonably large and complex simulation model since the number of those paths could be in the millions. Therefore, the objective of functional testing is to increase our confidence in model input–output transformation accuracy as much as possible rather than trying to claim absolute correctness.

Generation of test data is a crucially important but a very difficult task. The law of large numbers does not apply here. Successfully testing the model under 1000 input values (test data) does not imply high confidence in model input–output transformation accuracy just because 1000 is a large number. Instead, the number 1000 should be compared with the number of allowable input values to determine what percentage of the model input domain is covered in testing. The more the model input domain is covered in testing, the more confidence we gain in the accuracy of the model input–output transformation [48, 66].

In the case study, confidence intervals are constructed for each of the 14 performance measures by using actual observations collected on the vehicle waiting times. Confidence intervals are also constructed by using the simulation output data under the currently used traffic light timing. The actual and simulated confidence intervals with a confidence level of 95% are plotted corresponding to each performance measure. Lack of or little overlap between the actual and simulated confidence intervals revealed invalidity.

Graphical Comparisons. Graphical comparisons is a subjective, inelegant, and heuristic, yet quite practical approach, especially useful as a preliminary approach to model VV&T. The graphs of values of model variables over time are compared with the graphs of values of system variables to investigate characteristics such as similarities in periodicities, skewness, number and location of inflection points, logarithmic rise and linearity, phase shift, trend lines, and exponential growth constants [67–70].

Interface Testing. The interface testing (also known as *integration testing*) category consists of several techniques that are used to assess the accuracy of data use, (sub)model-to-(sub)model interface, and user–model interface:

1. *Data Interface Testing:* conducted to assess the accuracy of data inputted into the model or outputted from the model during execution. All data interfaces are examined to substantiate that all aspects of data input–output are correct. This form of testing is particularly important for those simulation models whose inputs are read from a database and/or the results of which are stored into a database for later analysis. The model's interface to the database is examined to ensure correct importing and exporting of data [51].

2. *Model Interface Testing:* conducted to detect model representation errors created as a result of (sub)model-to-(sub)model interface errors or invalid assumptions about the interfaces. It is assumed that each model component (submodel) or a model in distributed simulation is individually tested and found to be sufficiently accurate before model interface testing begins.

This form of testing deals with how well the (sub)models are integrated with each other and is particularly useful for object-oriented and distributed simulation models. Under the object-oriented paradigm (see Chapter 11), objects (1) are created with public and private interfaces, (2) interface with other objects through message passing, (3) are reused with their interfaces, and (4) inherit the interfaces and services of other objects.

Model interface testing deals with the accuracy assessment of each type of four interfaces identified by Sommerville [41]:

a. *Parameter Interfaces:* pass data or function references from one submodel to another.

b. *Shared Memory Interfaces:* enable submodels to share a block of memory where data are placed by one submodel and retrieved from there by other submodels.

c. *Procedural Interfaces:* used to implement the concept of encapsulation under the object-oriented paradigm. An object provides a set of services (procedures) that can be used by other objects and hides (encapsulates) how a service is provided to the outside world.

d. *Message-Passing Interfaces:* enable an object to request the service of another by way of message passing.

Sommerville [41] classifies interface errors into three categories:

a. *Interface Misuse:* occurs when a submodel calls another and uses its interface incorrectly. For submodels with parameter interfaces, a parameter being passed may be of the wrong type, may be passed in the wrong order, or the wrong number of parameters may be passed.

b. *Interface Misunderstanding:* occurs when submodel A calls submodel B without satisfying the underlying assumptions of submodel B's interface. For example, submodel A calls a binary search routine by passing an unordered list to be searched when in fact the binary algorithm assumes that the list is already sorted.

c. *Timing Errors:* occur in real-time, parallel, and distributed simulations that use a shared memory or a message-passing interface.

3. *User Interface Testing:* conducted to detect model representation errors created as a result of user-model interface errors or invalid assumptions about the interfaces. This form of testing is particularly important for testing human-in-the-loop, interactive, and training simulations. User interface testing deals with the assessment of the interactions between the user and the model. The user interface is examined from low-level ergonomic aspects to instrumentation and controls and from human factors to global considerations of usability and appropriateness for the purpose of identifying potential errors [37, 50, 51].

Object-Flow Testing. Object-flow testing is similar to *transaction-flow testing* [23] and *thread testing* [41]. It is used to assess model accuracy by way of exploring the life cycle of an object during model execution. For example, a dynamic object (aircraft) can be marked for testing in the VSE model shown in Figure 10.10. Every time the dynamic object enters a model component, the visualization of that component is displayed. Every time the dynamic object interacts with another object within the component, the interaction is highlighted. Examination of how a dynamic object flows through the activities and processes and interacts with its environment during its lifetime in model execution is extremely useful for identifying errors in model behavior.

Partition Testing. Partition testing is used for testing the model with the test data generated by analyzing the model's functional representatives (partitions). It is accomplished by (1) decomposing both model specification and implementation into functional representatives (partitions), (2) comparing the elements and prescribed functionality of each partition specification with the elements and actual functionality of corresponding partition implementation, (3) deriving test data to extensively test the functional behavior of each partition, and (4) testing the model by using the generated test data.

The model decomposition into functional representatives (partitions) is derived through the use of symbolic evaluation techniques that maintain algebraic expressions of model elements and show model execution paths. These functional representations are the model computations. Two computations are equivalent if they are defined for the same subset of the input domain that causes a set of model paths to be executed and if the result of the computations is the same for each element within the subset of the input domain [71]. Standard proof techniques are used to show equivalence over a domain. When equivalence cannot be shown, partition testing is performed to locate errors, or as Richardson and Clarke [72] state, to increase confidence in the equality of the computations due to the lack of error manifestation. By involving both model specification and implementation, partition testing is capable of providing more comprehensive test data coverage than other test data generation techniques.

Predictive Validation. Predictive validation requires past input and output data of the system being modeled. The model is driven by past system input data and its forecasts are compared with the corresponding past system output data to test the predictive ability of the model [73].

Product Testing. Product testing is conducted by the development organization after all submodels are successfully integrated (as demonstrated by the interface testing) and before acceptance testing by the client. No contractor wants to be in a situation where the product (model) fails the acceptance test. Product testing serves as a means of getting prepared for the acceptance testing. As such, the SQA group must perform product testing and make sure that all requirements specified in the legal contract are satisfied before delivering the model to the client organization [41]. As dictated by principle 12, testing each submodel successfully does not imply overall model credibility. Interface testing and product testing must be performed to substantiate overall model credibility.

Regression Testing. Regression testing is used to substantiate that correcting errors and/or making changes in the model do not create other errors and adverse side effects. It is usually accomplished by retesting the modified model with the previous test data sets used. Successful regression testing requires planning throughout the model development life cycle. Retaining and managing old test data sets are essential for the success of regression testing.

Sensitivity Analysis. Sensitivity analysis is performed by systematically changing the values of model input variables and parameters over some range of interest and observing the effect on model behavior [10]. Unexpected effects may reveal invalidity. The input values can also be changed to induce errors to determine the sensitivity of model behavior to such errors. Sensitivity analysis can identify those input variables and parameters to the values of which model behavior is very sensitive. Then model validity can be enhanced by assuring that those values are specified with sufficient accuracy [36, 45, 74, 75].

Special Input Testing. The special input testing category consists of the following techniques that are used to assess model accuracy by way of subjecting the model to a variety of inputs:

1. *Boundary Value Testing:* employed to test model accuracy by using test cases on the boundaries of input equivalence classes. A model's input domain can usually be divided into classes of input data (known as equivalence classes) which cause the model to function the same way. For example, a traffic intersection model might specify the probability of left turn in a three-way turning lane as 0.2, the probability of right turn as

0.35, and the probability of traveling straight as 0.45. This probabilistic branching can be implemented by using a uniform random number generator that produces numbers in the range $0 \leq RN \leq 1$. Thus three equivalence classes are identified: $0 \leq RN \leq 0.2$, $0.2 < RN \leq 0.55$, and $0.55 < RN \leq 1$. Each test case from within a given equivalence class has the same effect on the model behavior (i.e., produces the same direction of turn). In boundary analysis, test cases are generated just within, on top of, and just outside the equivalence classes [48]. In the example above, the following test cases are selected for the left turn: 0.0, $\pm$ 0.000001, 0.199999, 0.2, and 0.200001. In addition to generating test data on the basis of input equivalence classes, it is also useful to generate test data that will cause the model to produce values on the boundaries of output equivalence classes [48]. The underlying rationale for this technique as a whole is that the most error-prone test cases lie along the boundaries [76]. Notice that invalid test cases used in the example above will cause the model execution to fail; however, this failure should be as expected and meaningfully documented.

2. *Equivalence Partitioning Testing:* partitions the model input domain into equivalence classes in such a manner that a test of a representative value from a class is assumed to be a test of all values in that class [33, 41, 50, 51].

3. *Extreme Input Testing:* conducted by running/exercising the simulation model by using only minimum values, only maximum values, or arbitrary mixture of minimum and maximum values for the model input variables.

4. *Invalid Input Testing:* performed by running/exercising the simulation model under incorrect input data and cases to determine whether the model behaves as expected. Unexplained behavior may reveal model representation errors.

5. *Real-Time Input Testing:* particularly important for assessing the accuracy of simulation models built to represent embedded real-time systems. For example, different design strategies of a real-time software system to be developed to control the operations of the components of a manufacturing system can be studied by simulation modeling. The simulation model representing the software design can be tested by way of running it under real-time input data that can be collected from the existing manufacturing system. Using real-time input data collected from a real system is particularly important to represent the timing relationships and correlations between input data points.

6. *Self-Driven Input Testing:* conducted by running/exercising the simulation model under input data randomly sampled from probabilistic models representing random phenomena in a real or futuristic system. A probability distribution (e.g., exponential, gamma, Weibull) can be fit to collected data or triangular and beta probability distributions can be used in the absence of data to model random input conditions (Chapter 3; [11, 12]). Then, using random variate generation techniques, random values can be sampled from the probabilistic models to test the model validity under a set of observed or speculated random input conditions.

7. *Stress Testing:* intended to test the model validity under extreme workload conditions. This is usually accomplished by increasing the congestion in the model. For example, the model in Figure 10.10 can be stress tested by increasing the number of flights between two locations to an extremely high value. Such increase in workload may create unexpected high congestion in the model. Under stress testing, the model may exhibit invalid behavior; however, such behavior should be as expected and meaningfully documented [48, 63].

8. *Trace-Driven Input Testing.* conducted by running/exercising the simulation

TABLE 10.3 Statistical Techniques Proposed for Validation

Technique	References
Analysis of variance	[77]
Confidence intervals/regions	[10,11,30]
Factor analysis	[67]
Hotelling's T^2-tests	[10,26,78–80]
Multivariate analysis of variance	[81]
Standard MANOVA	
Permutation methods	
Nonparametric ranking methods	
Nonparametric goodness-of-fit tests	[77,82]
Kolmogorov–Smirnov test	
Cramer–Von Mises test	
Chi-square test	
Nonparametric tests of means	[10]
Mann–Whitney–Wilcoxon test	
Analysis of paired observations	
Regression analysis	[67,83,84]
Theil's inequality coefficient	[85–87]
Time-series analysis	
Spectral analysis	[45,84,88–90]
Correlation analysis	[91]
Error analysis	[92,93]
t-Test	[10,94]

model under input trace data collected from a real system. For example, a computer system can be instrumented by using software and hardware monitors to collect data by tracing all system events. The raw trace data is then refined to produce the real input data for use in testing the simulation model of the computer system.

Statistical Techniques. Much research has been conducted in applying statistical techniques for model validation. Table 10.3 presents the statistical techniques proposed for model validation and lists related references. The statistical techniques listed in the table require that the system being modeled be completely observable (i.e., all data required for model validation can be collected from the system). Model validation is conducted by using the statistical techniques to compare the model output data with the corresponding system output data when the model is run with the "same" input data that derive the real system. As dictated by principle 11, a comparison of model and system multiple outputs must be carried out by using a multivariate statistical technique to incorporate the correlations among the output variables. A recommended validation procedure based on the use of simultaneous confidence intervals is presented below.

Validation Procedure Using Simultaneous Confidence Intervals. The behavioral accuracy (validity) of a simulation model with multiple outputs can be expressed in terms of the differences between the corresponding model and system output variables when the model is run with the same input data and operational conditions that drive the real system. The range of accuracy of the jth model output variable can be represented by the jth confidence interval (CI) for the differences between the means of the jth model and system output variables. The simultaneous confidence intervals (SCIs) formed by these CIs are called the *model range of accuracy* (MRA) [30].

Assume that there are k output variables from the model and k output variables from the system, as shown in Figure 10.6. Let

$$(\boldsymbol{\mu}^m)' = [\mu_1^m, \mu_2^m, \ldots, \mu_k^m] \quad \text{and} \quad (\boldsymbol{\mu}^s)' = [\mu_1^s, \mu_2^s, \ldots, \mu_k^s]$$

be the k-dimensional vectors of the population means of the model and system output variables, respectively. Basically, there are three approaches for constructing the SCI to express the MRA for the mean behavior.

In approach I, the MRA is determined by the $100(1-\gamma)\%$ SCI for $\boldsymbol{\mu}^m - \boldsymbol{\mu}^s$ as

$$[\boldsymbol{\delta} - \boldsymbol{\tau}] \tag{3}$$

where $\boldsymbol{\delta}' = [\delta_1, \delta_2, \ldots, \delta_k]$ representing lower bounds and $\boldsymbol{\tau}' = [\tau_1, \tau_2, \ldots, \tau_k]$ representing upper bounds of the SCI. We can be $100(1-\gamma)\%$ confident that the true differences between the population means of the model and system output variables are simultaneously contained within (3).

In approach II, the $100(1-\gamma^m)\%$ SCI are first constructed for $\boldsymbol{\mu}^m$ as

$$[\boldsymbol{\delta}^m, \boldsymbol{\tau}^m] \tag{4}$$

where $(\boldsymbol{\delta}^m)' = [\delta_1^m, \delta_2^m, \ldots, \delta_k^m]$ and $(\boldsymbol{\tau}^m)' = [\tau_1^m, \tau_2^m, \ldots, \tau_k^m]$. Then the $100(1-\gamma^s)\%$ SCIs are constructed for $\boldsymbol{\mu}^s$ as

$$[\boldsymbol{\delta}^s, \boldsymbol{\tau}^s] \tag{5}$$

where

$$(\boldsymbol{\delta}^s)' = [\delta_1^s, \delta_2^s, \ldots, \delta_k^s] \quad \text{and} \quad (\boldsymbol{\tau}^s)' = [\tau_1^s, \tau_2^s, \ldots, \tau_k^s]$$

Finally, using the Bonferroni inequality, the MRA is determined by the following SCI for $\boldsymbol{\mu}^m - \boldsymbol{\mu}^s$ with a confidence level of at least $(1-\gamma^m-\gamma^s)$ when the model and system outputs are dependent and with a level of at least $(1-\gamma^m-\gamma^s+\gamma^m\gamma^s)$ when the outputs are independent [95]:

$$[\boldsymbol{\delta}^m - \boldsymbol{\tau}^s, \boldsymbol{\tau}^m - \boldsymbol{\delta}^s] \tag{6}$$

In approach III the model and system output variables are observed in pairs and the MRA is determined by the $100(1-\gamma)\%$ SCI for $\boldsymbol{\mu}^d$, the population means of the differences of paired observations, as

$$[\boldsymbol{\delta}^d, \boldsymbol{\tau}^d] \tag{7}$$

where $(\boldsymbol{\delta}^d)' = [\delta_1^d, \delta_2^d, \ldots, \delta_k^d]$ and $(\boldsymbol{\tau}^d)' = [\tau_1^d, \tau_2^d, \ldots, \tau_k^d]$.

The approach for constructing the MRA should be chosen with respect to the way the model is driven. The MRA is constructed by using the observations collected from the model and system output variables by running the model with the "same" input data and operational conditions that drive the real system. If the simulation model is self-driven, "same" indicates that the model input data are coming independently from the same populations or stochastic process of the system input data. Since the model and system input data are independent of each other but coming from the same populations, the model and system output data are expected to be independent and identically distributed. Hence approach I or II can be used. The use of approach III in this case would be less efficient. If the simulation model is trace driven, "same" indicates that the model input data are exactly the same as the system input data. In this case the model and system output data are expected to be dependent and identical. Therefore, approach II or III should be used.

Sometimes, the model sponsor, model user, or a third party may specify an acceptable range of accuracy for a specific simulation study. This specification can be made for the mean behavior of a stochastic simulation model as

$$L \leq \boldsymbol{\mu}^m - \boldsymbol{\mu}^s \leq \mathbf{U} \tag{8}$$

where $\mathbf{L}' = [L_1, L_2, \ldots, L_k]$ and $\mathbf{U}' = [U_1, U_2, \ldots, U_k]$ are the lower and upper bounds of the acceptable differences between the population means of the model and system output variables. In this case, the MRA should be compared against (6) to evaluate model validity.

The shorter the lengths of the MRA, the more meaningful is the information they provide. The lengths can be decreased by increasing the sample sizes or by decreasing the confidence level. However, such increases in sample sizes may increase the cost of data collection. Thus a trade-off analysis may be necessary among the sample sizes, confidence levels, half-length estimates of the MRA, data collection method, and cost of data collection. For details of performing the trade-off analysis, see ref. 30. The confidence interval validation procedure is presented in Figure 10.11.

Structural Testing. The structural testing (also known as *white-box testing*) category consists of the six techniques discussed below. Structural (white-box) testing is used to evaluate the model based on its internal structure (how it is built), whereas functional (black-box) testing is intended for assessing the input–output transformation accuracy of the model. Structural testing employs data flow and control flow diagrams to assess the accuracy of internal model structure by examining model elements such as statements, branches, conditions, loops, internal logic, internal data representations, submodel interfaces, and model execution paths.

1. *Branch Testing:* conducted to run/exercise the simulation model under test data so as to execute as many branch alternatives as possible, as many times as possible, and to substantiate their accurate operations. The more branches are tested successfully, the more confidence we gain in model's accurate execution with respect to its logical branches [23].

2. *Condition Testing:* conducted to run/exercise the simulation model under test data so as to execute as many (compound) logical conditions as possible, as many times as possible, and to substantiate their accurate operations. The more logical conditions are tested successfully, the more confidence we gain in model's accurate execution with respect to its logical conditions.

3. *Data Flow Testing:* uses the control flow graph to explore sequences of events related to the status of data structures and to examine data-flow anomalies. For example, sufficient paths can be forced to execute under test data to assure that every data element and structure is initialized prior to use or every declared data structure is used at least once in an executed path [23].

4. *Loop Testing:* conducted to run/exercise the simulation model under test data so

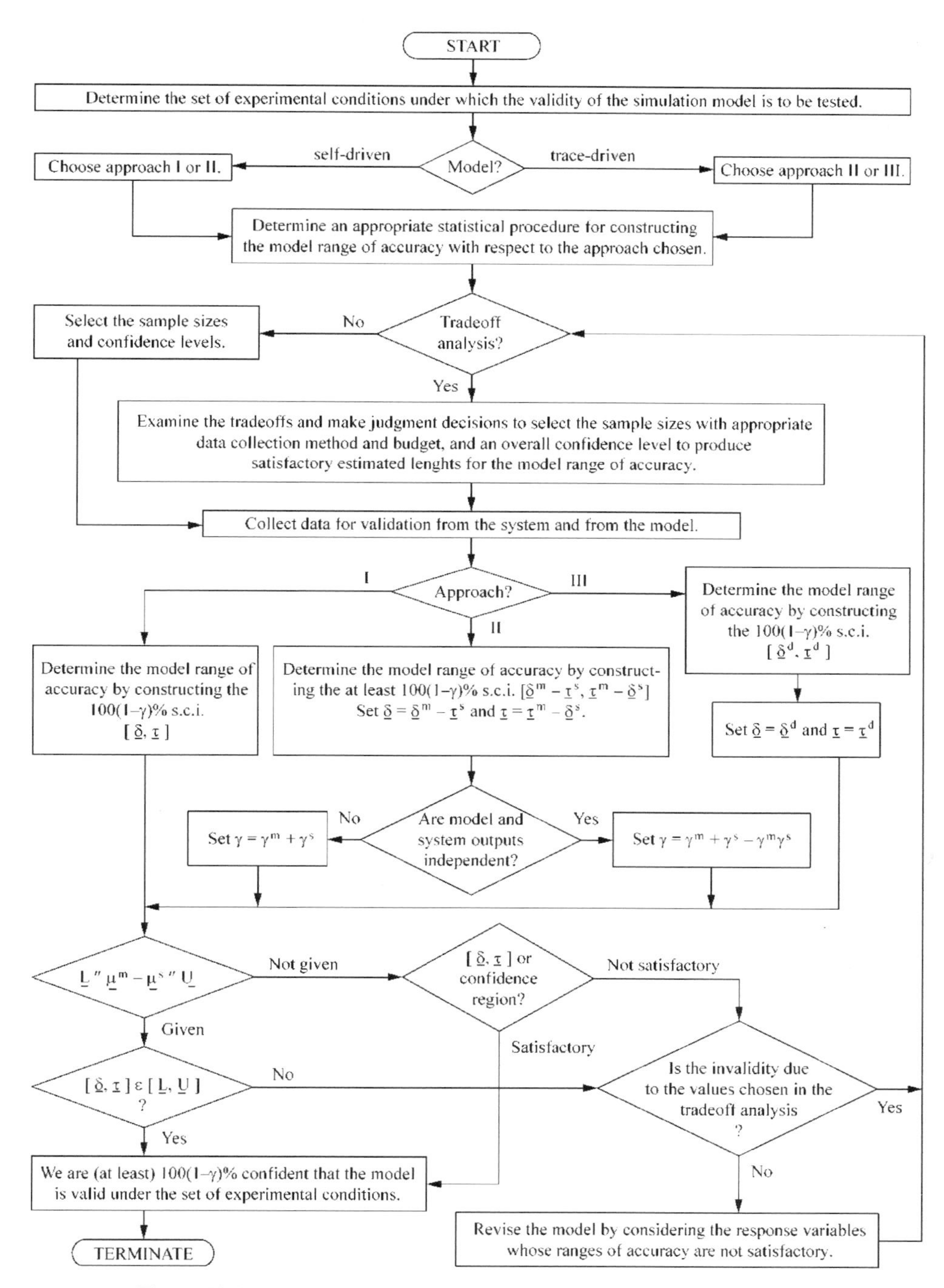

Figure 10.11 Validation procedure using simultaneous confidence intervals.

as to execute as many loop structures as possible, as many times as possible, and to substantiate their accurate operations. The more loop structures are successfully tested, the more confidence we gain in the model's accurate execution with respect to its loop structures [50].

5. *Path Testing:* conducted to run/exercise the simulation model under test data so as to execute as many control flow paths as possible, as many times as possible, and to substantiate their accurate operations. The more control flow paths are tested successfully, the more confidence we gain in model's accurate execution with respect to its control flow paths. However, 100% path coverage is impossible to achieve for a reasonably large simulation model [23].

Path testing is performed in three steps [71]. In step 1 the model control structure is determined and represented in a control flow diagram. In step 2 test data are generated to cause selected model logical paths to be executed. Symbolic execution can be used to identify and group together classes of input data based on the symbolic representation of the model. The test data are generated in such a way as to (1) cover all statements in the path, (2) encounter all nodes in the path, (3) cover all branches from a node in the path, (4) achieve all decision combinations at each branch point in the path, and (5) traverse all paths [96]. In step 3, by using the generated test data, the model is forced to proceed through each path in its execution structure, thereby providing comprehensive testing.

In practice, only a subset of all possible model paths are selected for testing, due to budgetary constraints. Recent work has sought to increase the amount of coverage per test case or to improve the effectiveness of the testing by selecting the most critical areas to test. The path prefix strategy is an "adaptive" strategy that uses previous paths tested as a guide in the selection of subsequent test paths. Prather and Myers [96] prove that the path prefix strategy achieves total branch coverage.

The identification of essential paths is a strategy that reduces the path coverage required by nearly 40% [97]. The basis for the reduction is the elimination of nonessential paths. Paths that are overlapped by other paths are nonessential. The model control flow graph is transformed into a directed graph whose arcs (called *primitive arcs*) correspond to the essential paths of the model. Nonessential arcs are called *inheritor arcs* because they inherit information from the primitive arcs. The graph produced during the transformation is called an *inheritor-reduced graph*. Chusho [97] presents algorithms for efficiently identifying nonessential paths and reducing the control graph into an inheritor-reduced graph and for applying the concept of essential paths to the selection of effective test data.

6. *Statement Testing:* conducted to run/exercise the simulation model under test data so as to execute as many statements as possible, as many times as possible, and to substantiate their accurate operations. The more statements are tested successfully, the more confidence we gain in the model's accurate execution with respect to its statements [23].

Submodel/Module Testing. Submodel/module testing requires a top-down model decomposition in terms of submodels/modules. The executable model is instrumented to collect data on all input and output variables of a submodel. The system is similarly instrumented (if possible) to collect similar data. Then each submodel behavior is compared with corresponding subsystem behavior to judge submodel validity. If a subsystem can be modeled analytically (e.g., as an M/M/1 model), its exact solution can be compared against the simulation solution to assess validity quantitatively.

Validating each submodel individually does not imply sufficient validity for the entire model as dictated by principle 12; each submodel is found sufficiently valid with some allowable error and the allowable errors can accumulate to make the entire model invalid. Therefore, after individually validating each submodel, the entire model itself must be subjected to overall testing.

Symbolic Debugging. Symbolic debugging assists in model VV&T by employing a debugging tool that allows the modeler to manipulate model execution while viewing the model at the source code level. By setting 'breakpoints" the modeler can interact with the entire model one step at a time, at predetermined locations, or under specified conditions. While using a symbolic debugger, the modeler may alter model data values or cause a portion of the model to be "replayed," that is, executed again under the same conditions (if possible). Typically, the modeler utilizes the information from execution history generation techniques, such as tracing, monitoring, and profiling, to isolate a problem or its proximity. Then the debugger is employed to understand how and why the error occurs.

Current state-of-the-art debuggers (or interactive run-time controllers) allow viewing the run-time code as it appears in the source listing, setting "watch" variables to monitor data flow, viewing complex data structures, and even communicating with asynchronous I/O channels. The use of symbolic debugging can greatly reduce the debugging effort while increasing its effectiveness. Symbolic debugging allows the modeler to locate errors and check numerous circumstances that lead up to the errors [42].

Top-Down Testing. Top-down testing is used in conjunction with top-down model development strategy. In top-down development, model construction starts with the submodels at the highest level and culminates with the submodels at the base level (i.e., the ones that are not decomposed further). As each submodel is completed, it is tested thoroughly. When submodels belonging to the same parent have been developed and tested, the submodels are integrated and integration testing is performed. This process is repeated in a top-down manner until the whole model has been integrated and tested. The integration of completed submodels need not wait for all "same level" submodels to be completed. Submodel integration and testing can be, and often is, performed incrementally [41].

Top-down testing begins with testing the global model at the highest level. When testing a given level, calls to submodels at lower levels are simulated using submodel stubs. A *stub* is a dummy submodel that has no other function than to let its caller complete the call. Fairley [65] lists the following advantages of top-down testing: (1) model integration testing is minimized, (2) early existence of a working model results, (3) higher-level interfaces are tested first, (4) a natural environment for testing lower levels is provided, and (5) errors are localized to new submodels and interfaces.

Some of the disadvantages of top-down testing are (1) thorough submodel testing is discouraged (the entire model must be executed to perform testing), (2) testing can be expensive (since the whole model must be executed for each test), (3) adequate input data are difficult to obtain (because of the complexity of the data paths and control predicates), and (4) integration testing is hampered (again, because of the size and complexity induced by testing the whole model) [65].

Visualization/Animation. Visualization/animation of a simulation model greatly assists in model VV&T [24, 98]. Displaying graphical images of internal (e.g., how customers are served by a cashier) and external (e.g., utilization of the cashier) dynamic behavior of a model during execution enables us to discover errors by seeing. For exam-

Figure 10.12 Traffic intersection simulation model animation.

ple, in the case study, we can observe the arrivals of vehicles in different lanes and their movements through the intersection as the traffic light changes, as shown in Figure 10.12. Seeing the animation of the model as it executes and comparing it with the operations of the real traffic intersection can help us identify discrepancies between the model and the system. In the case study, the animation was extremely useful for identifying bugs in the model logic. Many errors were reflected in the animation and were easily noticed.

Seeing the model in action is very useful for uncovering errors; however, seeing is not believing in visual simulation [99]. Observing that the animation of model behavior is free of errors does not guarantee the correctness of the model results.

10.4.4 Formal VV&T Techniques

Formal VV&T techniques are based on mathematical proof of correctness. If attainable, proof of correctness is the most effective means of model VV&T. Unfortunately, "if attainable" is the overriding point with regard to formal VV&T techniques. Current state-of-the-art proof of correctness techniques are simply not capable of being applied even to a reasonably complex simulation model. However, formal techniques serve as the foundation for other VV&T techniques and the most commonly known eight techniques are described briefly below: (1) induction, (2) inductive assertions, (3) inference, (4) λ-calculus, (5) logical deduction, (6) predicate calculus, (7) predicate transformation, and (8) proof of correctness [42, 100].

Induction, *inference*, and *logical deduction* are simply acts of justifying conclusions on the basis of premises given. An argument is valid if the steps used to progress from the premises to the conclusion conform to established *rules of inference*. Inductive reasoning is based on invariant properties of a set of observations (assertions are invariants since their value is defined to be true). Given that the initial model assertion is correct, it stands to reason that if each path progressing from that assertion can be shown to be correct and subsequently each path progressing from the previous assertion is correct, and so on, the model must be correct if it terminates. Formal induction proof techniques exist for the intuitive explanation just given.

Birta and Özmizrak [101] present a knowledge-based approach for simulation model validation based on the use of a validation knowledge base containing rules of inference.

Inductive assertions are used to assess model correctness based on an approach that is very close to formal proof of model correctness. It is conducted in three steps. In step 1 input-to-output relations for all model variables are identified. In step 2 these relations are converted into assertion statements and are placed along the model execution paths in such a way as to divide the model into a finite number of "assertion-bound" paths, that is, an assertion statement lies at the beginning and end of each model execution path. In step 3 verification is achieved by proving that for each path: If the assertion at the beginning of the path is true and all statements along the path are executed, the assertion at the end of the path is true. If all paths plus model termination can be proved, by induction, the model is proved to be correct [102, 103].

The λ-*calculus* [104] is a system for transforming the model into formal expressions. It is a string-rewriting system and the model itself can be considered as a large string. The λ-calculus specifies rules for rewriting strings (i.e., transforming the model into λ-calculus expressions). Using the λ-calculus, the modeler can formally express the model so that mathematical proof of correctness techniques can be applied.

The *predicate calculus* provides rules for manipulating predicates. A predicate is a combination of simple relations, such as *completed_jobs* > *steady_state_length*. A predicate will either be true or false. The model can be defined in terms of predicates and manipulated using the rules of the predicate calculus. The predicate calculus forms the basis of all formal specification languages [105].

Predicate transformation [106, 107] provides a basis for verifying model correctness by formally defining the semantics of the model with a mapping that transforms model output states to all possible model input states. This representation provides the basis for proving model correctness.

Formal *proof of correctness* corresponds to expressing the model in a precise notation and then mathematically proving that the executed model terminates and it satisfies the requirements specification with sufficient accuracy [37, 105]. Attaining proof of correctness in a realistic sense is not possible under the current state of the art. However, the advantage of realizing proof of correctness is so great that when the capability is realized, it will revolutionize the model VV&T.

10.5 CREDIBILITY ASSESSMENT STAGES

It is very important to understand the 15 principles of VV&T presented in Section 10.3 when applying more than 75 VV&T techniques described in Section 10.4 throughout the entire life cycle of a simulation study given in Figure 10.1. The principles help the researchers, practitioners, and managers better understand what VV&T is all about. These principles serve to provide the underpinnings for the VV&T techniques. Understanding and applying the principles is crucially important for the success of a simulation study.

Table 10.4 marks the VV&T techniques that are applicable for each major credibility assessment stage of the life cycle of a simulation study. The rows of Table 10.4 list the VV&T techniques in alphabetical order. The column labels correspond to the major credibility assessment stages in the life cycle:

- Formulated problem VV&T
- Feasibility assessment of simulation
- System and objectives definition VV&T
- Model qualification
- Communicative model VV&T
- Programmed model VV&T
- Experiment design VV&T
- Data VV&T
- Experimental model VV&T
- Presentation VV&T

It should be noted that the list above shows only the major credibility assessment stages and that many other VV&T activities exist throughout the life cycle.

10.5.1 Formulated Problem VV&T

Formulated problem VV&T deals with substantiating that the formulated problem contains the *actual* problem in its entirety and is sufficiently well structured to permit the derivation of a sufficiently credible solution [9]. Failure to formulate the *actual* problem results in a type III error. Once a type III error is committed, regardless of how well the problem is solved, the simulation study will either end unsuccessfully or with a type II error. Therefore, the accuracy of the formulated problem greatly affects the credibility and acceptability of simulation results.

In the case study, type III error may be committed if the problem domain boundary excludes the adjacent traffic intersections. It is possible that the traffic light timings of the adjacent intersections are set in such a way that they all turn green at the same time for the traffic traveling toward the intersection under study. Such light timings may be the root cause of congestion. Correcting the light timings at the adjacent traffic intersections may very well solve the congestion problem at the traffic intersection under study. Failure to identify such a cause may result in type III error.

Audit, cause–effect graphing, desk checking, face validation, inspections, reviews, and walkthroughs can be applied for conducting formulated problem VV&T. In applying cause–effect graphing, a causality network is created to analyze the potential root causes of the communicated problem [9]. The questionnaire developed by Balci and Nance [9] with 38 indicators can be used in applying audit, inspections, reviews, and walkthroughs.

TABLE 10.4 Applicability of the VV&T Techniques for the Credibility Assessment Stages

	FP VV&T	FA of Simulation	S&OD VV&T	Model Qualification	CM VV&T	PM VV&T	ED VV&T	Data VV&T	EM VV&T	Presentation VV&T
Acceptance testing									×	
Alpha testing									×	
Assertion checking						×	×	×	×	
Audit	×	×	×	×	×	×	×	×	×	×
Authorization testing						×			×	
Beta testing									×	
Bottom-up testing						×			×	
Boundary value testing						×			×	
Branch testing						×	×		×	
Calling structure analysis					×	×			×	
Cause–effect graphing	×				×	×			×	
Comparison testing									×	
Concurrent process analysis					×	×			×	
Condition testing						×	×		×	
Control flow analysis					×	×	×		×	
Data dependency analysis					×	×	×	×	×	
Data flow analysis					×	×	×	×	×	
Data flow testing						×	×	×	×	
Data interface testing								×	×	
Debugging						×	×		×	
Desk checking	×	×	×	×	×	×	×	×	×	×
Documentation checking	×	×	×	×	×	×	×	×	×	×
Equivalence partitioning testing						×			×	
Execution monitoring						×	×		×	
Execution profiling						×	×		×	

TABLE 10.4 (*Continued*)

	FP VV&T	FA of Simulation	S&OD VV&T	Model Qualification	CM VV&T	PM VV&T	ED VV&T	Data VV&T	EM VV&T	Presentation VV&T
Execution tracing						×	×		×	
Extreme input testing						×			×	
Face validation	×	×	×	×	×	×	×	×	×	×
Fault/failure analysis					×	×	×	×	×	
Fault/failure insertion testing						×	×	×	×	
Field testing									×	
Functional testing						×	×		×	
Graphical comparisons						×	×		×	
Induction						×			×	
Inductive assertions						×			×	
Inference						×			×	
Inspections	×	×	×	×	×	×	×	×	×	×
Invalid input testing						×			×	
Lambda calculus						×			×	
Logical deduction						×			×	
Loop testing						×	×		×	
Model interface analysis					×	×			×	
Model interface testing						×			×	
Object-flow testing						×			×	
Partition testing						×	×		×	
Path testing						×	×		×	
Performance testing									×	
Predicate calculus						×			×	
Predicate transformation						×			×	
Predictive validation									×	
Product testing									×	
Proof of correctness						×			×	
Real-time input testing						×			×	

Regression testing						×			×	
Reviews	×	×	×	×	×	×	×	×	×	×
Security testing									×	
Self-driven input testing						×			×	
Semantic analysis						×	×		×	
Sensitivity analysis						×	×		×	
Standards testing						×			×	
State transition analysis					×	×	×		×	
Statement testing						×	×		×	
Statistical techniques (Table 10.3)								×	×	
Stress testing						×			×	
Structural analysis					×	×	×		×	
Submodel/module testing						×	×		×	
Symbolic debugging						×	×		×	
Symbolic evluation					×	×	×		×	
Syntax analysis						×	×		×	
Top-down testing						×	×		×	
Trace-driven input testing						×			×	
Traceability assessment					×	×			×	
Turing test									×	
User interface analysis					×	×			×	
User interface testing						×			×	
Visualization/animation						×	×	×	×	×
Walkthroughs	×	×	×	×	×	×	×	×	×	×

10.5.2 Feasibility Assessment of Simulation

Audit, desk checking, face validation, inspections, reviews, and walkthroughs can be applied for assessing the feasibility of simulation with the use of indicators such as: (1) Are the benefits and cost of simulation solution estimated correctly? (2) Do the potential benefits of simulation solution justify the estimated cost of obtaining it? (3) Is it possible to solve the problem using simulation within the time limit specified? (4) Can all of the resources required by the simulation project be secured? and (5) Can all of the specific requirements (e.g., access to pertinent classified information) of the simulation project be satisfied?

10.5.3 System and Objectives Definition VV&T

For the purpose of generality, the term *system* is used to refer to the entity that contains the formulated problem. System and objectives definition VV&T deals with assessing the credibility of the system investigation process in which system characteristics are explored for consideration in system definition and modeling.

Audit, desk checking, face validation, inspections, reviews, and walkthroughs can be applied for conducting system and objectives definition VV&T by using indicators such as: (1) Since systems and objectives may change over a period of time, will we have the same system and objectives definition at the conclusion of the simulation study (which may last from six months to several years)? (2) Is the system's environment (boundary) identified correctly? (3) What counterintuitive behavior may be caused within the system and its environment? (4) Will the system significantly drift to low performance requiring a periodic update of the system definition? and (5) Are the interdependency and organization of the system characterized accurately? The objective here is to substantate that the system characteristics are identified and the study objectives are explicitly defined with sufficient accuracy. An error made here may not be caught until very late in the life cycle resulting in a high cost of correction or an error of type II or III.

10.5.4 Model Qualification

Model qualification is intended for assessing the credibility of the model formulation process. A model should be conceptualized under the guidance of a structured approach such as the conical methodology [5]. One key idea behind the use of a structured approach is to control the model complexity so that we can verify and validate the model successfully. The use of a structured approach is an important factor determining the success of a simulation project, especially for large-scale and complex models.

During the conceptualization of the model, one makes many assumptions in abstracting reality. Each assumption should be explicitly specified. Model qualification deals with the justification that all assumptions made are appropriate and the conceptual model provides an adequate representation of the system with respect to the study objectives. Audit, desk checking, face validation, inspections, reviews, and walkthroughs can be applied for conducting model qualification.

In the case study, many assumptions including the following, are made in abstracting the traffic intersection operation: (1) pedestrians are excluded; (2) bicycles and emergency vehicles are excluded; (3) the light timing cycle length is assumed constant and the sensor in lane 9 is ignored; (4) the yellow light is included in the green since most drivers pass on yellow; (5) all drivers obey the traffic laws; and (6) all vehicles have the same size. These assumptions were justified to be appropriate under the study objectives in the model qualification credibility assessment stage.

10.5.5 Communicative Model VV&T

Communicative model VV&T deals with confirming the adequacy of the communicative model to provide an acceptable level of agreement for the domain of intended application. *Domain of intended application* [25] is the prescribed conditions for which the model is intended to match the system under study. *Level of agreement* [25] is the required correspondence between the model and the system, consistent with the domain of intended application and the study objectives.

In the case study, the graphical model design specification, shown in Figure 10.3, is justified to be sufficiently accurate. Inspections are conducted to substantiate that all vehicle movements in the model design specification represent the real-life movements with sufficient accuracy. Specifications of all classes is found to be appropriate.

10.5.6 Programmed Model VV&T

Programmed model VV&T deals with the assessment of programmed (executable) model accuracy. Most of the techniques in Table 10.4 are applicable for conducting programmed model VV&T. In the case study, many of the applicable techniques in Table 10.4 were used to assess the executable model accuracy. Specifically, the animation was very helpful. In addition, tracing of message passing was instrumental in revealing some of the bugs.

10.5.7 Experiment Design VV&T

Experiment design VV&T deals with substantiating the sufficient accuracy of the design of experiments. The techniques marked in Table 10.4 can be applied for conducting experiment design VV&T with the use of indicators such as: (1) Are the algorithms used for random variate generation theoretically accurate? (2) Are the random variate generation algorithms translated into executable code accurately? (Error may be induced by computer arithmetic or by truncation due to machine accuracy, especially with order statistics (e.g., $X = -\log_e(1-U)$) [108]); (3) How well is the random number generator tested? (using a generator that is not rigorously shown to produce uniformly distributed independent numbers with sufficiently large period may invalidate the entire experiment design); (4) Are *appropriate* statistical techniques implemented to design and analyze the simulation experiments? How well are the underlying assumptions satisfied? (see ref. 109 for several reasons why output data analyses have not been conducted in an appropriate manner); (5) Is the problem of the initial transient (or the startup problem) [110] appropriately addressed? and (6) For comparison studies, are identical experimental conditions replicated correctly for each of the alternative operating policies compared?

10.5.8 Data VV&T

Data VV&T involves input data model VV&T and deals with substantiating that all data used throughout the model development phases of the life cycle in Figure 10.1 are accurate, complete, unbiased, and appropriate in their original and transformed forms. An input data model is the characterization of an input process (e.g., characterization of an arrival process by Poisson probability distribution). U.S. GAO [111] emphasizes the importance of input data model validation in credibility assessment of simulations. In those cases where data cannot be collected, data values may be determined through calibration. *Calibration* is an iterative process in which a probabilistic characterization for an input variable or a fixed value for a parameter is tried until the model is found to be sufficiently valid.

The techniques marked in Table 10.4 can be applied for conducting data VV&T with the use of indicators such as: (1) Does each input data model possess a sufficiently accurate representation? (2) Are the parameter values identified, measured, or estimated with sufficient accuracy? (3) How reliable are the instruments used for data collection and measurement? (4) Are all data transformations done accurately? (e.g., are all data transformed correctly into the same time unit of the model?) (5) Is the dependence between the input variables, if any, represented by the input data model(s) with sufficient accuracy? (blindly modeling bivariate relationships using only correlation to measure dependency is cited as a common error by Schmeiser [108]); and (6) Are all data up to date?

10.5.9 Experimental Model VV&T

Experimental model VV&T deals with substantiating that the experimental model has sufficient accuracy in representing the system under study consistent with the study objectives. All of the techniques listed in Table 10.4 can be applied for conducting experimental model VV&T. The applicability of the VV&T techniques depends on the following cases, where the system being modeled is (1) completely observable—all data required for model VV&T can be collected from the system, (2) partially observable—some required data can be collected, or (3) nonexistent or completely unobservable. The statistical techniques in Table 10.3 are applicable only for case 1.

In the case study, many of the applicable techniques in Table 10.4 were used to assess the experimental model accuracy. Some of the statistical techniques in Table 10.3 were also used.

10.5.10 Presentation VV&T

Presentation VV&T deals with justifying that the simulation results are interpreted, documented, and communicated with sufficient accuracy. Since all simulation models are descriptive, simulation results must be interpreted. A descriptive model describes the behavior of a system without any value judgment on the "goodness" or "badness' of such behavior. In the simulation of an interactive computer system, for example, the model may produce a value of 20 seconds for the average response time. But it does not indicate whether the value 20 is a "good" result or a "bad" one. Such a judgment is made by the simulation analyst depending on the study objectives. Under one set of study objectives the value 20 may be too high; under another, it may be reasonable. The project team should review the way the results are interpreted in every detail to evaluate interpretation accuracy. Errors may be induced due to the complexity of simulation results, especially for large-scale and complex models.

Gass [112] points out that "we do not know of any model assessment or modeling project review that indicated satisfaction with the available documentation." Nance [5] advocates the use of standards in simulation documentation. The documentation problem should be attributed to the lack of automated support for documentation generation integrated with model development continuously throughout the entire life cycle. The model development environment [16, 29, 113, 114] provides such computer-aided assistance for documenting a simulation study with respect to the phases, processes, and credibility assessment stages of the life cycle in Figure 10.1.

The simulation project team must devote sufficient effort in communicating technical simulation results to decision makers in a language they will understand. They must pay more attention to translating from the specialized jargon of the discipline into a form that is meaningful to the nonsimulationist and nonmodeler. Simulation results may be presented to the decision makers as integrated within a decision support system (DSS). With the help of a DSS, a decision maker can understand and utilize the results much better. The integration accuracy of simulation results within the DSS must be verified. If results are directly presented to the decision makers, the presentation technique (e.g., overheads, slides, films, etc.) must be ensured to be effective enough. The project management must make sure that the team members are trained and possess sufficient presentation skills. Audit, desk checking, face validation, inspections, reviews, visualization/animation, and walkthroughs can be applied for conducting presentation VV&T.

10.6 CONCLUDING REMARKS

The life-cycle application of VV&T is extremely important for successful completion of complex and large-scale simulation studies. This point must be clearly understood by the sponsor of the simulation study and the organization conducting the simulation study. The sponsor must furnish funds under the contractual agreement and require the contractor to apply VV&T *throughout* the entire life cycle of a simulation study.

Assessing credibility throughout the life cycle is an onerous task. Applying the VV&T techniques throughout the life cycle is time consuming and costly. In practice, under time pressure to complete a simulation study, the VV&T and documentation are sacrificed first. Computer-aided assistance for credibility assessment is required to alleviate these problems. More research is needed to bring automation to the application of VV&T techniques.

Integration VV&T with model development is crucial. This integration is best achieved within a computer-aided simulation model development environment [16, 29, 113, 114]. More research is needed for this integration. The question of which of the applicable VV&T techniques should be selected for a particular VV&T activity in the life cycle should be answered by taking the following into consideration: (1) model type, (2) simulation type, (3) problem domain, and (4) study objectives.

How much to test or when to stop testing depends on the study objectives. The testing should continue until we achieve sufficient confidence in credibility and acceptability of simulation results. The sufficiency of the confidence is dictated by the study objectives. Establishing a simulation quality assurance (SQA) program within the organization conducting the simulation study is extremely important for successful credibility assessment. The SQA management structure goes beyond VV&T and is also responsible for assessing other model quality characteristics such as maintainability, reusability, and usability (human–computer interface). The management of the SQA program and the management of the simulation project must be independent of each other and neither should be able to overrule the other [37].

Subjectivity is, and always will be, part of the credibility assessment for a reasonably complex simulation study. The reason for subjectivity is twofold: modeling is an art and credibility assessment is situation dependent. A unifying approach based on the use of indicators measuring qualitative as well as quantitative aspects of a simulation study should be developed.

REFERENCES

1. Ören, T. I. (1981). Concepts and criteria to assess acceptability of simulation studies: a frame of reference, *Communications of the ACM*, Vol. 24, No. 4, pp. 180–189.
2. Ören, T. I. (1986). Artificial intelligence in quality assurance of simulation studies, in *Modelling and Simulation Methodology in the Artificial Intelligence Era*, M. S. Elzas, T. I. Ören, and B. P. Zeigler, eds., North-Holland, Amsterdam, pp. 267–278.
3. Ören, T. I. (1987). Quality assurance paradigms for artificial intelligence in modelling and simulation," *Simulation*, Vol. 48, No. 4, pp. 149–151.
4. Balci, O. (1990). Guidelines for successful simulation studies, in *Proceedings of the 1990 Winter Simulation Conference*, O. Balci, R. P. Sadowski, and R. E. Nance, eds., IEEE, Piscataway, N.J., pp. 25–32.
5. Nance, R. E. (1994). The conical methodology and the evolution of simulation model development, *Annals of Operations Research*, Vol. 53, pp. 1–46.
6. Banks, J., D. Gerstein, and S. P. Searles (1987). Modeling processes, validation, and verification of complex simulations: a survey, in *Methodology and Validation*, O. Balci, ed., Society for Computer Simulation, San Diego, Calif., pp. 13–18.
7. Knepell, P. L., and D. C. Arangno (1993). Simulation validation: a confidence assessment methodology, *Monograph 3512-04*, IEEE Computer Society Press, Los Alamitos, Calif.
8. Woolley, R. N., and M. Pidd (1981). Problem structuring: a literature review, *Journal of the Operational Research Society*, Vol. 32, No. 3, pp. 197–206.
9. Balci, O., and R. E. Nance (1985). Formulated problem verification as an explicit requirement of model credibility," *Simulation*, Vol. 45, No. 2, pp. 76–86.
10. Shannon, R. E. (1975). *Systems Simulation: The Art and Science*, Prentice Hall, Upper Saddle River, N.J.
11. Law, A. M., and W. D. Kelton (1991). *Simulation Modeling and Analysis*, 2nd ed., McGraw-Hill, New York.
12. Banks, J., J. S. Carson, and B. L. Nelson (1996). *Discrete-Event System Simulation*, 2nd ed., Prentice Hall, Upper Saddle River, N.J.
13. Vincent, S., and A. M. Law (1995). ExpertFit: total support for simulation input modeling," in *Proceedings of the 1995 Winter Simulation Conference*, C. Alexopoulos, K. Kang, W. R. Lilegdon, and D. Goldsman, eds., IEEE, Piscataway, N.J., pp. 395–400.
14. Overstreet, C. M., and R. E. Nance (1985). A specification language to assist in analysis of discrete event simulation models, *Communications of the ACM*, Vol. 28, No. 2, pp. 190–201.
15. Martin, J., and C. McClure (1985). *Diagramming Techniques for Analysts and Programmers*, Prentice Hall, Upper Saddle River, N.J.
16. Balci, O., A. I. Bertelrud, C. M. Esterbrook, and R. E. Nance (1995). A picture-based object-oriented visual simulation environment, in *Proceedings of the 1995 Winter Simulation Conference*, C. Alexopoulos, K. Kang, W. R. Lilegdon, and D. Goldsman, eds., IEEE, Piscataway, N.J., pp. 1333–1340.
17. Orca Computer (1996). *Visual Simulation Environment User's Guide*, Orca Computer, Inc., Blacksburg, Va.

18. Orca Computer (1996). *Visual Simulation Environment Reference Manual*, Orca Computer, Inc., Blacksburg, Va.

19. Banks, J. (1996). Software for Simulation, in *Proceedings of the 1996 Winter Simulation Conference*, J. M. Charnes, D. J. Morrice, D. T. Brunner, and J. J. Swain, eds., IEEE, Piscataway, N.J., pp. 31–38.

20. Balci, O. (1988). The implementation of four conceptual frameworks for simulation modeling in high-level languages, in *Proceedings of the 1988 Winter Simulation Conference*, M. A. Abrams, P. L. Haigh, and J. C. Comfort, eds., IEEE, Piscataway, N.J., pp. 287–295.

21. Fishman, G. S. (1978). *Principles of Discrete Event Simulation*, Wiley-Interscience, New York.

22. Hetzel, W. (1984). *The Complete Guide to Software Testing*, QED Information Sciences, Wellesley, Mass.

23. Beizer, B. (1990). *Software Testing Techniques*, 2nd ed., Van Nostrand Reinhold, New York.

24. Sargent, R. G. (1996). Verifying and validating simulation models, in *Proceedings of the 1996 Winter Simulation Conference*, J. M. Charnes, D. J. Morrice, D. T. Brunner, and J. J. Swain, eds., IEEE, Piscataway, N.J., pp. 55–64.

25. Schlesinger, S., et al. (1979). Terminology for model credibility, *Simulation*, Vol. 32, No. 3, pp. 103–104.

26. Balci, O., and R. G. Sargent (1981). A methodology for cost-risk analysis in the statistical validation of simulation models, *Communications of the ACM*, Vol. 24, No. 4, pp. 190–197.

27. Nance, R. E., and C. M. Overstreet (1987). Diagnostic assistance using digraph representations of discrete event simulation model specifications, *Transactions of the SCS*, Vol. 4, No. 1, pp. 33–57.

28. Balci, O., and R. E. Nance (1992). The simulation model development environment: an overview, in *Proceedings of the 1992 Winter Simulation Conference*, J. J. Swain, D. Goldsman, R. C. Crain, and J. R. Wilson, eds., IEEE, Piscataway, N.J., pp. 726–736.

29. Derrick, E. J., and O. Balci (1995). A visual simulation support environment based on the DOMINO conceptual framework, *Journal of Systems and Software*, Vol. 31, No. 3, pp. 215–237.

30. Balci, O., and R. G. Sargent (1984). Validation of simulation models via simultaneous confidence intervals, *American Journal of Mathematical and Management Sciences*, Vol. 4, No. 3–4, pp. 375–406.

31. Johnson, M. E., and M. Mollaghasemi (1994). Simulation input data modeling, *Annals of Operations Research*, Vol. 53, pp. 47–75.

32. Watson, C. E. (1976). The problems of problem solving, *Business Horizons*, Vol. 19, No. 4, pp. 88–94.

33. Perry, W. (1995). *Effective Methods for Software Testing*, Wiley, New York.

34. Hollocker, C. P. (1987). The standardization of software reviews and audits, in *Handbook of Software Quality Assurance*, G. G. Schulmeyer and J. I. McManus, eds., Van Nostrand Reinhold, New York, pp. 211–266.

35. Adrion, W. R., M. A. Branstad, and J. C. Cherniavsky (1982). Validation, verification, and testing of computer software, *Computing Surveys*, Vol. 14, No. 2, pp. 159–192.

36. Hermann, C. F. (1967). Validation problems in games and simulations with special reference to models of international politics, *Behavioral Science*, Vol. 12, No. 3, pp. 216–231.

37. Schach, S. R. (1996). *Software Engineering*, 3rd ed., Richard D. Irwin, Homewood, Ill.

38. Ackerman, A. F., P. J. Fowler, and R. G. Ebenau (1983). Software inspections and the industrial production of software, in *Software Validation: Inspection, Testing, Verification, Alternatives, Proceedings of the Symposium on Software Validation*, Darmstadt, Germany, September 25–30, H.-L. Hausen, ed., pp. 13–40.

39. Dobbins, J. H. (1987). Inspections as an up-front quality technique. in *Handbook of Software Quality Assurance*, G. G. Schulmeyer and J. I. McManus, eds., Van Nostrand Reinhold, New York, pp. 137–177.
40. Knight, J. C., and E. A. Myers (1993). An improved inspection technique, *Communications of the ACM*, Vol. 36, No. 11, pp. 51–61.
41. Sommerville, I. (1996). *Software Engineering*, 5th ed., Addison-Wesley, Reading, Mass.
42. Whitner, R. B., and O. Balci (1989). Guidelines for selecting and using simulation model verification techniques, in *Proceedings of the 1989 Winter Simulation Conference*, E. A. MacNair, K. J. Musselman, and P. Heidelberger, eds., IEEE, Piscataway, N.J., pp. 559–568.
43. Schruben, L. W. (1980). Establishing the credibility of simulations, *Simulation*, Vol. 34, No. 3, pp. 101–105.
44. Turing, A. M. (1963). Computing machinery and intelligence, in *Computers and Thought*, E. A. Feigenbaum and J. Feldman, eds., McGraw-Hill, New York, pp. 11–15.
45. Van Horn, R. L. (1971). Validation of simulation results, *Management Science*, Vol. 17, No. 5, pp. 247–258.
46. Deutsch, M. S. (1982). *Software Verification and Validation: Realistic Project Approaches*, Prentice Hall, Upper Saddle River, N.J.
47. Myers, G. J. (1978). A controlled experiment in program testing and code walkthroughs/inspections, *Communications of the ACM*, Vol. 21, No. 9, pp. 760–768.
48. Myers, G. J. (1979). *The Art of Software Testing*, Wiley, New York.
49. Yourdon, E. (1985). *Structured Walkthroughs*, 3rd ed., Yourdon Press, New York.
50. Pressman, R. S. (1996). *Software Engineering: A Practitioner's Approach*, 4th ed., McGraw-Hill, New York.
51. Miller, L. A., E. H. Groundwater, J. E. Hayes, and S. M. Mirsky (1995). Survey and assessment of conventional software verification and validation methods, *Special Publication NUREG/CR-6316*, Vol. 2, U.S. Nuclear Regulatory Commission, Washington, DC.
52. Rattray, C., ed. (1990). *Specification and Verification of Concurrent Systems*, Springer-Verlag, New York.
53. Moose, R. L., and R. E. Nance (1989). The design and development of an analyzer for discrete event model specifications, in *Impacts of Recent Computer Advances on Operations Research*, R. Sharda, B. L. Golden, E. Wasil, O. Balci, and W. Stewart, eds., Elsevier, New York, pp. 407–421.
54. Dunn, R. H. (1984). *Software Defect Removal*, McGraw-Hill, New York.
55. Allen, F. E., and J. Cocke (1976). A program data flow analysis procedure, *Communications of the ACM*, Vol. 19, No. 3, pp. 137–147.
56. Yücesan, E., and S. H. Jacobson (1992). Building correct simulation models is difficult, in *Proceedings of the 1992 Winter Simulation Conference*, J. J. Swain, D. Goldsman, R. C. Crain, and J. R. Wilson, eds., IEEE, Piscataway, N.J., pp. 783–790.
57. Yücesan, E., and S. H. Jacobson (to appear). Computational issues for accessibility in discrete event simulation, *ACM Transactions on Modeling and Computer Simulation*, Vol. 6, No. 1, pp. 53–75.
58. King, J. C. (1976). Symbolic execution and program testing, *Communications of the ACM*, Vol. 19, No. 7, pp. 385–394.
59. Dillon, L. K. (1990). Using symbolic execution for verification of Ada Tasking programs, *ACM Transactions on Programming Languages and Systems*, Vol. 12, No. 4, pp. 643–669.
60. Ramamoorthy, C. V., S. F. Ho, and W. T. Chen (1976). On the automated generation of program test data, *IEEE Transactions on Software Engineering*, Vol. SE-2, No. 4, pp. 293–300.
61. Stucki, L. G. (1977). New directions in automated tools for improving software quality, in

Current Trends in Programming Methodology, Vol. 2, R. Yeh, ed., Prentice Hall, Upper Saddle River, N.J., pp. 80–111.

62. Department of Defense (1995). Modeling and simulation (M&S) master plan, *DoD 5000.59-P*, October.
63. Dunn, R. H. (1987). The quest for software reliability, in *Handbook of Software Quality Assurance*, G. G. Schulmeyer and J. I. McManus, eds., Van Nostrand Reinhold, New York, pp. 342–384.
64. Fairley, R. E. (1975). An experimental program-testing facility, *IEEE Transactions on Software Engineering*, Vol. SE-1, No. 4, pp. 350–357.
65. Fairley, R. E. (1976). Dynamic testing of simulation software, in *Proceedings of the 1976 Summer Computer Simulation Conference*, Washington, D.C., July 12–14, Simulation Councils, La Jolla, Calif., pp. 708–710.
66. Howden, W. E. (1980). Functional program testing, *IEEE Transactions on Software Engineering*, Vol. SE-6, No. 2, pp. 162–169.
67. Cohen, K. J., and R. M. Cyert (1961). Computer models in dynamic economics, *Quarterly Journal of Economics*, Vol. 75, No. 1, 112–127.
68. Forrester, J. W. (1961). *Industrial Dynamics*, MIT Press, Cambridge, Mass.
69. Miller, D. K. (1975). Validation of computer simulations in the social sciences, in *Proceedings of the 6th Annual Conference on Modeling and Simulation*, Pittsburgh, Pa., pp. 743–746.
70. Wright, R. D. (1972). Validating dynamic models: an evaluation of tests of predictive power, in *Proceedings of the 1972 Summer Computer Simulation Conference*, San Diego, Calif., July 14–16, Simulation Councils, La Jolla, Calif., pp. 1286–1296.
71. Howden, W. E. (1976). Reliability of the path analysis testing strategy, *IEEE Transactions on Software Engineering*, Vol. SE-2, No. 3, pp. 208–214.
72. Richardson, D. J., and L. A. Clarke (1985). Partition analysis: a method combining testing and verification, *IEEE Transactions on Software Engineering*, Vol. SE-11, No. 12, pp. 1477–1490.
73. Emshoff, J. R., and R. L. Sisson (1970). *Design and Use of Computer Simulation Models*, Macmillan, New York.
74. Miller, D. R. (1974). Model validation through sensitivity analysis, in *Proceedings of the 1974 Summer Computer Simulation Conference*, Houston, Texas, July 9–11, Simulation Councils, La Jolla, Calif., pp. 911–914.
75. Miller, D. R. (1974). Sensitivity analysis and validation of simulation models, *Journal of Theoretical Biology*, Vol. 48, No. 2, pp. 345–360.
76. Ould, M. A., and C. Unwin (1986). *Testing in Software Development*, Cambridge University Press, Cambridge.
77. Naylor, T. H., and J. M. Finger (1967). Verification of computer simulation models, *Management Science*, Vol. 14, No. 2, pp. B92–B101.
78. Balci, O., and R. G. Sargent (1982). Some examples of simulation model validation using hypothesis testing, in *Proceedings of the 1982 Winter Simulation Conference*, H. J. Highland, Y. W. Chao, and O. S. Madrigal, eds., IEEE, Piscataway, N.J., pp. 620–629.
79. Balci, O., and R. G. Sargent (1982). Validation of multivariate response models using Hotelling's two-sample T^2 test, *Simulation*, Vol. 39, No. 6, pp. 185–192.
80. Balci, O., and R. G. Sargent (1983). Validation of multivariate response trace-driven simulation models, in *Performance '83*, A. K. Agrawala and S. K. Tripathi, eds., North-Holland, Amsterdam. pp. 309–323.
81. Garratt, M. (1974). Statistical validation of simulation models, in *Proceedings of the 1974 Summer Computer Simulation Conference*, Houston, Texas, July 9–11, Simulation Councils, La Jolla, Calif., pp. 915–926.

82. Gafarian, A. V., and J. E. Walsh (1969). Statistical approach for validating simulation models by comparison with operational systems, in *Proceedings of the 4th International Conference on Operations Research*, Wiley, New York, pp. 702–705.
83. Aigner, D. J. (1972). A note on verification on computer simulation models, *Management Science*, Vol. 18, No. 11, pp. 615–619.
84. Howrey, P., and H. H. Kelejian (1969). Simulation versus analytical solutions, in *The Design of Computer Simulation Experiments*, T. H. Naylor, ed., Duke University Press, Durham, N.C., pp. 207–231.
85. Kheir, N. A., and W. M. Holmes (1978). On validating simulation models of missile systems, *Simulation*, Vol. 30, No. 4, pp. 117–128.
86. Rowland, J. R., and W. M. Holmes (1978). Simulation validation with sparse random data, *Computers and Electrical Engineering*, Vol. 5, No. 3, pp. 37–49.
87. Theil, H. (1961). *Economic Forecasts and Policy*, North-Holland, Amsterdam.
88. Fishman, G. S., and P. J. Kiviat (1967). The analysis of simulation generated time series, *Management Science*, Vol. 13, No. 7, pp. 525–557.
89. Gallant, A. R., T. M. Gerig, and J. W. Evans (1974). Time series realizations obtained according to an experimental design, *Journal of the American Statistical Association*, Vol. 69, No. 347, pp. 639–645.
90. Hunt, A. W. (1970). Statistical evaluation and verification of digital simulation models through spectral analysis, Ph.D. dissertation, University of Texas at Austin, Austin, Texas.
91. Watts, D. (1969). Time series analysis, in *The Design of Computer Simulation Experiments*, T. H. Naylor, ed., Duke University Press, Durham, N.C., pp. 165–179.
92. Damborg, M. J., and L. F. Fuller (1976). Model validation using time and frequency domain error measures, *ERDA Report 76-152*, NTIS, Springfield, Va.
93. Tytula, T. P. (1978). A method for validating missile system simulation models, *Technical Report E-78-11*, U.S. Army Missile R&D Command, Redstone Arsenol, Ala., June.
94. Teorey, T. J. (1975). Validation criteria for computer system simulations, *Simuletter*, Vol. 6, No. 4, pp. 9–20.
95. Kleijnen, J. P. C. (1975). *Statistical Techniques in Simulation*, Vol. 2, Marcel Dekker, New York.
96. Prather, R. E., and J. P. Myers, Jr. (1987). The path prefix software testing strategy, *IEEE Transactions on Software Engineering*, Vol. SE-13, No. 7, pp. 761–766.
97. Chusho, T. (1987). Test data selection and quality estimation based on the concept of essential branches for path testing, *IEEE Transactions on Software Engineering*, Vol. SE-13, No. 5, pp. 509–517.
98. Bell, P. C., and R. M. O'Keefe (1994). Visual interactive simulation: a methodological perspective, *Annals of Operations Research*, Vol. 53, pp. 321–342.
99. Paul, R. J. (1989). Visual simulation: seeing is believing? in *Impacts of Recent Computer Advances on Operations Research*, R. Sharda, B. L. Golden, E. Wasil, O. Balci, and W. Stewart, eds., Elsevier, New York, pp. 422–432.
100. Khanna, S. (1991). Logic programming for software verification and testing, *The Computer Journal*, Vol. 34, No. 4, pp. 350–357.
101. Birta, L. G., and F. N. Özmizrak (1996). A knowledge-based approach for the validation of simulation models: the foundation, *ACM Transactions on Modeling and Computer Simulation*, Vol. 6, No. 1, pp. 76–98.
102. Manna, Z., S. Ness, and J. Vuillemin (1973). Inductive methods for proving properties of programs, *Communications of the ACM*, Vol. 16, No. 8, pp. 491–502.
103. Reynolds, C., and R. T. Yeh (1976). Induction as the basis for program verification, *IEEE Transactions on Software Engineering*, Vol. SE-2, No. 4, pp. 244–252.

104. Barendregt, H. P. (1981). *The Lambda Calculus: Its Syntax and Semantics*, North-Holland, New York.

105. Backhouse, R. C. (1986). *Program Construction and Verification*, Prentice Hall International, London.

106. Dijkstra, E. W. (1975). Guarded commands, non-determinacy and a calculus for the derivation of programs, *Communications of the ACM*, Vol. 18, No. 8, pp. 453–457.

107. Yeh, R. T. (1977). Verification of programs by predicate transformation, in *Current Trends in Programming Methodology*, Vol. 2, R. Yeh, ed., Prentice Hall, Upper Saddle River, N.J., pp. 228–247.

108. Schmeiser, B. (1981). Random variate generation, in *Proceedings of the 1981 Winter Simulation Conference*, T. I. Ören, C. M. Delfosse, and C. M. Shub, eds. IEEE, Piscataway, N.J., pp. 227–242.

109. Law, A. M. (1983). Statistical analysis of simulation output data, *Operations Research*, Vol. 31, No. 6, pp. 983–1029.

110. Wilson, J. R., and A. A. B. Pritsker (1978). A survey of research on the simulation startup problem, *Simulation*, Vol. 31, No. 2, pp. 55–58.

111. U.S. GAO (1987). DOD simulations: improved assessment procedures would increase the credibility of results, *GAO/PEMD-88-3*, U.S. General Accounting Office, Washington, D.C., Dec.

112. Gass, S. I. (1983). Decision-aiding models: validation, assessment, and related issues for policy analysis, *Operations Research*, Vol. 31, No. 4, pp. 603–631.

113. Balci, O. (1986). Requirements for model development environments, *Computers and Operations Research*, Vol. 13, No. 1, pp. 53–67.

114. Balci, O., and R. E. Nance (1987). Simulation model development environments: a research prototype, *Journal of the Operational Research Society*, Vol. 38, No. 8, pp. 753–763.

PART III

RECENT ADVANCES

CHAPTER 11

Object-Oriented Simulation

JEFFREY A. JOINES AND STEPHEN D. ROBERTS
North Carolina State University

11.1 INTRODUCTION

An Object-Oriented Simulation (OOS) models the behavior of interacting objects over time. Object collections, called classes, encapsulate the characteristics and functionality of common objects. A set of object classes has been written in C++ which can be used to create simulation models and simulation packages. The simulations built with these tools possess the benefits of an object-oriented design, including the use of encapsulation, inheritance, polymorphism, run-time binding, and parameterized typing. These concepts are illustrated by creating a set of object frames to describe various simulation requirements. Simulation modeling is encapsulated within a set of modeling frameworks. From this set of modeling frameworks, a network queueing simulation language is developed which has several notable features not available in other non-OOS simulation languages. The extensibility and reusability of the simulation modeling concepts are demonstrated with examples. Object-oriented simulations provide full accessibility to the base language, faster executions, portable models and executables, a multi-vendor implementation language, and a growing variety of complementary development tools.

The idea of an object-oriented simulation has great intuitive appeal because it is very easy to view the real world as being composed of objects. In a manufacturing cell, the physical objects include the machines, the workers, the parts, the tools, and the conveyors. However, the part routings, the schedule, the work plan, and other information items could be also viewed as objects. All these objects interact to produce system behavior. A simulation simply viewed manipulates these objects over time.

It is quite easy to describe many existing simulation languages using object terminology. A simulation language or simulation package provides a user with a set of predefined *object classes* from which the simulation modeler can create needed *objects*. For example, a network-based queueing language will typically view a system as having *entity* objects that travel through a network of queue objects, being served by resource objects. Using the simulation language (object classes), the modeler would

Handbook of Simulation, Edited by Jerry Banks.
ISBN 0-471-13403-1

declare the network by defining the *node* objects and their connecting *branch* objects. The node objects would be described as *sources* (where entities arrive to the network), *queues* (where entities wait) and *activities* (where entities are served), with and without resources, and *sinks* (where entities leave the network). Predefined entity objects, sometimes called *transactions*, can be made to arrive to the network through source nodes. Most languages allow attributes to be assigned to the transactions. *Resource* objects and their behavior may be defined. Simulation support objects would include probability distributions, global variables, and statistical tables and histograms. The simulation modeler creates objects and specifies their behavior through the parameters available. The objects communicate with each other through *messages* and functions. The integration of all the objects into a single package provides the complete simulation model.

Some simulation packages/languages provide for special functionality, such as that needed for manufacturing simulations. Object classes may be defined for machines, conveyors, transporters, cranes, robots, and so on. These special objects have direct usefulness in particular situations. Simulation packages centered around such objects may be directed at specific *vertical* application areas such as automated guided vehicles, robots, flexible manufacturing systems, finite-capacity planning, and so on.

In this chapter we describe an object-oriented simulation platform and show how the platform is implemented and subsequently used. Although the description will concentrate on one platform in one programming language, its discussion should be sufficiently general to be representative of the greater interest in object-oriented simulation. We focus initially on the fundamental class structure (Section 11.2) and the design of a complete object-oriented simulation system (Section 11.3). A network simulation language is then developed within the context of this design as an illustration (Section 11.4). Finally, we demonstrate the reusability and extensibility of an object-oriented simulation through examples (Section 11.5). Everything is implemented in C++ which affects the implementation of the object-oriented simulation package. C++ is an object-oriented extension to the C programming language [1].

11.1.1 Object-Oriented Thinking

The general conceptual design of an object-oriented simulation could employ the hierarchical approach illustrated in Figure 11.1. At the outer-level users, specific simulation models can be directly parameterized by model. At some point during the simulation study the specific model may be insufficient for the application and the modeler will need to resort to more fundamental modeling and simulation concepts and features. At the inner level in the hierarchy (i.e., the inside circle), the user can employ the C++ general programming language to implement any programming concept or feature. Thus in an object-oriented simulation environment, the user can relate to the design in different ways. The common notion of a simulation language falls somewhere in the middle of this design with limited opportunity to travel in or out. Most simulation languages simply provide a programming interface when the simulation concepts and features are insufficient.

Users may relate to an object-oriented simulation design at several design levels and in several ways. Persons interested only in the simulation results may simply execute the various simulation models, while very knowledgeable persons may employ raw C++ and develop new features. The concepts at each level are encapsulated so that simulation model users, for instance, need not be concerned about the concepts at a lower level. The more sophisticated user, however, can delve deeper into the design, eventually reaching

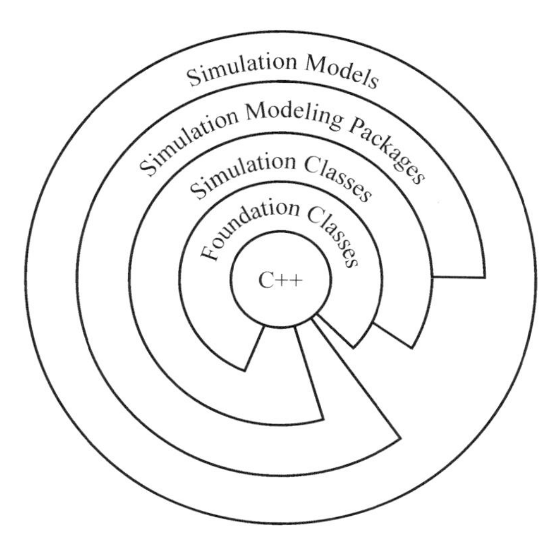

Figure 11.1 General conceptual OOS design.

the C++ level. Implicit in this design is a hierarchy of information, ranging from specific behavior of specific models to general program and simulation behavior.

In comparison, general computer users vary in their use of computer software. Many computer users simply execute software applications, like a spreadsheet or word processor, without paying much attention to the details. Some users may write computer programs or develop add-ins that perform important needed tasks which other people may execute. More knowledgeable persons may write fundamental software like compilers that other programmers use. Sometimes, those who write software for others are called *system software programmers* while those who write programs to solve specific problems are called *application programmers*. In contrast, simulation users are generally confined to be application users and only people in the vendor organizations can do simulation software development.

Thus one perspective on the contribution of object-oriented simulation to the general area of simulation is that simulation software engineering is now being added to simulation applications engineering. This addition provides some simulation users not only with a full array of simulation tools but with also the means to add new tools. Previously, the addition of new tools or new products rested solely in the domain of vendor organizations.

11.1.2 Appeal of Object-Oriented Simulation

The object-oriented simulation (OOS) concept has great intuitive appeal in applications because the notion of objects interacting with each other exhibits behavior similar to real-world experiences. It is also easy to accept the visual interpretation of objects and understand their potential for computer animation of simulations. Since object-oriented simulations focus on objects, there is the possibility of dividing the simulation computation among objects. Objects provide a natural means of organizing the simulation and offer the potential of delegating portions of the execution to different processors, either parallel or distributed (as shown in Chapter 12). Finally, since objects are often themselves made up of other objects, it is natural to decompose a system by its objects and view its behavior in terms of interacting objects.

Why Extensibility and Re-use. Because many simulation languages offer prespecified functionality produced in another language (assembly language, C, FORTRAN, etc.), the user cannot access the internal mechanisms within the language. Instead, only the vendor can make modifications to the internal functionality. Reusing language features requires that the user code any new features as though they were a completely separate package. Therefore, full integration with the existing language is not possible.

Also, users have only limited opportunity to extend an existing language feature. Some simulation languages allow for certain programminglike expressions or statements, which are inherently limited. Most languages allow the insertion of procedural routines written in other general-purpose programming languages. None of these procedures can, in any way, become an inherent part of the preexisting language.

Thus none of these approaches is fully satisfactory because, at best, any procedure written cannot use and change the behavior of a preexisting object class. Also, any new object classes defined by a user in a general programming language do not coexist directly with vendor code. The Arena software [2] provides some compositional extensibility by a template approach to representing collections of SIMAN statements (which may include the graphical representation). SLX [3] will provide extensibility within a general-purpose simulation language. However, neither of these approaches should be considered object oriented, due to lack of true extensibility (inheritance and parameterization). These topics are discussed in Section 11.2.

Simulation Software Engineering. Object-oriented simulation deals directly with the limitation of extensibility by permitting full data abstraction as well as procedural abstraction. Data abstraction means that new data types with their own behavior can be added arbitrarily to the programming language (abstract data types). When a new data type is added, it can assume a role as important as any implicit data types. For example, a user-defined data type that manages complex numbers can be as fundamental to a user's language ("first class") as the implicitly defined integer data type. In the simulation language context, a new user-defined robot class can be added to a language that contains standard resources without compromising any aspect of the existing simulation language and the robot may be used as a more complex resource.

11.1.3 Object-Oriented Simulation Software

Much of the interest in object-oriented programming was stimulated by the Simula language [4]. It introduced many of the object-oriented concepts including classes, inheritance, polymorphism, and run-time binding, all in the context of discrete-event simulation. These concepts are described in the next section. Although Simula was never fully appreciated as a simulation language, it was the basis for many of the prominent object-oriented programming languages that followed, such as Smalltalk [5], Eiffel [6], and C++. A sometimes overlooked fact is that Smalltalk continued the simulation heritage and contains an entire framework for simulation [7]. The growth of general object-oriented software and any attempt at enumerating them is well beyond the scope of this chapter; however, it is fair to observe that object orientedness has permeated almost every area of computer software development. In the broad spectrum of software development, computer simulation is a fairly narrow interest. Simulation developers are, however, discovering (or rediscovering) the benefits of the object-oriented approach and incorporating many object-oriented concepts into their simulation languages and packages. In the noncommercial arena, SimPack [8] and C++SIM [9] are two C++-based simulation packages; commercial packages include ModSimIIIa [10], Sim++ [11], C++/CSIM17 [12], and Simple++ [13]. Just how much and in what way object-oriented concepts are included depends on the developer's point of view and acceptance of the concepts. Suffice it to say, the claim of being object-oriented has many meanings, and users need to be alert to the various differences. In this chapter we attempt to clarify some of the fundamental points of view in object-oriented simulation.

11.2 OBJECTS AND CLASSES: FUNDAMENTAL CONCEPTS

The *class* concept is fundamental to object-oriented software. A class provides a pattern for the content of objects and defines their type. An example (as it appears in C++) is the `Exponential` class in Table 11.1, which is used to obtain exponential random variates (objects). The class definition determines the object's characteristics or properties. Table 11.1 is explained in the context of the following subsections.

11.2.1 Class Properties

The class definition specifies the object's properties, namely the data objects and the member functions that manipulate. These properties are generally grouped into *public* and *private* sections (C++ also permits another grouping). When the object is created, the *public* properties can be accessed from outside the object. The *private* properties are information kept strictly locked within an object and are available only to the object's member functions. For example, the object `mu` (exponential mean) is declared as a private data member of type double and cannot be directly accessed. However, a public function called `getMu()` does return the value of `m`, while `setMu()` allows the user

TABLE 11.1 `Exponential` **Class**

```
#include "random.h"

/* expon.h contains Class Exponential. This class describes an
inverse transformation generator for Exponential variables. */

class Exponential: public Random {
private:
  double mu;
public:
  Exponential ( double mean, unsigned int control=0, long seed=0 );
  Exponential ( int mean, unsigned int control=0, long seed=0 );
  virtual double sample();
  void setMu ( double initMu ) { mu = initMu; }
  double getMu() { return mu; }
};
```

to change the value of `mu`. Making a property private restricts unauthorized use and encapsulates the object's properties.

11.2.2 Inheritance

The `Exponential` class was not defined from scratch. For instance, it doesn't say anything about the use or origin of random numbers. Because the random number generator establishes the source of randomness for all random processes, it is defined in its own class to provide reusability. Hence, the `Exponential` class is *derived* from the `Random` class so that the `Exponential` class has access to all the public properties of the `Random` class without having to recode them.

This use of previously developed classes is called *inheritance*; for example, `Exponential` inherits the random generator properties from `Random`. In fact, this inheritance makes the `Exponential` class a kind of `Random` class, and in object-oriented terminology a "kind of" is considered an "is-a" relationship. The other major type of relationship between two classes is the "has-a" relationship. In the case of the exponential class, the `Exponential` has an object called `mu` of type double. A "has-a" relationship is the result of rather than inheritance (it is used to compose the object). The Arena software only provides the means to produce compositions and therefore has limited extensibility. The ability to inherit or specialize/extend objects is a fundamental feature of a true object-oriented language.

11.2.3 Construction and Initialization of Objects

When a class object is needed, the creation and initialization of it is provided by a special member function called a *constructor*. Constructors are member functions whose name is the same as the class. C++ will provide one if it isn't included in the class definition. In the case of the `Exponential` class, there are two constructors. One constructor requires a parameter of type double while the other needs an integer. Notice that some of the arguments have specified defaults, so the user doesn't have to specify all the potential features of an exponential object (these additional arguments pertain to the control of the random number stream). Within the constructors (details not shown), space is allocated for the `Exponential` object and data members are assigned.

Although not used in the `Exponential` class, C++ permits user-specified destructors. A destructor will clean up any object responsibilities (e.g., collecting statistics) and deallocate any acquired memory.

11.2.4 Run-Time Binding

The `sample()` function is specified as a virtual function in both the `Random` (not shown) and `Exponential` class so that at run time the program will decide from which random variate to sample. This approach of tying the variate to the sample at run time is also called *delayed* or *run-time binding*. Run-time binding may extract a small run-time computational penalty but makes this entire specification of sampling from variates much easier to write, maintain, and use. With run-time binding, new variate types can be added through inheritance without altering the existing simulation code. Without run-time binding, the object designer must anticipate every potential combination of future uses.

11.2.5 Polymorphism

The `Exponential` class has two constructors, so users may specify either floating-point or integer arguments for the mean interarrival time. Although it is not necessary in this case (C++ will make the right conversion), it does illustrate the use of polymorphism—where the same property applies to different objects. Thus the exponential object is appropriately specified, regardless of whether an integer or double precision is given. This encapsulation of the data makes the addition of new types for parameters very easy and localized. Under other circumstances, polymorphism allows users to produce the same behavior with different objects. For example, one message "request" can be used for AGV or Trucks rather than a message for each type; or specifying a resource requirement at an activity where this requirement might be a single resource, a *team* of resources, or a *group* of resources.

11.3 SIMULATION CLASS HIERARCHY AND FRAMES FOR OOS

A key to the creation of a fully integrated simulation package is the use of a *class inheritance hierarchy* (introduced in Section 11.2). With C++ being the most abstract form (lowest level) of a simulation package, more concrete elements are added so that at the highest level, the final product may be a specific simulation model. A specific simulation is also a kind of simulation model, which is a kind of simulation, which is a kind of programming project, which is a kind of C++ program.

An inheritance hierarchy can be viewed as a tree. The base of the tree is the most abstract class and the leaves present the most specific class. Thus the convention is that the base is known as the lower level, whereas the leaves are considered the higher level. In the figures that follow, we adopt the standard convention that the base is given at the top of the hierarchy and the leaves are at the bottom (opposite to the way that a real tree grows).

In order to collect classes into levels of abstraction, we introduce the notion of object-based frames. A *frame* is a set of classes that provide a level of abstraction in the simulation and modeling platform. A frame is not a C++ construct and therefore must be viewed conceptually. It is a convenient means for describing various levels within the simulation class hierarchy.

11.3.1 Foundation Frame

The foundation classes provide a base structure from which more simulation-specific classes may be created. These foundation classes are not simulation specific. The hierarchy for the foundation classes used in the simulation package is given in Figure 11.2. Classes in Figure 11.2 provide a variety of general support which are useful in building simulation languages and simulation packages. Although not specific to simulation, they provide a foundation framework from which more simulation-specific classes may be created. Many of these support classes can now be found in the standard template library (STL).

Abstract Objects. The `AbstractObject` forms the *fundamental base class* for the entire design and all other classes are derived from this base class. The `Abstract Object` class defines and characterizes all the essential properties every class in this

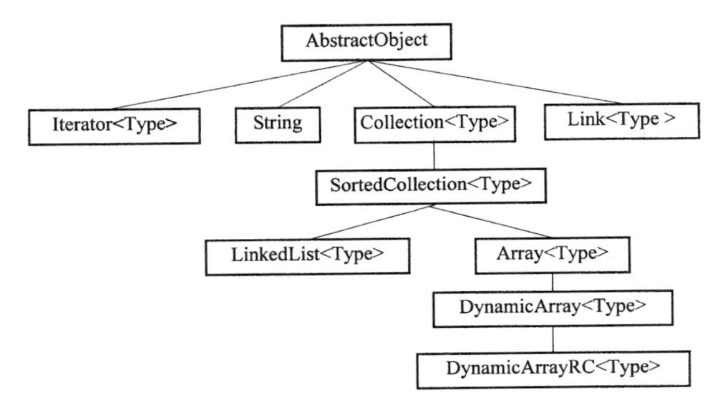

Figure 11.2 Hierarchy for foundation classes.

design should possess. No instances or objects of `AbstractObject` can be created since its primary purpose is to ensure that all classes have the same basic form. Such a common form gives uniform character to the design and allows all classes to share common desirable properties.

The `AbstractObject` class provides for the following general common properties in all derived classes, making them *nice* classes (see Carroll and Ellis [14]):

1. *Default Constructor:* constructs objects without user-specified parameters.
2. *Copy Constructor:* establishes a mechanism for creating a new object as a copy of another within this class.
3. *Assignment Operator:* allows objects to be the target of an assignment using the assignment operator.
4. *Equality Operator:* tests the "equality" of one object with another using the equality operators.
5. *Destructor:* provides for the orderly destruction of an object.

Nice classes promote reusability and classes derived from `AbstractObject` will need to either inherit these properties or provide them within the class.

Foundation Support Classes. The foundation support classes provide useful classes for the general manipulation of objects important in the creation of simulation languages/packages. These include the classes for strings, arrays, and linked lists. Arrays may be dynamically dimensioned and may have their index range checked automatically.

These classes augment the C++ language with container classes (e.g., linked lists, dynamic arrays) that hold multiple objects. Similar classes are now widely available in the new standard template library (STL) [15]. These libraries make quick work of many other elements that may be needed to build a simulation language or package.

11.3.2 Simulation Frame

The simulation frame classes provide basis simulation functionality, including random number and random variate generation, statistics collection, and base simulation ele-

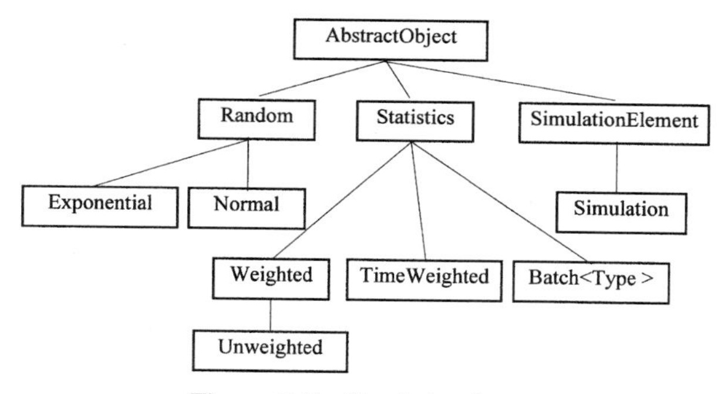

Figure 11.3 Simulation frame.

ments. The simulation class hierarchy is shown in Figure 11.3. As can be seen, all the classes are derived from `AbstractObject` to maintain a common class design throughout any simulation project.

Random Numbers and Random Variates. Random number generation is obtained from the `Random` class. Random variate generators are derived from the `Random` class so that each source of variate generation has its own random number generator (or generators). This design has two benefits: it facilitates the use of inverse transform method of random variate generation, and by associating each variate generator with its own random number stream, variance reduction through correlated sampling is possible. Random number and random variate generation properties include (1) setting and getting generator parameters, (2) obtaining random numbers/variates, and (3) creating antithetic sampling.

Statistics Collection. Basic statistics can be collected on `Weighted`, `Unweighted`, and `TimeWeighted` variables. Also, statistics may be batched from any of the basic statistic types. Tables, plots, and histograms may be displayed for basic or batched statistics. Statistics collection properties include (1) stopping and starting statistics collection, (2) clearing the statistics, and (3) reporting statistics. Basic statistics are collected during the simulation and provide (1) observation base of (weighted) observations or time, (2) mean and standard deviation, and (3) minimum and maximum observations. Batched statistics are also collected during the simulation and provide both overbatch and current batch results. Batches can be based on time intervals or numbers of observations.

Simulation Component Classes. `SimulationElement` contains the simulation time and manages the event calendar. It provides for event and time management by being capable of (1) scheduling events, (2) getting the next event, and (3) getting and setting the current time. This class provides an important base class from which modeling classes are derived.

The `Simulation` class has the run control properties which manage the complete simulation and include (1) getting the current replication number, (2) setting the number of replications, (3) setting the length of the run, (4) stopping the simulation or current replication, and (5) printing summary and individual output reports.

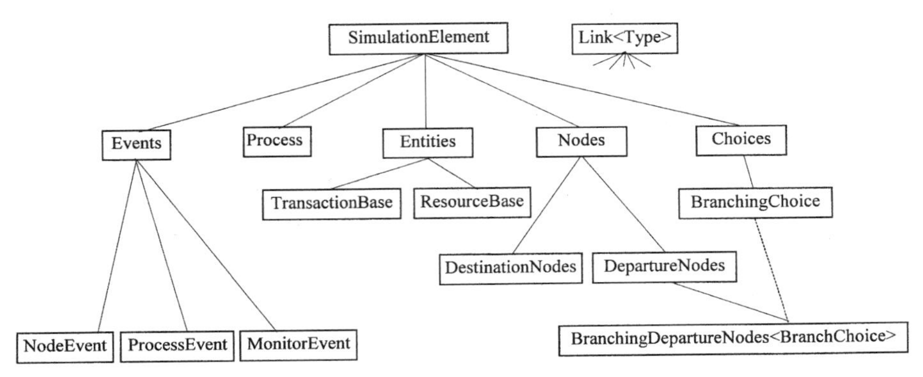

Figure 11.4 Simulation modeling frame.

11.3.3 Simulation Modeling Frame

To aid in the construction of simulation languages and packages, several simulation modeling classes have been designed and implemented. The components of the modeling frame are events, entities, processes, nodes, and choices. These components are derived from both the `SimulationElement` and the `Link` classes (see Figure 11.4). In Figure 11.4 and subsequent figures, the solid lines are inherited relationships ("is-a") while the dashed lines are composition relationships ("has-a").

`Entities` provide active elements for the simulation, whether permanent or temporary. The properties of entities include (1) getting the entity's creation time, (2) obtaining its status, (3) getting its current location, (4) obtaining the entry time of the entity's current state, and (5) getting the entity's time in the system. `TransactionBase` and `ResourceBase` classes are derived from `Entities` and extend the entities for use in general networks. The `TransactionBase` class provides entities that may need service and has properties for (1) getting and setting the node entry time, (2) getting the creation node, and (3) getting and setting the identification number. The `ResourceBase` class provides entities that can provide service and has properties for (1) getting and setting the resource name, (2) getting and setting resource states, and (3) defining the resource states.

`Processes` provide an encapsulated means for describing simulation processes (not computer tasks) such as seizing and releasing resources and reneging at queues. The process class is generally used to provide a means of decomposing a complex simulation activity, such as preempting a resource, and is a form of helper class for the simulation.

`Nodes` are used for network modeling and contain properties which include (1) getting and setting the node count, (2) getting node identification number, (3) obtaining the node type, (4) accessing a list of all nodes in the network, and (5) finding the entities at the current node. `Nodes` are derived from the `Destination` and `Departure` nodes. A destination node can be entered while a departure node may be exited. Often, departure nodes have branches connected to them and therefore need a means to choose one among several branches. These departure nodes are called `BranchingDeparture` nodes, and means of choosing the branch for the departure is its *branching choice*. The properties of the destination node include the *entering process*, while the departure nodes provide the *exiting process*. The `BranchingDeparture` nodes obtain their branching choice and related branching functions through the *class parameters* found within the <>brackets (parameterized classes are described in more detail in Section 11.4). These specifications help to compose the node and provide a parameterized "has-a" relationship.

The properties of the destination node include the entering process, while the departure nodes provide the exiting process. These two activities are handled through the virtual functions `executeEntering()` and `executeLeaving(1)`, respectively. All destination nodes need to either provide the `executeEntering()` function or inherit this property from a base class. This same is required of the `DepartureNodes` and the `executeLeaving()` function. This mechanism allows for new node types to be added directly and used with the existing simulation language.

`Choices` are used to give the simulation model "intelligence." Routes, rules, and policies may be modeled through the various choices. Of direct relevance to the `BranchingDeparture` node is the branching decision made upon exit from the node.

`Events` contain the properties related to simulation event management and provide (1) the means for setting and getting the event time, (2) setting and getting other event information (e.g., the transaction associated with the event), and (3) processing the event. `Node`, `Process`, and `Monitor` events provide specialized properties that are needed when events occur within a node, a process, or are independently specified. When `Events` are pulled off the event calendar, the virtual function `processEvent()` is invoked. Therefore, at run time, the program decides which type of event to process. This allows for other types of events not yet envisioned. `NodeEvents` are associated with `DepartureNodes` (e.g., an arrival event occurs at a source node, while an end-of-service event occurs at an activity). In both instances, transactions are leaving or departing from the particular node. When processing a `NodeEvent` (i.e., invoking the `processEvent()` function), the event invokes the departure nodes `executeLeaving()` function. Since the `executeLeaving()` is *virtual*, it allows for the addition of new types of departure node events without altering existing code.

11.3.4 Frames and Frameworks

While frames provide a convenient means to describe the levels of abstraction within the entire object-oriented simulation platform, another means of encapsulation is needed to deal with the broad simulation modeling concepts and features contained within the design. In a sense, the frames are quite similar to class libraries which can be called upon in the development of an actual simulation modeling language or package. However, for the higher-level modeling classes, these librarylike collections of classes are too complexly interrelated to be represented simply as a single level of abstraction. At the lower levels, users exercise complete control over the "flow of control" and direct the interaction among objects. At the higher level, more of the flow of control exists among the classes and users tend to finish the implementation details rather than providing the complete design. A better approach to the description of these higher-level complex interactions is the notion of *frameworks*. For our *purposes*, *frameworks* are used to describe those collections of classes that provide a set of specific modeling facilities. The frameworks may consist of one or more class hierarchies. These collections make the use and reuse of simulation modeling features more intuitive and provide for greater extensibility.

There are three important distinctions between frameworks and class libraries: (1) class libraries embody behavior while frameworks not only embody behavior but also specify the rules or protocol that govern how the behaviors are to be used; (2) class libraries are used to instantiate and call member functions, but depend fundamentally

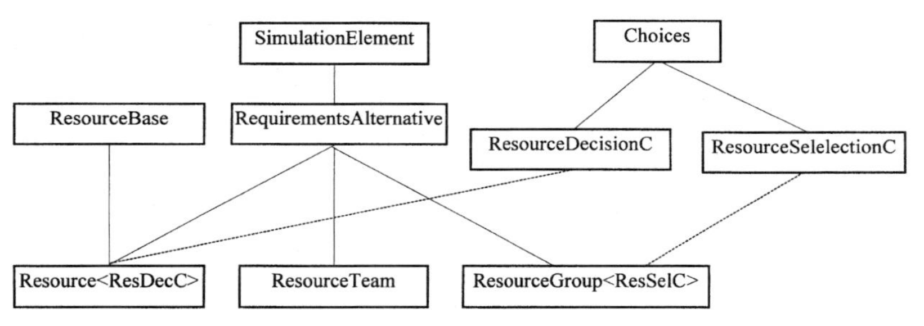

Figure 11.5 Resource framework.

on an external flow of control, whereas frameworks not only perform the same functions as libraries but also manage the flow of control among objects; and (3) class libraries largely incorporate reuse of design implementation, whereas frameworks provide for design reuse. It is in the spirit of these concerns that we provide a set of frameworks for simulation modeling.

To illustrate the development of a framework with the simulation modeling frame, consider the resources framework. The class hierarchy within the resource framework is shown in Figure 11.5. Notice that, again, the choices are used in composition rather than in inheritance. The resource framework, like all frameworks, provides *two* interfaces. The first and most common interface [also called an *applications programmer interface* (API)] is the direct use of the resource framework. Using the resource framework, the user can create individual resources, resource teams, and arbitrary resource groups. These "resources" make decisions about what to do (through resource decisions) and become targets for requests for service (by resource selections) to satisfy a service requirement. Modeling with the resource framework, the user simply instantiates and manipulates resource objects through various prespecified functions. This use is similar to what a typical user of a simulation language would do, except that the object base provides a simpler and more consistent use.

The second interface is the *extension interface* to the resource framework, which is needed for extensibility. For example, the resource selection may require some type of "look ahead" feature that selects resources based on how busy they are expected to become. If a means to look ahead in the network is needed, a new kind of resource selection may be designed that "extends" those currently available. Such extendibility is essential if users are to exploit the framework fully. There are several frameworks that compose the modeling frame. These include:

1. *Transaction Framework:* establishes the basic properties of the transaction and provides a means to create transactions, to bring them into the network, to branch them from node to node in the network, to cause them to exit the network, and to destroy them.
2. *Resource Framework:* establishes the basic properties of resources, provides for resource teams and resource groups, provide for preemption of resources, seize resources, and release resources.
3. *Queuing Framework:* stores transactions awaiting resources, ranking transactions in queue, conditional and unconditional reneging from queues, and gating transactions in queues until conditions are appropriate for their future movement.

4. *Activity Framework:* delays transactions for some specified time, may free resources and chose resource alternatives, and abort transactions from the activity.

More will be said about the various frameworks during the development of the example simulation language (Section 11.4) and its embellishments (Section 11.5).

11.4 CREATING A SPECIFIC OBJECT-ORIENTED SIMULATION

Special simulation languages and packages may be created from the object classes within modeling frameworks. In this section we present the YANSL network queueing simulation language. YANSL is an acronym for "Yet Another Network Simulation Language." YANSL is not intended to be another network queuing simulation language. Instead, it is just one *instance* of the kind of simulation capability that can be developed within an object-oriented simulation environment.

11.4.1 Basic Concepts and Objects in YANSL

YANSL was developed to illustrate the importance of object-oriented simulation. YANSL is a network queueing simulation package similar to the style of GPSS/H [16], SLAM [17], SIMAN [18], or INSIGHT [19], but without the "bells and whistles." Users familiar with any of these languages should recognize, however, that it is a very powerful alternative.

Classes Used to Create YANSL. Several classes are chosen from the modeling frameworks to create the YANSL modeling package. These classes are collected together to form a simple modeling/simulation language. In the next section (Section 11.5), we will show how to use other classes to create more complicated features and we will extend the language. The general simulation support classes, such as the variate generation, statistics collection, and time management, are used indirectly through the modeling frameworks. The network concepts are somewhat enhanced, but taken from the modeling framework. A choices class is introduced.

The YANSL network consists of transaction and resource entities moving through a network of nodes. The transactions are the entities that flow through the network while resources serve transactions at activities. What transactions and resources represent depends on the system being modeled. Both the transactions and resources are used directly from the modeling framework. The YANSL network nodes are used largely from the modeling framework but are specialized as follows:

1. *Source Node:* creates transactions and branches them into the network.
2. *Queue Node:* causes transactions to wait until resources are available at the associated activity.
3. *Activity Node:* place where transactions are delayed or possibly serviced by one or more required resources.
4. *Sink Node:* where transactions leave the network.

The YANSL node derivation hierarchy is shown in Figure 11.6. The higher-level nodes (`Assign`, `Activity`, `Queue`, `Source`, and `Sink`) are used directly by the

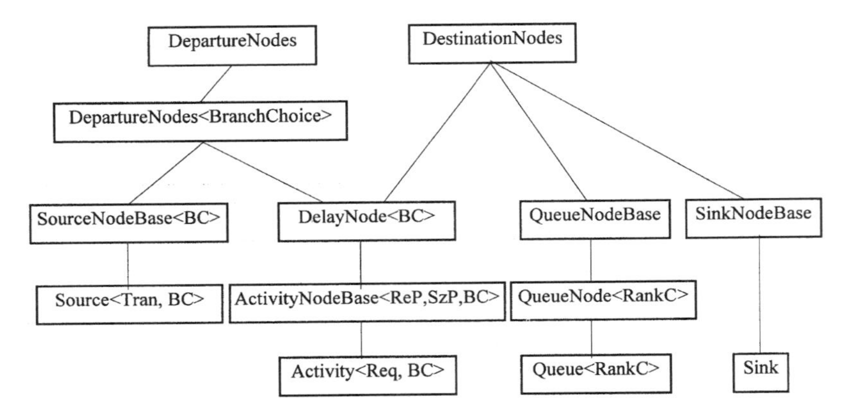

Figure 11.6 YANSL node derivation hierarchy.

YANSL modeler. Lower-level nodes provides abstractions where are less specific, thus allowing specialization for other simulation constructs (e.g., the `QueueNodeBase` class excludes ranking and statistics). Sink and queue nodes can have transactions branched to them and are therefore destination nodes. A delay node is both a departure and a destination node, so it inherits from both the departure and destination node classes. This inheritance from several parents (base classes in C++) is called *multiple inheritance*. An activity node is a kind of delay node that contains resource requirements. The properties of the YANSL nodes allow transactions to be created at source nodes, wait at queue nodes, be delayed at activity nodes, and exit the network at sink nodes.

Resources may service transactions at activity nodes. The resource hierarchy for YANSL uses the resource framework shown in Figure 11.5. The resource classes allow resources to be identified as individuals, as members of alternative groupings at an activity, or as members of teams. When there is a choice of resources at an activity, a resource selection method is employed. The ability to request a resource choice at run time without specifying it explicitly is another example of polymorphism.

The `Choices` available in YANSL extend those in the modeling frameworks and are given in Figure 11.7. The choices available add broad flexibility to the decision-making functions in the simulation without needing different extra classes for each different extra function. Instead, classes are parameterized with these choice classes and the choices consist of several methods. Parameterized types provide for another basic extensibility characteristic necessary for true object-oriented simulation language. Specifically in YANSL, they allow for the selection of alternative branches from a departure node, selection among alternative resources in requirements at an `Activity`, as well as providing the decision-making ability for resources to choose what to do next, and ranking choices among transactions at a `Queue`. The choices are used to represent the time-dependent and changing decisions that need to be modeled.

Modeling with YANSL. When modeling with YANSL, the modeler views the model as a network of elemental queueing processes (graphical symbols could be used). Building the simulation model requires the modeler to select from the predefined set of node types and integrate these into a network. Transactions flow through the network and have the same interpretation they have in the other simulation languages. Transactions

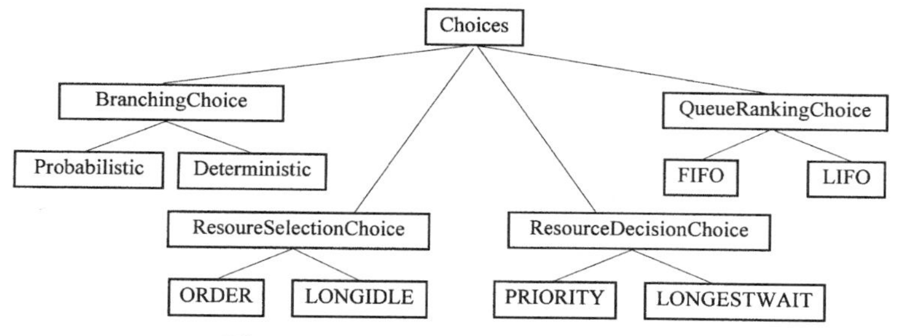

Figure 11.7 YANSL choices hierarchy.

may require resources to serve them at activities and thus may need to queue to await resource availability. Resources may be fixed or mobile in YANSL, and one or more resources may be required at an activity. Unlike some network languages, resources in YANSL are active entities, like transactions, and may be used to model a wide variety of real-world items (notice that this feature is, by itself, more powerful than some existing languages).

11.4.2 Case Study: TV Inspection and Repair

As a portion of their production process, TVs are sent to a final inspection station (refer to refs. 20 to 22 for more information as well to the harbor problem in ref. 23). Some TVs fail inspection and are sent for repair. After repair they are returned for reinspection. Transactions are used to represent the TVs. The resources needed are the inspector and the repairperson. The network is composed of a source node that describes how the TVs arrive, a queue for possible wait at the inspect activity, the inspect activity and its requirement for the inspector, a sink where good TVs leave, a queue for possible wait at the repair activity, and the repair activity. Figure 11.8 displays a visual network interpretation.

Figure 11.9 displays an updated hierarchy of Figure 11.1. Notice that YANSL is just one of many modeling languages that could be developed and that the TV inspection model is just one many models that can be created. Transactions branch from the source to the inspect queue, are served at the inspect activity, branch to either the sink or to the repair queue, are possibly served at the repair activity, and return to the inspect queue. The data used in the simulation is that the interarrival time of TVs is exponentially distributed with a mean interarrival time of 5.0 minutes, the service time is exponentially distributed with a mean of 3.5 minutes, the probability that a TV is good after being inspected in 0.85, and a repair time that is exponentially distributed with a mean of 8.0 minutes.

11.4.3 YANSL Model

The YANSL network has all the graphical and intuitive appeal of any network-based simulation language. A graphical user interface could be built to provide convenient modeling with error checking and help offered to the user. Whatever the modeling system used, the ultimate computer readable representation of the model might appear as shown in Table 11.2. The model in Table 11.2 has the character of many network simu-

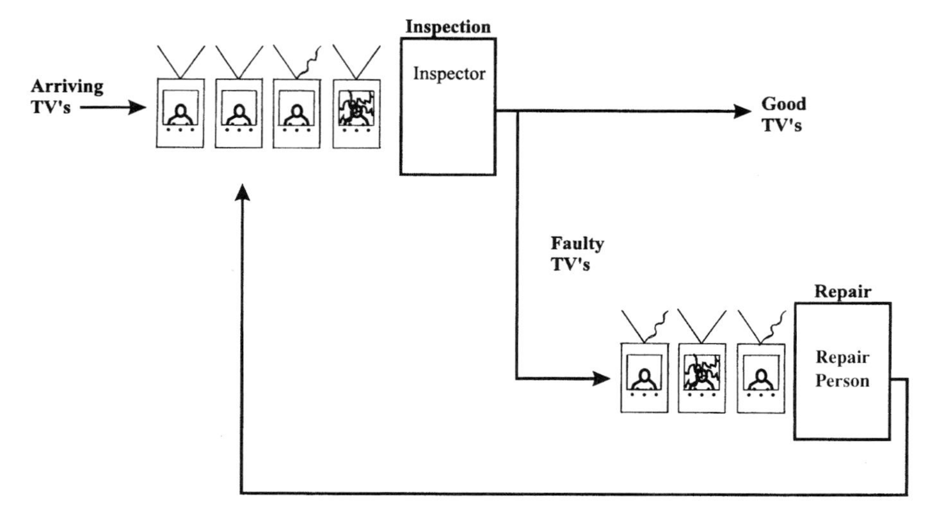

Figure 11.8 Extending the transaction framework.

lation languages. There is almost a one-to-one correspondence between the model component and the problem elements. No more information is specified than necessary. The statements are highly readable and follow a simple format. The predefined object classes grant the user wide flexibility. While the statements in the YANSL model are very similar to those in SIMAN, SLAM, or INSIGHT, they are all legitimate C++ code. Also, this model runs in dramatically less time than a similar SIMAN V model runs on the same machine! But the real advantage of YANSL is its extensibility.

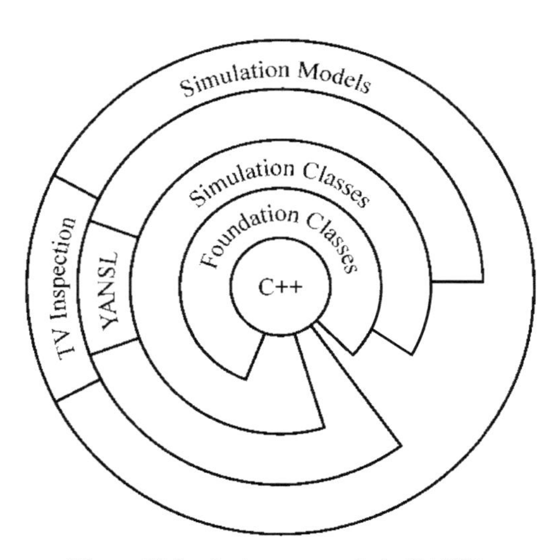

Figure 11.9 Assignment node in YANSL.

TABLE 11.2 YANSL Statement Model

```
#include "simulation.h"

main( ) {
// SIMULATION INFORMATION
Simulation tvSimulation( 1 ); // One replication

// DISTRIBUTIONS
Exponential  interarrival( 5 ),
             inspectTime( 3.5 ),
             repairTime( 8.0 );

// RESOURCES
Resource< PRIORITY > inspector, repairperson;

// NETWORK NODES

  /** Transactions Arrive **/              // Begin at 0.0 and quit at 480.0
Source< Transaction, DETERMINISTIC >  tvSource( "TV Source",
                                           interarrival, 0.0, 480 );

/** Inspection **/
Queue< FIFO > inspectQueue ( "Inspection Queue" );
   inspector.addNextDecision( inspectQueue );
Activity<RequirementSet, PROBABILITY> inspect
                              ("Inspection Station", inspectTime );
   inspect.addRequirement( inspector );
   inspectQueue.addActivity( inspect );

  /** Repair **/
Queue< FIFO > repairQueue( "Repair Queue" );
   repairperson.addNextDecision( repairQueue );
Activity<RequirementSet, DETERMINISTIC> repair( "Repair Station",
                                                      repairTime );
   repair.addRequirement( repairperson );
   repairQueue.addActivity( repair );

  /** Transactions Leave **/
Sink finish( "Leave");

//NETWORK BRANCHES
tvSource.addNexBranch( inspectQueue );
inspect.addNextBranch( finish, .85 );      // 85% are good and leave
inspect.addNextBranch( repairQueue, .15 ); // 15% need repair
repair.addNextBranch( inspectQueue );

//RUN the Simulation
tvSimulation.run( );
}
```

11.4.4 Objects and Their Specification

Lets take a closer look at the YANSL statements in Table 11.2. The model is enclosed in a recognizable C/C++ format, namely having a `#include` statement that includes all the simulation objects, a `main()` function header, and {} which enclose the block of code (YANSL statements). This format is left only to reveal it is C++ code. This format could be eliminated by the C preprocessor commands that would take a `Begin` and `End` and `StartSimulation` for the conventional C tokens. Also, the clever programmer could accept other more intuitive information and convert it to the YANSL (C++) format.

There are two types of YANSL statements. The first is the declaration of objects in the model. These statements describe the elements in the simulation. The second type of statement is member function calls or messages to structure the model. The same division of statements occurs in existing simulation languages. The only order requirement for statements is that an object must be declared before it is used (determined by C++). Thus the statements are ordered by declaring first the general information needed (like the distributions) and then we specify the network entities (resources, nodes, and branches).

Object Declarations. The objects in YANSL are declared in a form consistent with C++. The object class is specified first, then the objects are named. Initialization of specific objects is done in parentheses. For instance,

```
Exponential interarrival (5),
inspectTime (3.5),
repairTime (8.0);
```

creates three exponential distributions whose names are `interarrival`, `inspectTime`, and `repairTime` and whose initialization parameters are given in parentheses. It is important to note that the mean interarrival time is specified as an integer 5, but in fact it is assumed to be a floating point 5.0 (recall discussion in Section 11.2). This illustrates a simple case of overloading. Here, initialization of the interarrival object can take either an integer or a floating-point parameter. In object-oriented terminology, exponential objects are initialized by either an integer or floating-point object.

Some object declarations appear more complex because the object class is also parameterized by information in <>. In object-oriented terminology, these are called *parameterized types*. Parameterized types are created by class *templates* so that the ultimate specification of a class is not known until that class is declared in the model to create the object (both the class and the object are created). Templates make it easy for a user to specify a kind of class rather than having a whole bunch of classes whose similarities are greater than their differences. Thus they provide for another form extensibility. A parameterized type is used when the object class needs some information. Class parameterization should not be confused with initialization of objects where the object needs some information. As an example, consider

```
Queue <FIFO> inspectQueue ("Inspect Queue");
```

where the `Queue` class needs some ranking choice class called `FIFO`, while the object `inspectQueue` is initialized with the string "Inspect Queue," the name of the queue.

Because all queues are similar except for the ranking of the transaction, this eliminates the need to have different queue class for each kind of ranking. Notice that a class will be parameterized with another class, while an object is parameterized with another object.

Because YANSL is really C++, all the built-in classes from C++ are directly available to the YANSL user. These include `integer`, `float`, `char`, and so on. Because an object-oriented language does not distinguish any differently between these C++ classes and the ones we have added, use of all the classes is very similar. In the programming language literature, this property of having user objects treated like built-in objects means everything is treated as a first-class object.

Using the Objects. The other statements in YANSL provide direct use of the objects. These are actual member function calls in C++. In general object-oriented terminology, it is also called *message passing*. For example, the message `addNextDecision()` with `inspectQueue` object as a parameter is sent to the `inspector` object, as follows:

```
inspector.addNextDecision (inspectQueue);
```

In C++ terminology, the `addNextDecision()` member function for the `inspector` object is passed the `inspectQueue` object. The purpose of this message/function is to allow the inspector to service the queue of the inspection activity when this inspector is free to choose what to do next.

Notice the encapsulation of functionality. The resource class obviously has the ability to accept information about what a resource is to do when it is available. All this is contained in the resource class. If you want some different functionality of resource behavior, all the changes would be confined to the code in the resource class.

The YANSL functions are used to specify the behavior of the objects in the simulation. The `addNextDecision()` message specifies what queues the resources serve, the `addNextBranch()` specifies how transactions branch from the departure nodes, the `addActivity()` associates the activity with the queue, and the `addRequirement()` specifies the resource requirements at the activities. Finally, the `tvSimulation.run()` causes the simulation execution to begin. The simulation will continue until no events remain to be processed or some other criterion satisfied.

11.4.5 Running the Simulation

The prior model is compiled under a C++ compiler (a compiler should be AT& T version 3.0–compatible), linked with the YANSL simulation library, and executed. Currently, the YANSL simulation library has been compiled under Borland C++ 4.0 [24]. C++ is strongly typed, so error checking is very good. The current version of the software may be obtained via anonymous FTP from ftp.eos.ncsu.edu/pub/simul.

Also, the simulation is easily linked into other C++ libraries which may be used for graphics and statistical analysis. In a sense, YANSL has the same relationship to C++ that GASP IV [25] has to FORTRAN. The major difference is that whereas GASP was a set of FORTRAN functions (a library) that the model builder called, YANSL is both a set of data and the functions that manipulate these data organized into simulation objects (rather than simulation functions). As such, YANSL is a modeling language more like SLAM but fully compatible with the entire C++ language. Thus YANSL is fully extensible and not limited simply to permit general procedures to be "inserted" into a specific simulation structure.

11.4.6 Embellishments

There is no distinction between the base features of YANSL and its extensions, illustrating the seamless nature of user additions. Many embellishments are simply parallel application of the approaches used in the prior sections. For example, the embellishments shown in the earlier papers [20,22,23] could be applied here. These embellishments can be added for a single use or they can be made a permanent part of YANSL, say YANSL II. In fact, a different kind of simulation language, say for modeling and simulating logistics systems, might be created and called LOG-YANSL for those special users. Perhaps the logistics users would get together and share extensions and create a more general LOG-YANSL II. And so it goes! For those familiar with some existing network simulation language, consider the difficulty of doing the same.

11.5 EXPLORING REUSE AND EXTENSIBILITY WITHIN OOS

The interface to an object-oriented simulation provides for both the use and extension to the existing simulation code. In this section we explore some of the use and the extension of the simulation/modeling framework in expanding YANSL to a wider variety of modeling situations. These embellishments are intended to demonstrate the fundamental contributions of an object-oriented simulation. Initially, we use the existing modeling framework to demonstrate some very powerful simulation features. The later examples extend the modeling concepts and incorporate new features into the language in a way that appears to extend the original language.

11.5.1 Working at More Than One Activity

The resource framework (see Figure 11.5) is capable of representing resources that can service more than one activity. To illustrate this feature, suppose that we add a third worker who can inspect TVs and in addition will repair TVs when there is nothing to inspect. The additions and changes in Table 11.3 are made to the original model in Table 11.2 which adds the worker, specifies the worker's decision process when the worker finishes a job, and specifies the selection among alternative resources at the activity nodes (inspection and repair stations).

A new resource called `inspectRepairperson` is now declared and the `addNextDecision()` function states that this person will serve, in `PRIORITY` order, the `inspectQueue` and then the `repairQueue`. `PRIORITY` is a resource selection choice. Since both the inspection and the repair activities now have a choice of resources, two resource group objects called the `inspectList` and the `repairList` are created which will be used to specify how the activity requirement is chosen from the alternatives. In this case the resource is selected on the basis of `ORDER`, which is a resource decision choice. Finally, at the two activities, the `addRequirement()` function specifies the resource selection object rather than the resource object. This overloading of the `addRequirement()` function argument is another example of polymorphism applied to the user-defined classes. Therefore, a user of YANSL now may

TABLE 11.3 Floating Resources Example

```
//Add the new Resource, specify served queues
Resource< PRIORITY >    inspectRepairperson;
  inspectRepairperson.addNextDecision( inspectQueue );
  inspectRepairperson.addNextDecision( repairQueue );

//Add the Resource Group for activities
ResourceGroup< ORDER >  inspectList;
  inspectList.addRequirementAlternative( inspector );
  inspectList.addRequirementAlternative
    ( inspectRepairperson );

ResourceGroup< ORDER >  repairList;
  repairList.addRequirementAlternative( repairperson );
  repairList.addRequirementAlternative
    ( inspectRepairperson );

//Add at the Inspect Activity
inspection.addRequirement( inspectList );

//Add at the Repair Activity
repair.addRequirement( repairList );
```

specify a requirement using several resource alternatives with the exact same form used to specify a single resource and new decision rules may be easily included.

Smarter Resources. Resources within YANSL actually inherit more intelligent behavior. Suppose that we add a third queue by separating the repaired TVs from the ones newly arriving. Now the `inspectRepairperson` has a choice of three queues to service. Instead of choosing directly among them, suppose that we consider a more complex ("smarter") decision process. Lets assume that inspection is preferred to repair but that the choice among the two repair queues will be based on the length of wait for the TVs who are first in their respective queues. The additional YANSL statements necessary to achieve this result are given in Table 11.4.

First the new queue, `repairedInspectQueue`, is declared and added to the `inspectActivity`. The branch from the `repairActivity` is now directed to this new queue. In addition, we declared a new node from the resource framework called the `DecisionNode`. `DecisionNodes` model the +decision choices of the resources. The first decision node, called the `inspectNode`, is used to choose at the inspection station. The choice is `LONGESTWAIT` and the `repairedInspect Queue` and the `inspectQueue` are the choice objects. The next decision node, called the `serviceNode`, is a decision choice based on `ORDER` with the `repairQueue` and the `inspectNode` decision node as the choice objects. Again the polymorphism of the language permits the choice objects to be either a queue or another decision node. Finally, the `inspectRepairperson` now uses a `DECISIONNODE` as the primary decision choice mechanism and begins the choice with the `serviceNode`.

Working in Teams. A further inherited feature of the resource framework is the ability to model combinations of resource requirements at activities. One of the more interesting uses is the notion of resource teams. A *resource team* is a specific combination of resources that can satisfy a resource requirement. In this case the notion of a require-

TABLE 11.4 Resource Decision Nodes in YANSL

```
//Add the different Queues for repaired TVs at the inspection
Queue<SORT> repairedInspectQueue("Repaired Insp.Q");
  repairedInspectQueue.addActivity( inspect );

//Branch to the Queue from the repair Station replaces previous
repair.addNextBranch( repairedInspectQueue )

//Add the new Decision Nodes, specify served queues
DecisionNode< LONGESTWAIT >    inspectNode;
  inspectNode.addNextDecision(repairedInspectQueue);
  inspectNode.addNextDecision( inspectQueue );

DecisionNode< PRIORITY >    serviceNode;
  serviceNode.addNextDecision( inspectNode );
  serviceNode.addNextDecision( repairQueue );

Resource< DECISIONNODE >  inspectRepairperson;
  inspectRepairperson.addNextDecision( serviceNode );

... Include the Previous Two ResourceGroups in Table 11.3
```

ment is extended from individual resources to a specific set of groupings. With respect to the TV example, suppose that the inspection activity also requires an inspection tool in addition to either the `inspector` or the `inspectRepairperson`. There are two inspect tools, `inspectToolA` and `inspectToolB`. However, while the inspector can work with either of the two inspect tools, the `inspectRepairperson` can only work with `inspectToolB`. Now the choice of resources at the inspect activity must be the `inspector` using either `inspectToolA` or `inspectToolB` or the `inspectRepairperson` using `inspectToolB`. These constitute two *teams*, either of which can satisfy the resource requirement at the inspect activity. Table 11.5 provides the new additional statements.

We return the `inspectRepairperson` to servicing the `inspectQueue` and the `repairQueue`. Next we add the two inspection tools, `inspectToolA` and `inspectToolB`, and specify that they serve the `inspectQueue`. A `toolList` is created as a resource group (a resource group is a feature within the resource framework). The `toolList` consists of the two inspect tools. Now the two teams are declared, the `inspectTeam` and the `inspectRepairTeam`. The `inspector` and the `toolList` is added as requirement alternatives for the `inspectTeam` and the `inspectRepairperson` and the `inspectToolA` is added as the requirement alternatives for the `inspectRepairTeam`. Next the resource groups are formed for the `inspectList` and the `repairList`. Notice that the list for the inspect consists of the two teams, whereas the lists for the repair consists of the two resources.

11.5.2 Deriving a New Type of Transaction

So far we have used the transaction class to represent TVs but now we would like some way to distinguish the TVs that are newly arrived from those that have been inspected to those that have been repaired. In a network simulation language, this distinction would be obtained by assigning attributes to the transaction. The same can be done by extending the transaction framework in YANSL (seen in Figure 11.10 and described in Section 11.4).

TABLE 11.5 Using Resource Teams

```
//Add the new Resource, specify served queues
Resource< PRIORITY >  inspectRepairperson;
  inspectRepairperson.addNextDecision( inspectQueue );
  inspectRepairperson.addNextDecision( repairQueue );

//Add the two inspection tools, specify served queues
Resource< PRIORITY >  inspectToolA, inspectToolB;
  inspectToolA.addNextDecision( inspectQueue );
  inspectToolB.addNextDecision( inspectQueue );

//Add the Resource Group List for selection among the tools
ResourceGroup< ORDER >  toolList;
  toolList.addRequirementAlternative(inspectToolA );
  toolList.addRequirementAlternative(inspectToolB );

//Add the Two Resource Teams
ResourceTeam  inspectTeam, inspectRepairTeam;
  inspectTeam.addRequirementAlternative(inspector);
  inspectTeam.addRequirementAlternative(toolList);
  inspectRepairTeam.addRequirementAlternative
    (inspectRepairperson);
  inspectRepairTeam.addRequirementAlternative(inspectToolA);

//Add the Resource Groups for the various Activities
ResourceGroup< ORDER >  inspectList;
  inspectList.addRequirementAlternative( inspectorTeam );
  inspectList.addRequirementAlternative( inspectRepairTeam );

ResourceGroup< ORDER >  repairList;
  repairList.addRequirementAlternative( repairperson );
  repairList.addRequirementAlternative( inspectRepairperson );

//Add at the Inspect Activity
inspection.addRequirement( inspectList );

//Add at the Repair Activity
repair.addRequirement( repairList );
```

First, we derive a new type of transaction called a "TV" (see Table 11.6) and give the TV two private properties corresponding to the number of repairs and the color of the TV. The public functions set the value of color, increment repairs, and get the value of the private data containing the number of repairs. Although this is a small change, the TV could be given more complex properties, such as some kind of repair order

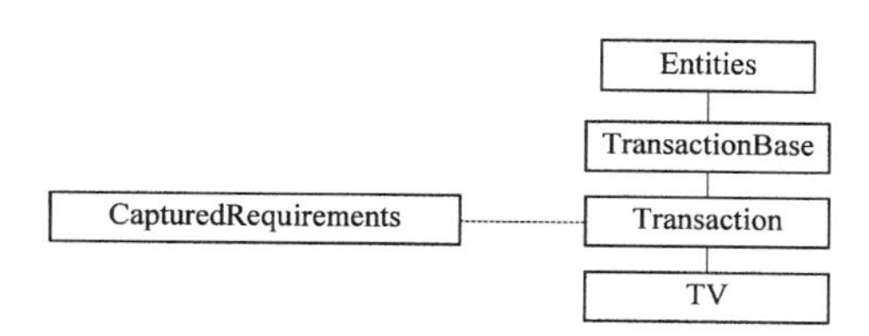

Figure 11.10 YANSL queueing framework.

TABLE 11.6 New Type of Transaction

```
#include "transact.h"

class TV : public Transaction{
private:
  int numRepairs;
  bool color;

public:
  TV( ) { numRepairs = 0; color = TRUE}
  void setcolor( bool cr ) { color = cr; }
  void incrementRepairs( ) } numRepairs++; }
  int getNumRepairs( ) { return numRepairs; }
  int compare( TV * );
};
```

object (a "has-a" relationship). `TV` is a derived class from Transaction (an "is-a" relationship) as seen in the transaction framework in Figure 11.10. The `Transaction` class extends the characteristics of the `TransactionBase` class (described in Section 11.3) with the ability to store and capture resources and has properties for getting and setting these captured resources. The captured resources are stored inside the CapturedRequirements class, which has properties that include (1) getting and setting these resources, (2) changing the state of these resources, and (3) telling any released resources to look for work or make a service decision. Because a TV is a kind of transaction, all the things transactions can do, TVs can do. Thus there is no need to write any special code or do anything special for TVs as they inherit all the functionality of the transactions.

Add an Assignment Node. Now that there are attributes associated with TVs designating their repairs and color, there needs to be some kind of assignment node that can cause the attribute to be changed. Thus we need to add a new node to YANSL. Node classes are formed in a class hierarchy, as shown in Figures 11.4 and 11.6. This hierarchy starts with a broad division of nodes and defines more specific nodes higher in the hierarchy. Nodes higher in the hierarchy inherit the properties of the nodes above them. A portion of that hierarchy is given in Figure 11.11.

In the hiearchy, nodes are broadly defined as departure and destination nodes. Departure nodes may have branches connected to them and therefore need a branching choice (BC). Recall that sink, queue, and activity nodes can have transactions branched to them and are therefore destination nodes. An *assign* node is both a departure and a destination

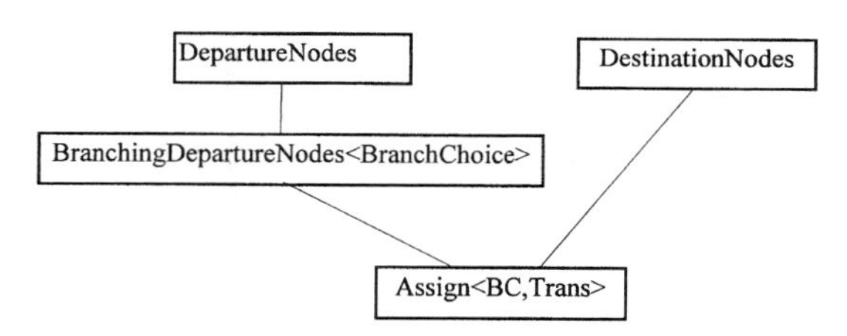

Figure 11.11 Extension of the activity framework in YANSL.

TABLE 11.7 Assignment Node in YANSL

```
#include "node.h"

template< class BC, class TransactionType>
class Assign : virtual public DestinationNode,
               virtual public BranchingDepartureNode< BC > {
protected:
  void (TransactionType :: *ptrFun) ( );
public:
  Assign(string str, void (TransactionType :: *pFun) ( ),
                                        bool statis=FALSE);
  virtual bool executeEntering( Transaction* tptr ){
    (((TransactionType*) (tptr))->*ptrFun) ( );
    executeLeaving(tptr);
    return true;
  }
  virtual void executeLeaving( Transaction* tptr ){
    branch.nextNode( )->executeEntering( tptr );
  }
};
```

node, so it inherits from both the departure and branching destination node classes. This inheritance from multiple parents is another example of multiple inheritance. Note that the delay node is also derived through multiple inheritance, since it is also a destination node. Not all object-oriented languages permit multiple inheritance like C++. Portions of the new assignment node class are shown in Table 11.7. The assignment node is also parameterized with the type of transaction.

Multiple inheritance is specified in the header of the class definition. The `executeEntering()` and `executeLeaving()` are virtual functions in departure and destination classes that act as placeholders, permitting the assignment node special functionality as transactions enter and leave (remember that TVs are simply a kind of transaction and thus they can use the assignment node). The following statement declares a TV assignment node (`tvAssign`) that will increment the number of repairs.

```
Assign<DETERMINISTIC, TV> tvAssign("TV Assign Node",
                                   TV::incrementRepairs);
```

In this case the assign node simply increments the number of repairs. Other assignments could be handled similarly.

Add a New Queue Ranking Choice. Now that TV objects remember their repairs, let us show how to extend the queueing framework (seen in Figure 11.12) to rank TVs according to the number of times that they have been repaired. Since transactions branch to queue nodes, the `QueueNodeBase` is a "kind of" `DestinationNode`. The `QueueNodeBase` class provides destination nodes with the ability to be associated with an activity and has properties for getting and setting the activity node that services the queue. YANSL queues can be associated with only one activity, while activities can be served by many different queues. The `QueueNode` provides the basic means for storing and ranking transactions awaiting resource requirements. Recall from the original

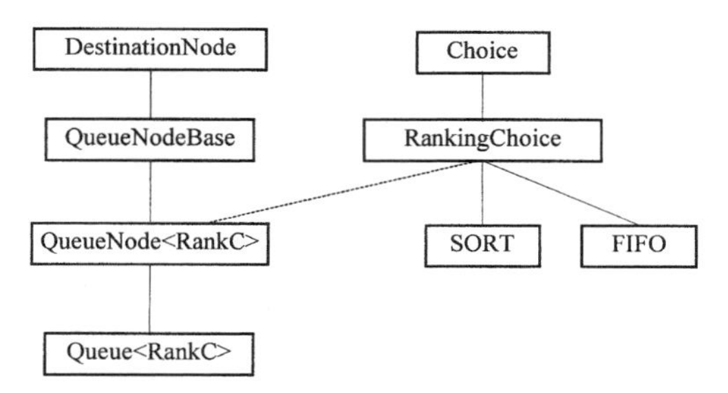

Figure 11.12 YANSL queueing framework.

model that queue nodes are parameterized by a `RankingChoice` that specifies how transactions are ranked at the queue.

So far, all that has been specified is the FIFO class. Because the queue class is parameterized, we can easily add a new ranking choice. Ranking choices are encapsulated as classes so that they can easily be modified. The new ranking choice class (`SORT`) seen in Table 11.8 uses the comparison function of the transaction class (`compare`) as seen in Table 11.6 for the `TV` class). The virtual functions in this class must be completed to perform the sort. The `rankInQueue()` member function would use the transaction's comparison function to determine its rank in the queue. Now the queue at the inspection activity would be specified by

```
Queue<SORT> inspectQueue("Inspection Queue");
```

Parameterized types create templates for classes so that the ultimate specification of a class is not known until that class is declared to create the object. *Templates* make it easy for a user to specify a kind of class rather than having a whole bunch of classes whose similarities are greater than their differences. Some network simulation languages approach this issue by having more general node types, like an "operation" node, but these general types cannot, in general, yield specific objects—only their subtypes create objects (in C++, such a class would be a pure virtual class).

Make Activity Time Depend on Transaction. Another interesting change in the basic model is to make the inspection time depend on whether or not it has been repaired. In Table 11.9 we add a new type of TV inspection activity to the activity framework shown in Figure 11.13. The activity framework provides for the basic means of delaying transactions and the ability to seize and release resources. The `DelayNode`, both a

TABLE 11.8 New Queue Ranking Choice

```
class SORT : virtual public RankingChoice{
public:
  virtual void addtoQueue( Transaction *tptr );
  virtual Transaction* removeFromQueue( );
  virtual int rankInQueue( Transaction *tptr );
};
```

TABLE 11.9 Inspection Time Based on the Number of Repairs

```
template< class REQ, class BC >
class InspectionActivity : public virtual Activity<REQ, BC >{
  public:
    InspectionActivity( string str, Random*, Random *);
    virtual BOOL executeEntering( Transaction * );

  protected:
    Random  *repairVariate;
};

template< class REQ, class BC >
InspectionActivity<REQ, BC>::InspectionActivity(Random
          *actTime, Random *repairTime) : Activity< BC >
            (actTime) {
  repairVariate = repairTime;
}

template< class REQ, class BC >
bool InspectionActivity<REQ, BC>::
          executeEntering(Transaction *tptr){
  //... same as activity class

  /* If the TV has been repaired Inspection
    time is different */

  double eventTime = ( ! ( (TV*)tptr)->getNumRepairs( ) ) ?
          currentTime+actVariate->sample( ) :
          currentTime+repairVariate->sample( );

  scheduleEvent(new NodeEvent(tptr,eventTime, this,
    NODEEVENT::ENDOFSERVICE));
  return TRUE;
}
```

departure and destination node, causes transactions to be delayed for a specified amount of time and has properties that include getting and setting a random service or delay-time variate. The `ActivityNodeBase` class extends the properties of the `DelayNode` by allowing transactions to capture and release resources. This procedure is provided by specifying two process classes, `SeizeProcess` and `ReleaseProcess`. As stated in Section 11.3, `Processes` provide an elegant way to define various simulation procedures. These process classes provide the activity with the basic ability to allow transactions to seize and capture resource requirements and a means for these resources to be released upon completion of the activity. The `RequirementSet` class, a helper class, contains the requirement alternatives specified for an activity and includes the properties for (1) setting and getting the various requirements, and (2) determining which requirements are available and necessary. The `Activity` class is extended in Table 11.9 to allow for the determination of the service times depending on whether or not the TV set has been repaired. This new kind of activity, called the `InspectionActivity`, uses all the properties of the activity but provides a different activity time if the TV has been repaired. Notice that only a placeholder for the new time variate is needed, along with a definition of the `executeEntering()` function which determines the

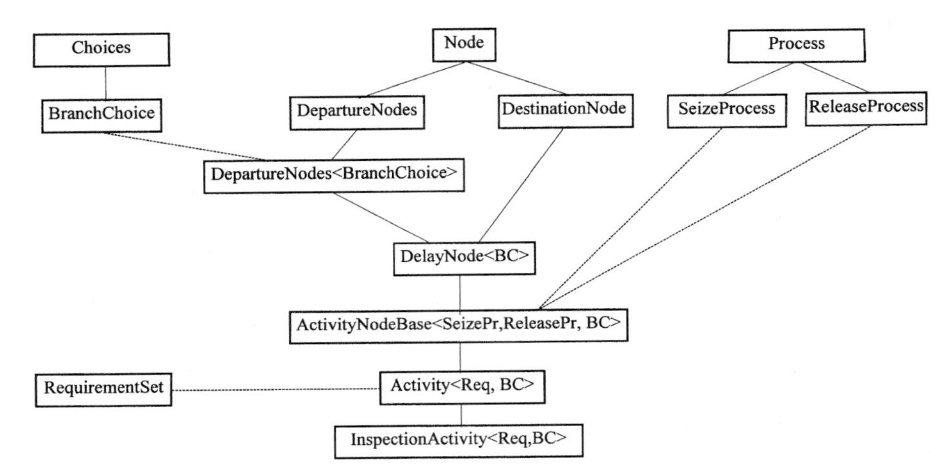

Figure 11.13 Extension of the activity framework in YANSL.

time of the event and schedules this end of service `NodeEvent` to happen. The only changes to the original model (see Table 11.2) is the `inspect` activity is now declared as an `InspectionActivity` type. Also, a new random variate associated with the inspection time if a TV has been repaired is created and passed to the `inspect` object.

```
Exponential inspectRepairTime(2.5);
InspectionActivity<RequirementSet,PROBABILITY>
inspect ("Inspection Station", inspectTime, inspectRepairTime);
```

Now the new `InspectionActivity` is used just like the original activity within the YANSL network.

11.5.3 Grouping Transactions Together

Suppose that good TVs are accumulated on a conveyor in front of a palletizer where eight are gathered into a group and palletized. Now a single pallet leaves the palletizing activity rather than eight TVs. This problem requires that we specifically accumulate and then combine eight TVs into a single object. Also, the activity should not process eight objects but only one, and only one object should leave the palletizer. A new kind of node for grouping and managing transactions may be defined, as in Table 11.10. C++ provides a simple means to create and destroy objects through its `new` and `delete` memory management operators. These operators can be overloaded to apply to specific objects, if needed.

11.6 CONCLUSIONS

Modeling and simulation in an object-oriented language possess many advantages. We have shown how internal functionality of a language now becomes available to a user

TABLE 11.10 Grouping Node in YANSL

```
template< class TransactionGroupType, class BC >
class Group : virtual public DestinationNode,
        virtual public BranchingDepartureNode< BC > {
  private:
    int target;
    int current;

  public:
    Group( int max ) { current = 0; target = max; }
    virtual bool executeEntering( Transaction* tptr ) {
      if( ++current == target ) {
        TransactionGroupType* tnew = new TransactionGroupType;
        current = 0;
        executeLeaving(tnew);
      }
      delete tptr;
      return TRUE;
    }
    virtual void executeLeaving( Transaction* tptr ) {
      branch.nextNode( )->executeEntering( tptr );
    }
};
```

(at the discretion of the class designer). Such access means that existing behavior can be altered and new objects with new behavior introduced. Furthermore, users may benefit directly from the work of others through inheritance and polymorphism. The object-oriented approach provides a consistent means of handling these problems (other general object-oriented languages in addition to C++ include Smalltalk [5] and Eiffel [6].

The user of a simulation in C++ is granted lots of speed in compilation and execution. The C language has been the language of choice by many computer users and now C++ is supplanting it. With the C++ standard [26] adopted, all C++ compilers are expected to accept the same C++ language. We can build an executable simulation on one machine and run it on another only as long as the operating systems are compatible—you don't need a C++ compiler on both machines. Most commercial simulation languages require some proprietary executive.

Because C++ has many vendors, the price of compilers is low while the program development environments are excellent. For example, the Borland package includes a optimizing compiler, a fully interactive debugger, an object browser, a profiler, and an integrated environment that allows you to navigate between a code editor and the other facilities. Also numerous class libraries for windowing, graphics, and so on, are appearing that are fully compatible with C++. Graphical user interfaces for simulation modeling, animation of simulation, and statistical analysis of simulation results could be offered by individual vendors. Their interoperability would be ensured by their use of a common means for defining and using objects.

The object-oriented framework offers great potential because of its extensibility. To take full advantage of object-oriented simulation will require more skill from the user. However, that same skill would be required of any powerful simulation modeling package, but with greater limitations.

REFERENCES

1. Lippman, S. B. (1991). *C++ Primer*, 2nd ed., Addison-Wesley, Reading, Mass.
2. Hammann, J. E., and N. A. Markovitch (1995). Introduction to Arena, in *Proceedings of the 1995 Winter Simulation Conference*, C. Alexopoulos, K. Kang, W. R. Lilegdon, and D. Goldsman, eds., IEEE, Piscataway, N.J.
3. Henriksen, J. O. (1995). An introduction to SLX, in *Proceedings of the 1995 Winter Simulation Conference*, C. Alexopoulos, K. Kang, W. R. Lilegdon, and D. Goldsman, eds., IEEE, Piscataway, N.J.
4. Birtwistle, G. M., et al. (1973). *SIMULA Begin*, Petrocelli/Charter, New York.
5. Goldberg, A., and D. Robson (1989). *Smalltalk-80: The Language*, Addison-Wesley, Reading, Mass.
6. Meyer, B. (1992). *Eiffel: The Language*, Prentice Hall, Upper Saddle River, N.J.
7. Fleishman, E. A., and W. E. Hemple (1994). Design of object oriented simulations in SmallTalk, *Simulation*, Vol. 49, pp. 239–252.
8. Fishwick, P. A. (1995). *Simulation Model Design and Execution*, Prentice Hall, Upper Saddle River, N.J.
9. Little, M. C., and D. L. McCue (1994). Construction and use of a simulation package in C++, *C User's Journal*, Vol. 12, No. 3.
10. Belanger, R., and A. Mullarney (1990). *Modsim II Tutorial*, CACI Products Company, La Jolla, Calif.
11. Lomow, G., and D. Baezner (1991). A tutorial introduction to object-oriented simulation and Sim++, in *Proceedings of the 1991 Winter Simulation Conference*, B. L. Nelson, W. D. Kelton, and G. M. Clark, eds. IEEE, Piscataway, N.J.
12. Schwetman, H. (1995). Object-oriented simulation modeling with C++/CSIM17, in *Proceedings of the 1995 Winter Simulation Conference*, C. Alexopoulos, K. Kang, W. R. Lilegdon, and D. Goldsman, eds., IEEE, Piscataway, N.J.
13. Geuder, D. (1995). Object-oriented modeling with Simple++, in *Proceedings of the 1995 Winter Simulation Conference*, C. Alexopoulos, K. Kang, W. R. Lilegdon, and D. Goldsman, eds., IEEE, Piscataway, N.J.
14. Carroll, M. D., and M. A. Ellis (1995). *Designing and Coding Reusable C++*, Addison-Wesley, Reading, Mass.
15. Plauger, P. (1995). *The Draft C++ Library*, Prentice Hall, Upper Saddle River, N.J.
16. Schriber, T. J. (1991). *An Introduction to Simulation Using GPSS/H*, Wiley, New York.
17. Pritsker, A. A. B. (1995). *Introduction to Simulation and SLAM II*, 4th ed., Halsted Press, New York.
18. Pegden, C. D., R. E. Shannon, and R. P. Sadowski (1995). *Introduction to Simulation Using Siman*, 2nd ed., McGraw-Hill, New York.
19. Roberts, S. D. (1983). *Modeling and Simulation with INSIGHT.* Regenstrief Institute, Indianapolis, Ind.
20. Joines, J. A., K. A. Powell, and S. D. Roberts (1992). Object-Oriented Modeling and Simulation with C++, in *Proceedings of the 1992 Winter Simulation Conference*, J. J. Swain, D. Goldsman, R. C. Crain, and J. R. Wilson, eds., IEEE, Piscataway, N.J.
21. Joines, J. A., and S. D. Roberts (1994). Design of object-oriented simulations in C++, in *Proceedings of the 1994 Winter Simulation Conference*, J. D. Tew, S. Manivannan, D. A. Sadowski, and A. F. Seila, eds., IEEE, Piscataway, N.J.
22. Joines, J. A., and S. D. Roberts (1995). Design of object-oriented simulations in C++, in *Proceedings of the 1995 Winter Simulation Conference*, C. Alexopoulos, K. Kang, W. R. Lilegdon, and D. Goldsman, eds., IEEE, Piscataway, N.J.

23. Joines, J. A., K. A. Powell, and S. D. Roberts (1993). Building object-oriented simulations with C++, in *Proceedings of the 1993 Winter Simulation Conference*, G. W. Evans, M. Mollaghasemi, E. C. Russell, and W. E. Biles, eds., IEEE, Piscataway, N.J.
24. Borland (1993). *Borland C++ Version 4.0*, Borland International, Scotts Valley, Calif.
25. Pritsker, A. A. B. (1974). *The GASP IV Simulation Language*, Wiley, New York.
26. Koenig, A. (1995). *Working Paper for the Draft Proposed International Standard for Information Systems—Programming Language C++*, NIST, Springfield, Va.

CHAPTER 12

Parallel and Distributed Simulation

RICHARD M. FUJIMOTO
Georgia Institute of Technology

12.1 INTRODUCTION

Any simulation tool will be of limited value if each run requires many days or weeks to complete. Unfortunately, many simulations of complex systems require this much time, greatly restricting the number and scale of experiments that can be performed. For example, simulations of modest-sized telecommunication networks may require processing 10^{11} or more simulated events, typically requiring a day or more of CPU time with existing simulation tools executing on a high-performance workstation. Simulations of large networks simply cannot be performed because the computation time is prohibitive.

The high-performance discrete-event simulation community has been developing technologies to execute large simulation programs in as little time as possible, thereby improving the productivity of engineers and scientists using these tools. Specifically, a considerable amount of attention has been focused on utilizing multiple processors, each working on a different portion of the simulation computation, to reduce model execution time. Researchers in the parallel and distributed simulation (PADS)* community have demonstrated orders-of-magnitude reductions in execution time for large-scale simulation problems using parallel computers.

Simultaneously, the needs of the military establishment to have more effective and economical means to train personnel has driven a large body of work in developing virtual environments where geographically distributed participants can interact with each other as if they were in actual combat situations. For example, tank commanders in Texas might be fighting helicopter pilots in California in simulated military exercises, with each operating in a virtual environment providing realistic imagery of a battlefield in Iraq. Work in the distributed interactive simulation (DIS) community has expanded to encompass other uses in the military [e.g., testing and evaluation (T&E) of new weapons systems to identify potential problems before the systems are deployed]. Application of distributed simulation to commercial applications, such as entertainment, air traffic con-

*We adopt the acronym PADS, borrowing from the name of the annual workshop that publishes much of the research results produced in this field.

Handbook of Simulation, Edited by Jerry Banks.
ISBN 0-471-13403-1

troller training, and emergency planning exercises to prepare for earthquakes or other disasters are also increasing. In contrast to PADS research, the principal goal of DIS work has historically been to develop cost-effective, realistic virtual environments utilized by geographically distributed personnel to prepare them for situations they could encounter later on the battlefield.

Although their motivations and techniques are different, both PADS and DIS research share one common theme: execution of simulation programs on multiple CPUs interconnected through a network. This chapter is concerned with technologies to enable a single discrete-event simulation, possibly composed of many autonomous simulation programs, to be executed on platforms containing, potentially, thousands of computers. Distributed execution of discrete-event simulation programs has spawned a considerable amount of interest over the last two decades, with interest in this technology rapidly growing in recent years.

The remainder of the chapter is organized as follows. First, the hardware platforms commonly used for parallel and distributed simulation are surveyed and contrasted, emphasizing the distinctions that have the greatest impact on parallel/distributed execution. Second, techniques used to achieve high performance through parallel/distributed execution are described, with particular emphasis on synchronization, which is at the core of much of the work in this area. The two principal approaches, termed *conservative* and *optimistic* synchronization, are described. Next, an approach to achieving high performance using a completely different technique called *time parallel execution* is described and illustrated through examples in simulating caches and queues. In the last section we describe work in the DIS community and basic underlying principles used in these distributed simulation systems. The objectives and approaches used in this work are contrasted with that used in PADS research.

12.2 HARDWARE PLATFORMS

The most prevalent multiple-processor hardware platforms for executing parallel or distributed simulations today are:

1. *Shared Memory Multiprocessors.* The distinguishing characteristic of these machines is that the computing platform supports memory (program variables) that may be shared among software executing on different processors. Thus one can define a variable X that can be read or modified autonomously by one processor without the intervention of another. In nonuniform memory access (NUMA) multiprocessors the machine differentiates between local and nonlocal memory and provides faster access to variables stored in local memory. This is typically realized by attaching memory components to each processor and providing the ability of one processor to read or write into the memory attached to another processor. Another approach is to provide separate hardware modules to hold the memory that are *not* associated with any particular processor and to provide local "cache memory" with each processor to hold frequently accessed instructions and data. This provides a simpler programming model than the NUMA approach because programmers need not distinguish between local and nonlocal memory in writing their programs. The shared memory approach contrasts with message-passing machines, where processors communicate only by exchanging messages. Message passing can be implemented using shared memory by defining shared data structures (queues) to hold incoming and outgoing messages for each processor.

2. *Distributed Memory Multicomputers.* These machines do not support shared variables, but rather, all communications between processors must occur via message passing. Unlike shared memory machines, a processor cannot directly access memory that belongs to another processor. In principle, shared memory operations could be implemented in software on these machines (e.g., by sending and receiving "read" and "write" messages); however, the overheads associated with these operations may be prohibitive. Like shared memory multiprocessors, all processors in distributed memory machines are in close physical proximity, typically within a single cabinet and almost always within a single room.

3. *SIMD Machines.* SIMD stands for *single instruction stream, multiple data stream.* The central characteristic of these machines is that all processors must execute the same instruction (but using different data) at any instant in the program's execution. Typically, these machines execute in "lock step," synchronous to a global clock; that is, *all* processors must complete execution of the current instruction (some may choose not to execute that instruction) before any is allowed to proceed to the next instruction. Lock-step execution and the constraint that all processors must execute the same instruction distinguishes these machines from the other types of machines that are described here.

4. *LAN-Based Distributed Computers.* LAN stands for *local area network.* These machines consist of a collection of workstations or personal computers in a limited geographic area (e.g., within a single building or a university campus) interconnected through a high-speed network.

5. *Geographically Distributed Computers.* These machines are similar to LAN-based machines except that the machines are distributed over much larger geographic distances (e.g., a metropolitan area, across a nation, or even worldwide). The interconnection network is referred to as a MAN (metropolitan area network) if the extent is a city, or a WAN (wide area network) if the extent is national or global.

Important characteristics distinguishing these platforms are the latency associated with communications between processors and the degree of heterogeneity among the processors in the system. Shared memory machines, multicomputers, and SIMD machines have low communication latency (typically, a few tens of microseconds or less to transmit a message from one processor to another) and utilize processors from the same manufacturer. LAN-based machines typically have communication latencies on the order of a millisecond, although this is gradually being reduced to approach multiprocessor performance with new switching techniques that bypass traditional communication protocols. Geographically distributed machines typically have latencies of tens or hundreds of milliseconds or more; the speed of light (approximately 5 ms per 1000 km in optical fiber, or about the time to access a mass storage device such as a moving head disk in today's technology) places a fundamental lower bound on communication latency in these machines. LAN- and WAN-based machines often contain computers from different manufacturers. Communication latency is important because it has a large impact on performance; if latencies are large, the computers may spend much of their time waiting for messages to be delivered. A modern personal computer can execute tens to hundreds of thousands of machine instructions (e.g., simple arithmetic operations) in 1 ms. Heterogeneity is important if one is interconnecting federations of existing simulations that execute on machines from different manufacturers (e.g., DIS).

The distinction between shared memory multiprocessors and multicomputers is important because the common address space provided by shared memory machines allows global data structures referenced by more than one processor to be used; these are not so easily implemented in distributed memory multicomputers. Also, different memory management techniques may be used in shared memory machines. For example, it is possible to define memory management protocols in shared memory computers that use, to within a constant factor, the same amount of memory as a sequential execution of the program; however, such techniques have not yet been developed for distributed memory machines.

Simulations that execute on shared memory multiprocessors, multicomputers, or SIMD machines are typically referred to as *parallel discrete-event simulation* (PDES) (or simply parallel simulation) programs. Simulations executing on LAN-, MAN-, or WAN-based distributed computers (which may include one or more multiprocessors, multicomputers, and/or SIMD machines) are usually referred to as *distributed simulations*.

The focus here is on the execution of a *single* simulation model on multiple computers. An alternative approach using parallel and distributed computers is when many independent runs of a simulation program are required, where each run uses (for example) different parameter settings or random number generator seeds. The latter approach is often referred to as the *replicated trials* method. Replicate trials is a simple, effective approach that can be used when the entire simulation program fits into the memory of a single machine and the execution time of each run is not an issue.

12.3 BASIC CONCEPTS AND AN EXAMPLE

In the next sections we focus on techniques developed in the PADS community to speed up the execution of discrete-event simulation programs. To date, most of this work has focused on parallel computers (multiprocessors, multicomputers, and SIMD machines) and LAN-based platforms. Basic concepts and terminology used in discrete-event simulation are first reviewed and then extended for parallel/distributed execution.

In a discrete-event simulation program the system is modeled as if the state of the system changes only at discrete points in simulated time. The simulation model "jumps" from one state to another upon the occurrence of an *event*. For example, a simulation of a store-and-forward communication network might include state variables to indicate the length of message queues, the status of communication links (busy or idle), and so on. Typical events might include arrival of a message at some node in the network, forwarding a message to another network node, component failures, and so on.

Sequential simulation programs typically utilize three data structures: (1) the *state variables* that describe the state of the system, (2) an *event list* containing all pending events that have been scheduled but have not yet taken effect, and (3) a global *clock* variable to denote how far the simulation has progressed. Each event contains a time stamp and usually denotes some change in the state of the system being simulated. The time stamp indicates when this change occurs in the actual system. The main loop of the simulation program repeatedly removes the smallest time-stamped event from the list and processes that event. Processing an event involves executing software (an event procedure) to effect the appropriate change in state and scheduling zero or more new events into the simulated future to model causal relationships in the system under investigation. Modern simulation languages often contain higher-level constructs such as processes where a set of events defining the behavior of a particular entity is represented as a sequence of actions in the simulation code (e.g., the aircraft took off, then flew to its next destination, then waited to land, etc.) rather than as a set of separate event procedures. Process abstractions are usually built on top of the event list mechanism described above, however, so they are not discussed further here.

For example, consider a simulation of air traffic in the United States. Only three airports—SFO in San Francisco, ORD in Chicago, and JFK in New York—will be considered. A (simplified) model for the operation of these airports might include two types of events: (1) an arrival event denoting an aircraft arriving at an airport and (2) a departure event denoting an aircraft leaving to travel to another airport. Upon arrival, each aircraft must (1) wait for a runway and land the aircraft (assume that the aircraft uses the runway for R units of time), (2) travel to the gate and unload and load new passengers (assume that this requires G units of time), and (3) depart and travel to another airport (assume that this requires F units of time). This scenario is repeated with the aircraft moving successively from one airport to another. The time at which the aircraft lands depends on the number that are waiting to use the runway. Assume that aircraft land in the order in which they arrive at the airport and R, G, and F are fixed, known quantities. Queueing at the runway for *departing* aircraft will not be considered here.

The simulation program for this system utilizes two different types of events: an *arrival event* denotes the arrival of a new aircraft at an airport, and a *departure event* denotes the departure of an aircraft destined for another airport. The simulation program modeling a single airport (SFO) in this system is shown in Figure 12.1*a*. A procedure is defined for each type of event. This procedure is executed each time an event of this type is removed from the pending event list for processing. Here *Now* indicates the time stamp of the event being processed. A variable called *Runway_Free* indicates when the runway will become available for the next aircraft to land and is used to determine when the aircraft may begin to land. If *Runway_Free* is smaller than *Now*, the aircraft may begin to land immediately. After landing, the aircraft moves to the gate and schedules a departure event (G time units after landing) at the airport where it landed. The second procedure for the departure event then schedules a new arrival event F time units into the future at the next airport that the aircraft visits. An actual air transportation model would be more complex, as it would include queueing departing aircraft and other effects, such as weather affecting the mean time between aircraft; however, this simple model will suffice here.

Now consider a *distributed* simulation program to model this system. The system being simulated, referred to as the *physical system*, is viewed as a collection of *physical processes* (here, airports) that interact by "sending" aircraft between each other. An actual air traffic system would also include other interactions, such as radio transmissions between aircraft and airports, but these interactions will be ignored here to simplify the discussion. In the distributed simulation program, the physical system is represented by a collection of *logical processes* (or LPs), each modeling a single physical process. Thus in the air traffic control example, each airport is represented by a single logical process (see Figure 12.1*b*). Interactions in the physical system are represented by time-stamped messages (or events; we use these two terms synonymously here), exchanged between the logical processes.

For example, an event with time stamp 9.0 might be sent from the JFK to the SFO process to represent an aircraft flying from New York to San Francisco, arriving at SFO at 9:00 A.M. Upon receiving this event, the SFO process would then schedule a departure event for itself, as described earlier in the sequential simulation. When this departure event is processed, it sends a new arrival event to (say) the ORD LP to model a flight from SFO to ORD. An extension of this approach is to model each aircraft as

```
/*
 * Now = current simulation time
 * R = time runway in use to land aircraft (constant)
 * G = time required at gate (constant)
 * F = time to fly to next airport (constant)
 *
 * Runway_Free = time runway becomes free (state variable, initialized to 0)
 */

Arrival Event at SFO:
/* compute time aircraft done using runway */
Runway_Free = max(now,Runway_Free) + R;
Schedule Departure Event at SFO at time Runway_Free + G;

Departure Event at SFO:
Schedule Arrival Event at next destination at time now + F;
```

(a)

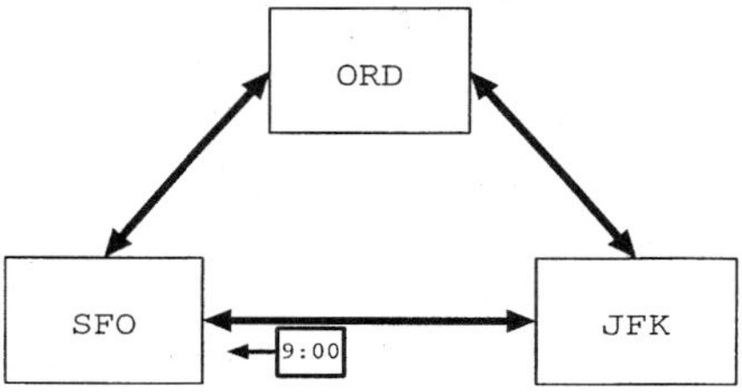

ORD

SFO

JFK

9:00

(b)

Figure 12.1 Air traffic simulation example: (*a*) simulation program for a single airport (SFO); (*b*) networked distributed simulator.

a logical process; however, we will not pursue this approach here. In this example the distributed simulation program is identical to that shown in Figure 12.1*a* except that "scheduling an event" is replaced by sending a message to the relevant logical process.

To summarize, a parallel simulation program may be viewed as a collection of sequential simulation programs (logical processes) that communicate by scheduling time-stamped events (sending time-stamped messages) to each other. It will be seen momentarily that it is important that *all* interactions must occur via this message-passing mechanism.

12.4 SYNCHRONIZATION

Recall that in the sequential simulation it is crucial that one always select the smallest time-stamped event (say E_{10}) from the event list as the one to be processed next. This is because if one were to select some other event containing a larger (later) time stamp, say E_{20}, it would be possible for E_{20} to modify state variables used by E_{10}. This would amount to simulating a system where the future could affect the past! This is clearly unacceptable; errors of this nature are called *causality errors*. Although it is easy to avoid causality errors in sequential simulation programs by using a centralized list of unprocessed events, this is much more problematic in distributed simulations, as will be discussed momentarily. This gives rise to the *synchronization problem*.

Consider parallelization of a simulation program that is based on the foregoing paradigm using logical processes. The greatest opportunity for parallelism arises from processing events concurrently on different processors. However, a direct mapping of this paradigm onto (say) a shared memory multiprocessor quickly runs into difficulty. Consider the concurrent execution of two arrival events, E_{10} at the LP for ORD and E_{20} at SFO, with time stamps 10 and 20, respectively. If E_{10} writes into a state variable that is read by E_{20}, then E_{10} must be executed before E_{20} to ensure that no causality error occurs.* In other words, for the computation to be correct, certain *sequencing constraints* must be maintained.

To avoid scenarios such as this, the restriction is made that there *cannot be any state variables that are shared between logical processes* (exceptions that do allow shared states are described in refs. 1 to 3). The state variables of the entire simulation program must be partitioned into state vectors, with one state vector per LP. Each logical process contains a portion of the state corresponding to the physical process it models, as well as a local clock that denotes how far the process has progressed.

One can ensure that no causality errors occur if one adheres to the following constraint:

Local Causality Constraint A discrete event simulation, consisting of logical processes (LPs) that interact by exchanging time-stamped messages, obeys the local causality constraint if and only if each LP processes events in nondecreasing time-stamp order.

Assuming that LPs interact exclusively by exchanging messages, this constraint is sufficient to guarantee that no causality errors occur. It is not a necessary constraint, however, because two events within a single LP may be independent of each other, in which case processing them out of time-stamp sequence does not lead to causality errors.

Although the exclusion of shared state in the logical process paradigm helps prevent many types of causality errors, by itself it does not guarantee adherance to the local causality constaint. Again consider two events, E_{10} at logical process LP_{ORD} with time stamp 10 and E_{20} at LP_{SFO} with time stamp 20 (see Figure 12.2). If E_{10} schedules a new event E_{15} for LP_{SFO} which contains the time stamp 15, E_{15} could affect E_{20}, necessitating sequential execution of all three events. If one had no information regarding what events could be scheduled by what other events, one would be forced to conclude that the only event that is "safe" to process is the one containing the smallest time stamp, limiting concurrent execution to events containing the same time stamp.

Consider this situation from the perspective of the physical system. There, the "cause" must always precede the "effect." These cause-and-effect relationships in the physical system become sequencing constraints in the simulation.† It is the simulation

*To simplify the discussion, ignore concurrent execution of *portions* of E_{10} and E_{20} that still satisfy this sequencing constraint.

†The simulation may actually have more constraints that arise as an artifact of the way it was programmed (e.g., constraints arising from updating statistics variables).

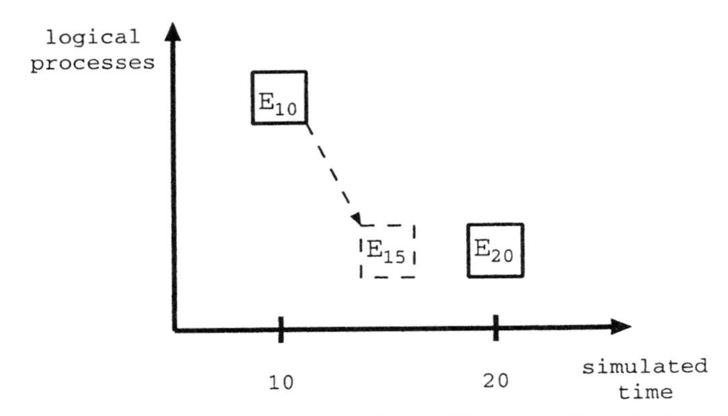

Figure 12.2 Event E_{10} affects E_{20} by scheduling a third event, E_{15}, which modifies a state variable used by E_{20}. This necessitates sequential execution of all three events.

mechanism's responsibility to ensure that these sequencing constraints are not violated when the simulation program is executed on the parallel computer.

Operationally, one must decide whether or not E_{10} can be executed concurrently with E_{20}. But how does the simulation determine whether or not E_{10} affects E_{20} without actually performing the simulation for E_{10}? This is the fundamental dilemma that must be addressed. The scenario in which E_{10} affects E_{20} can be a complex sequence of events and is critically dependent on event time stamps.

Thus a synhronization mechanism is required to ensure that event computations are performed in time-stamp order. This is nontrivial because the sequencing constraints that dictate which computations must be executed before which others are, in general, quite complex and highly data dependent. This contrasts sharply with other areas where parallel computation has had a great deal of success (e.g., vector operations on large matrices of data). There, much is known about the structure of the computation at compile time.

Synchronization mechanisms broadly fall into two categories: *conservative* and *optimistic*.* A more detailed taxonomy of simulation mechanisms is described in ref. 4. Conservative approaches strictly *avoid* the possibility of any causality error ever occurring. These approaches rely on some strategy to determine when it is safe to process an event (i.e., they must determine when all events that could affect the event in question have been processed). On the other hand, optimistic approaches use a *detection and recovery* approach: Causality errors are detected, and a *rollback* mechanism is invoked to recover. The following sections describe details and the underlying concepts behind several conservative and optimistic simulation mechanisms that have been proposed.

Assume that the simulation consists of N logical processes, $LP_0, \ldots, LP_{N-1}$. $Clock_i$ refers to the simulated time up to which LP_i has progressed: when an event is processed, the process's clock is advanced automatically to the time stamp of that event. If LP_i may send a message to LP_j during the simulation, a *link* is said to exist from LP_i to LP_j.

*It may be noted that this taxonomy applies more broadly to other areas of parallel and distributed computing (e.g., transaction processing systems).

12.5 CONSERVATIVE MECHANISMS

Historically, the first distributed simulation mechanisms were based on conservative approaches. As discussed earlier, the basic problem that conservative mechanisms must solve is to determine when it is safe to process an event. More precisely, if a process contains an unprocessed event E_{10} with time stamp 10 (and no other with smaller time stamp), and that process can determine that it is impossible to later receive another event with time stamp smaller than 10, it can safely process E_{10} because it can guarantee that doing so will not later result in a violation of the local causality constraint. Processes containing no safe events must block; this can lead to deadlock situations if appropriate precautions are not taken.

12.5.1 Deadlock Avoidance

Independently, Chandy and Misra [5] and Bryant [6] developed some of the first synchronization algorithms. These approaches require that one *statically* specify the links that indicate which processes may communicate with which other processes. To determine when it is safe to process a message, it is required that the sequence of time stamps on messages sent over a link be nondecreasing and that the communications facility guarantee that messages are received in the same order in which they were sent (software to reorder messages is necessary if the network does not guarantee this). This guarantees that the time stamp of the last message received on an incoming link is a lower bound on the time stamp of any subsequent message that will be received later.

Messages arriving on each incoming link are stored in first in, first out (FIFO) order, which is also time-stamp order because of the foregoing restrictions. Each link has a clock associated with it that is equal to the time stamp of the message at the front of that link's queue if the queue contains a message, or the time stamp of the last message received if the queue is empty. The process repeatedly selects the link with the smallest clock and if there is a message in that link's queue, processes it. If the selected queue is empty, the process blocks. This protocol guarantees that each process will only process events in nondecreasing time-stamp order, thereby ensuring adherence to the local causality constraint.

For example, consider the air traffic simulation described earlier. Each airport LP will have one queue to hold incoming messages from each of the other airports that are simulated, as well as a queue to hold messages it has sent to itself. Again, assume that there are only three airports: SFO, ORD, and JFK. Consider the queues in the JFK process. Suppose that the queue holding messages from ORD contains messages with time stamps 20, 27, and 35, and the queue for SFO has messages with time stamps 25, 31, and 32. Assume that the queue holding messages sent by JFK to itself has a single message with time stamp 40. This implies that the next arrival event sent from ORD to JFK must have time stamp at least 32, and the next message sent by JFK to itself must have a time stamp of at least 40. The JFK process will now process arrival messages in the following order, assuming that no new messages arrive: 20 (ORD), 25 (SFO), 27 (ORD), 31 (SFO), 32 (SFO). Assuming that no new messages have been received, the JFK process will block at this point, even though there is an unprocessed message with time stamp 35 from ORD. At this point there will also be several messages that it has sent to itself, but all of these have time stamps of 40 or larger. The LP must block because it cannot guarantee that a new message won't later arrive from SFO with time stamp less than 35 (as mentioned earlier, it must have a time stamp of at least 32), so none of the messages buffered in its queues are safe to process.

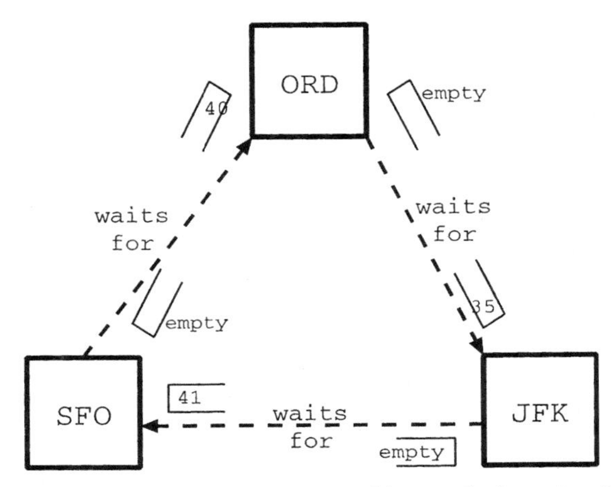

Figure 12.3 Deadlock situation. Each process is waiting on the incoming link containing the smallest link clock value because the corresponding queue is empty. All three processes are blocked, even though there are event messages in other queues that are waiting to be processed. Queues holding messages sent by a process to itself are not shown in this figure.

If a cycle of empty queues arises that has sufficiently small clock values, each process in that cycle must block and the simulation deadlocks. Figure 12.3 shows one such deadlock situation. It can be seen that JFK is waiting for SFO, SFO is waiting for ORD, and ORD is waiting for JFK. In general, if there are relatively few unprocessed event messages compared to the number of links in the network, or if the unprocessed events become clustered in one portion of the network, deadlock may occur very frequently.

Null messages are used to avoid deadlock situations. Null messages are used only for synchronization purposes and do not correspond to any activity in the physical system. A null message with time stamp T_{null} that is sent from LP_A to LP_B is essentially a promise by LP_A that it will not send a message to LP_B carrying a time stamp smaller than T_{null}. How does a process determine the time stamps of the null messages it sends? The clock value of each *incoming* link provides a lower bound on the time stamp of the next unprocessed event that will be removed from that link's buffer. When coupled with knowledge of the simulation performed by the process (e.g., a minimum time stamp increment for any message passing through the logical process), these incoming bounds can be used to determine a lower bound on the time stamp for the next *outgoing* message on each output link. Whenever a process finishes processing an event, it sends a null message on each of its outgoing links indicating this bound; the receiver of the null message can then compute new bounds on its outgoing links, send this information on to its neighbors, and so on. It is up to the application programmer to determine the time stamps assigned to null messages.

In the airport example, assume that the minimum time for an aircraft to land, exchange passengers and depart, and then fly from SFO to JFK is 5 units of time. Further, suppose SFO is currently at simulated time 34 (i.e., the last event it processed contained a time stamp of 34). Using the algorithm above, SFO would send a null message to JFK with a time stamp of 39, indicating that it will not schedule any new arrival events at JFK with time stamp smaller than 39. Upon receipt of this message, the JFK process can now process the arrival message from ORD with time stamp 35 without fear of violating the local causality constraint.

The algorithm proposed by Chandy and Misra requires that each LP send a null message on each outgoing link after processing each event.* This guarantees that processes always have updated information on the time stamp of future messages that can be received from each of the other processes. It can be shown that this mechanism avoids deadlock as long as one does not have any cycles in which the collective time-stamp increment of a message traversing the cycle *could* be zero. A necessary and sufficient condition for deadlock using this scheme is that a cycle of links must exist with the same link clock time [8]. This implies that certain types of simulations cannot be performed (e.g., queueing network simulations in which the minimum service time for jobs passing through a server is zero). One way to circumvent this problem is to assume that a small, minimum, positive service time is always used.

12.5.2 Deadlock Detection and Recovery

One potential problem with the deadlock avoidance approach is that an excessive number of null messages may have to be sent. Chandy and Misra [9] developed a second synchronization algorithm that eliminates the use of null messages. The mechanism is similar to that described above except that no null messages are created. Instead, the computation is allowed to deadlock. A separate mechanism is used to detect when the simulation is deadlocked, and still another mechanism is used to break the deadlock. Deadlock detection mechanisms are described in refs. 10 and 11 and are beyond the scope of the current discussion. The deadlock can be broken by observing that the message(s) containing the smallest time stamp is (are) always safe to process. Alternatively, one may use a distributed computation to compute lower-bound information (not unlike the distributed computation using null messages described above) to enlarge the set of safe messages. Unlike the deadlock avoidance approach, this mechanism does not prohibit cycles of zero time-stamp increment, although performance may be poor if many such cycles exist.

The mechanism described above only attempts to detect and recover from *global* deadlocks. One can modify this approach to detect and recover from *local* deadlocks (i.e., situations where only a portion of the network has deadlocked) [10]. In particular, a preprocessing step can be used that identifies all subnetworks that are prone to deadlock, then applying these techniques on individual subnetworks. The overhead to implement this approach may be large, however, if the network topology contains many cycles. An alternative approach based on detecting specific types of cycles of blocked processes is described in ref. 12.

Several other conservative synchronization algorithms have been developed. Some of the key ideas used by these mechanisms are described next.

12.5.3 Synchronous Operation

Several researchers have proposed algorithms where the parallel simulation as a whole repeatedly cycles through phases of (1) determining which events are safe to process, and (2) processing these safe events [13–16]. Barrier synchronizations are used to keep the phases from interfering with each other.

*A variation on this approach is to have processors explicitly request null messages rather than always blindly sending them out; this can reduce the number of null messages somewhat [7].

It is instructive to compare the synchronous style of execution with the deadlock detection and recovery approach described earlier. Both share the characteristic that the simulation moves through phases of (1) processing events, and (2) performing some global synchronization function to decide which events are safe to process. The two methods differ in the way they enter into the synchronization phase.

In the best case, the detection and recovery strategy will never deadlock, eliminating most of the clock synchronization overhead. In contrast, synchronous methods will continually block and restart throughout the simulation. On the other hand, the synchronous methods do not require a deadlock detection mechanism. However, an important disadvantage of the detection and recovery method is that during the period leading up to a deadlock when the computation is grinding to a halt, execution may be largely sequential. This can lead to limited speed-up, in accordance with Amdahl's law, which states that no more than k-fold speed-up is possible if $1/k$th of the computation is sequential. Synchronous methods have some control over the amount of computation that is performed during each iteration, so, at least in principle they offer a mechanism for guarding against such behavior.

The feature that separates different synchronous approaches is principally the method used to determine which events are safe to process. A common thread that runs through many techniques is the minimum time-stamp increment function used in the original deadlock avoidance approach. For example, in the air traffic example, if the minimum time to fly between two airports is 3 units of simulated time, an event in one airport LP must schedule a new event in another LP at least 3 units of time into the simulated future. A simple extension of this concept leads to the notion of *distance* between processes; distance provides a lower bound in the amount of simulated time that must elapse for an unprocessed event on one process to propagate (and possibly affect) another process. It is clear that the physical distance between airports and the maximum speed of aircraft translate directly into the simulated time distance between the LPs modeling those airports. Later, a more general principle called look-ahead is discussed that encompasses both minimum time-stamp increments and distance between objects.

12.5.4 Conservative Time Windows

Lubachevsky proposed using a moving simulated time window to reduce the overhead associated with determining when it is safe to process an event [15]. The lower edge of the window is defined as the minimum time stamp of any unprocessed event. The upper edge depends on the window size, as will be discussed momentarily. Only those unprocessed events whose time stamp resides within the window are eligible for processing.

The purpose of the window is to reduce the *search space* one must traverse in determining if an event is safe to process. For example, if the window extends from simulated time 10 to time 20, and the application is such that each event processed by an LP generates a new event with a minimum time-stamp increment of 8 units of simulated time, each LP need only examine the unprocessed events in neighboring LPs to determine which events are safe to process. No unprocessed event two or more hops away can affect one in the 10-to-20 time window because such an event would have to have a time stamp earlier than the start of the window.

An important question is the method to be used for determining the size of the time window. If the window is too small, there will be too few events available for concurrent execution. On the other hand, if the window is too large, the simulation mechanism behaves in much the same way as if no time window were used at all (such mechanisms implicitly assume an infinitely large time window), implying that the overhead to manage the window mechanism is not justified. Setting the window to an appropriate size requires application specific information that must be obtained either from the programmer, the compiler, or from monitoring the simulation at run time.

12.5.5 Improving Look-ahead by Precomputing Service Times

Look-ahead refers to the ability to predict what will happen, or more important, what will *not* happen, in the simulated future. If a process at simulated time *Clock* can predict with complete certainty all events it will generate up to simulated time $Clock + L$, the process is said to have lookahead L. In general, a process may have different look-aheads on links to different processes. For instance, as was seen earlier, if the minimum time to fly from JFK to ORD is 3 units of time and the minimum time to SFO is 5, JFK has a look-ahead of 3 on its link to ORD and 5 to SFO.

Nonzero minimum time-stamp increments are the most obvious form of look-ahead: A minimum time-stamp increment of M translates directly into a look-ahead of (at least) M because the process can guarantee that no new event messages will be created with time stamp smaller than $Clock + M$. Look-ahead enhances one's ability to predict future events, which in turn can be used to determine which other events are safe to process. It is used in the deadlock avoidance approach to determine the time stamps that are assigned to null messages. It is also used to some extent in the deadlock detection and recovery algorithm because whenever a process sends a message with a time-stamp increment of T to another process, it is guaranteeing that no other messages will follow on that link that contain a time stamp smaller than $Clock + T$.

Nicol proposed improving the look-ahead ability of processes by precomputing portions of the computation for future events [17]. For example, in queueing network simulation using first come, first served queues without preemption, one can precompute the service time of jobs that have not yet been received. If the server process is idle and its clock has a value of 100 and the service time of the next job has been precomputed to be 50, the lower bound on the time stamp of the next message it will send is 150 rather than 100. If the average service time is much larger than the minimum, this will provide a better lower bound on the time stamp of the next message.

Interestingly, the ability to use precomputation to improve look-ahead *itself* requires look-ahead ability. Precomputation is possible if one can predict aspects of future event computations without knowledge of the event message that causes that computation, or the state of the process when that future event computation would take place. For example, if the service time depends on a parameter in the message (e.g., a message length for a communication network simulation), precomputation would not be so simple. Nevertheless, precomputation appears to be a useful technique when it can be applied.

Other conservative protocols have been proposed; however, this subject is beyond the scope of the current discussion. See refs. 18 and 19 for surveys of work in this area.

12.5.6 Performance of Conservative Mechanisms

The degree to which processes can look ahead and predict future events plays a critical role in the performance of conservative strategies. Actually, what is more important than predicting future events is the fact that a process with look-ahead L can guarantee that *no events* other than the ones that it can predict will be generated up to time *Clock* +

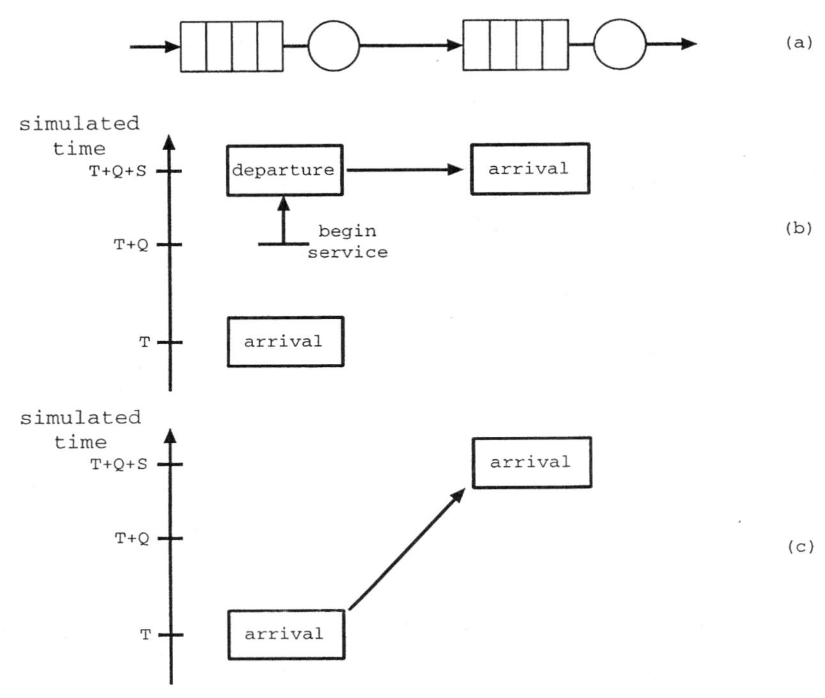

Figure 12.4 Two approaches to simulating a queueing network [18]: (*a*) two queues connected in tandem, each using a first come, first served discipline and no preemption; (*b*) history of events when using the classical approach that does not exploit look-ahead; (*c*) history for approach that does exploit look-ahead.

L. This may enable other LPs to safely process pending event messages that they have already received.

To illustrate this point, let us consider the simulation of a queueing network consisting of two stations connected in tandem as shown in Figure 12.4*a*. Queueing networks are a more abstract representation of systems such as the air traffic example developed earlier, where each station models an airport and each job moving between stations models an aircraft. The first station is modeled by logical process LP_1 and the second by LP_2. Each station contains a server and a queue that holds jobs (customers) waiting to receive service. Assume that incoming jobs are served in first come, first served order.

The classical textbook approach to programming the simulation is to use two types of events: (1) an arrival event denotes a job arriving at a station, and (2) a departure event denotes a job completing service and moving on to another server. As shown in Figure 12.4*b*, a job J arriving at the first station at time T will, in general, (1) spend Q units of time ($Q \geq 0$) in the queue, waiting to be served, and (2) an additional S units of time being served, before it is forwarded to LP_2.

The simulation program described above has *poor* look-ahead properties. In particular, LP_1 must advance its simulated time clock to $T + Q + S$ before it can generate a new arrival event with time stamp $T + Q + S$. It has zero look-ahead with regard to generating new arrival events.

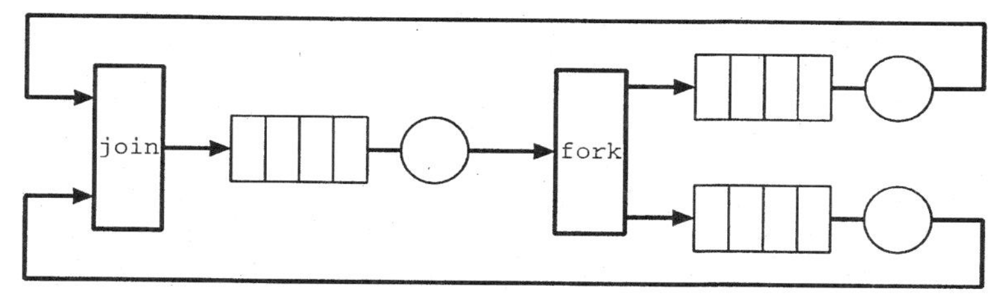

Figure 12.5 Central server queueing network [21]. The fork process routes incoming jobs to one of the secondary servers. Here the fork process is equally likely to select either server.

An alternative approach to programming this simulation is depicted in Figure 12.4*c*. Here the departure event has been eliminated, and processing one arrival event causes a new arrival event to be scheduled immediately. This is possible because first come, first serve queues are used and no preemption is possible. The event at time T can predict the arrival event at $T + Q + S$ because both Q and S can be computed at simulated time T. In particular, Q is the remaining service time for the job being served at time T, plus the service times of all jobs preceding J in the queue. Similarly, S can be computed at simulated time T because it does not depend on the state of the process at a time later than T. The look-ahead using this alternative approach is $Q + S$.

Programming the simulation to exploit look-ahead can improve performance dramatically. Figure 12.5 shows performance measurements of simulating a central server queueing network (shown in Figure 12.6) using Chandy and Misra's deadlock detection and recovery algorithm on a BBN Butterfly multiprocessor. Each logical process executes on a separate processor. A closed network is simulated with a fixed number of circulating jobs (referred to as the *message population*). Figure 12.5*a* shows the average number of messages processed between deadlocks, and Figure 12.5*b* shows speed-up relative to a sequential event list implementation where the event list is implemented using a splay tree [20]. Each graph indicates measurements of the classical approach to programming the simulation illustrated in Figure 12.4*b* and the approach optimized to exploit look-ahead (Figure 12.4*c*). As can be seen, the version that exploits look-ahead far outperforms the version that does not. Similar speed-up curves were observed for the deadlock avoidance approach using null messages [21].

Before continuing we should note that the foregoing situation is one where the application contains good look-ahead, and the simulation program could easily be modified to exploit it. This is not always the case, however. For example, consider a queueing network where the service time distribution has a minimum of zero (e.g., an exponential distribution) and preemption may occur. A high-priority job that has been simulated up to time T could (albeit very unlikely) affect every station in the network at time T, so no station can look ahead beyond T. Exploiting look-ahead in simulations such as these is much more challenging.

Much has been learned concerning the performance of conservative mechanisms. In general, conservative mechanisms must be adept at predicting what will *not* happen in order to be successful. It is the fact that "*no* smaller time-stamped event will later be received" is the firing condition that allows an event to be safely processed. Effectively exploiting the look-ahead properties of the simulation is the key to achieving good performance with these methods.

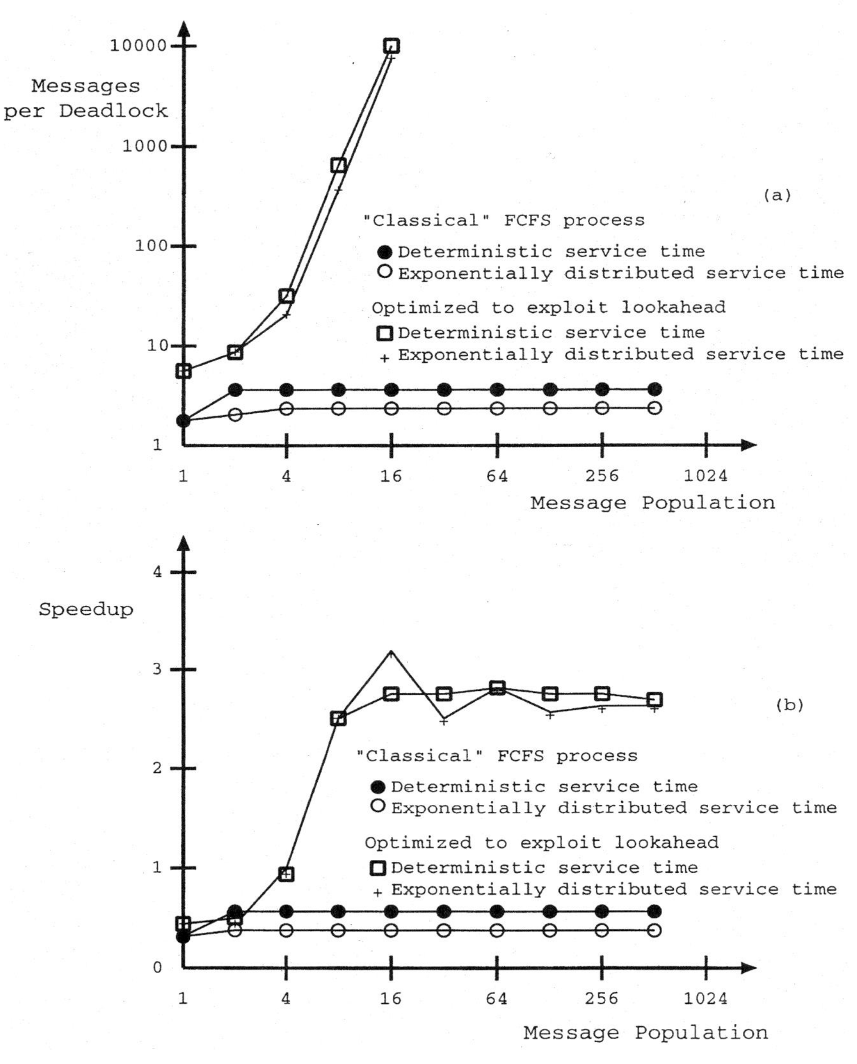

Figure 12.6 Performance of deadlock detection and recovery simulation program for central server queueing network [18]: (*a*) average number of messages processed between deadlocks as a function of the message population; (*b*) speed-up over a sequential event list implementation.

12.5.7 Critique of Conservative Mechanisms

Table 12.1 summarizes key advantages and disadvantages of conservative protocols. The column in this table concerning optimistic protocols is discussed later. Perhaps the most obvious drawback of conservative approaches is that they cannot fully exploit the parallelism available in the simulation application. If it is possible that event E_A *might* affect E_B either directly or indirectly, conservative approaches must execute E_A

TABLE 12.1 Comparing Conservative and Optimistic Synchronization

	Conservative	Optimistic
Parallelism	Limited by worst-case scenario	Not limited by worst case
Performance challenges	Low look-ahead	State saving/rollback overhead
Memory use	No logs needed	Need state and message logs
Protocol development	Straightforward	Must consider rollbacks
Application development	Must exploit look-ahead	Greater protocol transparency
	Potentially complex/fragile code	More robust to model changes
		Potential for unexpected errors

and E_B sequentially. If the simulation is such that E_A seldom affects E_B, these events could have been processed concurrently most of the time. In general, if the worst-case scenario for determining when it is safe to proceed is far from the typical scenario that arises in practice, the conservative approach will usually be overly pessimistic and force sequential execution when it is not necessary.

Another way of stating this fact is to observe that conservative algorithms rely on look-ahead to achieve good performance.* If there were no look-ahead, the smallest time-stamped event in the simulation *could* affect every other pending event, forcing sequential execution no matter what conservative protocol is used. Characteristics such as preemptive behavior or dependence of an output message with time stamp T on the state of an LP at time T diminish the look-ahead properties of the simulation. Conservative algorithms appear to be poorly suited for simulating applications with poor look-ahead properties, even if there is a healthy amount of parallelism available.

A related problem faced by conservative methods concerns the question of robustness; it has been observed that seemingly minor changes to the application may have a catastrophic effect on performance [22]. For example, adding short, high-priority messages that interrupt normal processing in a computer network simulation can destroy the look-ahead properties on which the mechanism relies and lead to severe performance degradations. This is problematic because experimenters often do not have advance knowledge of the full range of experiments that will be required, so it behooves them to invest substantial amounts of time parallelizing the application if an unforeseen addition to the model at some future date could invalidate all of this work.

Most conservative schemes require knowledge concerning logical process behavior to be explicitly provided by the simulation programmer for use in the synchronization protocol. The deadlock detection and recovery algorithm is perhaps the only existing conservative approach that does not explicitly require such knowledge from the user. Information such as minimum time-stamp increments or the guarantee that certain events really have no effect on others (e.g., an arrival event in a queueing network simulation may not affect the job that is currently being serviced) may be difficult to ascertain for complex simulations. Users would be ill advised to give overly conservative estimates (e.g., a minimum time-stamp increment of zero) because very poor performance may result. Overly optimistic estimates can lead to violations of causality constraints and erroneous results.

Perhaps the most serious drawback with existing conservative simulation protocols is that the simulation programmer must be concerned with the details of the synchro-

*That is, except in a few special instances such as feedforward networks that do not contain cycles.

nization mechanism in order to achieve good performance. Proponents of optimistic approaches argue that the user should not have to be concerned with such complexities, just as programmers of *sequential* simulations need not be concerned with the details of the implementation of the event list. Of course, certain guidelines that apply to *all* parallel programs must be followed when developing parallel simulation code (e.g., ensuring there is sufficient computation between interprocessor communications to prevent message passing overhead from severely degrading performance). However, also requiring the programmer to be intimately familiar with the synchronization mechanism and to program the application to maximize its effectiveness will often lead to "fragile" code that is difficult to modify and maintain. One potential solution to this problem is to define and utilize a simulation language where the essential information needed by the simulation mechanism can be automatically extracted from the simulation primitives [23,24]. It remains to be seen to what extent this approach can be effective.

Despite these drawbacks, conservative algorithms can be effective in speeding up the execution of parallel simulation programs, and useful systems have been deployed successfully using these techniques. It is clear that much additional work is required to enable widespread use of this approach, however.

12.6 OPTIMISTIC MECHANISMS

Optimistic methods detect and recover from causality errors rather than strictly avoiding them. In contrast to conservative mechanisms, optimistic strategies need not determine when it is safe to proceed; instead, they determine when an error has occurred and invoke a procedure to recover. One advantage of this approach is that it allows the simulation program to exploit parallelism in situations where it is possible causality errors *might* occur but in fact do not. Also, dynamic creation of logical processes can be easily accommodated [25].

The *Time Warp mechanism*, based on the *Virtual Time paradigm*, is the best known optimistic protocol [26]. Here virtual time is synonymous with simulated time. In Time Warp, a causality error is detected whenever an event message is received that contains a time stamp smaller than that of the process's clock (i.e., the time stamp of the last message processed). The event causing rollback is called a *straggler*. Recovery is accomplished by undoing the effects of all events that have been processed prematurely by the process receiving the straggler (i.e., those processed events that have time stamps larger than that of the straggler).

An event may do two things that have to be rolled back: It may modify the state of the logical process, and/or it may send event messages to other processes. Rolling back the state is accomplished by saving the process's state periodically and restoring an old state vector on rollback. "Unsending" a previously sent message is accomplished by sending a negative or *antimessage* that annihilates the original when it reaches its destination. Messages corresponding to events in the simulation are referred to as *positive* messages. If a process receives an antimessage that corresponds to a positive message that it has already processed, that process must also be rolled back to undo the effect of processing the soon-to-be-annihilated positive message. Recursively repeating this procedure eventually allows all of the effects of the erroneous computation to be canceled. It can be shown that this mechanism always makes progress under some mild constraints.

As noted earlier, the smallest time-stamped, unprocessed event in the simulation will always be safe to process. In Time Warp, the smallest time stamp among all unprocessed event messages (both positive and negative) is called *global virtual time* (GVT). No event with time stamp smaller than GVT will ever be rolled back, so storage used by such events (e.g., saved states) can be discarded.* Also, irrevocable operations (such as I/O) cannot be committed until GVT sweeps past the simulated time at which the operation occurred. The process of reclaiming memory and committing irrevocable operations is referred to as *fossil collection*. Several algorithms have been proposed for computing GVT. Detailed discussion of this topic is beyond the scope of the present discussion, but is discussed elsewhere [27–29].

12.6.1 Lazy Cancellation

Optimizations have been proposed to repair the damage caused by an incorrect computation rather than to repeat them completely. For instance, it may be the case that a straggler event does not sufficiently alter the computation of rolled-back events to change the (positive) messages generated by these events. The Time Warp mechanism described above uses *aggressive* cancellation (i.e., whenever a process rolls back to time T, antimessages are immediately sent for any previously sent positive message with a time stamp larger than T). In lazy cancellation [30], processes do not immediately send the antimessages for any rolled-back computation. Instead, they wait to see if reexecution of the computation regenerates the *same* messages; if the same message is recreated, there is no need to cancel the message. An antimessage created at simulated time T is only sent after the process's clock sweeps past time T without regenerating the same message.

Depending on the application, lazy cancellation may either improve or degrade performance. Lazy cancellation does require some additional overhead whenever an event is processed. The simulation executive must check on each message send to determine if the message now being sent has already been transmitted; one or more message comparisons with antimessages may be required if one is reexecuting previously rolled back events. Also, lazy cancellation may allow erroneous computations to spread further than they would under aggressive cancellation, so performance may be degraded if the simulation program is forced to execute many incorrect computations. One can construct cases where lazy cancellation executes a computation with N-fold parallelism N times *slower* than aggressive when N processors are used [31].

On the other hand, lazy cancellation has the interesting property that it can allow the computation to be executed in less time than the critical path execution time [32,33]. The explanation for this phenomenon is that computations with incorrect or only partially correct input may still generate correct results!† Therefore, one may execute some computations prematurely, yet still generate the correct answer. This is not possible using aggressive cancellation because rolled-back computations are discarded immediately, even if they did generate the correct result. One can construct a case where lazy cancellation can execute a sequential computation with N-fold speed-up using N processors, while aggressive cancellation requires the same amount of time as the sequential execution [31]. The conclusion one can make from this analysis is that while aggressive

*Actually, one state vector older than GVT is required to restore a process's state in case a rollback to GVT occurs.

†For example, suppose that the event computes the minimum of two variables, A and B, and executes prematurely using an incorrect value for A. If both the correct and incorrect values of A are greater than B, the incorrect execution produces exactly the same result as the correct one.

cancellation will not perform better than the critical path execution time, lazy cancellation can perform arbitrarily better or worse depending on details of the application and the number of available processors.

Although it is instructive to construct best- and worst-case behaviors for lazy and aggressive cancellation, it is not clear that such extreme behaviors arise in practice. Empirical evidence suggests that lazy cancelation tends to perform as well as, or better than, aggressive cancellation in practice [31,34].

12.6.2 Lazy Reevaluation

The *lazy reevaluation* optimization (also sometimes called *jump forward*) is somewhat similar to lazy cancellation but deals with state vectors rather than messages [35]. Consider the case where the state of the process is the same after processing a straggler event message as it was before. If no new messages arrived, it is clear that the reexecution of rolled-back events will be identical to the original execution. Therefore, one need not reexecute the rolled-back events, but instead, "jump forward" over them. This requires a comparison of state vectors to determine if the state has changed.

One situation where one could derive significant benefit from lazy reevaluation is "read-only" or query events. Here lazy reevaluation avoids the expense of regenerating states when a query event causes a rollback.* It is worth mentioning, however, that lazy reevaluation may significantly complicate the Time Warp code, detracting from its maintainability. It was implemented in a Time Warp executive developed at the Jet Propulsion Laboratory [38], but later removed for this reason.

Paralleling the work in conservative protocols, a variety of other optimistic protocols have been developed [18,19]. Most define methods to limiting the amount of optimism (i.e., the degree that some processes can advance ahead of others).

12.6.3 Performance of Optimistic Mechanisms

Several successes have been reported in using Time Warp to speed up real-world simulation problems. Substantial speed-ups have been reported by researchers at JPL in simulations of battlefield scenarios [39], communication networks [40], biological systems [41], and simulations of other physical phenomena [42]. Typical speed-ups on the JPL Mark III hypercube (a 68020-based, message-passing machine) ranged from 10 to 20 using 32 processors. Fujimoto also reports good performance using another, independently developed version of Time Warp for queueing network simulations [43] and various synthetic workloads [44] where speed-ups as high as 57 using a 64-processor BBN Butterfly were reported.

We earlier observed that look-ahead appears to be essential to obtain significant speed-ups using *conservative* algorithms for most problems of practical interest. Is the same true for optimistic methods? Empirical evidence indicates that while look-ahead improves the performance of optimistic algorithms, it is not a prerequisite for obtaining good performance.

For example, Figure 12.7 compares the performance of Time Warp with the conservative deadlock avoidance and deadlock detection and recovery algorithms for a closed queueing network simulation. Speed-up over a sequential event list simulation program is shown as the message density (the message population divided by the number of logi-

*See refs. 36 and 37 for a discussion of other mechanisms to handle queries.

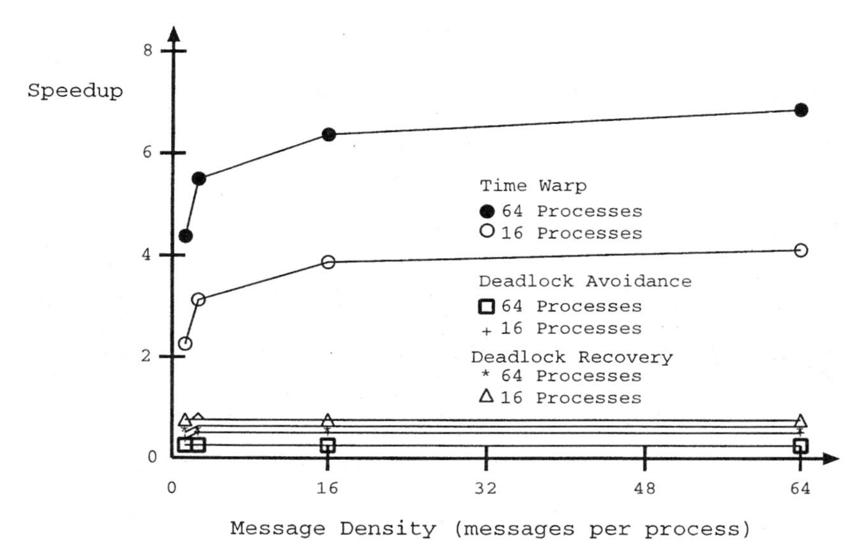

Figure 12.7 Speedup of time warp and conservative algorithms for a queueing network simulation where the service time distribution is exponential (minimum service time is zero), and preemption is allowed [18]. One percent of the jobs in the network have high priority.

cal processes) is varied. An eight-processor BBN Butterfly multiprocessor was used in these experiments. An exponential service time distribution with a minimum value of zero is used.* Further, some fraction of the jobs (here 1%) are designated as high priority, while the rest are low priority. High-priority jobs *preempt* service from low-priority jobs. As noted earlier, this simulation contains very poor look-ahead characteristics and cannot be optimized as was done earlier for the simulation program using first come, first served queues. As can be seen, Time Warp is able to obtain a significant speed-up for this problem, while the conservative algorithms have difficulty.

12.6.4 Critique of Optimistic Methods

Table 12.1 summarizes key advantages and disadvantages of optimistic protocols and compares them with conservative protocols. A critical question faced by optimistic systems such as Time Warp is whether the system will exhibit thrashing behavior where most of its time is spent executing incorrect computations and rolling them back. Here the concern is that incorrect computations will be executed at the expense of correct ones; indeed, if the application contains only limited parallelism relative to the number of available processors, a significant degree of rollback is inevitable, and in fact, may be perfectly acceptable. As discussed earlier, a variety of techniques are available to avoid overly optimistic execution and excessive rollbacks.

A more serious problem with the Time Warp mechanism is the need to save the

*Strictly speaking, the deadlock avoidance algorithm cannot simulate this network because cycles containing zero look-ahead exist. A "fortified" version of the deadlock avoidance approach was used that is supplemented with a deadlock detection and recovery mechanism to circumvent this problem. As mentioned earlier, minimum service times could also have been used to solve this problem.

state of each logical process periodically. As mentioned earlier, state-saving overhead can seriously degrade the performance of many Time Warp programs, even if the state vector is relatively modest in size. The state-saving problem is further exacerbated by applications requiring dynamic memory allocation because one may have to traverse complex data structures to save the process's state. The amount of time in each event to save state must be small relative to the amount of simulation computation per event to avoid serious performance degradations. This may be difficult to achieve for certain applications. Incremental state-saving techniques are essential for programs with large numbers of state variables, although as mentioned earlier, care must be taken to minimize rollback overheads. Alternatively, hardware support has also been proposed to solve the state-saving problem [1,45].

Optimistic algorithms tend to use much more memory than their conservative counterparts to store old messages and state information. One can implement Time Warp using no more memory than is required by the corresponding sequential simulation [46,47]; however, performance will be poor if one attempts to run Time Warp simulations using this little memory. Protocols have been defined to roll back processes, if necessary, to reclaim memory resources as needed. This provides a mechanism that allows Time Warp to live gracefully with whatever memory is provided to it. Perhaps more surprisingly, Jefferson also shows that existing *conservative* synchronization algorithms are *not* storage optimal. For example, an LP that does not receive messages from any other LP may not be constrained by the conservative algorithm from executing ahead of the others and generating messages (which consume memory) before they would have been created during a sequential execution. Adding a similar mechanism to existing asynchronous protocols to guarantee execution using the minimum amount of memory without introducing a significant performance degradation is an open question.

Unlike conservative approaches, optimistic systems need to be able to recover from arbitrary program execution errors (e.g., the simulation program performing a divide-by-zero operation) because such errors may be erased by a subsequent rollback. Erroneous computations may enter infinite loops, requiring the Time Warp executive to interact with the hardware's interrupt system. In certain languages, pointers may be manipulated in arbitrary ways; Time Warp must be able to trap illegal pointer uses that result in run-time errors and prevent incorrect computations from overwriting non-state-saved areas of memory. Although such problems are, in principal, not insurmountable, they may be difficult to circumvent in certain systems without appropriate hardware support. The alternative taken by most existing Time Warp systems is to leave the task of analyzing incorrect execution sequences to the user (e.g., by always checking array indices at run time and explicitly testing to ensure that loops will terminate).

Finally, the Time Warp mechanism is much more complex to implement than conservative approaches, particularly if one attempts to catch errors such as those described above. Although the actual Time Warp code is not very complex if one ignores the error-handling aspects, inexperienced implementors may make seemingly minor design mistakes that lead to extremely poor performance. For example, use of an inappropriate policy for determining which logical processes are executed when can be catastrophic. Debugging Time Warp implementations is time consuming because it may require detailed analysis of complex rollback scenarios. A certain amount of design experience (or luck) with optimistic execution is often required to obtain a good, robust implementation of Time Warp. On the other hand, this development cost need only be paid once when developing the Time Warp kernel rather than with each new application.

Because of its relaxed dependence on look-ahead, optimistic simulation techniques such as Time Warp are believed to offer the best hope for providing a general-purpose parallel simulation executive. Time Warp does entail additional overheads that are not incurred in conservative simulation approaches (e.g., for state saving and rollback). In general, it is difficult to predict without experimental evaluation whether an optimistic or conservative mechanism will yield better performance for a specific application, but a general rule is that if the application contains very poor look-ahead properties, optimistic methods are more likely to achieve significant speed-ups. On the other hand, if state-saving overheads are expected to dominate, conservative methods are preferred.

12.7 TIME PARALLEL SIMULATION

The approaches to distributed simulation discussed thus far use a *spatial* decomposition of the simulation model into a collection of logical processes. We now turn our attention to another approach that has been utilized successfully for certain simulation problems: *temporal decomposition*.

One can view a simulation as a computation that must determine the values of certain *state variables* across simulated time. The state variables capture the state of the system (e.g., the number of customers waiting for service in a simulation of a queue). Changes in the state of the system occur at discrete points in simulated time (e.g., when a new customer enters the system). This space-time view of the simulation [48] is depicted in Figure 12.8 using a two-dimensional graph, where the vertical axis represents the state variables and the horizontal axis represents simulated time. The goal of the simulation program is to "fill in" the graph by computing the values of each of the state variables across simulated time. A parallel simulator attempts to use multiple processors that simultaneously fill in different portions of the space-time graph.

In Figure 12.8*a* the graph is divided into horizontal strips, with a logical process responsible for the computation within each strip. This is the *space-parallel* approach, which has been the basis for the synchronization algorithms discussed earlier. An alternative approach called the *time-parallel* method partitions the space-time graph into

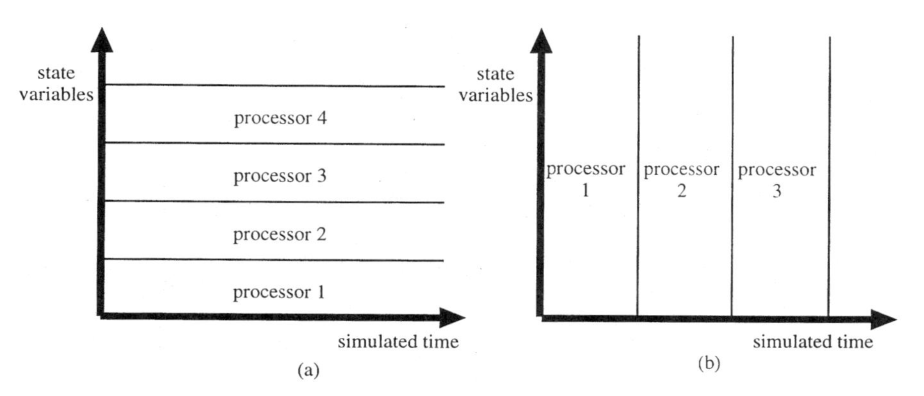

Figure 12.8 Space-time diagram depicting simulated time on the horizontal axis and state variables on the vertical axis: (*a*) approach exploiting space parallelism; (*b*) approach exploiting time parallelism.

vertical strips as shown in Figure 12.8*b* and assigns a separate processor to each strip. The simulated time axis is divided into intervals $[T_1, T_2]$, $[T_2, T_3]$, ... , $[T_i, T_{i+1}]$, ... with processor i assigned the task of computing the portion of the space-time graph within the interval $[T_i, T_{i+1}]$.

Time-parallel simulation methods have been developed for attacking a handful of specific simulation problems (e.g., measuring the loss rate of a finite-capacity queue). Because the amount of parallelism available in time-parallel algorithms is proportional to the length of the simulation run, this technique allows for massively parallel execution for many simulation problems.

A central question that must be addressed by time-parallel simulation programs is ensuring the states computed at the "boundaries" of the time intervals "match." Specifically, it is clear that the state computed at the end of the interval $[T_{i-1}, T_i]$ must match the state at the beginning of interval $[T_i, T_{i+1}]$. Thus this approach relies on being able to perform the simulation corresponding to the ith interval without first completing the simulations of the preceding ($i - 1$, $i - 2$, ... , 1) intervals.

Because of the state-matching problem, time-parallel simulation is really more a methodology for developing massively parallel algorithms for specific simulation problems than a general approach for executing arbitrary discrete-event simulation models on parallel computers. Time-parallel algorithms are currently not as robust as space-parallel approaches because they rely on specific properties of the system being modeled (e.g., specification of the system's behavior as recurrence equations and/or a relatively simple state descriptor). This approach is currently limited to a few important applications (e.g., queueing networks, Petri nets, cache memories, and statistical multiplexers in telecommunication networks). Space-parallel simulations offer greater flexibility and wider applicability, but concurrency is limited to the number of logical processes. In some cases, both time and space parallelism can be used [49].

One approach to solving the state-matching problem is to have each processor "guess" the initial state of its simulation and then simulate the system based on this guessed initial state [50]. In general, the initial state will not match the final state of the previous interval. After the initial simulation of each interval has been completed, a fix-up computation is performed to account for the fact that the wrong initial state was used in each interval simulation. This might be performed, for instance, simply by repeating the simulation, using the final state computed in the previous interval as the new initial state. This "fix-up" process is repeated until the initial state of each interval matches the final state of the previous interval. In the worst case, N such iterations are required when there are N intervals. However, if the final state of each interval simulation is seldom dependent on the initial state, far fewer iterations will be needed.

12.7.1 Time-Parallel Cache Simulation

To illustrate this approach, consider the simulation of a cache memory in a computer memory system employing a least-recently-used replacement policy. Briefly, a cache is a high-speed memory that holds recently referenced data and instructions. The goal is to store frequently used data and instructions in the cache that can be accessed very quickly, typically an order of magnitude faster than main memory. Because the cache memory has very high speed, it is expensive, so the computer system may contain only a limited amount of cache memory. If data or instructions are referenced by the CPU that do not reside in the cache, the information must be loaded into the cache, displacing other data/instructions from the cache if there is no unused memory in the cache. Main memory is partitioned into a collection of fixed-sized blocks (a typical block size is 64 bytes), and some set of blocks are maintained within the cache. The cache management hardware includes tables that indicate which blocks are currently stored in the cache.

The cache *replacement policy* is responsible for determining which block to delete from the cache when a new block must be loaded. A commonly used policy is to replace the block that hasn't been referenced in the longest time, based on the premise that recently referenced blocks are likely to be referenced again in the near future. This approach is referred to as the *least recently used* (LRU) replacement policy. Due to implementation constraints, cache memories typically subdivide the cache into *sets* of blocks and use LRU replacement within each set.

Heidelberger and Stone were among the first to propose the time-parallel simulation approach, in the context of simulating cache memory systems using LRU replacement [51]. This approach is effective for this application because the final state of the cache is usually not dependent on the cache's initial state. The input to the simulation is a sequence of memory references, with each reference denoting a read or write to a location in memory. The sequence is partitioned into N subsequences, one for each processor participating in the simulation. The parallel simulation proceeds as follows:

1. Each processor (CPU) executes a simulation of the subsequence assigned to it, assuming that the cache is initially empty. This simulation is actually incorrect in all processors except the first because in reality, the cache is not initially empty except for the processor assigned the first subsequence.
2. Each processor (except the first) begins to repeat the simulation assuming the initial state of the simulation for subsequence i is the same as the final state of the cache at the end of the simulation of subsequence $i - 1$.
3. The simulation in step 2 continues until either the state of the cache becomes the same as (matches) that of the previous simulation run at that point in the subsequence or the end of the subsequence is reached.
4. Steps 2 and 3 are repeated until each subsequence simulation terminates by converging to a prior state rather than reaching the end of the subsequence.

When using an LRU replacement policy, the final state of the simulation of a subsequence is guaranteed to be independent of the initial state if the subsequence fills an empty cache with new data elements. If each subsequence is sufficiently long to have this property, only two passes through the subsequences (one initial pass, and one to fix up the initial portions of the first pass) will be required to complete the simulation.

A variation on this approach devised in the context of simulating statistical multiplexers for asynchronous transfer mode (ATM) switches precomputes certain points in time where one can guarantee that a buffer overflow (full queue) or underflow (empty queue) will occur [52,53]. Because the state of the system (i.e., the number of occupied buffers in the queue) is known at these points, independent simulations can be begun at these points in simulated time, thereby eliminating the need for a fix-up computation. Ammar and Deng also use a related approach for simulating Petri networks [54].

12.7.2 Time-Parallel Simulation of Queues Using Parallel Prefix

Another interesting approach to time-parallel simulation is described in ref. 55. This algorithm simulates a G/G/1 queue (a queue with a single server and general distributions for service and interarrival times) where service times are not dependent on the state of the queue. The approach is to represent the queue's behavior as a set of recurrence equations which are then solved using well-known parallel prefix algorithms. The parallel prefix computation enables the state of the system at various points in simulated time to be computed concurrently.

A prefix computation is one that computes the N initial products of $P_i = X_1 * X_2 * \ldots * X_i$ for $i = 1, 2, \ldots, N$ where $*$ is an associative operator. Well-known algorithms exist for performing this computation in parallel. The key observation is that the simulation of a G/G/1 queue can be recast into this form. Specifically, let r_i denote the interarrival time preceding the ith job and s_i denote the service time for the ith job. r_i and s_i can be precomputed, in parallel, since they are independent random numbers. The simulation must compute the arrival and departure times of the ith job, A_i and D_i, respectively. Because the arrival time of the ith job is simply the arrival time of the $(i-1)$st job plus the interarrival time for the ith job ($A_i = A_{i-1} + r_i$), computation of arrival times is already in the proper form and a parallel prefix computation can be applied directly.

The departure times can also be written in the form of a parallel prefix computation because $D_i = \max(D_{i-1}, A_i) + s_i$, which can be rewritten as

$$D_i = \begin{bmatrix} d_i \\ 0 \end{bmatrix} \qquad M_i = \begin{bmatrix} s_{i-1} - r_i & 0 \\ -\infty & 0 \end{bmatrix}$$

where matrix multiplication is performed using max as the additive operator (with identity $-\infty$) and + as the multiplicative operator (identity 0). Parallel prefix algorithms can now be used to solve for D_i.

12.8 DISTRIBUTED INTERACTIVE SIMULATION

The algorithms discussed thus far evolved from the high-performance computing community. Independently and in approximately the same time frame, a separate community has evolved that has been driven by the Department of Defense to interconnect separately developed simulators (e.g., manned tank simulators) on distributed computing platforms to create synthetic environments for training military personnel. The set of standards and approach to enable interoperability among separately developed simulators have become known as *distributed interactive simulation* (DIS).

While the foundations for PADS research lies in early research concerning synchronization, the precursor to DIS was the SIMNET (SIMulator NETworking) project (1983–1989), which demonstrated the feasibility of interconnecting several autonomous simulators in a distributed environment for training exercises [56]. SIMNET was used as the basis for the initial DIS protocols and standards, and many of the fundamental principles defined in SIMNET remain in DIS today. SIMNET realized over 250 networked simulators at 11 sites in 1990.

From a model execution standpoint, a DIS exercise can be viewed as a collection of autonomous simulators (e.g., tank simulators), each generating a virtual environment representing the battlefield as seen from the perspective of the entities that it models. Each simulator sends messages, called *protocol data units* (PDUs), whenever its state changes in a way that might affect another simulator. Typical PDUs include movement to a new location or firing at another simulated entity.

To achieve interoperability among separately developed simulators, a set of evolving standards have been developed [57]. The standards specify the format and contents of PDUs exchanged between simulators as well as when PDUs should be sent. Communication protocols and requirements are specified in a separate document [58]. DIS is based on several underlying design principles [59]:

- *Autonomy of Simulation Nodes.* Autonomy facilitates ease of development, integration of legacy (preexisting) simulators, and simulators joining or leaving the exercises while it is in progress. Each simulator advances simulation time according to a local real-time clock. Simulators are *not* required to determine which other simulators must receive PDUs; rather, PDUs are broadcast to all simulators and the receiver must determine those that are relevant to its own environment.
- *Transmission of "Ground Truth" Information.* Each node sends absolute truth about the state of the entities it represents. Degradations of this information (e.g., due to environment or sensor limitations) are performed by the receiver.
- *Transmission of State Change Information Only.* To economize on communications, simulation nodes only transmit changes in behavior. If a vehicle continues to "do the same thing" (e.g., travel in a straight line with constant velocity), the rate at which state updates are transmitted is reduced. Simulators do transmit "keep alive" messages (e.g., every 5 seconds), so new simulators entering the exercise can include them in their virtual environment.
- *"Dead-Reckoning" Algorithms.* All simulators use common algorithms to extrapolate the current state (position) of other entities between state updates. More will be said about this later.
- *Simulation Time Constraints.* Because humans cannot distinguish differences in time of less than 100 ms, an application-to-application communication latency of no more than this amount is required. Lower latencies are needed for other, non-training, simulators (e.g., testing of weapons systems).

12.8.1 Contrasting DIS and Parallel Simulation

It is instructive to contrast the work in the PADS community described earlier with that in DIS. Table 12.2 summarizes some key technical differences between PADS and DIS research to date. One of the most important features distinguishing these communities is that DIS has been focused primarily on real-time environments, while PADS work has focused on non-real-time "as fast as possible" simulations. This is because DIS has evolved from virtual training environments, while PADS has evolved from analytic simulation tools for engineering design.

In PADS, performance is paramount, and speed-up relative to a sequential execution is used as the primary metric. By contrast, realism of the virtual environment is of principal importance in DIS, with *scalability* as a second, important goal. Intuitively, a distributed simulation is said to be scalable if it can be expanded to include more simulated entities executing on proportionately larger hardware configurations and the simulator is still able to meet its stated objectives (e.g., value as a training mechanism), which in turn translates to real-time performance. While scalability is also important in PADS research, the issue has become a more pressing concern in DIS because of

TABLE 12.2 Contrasting PADS and DIS Research

	PADS	DIS
Speed requirement	As fast as possible	Real time
Typical applications	VLSI circuits, telecomm, wargaming, transportation	Military training, entertainment, air traffic control, emergency planning
Performance metric	Speed-up	Realism, scalability
Simulation model	Single model	Federation of models
Distribution	Single site	Geographically distributed
Communication	Arbitrary latency; reliable	100–300 ms latency; unreliable
Network	Multiprocessor or LAN	LAN and WAN

the military's desire to expand DIS demonstration exercises to include more simulated entities and sites.

A second distinction between PADS and DIS simulations concerns the models themselves. To date, PADS work has been largely concerned with execution of a single large simulation model. The components of the model are usually developed in a single simulation environment using one language and are developed from the start as a single integrated system. Research has focused on simulation languages and tools for rapid development of large simulation models. There is no question of interoperability among the different components of the simulation because they are designed to do so from the start. The question of retrofitting an existing model to execute on a parallel platform has received only a modest amount of attention (see, e.g., refs. 60 and 61). By contrast, achieving interoperability among existing and new simulation models is central to much of the work in the DIS community and is one of the most difficult technical problems being attacked. The aggregate-level simulation protocol (ALSP) project is a second effort to bridge this gap by combining separately developed constructive simulators using a PADS synchronization protocols [62].

To date, most PADS research utilizes tightly coupled multiprocessors (using shared memory or message passing for communications) or LAN (local area network)-based distributed computing environments, while DIS exercises usually utilize LAN and WAN (wide area network) interconnects. Different assumptions are made by these communities concerning the network. PADS research generally assumes *reliable communications* but *arbitrary communications latency*. This originates from the analytic nature of typical PADS applications. By contrast, most DIS work assumes unreliable communications but bounded maximum latencies for message delivery. Unreliable communications are used because message acknowledgment protocols that are used to ensure reliable delivery may compromise real-time performance. Message losses can often be tolerated because many of the messages that are transmitted are simply periodic updates of state information (e.g., the position of vehicles), so subsequent messages allow the simulator to recover gracefully. Moreover, glitches in DIS simulations can often be tolerated as long as they do not happen too frequently and do not have lasting effects. Maximum latencies also stem from the real-time nature of early DIS applications. Because of human limitations to perceive nearly simultaneous events, latencies of up to 100 or 300 ms are acceptable [59]; Correct modeling of more tightly coupled systems [e.g., components of a weapons system (say, a simulated missile and its guidance system)] require lower latencies [63].

12.8.2 Dead Reckoning

DIS simulations use a technique called *dead reckoning* to reduce interprocessor communication to distribute position information. This reduction is realized by observing that rather than sending new position coordinates of moving entities at some predetermined frequency, processors can estimate the location of other entities through a local computation. In principle, one could duplicate a remote simulator in the local processor so that any dynamically changing state information is readily available. When applied to computing position information of moving entities, this local computation is referred to as the *dead-reckoning model* (DRM).

In practice, the DRM is only an approximation of the true simulator. An approximation is used because (1) the DRM does not receive inputs received by the actual simulator (e.g., a pilot using a flight simulator decides to travel in a new direction), and (2) to economize on the amount of computation required to execute the DRM. The DRM is realized as a simplified, lower-fidelity version of the true model. To limit the amount of error between the true and DRM, the true simulator maintains its own copy of the DRM to determine when the divergence between them has become "too large" (i.e., the difference between the true position and the dead-reckoned position exceeds some threshold). When this occurs the true simulator transmits new, updated information (the true position) to "reset" the DRM. To avoid "jumps" in the display when the DRM is reset, simulators may realize the transition to the new position as a sequence of steps [64].

12.8.3 Communications in DIS

Work in DIS is now attempting to scale exercises to include more entities and sites (locations). Current goals call for exercises including 50,000 entities at 30 sites by 1997 and 100,000 entities at 50 sites by 2000. Significant changes to DIS are required to enable simulations of this size, particularly with respect to the amount of communication that is required. Because DIS uses broadcasts, the number of messages that must be transmitted is proportional to N^2, where N is the number of entities.

Even with dead reckoning, the DIS protocol described above does not scale to such large simulations. As mentioned above, an obvious problem is the reliance on broadcasts. There are two problems here: (1) realizing the communication bandwidth required to perform the broadcasts, estimated to be 375 Mbits per second per platform for a simulation with 100,000 players [65], is too costly, and (2) the computation load required to process incoming PDUs is excessive and wasteful, particularly as the size of the exercise increases because a smaller percentage of the incoming PDUs will be relevant to each simulator.

Several techniques have been developed to address this problem [66]:

- *Relevance Filtering.* Rather than using broadcasts, messages are sent to only a subset of the simulation entities [65,66]. For example, the battlefield can be divided into a grid, and entities need only send state-update PDUs to entities in grid sectors in or near that generating the PDU. Relevance filters can be used for other information as well (e.g., radio communications).
- *Distributed Representation.* Information concerning remote entities can be stored locally. This is particularly effective for information that seldom changes. Like dead reckoning, remote computations can be replicated locally to generate dynamically changing data.

- *Compression.* Redundant information can be eliminated from the PDU. The *protocol independent compression algorithm* (PICA) uses a reference state PDU that is known to the communicating entities and only transmits differences from the reference state [67]. PICA has been reported to yield fourfold compression of entity state PDUs [66].
- *Bundling.* Several PDUs may be bundled into larger messages to reduce overheads.
- *Overload Management.* These mechanisms reduce the communications load during periods of high utilization. For example, dead-reckoning thresholds may be adjusted to generate less traffic when the network is loaded.
- *Fidelity Management.* Different degrees of detail can be sent to different entities. For instance, less frequent state updates can be sent to distant receivers than those in close proximity. This is particularly useful for *wide area viewers* such as aircraft that can "see" large areas at one time.

Relevance filtering and multicast communications go hand in hand, although it should be noted that relevance filtering still has merit even if multicast is not available. Multicast communications mechanisms provide for efficient transmission of messages where identical copies must be sent to many destinations. A multicast *group* refers to a set of destinations that should all receive a copy of any message sent to the group. Multicast is more challenging in DIS than other applications (e.g., teleconferencing or video on demand) because of the need for a large number of multicast groups and the dynamic nature of the groups. It is estimated that from 1000 to 10,000 active multicast groups (different sets of destinations) will be needed, with entities joining or leaving groups at a rate of hundreds per second [68]. Changes to multicast groups should occur with low latency (e.g., 1 ms).

12.8.4 Synchronization and Time Management

In DIS, the term *synchronization* usually refers to the problem of ensuring that the real-time clocks distributed throughout the network advance in synchrony with each other [63]. *Time management* refers to the method used to advance simulated time in each simulator (i.e., what the PADS community refers to as *synchronization*). The synchronization and time management mechanisms are responsible for ensuring that *temporal correlation* is achieved (i.e., temporal aspects of the simulation exercise correspond to real-world behavior). While PADS simulation protocols guarantee that all logical processes observe the same, time-stamped ordered sequence of events, DIS makes no such guarantees. This is a well-known problem in DIS today. Correlation problems can occur because:

- Messages may be lost. While DIS can tolerate some lost PDUs, as discussed earlier, loss of certain events such as detonation of ordinances could be more problematic.
- No mechanism is provided to ensure that events are processed in time-stamp order. PDUs may arrive out of order because of communication delay variations or differing (real-time) clocks in different simulators. Different simulators may perceive the same set of events in different orders, possibly resulting in different observed outcomes in different parts of the network. This is clearly undesirable.

In DIS, each PDU contains a time stamp with the *current* time of the simulator (obtained from the simulator's real-time clock) generating the PDU. This is in contrast to PADS simulations that typically generate events into the simulated future [i.e., with (simulated time) time stamp greater than the current time of the entity scheduling the event]. Thus events in DIS always arrive "late." Receivers can compensate by determining the communication delay in transmitting the message. *Relative time-stamp* schemes do this based on past message transmission times, and *absolute time-stamp* schemes assume synchronized real-time clocks in the sender and receiver to determine the latency simply by computing the difference between the send and receive times [69].

Much of the work in the DIS community to address the temporal correlation problem has been concerned with maintaining real-time clocks that are synchronized to a standard clock called *coordinated universal time* (UTC). Several approaches have been used for this task (see ref. 63). Methods include broadcasting UTC on radio services, use of a U.S. National Institute of Standards and Technology (NIST) dial-up time service, use of a global positioning system (GPS) used by radio navigation systems [70,71], and network protocols such as Network Time Protocol (NTP) to distribute clock information over the network [72]. The relationship between clock synchronization and temporal correlations is discussed in ref. 73.

In addition to timing and synchronization errors, other correlation problems may arise due to difference in the virtual environments perceived by different simulator nodes. A tank that believes it is hiding behind a tree may actually be visible to other simulators because of differences in spatial computations. This can lead to "unfair" scenarios that reduce the realism of the exercise.

Work in DIS encompasses a variety of other topics that are related to interoperability and producing realistic synthetic environments, as opposed to distributed execution. *Computer-generated forces* are artificial intelligence techniques to generate automated or semiautomated models for forces, enabling the number of simulation participants to be much larger than the number of personnel participating in the exercise. *Aggregation and deaggregation* algorithms enable interoperability between virtual simulators representing individual, deaggregated entities (e.g., individual tanks) and constructive simulators with aggregated entities (a column of tanks) by aggregating and deaggregating entities as needed. Work in *validation, verification, and accreditation* (VV&A) is concerned with defining appropriate performance metrics and measurement mechanisms to ascertain the extent that simulation exercises meet their goals. *Physical environment representation* is concerned with providing entities with common views of the battlefield in an environment changing because of constructed (e.g., introduction of craters when shells explode) and natural (e.g., roads washed out by thunderstorms) causes.

12.9 CONCLUSIONS

The goal of this chapter is to provide insight into the problem of executing discrete-event simulation programs on parallel and distributed computers. Conservative and optimistic simulation techniques offer a general approach requiring a spatial decomposition of the simulation program into logical processes that may execute concurrently on different processors. Optimistic methods offer the greatest potential as a general-purpose simulation engine where the application can be cleanly separated from the underlying execution mechanism. Conservative methods require specification of look-ahead which is in general difficult to accomplish without help from the programmer, but avoid some of the overheads and implementation complexities associated with optimistic mechanisms. Parallel simulation software executives (e.g., those developed in research labs and universities) are widely available (e.g., the Georgia Tech Time Warp software is available from the author of this chapter and is being used extensively in modeling air transportation systems and telecommunication networks). However, few commercial versions of parallel simulation executives are available at the time of this writing. Time-parallel simulation techniques offer massively parallel execution for simulation problems with little spatial parallelism (e.g., simulation of a single queue), but are currently limited in applicability to a handful of problems. Finally, simulation techniques developed in the DIS community relax causality constraints and focus on interoperability issues in constructing federations of autonomous simulations.

The state of the art in parallel and distributed simulation has advanced rapidly in recent years. The technology is now having a major impact in both the commercial and military simulation communities. As networking infrastructure improves to provide greater connectivity among previously isolated machines, one can expect distributed simulation, especially those embraced by the DIS community, to be much more common and have a much broader impact in the years ahead.

REFERENCES

1. Fujimoto, R. M. (1989). The virtual time machine, in *Proceedings of the International Symposium on Parallel Algorithms and Architectures*, pp. 199–208.
2. Jones, D. W. (1986). Concurrent simulation: an alternative to distributed simulation, in *Proceedings of the 1986 Winter Simulation Conference*, J. R. Wilson, J. O. Henriksen, and S. D. Roberts, eds., IEEE, Piscataway, N.J., pp. 417–423.
3. Jones, D. W., C.-C. Chou, D. Renk, and S. C. Bruell (1989). Experience with concurrent simulation, in *Proceedings of the 1989 Winter Simulation Conference*, E. A. MacNair, K. J. Musselman, and P. Heidelberger, eds., IEEE, Piscataway, N.J., pp. 756–764.
4. Reynolds, P. F., Jr. (1988). A spectrum of options for parallel simulation, in *Proceedings of the 1988 Winter Simulation Conference*, M. Abrams, P. Haigh, and J. Comfort, eds., IEEE, Piscataway, N.J., pp. 325–332.
5. Chandy, K. M., and J. Misra (1979). Distributed simulation: a case study in design and verification of distributed programs, *IEEE Transactions on Software Engineering*, Vol. SE-5, No. 5, pp. 440–452.
6. Bryant, R. E. (1977). Simulation of packet communication architecture computer systems, *MIT-LCS-TR-188*, Massachusetts Institute of Technology, Cambridge, Mass.
7. Su, W. K., and C. L. Seitz (1989). Variants of the Chandy–Misra–Bryant distributed discrete-event simulation algorithm, in *Proceedings of the SCS Multiconference on Distributed Simulation*, Vol. 21, pp. 38–43, SCS Simulation Series, March.
8. Peacock, J. K., J. W. Wong, and E. G. Manning (1979). Distributed simulation using a network of processors, *Computer Networks*, Vol. 3, No. 1, pp. 44–56.
9. Chandy, K. M., and J. Misra (1981). Asynchronous distributed simulation via a sequence of parallel computations, *Communications of the ACM*, Vol. 24, No. 4, pp. 198–205.
10. Misra, J. (1986). Distributed-discrete event simulation, *ACM Computing Surveys*, Vol. 18, No. 1, pp. 39–65.

11. Dijkstra, E. W., and C. S. Scholten (1980). Termination detection for diffusing computations, *Information Processing Letters*, Vol. 11, No. 1, pp. 1–4.
12. Liu, L. Z., and C. Tropper (1990). Local deadlock detection in distributed simulations, in *Proceedings of the SCS Multiconference on Distributed Simulation*, Vol. 22, pp. 64–69, SCS Simulation Series, January.
13. Ayani, R. (1989). A parallel simulation scheme based on the distance between objects, in *Proceedings of the SCS Multiconference on Distributed Simulation*, Vol. 21, pp. 113–118, SCS Simulation Series, March.
14. Chandy, K. M., and R. Sherman (1989). The conditional event approach to distributed simulation, in *Proceedings of the SCS Multiconference on Distributed Simulation*, Vol. 21, pp. 93–99, SCS Simulation Series, March.
15. Lubachevsky, B. D. (1989). Efficient distributed event-driven simulations of multiple-loop networks, *Communications of the ACM*, Vol. 32, No. 1, pp. 111–123.
16. Nicol, D. M. (1989). The cost of conservative synchronization in parallel discrete event simulations, *Technical Report 90-20*, ICASE, June.
17. Nicol, D. M. (1988). Parallel discrete-event simulation of FCFS stochastic queueing networks, *SIGPLAN Notices*, Vol. 23, No. 9, pp. 124–137.
18. Fujimoto, R. M. (1990). Parallel discrete event simulation, *Communications of the ACM*, Vol. 33, No. 10, pp. 30–53.
19. Nicol, D. M. and R. M. Fujimoto (1994). Parallel simulation today, *Annals of Operations Research*, Vol. 53, pp. 249–286.
20. Sleator, D. D. and R. E. Tarjan (1985). Self-adjusting binary search trees, *Journal of the ACM*, Vol. 32, No. 3, pp. 652–686.
21. Fujimoto, R. M. (1989). Performance measurements of distributed simulation strategies, *Transactions of the Society for Computer Simulation*, Vol. 6, No. 2, pp. 89–132.
22. Leung, E., J. Cleary, G. Lomow, D. Baezner, and B. Unger (1989). The effects of feedback on the performance of conservative algorithms, in *Proceedings of the SCS Multiconference on Distributed Simulation*, Vol. 21, pp. 44–49, SCS Simulation Series, March.
23. Bagrodia, R. L., and W.-T. Liao (1990). Maisie: a language and optimizing environment for distributed simulation, in *Proceedings of the SCS Multiconference on Distributed Simulation*, Vol. 22, pp. 205–210, SCS Simulation Series, January.
24. Cota, B. A., and R. G. Sargent (1990). A framework for automatic lookahead computation in conservative distributed simulations, in *Proceedings of the SCS Multiconference on Distributed Simulation*, Vol. 22, pp. 56–59, SCS Simulation Series, January.
25. Tinker, P. A., and J. R. Agre (1989). Object creation, messaging, and state manipulation in an object oriented Time Warp system, in *Proceedings of the SCS Multiconference on Distributed Simulation*, Vol. 21, pp. 79–84, SCS Simulation Series, March.
26. Jefferson, D. R. (1985). Virtual time, *ACM Transactions on Programming Languages and Systems*, Vol. 7, No. 3, pp. 404–425.
27. Lin, Y.-B., and E. D. Lazowska (1989). Determining the global virtual time in a distributed simulation, *Technical Report 90-01-02*, Department of Computer Science, University of Washington, Seattle, Wash.
28. Preiss, B. R. (1989). The Yaddes distributed discrete event simulation specification language and execution environments, in *Proceedings of the SCS Multiconference on Distributed Simulation*, Vol. 21, pp. 139–144, SCS Simulation Series, March.
29. Samadi, B. (1985). Distributed simulation, algorithms and performance analysis, Ph.D. thesis, University of California, Los Angeles.
30. Gafni, A. (1988). Rollback mechanisms for optimistic distributed simulation systems, in *Proceedings of the SCS Multiconference on Distributed Simulation*, Vol. 19, pp. 61–67, SCS Simulation Series, July.

31. Reiher, P. L., R. M. Fujimoto, S. Bellenot, and D. R. Jefferson (1990). Cancellation strategies in optimistic execution systems, in *Proceedings of the SCS Multiconference on Distributed Simulation*, Vol. 22, pp. 112–121, SCS Simulation Series, January.

32. Berry, O. (1986). Performance evaluation of the Time Warp distributed simulation mechanism, Ph.D. thesis, University of Southern California, May.

33. Som, T. K., B. A. Cota, and R. G. Sargent (1989). On analyzing events to estimate the possible speedup of parallel discrete event simulation, in *Proceedings of the 1989 Winter Simulation Conference*, E. A. MacNair, K. J. Musselman, and P. Heidelberger, eds., IEEE, Piscataway, N.J., pp. 729–737.

34. Lomow, G., J. Cleary, B. Unger, and D. West (1988). A performance study of Time Warp, in *Proceedings of the SCS Multiconference on Distributed Simulation*, Vol. 19, pp. 50–55, SCS Simulation Series, July.

35. West, D. (1988). Optimizing Time Warp: lazy rollback and lazy re-evaluation, M.S. thesis, University of Calgary, January.

36. Gates, B., and J. Marti (1988). An empirical study of Time Warp request mechanisms, in *Proceedings of the SCS Multiconference on Distributed Simulation*, Vol. 19, pp. 73–80, SCS Simulation Series, July.

37. Puccio, J. (1988). A causal discipline for value return under Time Warp, *Proc. SCS Multiconference on Distributed Simulation*, Vol. 19, pp. 171–176, SCS Simulation Series, July.

38. Jefferson, D. R., B. Beckman, F. Wieland, L. Blume, M. DiLorento, P. Hontalas, P. Reiher, K. Sturdevant, J. Tupman, J. Wedel, and H. Younger (1987). The Time Warp Operating System, *Proceedings of the 11th Symposium on Operating Systems Principles*, Vol. 21, No. 5, pp. 77–93.

39. Wieland, F., L. Hawley, A. Feinberg, M. DiLorento, L. Blume, P. Reiher, B. Beckman, P. Hontalas, S. Bellenot, and D. R. Jefferson (1989). Distributed combat simulation and Time Warp: the model and its performance, in *Proceedings of the SCS Multiconference on Distributed Simulation*, Vol. 21, pp. 14–20, SCS Simulation Series, March.

40. Presley, M., M. Ebling, F. Wieland, and D. R. Jefferson (1989). Benchmarking the Time Warp Operating System with a computer network simulation, in *Proceedings of the SCS Multiconference on Distributed Simulation*, Vol. 21, pp. 8–13, SCS Simulation Series, March.

41. Ebling, M., M. DiLorento, M. Presley, F. Wieland, and D. R. Jefferson (1989). An ant foraging model implemented on the Time Warp Operating System, in *Proceedings of the SCS Multiconference on Distributed Simulation*, Vol. 21, pp. 21–26, SCS Simulation Series, March.

42. Hontalas, P., B. Beckman, M. DiLorento, L. Blume, P. Reiher, K. Sturdevant, L. Van Warren, J. Wedel, F. Wieland, and D. R. Jefferson (1989). Performance of the colliding pucks simulation on the Time Warp Operating System, in *Proceedings of the SCS Multiconference on Distributed Simulation*, Vol. 21, pp. 3–7, SCS Simulation Series, March.

43. Fujimoto, R. M. (1989). Time Warp on a shared memory multiprocessor, *Transactions of the Society for Computer Simulation*, Vol. 6, No. 3, pp. 211–239.

44. Fujimoto, R. M. (1990). Performance of Time Warp under synthetic workloads, in *Proceedings of the SCS Multiconference on Distributed Simulation*, Vol. 22, pp. 23–28, SCS Simulation Series, January.

45. Fujimoto, R. M., J. Tsai, and G. Gopalakrishnan (1988). Design and performance of special purpose hardware for Time Warp, in *Proc. 15th Annual Symposium on Computer Architecture*, pp. 401–408, June.

46. Jefferson, D. R. (1990). Virtual time II: storage management in distributed simulation, in *Proceedings of the 9th Annual ACM Symposium on Principles of Distributed Computing*, pp. 75–89, August.

47. Lin, Y.-B., and B. R. Preiss (1991). Optimal memory management for time warp parallel simulation, *ACM Transactions on Modeling and Computer Simulation*, Vol. 1, No. 4.

48. Bagrodia, R., W.-T. Liao, and K. M. Chandy (1991). A unifying framework for distributed simulation, *ACM Transactions on Modeling and Computer Simulation*, Vol. 1, No. 4.
49. Gaujal, B., A. G. Greenberg, and D. M. Nicol (1993). A sweep algorithm for massively parallel simulation of circuit-switched networks, *Journal of Parallel and Distributed Computing*, Vol. 18, No. 4, pp. 484–500.
50. Lin, Y.-B., and E. D. Lazowska (1991). A time-division algorithm for parallel simulation, *ACM Transactions on Modeling and Computer Simulation*, Vol. 1, No. 1, pp. 73–83.
51. Heidelberger, P., and H. Stone (1990). Parallel trace-driven cache simulation by time partitioning, in *Proceedings of the 1990 Winter Simulation Conference*, O. Balci, R. P. Sadowski, and R. E. Nance, eds., IEEE, Piscataway, N.J., pp. 734–737.
52. Nikolaidis, I., R. M. Fujimoto, and A. Cooper (1993). Parallel simulation of high-speed network multiplexers, in *Proceedings of the IEEE Conference on Decision and Control*, December.
53. Andradóttir, S., and T. Ott (1995). Time-segmentation parallel simulation of networks of queues with loss or communication blocking, *ACM Transactions on Modeling and Computer Simulation*, Vol. 5, No. 4, pp. 269–305.
54. Ammar, H., and S. Deng (1992). Time warp simulation using time scale decomposition, *ACM Transactions on Modeling and Computer Simulation*, Vol. 2, No. 2, pp. 158–177.
55. Greenberg, A. G., B. D. Lubachevsky, and I. Mitrani (1991). Algorithms for unboundedly parallel simulations, *ACM Transactions on Computer Systems*, Vol. 9, No. 3, pp. 201–221.
56. Kanarick, C. (1991). A technical overview and history of the SIMNET project, in *Advances in Parallel and Distributed Simulation*, Vol. 23, pp. 104–111, SCS Simulation Series, January.
57. IEEE (1993). Standard for information technology: protocols for distributed interactive simulation applications, *IEEE Standard 1278*, IEEE, Piscataway, N.J.
58. IST (1994). Standard for distributed interactive simulation: communication architecture requirements, Institute for Simulation and Training, Orlando, Fla.
59. DIS Steering Committee (1994). The DIS vision, a map to the future of distributed simulation, *Technical Report IST-SP-94-01*, Institute for Simulation and Training, Orlando, Fla.
60. Tsai, J. J., and R. M. Fujimoto (1993). Automatic parallelization of discrete event simulation programs, in *Proceedings of the 1993 Winter Simulation Conference*, G. W. Evans, M. Mollaghasemi, E. C. Russell, and W. E. Biles, eds., IEEE, Piscataway, N.J., pp. 697–705, December.
61. Nicol, D. M., and P. Heidelberger (1995). On extending parallelism to serial simulators, in *Proceedings of the 9th Workshop on Parallel and Distributed Simulation*, pp. 60–67, June.
62. Wilson, A. L., and R. M. Weatherly (1994). The aggregate level simulation protocol: an evolving system, in *Proceedings of the 1994 Winter Simulation Conference*, J. D. Tew, S. Manivannan, D. A. Sadowski, and A. F. Seila, eds., IEEE, Piscataway, N.J., pp. 781–787.
63. Cheung, S., and M. Loper (1994). Synchronizing simulations in distributed interactive simulations, in *Proceedings of the 1994 Winter Simulation Conference*, J. D. Tew, S. Manivannan, D. A. Sadowski, and A. F. Seila, eds., IEEE, Piscataway, N.J., pp. 1316–1323.
64. Fishwick, P. A. (1994). *Simulation Model Design and Execution: Building Digital Worlds*, McGraw-Hill, New York.
65. Macedonia, M. R., M. J. Zyda, D. R. Pratt, D. P. Brutzman, and P. T. Barham (1995). Exploiting reality with multicast groups: a network architecture for large-scale virtual environments, in *Proceedings of the 1995 IEEE Virtual Reality Annual Symposium*, pp. 11–15, March.
66. Van Hook, D. J., J. O. Calvin, M. Newton, and D. Fusco (1994). An approach to DIS scalability, in *Proceedings of the 11th Workshop on Standards for the Interoperability of Distributed Simulations*, Vol. 2, pp. 347–356.
67. DiCaprio, P. N., C. J. Chiang, and D. J. Van Hook (1994). PICA performance in a lossy

communications environment, in *Proceedings of the 11th Workshop on Standards for the Interoperability of Distributed Simulations*, Vol. 2, pp. 363–366.

68. Miller, D. C. (1995). A brief history of distributed interactive simulation (presentation notes), in *Distributed Interactive Simulation IP/ATM Multicast Symposium*, May.
69. Golner, M., and E. Pollak (1994). The application of network time protocol (NTP) to implementing DIS absolute timetamps, in *Proceedings of the 11th Workshop on Standards for the Interoperability of Distributed Simulations*, Vol. 2, pp. 431–440.
70. Kress, J., J. R. Phipps, and D. Carver, Jr. (1994). Synchronization of large scale distributed simulations and programs, in *Proceedings of the 10th Workshop on Standards for the Interoperability of Distributed Simulations*, Vol. 2, pp. 611–623.
71. Forbes, J. (1994). Synchronization and absolute time stamping in the DIS environment, the BFTT method, in *Proceedings of the 10th Workshop on Standards for the Interoperability of Distributed Simulations*, Vol. 2, pp. 625–627.
72. Mills, D. L. (1991). Internet Time Synchronization: The Network Time Protocol, IEEE Transactions on Communications, Vol. 39, No. 10, pp. 1482–1493.
73. Katz, A. (1994). Synchronization of networked simulators, in *Proceedings of the 11th Workshop on Standards for the Interoperability of Distributed Simulations*, Vol. 2, pp. 81–87.

CHAPTER 13

On-Line Simulation: Need and Evolving Research Requirements

WAYNE J. DAVIS
University of Illinois

13.1 INTRODUCTION

In this chapter we explore the future evolution of simulation modeling and analysis techniques. It is obvious that no one can predict the future with certainty. However, it is immediately evident that today's systems are more complex than their predecessors and that these systems will affect future simulation tools and the associated methods that are employed for simulation analyses. One class of systems that has significantly influenced the development of simulation tools is the flexible manufacturing system (FMS). Since FMSs have been introduced, several simulation tools, (e.g., AweSim, ARENA, and WITNESS, among others), now include provisions for modeling material handling systems (see Chapter 14). As new classes of systems are developed, this evolutionary trend will continue.

The flexible manufacturing system is only one example of a new class of large-scale discrete-event systems that are being developed or are in the conceptual design stage. Other systems include advanced air or vehicular traffic control systems (Chapter 16), real-time battle management systems (Chapter 19), emergency response systems, and so on. In fact, one goal within the manufacturing arena is to integrate several flexible manufacturing systems into a much larger enterprise system to provide an agile manufacturing environment.

The desire to design and operate large-scale, discrete-event systems has evolved based on two distinct realities. First, there are the real-world needs. For example, in most metropolitan areas, vehicular traffic (especially at rush hour) has become a major problem. The general consensus is that a system must be developed to coordinate the traffic flow rather than relying solely on the combined actions of the individual vehicle operators, who can only make decisions within their immediate domain while attempting to optimize their individual performance objectives. Someone, or something, must consider the performance of the entire traffic system.

Handbook of Simulation, Edited by Jerry Banks.
ISBN 0-471-13403-1

Second, computer and networking technologies have now provided new opportunities to explore such systems. As recently as a decade ago, such systems would have been much more difficult, if not impossible, to construct. Even now, some questions remain about whether the current information-processing capabilities are sufficient. However, few believe that current capabilities have reached their ultimate potential. In fact, these capabilities continue to increase at an exponential rate, and their ultimate limits (if they exist) are unknown.

Desire plus information-processing capability do not solve the problems, however. A case in point is the development of an air traffic control system. The system currently employed in the United States was developed in the 1970s and is long overdue for replacement. Yet even given advancements in both computer and software technologies, engineers and programmers have spent years and millions of dollars trying to provide a replacement system. When will a new system be available? No one is really certain.

A key element in the implementation of these systems is the specification of the control architecture that manages their operation. The complexity of these systems will never permit the definition of a single monolithic controller that can manage the entire system. Could a single controller ever manage a Fortune 500 corporation or the vehicular traffic flow pattern for the Chicago metropolitan highway system? The overall control requirements must be distributed within a control architecture. There is still another major concern. In the real-time management setting, planning and control cannot be addressed independently. That is, in the real-time environment, a controller cannot rely upon another, separate planning entity to perform its planning. Each controller must plan its own strategy, which it then implements.

Researchers have already recognized the need for integrated planning and control in the real-time management of simple systems such as robotics and have developed a new class of intelligent controllers. (The interested reader may refer to the annual *Proceedings of the IEEE International Symposium on Intelligent Control* or the numerous other control journals.) Significant advances have been made in the intelligent control technologies, but these results are not sufficient to address the emerging class of large-scale discrete-event systems. The present technology for intelligent control addresses primarily the management of a single subsystem only. The coordinated operation of more than one subsystem has been considered only in a very limited sense. Most systems under analysis are continuous state in nature (i.e., they can be modeled by differential equations), and there has only been minimal consideration of discrete-event systems.

The design, implementation, and real-time management of large-scale systems will require the coordinated operation of numerous (mostly discrete-event) subsystems. This coordinated approach will require that sophisticated intelligent controllers be defined for each subsystem to address planning and control requirements in real time. Because every intelligent controller must plan its own actions, each controller will be able to plan only to the extent that it can implement. Furthermore, each controller's actions must also be coordinated with the actions of the other intelligent controllers in order to provide a coordinated response for the overall system. Because of the need for each subsystem to interact with other subsystems, we refer to this emerging class of intelligent controllers as *coordinators*.

Once an intelligent control (coordination) architecture is proposed, the consequences on the overall system behavior must be tested before the system can be implemented. That is, the response of the system must be simulated under a variety of operating scenarios. Unfortunately, today's simulation tools are not capable of modeling the interactions among the controllers, and therefore, they cannot accurately access the constraints that a proposed control hierarchy imposes on operation of the system. In this respect, the system designer cannot verify a proposed design before it is implemented. The consequences derived from this limitation have already been experienced in the design and operation of FMSs. Few, if any, FMSs have achieved their anticipated performance goals, and many have operational defects, such as a tendency to deadlock. Mize et al. (1992) have concluded that the current simulation tools should not be used to predict the absolute performance characteristics of FMSs. Rather, their use should be restricted to comparing the relative performance of alternative systems only. In short, Mize is saying that current simulation tools *cannot accurately predict* the performance of this class of systems.

Over the last decade, our research laboratory has had an opportunity to work with several real-world FMSs. Our efforts to model these systems have uncovered several deficiencies in the modeling approaches that are being employed by the current simulation tools (see Davis et al., 1993). In the next section we attempt to describe these deficiencies. We also define a new modeling approach that we believe can address these deficiencies. In developing this modeling approach, we define a possible intelligent control architecture for managing these systems. We develop the functions that must be addressed by each controller and demonstrate the evolving need for on-line simulation analyses. An overview of the integrated solution approach for the modeling planning and control of these systems which serves as the basis for this chapter will appear in Davis, Macro and Setterdahl (1998).

On-line simulation is an evolving technology (see Chapter 21) for which there is currently minimal theoretical guidance. In Section 13.3 we contrast two types of planning: off-line planning, which employs the current simulation technologies; and on-line planning, which requires on-line simulation analysis. Based on this discussion, several individual facets of the on-line simulation approach are then discussed, and future research requirements are outlined. In particular, we explore the nature of the advancements that must be made in the areas of input analysis (Chapter 3), output analysis (Chapter 7), experimental design (Chapter 6), optimization (Chapter 9), and validation and verification (Chapter 10). The chapter ends with a brief description of our current research efforts toward implementing on-line simulation analyses.

13.2 EVOLUTION OF NEW SIMULATION MODELING APPROACHES

To provide an illustrative example of a complex discrete event, in this section we focus on modeling of a flexible manufacturing system (FMS). This section begins by describing a generic FMS using a conventional modeling approach. Then a more descriptive model is formulated in order to demonstrate the limitations of the current modeling approach. Next, an alternative modeling approach is proposed. Finally, using this new modeling approach, we speculate on what future modeling capabilities exist for these systems.

13.2.1 Observations Pertaining to the Capabilities of Conventional Simulation Approaches

In Figure 13.1 we provide a schematic for a generic FMS within which job entities of several part types are processed. For this generic FMS, we assume that there is an entry mechanism that introduces the job entities into the system and an exit mechanism where

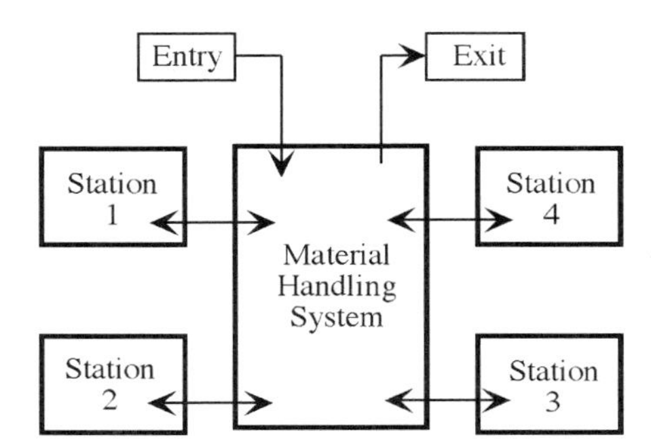

Figure 13.1 Schematic diagram for a typical manufacturing cell.

these depart from the system. The entry mechanism for most simulation models specifies a probabilistic distribution for interarrival times between successive job entities. This entry mechanism also assigns a given part type to the arriving job entity. Other attributes pertaining to job entity can also be initialized. For example, the arrival time for the entity is often recorded.

While the entity resides within the FMS, it will visit one or more of the four workstations included within the FMS. The stations that a given job entity will visit and the order in which these visits will occur depend on the part type to which the job entity belongs. Today, most simulation tools permit the modeler to specify the sequence of stations that will be visited by each entity of a given part type. Typically, the definition of the sequence requires that the modeler first specify a label for the sequence that corresponds to the part type. Then the modeler specifies the sequence of stations that will be visited. For each station that is visited, the modeler may also be permitted to specify a set of attribute values that will be assigned to an entity before it arrives at the next station in its visitation sequence. Based on these observations, these sequences can be viewed as a simplified specification for the process plan to manufacture a given part type.

Once an entity arrives at a given station, it is typically placed into a queue at the beginning of a subnetwork of nodes that the entity will traverse while it resides at the station. Consider a typical station subnetwork as shown in Figure 13.2. After the entity arrives at the station, it joins the initial queue, where it waits until an operator and the machine become available. When both resources are available, a setup operation then occurs with a duration equal to the value stored in the entity's SetupTime attribute. The value of the entity's SetupTime attribute is typically initialized before the entity arrives at the station using the specifications made within the sequence statement for the part type to which the job entity belongs. After the setup operation is completed, the operator could be freed and the actual processing would occur. This would be modeled by a delay equal to the value stored in the entity's RunTime attribute. After this delay occurs, the machine would be freed and the entity would join the output queue for the workstation, where it would wait until the material handling system transfers the entity to the next workstation.

Most modern simulation tools provide a significant capability for specifying the material handling system's characteristics. These capabilities include options to model transport systems such as automated guided vehicle systems or conveyors. Most simulation languages also include basic modeling elements which permit an entity that is

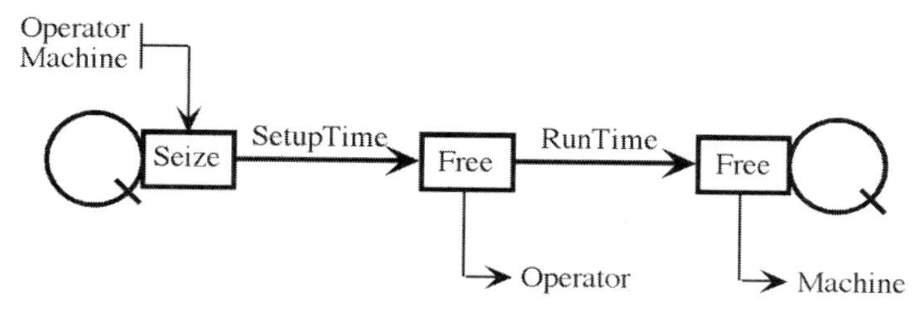

Figure 13.2 Example network for job entity flow within a workstation.

waiting at a given station to request the services of the material handling system in order to transport it from the current station where it resides to the next station in its visitation sequence. When the entity gains access to the material handling system, the station index, which is stored as one of the entity's attributes, is incremented to the next station that is to be visited and the transporter then begins to transport the entity to that next station. At the same time, the entity's attributes that are to be employed at the next station, such as SetupTime and RunTime, are updated using the specifications for the assignment of attribute values that were included by the modeler within the station visitation sequence.

We have now reached a point where we can begin discussing the deficiencies of the current modeling approaches. First, even though the discussed system is a flexible manufacturing system, the modeler is typically limited to specifying a single visitation sequence for each part type. That is, the modeler usually cannot specify alternative sequences for the same part type. The reason for this limitation is that the decision about what happens next is dependent on the current state of the entity and of the node at which it resides.

As example of this limitation, consider the FMS discussed in Flanders and Davis (1995). In this FMS, there were seven milling machines. Five of the milling machines were dedicated to machining steel parts and the remaining two milling machines were dedicated to machining aluminum parts. The tooling employed in the machining processes was optimized so that any machine dedicated to a given material type (i.e., steel or aluminum) could perform any required machining operation on any part of a given material type. Each part typically required three distinct fixturings (i.e., mounting the part on a fixture) to be employed. After a given part was fixtured, there was a choice of either five milling machines if the part was steel, or two milling machines if the part was aluminum, where the part could be assigned for machining. The decision to which machine the part was to be assigned depended on the loading of all the machines dedicated to machining the material type for the requesting part and the remaining tool life for the tools that resided at a given machine. In this example there simply was no way that the modeler could specify the requirements for assigning the part to a machine using a station visitation sequence.

With current modeling technologies it is very difficult to manage the behavior of the entity by considering the state of a given subsystem of nodes and the entities that reside in those nodes. Simply stated, specifying the controllers that can make decisions on what to do next is nearly impossible using current modeling technologies. Most languages do permit the modeler to introduce programming patches that will allow more complex routing logic to be considered, but employment of these patches usually makes the model much more complex to debug and validate.

When the model includes a material handling system, the simulation tool employed provides the controller for managing that material handling system. This controller is typically coded within the simulation software and is not accessible to the modeler. What the modeler provides is the specification for basic layout of the material handling system and the points in the model where an entity can either access or egress from the material handling system. While the entity is under the control of the material handling system, the modeler has little or no access to the control logic that governs the behavior of the entity. There are several situations where this limitation can be a significant problem. One particular situation arose when we desired to manage the placement of entities on a conveyor. Most simulation tools will simply place an entity on a conveyor whenever the first conveyor position of sufficient width reaches the entity that is waiting to access the conveyor. In certain cases it may be desirable if the conveyor position is not given to first entity that is reached, as it may be more beneficial for the conveyor position to be allocated to an entity that will be reached later.

The ability to model material handling is also limited by the types of material handling devices provided by the simulation tool. Few, if any, simulation tools provide the capability to model automated storage and retrieval systems. In fact, the capabilities for modeling material handling systems within most current simulation tools will not permit an entity to be stored within the material handling system. Rather, the modeling capabilities are provided only as a means for transferring an entity from one location to another while operating under the control logic provided by the vendor of the simulation tool.

Now, let us assume that the generic FMS depicted in Figure 13.1 is dedicated to machining discrete parts. Let us assume further that the system contains dedicated material handling systems for both the job entities, representing parts to be machined, and the tooling required to perform the machining at a given station. In the literature, few simulation models for FMSs of this complexity are discussed, yet most real-world flexible manufacturing systems for discrete-part machining require some mechanism for delivering tools to the workstations. If one attempts to model an FMS of this complexity using current simulation tools, he/she will immediately observe additional deficiencies with these tools.

The first deficiency is that most current simulation tools are focused primarily on modeling a single basic entity type. For most manufacturing applications, this is the job entity. In the scenario proposed, there are two basic entity types arriving at a station: jobs and tools. The desire to consider two types of entities will typically require that the modeler write two sets of network logic for each station, one for the job entity and one for a tool entity. When an entity arrives, the modeler must first test the entity's attributes to determine what type of entity has arrived and then invoke the proper network logic to handle that basic entity type. Second, the modeler must specify the order in which a given tool will be used in the processing of a given part type at the given station. The capabilities for specifying the station visitation sequences will not permit this specification to be made. Again, the modeler will have to resort to programming patches to store this more detailed processing information. Third, the flow of the job entity along its dedicated station's subnetwork must be synchronized with the flow of the tool entity along its subnetwork at the given station. That is, the overall run-time delay for the job entity is dependent on the delays associated with loading each tool into the machine spindle's chuck, executing the NC-machine code for cutting to be performed with the tool, and finally, unloading the tool from the machine spindle's chuck. Fourth, the modeler must provide a mechanism for modeling the systemwide tool management system. Tools have a finite life, after which they must be replaced. In addition, most real-world FMSs will permit the sharing of tools among machines. That is, the tools represent resources that can be moved from one machine to another, depending on where they are currently needed. Furthermore, a given tool can be used at several different points in a given processing plan, and may also be used in several different process plans. This implies that the tools can also be employed by more than one machine. In fact, a goal in designing the collective set process plans for the parts to be manufactured within a given cell is often to minimize the total number of tool types required to manufacture entire set of part types within the cell. In general, tooling is very expensive. Hence tool management is a critical concern, and its consequences upon the production flow must be considered within the model.

To make the foregoing deficiencies more concrete, consider the schematic for a generic automated machining workstation depicted in Figure 13.3. This automated workstation is comprised of several processes each with its own controller. The overall coordination of the actions occurring within the workstation is managed by the workstation controller. When a job entity arrives at the machining workstation, the workstation controller issues two commands to separate processes. First, it tells the part carousel to move an empty storage position to a location where the incoming job entity can be stored. It then tells the part loader to remove the part from the material handling system and place it on the empty part storage position. A similar situation occurs when an incoming tool entity arrives at the workstation. Parts and tools can be returned to their respective material handling systems at the FMS's cell level by reversing the procedure.

When a job entity is to be processed, the part carousel is first instructed to position the job entity at a location so that it can be loaded into the work area using the part exchanger. The part exchanger is then instructed to load the job entity into the machine's work area. After the processing plan has been downloaded into the station controller and after the location of the part within the work area has been calibrated (which represents

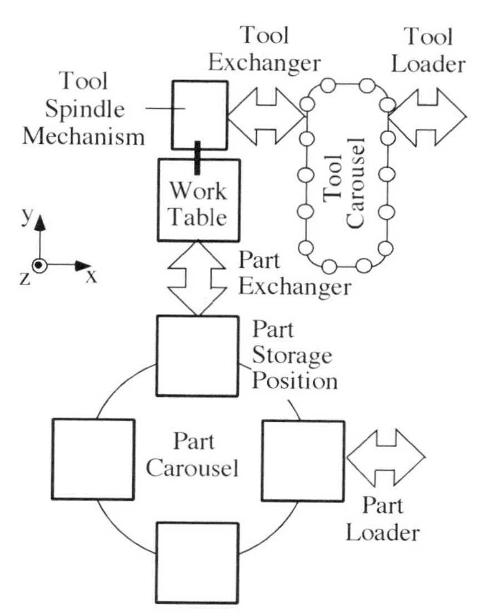

Figure 13.3 Schematic for a typical automated workstation with a discrete-machining FMS.

an element of the set up procedure), the workstation controller rotates the tool carousel to the position that will permit the first tool to be employed in the process plan to be loaded into the tool mechanism by the tool exchanger. The NC-machining instructions to be implemented with that tool are then downloaded into the controller that manages the worktable and tool spindle mechanism. After this instruction is executed, the tool exchanger is instructed to remove the current tool from the spindle and to return it to a given position on the tool carousel. The next tool is then loaded into the spindle mechanism and the process is repeated.

After all the processing steps have been implemented for the given job entity, the part exchanger is then instructed to remove the job entity from the work area into a given position on the part carousel. The workstation controller then notifies the cell controller within the FMS that the job entity has finished its assigned processing at the given station and that the job entity is now ready for removal from the station.

It should be noted that several tasks can occur concurrently at the workstation. For example, both a tool and a part can be loaded into the workstation while another part is being processed in the work area. In addition, there can be contention for a processes. For example, a tool loading and tool exchange both require access to the tool carousel. In short, the workstation controller must perform real-time planning in order to schedule the order in which pending tasks will be implemented.

Although the schematic for the workstation is fairly simple, its complexity cannot be modeled by most simulation languages. First, there are no provisions for modeling the workstation controller that is managing the individual processes within the workstation. Second, there are no modeling mechanisms for handling the sequence of tools required in a given processing plan for a given part type residing at the workstation. Third, there are no mechanisms for modeling the carousels. Recall that programming elements provided by most simulation languages will not permit the storage of entities within the material handling system.

The operations of the processes for the generic workstation are nevertheless discrete-event in nature. These discrete events represent the start and finish events associated with the execution of each task at a given process. Furthermore, the duration of most of these tasks is nearly deterministic. For automated equipment we can typically compute the time required to move a carousel from one location to another with very high accuracy. Similarly, we can typically compute the time required to implement a segment of a machine's NC code within a few milliseconds.

However, we cannot model the dynamics of this generic workstation today with conventional simulation languages and we certainly cannot incorporate its dynamics into the model for the overall FMS. Instead, we ignore the internal dynamics of the workstation and make probabilistic approximations for the time that will be required to implement any given aggregate processing task, such as those illustrated in Figure 13.2. Perhaps this modeling approach is acceptable when we are attempting to make steady-state estimates for the system performance, but it simply is not sufficiently accurate when we must address on-line planning and control concerns.

Even the use of this current modeling approach in order to make steady-state approximations is suspect. As stated earlier, most simulation models of FMSs ignore the flows of presumed secondary entities, such as tooling, fixturing, and even processing information. This assumption to ignore the flow of secondary entities is valid only if these flows do not represent a constraint on the overall system performance. Yet in numerous cases where we have modeled real-world discrete-part machining FMSs, the assumption to ignore the flow of secondary entities has not been valid. In every case the flow of the presumed secondary resources has imposed a significant constraint on the overall production flow (see Hedlund et al., 1990; Dullum and Davis, 1992; Flanders and Davis, 1995).

Additionally, when we have used detailed models to develop statistical estimates of the steady-state performance of an FMS and compared them to the state-state estimates associated with simplified models employing the conventional modeling approach, we have discovered that the steady-state estimates are not the same. Invariably, the approximate model significantly overestimates the projected performance of the system. This observation confirms the statement made by Mize et al. (1992), which concluded that current simulation tools should not be used to project the absolute performance of a FMS. Their observations also confirm another finding that we have made in our interaction with real-world FMSs. To date, we have not found an operational FMS that has achieved the performance projections based on its preliminary simulation analysis. This fact may be a major reason for manufacturers' current disinterest, if not distrust, of FMSs.

Finally, the development of steady-state performance statistics for a FMS is an oxymoron. If we are to operate the system in a flexible manner, a steady-state operation condition can never be achieved. In short, we are trying to estimate statistics for a situation that can never occur. This situation within a flexible manufacturing scenario must be contrasted against that of the assembly line. For an assembly line, steady-state operating conditions can be specified, and here it makes sense to estimate steady-state performance.

13.2.2 Toward a New Modeling Approach

In Figure 13.4 we attempt to summarize the interaction among subsystems and the flow of entities that arise in the operation of an FMS. In this figure we have established two basic flow planes for the interactions among the controllers that are managing the system. The vertical plane deals with the controller interactions associated with the flow of job entities; the horizontal plane addresses the flow of tool entities. For each entity type, we have indicated the basic controllers that are involved, including the cell-level and machine-level controllers for both the material handling and tool handling systems. When a job entity arrives at the station, it is initially under the control of the cell material handling system (MHS). When the workstation accepts control of the job entity at its arrival event (A), the control of that entity is passed from the cell controller to the workstation controller and then to the machining station's material handling system. The control of the entity remains there until the processing of the job entity is initiated. At that point, the control of the entity is returned to workstation and the start job (S_J) event occurs. After the station controller downloads its basic task instructions to the process controller, the start task (S_T) event occurs and the process controller assumes control of the job entity.

Arrival events also occur for tools, at which point control of the tool is passed from the cell tool handling system (THS) to the machining station THS via the cell and the workstation controllers. The control of the tool remains with the station THS until it is needed to execute a processing instruction. When this occurs the tool is loaded into the process, and the process controller assumes control of the tool entity. At this point, a start instruction (S_I) occurs. After each instruction is completed, a finish instruction event occurs (F_I), and control of the tool is returned to the station THS when it is removed from the machine's spindle. This process is repeated for all the remaining tools that are needed to complete the set of processing instructions within a processing task.

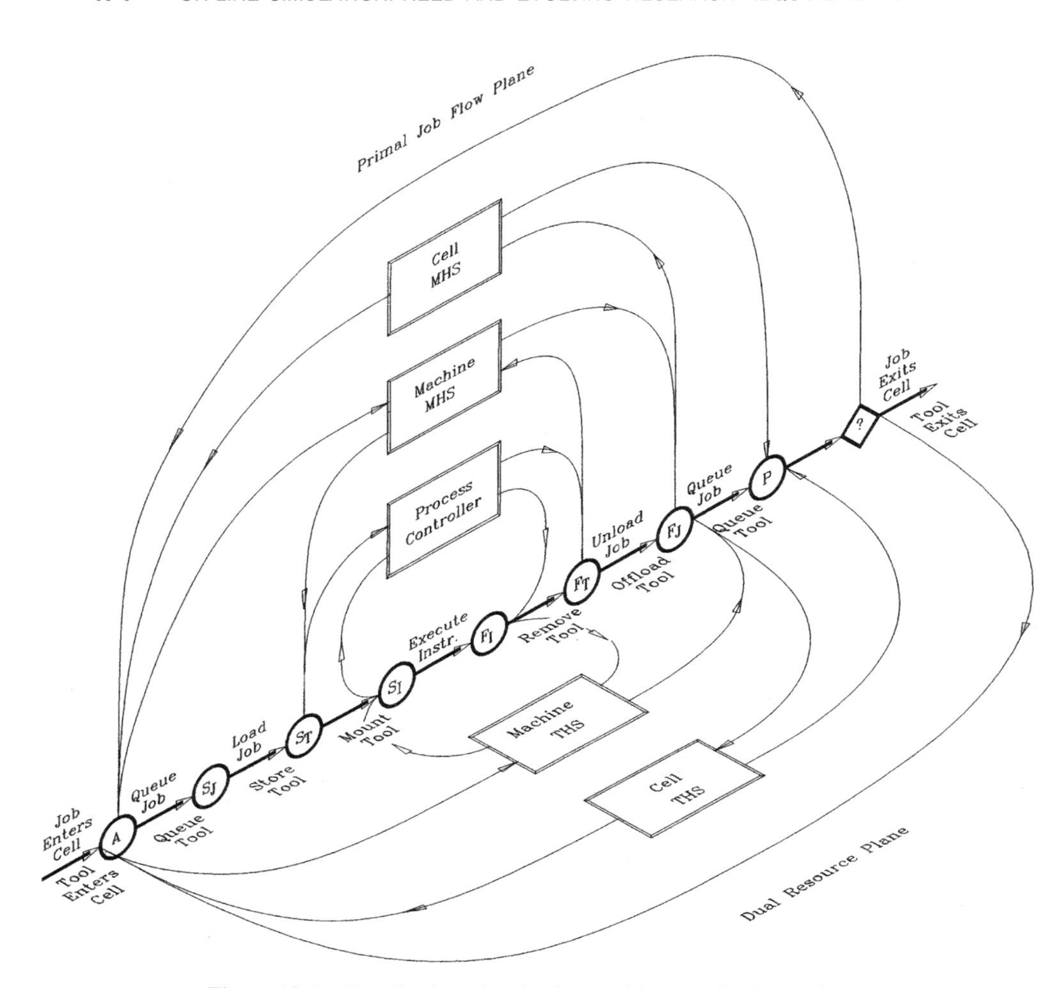

Figure 13.4 Coordination of entity flows with controller interactions.

After all the instructions within a task are completed, a finish task (F_T) event occurs. At this point, the control of the job entity is returned through the station controller to the station's MHS. However, since the processing on this job entity is completed, the control of the job disposition is actually returned to the cell controller, which then passes that control to the cell MHS, which then schedules its pickup (P) event. When the pickup event occurs, the cell MHS takes physical control of the job entity. The cell MHS is then instructed by the cell controller to deliver the job entity to the next station. This task assignment will lead to the next arrival event at the next station that is to be visited by the job entity. The tool entities may become worn or no longer needed at a given workstation. To this end, pickup events occur for tools also.

In reality, Figure 13.4 represents a significant simplification of the real-world scenario. There are many more entity-type planes that could be considered. For example, we could provide a plane for the flow of task description (i.e., process plans) information. We have already encountered FMSs where it can take several minutes to download the detailed processing information to a given workstation.

Despite this simplification, however, we can make several important generalizations based on this figure. First, the events associated with the operation of these systems actually represent moments in time where controllers interact. Second, when controllers interact, there is typically a reassignment of an entity's ownership from one controller to another. While a controller has ownership of an entity, typically one or more tasks will be assigned to the controller for implementation upon or with the entity by the controller's supervisor. After the execution of assigned tasks has been completed, the physical control of the entity remains with the subordinate controller to which the task was assigned, but the control for the disposition of the entity is returned to the supervisor, who may then ask the subordinate to perform another task on the entity or arrange the necessary material handling tasks which will allow the subordinate controller to return the physical entity to the supervisor when the resulting pickup event occurs.

Based on this discussion, we can now define three sets of commands that a supervisory control can provide to a subordinate controller. The first command is for the subordinate to accept an entity into its control domain. The second is for the subordinate to perform one or more tasks upon or with an entity within its control domain. The third command is for the subordinate to return an entity or remove it from its control domain. Using the principle of feedback control, each of these commands will require that the subordinate controller provide the supervisory controller a feedback response in order to update the supervisor about the subordinate's success in fulfilling the requirements of the command.

By now it is certainly clear that we must model the interactions among the controllers that are managing these complex discrete-event systems in order to accurately portray the state evolution interactions. Yet the modeling elements provided by the current simulation tools do not permit us to address these controller interactions. Rather, we must seek a new modeling element, template if you like, which will permit the modeler to decompose the overall system into a hierarchical collection of subsystems where each subsystem will be managed by an intelligent controller. This hierarchical architecture assumes that the intelligent controller within a given subsystem will have a supervisory controller that assigns tasks to it. The intelligent controller for the subsystem will also serve as the supervisory controller to the intelligent controllers for the subsystems that reside within the considered subsystem.

Based on the observation that a given subsystem can be contained within another supersystem and that the given subsystem itself can contain internal subsystems, the desire is to develop the subsystem modeling template such that it can be recursively employed in order to define the necessary levels of hierarchical subsystems that are needed to model the overall system. This template now exists. The first formulation of this modeling template was developed in collaboration with Motorola Corporation and published in Tirpak et al. (1992a). Since its original publication, the modeling template has been further generalized and is now termed the *coordinated object*. The schematic for the coordinated object is depicted in Figure 13.5.

Let us first compare this new modeling template to the structure of the generic FMS pictured in Figure 13.1. In Figure 13.1 the generic FMS has an entry mechanism. In Figure 13.5 the coordinated object has an input inhibit flag and input queue. The input port through which the entity enters the coordinated object belongs to the coordinated object's supervisor and has not been included in Figure 13.5. When an entity arrives at a given

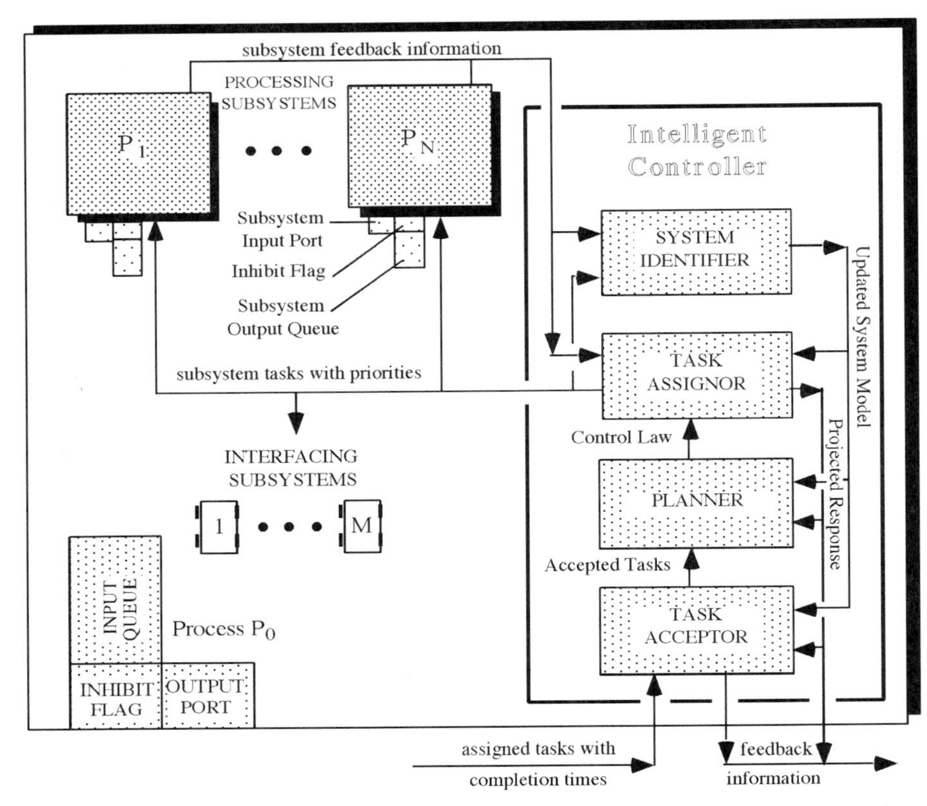

Figure 13.5 Schematic of the coordinated object: basic module for planning and control.

coordinated object, it is still under the physical control of the supervisor and resides in the input port. The supervisor of the coordinated object then instructs the controller within the coordinated object to accept the entity into its control domain. After the coordinated object's controller accepts the entity into its control domain, the entity then joins the input queue, which belongs to the coordinated object. The input inhibit flag is also controlled by the coordinated object's controller. It can prevent entities from entering the coordinated object. Its inhibiting actions can be comprehensive or selective. For example, if a given processing subsystem within the coordinated object is inoperative, the coordinated object may prevent any entity requiring the services of that processing subsystem from entering the coordinated object. On the other hand, if the input queue is full, the coordinated object may prevent any other entity from entering the system.

In Figure 13.1 there is also an output mechanism for entities exiting the FMS. The coordinated object contains an output port, an output inhibit flag, and an output queue. The output port belongs to the coordinated object. When a coordinated object finishes its assigned tasks upon a given entity, it is placed into the output port and the supervisor is notified. When the supervisor requests that the entity be returned, the entity joins the output queue, which belongs to the supervisor. The supervisor also manages the output inhibit flag, which can prevent entities from being returned to it.

An inspection inside the coordinated object indicates that it possesses N processing subsystems. These correspond to the workstations in Figure 13.1 and represent subsystems (processes) where processing tasks will be executed. The coordinated object also contains interfacing subsystems. The material handling system included in Figure 13.1 represents one type of interfacing subsystem. The interfacing subsystems perform tasks that support the production, such as transporting entities. The distinction between interfacing and processing subsystem will be further delineated shortly.

An important feature of the coordinated object is that the intelligent controller managing the collection of subsystems within the coordinated object is included as a critical element of the coordinated object. The intelligent controller is responsible for accepting tasks to be performed on or with entities within its control domain from its supervisory controller. These tasks are then decomposed into a set of subtasks that can be executed at the subsystems. When these subtasks are defined, the intelligent controller must schedule the execution of each subtask at one of its subordinate subsystems. These subtasks are then assigned to the intelligent controller within each subsystem after the entities are moved into the appropriate subsystem's control domain.

The overall specification of the tasks that can be executed by the processing subsystems is contained within the process plan database for the part type of the entity upon which a processing task is defined. The process plans are thus employed to decompose the tasks that have been assigned to the coordinated object into subtasks that can be scheduled for execution at its subordinate processing subsystems. However, the execution of processing tasks at a processing subsystem usually requires that the entity, as well as other entities that support its execution such as tooling, be delivered to the assigned processing subsystem's control domain. To implement the basic processing subtasks, additional supporting subtasks must be defined that will be executed by the interfacing subsystems. The task descriptions for executing the supporting tasks are typically not dependent on the information contained within the processing plan. Rather, the manner in which an interfacing subsystem will execute a task is already known by the controller. For example, the material handling system will typically move any job entity, regardless of its part type, from one location to another using the same basic instruction set. The same situation is true for the tool handling system in its delivery of tools. The local area network also transfers information contained within the detailed processing plan from the cell controller to a workstation controller. All of these subsystems are interfacing subsystems.

As stated before, it is our desire to employ our modeling template recursively, *the coordinated object*, in a manner that will permit us to define the complete hierarchical control architecture for the system. To understand this approach, consider now a workstation in Figure 13.1 which represents a processing subsystem within our cell-level coordinated object. Using the schematic of a generic machining workstation as pictured in Figure 13.3, we continue our decomposition using the coordinated object template. First, the workstation is also a coordinated object. Let us assume that it contains a single processing subsystem that includes the worktable and machine spindle and two workstation-level interfacing subsystems: the workstation material handling system and the workstation tool handling system. All three of these subsystems are also coordinated objects. The station-level processing subsystem contains a single controller that coordinates the individual controllers that are managing the worktable and the machine spindle. The last individual process controllers are not coordinated objects because they do not have any subordinate subsystems. The station-level material handling system is also a coordinated object that manages the individual controllers for the part carousel, part loader, and part exchanger. Similarly, the station-level tool handling system is a coordinated object whose controller manages the tool carousel, the tool loader, and the tool exchanger.

As we have recursively applied our modeling template, the coordinated object, to the generic cell in Figure 13.1 and the automated machining station in Figure 13.3, we now see that at least four levels of hierarchical control have emerged. At the lowest level of this hierarchy are the basic processes that execute the basic primitive tasks resulting from the overall task decomposition process. For example, the worktable may be instructed to move to a given *x-y* position while traversing along a specified trajectory while the machine spindle process may be instructed to rotate with a given angular velocity. At the higher levels, the intelligent controller is performing the necessary planning, which eventually results in the primitive tasks that are being executed at the lowest-level processes.

We call the resulting control hierarchy, the recursive object-oriented coordination hierarchy (ROOCH). The development of the ROOCH for the overall system results in two important specifications. First, it specifies a set of hierarchically nested subsystems, which includes every process contained within the overall system. Second, it specifies the controller for each subsystem and delineates for each controller a controller that will serve as its supervisor and the controllers that are its subordinates.

With respect to system dynamics for the overall system modeled under the ROOCH paradigm, there are two essential sets of dynamics to be considered. First, there are the dynamics associated with the entities moving through the system. Second, there is the task decomposition and execution process, which results from interactions among the controllers. Current simulation tools focus on modeling the first set of dynamics while ignoring the second set of dynamics entirely. It is our contention, however, that the first set of dynamics, the movement of entities, is entirely dependent on the second set of dynamics resulting from the controller interactions. Therefore, we believe that to model effectively a complex discrete-event system such as an FMS operating under a sophisticated control architecture, we must focus on modeling the interactions among the intelligent controllers.

The development of the specifications for the functional operations of the intelligent controllers as they interact with each other is still in its infancy. We are now researching these architectures in our Manufacturing Systems Laboratory. Similar research is being conducted by the Intelligent Systems Division at the National Institute of Standards and Technology (see Albus and Meystell, 1995). We can, however, discuss the basic functionality that must be addressed by this controller. In Figure 13.6 the intelligent controller has been decomposed into four basic functional elements. The Task Acceptor is responsible for interacting with its supervisor's Task Assignor in order to accept new tasks that will be executed within the coordinated object. One element of the task acceptance process is the determination of a time for the completion of the assigned task. This completion time represents the time at which the supervisor will regain control of the entity upon which new task(s) will be performed.

The Planner has several responsibilities. First, it must decompose the assigned tasks into subtasks that can be implemented at its subordinate processing subsystems. Second, it must also define the subtasks that must be executed by the interfacing subsystems in order to support the execution of the processing subtasks at the processing subsystems. Next, it must assign a subordinate subsystem to execute each subtask. Finally, it must schedule when an assigned subtask will be executed at the subordinate subsystem. Recall, however, that planning can also occur at the subordinate subsystem, which may

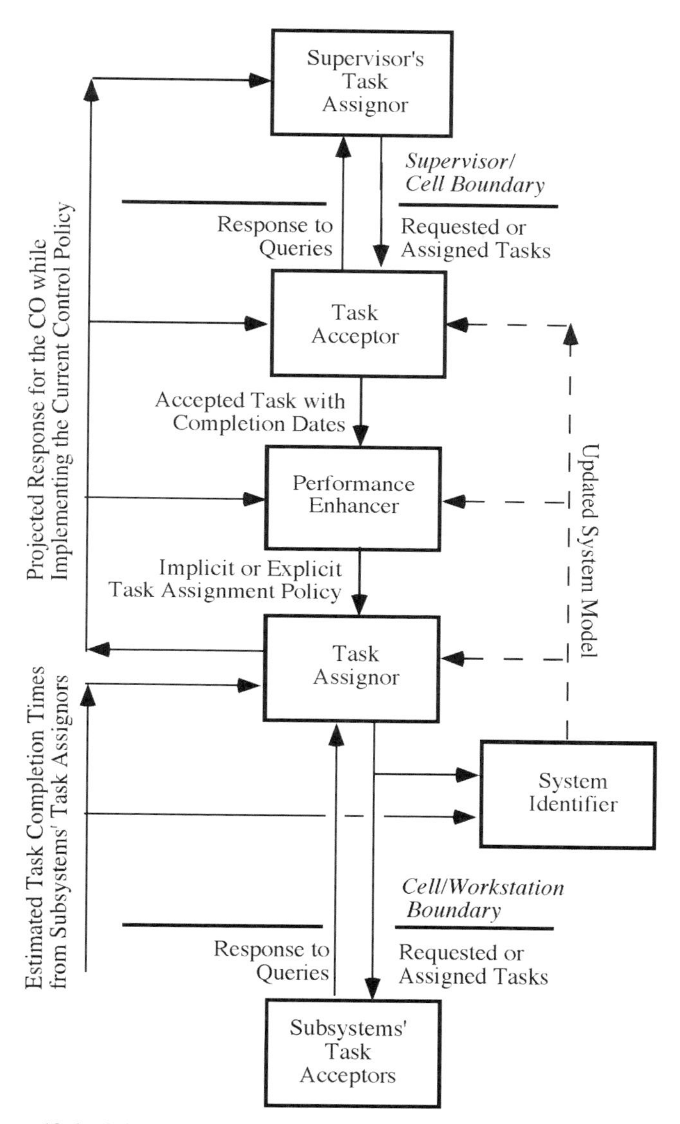

Figure 13.6 Schematic for one possible realization of an intelligent controller.

also possess an intelligent controller. Therefore, a given intelligent controller can plan only to the extent that it can implement and must rely on subordinates to plan for the execution of tasks that it cannot implement. Hence, the planning executed by the Planner within a given intelligent controller is by its very nature imprecise. Any schedule that it develops for completing its assigned tasks cannot be fully specified. To this end, we must rely on the principle of feedback control in order to permit the controller to update its schedule based on feedback information reported by its subordinate subsystems as they implement their currently assigned tasks. For the required feedback control to occur, the schedule developed by the Planner must be converted into a control law that will permit the schedule to be updated based on feedback information received from the coordinated object's subordinate subsystems while they are executing their assigned tasks.

The Task Assignor employs the current control law to assign tasks to its subordinate subsystems. The assignment process requires the Task Assignor to interact with the Task Acceptor within the subordinate subsystem where a given task is to be assigned. The outcome of this process is the assignment of the task to the subordinate subsystem and a negotiated completion time at which the control of the entity will be returned to the Task Assignor.

The Task Assignor is also responsible for collecting feedback information from each of the subordinate subsystems and organizing the information in order to describe the current state of the collective set of subsystems. The feedback information will contain not only the subordinate subsystems' current state, but will also provide an estimate of the subsystems' future performance in executing the tasks that have already been assigned to them. Please note that to insure continuous activity at the subordinate subsystem level, there must be planning for future tasks as well as control of the tasks that are currently being executed.

Once the current state information is collected, the Task Assignor then employs this information to project the overall behavior of the coordinated object while it operates under the control law that has been selected by the Planner. This projected performance for the coordinated object is then transmitted to the remaining functions within the intelligent controller and to the Task Assignor within the supervisor's intelligent controller. The Planner employs the projected response for the coordinated object operating under the currently selected control law as a standard against which the performance of alternative control laws are to be compared. If an alternative control law is demonstrated to provide better performance than the control law that is being implemented, the new control law will be submitted to the Task Assignor for implementation.

The Task Acceptor employs the projected response for the coordinated object in order to provide a nominal system trajectory which it uses to negotiate the operational constraints such as the projected completion date for new tasks with its supervisor's Task Assignor. The System Identifier uses this projected trajectory to update its model of the coordinated object as it interacts with its subordinate subsystems. We assume that these subsystems will be time variant and that the need to update subsystem models is constant.

The supervisor's Task Assignor uses the projected system trajectory for the coordinated object in a manner that is similar to the manner in which the Task Assignor employs the feedback information from its subordinate subsystems. In fact, the corresponding behavior for each element in the supervisor's intelligent controller is similar to that of the corresponding function within the coordinated object's intelligent controller. If a subordinate subsystem is also a coordinated object, its intelligent controller's function also addresses similar responsibilities. Such is the beauty of recursion.

Although we do not discuss the detailed operation of each function within the intelligent controller, it is clear that most of these functions must be able to predict the effect of their functional responsibilities on the operation of the system. It is evident, based on the complexity of these systems, that the future system trajectory cannot be described analytically. Simulation will be essential. However, the required simulation analysis is significantly different from the conventional off-line simulations that most simulation practitioners consider. These simulation analyses must consider the current state of the involved systems. The analysis must be performed in real time. In fact, the simulations themselves must be performed much faster than real time in order to provide sufficient simulation trials for the real-time analysis. We term this form of simulation analysis *on-line analysis*. We discuss on-line analysis extensively in the next section.

13.2.3 Toward New Simulation Capabilities

Using the recursive object-oriented coordination hierarchy (ROOCH), we have developed a new simulation language called the hierarchical object-oriented programmable logic simulator (HOOPLS). A HOOPLS-based model of a FMS consists of four primary frames. The first simulation frame is the model frame, which contains the specifications for the ROOCH associated with the FMS modeled.

The second simulation frame is the control frame, which provides for the exhaustive definition of the control messages that will be issued or received by each controller contained within the ROOCH. The control frame also defines the state transition mechanisms which are executed on the receipt of a control message and the subsequent control message(s) that will be issued. Specification of these control messages and state transition mechanisms is a challenging but essential task. Given that HOOPLS explicitly models the interaction among the controllers within the ROOCH and considers the flow of all entities to be a consequence of these interactions, HOOPLS has abandoned the use of a traditional event calendar. Instead, HOOPLS employs a message relay which stores the control messages that will be passed among the controllers and orders them chronologically, based on their delivery times. Each control message designates the controller that issued the message, the recipient controller for the message, the message content, and the scheduled time delay until delivery. The message relay is responsible for delivering the control message to the recipient controller at the appropriate simulated time. It is obvious that the prescribed operation of the message relay physically mimics the communication network that links the controllers in an actual FMS.

The third simulation frame in the HOOPLS-based model is the processing plan frame. The processing plan details not only which manufacturing processes will be required, and in which order, but also details which supporting resources (e.g., tooling) will be required, and for what duration in order to complete a processing task.

The fourth simulation frame is the experiment frame, which specifies the experimental parameters governing the simulation study. HOOPLS expands the requirement for the experiment frame, however, by requiring that it include extensive capabilities for initializing the simulation to a known system state. The need for this provision will be explained in Section 13.3. However, its needs has already been recognized when we discussed the four basic functions that are contained within the intelligent controller.

The specifications for the HOOPLS language are currently under development. Before finalizing these specifications, we hope to model a wide variety of flexible manufacturing environments and to expand the scope of applicability for the proposed language to the modeling of other classes of discrete-event systems.

To demonstrate the descriptive power of HOOPLS-based modeling approach, we will now discuss a recent application of HOOPLS to develop a real-time emulator (we discuss this term shortly), for the Rapid Acquisition of Manufactured Parts (RAMP) FMS, operated by the Department of Defense. The RAMP FMSs (there are several configurations) were developed by the U.S. Department of Defense's Flexible Computer Integrated Manufacturing program as a means for manufacturing spare parts for

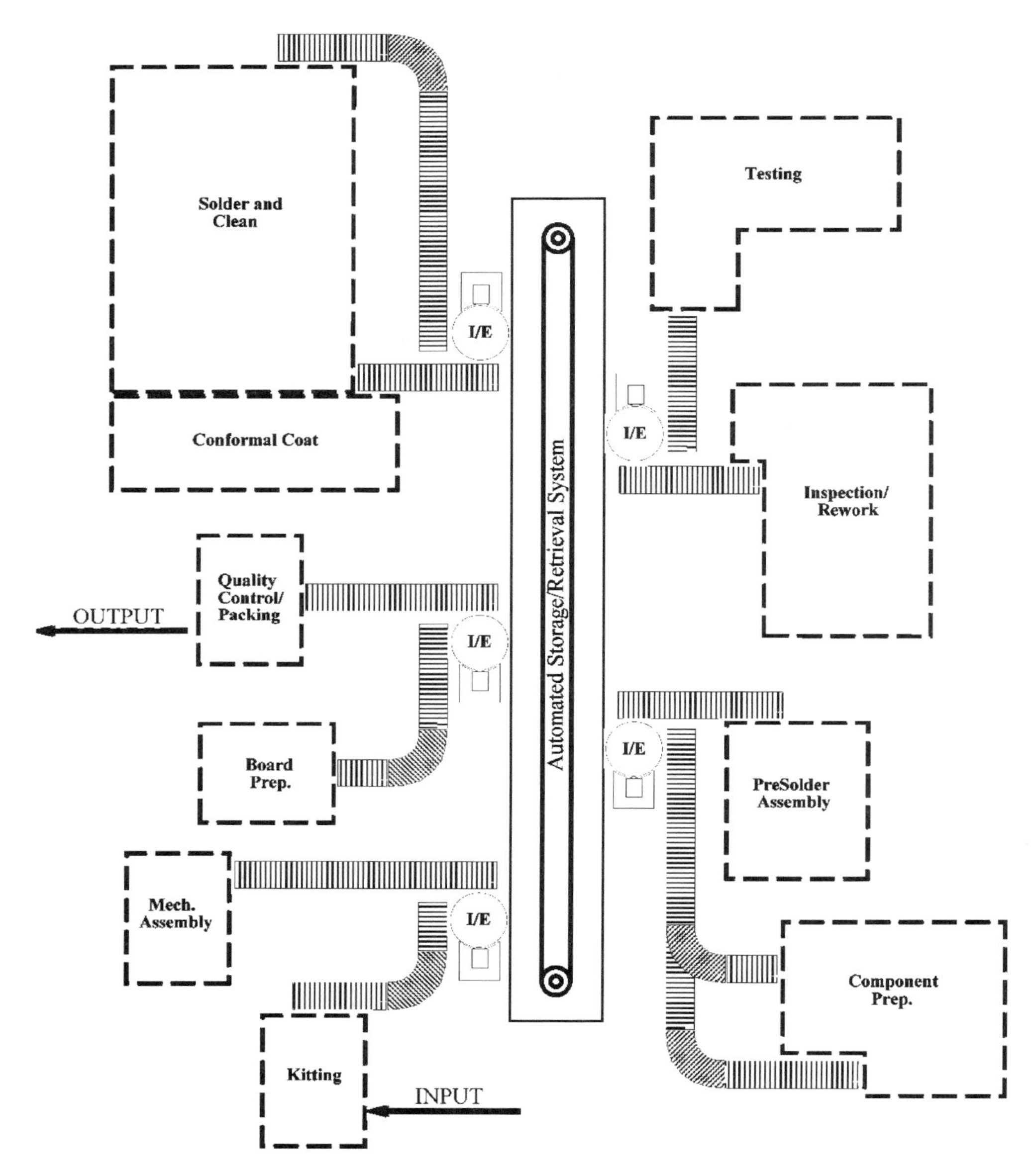

Figure 13.7 Schematic layout for the RAMP FMS for manufacturing circuit boards.

defense systems in small lot sizes. (See Davis et al., 1994.) As a collection, these FMSs are some of the most complex manufacturing systems that our research laboratory has encountered. The particular RAMP being considered here has been designed to manufacture replacement circuit boards using through-the-hole assembly technologies where the component leads are actually inserted into holes contained on the circuit board. A

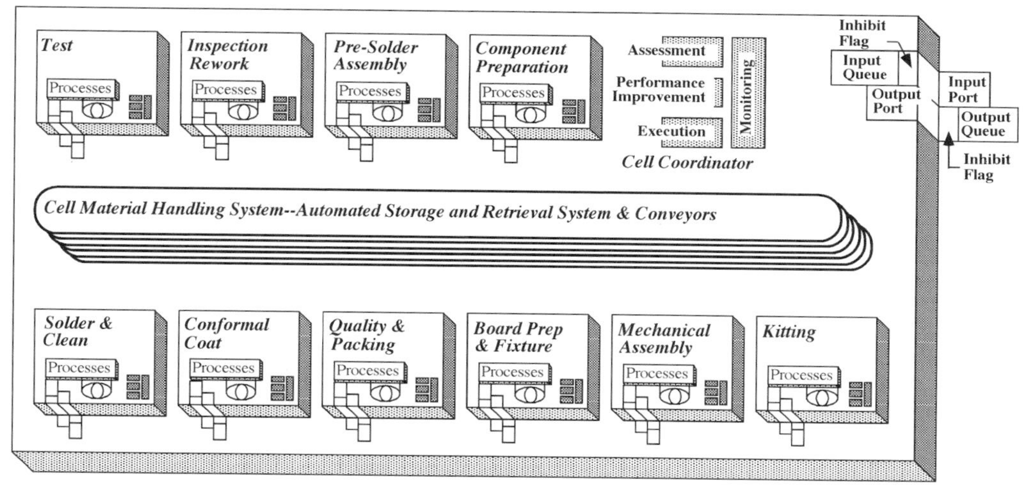

Figure 13.8 ROOCH for the RAMP FMS.

schematic for the this RAMP FMS is provided in Figure 13.7. The ROOCH developed for modeling this system is given in Figure 13.8.

In Table 13.1 we provide a basic description for the functions that are being addressed at each of its included workstations. In Table 13.2 we provide a list of the controllers that are managing this system. The reader will note that human beings perform a significant portion of the manufacturing responsibility. We have separated these controllers into three classes. A coordinate node indicates a controller within a coordinated object. Coordinate nodes in this case are defined at the RAMP cell level, at each workstation, and at the central automated storage and retrieval system (AS/RS), which is the primary material handler at the cell level. A unit process node represents a controller at a basic process that can implement a processing task. A transport process node represents a controller at a basic process that can implement a transport task. There are over 70 controllers in this system. In fact, the AS/RS system alone contains 31 controllers.

Numerous types of entities flow through this system. In fact, a given job entity for a given part type must be further delineated into several different entity types during its production process. Figure 13.9 provides a more detailed schematic for the manufacturing process. All the parts needed to assemble an order of a given board type arrive in an order kit tote at the kitting station. The bare printed circuit boards are placed into a bare-board tote. Each of the smaller components to be mounted on the board is placed into a part carousel, which is then placed into part tote. If needed, several part totes can be generated for a given order. The larger parts are placed into either a mechanical parts tote or a post-solder assembly tote, depending on whether they are placed on the board before or after wave soldering occurs. All of these totes are then returned to the AS/RS. At the board fixturing station, the bare boards in the bare-board tote will be individually placed into a fixture, and each fixtured board will be placed into an individual fixtured board tote. As indicated in Figure 13.8, various totes containing different types of parts for each job entity are needed at various stations. Based on this discussion, it is obviously impossible to model a job as a single entity moving through the system.

TABLE 13.1 Processing Steps and Entity Flow for Production in the RAMP FMS

Kitting: Generate the various parts' kits associated with the production order.
- Input: KIT tote containing all components for the order
- Output: Bare board tote containing up to 10 boards
  Pre-solder assembly tote(s) holding parts to be mounted before soldering
  Post-solder assembly tote holding parts to be mounted after soldering
  Mechanical tote containing large mechanical parts such as transformers

Board preparation: Place board into fixture.
- Input: Bare-board tote with bare boards
- Output: One-pallet tote for each fixtured board

Component preparation: Bend and tin the leads for the components.
- Input: Pre-solder assembly tote(s) and post-solder tote
- Output: Same

Pre-solder assembly: Place components on board prior to solder.
- Input: Pre-solder assembly tote(s) and all pallet totes for given order
- Output: Same

Pre-solder inspect: Check board prior to wave soldering.
- Input: Pallet totes for order and pre-solder assembly tote(s) if rework is needed
- Output: Same

Wave solder and clean
- Input: All pallet totes associated with order.
- Output: Same

Post-solder assembly: Mount smaller parts after wave soldering.
- Input: All pallet totes associated with order and post-solder assembly tote
- Output: Same

Post-solder inspection
- Input: All pallet totes and pre- and post-solder assembly totes if reworked
- Output: Same

Mechanical assembly: Remove board from fixture and mount larger mechanical parts.
- Input: All pallet totes for order and mechanical part tote
- Output: Printed wiring assembly tote for each board and mechanical part tote

Test: Perform bed-of-nails fixture test and burn-in test.
- Input: Printed wiring assembly totes for job and various part totes if reworked
- Output: Same

Conformal coat: Apply protective coating to board.
- Input: Printed wiring assembly totes for order
- Output: Same

Final quality control and inspection: Pack boards for shipping. Remove remaining parts.
- Input: Post-wiring assembly totes for order and various part totes
- Output: Packaged boards and unused parts.

Note: The operator is required to remove parts from, and place parts into, the totes at each processing station. The totes come in two sizes—tall and short. The ASRS has six levels of storage, with each level capable of holding 96 tall totes and 96 short totes.

There are also numerous supporting resources which are employed during the production processes that must be tracked. For example, there are numerous types of totes. In fact, there are nearly 1200 totes of different types that must be constantly tracked within the system. There are also various fixtures that must be tracked. Some of these fixtures are used to hold the boards and are stored with a fixtured board tote. Other are used during testing operations. The human operators in this system are also cross-trained. In fact, each operator has his or her own unique qualifications. During the oper-

TABLE 13.2 Controllers in the RAMP FMS

- **Cell Controller** (Coordinate Node)
  - Subordinate controllers:
    - Kitting station controller (coordinate node)
      - Subordinate Controllers:
        - Processor: human operator (unit process node)
        - Material handling controller: human operator (transport process node)
    - Board preparation controller (coordinate node)
      - Subordinate controllers:
        - Processor: human operator (unit process node)
        - Material handling controller: human operator (transport process node)
    - Component preparation controller (coordinate node)
      - Subordinate controllers:
        - Processor: human operator (unit process node)
        - Material handling controller: human operator
        - Part carousel controller (transport process node)
        - Tinning robot controller (unit process node)
    - Pre-solder assembly controller (coordinate node)
      - Subordinate controllers:
        - Processor: human operator (unit process node)
        - Material handling controller: human operator
        - Part carousel controller (transport process node)
        - Part location indicator controller (unit process node)
    - Test facility controller (coordinate node)
      - Subordinate controllers:
        - Processor: human operator (unit process node)
        - Material handling controller: Human Operator (transport process node)
        - Burn-in test controller (unit process node)
        - Conductivity test controller (unit process node)
    - Inspection/rework controller (coordinate node)
      - Similar to board preparation
    - Mechanical assembly controller (coordinate node)
      - Similar to board preparation
    - Conformal coat controller (coordinate node)
      - Similar to board preparation
    - Quality and packaging controller (coordinate node)
      - Similar to board preparation
    - Wave solder and clean controller (coordinate node)
      - Similar to board preparation
    - Automated storage and retrieval system controller (coordinate node)
      - Subordinate controllers:
        - Storage layer controller (6) (transport process node)
        - Inserter/extractor controller (5) (transport process node)
        - Station input/output conveyors controllers (20) (transport process node)

ation, these operators will be assigned to the station where their services are best needed. The Department of Defense did attempt to construct a simulation model of this facility and was unsuccessful. The contractor who performed the simulation should not be blamed, however. It is our contention that it is simply impossible to model this system using conventional tools.

Under a small contract, our laboratory undertook the challenge of modeling this sys-

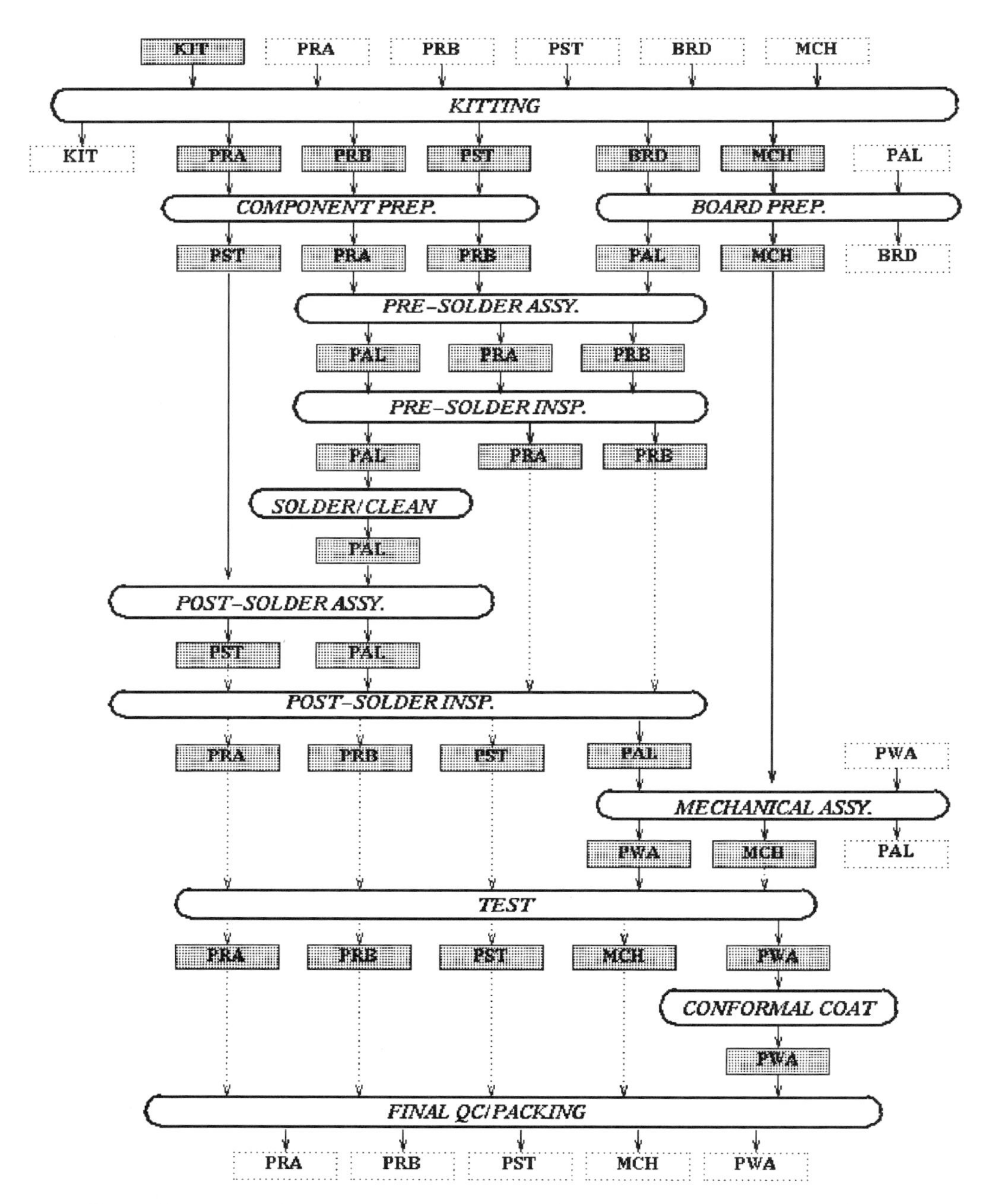

Figure 13.9 Process/resource flow diagram for circuit board production in the RAMP FMS.

tem using the HOOPLS paradigm. Given the size of the contract, we were forced to employ the services of two undergraduate students in computer science who had no prior knowledge of simulation. We chose to program the model in C++. Instead of developing a standard simulation model for the RAMP FMS, however, we elected to demonstrate the additional capabilities that a HOOPLS-based model could provide. To this end we set out to construct a real-time emulation for this RAMP FMS.

Under the HOOPLS paradigm, we programmed the control logic for each controller in C++. We next developed our own messaging service that would permit these controllers to be distributed across the Internet. Using this messaging service, each control object communicated with the other control objects over the Internet. That is, when one controller desired to send a message to another controller, it sent the message to the messaging service, which then forwarded the message to the Internet IP address where the recipient controller object resided.

For each controller, we also developed a detailed controller window which displayed the real-time state information for the subsystem being managed by that controller. However, each controller and its control window were retained as distinct programmed objects. The control window objects were programmed as Java applets. This feature permits the real-time emulation to be viewed at any site on the World-Wide Web using a standard webbrowser. What evolved was real-time emulation capability for the RAMP FMS that could be virtually viewed from anywhere on the World-Wide Web.

The real-time emulation operates in the following manner. First, the controller objects for each controller contained within the RAMP FMS are assigned to a given computer whose IP address is registered with the message service. After all the locations of all the controllers are known, the emulation can begin. The message service is responsible for maintaining the real-time clock that is to be employed by all the control objects. At each second, it sends a time message to each emulated control object so that the overall actions of entire collection is synchronized to a single clock. This feature also permits the emulation to operate faster or slower than real time by employing time scaling at the messaging service object. In the future, the viewer of the emulator will be able to speed up the emulation when he or she desires to move quickly through a given phase of the system's operation or slow the emulation down when the system behavior is to be observed in minute detail.

Each control object is responsible for managing the events that occur within its domain. Events are delayed from occurring until the real-time moment that they are scheduled to occur. When an event occurs, one or more message may need to be sent to another controller. These messages are then first sent to the message service and then forwarded to the recipient controller. Upon receiving the message, the recipient controller makes the appropriate state transitions and schedules when future events (or control messages) will be transmitted to other controllers.

The control window's object is distinct from the controller object and can be executed upon different computers. The entire ensemble of controller window objects are typically bundled into a single Java applet which can be downloaded to any computer with a web browser. When the control window's Java applet is executed, it first registers itself with the messaging service. At this point a control window for the overall cell controller is presented on the screen. The contents of this window are very similar to the standard animation for a FMS depicting entities flowing through the various stations. When the control window applet has registered itself with the message service, the message service requests that the emulated control object start sending updated state information at each second interval. The messaging service then forwards this state information with the current time stamp to the requesting control window applet. The control window applet then updates the contents within the control window to reflect the new state information with each state information update that it receives from the messaging service.

The remote viewer of the control window can select any element within the cell controller window in order to access more detailed state information for that subsystem. When the user selects a given subsystem, another control window applet is opened and the message service is requested to start sending detailed state information for that subsystem. Note that windows for several subsystems can be open simultaneously at one viewing site. It is also possible that several individuals can view the real-time emulation form from several different computer concurrently.

There are several differences between the real-time emulation and a conventional animated simulation. First, the evolution of the system is being synchronized to a real-time clock. For most animated simulations, the speed at which the animation evolves is dependent on the speed of the computer processor that is performing the simulation. The second difference is the detail of the state information displayed. For most animated simulations, a viewer is limited to a single window. In our real-time emulation, every subsystem has its own control window which can be accessed for more detailed state information. Eventually, we hope to make the icons for the entities selectable also. That is, by clicking a given entity's icon, another window will open that will permit one to view the real-time detailed state information (attributes) for that entity. We believe that this feature would be very useful for the real-time debugging and validation of the model. The real-time emulation was demonstrated at the 1996 Winter Simulation Conference (see Davis et al., 1996).

In the development of real-time emulation capability, we wished to demonstrate the intrinsic ability that a HOOPLS-based model will have for managing the system it models. That is, the control objects developed for a HOOPLS-based model of a given system can actually control the modeled subsystem within the real system. We have already demonstrated this fact on a physical emulator for an FMS that we have constructed in our laboratory (see Gonzalez, 1997; Gonzalez and Davis, 1997). Here the control objects within the HOOPLS-based model for the physical FMS emulator have been placed on separate computers and attached to the physical equipment that they were designed to manage. The same programmed control objects are also being executed concurrently on another computer to perform an on-line simulation of the physical FMS emulator in order to project its future behavior in real-time while operating under the current policy that is being employed to schedule tasks.

Our goals in the development of HOOPLS are the same goals that we feel should guide the development of future simulation languages. First, future languages should provide a computer-aided tool for designing the systems that they are modeling. When the model is completed, the language should provide a model of sufficient fidelity that it can actually manage the system that it models. Second, the same model should have multiple uses. Certainly, it should be useful for off-line planning analyses that are addressed during the design of the system. As stated above, it should also be able to manage the systems operations in real time. Finally, the same model should be available to perform within on-line planning analyses to assess the future behavior of the system in real time.

Third, we should move toward a standard model description language. We believe that HOOPLS has taken a major step in this respect by modeling the controller interactions. Currently, we are attempting to model several different types of manufacturing systems. Our goal is to develop a set of basic controller types that are applicable across most manufacturing scenarios. We are also attempting to define a generalized structure for the process plan database. These generic specifications will govern our eventual development of the HOOPLS-based modeling and design tool for these complex discrete-event systems.

By modeling controller interactions, we also believe that the maintainability and reusability of the model can be enhanced. We are currently demonstrating this feature in our modeling of the RAMP FMS. At one RAMP site, the operator has added a new station which itself is an automated line for assembling circuit boards using surface-mount technologies. As we integrate this new station into our current real-time emulation model for the RAMP FMS, nearly all the existing code will be reusable. Our focus will be simply on programming the capabilities for the new station and expanding the current cell controllers dynamics in order to permit it to communicate with the new station. We must also add a new entry point for the automated storage and retrieval system so that totes can be delivered to the new station. We must describe the entities and process plans that can be executed on in the expanded FMS. Note, however, that all of these modifications are additions to existing code (not major rewriting of the existing code), most of which will be completely retained within the emulation model for the expanded system.

We believe that by modeling controller interactions this reusability can be expanded in ways that most cannot imagine. Consider a given piece of equipment. Its controller is typically specified by the vendor. Now assume that the vendor of this equipment provides a detailed model of its system which addresses all the control input and output messages. This piece of code should be useable by any simulation model which employs the controller interactions as a basis for modeling the system. The vendor can place the control object for its equipment into a national library which is assessable via the internet. When the modeler is designing a system, he or she can then download the model for the given piece of equipment and integrate it quickly into the overall system that the designer is investigating. In this manner, the designer should be able quickly to test alternative configurations of equipment within the proposed system before it is implemented. In fact, if an agency such as the National Institute of Standards and Technology can take an initiative in this area, a standardized set of commands for basic equipment types can be defined. We know that this is feasible because large manufacturers have already specified a set of command instructions which they will employ when they seek to purchase equipment from a given vendor. In this manner, all equipment of a given type employs the same instructional set, independent of its vendor. If a single corporation can accomplish this standardization within its manufacturing facilities, it can also be done at the national/international level.

Furthermore, since the controllers interactions can be standardized, any simulation model that uses these standardized commands should be able to integrate with any other modeled system which also employs these commands. That implies further that models generated by one vendor's simulation tool should be able to integrate with another model produced with another simulation tool. Again, we believe that it will take the leadership of a national agency such as the National Institute of Standards and Technology to make this standardization. The important observation, however, is that this standardization is feasible.

Other researchers are also considering the modeling of controller interactions. Researchers at Georgia Tech have developed OOSIM, which shares many modeling concepts and capabilities with HOOPLS (see Narayanan et al., 1992; Govindaraj et al.,

1993). Researchers at Texas A& M University are also attempting to employ a specialized version of the ARENA simulation language, to control their demonstration FMS (see Smith et al., 1994; Drake et al., 1995; Peters et al., 1996).

The reader should be cautioned from believing, however, that modeling controller interactions solves every issue pertaining to the modeling, planning, and control issues associated with the design and management of these complex discrete-event systems. It does not. Once we accept the fact that there are alternative approaches for modeling discrete-event systems beyond the conventional approaches employing entity flows through stochastic queuing networks, the entire area of discrete-event simulation is open to development of new modeling approaches. In addition to HOOPLS, we are developing a new simulation tool which is being explicitly designed to support scheduling. One version of this new language, called the flexible object-oriented production planning simulator (FOOPPS), has been designed to support the integrated tasks of master production scheduling and manufacturing resource planning in a flexible manufacturing environment (see Seiveking and Davis, 1995; Davis et al., 1997).

We believe that the conventional modeling approaches do address a broad class of discrete-event systems. However, they have been shown on several occasions to be inadequate for some systems. In the near future, major new approaches for modeling complex discrete-event systems will appear. We further believe that with their introduction, the conventional uses of simulation will be significantly expanded. One particular new use will be the development of on-line simulation technologies which will be needed to support the on-line planning and control requirements within the intelligent control architectures that manage these systems. In light of this observation, we now discuss on-line simulations.

13.3 EVOLVING SIMULATION TECHNOLOGIES FOR ON-LINE PLANNING AND CONTROL

To understand the role that simulation plays in on-line planning and control, we need to develop a basis for comparison of the new technologies with existing simulation and planning technologies. We begin by providing a brief overview of off-line planning with simulation, which is addressed in detail in Chapters 3, 6, 7, 9, and 10. A schema for on-line planning is then provided. Two approaches to on-line planning, reactive and proactive, are discussed. Finally, we discuss technologies that are needed to implement on-line simulation.

13.3.1 Overview of Off-Line Planning Using Conventional Simulation Technology

The off-line approach to simulation analysis begins by assuming that there is a system to be modeled, as is depicted at the top center of Figure 13.10. This system is characterized by its *state variables*, which are assumed to change or evolve with time. Using classical system theory, there must be a set of *inputs* to the system that represents the environmental interactions on the system. The behavior of the system is characterized by its *state transition function*, which determines the evolution of the system state variables given the system's current state and the sequence of inputs that the system experiences. As the system state evolves, the system will generate *outputs* that represent the interaction of the system on its environment. Again, it is assumed that there exists an *output*

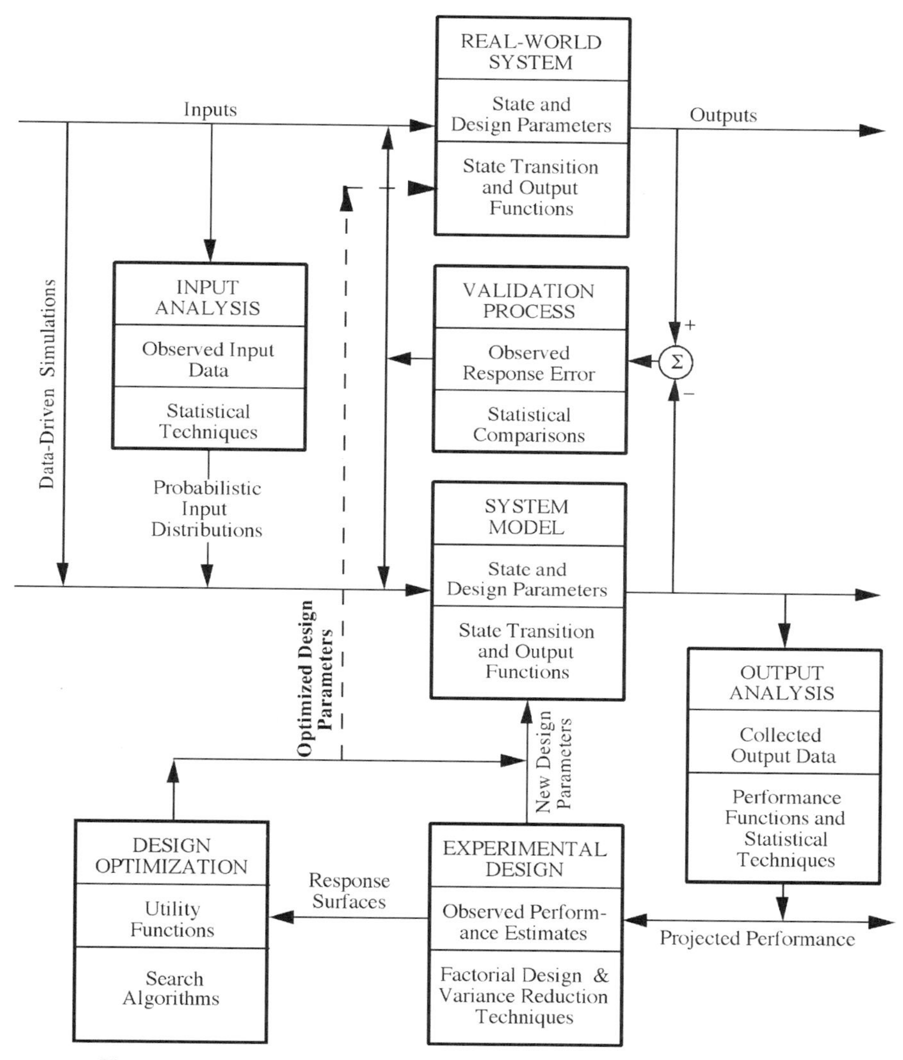

Figure 13.10 Schematic for the off-line planning process using simulation.

function which defines the generated output based on the current system state and the system inputs. Thus the state transition function and the output functions are dependent on both the current system state and the input stream. The role of the modeler is to translate the state transition and the output functions for the real system into a model that can then be exercised, or simulated, to predict the behavior of the system. As in Figure 13.10, the system model also receives inputs and generates outputs just as in the real system.

Since both the system and its model accept inputs and generate outputs, the modeler has a method for determining the accuracy of the created model, and the *validation process* is introduced (see Chapter 10). As illustrated in Figure 13.10, the validation process begins with the assumption that both the real system and its model are initialized to the same state and are subjected to the same input stream. It is often assumed that the modeler will have control over the input stream used for the validation process. To this end, Figure 13.10 depicts an input stream that is being generated by the validation process itself. However, when the actual system is employed, it is often expensive, if not impossible, to provide an arbitrary input stream. Usually, the modeler carefully records an input stream as it affects the real system and the resulting output stream that the system generates. The same input stream is then applied to the model and its output stream is recorded.

If the model is correct, the output streams for the real system and the model should be the same. To verify this, the validation process computes the error term between the two output streams as a function of time. The computed stream of output errors is then used to correct the model. In most cases, the validation process is actually comprised of two distinct steps, *validation* and *verification*. The validation step is concerned with developing the correct model for the system, while the verification step seeks to ensure that the model has been executed faithfully by the software program. Law and Kelton (1991) note that verification is often easier when one employs a commercial simulation tool. This statement assumes, however, that the simulation language employs the correct modeling philosophy. Unfortunately, as discussed in Section 13.2.1, current simulation languages have not been designed for modeling systems where controller interactions must be considered. In this case, the benefits of using a commercial simulation tool may not be realized during the validation process. In fact, its use may even hinder the validation process.

If the use of a commercial simulation tool is appropriate, modern capabilities for these tools, such as animation and interactive run controllers (debuggers), are often useful in the validation process. Still, it is nearly impossible to design a set of procedural steps that ensure a valid model in every situation. The validation process becomes even more complicated when one considers a time-variant system where the characteristics for essential elements of the modeled system change with time. Today, one usually assumes that the system is stationary, and validation is usually conducted in an off-line manner. For time-variant systems, validation must be conducted on-line while the system is in operation.

Currently, simulation is used primarily to support off-line planning, not on-line planning and control. Off-line planning usually attempts to improve the design of the real system. To facilitate the system design process, it is usually assumed that both the real-world system and its model are characterized by a set of design parameters that affect both the state transition and the output functions (see Figure 13.10). The primary purpose for simulation within the off-line planning scenario is to assess the performance of the system while it operates under a given set of values for the design parameter. The overall goal for off-line planning is to define the optimal set of values for these design parameters. However, the ability to specify the true optimal set of parameters can seldom be achieved for real-world systems. The model is always an approximation to the real-world system. Furthermore, it is virtually impossible to explore all feasible values for the design parameters.

The first step of the off-line planning approach is to specify the input stream that will be employed during the planning analysis. The most commonly used approach is to perform a statistical analysis of prior or potential inputs to the system and then to develop probabilistic distributions that characterize the recorded or predicted input streams. This process is often referred to as *input analysis* (see Chapter 3). A second, less common approach is simply to apply the recorded input streams to the real-world system as input streams to the model. The latter approach is referred to as a *data-driven simulation*. When either approach is adopted, the resulting simulation analyses are still conducted off-line (i.e., they are detached from the real-world system).

Using either a generated or recorded input stream, a simulation run (experiment) is then conducted for a given parameter setting. The output stream generated is then subjected to *output analysis* (see Chapter 7), which consists of two major steps. First, a set of performance measures is selected for the analysis. Then, using the generated output streams, the values of these performance measure are statistically quantified.

In performing this statistical analysis, several additional factors must be considered. First, the length of the employed simulation run and the number of replications that must be performed are determined. Second, measures are taken to ensure that transient effects arising from the initial conditions of the simulation runs are eliminated. In most cases the desire is to provide a steady-state analysis of the projected system performance.

As discussed in most simulation texts, there are two basic types of output analysis. The first type predicts the steady-state values for the performance measures using simulation trials conducted over an extended-time horizon. Here more than one replication (trial) may be performed, or a given simulation run may be partitioned into several component trials using batching techniques. The second type of simulation analysis, the *terminating simulation*, attempts to estimate the expected performance of the system while it operates over a repeated cycle. Here numerous simulation trials for the cycle being investigated are needed so that an estimate of the system state and performance can be made at any point in the cycle. This is accomplished by averaging the temporal values of the recorded state and performance measures across several simulation runs. As will be shown shortly, the terminating simulation analysis does have some similarities to the statistical analysis required for on-line simulations. However, there are also critical differences.

In both instances, the primary concern of the simulation analyst is to develop statistical estimates of the expected values of one or more performance measures. These values are assumed to depend on the values of the design variables associated with the simulation experiment. These, in general, should be independent of the initial system state for the simulation, except in the case of the terminating simulation, where the initial system state is assumed to be the same for every simulated operational cycle.

As mentioned previously, the overall goal of off-line planning is to specify an optimal set of design parameters within the limitations of the adopted modeling and search processes. To this end, the modeler typically seeks to describe performance measures as a function of the design parameters. This estimated function is often referred to as the *response surface*. The design parameters (which serve as the variables in the function) are typically referred to as the *factors*. *Experimental factorial design techniques* as well as other statistical approaches (see Chapter 6) can be employed to assist in defining which experimental settings for the factors (design parameters) should be investigated to maximize the information that can be obtained about the response surface. In addition, *variance reduction techniques* (see Law and Kelton, 1991), can be employed to further refine the statistical estimates resulting from simulation results for a given assignment of values for the factor (design parameters).

Given the response function, the modeler seeks the optimal assignment of the factors. If more than one performance measure is employed, the modeler might use utility function techniques or other multicriteria decision-making techniques to develop the best compromise among the performance measures (see Fishburn, 1987; Wolfram, 1988). In any case, search techniques are essential to establish the optimal parameter settings (see Chapter 9). Several types of search techniques might be employed. These include derivative-free search methods such as the Box method or the Nedler–Meade algorithm (see Himmelblau, 1972), or more recent stochastic search techniques such as genetic algorithms (see Goldberg, 1989), or simulated annealing (see Davis, 1987). Using perturbation analysis techniques (see Cassandras, 1993), it is now also possible to make estimates of the derivatives for the performance measures with respect to the design parameters. Hence derivative-based searches such as the conjugate gradient algorithms may also be considered (see Bazaraa et al., 1993).

These optimizations, however, are often very "noisy." In other words, the performance measures are typically random variables whose values must be estimated statistically. The parameters are often discrete-valued variables that can take on integer values only. For example, there can be only an integer number of servers at a given station, or the capacity of a queue must be integer. In short, determining the optimum set of values is not a trivial task, and optimality often represents a property that simply cannot be established or verified.

Once the desired set of design parameter values has been determined (see Figure 13.6), these are then implemented within the real-world system. At this point the modeler can return to the validation process to determine if the anticipated performance estimated through the off-line planning process is being achieved by the real-world system. This validation effort may lead to further improvements in the model, and a new off-line planning cycle may be warranted to further refine the selected values of the design parameters.

13.3.2 Overview of On-Line Planning and Control Using On-line Time Simulation Technology

On-line (real-time) planning and control begin with the assumption that a simulation model for the real system exists and has withstood the validation process. It is also assumed that off-line planning has been performed to determine the optimum set of design parameters for operation of the system. Like the off-line planning scenario, the real-time planning scenario presumes that both the real-world system and its model will be subjected to inputs from the environment. However, the on-line planning scenario necessitates that the modeler consider the same real-time inputs that affect the real system (see Figure 13.11). In this respect, the real-time simulation is necessarily data driven.

A critical element of both the real-world system and its model is the control policy that has been selected for implementation. In Figure 13.11 there is no mention of planning per se. In the real-time operation of the system, a selected plan must be implemented immediately. This implies that choosing a plan necessarily requires specification of the control policy that will implement the selected plan. Therefore, the plan selected and its control policy become intrinsically linked. As implied in Figure 13.11, the implementation of the control policy is the only element that needs to be considered.

The control policy selected, in conjunction with the state transition function, determines the evolution of the system state as a function of time. One may, in fact, view the system inputs as consisting of two components. The input to the system depicted in Figure 13.11 represents the *exogenous* input component over which the system has no control. On the other hand, the second component is the *endogenous control* input,

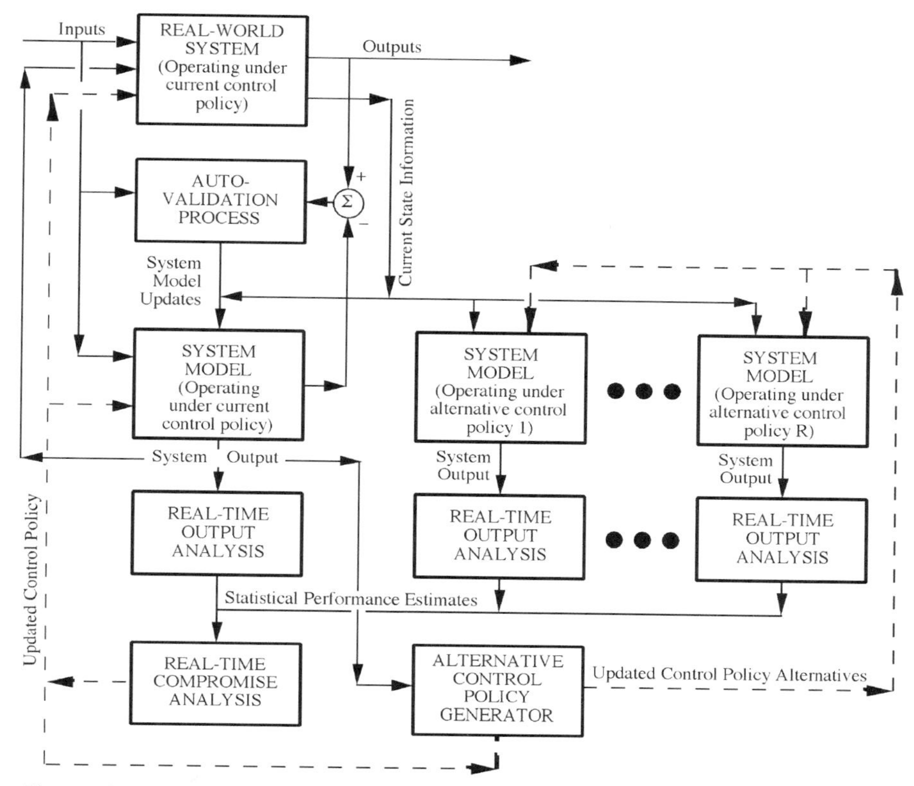

Figure 13.11 Schematic for the on-line planning/control process using real-time simulation.

which depends on the selected control policy. This control policy can generate the control inputs in either of two ways. If an open-loop control policy is employed, the control inputs generated under that policy are determined on an *a priori* basis and will not depend on the evolution of the system state. For a closed-loop control policy, however, the issued control inputs depend on both the selected control policy and the current state of the system.

The current system state for a on-line simulation provides another critical input to the system model(s) included within the on-line planning/control scenario. The need to employ the current state information as an input to the system model was a primary reason for including the extensive initialization capabilities with the experiment frame of the HOOPLS (discussed in Section 13.2.3).

Although the system model has presumably been validated, the validation process must continue under the on-line planning/control scenario. In most cases, off-line planning is performed once (usually when the system is initially configured) and not addressed again until the system is modified. Since off-line planning does not consider the real-time operation of the system, there is usually no need to update the system model between subsequent, off-line planning episodes.

Real-world systems are seldom stationary. That is, the operating characteristics change constantly with time. Typically, these changes are slow, but they can also be abrupt. For example, a new type of cutting tool with an extended tool life is introduced into a machining cell and the number of tooling replacements may be reduced significantly. Improved processing techniques could also significantly modify processing times.

Whether the system characteristics are changed slowly or abruptly, the system model must continually be updated to reflect these changes. To this end, Figure 13.11 includes an *autovalidation* process that compares the output projected by the model against the measured output from the system and updates the model to improve its accuracy. This autovalidation process functions in a manner similar to the system identification element which is contained in most conventional, self-organizing controllers for time-varying, continuous-state systems (see Astrom and Wittenmark, 1989; Eykhoff, 1974; Ljunj, 1987). Hence, the autovalidation process can be viewed as a system identifier for a discrete-event system. Unfortunately, the technology needed to construct this autovalidation capability does not currently exist. Because it does not exist, however, does not imply that it is not needed. Such technology is essential for the on-line simulation of real-world systems (see Section 13.3.6).

Under a true on-line planning/control scenario, it is essential that a on-line simulation of the implemented control policy be conducted concurrently with the implementation of the control policy by the real-world system. In general, the real-world system is stochastic. As the real-world system evolves in time, while operating under a selected control policy, it will realize only one of the potential-state trajectories that could occur. That is, the current operation of the system can be viewed as a single statistical experiment. The on-line simulation must necessarily be initialized to the current system state for the real-world system. Its statistics must characterize the future response of the real-world system given the current system state. Hence on-line simulation must employ simulation models that can be executed much faster than real time. Certainly, the evolution of the simulation will not be tied to a real-time clock, which was the case for the real-time emulator developed for the RAMP FMS (see Section 13.2.3). Rather, the simulation trials that analyze the system performance over a given planning horizon must be repeated numerous times to generate a set of output trials that can then be employed to statistically quantify future performance of the real system. *To provide the essential number of simulation trials for on-line analysis, the trials employed must actually be generated at a rate that is significantly faster than real time!*

While operating under current control law, statistical estimates for the future performance of the real system have three immediate uses. First, these provide another input for the real system and can be viewed as feedforward information to be employed by the controller of the real system. Second, the projected response of the system operating under the control policy selected can assist in the generation of alternative control policies for possible implementation. This use is discussed in Section 13.3.7. Finally, these projections provide a reference level of performance against which the predicted performance of other potential control policies can be compared.

Returning to Figure 13.11, the reader will note that R additional instantiations of the system model have been included. Each of these R instantiations considers an alternative control law for possible implementation. These simulations receive the current state of the real system as an input and any updates to the system model derived from the autovalidation process. Again, these real-time simulations generate trials as quickly as possible in order to statistically characterize the future performance of the real system as it operates under the alternative control law.

To compute the statistical estimates, the outputs generated by the on-line simulation trials for each system model are passed immediately to an on-line output analysis process. Obviously, there is a correspondence between the on-line output analysis function included in Figure 13.10 and the output analysis function included under the off-line planning scenario in Figure 13.9. There are, however, marked differences in the statistical technologies needed to implement the respective functions, as discussed in Section 13.3.4.

The actual statistical comparison of the future performance of a real-world system operating under current and alternative control laws is performed by the on-line compromise analysis function. It is assumed from the outset that multiple performance criteria will be considered and that a compromise solution must be defined. Again, the procedures for performing the compromise analysis are not currently known. One approach might be to develop a utility function for real-time performance criteria in order to provide a single, aggregate performance criterion that can then be used in the comparison. There are, of course, limitations to this approach as the contribution of the individual criterion becomes obscured in the aggregation process. There are also other approaches to multicriteria that have been considered in the operations research literature (see Fishburn, 1987; Wolfram, 1988). However, these approaches have not presently been extended to consider the stochastic performance measures that are to be evaluated using simulation.

Comparison approaches are also a research issue. For the comparison of stochastic entities, the notion of stochastic dominance has been developed. However, the requirements that are necessary to assert stochastic dominance are so restrictive that these are difficult to demonstrate. Stochastic dominance requires that (empirical) cumulative density functions for the compared quantity must not intersect. That is, the cumulative density function for the performance criterion arising from the operation of the system under one alternative control strategy must be consistently greater than the cumulative density function for the same performance measure operating under another control strategy. To assert complete dominance in the on-line planning situation, one control strategy must stochastically dominate all the others with respect to all performance criteria. This situation seldom occurs, and other forms of statistical comparison are needed. Issues arising from on-line compromise analysis are discussed in Section 13.3.5.

If an alternative control policy is demonstrated to provide an improved performance over the control policy that is currently being implemented, the new control policy is transmitted to the real-world system for immediate implementation. It is also transmitted to the reference simulation model, which then begins to estimate the future performance of the system as it operates under the new control policy. The control policy implemented is also transmitted to the alternative control policy generator to provide a starting point for generating new control policy alternatives. These alternatives subsequently will be subjected to on-line simulation at the other R instantiations of the system model (included in Figure 13.11). Like many of the other elements discussed above, the procedures for generating these alternative control policies are a research concern (see Section 13.3.7).

Together, the on-line compromise analysis function and the alternative control policy generator implement a planning algorithm that is similar to the design optimization function for the off-line planning scenario (see Figure 13.10). In particular, the alternative control policy generator function can be viewed as one of implementing the search algorithm. There is, however, no direct correspondence to the experimental design function (in Figure 13.10) for the on-line planning scenario (illustrated in Figure 13.11).

Given the discrete nature of the control policies to be evaluated, it is unlikely that a response function can be defined over a control policy space. That is, the definition of a response surface requires that a design variable space be defined such that the value of the performance criteria can be functionally related to the potential values of the design variables (factors). For control policies, this functional relationship is virtually impossible to define.

13.3.3 Proactive Versus Reactive On-Line Planning

The schematic for on-line planning/control provided in Figure 13.11 is actually a generalization upon, and an enhancement of, most of the real-time simulation, planning, and control algorithms contained in the literature (see Chapter 21). In fact, most published research in this area still has not provided an integrated approach. Most papers report experimental efforts to construct real-time schedulers only. Furthermore, the reported algorithms are usually not true on-line algorithms. These begin by choosing a scheduling/control policy that is to be implemented. This is accomplished by recording the current state of the real-world system and then performing the planning in an off-line fashion. Once the schedule is defined, it is executed in an open-loop manner. While this schedule is being executed, planning is usually suspended. The system performance is simply monitored until the planned and realized response deviates to such an extent that the current schedule is no longer valid. At this point, the system state is recorded and a new schedule is sought. The situation that triggers rescheduling is sometimes referred to as a *decision point* (see Harmonosky, 1990, 1995; Harmonosky and Robohn, 1991).

As stated above, the system usually operates in an open-loop control manner under most reported algorithms. That is, the feedback information realized from the measured response of the system while it operates under the selected schedule is ignored until the decision point occurs. When the decision point occurs, rescheduling is implemented. There is no attempt to update the schedule during the interim between decision points where rescheduling occurs. With this strategy it is clear that the scheduling becomes *reactive*, triggered primarily by system disruptions and/or unexpected system behavior.

Reactive planning/control is, by its very nature, inefficient. By analogy, reactive planning/control is tantamount to driving a vehicle down a highway and changing the direction of steering and/or speed only when one encounters a guardrail or another vehicle. On the other hand, reactive planning and control is much easier to implement. As we will see shortly, reactive planning and control relieves the modeler of having to provide many of the new technologies required to implement proactive on-line planning and control. In many respects, reactive planning may be viewed simply as performing off-line planning more frequently over shorter planning horizons.

A *proactive*, on-line scheduling and control methodology has been depicted in Figure 13.10. The functions in Figure 13.10 are designed to operate concurrently with the real-world system. That is, the task of seeking an improved schedule is constant and never-ending. Furthermore, most proposed reactive planning schemes assume that the system is deterministic. Such schemes perform a single simulation run for each scheduling alternative. The proactive planning/control framework attempts to provide a full assessment of the uncertainties that exist in future system performance. In addition, statistical estimates are constantly updated to account for the current state of the real-world system as it evolves in real time. As this information is formulated, it is forwarded to the real-world system to be employed as feedforward information for controlling that system. As in our driving analogy, the statistical information derived from real-time simulations is our forward-looking vision that permits us to see oncoming cars and turns so that we can modify our speed and direction appropriately. The schedule is constantly updated to reflect the current state of the system and the projected system performance. On the other hand, the closed-loop control policy that monitors the system state and precipitates the modification of the schedule remains in place until it is demonstrated that a better control policy exists.

Under the proactive planning approach, alternative control policies are always being evaluated, and a new control policy can be implemented whenever it is beneficial to do so. That is, we do not have to wait until a decision point occurs to change the control policy. The proactive approach always seeks a better plan for its associated implementing policy and can execute a new policy immediately. Meanwhile, the overall planning and control algorithm constantly assesses the outcome of implementing the current control policy based on the current state of the system.

The algorithm defined in Figure 13.11 has been specially developed for implementation in a concurrent computing environment. This algorithm is asynchronous in nature. That is, it does not require handshaking or two-way communication among the included functions. Thus the algorithm can be implemented in the simplest of concurrent computing environments, including a network of independent workstations. Of course, an advanced concurrent computing environment would permit enhanced communication among the included functions, but it is not essential.

Despite the complexity of the schematic in Figure 13.11, it still represents a simplification of the on-line planning/control situation. Our research in the on-line arena has demonstrated conclusively that on-line planning and control requires that both the planning and the subsequent implementation of the plan must be performed by the same agent to be effective. That is, on-line planning necessitates that an agent plan only to the extent that it can implement. Planning and control can be distributed across several intelligent controllers. The need for a whole new set of distributed planning and control algorithms thus evolves.

These observations led initially to the conceptualization of the coordinated object and the recursive, object-oriented coordination hierarchy, as discussed in Section 13.2. Recall that each of the coordinated objects contains a intelligent controller to perform its on-line planning and control. Looking carefully within the intelligent controller of the coordinated object pictured in Figure 13.6, four functions have been included. Figure 13.11 provides a schematic of the Planning Function that is responsible for determining an improved control law. The Task Assignor Function is responsible for implementing that control law. It performs this task by assigning tasks to its subordinate subsystems. In making this assignment, the Task Assignor Function interacts with the Task Acceptor Function within the subordinate subsystem's intelligent controller. The Task Acceptor Function's role is one of accepting new tasks to be executed by the coordinated object and establishing their completion dates. If new tasks are not accepted, the coordinated object will eventually have no new tasks to perform. Since new tasks are always being accepted even while currently assigned tasks are executed, the scheduling problem as addressed by the coordinated object is constantly being modified. Hence the need for on-line planning is constant. Meanwhile the System Identifier Function is constantly attempting to improve its model of the subordinate subsystem to permit the other functions to better address the on-line planning and control requirements.

The complexity of the overall operation of the intelligent controller is well beyond the scope of this chapter. Like the Planner, each of the other functions consists of several integral elements. These also employ their own on-line simulations. Furthermore, the overall operation of a given intelligent controller cannot be addressed without considering its coordinated interactions with the intelligent controllers at adjacent levels of the ROOCH. Davis (1992) and Davis et al. (1992) provide a more comprehensive formulation for the intelligent controller within this distributed planning and control framework. Davis (1992) and Davis et al. (1993, 1996) further develop the architecture for the coordinated objects and the associated algorithms needed to distribute planning and control. This work can also be viewed as an application of the more general reference model for the intelligent control of large-scale systems that is being developed by Albus and Meystel (1995). This reference model has now been extended to a framework for semiotic modeling (see Meystel, 1995). On the other hand, the control of discrete-event systems is still in its embryonic stage. Ho (1989) may be viewed as one of the pioneers in formalizing the control of discrete-event systems (see also Cassandras, 1993).

To be effective, however, on-line planning (scheduling) and control must also be coordinated with the other organizational functions that address strategic planning and control issues. The comprehensive set of interactions among all the agents that are addressing the entire set of planning and control functions gives rise to what we have termed *hyperlinked object-oriented architectures* (see Davis et al., 1995a). The evolving simulation requirements for these advanced architectures have also been published in Davis et al. (1995b). The important observation here is that although the primary current use for on-line simulation is to enable an on-line scheduling capability, there will probably be many new uses for on-line simulation in the future. A primary purpose of the semiotic modeling (as discussed above) is to provide a framework for modeling and to incorporate the supporting technologies for real-time data analysis. On-line simulation is one such technology.

We now investigate the new simulation-based technologies that are needed to support proactive on-line planning and control. We outline the needs for new simulation tools, new on-line statistical output analysis tools, procedures for on-line compromise analysis, and the selection of scheduling alternatives.

13.3.4 New Simulation Tools and On-Line Simulation Requirements

Now that an overview of on-line simulation and planning has been presented, new requirements for future simulation modeling tools may be defined. First, the need for a controller-based simulation approach becomes even more obvious. The output of the on-line planning process must be the control law by which the developed plan is to be implemented. Unless the simulation model can incorporate this law and model the interactions among the controllers that result from its implementation, there will be no way to evaluate the plan. Second, the capability to initialize the simulation to a measured system state has been established as a critical requirement for on-line simulation. As stated in Section 13.2.3, most current simulation tools are limited in this capacity.

There is also a need for the simulation modeling approach to produce hierarchical models. Recall that a primary assertion for on-line planning is that the planning must be distributed. Each coordinated object must plan its own schedule and implement it. On the other hand, the coordinated object should not plan for actions beyond its control authority to implement. Thus it does not make sense for the on-line planner (e.g., coordinated object) to consider the complete model for the overall system in its planning. Rather, it needs only to consider its behavior as it interacts with immediate subordinate subsystems. Simply stated, the span of control is limited for each on-line planner. It does not have direct control over the behavior of its supervisor, nor does it have direct control over the manner in which its subordinates interact with their own subordinates.

Furthermore, if the on-line planner simulates only the behavior that is derived from its subordinate interactions, the computational requirements for each on-line simulation trial will be reduced. The simulated model will typically be much smaller than the overall system model. The net effect will be that the planner can generate simulation trials at a far faster rate, which, in turn, has consequences for on-line output analysis.

13.3.5 Evolving Needs for On-Line Output Analysis Techniques

Hundreds of papers and numerous books and monographs have been written on the subject of output analysis for off-line planning. On the other hand, little or no research on on-line output analysis has been reported. Unfortunately, most of the literature pertaining to off-line output analysis is not (or has not been shown to be) relevant to the on-line counterpart. For practical purposes this is a virgin research area.

There is a singular concern that seems to render the off-line approaches to output analysis inappropriate to on-line simulation analyses. This is the need to initialize simulation trials to the current system state. As discussed earlier, off-line output analysis is concerned primarily with estimating the steady-state performance of the system. In most cases, explicit measures are taken to eliminate the effects of the transient response which arise after starting the system from a particular initial state. Even terminating simulation analyses assume that the same initial state and operational cycle will be investigated.

For on-line simulations, the focus is on the transient phase of operation. The system never reaches steady state over the short planning horizon. The modeler must analyze the performance of the system in the near term given the current system-state information. Thus, each moment's planning constraints are unique and depend on the current system's state. The same planning problem, as well as the same system state, may never again be observed over the operational lifetime of the system.

Figure 13.12 provides an overview of the ideal situation that evolves when one performs on-line simulation. First, one must separate the output analysis into two regions. The retrospective analysis considers the measured system response that has occurred

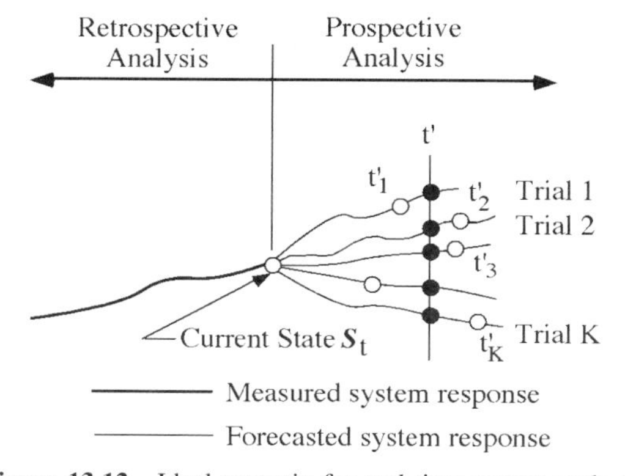

Figure 13.12 Ideal scenario for real-time output analysis.

prior to the current moment. Here, one might ask a question such as: What was the average sojourn time for job entities residing in the system during the last month? The answer to this retrospective question can easily be answered using the output data that were collected while the system operated.

In on-line output analyses, however, one is concerned with providing statistical estimates of the future response of the system. As indicated in Figure 13.12, the current state S_t is known and one desires to look at the evolution of the system-state trajectory beyond that state. In the ideal situation, the essential number of simulation trials (K) is generated immediately and a prospective output analysis can be implemented. If this ideal scenario could be implemented in a manner depicted in Figure 13.12 (i.e., the K simulation trials could be instantaneously generated), the output analysis would be similar to that for a terminating simulation, since all trials are being initiated to the same state S_t. For the moment, let us assume that this ideal scenario can be achieved.

For the prospective output analysis, several types of statistical estimates can be made. One type of analysis would attempt to predict the state at some future time, say t' (see Figure 13.12). This type of analysis is similar to state estimation and is addressed by most terminating simulations. A second type of analysis might focus on the occurrence of a specified event (e.g., when a given job will be completed). In Figure 13.12 we have depicted the occurrence of the selected event as the projected times t'_1 through t'_K for the trials 1 through K, respectively. Given these projected times, one could construct an empirical probability density function for the projected event time and compute its associated statistics. The third type of output analysis would first specify one or more performance measures, which would then be evaluated for each projected trajectory generated on a given simulation trial. These computed values would then be used to compute empirical probability density functions for the projected performance index and its associated statistics. For on-line planning situations, most output analysis will probably be of the third type. However, it is still possible to make the other types of statistical estimates.

The problem is, however, that the ideal situation depicted in Figure 13.12 cannot be realized. It requires computational time to generate the required simulation trials. Figure 13.13 is a more realistic depiction of the situation. In this figure, S^k denotes the initial state to which the kth simulation trial is initiated. While the simulation trial is being executed, the system continues to evolve. Let Δ represent the average time required to generate the simulation trial. If trial k was initiated at t, trial $k+1$ will be initiated at $t+\Delta$. During the Δ time interval, the system will evolve to S^{k+1}. This state may not have been projected by the previous trial k because we are assuming that the system is stochastic.

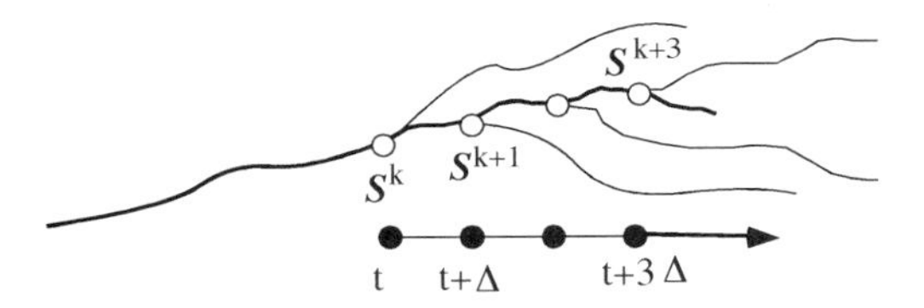

Δ = Computation time period to generate a simulation trial

Figure 13.13 Realistic scenario for addressing real-time output analysis.

Should one initialize the next simulation trial $k + 1$ to the now known state S^{k+1}? A paradox has arisen. Ideally, one would prefer to initialize every simulation trial to the same initial state, as depicted in Figure 13.12. If one adopts this approach, one can, with some statistical justification, treat the on-line output analysis in a manner similar to that of a terminating simulation analysis. However, if one adopts this approach, one must also ignore known information about the system performance that was measured while computing the essential number of simulation trials.

The second approach (depicted in Figure 13.13) is simply to initialize each simulation trial to the most recent, recorded state. However, if one adopts this approach, a set of simulation trials will be generated where each trial has been initialized to a different state. Hence a statistical analysis based on the resulting set of trials is certainly questionable. Using this approach, however, one employs all known information pertaining to the measured system response. In our research, we have adopted the second approach so that we may employ all known information. However, we are certainly aware of the statistical concerns that arise.

There are other statistical concerns pertaining to the number of simulation trials that will be used to perform output analysis. When one investigates the real-time dynamics of a system, the nature of its operation changes with time. At certain times, system behavior may be very transient (i.e., fast dynamics). At other times, it may be very slow (i.e., nearly steady state). If the system is operating with nearly steady-state dynamics, the desire is to increase the number of simulation trials employed in the statistical analysis so that one might achieve more confidence in the statistical estimates. On the other hand, if the system is operating in a highly transient regime, employment of a large number of simulation trials may utilize trials that were initialized to a starting state that is no longer relevant. It is our belief that during periods of highly transient behavior, it is necessary only to make some estimate of future performance. The retention of many old trials may provide an undesirable bias, a form of statistical inertia, in real-time estimates.

It is our observation that there are currently no solutions to the foregoing dilemma. However, after years of experimentation, we have adopted the following approach. For each real-time simulation, we maintain a fixed number of computed performance indices from prior simulation trials, usually the last 1000 to 2000 trials. These projected values for each performance index are held in a pushdown stack. As a new trial provides a projected performance value, the new value is added to the top of the stack and the oldest projected value is removed from the bottom of the stack. To compute statistical estimates for a performance index, we employ an adaptive sample size algorithm as investigated initially by Antonacci (1992). This algorithm begins by taking the 50 most recent projected values for a given performance index and computes its sample mean and associated confidence interval. This procedure is repeated for the 100, 250, 500, 1000, and 2000 most recent samples. Using the properties of statistical estimates, the confidence interval at a given confidence level for the 50 most recent projections should be larger than that of the 100 most recent at the same confidence level. The confidence interval for the 100 most recent samples should be larger than that of the 250 most recent, and so on. This is true, however, only if the system is operating in a nearly steady-state mode. If the system is operating in a transient mode, the sample variance can increase to a point where the confidence interval is actually larger for the larger sample sizes.

After computing the confidence intervals for the various sample sizes, we find the largest sample size employed to compute each performance index. We begin with a sample that includes the last 50 trials and incrementally increase that sample size to

100, 250, 500, 1000, or 2000 by requiring that the confidence interval for the next-largest size be fully contained within the confidence interval of the previous sample size. If the confidence interval for the larger sample size is not contained within the smaller sample size, we employ the smaller sample size to perform our estimates. If more than one performance index is considered, the smallest sample size determined for the analysis of any performance index is used for estimating all the performance indices.

Our approach attempts to maximize the sample size that can be employed in the statistical analysis. When transients occur in system performance, the sample size is automatically reduced to minimize the statistical bias that occurs when outdated simulation trials are employed. However, when the system performance is operating at nearly steady state, the sample size will automatically be increased to use all available data.

Returning to the on-line planning scenario depicted in Figure 13.10, we recall that R different control policy alternatives, in addition to the control policy currently being implemented, are being analyzed concurrently. To ensure that a consistent statistical analysis is performed for all control policies, the smallest acceptable sample size found for any scheduling policy is employed for the analyses of all policies.

The task of performing on-line statistical analysis for $R + 1$ control policy alternatives in real time is computationally challenging. Assuming that L performance measures are to be considered, there are $L(R + 1)$ empirical probability density functions, means, variances, and sets of confidence intervals to be updated in real time. In general, the performance indices are not entirely statistically independent of each other. To this end we typically compute $L(L - 1)(R + 1)/2$ covariances in real time.

The amount of data that can be generated in this manner is phenomenal. Therefore, graphic interfaces are essential. In Figures 13.14 and 13.15 we provide sample graphical outputs for an on-line simulation that we conducted for the Automated Manufacturing Research Facility's flexible manufacturing system that was previously operated by the National Institute of Standards and Technology. Figure 13.14 depicts the estimated average sojourn time (AvgTIS) against the average tardiness (AvgLt) for the next 100 jobs to be processed by the system. Each dot on the upper right screen plots the (AvgLt, AvgTIS) coordinates for the values of the indices computed on a given simulation trial. On this graphic the results of the last 1000 simulation trials are plotted. The dots on the computer screen are color-coded to represent the different scheduling/control policies that are being evaluated. For publication purposes, the color-coded graphic has been converted to a gray scale. In this study we compare three different strategies for allocating the two automated guided vehicles to the jobs that are waiting to be transferred from one station to another. The medium gray dots represent reported performance index values for a first in, first out (FIFO) strategy where the material handling requests are processed in the order in which they are received. The light gray dots represent the "LatestJob" strategy where jobs are given priority based on proximity to their due dates. Finally, the black dots represent the "SmartCart" control strategy, which attempts to minimize the distance that the carts must travel to process the next transfer request.

In Figure 13.14, the dots related to the FIFO strategy are of three distinct linear clusters which are angled at nearly 45°. The dots for the LatestJob strategy are also divided into three nearly horizontal clusters. One must look carefully to discern the dots from the SmartCart strategy. In Figure 13.14 they form stubby, vertical lines that lie primarily on top of the clusters associated with the LatestJob strategy. In Figure 13.15 there are again three primary clusters of dots for both FIFO strategies. In this figure the three clusters for LatestJob strategy blend into a single band. The dots for the

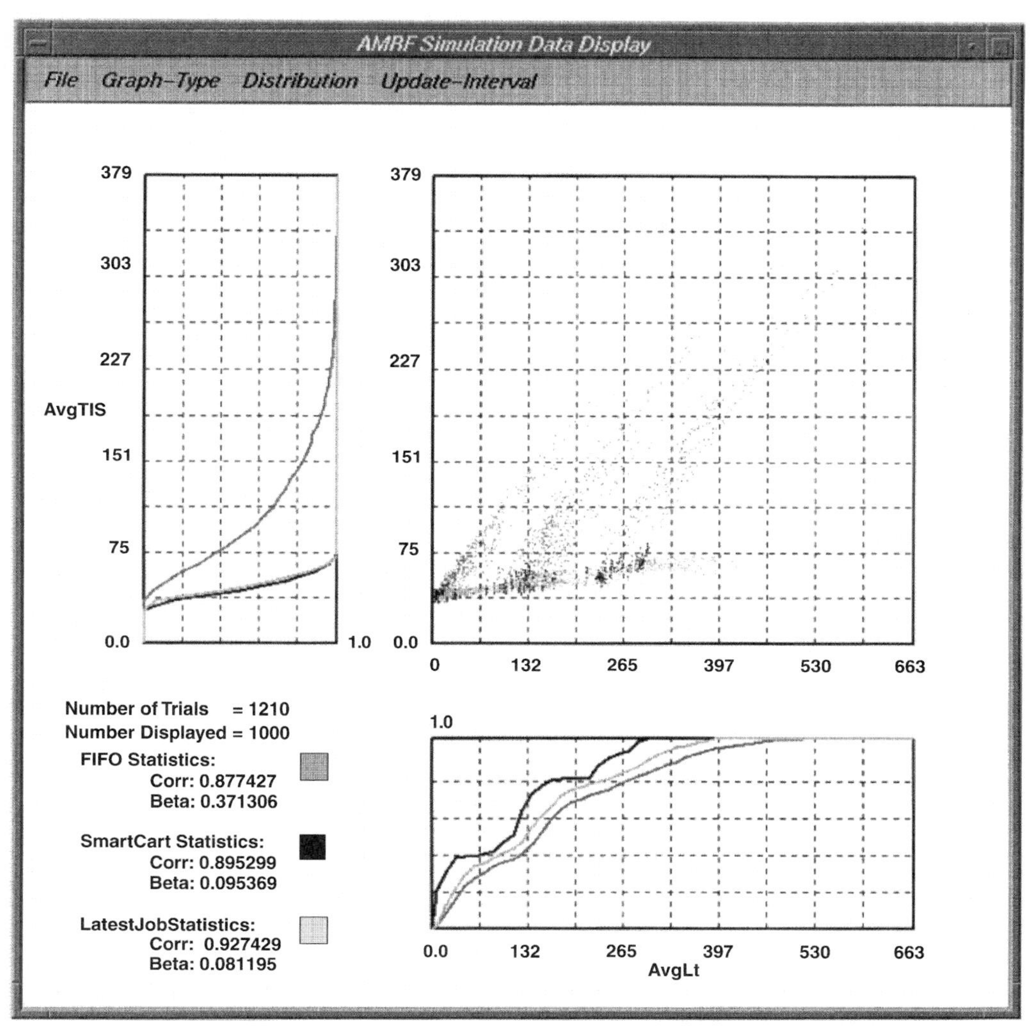

Figure 13.14 Sample graphic from an on-line simulation depicting the trade-offs between the average time in the system and the average lateness.

SmartCart strategy also form vertical lines within the LatestJob clusters, but the vertical lines are much longer in this graphic.

To the left and below the primary graphic, the empirical cumulative density functions are given for each performance index operating under each control policy. The functions are again coded by a gray scale with reference to the control policy being depicted. In the lower left-hand quadrant, the correlation coefficient is given for the two indices under each control policy. The slope of the best linear regression line for the data reported in the primary graphic is also given for each policy.

In Figure 13.14 we can see that the control policy chosen leads to significantly dif-

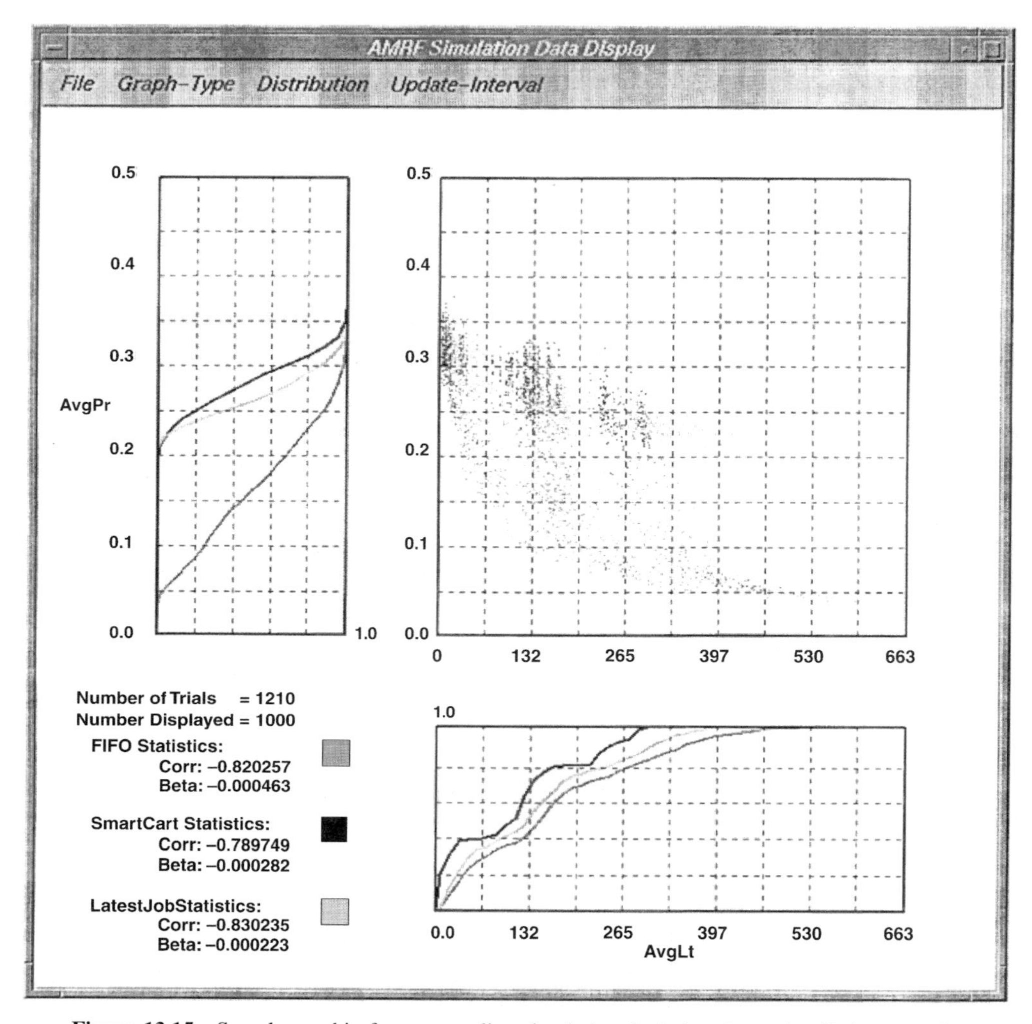

Figure 13.15 Sample graphic from an on-line simulation depicting the trade-offs between the average lateness and the average productivity.

ferent results. In particular, there is a strong linear relationship denoted between the two indices operating under the FIFO policy. This relationship is less pronounced under the other two policies. In Figure 13.14 another graphic for the same on-line simulation analysis is provided, but here the average productivity (AvgPr) is plotted against the AvgLt for the next 100 jobs to be processed. Note the remarkable difference between Figures 13.14 and 13.15 in the loci of the plotted dots. In particular, the relationships for the indices in Figure 13.14, under the FIFO strategy, were nearly linear, whereas the relationships in Figure 13.15 are certainly nonlinear. If L performance indices are being considered, there are $L(L-1)/2$ such graphics to be computed in real time.

Some additional comments regarding the clustered shapes of the dots are appropriate. First, these shapes can and do change with time. The shapes of the clusters are dependent on the current state of the system. Second, in Figure 13.14 we noted that the points in the main graphic for a given control policy do not form a continuous cloud or cluster of points. Rather, they separate into bands that are similarly shaped. We would expect the clusters not to have bands if the system were purely stochastic. We had conjectured that the bands might arise from abrupt changes in the initial state. However, for the graphics depicted in Figures 13.14 and 13.15, we held the initial state constant, and all the reported simulation trials use the same initial state. Although we have not yet proven it, we conjecture that these discrete-event systems may be chaotic and that the bands represent the existence of multiple attractors. In complex system theory it is known that such behavior can occur for even the simplest of nonlinear continuous state systems. Can such behavior also occur for these complex systems? This issue requires more research. We should note also that this concern has not arisen in past research since simulation analysis has not considered transient behavior. Attention to transient behavior may provide new insights into the true dynamics of discrete-event systems. There is still much to be learned about the dynamics of discrete-event systems.

The procedures outlined in this section have been developed from an empirical point of view. Our research in on-line simulation has discovered problems which we have discussed here and in earlier articles (see Davis et al., 1991; Tirpak et al., 1992b). We have attempted to engineer solutions to these problems, but there is limited statistical validity to such approaches for on-line output analysis, as discussed above. Much more research is needed.

13.3.6 Evolving Needs for Real-Time Compromise Analysis Techniques

The issue of selecting a control policy for implementation remains. A comparison of systems using simulation is discussed in Chapter 8. Typically, such analyses are performed in off-line fashion where one or more performance criteria are to be considered. In this section we address a similar task, but in an on-line fashion. We view the implementation of a given control policy as a distinct system configuration and our task is to select the system configuration operating under a particular control policy as the system configuration that is to be implemented at this current moment in time.

The discussion in Section 13.3.4 dealt with the on-line statistical characterization of the L considered performance indices for each of the possible control policy alternatives. The task remains to select the best control policy for implementation. To compare a given stochastic performance index across different control alternatives, we begin by considering the concept of stochastic dominance as discussed by Whitt (1979). Assume that $F_1(x)$ and $F_2(x)$ represent the cumulative probability density function for a given performance index as computed for control policies 1 and 2, respectively. The *principle of stochastic dominance* states that control policy 1 would dominate control policy 2 if the appropriate following conditions could be demonstrated:

$$F_1(x) > F_2(x) \text{ for every } x \qquad \text{(for a maximization)} \tag{1}$$

or

$$F_1(x) < F_2(x) \text{ for every } x \qquad \text{(for a minimization)} \tag{2}$$

In general, this is a very difficult result to achieve. However, if we return to Figure 13.13 and view the empirical cumulative density functions for both the AvgTIS and the AvgLt (both of which we desire to minimize), we see that SmartCart control strategy currently dominates both the FIFO and the LatestJob control strategies with respect to both indices. In Figure 13.14 we see that empirical cumulative density functions for the AvgPr under the SmartCart control strategy and LatestJob intersect. Hence, it cannot be asserted that the SmartCart policy dominates the LatestJob policy in this case.

Based on the data depicted in Figures 13.13 and 13.14, one would probably select the SmartCart policy for implementation. However, stochastic dominance cannot be strictly demonstrated. This is the usual case, as true stochastic dominance can seldom be demonstrated. In most cases, trade-offs must be considered for the development of a compromise solution.

In developing a compromise solution, it is desirable to employ many of the approaches to multicriteria decision making, or Pareto optimization, that have been developed in the planning literature. Unfortunately, most of these concepts have been developed for deterministic systems and have not been extended to stochastic systems. One might adopt a utility function approach, which aggregates the L individual performance criteria into a single criterion. If this approach were to be adopted, one could simply append another performance criterion, the utility function, to the current list of L considered performance criteria and evaluate it for each simulation trial. A graphic, such as the ones in Figures 13.13 and 13.14, could then be generated for the utility function in order to compare the utility function against each individual performance function that is considered within the utility function. Finally, one might attempt to use the principle of stochastic dominance to assert the control policy that should be employed, based on the statistics for the utility function.

Two problems arise when this approach is adopted. The first problem concerns the definition of the appropriate utility function. The statistics computed for each performance index under each alternative control strategy are time variant. Since the statistics for the performance indices are time variant, it follows that the statistical characterizations governing the compromise among the performance criteria are also time variant. Hence, in the on-line planning scenario, it is virtually impossible to define a utility function on an a priori basis for the entire period in which on-line planning is to be addressed.

The second problem concerns the ability to demonstrate and the benefits derived from demonstrating stochastic dominance. Even if a utility function could be defined, it is unlikely that one control policy would dominate the others stochastically. Furthermore, a demonstration of stochastic dominance is not a complete solution in itself. Stochastic dominance provides no information about the extent to which one control policy dominates another with respect to a given performance index, nor does it guarantee that the dominating control policy will generate a better performance index when implemented. In Figure 13.14, which includes the empirical probability density functions for the average sojourn time (AvgTIS), we see that the SmartCart strategy dominates the FIFO strategy to a much greater extent than it dominates the LatestJob Strategy. In the latter case the two empirical probability density functions almost lie on top of each other.

Stochastic dominance addresses the expected outcome resulting from applying the dominating control policy many times. However, under the on-line planning scenario, the current situation will probably be addressed at this time only. Hence, implementation of a control law provides an opportunity to conduct a single experiment. The outcome of the experiment cannot be guaranteed, as the system is assumed to be stochastic and, perhaps, chaotic.

Our laboratory is currently experimenting with new ways of computing dominance probabilities. We are now computing, in real time, the probability that one control policy will provide a performance value that is δ greater than the value generated by another control policy. This dominance probability density function is calculated for all possible combinations of performance indices and control alternatives and all possible values of δ. This is a computationally intensive operation, as there are $R + 1$ control alternatives and L performance criteria to be considered, requiring that a total of $LR(R+1)/2$ dominance probability distributions for δ are to be computed in real time. Furthermore, these probabilities are again dependent on the current state of the system and must be recomputed constantly. Even when these probabilistic dominance distributions are computed, we still do not know how to employ them efficiently to obtain the best compromise solution. This task remains a future research direction.

13.3.7 Evolving Needs for Autovalidation Procedures

As stated in Section 13.3.1, the validation process is often complex, and there are no standard procedures that apply for addressing this task over a broad spectrum of modeling situations. Even the suggestion to develop autovalidation procedures might appear to be folly. Nevertheless, it must be recognized that real-world systems are time variant, and the models that are employed in the on-line planning scenario must be adapted to reflect observed changes in the system dynamics. Again, there is a current need for which there is no solution.

A major factor that contributes to the complexity of the validation process is the underlying modeling philosophy of the simulation language chosen. Often, the modeler is forced to employ modeling tricks in order to address a particular element of the system's operation. When the modeler employs nonstandard modeling methods, model validation and verification processes are complicated. This deficiency can be addressed with new simulation languages that apply a modeling philosophy more congruent with the operation of the real-world systems. As stated above, we believe that it is essential to model controller interactions for most modern automated systems. We further believe that languages which adopt this philosophy will simplify modeling and permit the construction of models that are more easily validated.

A second factor that complicates the validation process evolves from the use of aggregation where the dynamics arising from the execution of several tasks is model as the execution of a single task. A common belief is that aggregation simplifies the validation process. This assertion has never been proven, however. In our years of modeling real-world systems, we have observed that a primary contribution to the stochastic nature of the models is derived from the aggregation process, not the system dynamics. In fact, one can take a deterministic system and construct a model that is stochastic due solely to aggregation. The fact is that when one ignores the detailed dynamics of a system, one reduces the ability to predict the dynamics of that system.

Most engineering disciplines require a testing of the consequences of making simplifying assumptions before application. For example, can one safely apply a linear model when the system has nonlinear characteristics? To test the simplifying assumptions, one must build two models, one with, and one without, the simplifying assumptions. Too often in simulation, however, we apply simplifying assumptions that are never tested. The required detailed models are never constructed.

It is our belief that we should model the system to the greatest detail possible to achieve the most accurate predictions of true system performance. With this approach we have found that the resulting models are typically less stochastic. In general, the remaining stochastic elements that are truly random variables associated with phenomena that are more readily measured. For example, the yield on a given process can easily be plotted as a function of time. The time required to perform a given instruction from a specific processing plan can be measured and statistically quantified.

It is also noted that many of the procedures employed in input analysis can be automated. This observation is, to a certain degree, substantiated by the fact that many of these procedures have already been programmed and are now included as standard elements of most commercial simulation packages. It is our belief that if we are concerned with statistical quantifying of simple random phenomena, there is a potential for the development of autovalidation procedures.

However, with the statistical characterization of stochastic phenomena that have evolved simply through the aggregation process, the validation process is typically much more complex. In this case the modeler must develop a probabilistic distribution for a phenomenon that represents the interaction of several random variables. These interacting random variables could have been investigated independently if an aggregation had not been adopted.

In summary, we challenge the inherent belief that the modeler should attempt aggregation in order to simplify the model. New, improved simulation tools would permit the modeler to address the detailed dynamics of the real system and should make validation a more exacting endeavor.

13.3.8 Selection of Scheduling Alternatives

As with the foregoing concerns, the selection of scheduling alternatives is also an area where considerable research is needed. A problem evolves with the nature of discrete-event systems because their dynamics are explicitly tied to the definition and the execution of tasks. Unfortunately, our current planning techniques do not adequately consider tasks. Most mathematical programming formulations for scheduling problems are tied to the definition of decision variables. These decision variables simply do not provide the information required to specify tasks. Too often, the modeler must make numerous simplifying assumptions in order to translate the scheduling problem into a mathematical formulation. In most cases the formulation considers the precedence relationships and required processing durations for the scheduled job at various machines only. These formulations ignore completely the constraints that arise from the provision of supporting resources such as tooling in a manufacturing setting. These formulations also typically do not consider alternative methods for implementing a given task. That which results is a formulation that has little correspondence to the real-world planning task that must be addressed.

Even if the decision can be expressed as a mathematical programming problem, the mathematical programming problem still must be solved. Here again, there are no guarantees. The complexity of the resultant formulation often requires heroic efforts to generate a solution. Heuristic approaches are often adopted. More recently, next-generation search procedures such as genetic algorithms have been applied. There are numerous articles from the genetic algorithm literature in which the scheduling problem is addressed as the classic traveling salesman problem (TSP). To transform the scheduling problem into a TSP, the modeler must once again make several simplifying assumptions which ultimately affect the utility of the solution.

If a solution can be achieved, the task remains to transfer this solution into a control policy that can be implemented. There are no well-tested procedures for making this transformation. Perhaps the solution tells us the order in which each job will be processed at each machine, but is this the control policy that we desire to implement? Recall that it is our desire to develop a closed-loop policy that will modify the schedule based on the current state of the system. The mathematical formulation of the problem, however, typically ignores the stochastic nature of the system. The constraints are deterministic. Seldom does the planner check the robustness of the derived solution against the uncertainties that exist in the real system. There simply is no basis for the generation of a closed-loop control law.

Such observations are not new. In the 1980s, there was a major initiative from several research groups to develop scheduling capabilities using expert systems. The attempt was to develop rule-based schedulers which would specify the control action based on the current state of the system. Hence, this approach did lead to a closed-loop control law.

One might argue that a major impetus for the development of expert-system, rule-based schedulers was the perceived consensus that mathematical programming could not address the scheduling problem. Complications again arose. First, there simply were no experts for many of these systems. This was especially true when new systems (e.g., flexible manufacturing systems) were brought on-line. Second, most expert systems assumed deterministic behavior when the systems were stochastic. Third, the quality of the schedules produced by these systems could not be tested. Expert systems do not guarantee optimal solutions. In fact, expert systems provide no capability at all for assessing the quality of the solutions they generate. Finally, the task of assessing the appropriate action for every possible state is simply insurmountable. Thus rule-based schedulers were unable to respond in all given situations.

In formulating the on-line planning schematic given in Figure 13.10 we did not make assumptions about how the scheduling/control policy will be generated. We assumed only that alternative control policies exist. On-line simulations are then conducted to project performance with respect to a set of criteria. (Note that the scheduling approaches above typically consider a single performance criterion.) We believe that the development of scheduling approaches for these discrete-event systems will remain a major research concern well into the future. Given the decades of scheduling research, this may appear to be an astonishing statement. The fact remains that our laboratory is still encountering numerous instances where operators of advanced manufacturing systems are turning off their expensive commercial schedulers and running their systems on an ad hoc basis (e.g., first in, first out basis) to increase their throughput and productivity.

Based on our research, we suggest that the following criteria be considered in developing a comprehensive scheduling approach:

- The scheduling approach should lead immediately to the definition of tasks that will be executed in a manner that satisfies the detailed processing plans for executing a job. The potential for alternative processing plans should also be considered.
- The scheduling approach should consider the provision of all supporting resources essential to task execution.
- The execution scheme for the planned tasks must be transferrable into a closed-loop control policy that can be implemented immediately at the controllers contained within the real-world system.
- The feedforward information derived from the on-line simulation of the currently implemented policy should be incorporated into the planning process.

Finally, on-line planning and control necessitates that both planning and control be distributed. Currently, there is little theoretical guidance to address this requirement, particularly in the case of discrete-event system management.

13.3.9 Implementation Considerations

Before leaving this section, perhaps it would be beneficial to mention a few concerns that arise during the implementation of the on-line simulation and planning analyses. This is certainly not a trivial concern. In developing on-line simulation analyses in our laboratory, we have attempted to implement the algorithms in a manner that will allow for asynchronous communication among the various implemented modules. In this regard, our attempt has been to define computational procedures that will minimize the communication among the various computational modules and to require one way communication only.

In the on-line simulation analysis used to generate Figures 13.13 and 13.14, we adopted the following approach. First, we began with a real-time emulation for the Automated Manufacturing Research Facility operated by the National Institute of Standards and Technology which was the subject FMS for the analysis. This emulation was programmed in C and wrote the current state information to a file on the network server's disk every 10 seconds of emulated time. The same C code that is performing the real-time emulation for the AMRF is also employed to generate the on-line simulation trials. We refer to this use of the simulation code as being the *simulation engine*.

In this study, three different simulation engines were employed, one for each rule considered for dispatching the material handling system. These rules included first in, first out, the latest job, and the shortest travel distance to determine which of the pending transportation requests would be handled next. Each engine first read the current state file stored on the network server and then generated 10 simulation trials using that current state information. On each simulation trial, the simulation engine also computed the value of the four considered performance indices for the next 100 jobs to be manufactured by the system. These indices included the average time in the system, the average tardiness, the average productivity and the average process utilization. The values of the performance indices computed from the 10 trials were then written into two other files, both with the same information, on the network server. At this point a given simulation engine reinitialized itself using the current state information stored on the network server and repeated the process of generating 10 more simulation trials.

As stated above, each simulation engine writes two files containing the project performance of the system for its 10 most recent trials. One of these files for each simulation engine serves as the input to the on-line statistical analysis module. When the on-line statistical module determines that one of its input files from one of the simulation engines is not empty, it inputs the 10 most recent trials and then updates its performance statistics for the particular dispatching rule that was employed to generate the latest estimates. It then checks its input files that were generated by the other simulation engines. It continues this checking and incorporates the inputted data wherever new data is available. In addition to generating the graphic output as depicted in Figures 13.13 and 13.14, the on-line statistical analysis module also outputs a file containing its current estimates for the empirical cumulative density function for each performance index while operating under each dispatching rule.

The on-line statistical compromise analysis function has access to two sets of input date. First it can use the second output file of performance estimates for the last 10 simulations as generated by a given simulation engine. It can also employ the output file of the cumulative density functions generated by the on-line statistical analysis module. The operation of the on-line compromise analysis module is very similar to that for the on-line statistical analysis module. As new data is inputted into the module, the statistics are updated and displayed to the user. The output of the statistical compromise analysis is the rule selected for dispatching the material handling system. This is written to a file which is then inputted by the real-time emulator for the AMRF. In this manner, the planning loop is closed.

Although the foregoing procedure works, it is very costly in terms of input and output requirements in its need to employ the services of the hard disk drive upon the network server. In fact, long-term execution of the on-line simulation analyses can be detrimental to the life of the hard disk drive. To overcome the current communication overhead, the entire suite of program has been reprogrammed in Java. A client-server has been developed to handle the communication among the various modules. The window interfaces to the various modules are now executable at any site on the world-wide web.

Within the next year, we hope to establish a home page for our laboratory on the World-Wide Web where both the real-time emulation and the on-line simulation analysis can be viewed by all. We also hope to provide detailed databases for several real-world FMSs which the public can download and employ in the development of simulation models for real-world FMSs. In the long term, we also hope to generate remote displays to our physical FMS emulator where the viewer will be able to monitor its real-time operation and state information, view its projected behavior via an on-line simulation analysis, and make control decisions that will affect its future performance.

13.4 CONCLUSIONS

For over three decades, significant research efforts have been devoted to the discrete-event modeling approaches and the simulation technologies needed to conduct off-line planning analyses. In the introduction we discussed a new class of large-scale system, one utilizing distributed control, for which we believe that current modeling, scheduling, and control techniques are inadequate. In Section 13.2 we addressed the limitation of current modeling approaches head on. We asserted that an essential requirement for modeling these systems is the ability to assess the impact that the controller interactions have on the dynamics of the system. We also asserted that the real-time management of these systems requires that on-line planning and control be considered in an integrated fashion and that the complexity of these systems requires that both planning and control be distributed across a hierarchy of intelligent controllers or coordinators.

To address these requirements, in Section 13.2 we introduced the notion of the coordinated object, which included an intelligent controller or coordinator to perform integrated on-line planning and control. The recursive object-oriented coordination hierarchy was then described as an architecture for assembling the coordinated objects into a real-time management structure such that planning and control are both distributed and coordinated. To model the performance of the ROOCH, a new simulation modeling methodology, the hierarchical object-oriented programmable logic simulator was introduced and our efforts to construct a real-time emulator for a real-world FMS were discussed.

In Section 13.3 we addressed the on-line planning and control problems that would probably be addressed by a coordinated object. First, the off-line and on-line planning scenarios using simulation were contrasted. Next, two forms of on-line planning (reactive and proactive) were discussed. Finally, we presented a series of technologies that must be developed to permit on-line simulation analyses to be conducted within on-line planning scenarios. The development of most of these technologies represents new research areas.

In summary, the writing of this futuristic assessment has presented a considerable challenge for several reasons. First, simulation is a very large topic in itself. The coupling of simulation with the evolving areas of on-line planning and control for a large-scale discrete-event system makes the task even more exciting and difficult. The writing of this chapter drew on years of experience with real-world systems. The conclusions and approaches are empirical; they were developed to address problems as the research team encountered them. More theoretical research is certainly justified.

REFERENCES

Albus, J., and A. Meystel (1995). A reference model architecture for design and implementation of semiotic control in large and complex systems, in *Architectures for Semiotic Modeling and Situation Analysis in Large Complex Systems: Proceedings of the 1995 ISIC Workshop*, J. Albus, A. Meystel, D. Pospelov, and T. Reader, eds., AdRem Press, Bala Cynwyd, Pa., pp. 21–32.

Antonacci, L. A. (1992). Experimentation in stochastic compromise analysis for real-time multicriteria decision making, unpublished master's thesis, W. J. Davis, advisor, Department of General Engineering, University of Illinois, Urbana, Ill.

Astrom, K. J., and B. Wittenmark (1989). *Adaptive Control*, Addison-Wesley, Reading, Mass.

Bazaraa, M. S., D. Sherali, and C. M. Shetty (1993). *Nonlinear Programming: Theory and Applications*, 2nd ed., Wiley, New York.

Booch, G. (1991). *Object Oriented Design with Applications*, Benjamin-Cummings, Redwood City, Calif.

Cassandras, C. G. (1993). *Discrete Event Systems: Modeling and Performance Analysis*, Aksen Associates.

Davis, L. (1987). *Genetic Algorithms and Simulated Annealing*, Pitman Publishing, London.

Davis, W. J. (1992). A concurrent computing algorithm for real-time decision making, *Proceedings of the ORSA Computer Science and Operations Research: New Developments in Their Interfaces Conference*, O. Balci, R. Sharda, and S. Zenios, eds., Pergamon Press, London, pp. 247–266.

Davis, W. J., H. Wang, and C. Hsieh (1991). Experimental studies in real-time, Monte Carlo simulation, *IEEE Transactions on Systems, Man and Cybernetics*, Vol. 21, No. 4, pp. 802–814.

Davis, W. J., A. T. Jones, and A. Saleh (1992). A generic architecture for intelligent control systems, *Computer Integrated Manufacturing Systems*, Vol. 5, No. 2, pp. 105–113.

Davis, W. J., D. L. Setterdahl, J. G. Macro, V. Izokaitis, and B. Bauman (1993). Recent advances in the modeling, scheduling and control of flexible automation, in *Proceedings of the 1993 Winter Simulation Conference*, G. W. Evans, M. Mollaghasemi, E. C. Russell, and W. E. Biles, eds., Society for Computer Simulation, San Diego, Calif., pp. 143–155.

Davis, W. J., J. J. Swain, G. Macro, and D. L. Setterdahl (1994). An object-oriented, coordination-based simulation model for the RAMP flexible manufacturing system, in *Proceedings of the Flexible Automation and Integrated Manufacturing Conference*, M. M. Ahmad and W. G. Sullivan, eds., Begell House, New York, pp. 138–147.

Davis, W. J., F. G. Gonzalez, J. G. Macro, A. C. Sieveking, and S. V. Vifquain (1995a). Hyperlinked architectures for the coordination of large-scale systems, in *Architectures for Semiotic Modeling and Situation Analysis in Large Complex Systems: Proceedings of 1995 ISIC Work-*

shop, J. Albus, A. Meystel, D. Pospelov, and T. Reader, eds., AdRem Press, Bala Cynwyd, Pa., pp. 21–32.

Davis, W. J., F. G. Gonzalez, J. G. Macro, A. C. Sieveking, and S. V. Vifquain (1995b). Hyper-linked applications and its evolving needs for future simulation, in *Proceedings of the 1995 Winter Simulation Conference*, C. Alexopoulos, K. Kang, W. R. Lilegdon, and D. Goldsman, eds., Society for Computer Simulation, San Diego, Calif., pp. 186–193.

Davis, W. J., J. J. Swain, G. Macro, A. L. Brook, M. S. Lee, and G. S. Zhou (1996). Developing a real-time emulation/simulation capability for the control architecture to the RAMP FMS, a state-of-the-art tutorial in *Proceedings of the 1996 Winter Simulation Conference*, J. M. Charnes, D. J. Morrice, D. T. Brunner, and J. J. Swaim, eds., Society for Computer Simulation, San Diego, Calif., pp. 171–179.

Davis, W. J., A. C. Brook, and M. S. Lee (1997). A new simulation methodology for master production scheduling, *Proceedings of the 1997 IEEE International Conference on Systems, Man and Cybernetics*.

Davis, W. J., J. Macro, and D. Setterdahl. (to appear). An integrated methodology for the modeling, scheduling and control of flexible automation, *Journal of Robotics and Intelligent Control*.

Drake, G. R., J. S. Smith, and B. R. Peters (1995). Simulation as a planning and scheduling tool for flexible manufacturing, in *Proceedings of the 1995 Winter Simulation Conference*, C. Alexopoulos, K. Kang, W. R. Lilegdon, and D. Goldsman, eds., Society for Computer Simulation, San Diego, Calif., pp. 805–812.

Dullum, L. M., and W. J. Davis (1992). Expanded simulation studies to evaluate tool delivery systems in a FMC, in *Proceedings of the 1992 Winter Simulation Conference*, J. J. Swain, D. Goldsman, R. C. Crain, and J. R. Wilson, eds., Society for Computer Simulation, San Diego, Calif., pp. 978–986.

Eykhoff, P. (1974). *System Identification: Parameter and State Identification*, Wiley, New York.

Fishburn, P. C. (1987). *Nonlinear Preference and Utility Theory*, Johns Hopkins University Press, Baltimore, Md.

Flanders, S. W., and W. J. Davis (1995). Scheduling a flexible manufacturing system with tooling constraints: an actual case study, *Interfaces*, Vol. 25, pp. 42–55.

Goldberg, D. E. (1989). *Genetic Algorithms in Search, Optimization and Machine Learning*, Addison-Wesley, Reading, Mass.

Gonzalez, F. G. (1997). Real-time distributed control and simulation of a physical emulator of a flexible manufacturing system, unpublished doctoral thesis, W. J. Davis, advisor, University of Illinois at Urbana-Champaign.

Gonzalez, F. G., and W. J. Davis (1997). A simulation-based controller for distributed discrete-event systems with application to flexible manufacturing, in *Proceedings of the 1997 Winter Simulation Conference*, S. Andradóttir, K. J. Healy, D. H. Withers, and B. L. Nelson, eds., Society for Computer Simulation, San Diego, Calif., pp. 845–852.

Govindaraj, T., L. F. McGinnis, C. M. Mitchell, D. A. Bodner, S. Narayanan, and U. Sreekanth (1993). OOSIM: a tool for simulating modern manufacturing systems, in *Proceedings of the 1993 National Science Foundation Grantees in Design and Manufacturing Conference*, pp. 1055–1062.

Harmonosky, C. M. (1990). Implementation issues using simulation for real-time scheduling, control, and monitoring, in *Proceedings of the 1990 Winter Simulation Conference*, O. Balci, R. P. Sadowski, and R. E. Nance, eds., Society for Computer Simulation, San Diego, Calif., pp. 595–598.

Harmonosky, C. M. (1995). Simulation-based real-time scheduling: review of recent developments, in *Proceedings of the 1995 Winter Simulation Conference*, C. Alexopoulos, K. Kang, W. R. Lilegdon, and D. Goldsman, eds., Society for Computer Simulation, San Diego, Calif., pp. 220–225.

Harmonosky, C. M., and S. F. Robohn (1991). Real-time scheduling in computer-integrated manufacturing: a review of recent research, *Journal of Computer Integrated Manufacturing*, Vol. 4, pp. 331–340.

Hedlund, E., W. Davis, and P. Webster (1990). Using computer simulation to compare tool delivery systems in an FMC, in *Proceedings of the 1990 Winter Simulation Conference*, O. Balci, R. P. Sadowski, and R. E. Nance, eds., Society for Computer Simulation, San Diego, Calif., pp. 641–645.

Himmelblau, D. M. (1972). *Applied Nonlinear Programming*, McGraw-Hill, New York, pp. 148–158.

Ho, Y. C. (1989). Dynamics of discrete-event systems, *Proceedings of the IEEE*, Vol. 77, No. 1, pp. 3–6.

Law, A. M., and W. D. Kelton (1991). *Simulation Modeling and Analysis*, 2nd ed., McGraw-Hill, New York.

Ljung, L. (1987). *System Identification: Theory for the User*, Prentice Hall, Upper Saddle River, N.J.

Meystel, A. (1995). *Semiotic Modeling and Situation Analysis: An Introduction*, AdRem Press, Bala Cynwyd, Pa.

Mize, J. H., H. C. Bhuskute, and M. Kamath (1992). Modeling of integrated manufacturing systems, *IIE Transactions*, Vol. 24, No. 3, pp. 14–26.

Narayanan, S. D., A. Bodner, U. Sreekanth, S. J. Dilley, T. Govindaraj, L. F. McGinnis, and C. M. Mitchell (1992). Object-oriented simulation to support operator decision making in semiconductor manufacturing, in *Proceedings of the 1992 International Conference on Systems, Man and Cybernetics*, IEEE, Piscataway, N.J., pp. 1510–1519.

Peters, B. A., J. J. Swain, S. Smith, J. Curry, and C. LaJimodiere (1996). Advanced tutorial: simulation based scheduling and control, in *Proceedings of the 1996 Winter Simulation Conference*, J. M. Charnes, D. J. Morrice, D. T. Brunner, and J. J. Swaim, eds., Society for Computer Simulation, San Diego, Calif., pp. 194–198.

Seiveking, A. C., and W. J. Davis (1995). An object-oriented simulation language for master production scheduling, in *Proceedings of the 1995 IEEE International Conference on Systems, Man and Cybernetics*, A. C. Sieveking and W. J. Davis, eds., pp. 189–194.

Smith, J. S., R. A. Wysk, D. T. Sturrock, S. E. Ramaswamy, G. D. Smith, and S. B. Joshi (1994). Discrete event simulation for shop floor control, in *Proceedings of the 1994 Winter Simulation Conference*, J. D. Tew, S. Manivannan, D. A. Sadowski, and A. F. Seila, eds., Society for Computer Simulation, San Diego, Calif., pp. 962–969.

Tirpak, T. M., S. M. Daniel, J. D. LaLonde, and W. J. Davis (1992a). A fractal architecture for modeling and controlling flexible manufacturing systems, *IEEE Transactions on Systems, Man and Cybernetics*, Vol. 22, No. 5, pp. 564–567.

Tirpak, T. M., S. J. Deligiannis, and W. J. Davis (1992b). Real-time scheduling of flexible manufacturing, *Manufacturing Review* (ASME), Vol. 5, No. 3, pp. 193–212.

Whitt, S. (1979). Comparing probability measures on a set with intransitive preference relation, *Management Science*, Vol. 25, No. 6, pp. 505–511.

Wolfram, S. (1988). Fundamentals of multicriteria optimization, in *Multicriteria Optimization in Engineering and in the Sciences*, S. Wolfram, ed., Plenum Press, New York.

PART IV

APPLICATION AREAS

CHAPTER 14

Simulation of Manufacturing and Material Handling Systems

MATTHEW W. ROHRER
AutoSimulations Incorporated

14.1 INTRODUCTION

Manufacturing and material handling systems provide a wealth of applications for simulation. Simulation has been used to solve manufacturing problems for many years, as indicated in the annual *Proceedings of the Winter Simulation Conference*. There are several reasons for this great use of simulation, including:

- Motivation is needed for manufacturers to stay competitive.
- A high level of automation is applied to manufacturing.
- Trends such as just-in-time manufacturing need to be tested.
- Manufacturing systems are quite well defined.
- Manufacturing and material handling systems are usually too complex for other analytic techniques.

In a global economy, successful manufacturers are constantly changing the way they do business in order to stay competitive. Companies are asking such questions as:

- When should the next piece of equipment be purchased?
- How many people will be needed next month to fill orders?
- Can a new order be accepted without delaying other work?
- How will the new plant operate five years from now?
- How can work-in-process inventory and cycle time be reduced while increasing throughput?

Simulation provides a method for finding answers to these and other questions about

Handbook of Simulation, Edited by Jerry Banks.
ISBN 0-471-13403-1

the behavior of a manufacturing system. Savolainen et al. (1995) indicate that simulation models are really formal descriptions of real systems that are built for two main reasons:

1. To understand conditions as they exist in the system today.
2. To achieve a better system design through performing what-if analysis.

Law and Kelton (1991) and Banks et al. (1997) give many benefits for simulation. Perhaps the most important benefit is that after the model is validated and verified, proposed changes can be investigated in the simulation model rather than in the real system.

As a result of material handling requirements and repetitive tasks, a major increase in automation has occurred in manufacturing. Robots perform many tasks that human beings used to perform. Ergonomic issues continue to drive the use of automation where repetitive tasks can cause injury to people. Also, automated material handling systems help move products quickly to the point of use, enabling manufacturers to increase their production. Automation projects require significant capital expenditures, and simulation can be used to test drive new equipment before purchasing it.

There have been many trends in manufacturing methods with names such as just-in-time (JIT), flexible work cells, and Kanban. Types of manufacturing systems as defined by Harrell and Tumay (1996) include, but are not limited to:

1. Project shop
2. Job shop
3. Cellular manufacturing
4. Flexible manufacturing systems
5. Batch flow shop
6. Line flow systems (production and assembly lines, transfer lines)

For many manufacturers, implementing a change in their operation can be risky. Once again, simulation can be used as the test bed for evaluation of new manufacturing methods and strategies.

One of the tasks of industrial and manufacturing engineers is to look for ways to improve manufacturing processes. Process improvement starts with measurement. The axiom "one can't improve what can't be measured" comes to mind. Using engineering discipline, manufacturing systems can be measured, data collected, and processes analyzed. Measurement efforts are the first step to better understanding of manufacturing systems. Where processes have been measured and data collected, simulation can be applied as a decision-making tool to enhance system understanding. When systems are not well defined or understood, it is difficult to build accurate models that are worthwhile.

Manufacturing and material handling systems can be arbitrarily complex and difficult to understand. The number of possible combinations of input variables can be overwhelming when trying to perform experimentation. Other methods of analysis, such as spreadsheet models or linear programs, may not capture all the intricacies of process interaction, downtime, queueing, and other phenomena observed in the actual system. After manufacturing system data have been collected and verified, simulation can be used to represent almost any level of detail to provide an accurate representation of a real-world system. From a model of the system, the behavior of the system and its components can be better understood.

When Should Simulation Be Used? Simulation is used most often in the planning and design of new or existing manufacturing facilities. Whenever a new system is being designed, and significant financial resources will be expended, simulation should be one of the steps in the design process. In the late 1990s, some companies use minimums from $50,000 to $100,000 as the capital expenditure level which requires that a simulation be performed. Most new facilities run into the millions or even billions of dollars in construction and equipment costs. The average simulation takes 4 to 10 person-weeks to complete, and with late 1990s prices of around $5000 per week, the cost extends from $20,000 to $50,000. If a manufacturer saves just one piece of equipment, or eliminates a potential bottleneck, the simulation will have paid for itself. Once the bricks and mortar are in place and the machines have been set on the floor, it is expensive to make changes, perhaps 10 times as expensive as making the change when the system was being designed.

While some models are used to plan and design, other models are used in the day-to-day operation of manufacturing facilities. These "as built" models provide manufacturers with the ability to evaluate the capacity of the system for new orders, unforeseen events such as equipment downtime, and changes in operations. Some operations models also provide schedules that manufacturers can use to run their facilities. Simulation can complement other planning and scheduling systems to validate plans and confirm schedules. Before taking a new order from a customer, a simulation model can show when the order will be completed and how taking the new order will affect other orders in the facility. Simulation can be used to augment the tasks of planners and schedulers to run the operation with better efficiency.

Experimentation with an actual system can be a costly endeavor. In a service industry such as banking, it is easy to add another teller to the operation without causing adverse effects in the operation. In a manufacturing facility, any disruption in the operation can cause problems that could be catastrophic. For example, an automotive supplier provides trim parts to an assembly line. The automaker keeps a minimum inventory of the parts at the line, following JIT principles and reducing work-in-process (WIP) inventory. The supplier attempts a process change that backfires and causes a delay in production of 8 hours. Because the automaker only has 4 hours worth of inventory on site, the assembly plant is forced to shut down, idling hundreds of workers. The automaker decides to find another supplier and takes its business elsewhere. With an accurate model, the supplier could have tested many different scenarios before making process changes on the shop floor. In this example the cost of idling workers as well as the lost business could easily have justified using simulation.

There are cases where simulation is not the appropriate tool. For example, when a closed-form solution is possible, simulation would be an unduly expensive alternative. Banks and Gibson (1997) give 10 rules when simulation is not the appropriate tool.

What Are the Steps in Getting Started with Simulation? Once the need to build a simulation model is recognized, the next step is to get a simulation project started. The size of the firm will determine how you proceed with a study. The following are some options for getting started:

- Use internal resources.
- Enlist the services of a qualified simulation consultant.
- Use a simulation consultant in combination with internal resources.

Regardless of the approach taken, be sure that objectives are clearly defined for the project. Also, when using internal resources, remember that the process of simulation is more than learning how to use a simulation tool. A course in simulation from a university or software vendor or other provider can help ensure the success of your first project. Banks and Gibson (1996) prescribe 12 steps in getting started with simulation.

What Software Should Be Used? The selection of simulation software depends on many considerations. These considerations and a brief description of numerous simulation software packages are covered in Chapter 25.

Chapter Outline. This chapter is organized into two sections, the first on manufacturing and the second on material handling. In each of the sections, issues specific to the topic are discussed. An example as well as case studies are provided to enhance understanding of the topic. Finally, a synopsis of recent articles from the *Proceedings of the Winter Simulation Conference* is given.

14.2 MANUFACTURING SYSTEMS

Manufacturing is the process of making a finished product from raw material using industrial machines. Almost everything in the home, at the workplace, or in moving between the two is the result of manufacturing. Examples include automobiles, airplanes, ships, home appliances, computers, and furniture. The list goes on. But how do these products get made, and what are the issues facing those who manufacture these goods?

One major issue for manufacturers is competition. In a free-market economy, anyone with a good idea and some money can start a manufacturing business. Competitive pressures force manufacturers to look at different ways of doing business so they can continue to produce. Manufacturing and industrial engineers are tasked with finding ways to improve operations through analysis.

Another issue is managing change. To stay competitive, manufacturers are changing their operations constantly. The companies that manage change most effectively come out on top. Change is inevitable, and those companies that resist it often find themselves out of business.

Since manufacturing covers a wide range of products and manufacturing methods, it would take an entire book to address every possible manufacturing operation and how simulation applies. This chapter gives an overview of the key issues in manufacturing operations as they relate to simulation. Manufacturing can be broken down into two types, discrete and batch. Discrete manufacturing involves individual pieces or parts. Batch manufacturing applies to work-in-process that is handled as a fluid or bulk solid. Simulation has been applied to both types of manufacturing; however, most applications have been in the discrete area.

Before manufacturing issues are discussed in detail, it is important to cover some "commonsense" guidelines for model level of detail. Although this topic has been discussed in other chapters, a brief review will help the reader decide which details are important to their application.

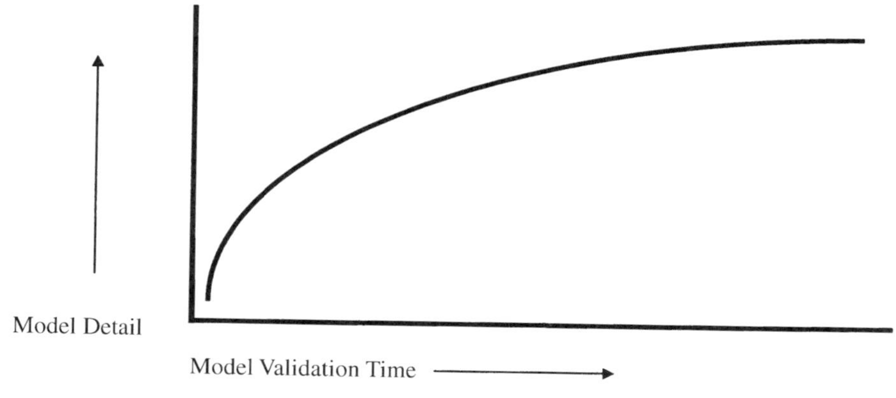

Figure 14.1 Model detail during validation.

14.2.1 Guidelines for Simulation Model Level of Detail

Every model is an approximation of the real world. It is a given that a modeler will leave out some details when building a model of the actual system. In the simulation community, this concept is referred to as the *level of abstraction*. The model will be an abstraction, or approximation, of the actual system. The important point to note is that some details will be omitted from a model, and choosing the right details to omit determines whether a simulation study will be successful. For example, do we include the janitor who sweeps around a drill in a machine shop? If the sweeping operation is not critical to the output of the machine shop, the answer is no.

Engineers are trained to be detail oriented. It goes against the grain to ask an engineer to omit information from a model. Simulation modelers often discuss the accuracy of their models in terms of a percent. The percent is usually how close the model gets to the results of the actual system. To get from a 95% accurate model to a 98% accurate model may take more effort than it takes to build the original model. Thus some compromise must be made or the simulation becomes impractical. A good rule is that it is easier to add detail later than it is to recoup time lost by adding unnecessary detail. Figure 14.1 shows how details are added as the model validation process proceeds. Note that during model validation, details are added as the model approaches an acceptable level of accuracy.

The process of validation is an iterative one. The modeler adds new details to the model, runs the model, and furnishes results to the project team. If the results are not sufficiently accurate, the project team identifies other details that should be included. The modeler adds these details, and the cycle starts anew. At some point, the project team must agree that the model is "close enough" to provide useful information, and the validation process gives way to experimentation.

14.2.2 Components of Manufacturing Systems

Although there are many types of manufacturing systems used to make the wide array of products available today, there are some common elements that describe most manufacturing operations. These common elements should be the basis for input data used

TABLE 14.1 Manufacturing Components

Product	Resources	Demand	Control
Parts/pieces	Equipment layout	Customer orders	Warehouse management
Routings	Number of machines	Start date	Inventory control
Process times	Downtime	Due date	Shop floor control
Setup times	Preventive maintenance	WIP inventory	WIP tracking
Bill of materials	Storage areas		PLCs
Yield	Tools/fixtures		Station rules
Rework	Labor—classification		
	Shift schedules		

by a simulation model. Table 14.1 shows these common elements in manufacturing systems. To build an accurate simulation model, the data in this table should be available and verified to be accurate. We briefly describe the data from this table in the following sections.

Product. Part, lots, or products are being manufactured. Products may move in manufacturing groups called *lots* that are made up of a number of pieces. Products may also proceed through the manufacturing process one at a time, like an automobile on an assembly line. Products have routings, which define the processing sequence. Product routing can be quite complex, as in a semiconductor wafer fabrication process, or it can be sequential, as in an automotive assembly line. The routing defines how product flows through the manufacturing process. For each operation, the processing and setup times determine how long a machine or operator will perform the given task. These times can be deterministic (constant) or stochastic (random), and can be machine/part determined or purely machine determined.

The bill of materials defines the subassemblies that comprise constituent final assembly. These subassemblies must be available for processing to proceed. Subassemblies also have a processing sequence of their own and must be either produced or purchased with sufficient lead time to be available for final assembly. Some simulation systems have features that allow the bill of materials to be modeled easily.

Yield and rework are found in many manufacturing operations. Because of imperfections in the manufacturing process, occasionally, products have to be scrapped or reworked. Both these factors affect the overall throughput of the operation, as well as other performance metrics. Yield and rework should be considered where their impact is noticeable.

Resources. Resources are used to manufacture products. Resources include machines and human beings as well as tools, fixtures, material handling systems, storage areas, and so on. The equipment layout affects the flow of operations, and if material handling is an issue, the layout should be included in a simulation.

Machines have both random failures as well as scheduled preventive maintenance. There are times when these machines are unavailable to do work, even though other machines are still working. Rerouting of work during machine down periods can have a major affect on the operation and should be included in a simulation model.

Human resources have skill levels and work on shift schedules. It is important to model shifts and downtimes when not all areas in a manufacturing facility are on the same schedule. If all machines and labor start and stop at the same time, and startup effects are negligible, it is not necessary to model shifts.

Demand. The demand on a manufacturing system is defined by customer orders. Customers usually order specified quantities of products and want them delivered on a particular date. The manufacturer is responsible for determining when to start products so that due dates can be met. Simulation can help determine the latest start date that will allow due dates to be met.

In a manufacturing operation, there is always some work-in-process. Most simulation models assume that the facility starts empty and idle. Often, simulation models need to be executed for a certain period of time to initialize the WIP inventory. If the model is large, this initialization time can be quite long, and it becomes necessary to find a way to get faster execution from the model. The answer is to read initial inventory from an external data source into the model. This allows the model to start just as the facility starts, with some WIP.

Control. Computer-based control systems make decisions about how product should be routed, collect information about current status of product, or maintain proper inventory levels. These control systems interface with simulation in two ways. First, they can provide input data to be used in the simulation. Examples are shop floor control systems that collect processing times and WIP tracking systems that can provide up-to-date inventory levels. Second, these systems often make operational decisions that should be represented in a simulation. Replication of control algorithms in a simulation model is one of the bigger challenges faced by model builders.

Local scheduling decisions are also called *station rules*. For each work center, the station rule is really an answer to the question "What do I work on next?" Station rules can be as simple as first in, first out (FIFO), or the rule can use a very complex decision tree with many steps. Making intelligent decisions at the machine level has been the topic of many a master's thesis, and there are many views on how best to make local scheduling decisions. For simulation it becomes important to be able to represent what happens in the real world with a high degree of accuracy. One should explore the complexities of manufacturing process decision making with the modeler to determine the right level of abstraction and the simulation tool that is most appropriate.

14.2.3 Downtime

Downtime is an aspect of manufacturing that is sometimes overlooked when building a simulation. Downtimes and failures can, however, have a significant effect on the performance of manufacturing systems. Banks et al. (1996) state that there are four options for handling downtime:

1. Ignore it.
2. Do not model it explicitly but adjust processing time appropriately.

3. Use constant values for time-to-failure and time-to-repair.
4. Use statistical distributions for time-to-failure and time-to-repair.

Of the four options, using statistical distributions for time-to-failure and time-to-repair is preferred. What this means to the manufacturer is that sufficient downtime data have to be collected to fit statistical distributions correctly. There are software packages available that help the modeler fit collected data to statistical distributions. One of the requirements for fitting distributions to data is that enough samples have been collected. In most cases, the more samples available, the better the chances that an accurate fit can be achieved. A practical starting point is at least 20 samples, but 50 to 100 is preferred. If at least 20 samples of downtime for a particular machine cannot be collected, that machine may not go down enough to have an effect on the operation.

In the absence of good downtime data, but where downtime is suspected to have an effect, the simulation model should be used to determine the effects of downtime on the system. Using the model, a sensitivity analysis can be performed with different time-to-failure and time-to-repair values and distributions. When analyzing the effects of downtime, the simulation analyst should not only look at the overall performance measures such as throughput and cycle time, but also at the local queueing and congestion found at work centers.

Preventive maintenance (PM) is also considered a type of downtime, but it is often scheduled at regular intervals. PMs can occur during a particular time of the day or after a specified number of operations have been completed by the work center. In some cases, PMs are scheduled during an off shift and thus do not affect the operation. When a PM interrupts normal operations at a work center, it should be included in a simulation.

One method used to validate models is to allow downtime to be toggled on and off for all equipment. By turning off all downtime, the model will produce the theoretical maximum value for throughput. As an example, consider an automotive assembly line that is expected to manufacture 55 jobs per hour (JPH). Downtime of all subsystems is included in the model, and with downtime turned on the model achieves only 50 JPH. If the downtime is turned off, the throughput is 53 JPH, 2 JPH short of the 55 JPH goal. Through careful analysis of the model, it was determined that the synchronization of two material handling devices was responsible for the 2 JPH shortfall. With downtime turned on, it would have been difficult to determine the exact cause of the production shortage. The use of simulation in the automotive industry is such an important application area that it is the focus of Chapter 15.

Catastrophic and rare events are downtime that can effect system performance. In most cases, rare events are ignored for one of several reasons:

- The effort to model the rare event is too great.
- The effect of the rare event is not significant.
- Not enough data exist to model the rare event accurately.

Events such as acts of nature, labor strikes, and power failures can literally shut down a manufacturing operation. Because they are not part of normal operation and are very difficult to predict, catastrophic events can be ignored for most simulation activities. A simulation model can be used to determine recovery strategies from rare and catastrophic downtimes.

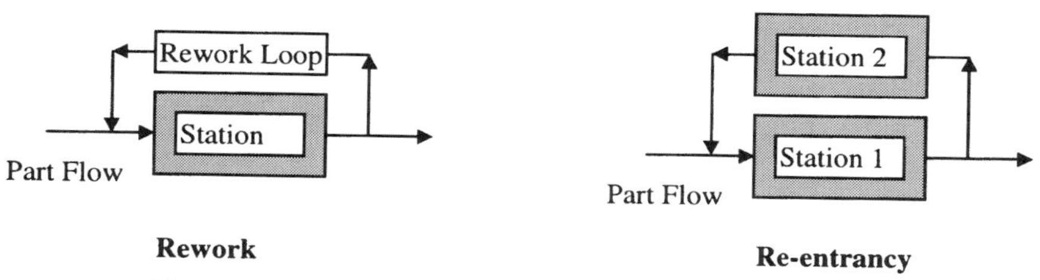

Figure 14.2 Rework and reentrancy. (From Mittler et al., 1995.)

14.2.4 Rework and Reentrancy

Reentrant process flow occurs when a particular station or work cell must be visited more than one time by the same part. Rework occurs when a part must be run through a work cell because the prior processing step was not completed successfully. Figure 14.2 shows the difference between rework and reentrancy. Using simulation it is possible to determine the effects of rework and reentrancy on a system. Rework is typically given as a percent of the parts processed, while reentrancy is provided in the part routing as explicit steps where the same machine(s) must be used. In either case, the true effects on queueing and congestion can be determined using simulation. Rework and reentrancy should not be overlooked. Additional information about the reentrancy phenomenon and the role of simulation is available from Banks and Dai (1997).

14.2.5 Handling Stochastic (Random) Events

For most manufacturing systems, one of the reasons to model is the presence of random events. Random number generation and random variate generation are discussed in Chapters 4 and 5, respectively. In this chapter, some of the common random events encountered in manufacturing systems are discussed. Random events in manufacturing systems are associated with:

- Processing time
- Setup time
- Downtime (time to fail and time to repair)
- Yield percentages
- Transportation time
- Truck arrivals at receiving docks

For all random events it is important to represent the distribution of randomness accurately in the simulation model. As mentioned earlier, there are several distribution-fitting software packages that can assist in finding the right distribution for collected data. Choosing the right distribution is a very important part of the simulation process. When a known distribution cannot be found for a set of data, an empirical distribution can be used. Two types of empirical distributions, discrete and continuous, are discussed in Banks et al. (1996).

An example illustrates the importance of fitting the correct distribution to a given data set. For time to repair on a lathe, 100 samples have been collected over a one-year

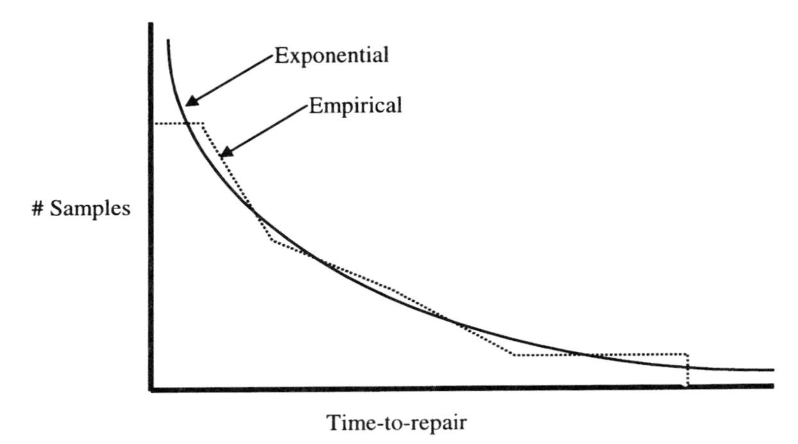

Figure 14.3 Empirical and exponential distributions.

period, the longest being 3 hours. Input data-fitting software show that the exponential distribution fits but only poorly. It is decided that an empirical distribution for the lathe mean time to repair (MTTR) is to be used. Figure 14.3 shows the empirical and exponential distributions for the data. Notice that the empirical distribution has a maximum while the exponential approaches the x-axis asymptotically. The empirical distribution does not generate values greater than the observed maximum, but the exponential distribution does. The events that occur on the tail of the exponential distribution could have a major effect on the system. Using the exponential distribution, it is possible to get values of 10 or even 20 hours for MTTR, although the probability is low. This example illustrates the importance of choosing the right distribution.

Bratley et al. (1987) recommend a mixed empirical and exponential distribution. This approach provides for the tail effects that are seen as rare events in the actual system. Like any distribution, the mixed distribution should be used with caution. Thorough sensitivity analysis should be performed to understand the effects of using an incorrect distribution.

An option to the empirical distribution is to use the actual data in a simulation model. An example would be trucks arriving at a receiving dock. Models that use actual data are called trace driven by Banks et al. (1996). Trace-driven models read actual data from files outside the simulation model. An important concern when using trace-driven models is that only the values observed historically will be repeated in the simulation.

14.2.6 Measures of Performance

The methods used to measure model performance should be the same as those used in the real system. Otherwise, it may be difficult to validate the model. With any of the performance measures discussed in this section, it is important to collect the average as well as the variability. Variability is usually indicated by the standard deviation, but maximum and minimum are also helpful in measuring performance. The following statistics are typically collected from manufacturing systems and should thus be provided by models of such systems:

- Throughput (aggregate as well as for different areas in the system)
- Cycle time (makespan)
- Queueing at work locations
- Response time of material handling equipment
- WIP (work in process)
- Utilization of resources (equipment and labor)
- System specific performance measures (scrap rate, waiting time at a process)

It is important to note that optimizing on one measure of performance can adversely affect another measure of performance. For example, if WIP is reduced, equipment utilization usually goes down. Understanding the relationships between measures of performance can help in the experimentation phase of a model.

14.2.7 Analysis Issues

Using the performance measures described in Section 14.2.6, model users (analysts and engineers) experiment with a model to understand the behavior of the system under varying conditions. The issues often encountered in analysis include:

- Determining the bottleneck
- Determining required staffing levels
- Evaluating the scheduling of tasks
- Evaluating the control system
- Recovery strategies for random events and surges

One of the most often asked questions of a simulation analyst is: "Where is my bottleneck?" Bottleneck identification can be an arduous task, as there are possibly many factors that contribute to a bottleneck. As changes are made to the model to find the bottleneck, the bottleneck itself can move from one work center to another. The analyst needs to look at performance measures at both the local and global level within the model to determine the bottleneck. Once the bottleneck is found, proposed solutions should be tested using the model.

Manufacturing system models can also be used to determine staffing requirements for an operation. The number of operators as well as operator certification levels should be used as input, with operator utilization used as the evaluation criteria. Order launch and resource scheduling can also be evaluated using simulation-based finite-capacity scheduling (FCS) systems, as discussed in Chapter 21. Scheduling decisions can be evaluated with a simulation model prior to implementing them on the shop floor.

In modern manufacturing systems, the computer controls many decisions affecting performance. No software system is without defects. Thus tools such as simulation can be used to test the control algorithms before they are connected to the physical equipment. The more intricate the control system, the greater the need to do simulation-based testing prior to commissioning.

14.2.8 Manufacturing Example: Wheel Shop

Let us now use a hypothetical machine shop to illustrate some of the simulation issues germane to manufacturing. This machine shop makes aftermarket specialty wheels for

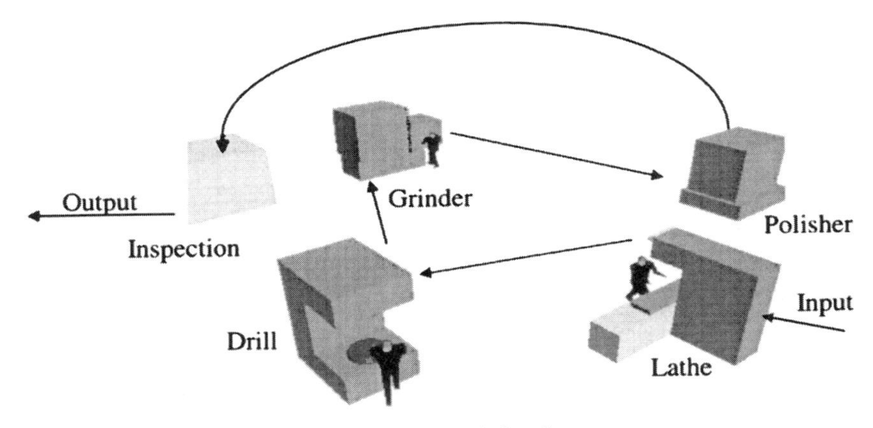

Figure 14.4 Wheel shop layout.

cars and light trucks. The wheels come in two types, custom and standard, and a variety of sizes are manufactured to accommodate different tire sizes. A picture of the machine shop layout is shown in Figure 14.4.

There are five machines in the shop; a grinder, polisher, lathe, mill, and inspection station. At present only three operators set up and process jobs at the machines. There are two different processing sequences, or routings, for the two types of wheels made. The routings are shown in Table 14.2. As shown in the table, there are two different routings. The machine order is the same; however, the processing times, setup requirements, and setup times are different for the two types of wheels. The format shown for the routing is generic. For a manufacturing application, it is good practice to read input data, like routings, in the user-supplied format. To facilitate experimentation and reduce errors, a model should read data as is, and modification of the data for simulation purposes should be kept to a minimum.

At this point we've described the equipment and logical routing of parts in the wheel shop. Table 14.3 shows one week of customer orders to be produced in the wheel shop.

TABLE 14.2 Typical Routing

Route Name	Step	Machine	Process Time	Units	Setup	Setup Time	Units
Custom	1	Lathe	6	hr	Custom	1	hr
	2	Drill	5	hr	Custom	1	hr
	3	Grind	3	hr	Custom	30	min
	4	Polish	4	hr	Custom	1	hr
	5	Inspect	1	hr	Custom	20	min
Standard	1	Lathe	5	hr	Standard	1	hr
	2	Drill	4	hr	Standard	1	hr
	3	Grind	3	hr	Standard	30	min
	4	Polish	4	hr	Standard	1	hr
	5	Inspect	1	hr	Standard	10	min

TABLE 14.3 Weekly Customer Orders

Order	Pieces	Start	Due
15 Custom	30	10/26/98 8:00	10/30/98 17:00
15 Standard	20	10/26/98 8:00	10/30/98 17:00
16 Custom	20	10/26/98 8:00	10/30/98 17:00
15 Standard	20	10/26/98 8:00	10/30/98 17:00
16 Custom	30	10/26/98 8:00	10/30/98 17:00
17 Standard	20	10/26/98 8:00	10/30/98 17:00
16 Custom	20	10/26/98 8:00	10/30/98 17:00
16 Standard	30	10/26/98 8:00	10/30/98 17:00
16 Custom	20	10/26/98 8:00	10/30/98 17:00
16 Standard	20	10/26/98 8:00	10/30/98 17:00

The orders are for three different-size wheels, 15, 16, and 17 in., in both standard and custom configurations. The orders will all be started on Monday, October 26 and must ship by the end of the week, which is October 30.

The questions to be answered are as follows:

1. How many operators are needed for the orders that are scheduled?
2. If the operating rule is changed for the machines, can the orders be completed in a more timely fashion?
3. What is the utilization of the equipment?
4. What is the utilization of the operators?

To answer the first question, the simulation will be run with two, three, and four operators, and the results compared. The number of orders completed and operator utilization will be used to measure the results. From Table 14.3 there are 10 orders to be processed in the week. Table 14.4 shows the simulation results of three scenarios. It appears that four operators are needed to handle the workload required by the current orders. Figure 14.5 shows a Gantt chart of the operation using four operators.

TABLE 14.4 Determining the Number of Operators Required

Number of Operators	Orders Completed	Operator Utilization
2	5	97
3	9	88
4	10	67

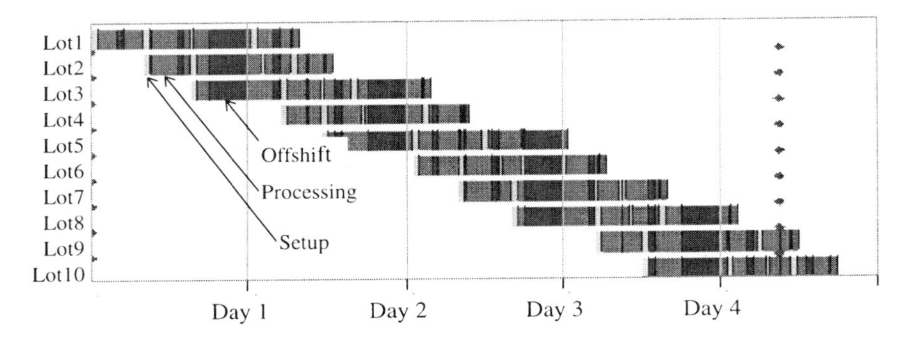

Figure 14.5 Gantt chart of wheel shop.

As can be seen from Figure 14.5, the orders are completed but not by their due date. A little overtime has to be worked to get the last two orders shipped. Also notice that there are many setups for each one of the wheel lots. In fact, almost every processing step has a setup before it. This raises the second question. There is an option of changing the operating rule for each work center. Currently, the shop uses a first in, first out (FIFO) rule, working on the jobs as they appear in the work queue. What if this FIFO rule is changed to a rule where jobs that require the same setup are taken first? This kind of a rule would minimize setup time at the machines. Figure 14.6 shows the results of running with a same setup operating rule. Note that the orders are now completed by their due date at the end of the week. Also, there are fewer setups. In this model the setup time as a percent of total time was reduced from 10% to 2% by changing the decision rule used at the work centers.

Equipment utilization is provided in Table 14.5. Note that the lathe and drill have the highest utilization, 82% and 73%, respectively. Most manufacturers consider a utilization of 85% maximum. Thus, to increase production, another lathe, and probably a drill, would be purchased. The simulation can determine where the bottleneck operation is located and can be used to determine when to purchase equipment.

Operator utilization was shown in Table 14.4. The number of operators needed to complete the orders on time was determined to be four. Operator utilization is about 67%

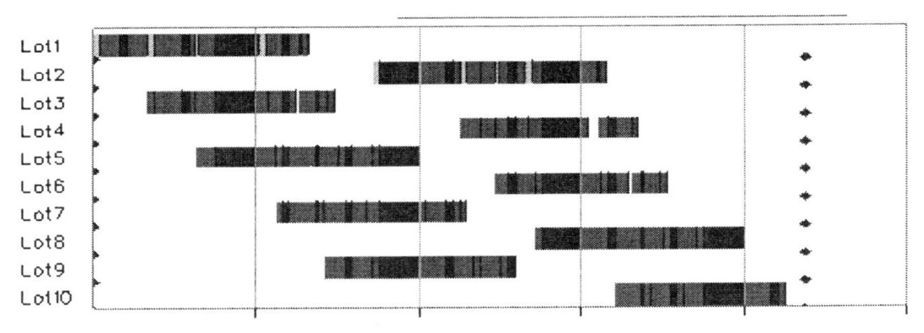

Figure 14.6 Improved processing with same setup rule.

TABLE 14.5 Work Center Utilization

Work Center	Utilization (%)
Lathe	82.0
Grinder	59.6
Polisher	68.7
Drill	72.9
Inspection	42.5

using four operators. This could be improved by decreasing the number of operators, but at the expense of delivery dates. Simulation can be used to support decisions such as the number of operators to employ.

As shown in this example, simulation can be used to make decisions about how to operate a manufacturing system. Through the judicious use of an accurate simulation model, manufacturers can run their operations more efficiently, and thus at greater profit.

14.2.9 Manufacturing Case Studies

Three case studies are given as examples of how simulation has been used in the manufacturing industry. Some details of these applications cannot be published because of the proprietary nature of the information.

Case Study 1: Appliance Manufacturing Line. A large appliance manufacturer in the midwest was introducing a new line of refrigerators. An assembly line already existed for older models, but a reconfiguration of the line was required for the new models being introduced. This assembly line used a closed-loop transport system to move units between about 100 assembly stations. At each assembly station there are several queue positions to compensate for the variability between processing at adjacent stations. It was determined through sampling that processing time fit a lognormal distribution. Once an operator finishes processing a unit, the unit is released to queue for the next station. The assembly line also contained a side loop where the refrigerator compressor must be evacuated before it can be charged with refrigerant. Units must stay in the evacuation loop for a set period of time to assure complete evacuation of the compressor. A simulation model was used to:

- Verify line balancing and queueing
- Determine the number of transporter carriers required
- Determine the maximum production possible for the line

The performance measures used were:

- Queueing at each assembly station
- Utilization of the operators at each station
- Throughput of the system at maximum input
- Congestion of the transport system

A line balance had been performed prior to the simulation. It was discovered that some of the processing times from the line balance needed to be adjusted because several processing stations were causing bottlenecks in the operation. Once processing times had been adjusted, the number of carriers on the transportation system was varied until the maximum line throughput was achieved. Adding more carriers caused more congestion on the transportation system, causing stations to be blocked, that in turn decreased throughput. At maximum throughput, the variability of the throughput was at the highest value observed. This was because when the system reaches its optimum, there are few constraints on throughput, and variations in random events translate directly to the performance metrics.

Simulation showed that the new refrigerator model could be accommodated on the existing assembly line with minor modifications. The line balancing was verified and corrected, and the number of carriers determined.

Case Study 2: Printer Semiconductor Wafer Fab. A major printer manufacturer in the Pacific Northwest was building a new semiconductor wafer fabrication facility. Semiconductor chips are used in the printer heads to control ink flow and other printer functions. The main goal of the simulation was to determine the level of automated material handling required to meet performance requirements. As with most semiconductor production facilities, other concerns are the release policy, cycle time, and the scheduling policy for the equipment. Lu et al. (1994) indicate that yield is affected directly by the time that product spends waiting to be processed, so reducing cycle time is nearly always an important goal in semiconductor manufacturing.

In the semiconductor processes, wafers are loaded into carriers called *cassettes*. Cassettes are transported between the processing equipment and storage areas called *stockers*, as shown in Figure 14.7. The processing equipment is usually organized into groups of similar machines, called *bays*.

In this example, a simulation model was developed to determine the extent to which automation would be required and the best management strategy for operating the pro-

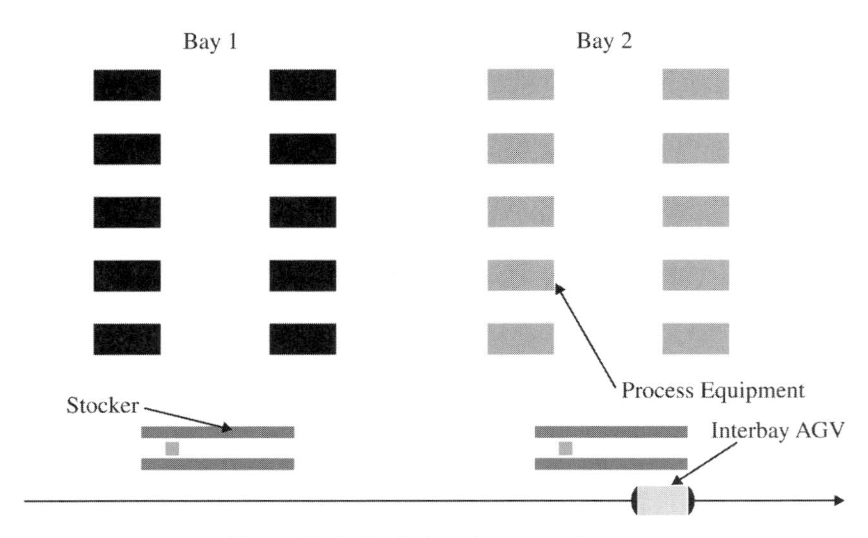

Figure 14.7 Typical semiconductor layout.

cessing equipment. Several alternatives for automation were evaluated, including different combinations of manual and automated material handling. Automated material handling equipment from several suppliers was also evaluated, and different track configurations were tested using the model.

Strategies for keeping bottleneck operations highly utilized were also evaluated using simulation. One such strategy, called *constraint-based WIP Management* (CBWM), was tested with the simulation model. In CBWM, the bottleneck process(es) are identified, and WIP levels are maintained at the bottleneck machines so that they never become idle. In CBWM, the process is managed from the very beginning to keep the bottleneck machines as busy as possible. The model helped determine the right level of automation for the facility. Additionally, scheduling strategies were evaluated to help run the facility as efficiently as possible.

Case Study 3: Aircraft Manufacturing. The Boeing Company was evaluating the manufacturing process for the new 777 aircraft to be built at their Everett, Washington assembly facility. The facility is currently used to manufacture the 747 and 767 aircraft as well. Large aircraft parts such as wings, fuselage sections, and engines are handled by an elaborate overhead crane system. Boeing needed a way to evaluate the crane handling capacity and to determine a realistic build rate for the new aircraft.

The model was used to choreograph crane movements on large parts. Visualization provided by three-dimensional animation allowed Boeing engineers to see how these large parts would flow between the assembly processes. Through the use of the model, Boeing was able to determine a realistic build rate to meet customer demands for the new aircraft.

14.3 MATERIAL HANDLING SYSTEMS

Material handling is one of the most important elements of manufacturing and one that is simulated often. For many operations, product can spend as much time moving around as it does being processed. Banks et al. (1996) say that material handling time can be as much as 85% of total manufacturing time. Additionally, material handling equipment involves large capital expenditures. Simulation provides the insurance that the material handling solution will work in the desired application, before any equipment is purchased.

In many applications of material handling systems, equipment from many suppliers is integrated into the total system. Since each supplier is responsible only for the equipment provided by them, it becomes necessary to have a way of testing the integrated solution. Simulation can help ensure that the system components work well together.

When evaluating the application of material handling systems, there are often many different solutions to the same problems. Should conveyors or fork trucks be used? Should the storage be manual or automated? The comparison of different alternatives can be accomplished using simulation. For these reasons, simulation has been used for material handling systems to help find the most efficient ways to move product. In this section the various types of material handling systems, both automated and manually operated, are described. The discussion includes both manufacturing and nonmanufacturing applications of material handling systems (e.g., airport baggage handling, warehousing, and distribution).

When modeling material handling systems, Norman (1984) says that equipment

TABLE 14.6 Input Parameters for Automated Material Handling Systems

Automated storage and retrieval systems	Power-and-free towline conveyors
Horizontal and vertical speeds	Speed
Acceleration and deceleration rates	Transfer time
Shuttle cycle time	Carrier spacing
Storage rack configuration	Carrier size
Storage depth (single or double)	Number of carriers
Random storage vs. storage zones	Area counters for carriers
Priorities for double-ended rack	Carrier bias banking requirements
Input–output locations	Bridge cranes
Multiple load handling capability	Speeds
Pallet conveyors	Acceleration and deceleration rates
Conveyor type (transport, queueing, accumulation)	Load and unload times
Speed	Area interference
Minimum load spacing	Scheduling rules
Load size (length, width)	Input–output locations
Right-angle transfer time	Automatic guided vehicles
Case conveyors	Guide-path layout
Speed	Control point location
Load size	Horizontal and vertical speeds
Transfer time	Acceleration and deceleration rates
Recirculation requirements	Load and unload times
	Vehicle blocking rules
	Empty vehicle management rules
	Battery charging rules
	Area counters and antideadlock prevention

Source: Norman (1984).

capacity, speed, and arrangement are the most important considerations. The capacity is the amount of product that can be handled by the equipment. Speed is the operating velocity of the equipment, which may include acceleration, deceleration, and different speeds, depending on the product begin carried. The arrangement is the layout of the material handling system. How do the fork trucks get from the receiving docks to the storage area? Should a new conveyor system be routed on the floor, or overhead?

Material handling systems can be classified as being either of vehicle type or nonvehicle type. Vehicle systems use a transporter to carry the load along a path that may or may not be predefined. Vehicle systems include fork trucks, pallet jacks, automated storage and retrieval systems (AS/RS), automated guided vehicles (AGV), and bridge cranes.

Conveyors are examples of nonvehicle systems. They have a fixed path and do not require that a vehicle be assigned to the load being transported. Conveyor systems can be made up of many individual conveyor sections, each with a drive motor that moves loads on the conveyor surface.

Each automated material handling system has its own set of input parameters that are important to simulation modeling. These parameters are shown in Table 14.6.

14.3.1 Conveyors

Conveyor systems can be classified by the type of conveyor as well as the size of the load moving on the conveyor. Conveyors can be either accumulating or nonaccumulating. On nonaccumulating conveyors, when one load stops, the entire conveyor stops, maintaining spacing for all loads. On accumulating conveyors, loads continue moving and can bunch up behind a load that has stopped. Conveyors can also have fixed windows that require all loads to be equally spaced, or they can have variable windows. There are larger conveyor systems that handle full pallets, and other conveyor types that handle smaller loads such as cases.

Another type of conveyor, the power-and-free conveyors, have carriers that attach to the load being transported. The conveyor has hooks called *dogs* that the carrier connects to in order to move. The dogs on a power and free conveyor are regularly spaced, like the fixed intervals on other conveyor types. Power-and-free conveyors are often seen in automotive paint applications.

For all conveyor types, the parameters indicated in Table 14.6 should be available to build an accurate model.

14.3.2 Vehicle Systems

Vehicle systems are defined as having a transporter that is assigned to one or more loads for movement. The transporter could be an AGV, fork truck, pallet jack, person, or a storage and retrieval machine (SRM). From Table 14.6, AS/RS, bridge cranes, and AGVs would all be considered vehicle systems.

Some vehicle systems, such as manually operated fork trucks, do not always follow a fixed path. These types of vehicles often have traffic patterns that can be defined by fixed paths. Most simulation software today requires that the user define the vehicle paths. Fork trucks and other manually operated vehicles can be modeled using the path-guided transporter modules available in simulation software. One difference between manually operated systems and automated systems is how deadlocks and collisions are avoided.

In most cases, human beings can resolve deadlock conditions "on the fly" for manually operated systems. For computer-controlled AGV systems, algorithms need to be implemented in the control software to avoid deadlocks. Most of the simulation software available today allows the user to define the control algorithms used in automatic systems. Modeling manual systems becomes more of a challenge. Representing human decisions in a simulation tool can be expensive and the additional effort may not bring improved accuracy in the results. Thus it may be necessary to make simplifying assumptions when modeling manual systems. One such assumption might be that fork trucks can pass through each other in the simulation as they would navigate around each other in the real world.

AS/RS systems are used for storage of all sizes of products, from cars to pallets to totes carrying small parts. In some cases there is a storage and retrieval machine (SRM) in every aisle. Other times one SRM may service multiple aisles. SRMs are vehicles, and the racks are a storage system. In most systems the SRM capacity is the main concern. Thus simulation is used to test the moving capacity of the cranes, and the individual storage locations are not tracked in detail. The size of an AS/RS rack can, in most cases, be determined without simulation. In some cases, such as semiconductor fabrication facilities, the stocker (AS/RS) storage capacity is determined using simulation.

Human beings are sometimes used to transport material and would thus fall into the category of vehicle systems. In this case one must consider the walking speed as well as pickup and setdown times. Additionally, whenever human beings are used for transport, or interact with other material handling equipment, the variability associated with human activities should be included in the model.

14.3.3 Other Material Handling Systems

Robots are sometimes used as material handling devices. Examples include palletizers, pick-and-place machines, package orientation devices, and so on. Robots can be modeled in detail to include rotation and translation speed of joint axes, or delay times can be taken for robot actions.

Carousels are a combination storage and material handling device. There are horizontal as well as vertical carousels. In a carousel, bins move along a closed loop to a picking point, usually at one end of the carousel. Most carousels are used for picking in distribution facilities. Carousels can be single or bidirectional and are usually grouped into a pod of several carousels with one human picker working between them. For simulation, the carousel speed and configuration are the main input parameters.

Vertical lifts are often used to move material from one elevation to another. Lifts can act like elevators, servicing several levels with bidirectional movement, or they can be moving continuously like a fixed window conveyor oriented vertically. Lift input usually includes the speed as well as the control algorithms for what the lift does when empty and full. Also, lifts usually have to coordinate with other material handling systems, such as fork trucks and conveyors. The details associated with lifts are important when queueing and lift capacity can have an effect on material flow.

14.3.4 Modeling Material Flow

Material flow between processes can be modeled in many ways. One common method used to evaluate material handling systems is the from/to method. For each source of material to each destination, a rate is given for the number of units that must be moved per unit time. A typical material flow diagram might look like Figure 14.8. These flow rates can be tabulated and used to drive a simulation model. If the model is built to be table driven, it is easy to experiment with different rates by changing the data in the from/to flow tab.

14.3.5 Modeling Random Events in Material Handling Systems

Ask a material handling equipment supplier about equipment downtime and you are likely to receive an optimistic estimate. Ask someone who operates a facility with automated material handling equipment and you are likely to hear a pessimistic estimate of the frequency with which a particular piece of equipment has gone down. Realistic equipment failure data are somewhere between what the vendors say and what the operators experience.

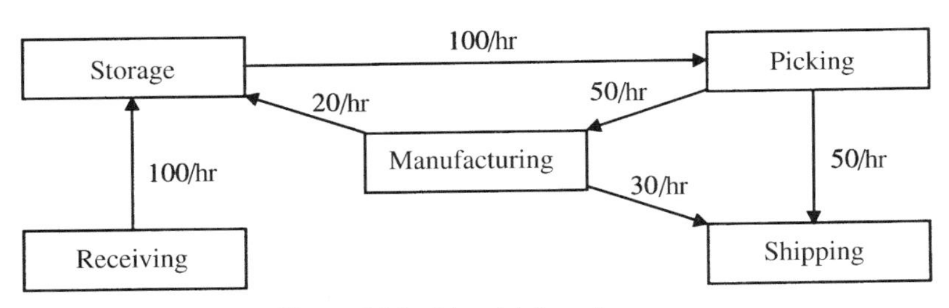

Figure 14.8 Material flow diagram.

Downtime for material handling equipment can exhibit itself in many ways, including motor or bearing failure, control system glitches, conveyor jams, battery failure for AGVs, and sensor malfunction. Usually, all the possible downtime types are distilled into a time-to-fail and time-to-repair description for a piece of equipment. If enough data exist, a probability distribution can be fit to the downtime data.

Where manual operations interact with automated equipment, the variability inherent in the manual operations should be included. Operator availability, cycle time, and break schedules can all have an adverse effect on the operation of automated equipment where the two interface. Care must be taken to capture and model manual variability.

Modeling downtime is important not only to understand system performance under downtime conditions. Simulation can also be used to formulate and test downtime recovery strategies. For example, when an SRM goes down in a storage system aisle, how can we continue to get to every type of product stored in the system? One strategy is to store like material in every aisle, so that if one aisle goes down, it is still possible to retrieve material stored in the other aisles.

14.3.6 Control Systems

Control systems are implemented in software that runs material handling systems. The control system can be as large and complex as a warehouse management system (WMS) or as simple as the programmable logic controller (PLC) that controls a set of conveyor sections. In either case the control system contains decision-making logic that should be tested as early as possible in the design of the system. Many simulation tools include languages that can be used to replicate control system algorithms. Some simulation tools can actually communicate with control system programs directly to help test the code. The earlier that control system defects can be found, the better the material handling system will operate.

14.3.7 Material Handling System Performance Criteria

In addition to the performance measures mentioned in the manufacturing section of this chapter, there are specific performance measures used to evaluate the performance of material handling systems. We describe performance measures for each type of material handling device.

Vehicles. Vehicle systems include automated as well as manual transportation devices. Their performance measures include:

- Utilization
- Time in different states (retrieving, delivering, parking)
- Number of trips made
- Congestion
- Response time
- Average number of loads on board
- Guide-path use and queueing
- Percent of time at top speed

Conveyors. Conveyor systems are made up of individual sections that link together to form a transportation network. Performance measures include:

- Entries per section
- Average time a load spends on a section
- Queueing (maximum, average number on a section)
- Recirculation events
- Number of start and stop events for the motor

AS/R and Bridge Crane Systems. AS/R and bridge crane systems are combination vehicle and storage devices. Performance measures include:

- Utilization (crane and storage area)
- Time in different states (retrieving, delivering, parking)
- Number of trips made
- Number of dual cycles made (full in and out trips)
- Average number of loads on board
- Percent of time at top speed

14.3.8 Material Handling Example: Packaging Line

We will use an example to show how material handling issues can be solved using simulation. The example is a packaging line that includes case sealing, inspection, repack, and shipping operations. Figure 14.9 shows a layout of the system. Conveyors are used to move cases of different sizes from the manufacturing area through the case sealers, inspection, and onto shipping. All conveyors run at 60 ft/min, and all are accumulation type. Cases come from the manufacturing area at the rate of 600 per hour. Table 14.7 shows the case sizes and frequencies per hour from manufacturing.

Case sealing time takes 18 seconds. Every 8 hours, sealing material and labels must be restocked on the sealing machines. Restocking takes 10 minutes per machine. Inspection takes 6 seconds, and if a package is rejected, it takes 1.5 minutes to repack.

The questions to be answered by the simulation are:

- Should four or five case sealers be used?
- What is the utilization of the inspectors?
- Is there enough accumulation space on the conveyors prior to the case sealers?

Upon running the simulation with four case sealers, it is found that the sealer handling the higher-volume boxes backs up after a short time, and utilization is very high (>95%). It is determined that five case sealers are needed to handle the proposed package volume. Inspector utilization is about 46%, so it is possible for the system to work with one inspector.

The maximum number of boxes on the conveyor section prior to the case sealer is eight, which is the maximum that the section can hold. This happens for sealer 1. If one more box arrives for this case sealer when the queue is full, cases stop along the main line. This causes the other case sealers to be blocked. So it looks like some length needs to be added to the accumulation conveyors prior to the case sealers.

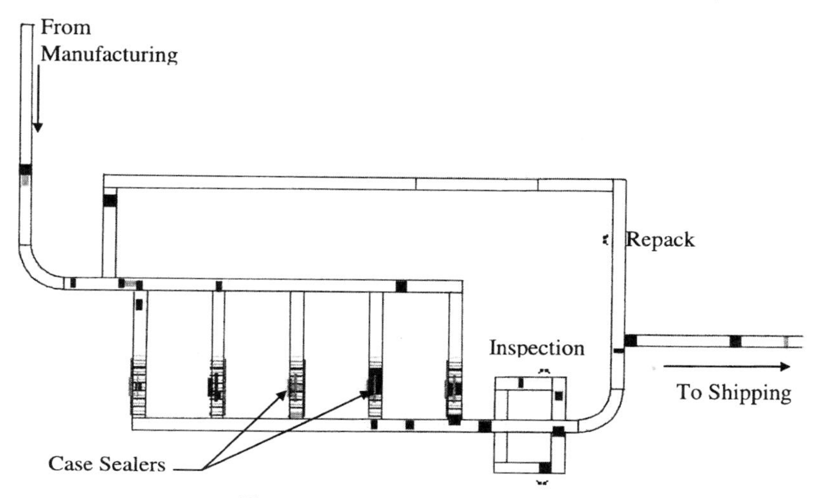

Figure 14.9 Packaging line layout.

14.3.9 Material Handling Case Studies

Once again we will use three case studies as examples of how simulation has been used in material handling system design.

Case Study 1: Chicago O'Hare Baggage Handling. In 1992 a new international terminal was added at O'Hare International Airport in Chicago. A baggage handling system was required for the new terminal. The system would handle luggage from ticket counters as well as connecting flights. A simulation was used to design the baggage handling conveyor system. The purpose of the model was to test several different conveyor layouts and strategies for storing bags. For some flights, passengers arrive up to 3 hours before departure, and their luggage must be stored until an aircraft has been assigned a gate and luggage can be loaded.

Each piece of luggage in the system was to be identified by a bar code label. Since some sections of the conveyor system moved at very high speed (>360 ft/min), identifying luggage became a key issue addressed by the simulation. If a bag could not be identified, it was sent to a manual encoding area where it was rescanned by an operator. Because of the high volume of bags being handled by the system, slight variations in the scanner no-read percent could flood the manual encoding area and shut down part

TABLE 14.7 Case Sizes and Frequencies

Case Number	Size (in.)	Number/Hour
1	18 × 30	120
2	24 × 30	120
3	30 × 30	120
4	28 × 28	240

of the system. The simulation was used to determine how many operators would be required in the manual encoding area.

Case Study 2: Cereal Material Handling Center. A major cereal producer in central Michigan was introducing some new products to an existing material handling system. In the system, cases of cereal are transported from production areas to a material handling center, where they are sorted, palletized, and loaded onto trucks. Conveyors are used to transport the cases, which have different sizes. Cases are accumulated into slugs for better control of the sorting process. Once pallets have been created, they are transported by an AGV system to queueing stands at the loading docks. Manually operated fork trucks load the pallets on the trucks.

A simulation model was built to design the material handling system, and the same model was used over a 5-year period to aid in the operation of the facility. For new products, the simulation was used to answer the following questions:

- How will the PLC logic need to change?
- Will there be enough AGVs to handle the higher-volume new products?
- Will there be enough sortation lanes?

The simulation found that some of the existing PLC logic would have to change to accommodate the new products. Also, it was determined that the existing number of AGVs would handle the increased volume of new products.

Case Study 3: Box Beef Handling System. A beef producer was evaluating a revolutionary new concept for box sortation and shipping for one of its plants in Texas. In this facility, beef products are boxed, sorted, and palletized for shipment. Some shipments are made using loose boxes, which is called *floor loading* of the trucks.

The proposed system contained sortation and transport conveyors, a unit load AS/RS for pallets, and two miniload AS/RS systems for cases. One of the miniload systems was for low-volume stock-keeping units (SKUs), while the other was for high-volume SKUs that need to be stored for short periods of time (<8 hours). The main questions addressed by the simulation were:

- Can the proposed system handle the current product volume and mix?
- Can the system handle a 20 to 30% increase in SKUs?
- Can the system handle an increase in production of 30%?
- What is the best ratio of floor and pallet loading at shipping?

The model found that current volume could be handled, but the proposed system could not handle an increase of 30%. Several design changes were made based on output from the simulation, and a new system configuration was discovered to handle the projected increase in volume. Increasing the number of SKUs caused problems in the sortation area, but these problems were solved with minor modifications in the control logic used at sortation. A ratio of about 80% pallet to 20% floor loading seemed to work best based on the results of the simulation.

14.4 ADDITIONAL EXAMPLES

An outstanding source of information about the application of simulation in manufacturing and material handling systems is the annual *Proceedings of the Winter Simulation Conference*. In 1996, there were 20 papers on the topic, including 19 papers in a track entitled 'Manufacturing Applications' and a tutorial on manufacturing applications. Many of the papers concern material handling. Four examples of these papers are given in the next paragraphs.

The first example is that by Jefferson et al. (1996), in which material transport in a semiconductor factory (wafer fab) is discussed. Movement of wafers in the fab creates risks, as the wafers are easily damaged. One solution to reduce handling is automated intrabay movement. A simulation model was developed to analyze the effects of using this type of movement system at Intel.

A second example, that by Bakst et al. (1996), concerns improving the towline material handling system at Random House, one of the largest publishing houses in the world. Even though a towline is a rather low-tech material handling system, the control issues that surround it are fertile grounds for simulation.

As a third example, Gunal et al. (1996) modeled chain conveyors and their equipment interfaces. These types of conveyors are prevalent in the manufacture of automobiles, particularly in the body and paint shops. Modeling chain conveyors requires much concern for detail. Of particular importance are the dimensional aspects of the interfaces, such as lift tables.

Finally, a fourth example of material handling is that by Takakuwa (1996), in which simulation of a large-scale material handling system is discussed. Takakuwa has expanded the details of this large-scale system over a series of papers beginning in 1989. The system under consideration consists of an AS/RS with stacker cranes, a looped track AGV system, aisle conveyors connecting these two systems, and outgoing conveyors. A modular simulation model was constructed, then experiments were performed. Analytical results are presented for major selected alternatives.

Many of the manufacturing examples in the WSC *Proceedings* concern semiconductor manufacturing systems. Sandell and Srinivasan (1996) evaluate lot release policies for semiconductor manufacturing systems using simulation. A full-factorial experiment with five factors was designed and executed. The result of the analysis is that no one release policy dominates for all scenarios.

Another semiconductor manufacturing application is that of Hallas et al. (1996). The study investigates a number of operational issues, such as lot size, test wafer proportion, and tool productivity on cost and production performance. Results indicate minimal differences between 24 and 48 wafer lot sizes, with significant differences in performance at 12 wafer lot sizes. The overall results indicate extreme sensitivity of a variety of fab performance measures to constraints determined by the operating characteristics, such as raw processing time and setups.

Kunnathur et al. (1996) describe a rule-based expert system driven by a simulation model that performs dynamic shop rescheduling. A heuristic was developed by the authors based on flow-time prediction strategy. Numerous experiments were conducted.

14.5 CONCLUSIONS

Simulation is an indispensable methodology for solving problems concerning manufacturing and material handling as these systems have become so complex. The literature is rich with successful examples. In modeling manufacturing and material handling systems, attention to details concerning the product, the resources, the nature of the demand, and the control system is important. Similarly, the way that downtime is treated, what to do about rework and reentrancy, and the modeling of random events must be treated properly. Selection of the appropriate measure or measures of performance is required. Model input data must be verified for accuracy. Examples of the application of simulation to specific problems help to understand the use of simulation in modeling manufacturing and material handling systems, but there is no substitute for experience. The use of simulation to solve manufacturing and material handling systems problems is based on the validity of the model that is being used.

REFERENCES

Bakst, J., J. Hoffner, and K. Jacoby (1996). Evaluation of a distribution center tow-line material handling system through simulation modeling, in *Proceedings of the 1996 Winter Simulation Conference*, J. M. Charnes, D. J. Morrice, D. T. Brunner, and J. J. Swain, eds., Association for Computing Machinery, New York, pp. 1099–1106.

Banks, J., and J. Dai (1997). Simulation studies of multiclass queueing networks, *IIE Transactions*, Vol. 29, No. 3, pp. 213–220.

Banks, J., and R. R. Gibson (1996). Getting started with simulation modeling, *IIE Solutions*, November.

Banks, J., and R. R. Gibson (1997). Don't simulate when: ten rules for determining when simulation is not appropriate, *IIE Solutions*, September.

Banks, J., J. S. Carson II, and B. L. Nelson (1996). *Discrete-Event System Simulation*, 2nd ed., Prentice Hall, Upper Saddle River, N.J.

Banks, J., M. Spearman, and V. Norman (1997). Uses of simulation: traditional and nontraditional, *Computer Integrated Manufacturing and Engineering*, Winter.

Bratley, P., B. L. Fox, and L. E. Schrage (1987). *A Guide to Simulation*, 2nd ed., Springer-Verlag, New York.

Gunal, A. K., S. Sadakane, and E. J. Williams (1996). Modeling of chain conveyors and their equipment interfaces, in *Proceedings of the 1996 Winter Simulation Conference*, J. M. Charnes, D. J. Morrice, D. T. Brunner, and J. J. Swain, eds., Association for Computing Machinery, New York, pp. 1107–1114.

Hallas, J. F., J. D. Kim, C. T. Mosier, and C. Internicola (1996). An investigation of operating methods for 0.25 micron semiconductor manufacturing, in *Proceedings of the 1996 Winter Simulation Conference*, J. M. Charnes, D. J. Morrice, D. T. Brunner, and J. J. Swain, eds., Association for Computing Machinery, New York, pp. 1023–1030.

Harrell, C., and K. Tumay (1996). *Simulation Made Easy: A Manager's Guide*, Industrial Engineering and Management Press, Norcross, Ga.

Jefferson, T., M. Rangaswami, and G. Stoner (1996). Simulation in the design of ground-based intrabay automation systems, in *Proceedings of the 1996 Winter Simulation Conference*, J. M. Charnes, D. J. Morrice, D. T. Brunner, and J. J. Swain, eds., Association for Computing Machinery, New York, pp. 1008–1013.

Kunnathur, A. S., S. Sampath, and P. S. Sundaraghavan (1996). Dynamic rescheduling of a job shop: a simulation study, in *Proceedings of the 1996 Winter Simulation Conference*, J. M. Charnes, D. J. Morrice, D. T. Brunner, and J. J. Swain, eds., Association for Computing Machinery, New York, pp. 1091–1098.

Law, A. M., and W. D. Kelton (1991). *Simulation Modeling and Analysis*, 2nd ed., McGraw-Hill, New York.

Lu, S. C. H., D. Ramaswamy, and P. R. Kumar (1994). Efficient scheduling policies to reduce mean and variance of cycle-time in semiconductor manufacturing plants, *IEEE Transactions on Semiconductor Manufacturing*, Vol. 7, No. 3, pp. 374–388.

Mittler, M., M. Purm, and O. Gihr (1995). Set management: minimizing syncronization delays of prefabricated parts before assembly, in *Proceedings of the 1995 Winter Simulation Conference*, C. Alexopoulos, K. Kang, W. R. Lilegdon, and D. Goldsman, eds., Association for Computing Machinery, New York, pp. 829–836.

Norman, V. B. (1984). *Simulation of Automated Material Handling and Storage Systems*, Auerbach, Princeton, N.J.

Sandell, R., and K. Srinivasan (1996). Evaluation of lot release policies for semiconductor manufacturing systems, in *Proceedings of the 1996 Winter Simulation Conference*, J. M. Charnes, D. J. Morrice, D. T. Brunner, and J. J. Swain, eds., Association for Computing Machinery, New York, pp. 1014–1022.

Savolainen, T., D. Beeckmann, P. Groumpos, and H. Jagdev (1995). Positioning of modeling approaches, methods and tools, *Computers in Industry*, Vol. 25, pp. 255–262.

Takakuwa, S. (1996). Efficient module-based modeling for a large-scale AS/RS-AGV system, in *Proceedings of the 1996 Winter Simulation Conference*, J. M. Charnes, D. J. Morrice, D. T. Brunner, and J. J. Swain, eds., Association for Computing Machinery, New York, pp. 1141–1148.

CHAPTER 15

Simulation in the Automobile Industry

ONUR ULGEN
Production Modeling Corporation and University of Michigan—Dearborn

ALI GUNAL
Production Modeling Corporation

15.1 INTRODUCTION

In this chapter we discuss the use of computer simulation in design and operation of car and truck assembly plants as well as automotive components manufacturing plants. Most of the automotive manufacturers worldwide and, in particular, the big three U.S.-based companies (General Motors Corporation, Ford Motor Company, and Chrysler Corporation) currently require that all new and modified manufacturing system designs be verified by simulation analysis before they are approved for final equipment purchases. In fact, there is a general push in the big three automotive companies that any new equipment purchase or manufacturing line modification costing more than several million dollars should be verified by simulation modeling before approval. Studies performed in the past are indicators of how useful simulation could be in the design and operation of production systems of all kinds, including vehicle manufacturing. Examples can be found in refs. 1 to 8.

In what follows, we discuss mainly the applications of discrete-event simulation in the automotive industry, with some discussion of the emerging role of robotics simulation. Applications of discrete-event simulation in the design and operation of vehicle manufacturing systems can be categorized in two different ways. The first classification is based on the stage of the development of the design of the system. Four categories are observed in this classification: conceptual design phase, detailed design phase, launching phase, and fully operational phase. The *conceptual phase* refers to the initial stage where new methods of manufacturing and material handling are tested by the engineers. Discrete-event simulation packages with three-dimensional animation capabilities are the popular simulation tools at this phase. The *detailed design phase* refers

Handbook of Simulation, Edited by Jerry Banks.
ISBN 0-471-13403-1

to the stage where detailed layout plans and equipment specifications are verified for the system. The principal factors considered here include equipment justifications (e.g., the number of hold tables, power and free carriers, the size of buffers), cycle-time verifications (e.g., conveyor speeds, line throughput), and line operational and scheduling issues (e.g., logic for evacuating ovens and paint booths, repairs, and product mix decisions). Discrete-event simulation packages with built-in detailed equipment features and three-dimensional animation features appear to be the most popular packages used at this stage. The *launching phase* refers to the stage where the plant operates below the designed operational conditions. In some cases it may take up to 6 months for the plant to operate under maximum-capacity conditions. Simulation studies performed at this stage are generally used to test operational policies (e.g., operate one of the two paint booths at a time, run each shop for one-half of the total available time, use different product mixes). Discrete-event simulation packages used at this stage do not typically require the detailed equipment features or the three-dimensional animation features. The simulation program generators with user-friendly features are the most popular packages used at this phase, as models tend to be at a macro level rather than a micro level. An explanation of macro and micro models, and an example of their interactions, appears in ref. 9. The *fully operational phase* refers to the stage where the plant is operating at its anticipated capacity. The simulation studies done at this phase consider product mix decisions, new product introductions, new operational policies, and line modifications. Simulation software used at this phase generally require the same capabilities as those used at the launching phase.

The second classification of the use of discrete-event simulation in automotive manufacturing plants is based on the nature of the problem to be investigated. Four major categories can also be identified in this classification: equipment and layout design issues, issues related to variation management, product-mix sequencing issues, and other operational issues. The equipment and layout design issues include typical problems such as number of machines required, cycle-time verification, identification of buffer storage locations, buffer size (strip conveyors and buffers for sequencing) analysis, and conveyor length and speed determination. Examples of typical problems in the *variation management* area are repair and scrap policy analysis, order-size variation, and paint booth scheduling. The *product-mix sequencing issues* typically include trim line and body shop sequencing, shift scheduling, and trim and final assembly line balancing. In the *other operational issues* area, typical applications involve priority assignment at traffic intersections, assembly-line sequencing, and shift and break scheduling. Table 15.1 summarizes various uses of simulation in vehicle assembly plants. The x-marks indicate typical phases(s) where simulation can play an essential role for the application area listed and where certain types of problems are more likely to be attacked by the designers or managers. For example, cycle-time verification problems are more likely to arise at earlier stages of the design and operation cycle. However, shift scheduling problems are likely to be solved once all equipment and layout design issues are finalized. It should be noted, however, that the table constitutes only a broad framework since, in reality, each type of problem area can be attacked in any phase of the design cycle.

In the following sections of this chapter, we discuss applications of discrete-event simulation in assembly plants, major-component plants, and small-components plants. Then we consider the nonmanufacturing applications of discrete-event simulation. In the following section we discuss the role that corporate groups, simulation service vendors, and equipment suppliers play in applying simulation in the automotive industry.

TABLE 15.1 Classification of the Applications of Simulation in the Automotive Industry

Application Category	Example Application	Phase			
		Conceptual Design	Detailed Design	Launch	Full Operation
Equipment and layout	Buffer size analysis	×	×	×	
	Surge bank locations	×	×	×	
	Cycle time verification	×	×	×	×
	Conveyor length and speed	×	×	×	
Variation management	Test-repair loop analysis		×	×	×
	Scrap analysis		×	×	×
	Paint gun spray purge scheduling		×	×	×
Product mix sequencing	Trim line sequencing		×	×	×
	Body shop sequencing	×	×	×	×
	Shift overlap	×	×	×	×
Detailed operational issues	Traffic priority management		×	×	×
	Assembly sequencing		×	×	×
	Shift and break scheduling		×	×	×

Simulation model life-cycle approaches are discussed next. In the final two sections of the chapter we review the emerging role of robotics simulation and discuss trends in the future of simulation in the automotive industry.

15.2 APPLICATIONS IN ASSEMBLY PLANTS

An automotive assembly plant typically has three major sections with respect to the stages of the assembly process: body shop, paint shop, and trim and final assembly (Figure 15.1). Each of these areas has different types of processes with unique features. There are many issues in an assembly plant that are addressed effectively through simulation. Following is a discussion of the typical issues.

The major components of a vehicle body are assembled in the body shop. The major components typically come from stamping plants. The inner and outer faces of doors, the inner and outer faces of body sides, the hood, and the trunk lid are some of those parts that go into body shop operations. The process to assemble body parts includes stations to bring components loosely together, stations to weld the components, and stations to inspect the structural integrity of the welded components. In the body shop, the emphasis is on the process more than on material movement. The operation times are very dependent on the model of the vehicle being made. Most operations require subassemblies to stop at a station for processing. The reliability of process machinery is the critical part of many problems that can be observed in a typical body shop. Consequently, adequate representation of the machine breakdowns and cycle times is an important part of a body shop simulation. Additional discussion of downtime modeling appears in ref. 10.

Once all major body parts are assembled, the body is sent to the paint shop as the second major phase in the assembly process. A typical paint shop will consist of several painting processes. In rough order, those processes include electrocoating, sealing, prime

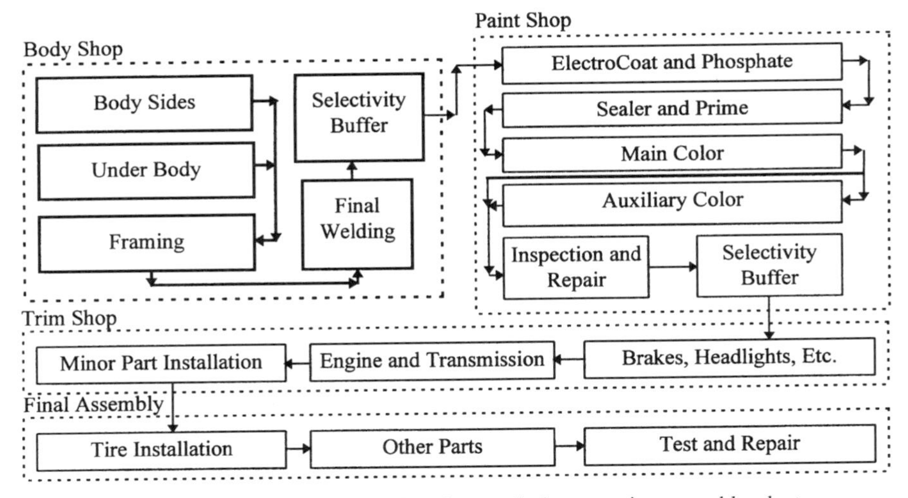

Figure 15.1 Parts of and job flow in a typical automotive assembly plant.

painting, main color painting, inspection, and auxiliary painting operations. The paint shop processes are such that many of the operations can be performed without stopping the vehicle at a station. Therefore, the capability and the reliability of the material handling equipment (conveyors in almost all cases) is important in a paint shop.

One of the typical problems in a paint shop is to sort the incoming sequence of vehicles to minimize the color changes in paint booths. Since changing colors involves a setup time, it is desired to sequence same color vehicles back to back. To achieve such *color blocking*, however, there is a need for a temporary storage place. Once a sufficient number of units are accumulated, vehicles of the same color can be pulled from this bank in order of color. Similar temporary storage banks are required to have the ability to empty the ovens that are used for baking different coats of paint. If for some reason, the processes ahead of a paint oven are blocked, there must be sufficient space to unload the units once they come out of an oven.

Another source of variation in a paint shop is the yield of the painting processes. Since the yield of processes is relatively low (in a typical paint shop, the first pass rate could be as low as 65%), the chances of rework are high. Because of the randomness in the process, it is always a challenge to keep vehicles in the same order as they arrived in the paint shop. Vehicles requiring rework would need additional passes through the paint booths. Keeping vehicles in a certain order is one of the critical problems in a typical assembly plant. The movement of material to stations is dependent on the type of vehicles to be assembled next. Consequently, there is a need to schedule a certain order of vehicles ahead of time. Once a schedule is made and issued to all workstations, changing the sequence may cause loss of production by creating material shortages at various points in the plant. Therefore, it is a common practice to accumulate the vehicle bodies in temporary storage areas after the paint process. Such temporary areas allow sufficient time for the late units to catch up to their position in the assembly sequence.

Once vehicles come out of a paint shop, they go into the trim and final assembly area. This area is where all the major and minor components of a vehicle are put together. In a typical setup, some of the minor components are assembled to the vehicle body in the trim shop. The brake system, headlights, and taillights are examples of those minor components. Then the major components are assembled: the engine, power train, and chassis. Finally, the vehicle is put on a final assembly line. Most of the operations in those areas are done by manual labor. Usually, vehicle bodies are not stopped for an operation except for major component assemblies. Consequently, the capability of the material handling equipment (almost all cases involve conveyors) is an important issue. Another issue is the ability to make available the required parts at the time they are needed at every operation.

15.2.1 Case Study: Body Shop Material Handling System Analysis

This simulation study was performed during the conceptual design phase of a new vehicle program for a major U.S. car manufacturer. The system studied consisted of the following components: (1) a car track system with several sections, (2) 90° turn tables between various sections of the car track system, (3) several robotic spot welding stations, (4) two load/unload stations for two different car models, and (5) a variable number of carriers for each car model. Figure 15.2 depicts a snapshot of a part of the model that shows the car track system. The objectives of the study were threefold:

1. Determine the best equipment configuration and the corresponding line throughput under a given set of operating parameters.
2. Determine the maximum allowable cycle time at the loading stations for either car model.
3. Determine the best number of carriers for each car model.

Some of the model assumptions were as follows:

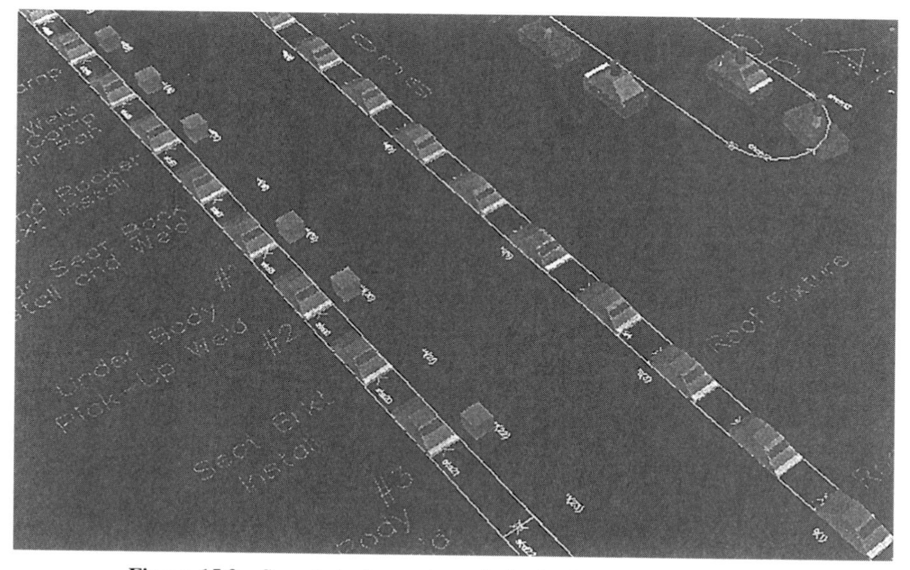

Figure 15.2 Snapshot of a section of a body shop simulation model.

TABLE 15.2 Results of Four Scenarios

	SCENARIO			
	1	2	3	4
Parameters				
RA station type	I	II	I	II
Main welding station type	Manual	Robotic	Robotic	Robotic
Flexible station configuration?	No	No	Yes	Yes
Results summary				
Number of model A completed/hour	32.47	33.29	33.44	33.57
Number of model B completed/hour	15.64	16.33	16.48	16.59
Average production rate/hour	48.11	49.62	49.92	50.16

- Target production rate is 38 model A and 18 model B cars.
- There will always be a vehicle waiting for loading at either load station.
- Each station has a randomly distributed downtime and a random repair time. The mean values and distributions of those random variables were known based on historical data from similar systems.
- The transfer time between two stations on the line is 6 seconds if the carriers do not stop at a station. If the vehicle stops at a station, the transfer time would be 8 seconds, taking acceleration and deceleration into account.
- A station is assumed to break down only after a full cycle of operation, not during a cycle.

The following parameters and variables (evaluated in the what-if scenarios) were used in the simulation model: (1) two different equipment configurations distinguished by cycle times and downtime data, (2) number of carriers allocated to each of two job types, (3) loading/unloading station cycles times. The performance statistics from the model included the line throughput for each type of car and the utilization of each station.

The system was modeled using the AutoMod software such that most operating parameters could be input from data files rather than by modifying the model. The system capacity was evaluated under equipment downtimes with different equipment configurations, ranging from labor-intensive to robotic processes. In particular, the type of equipment for the robotic alignment (RA) station was considered in detail. Table 15.2 summarizes the results from some of the scenarios investigated during the course of the study.

The results in the table indicate that the last alternative, with a flexible line configuration and a type II RA station, would provide the most output from the line. Other scenarios that are not displayed in the table also showed that the system performance was not highly sensitive to the number of carriers. In addition to those results, it was also found that the loading operation could be done manually for model A cars without affecting the throughput, although a longer cycle time was needed. Consequently, a line configuration with a best mixture of robotic and manual operations was determined and submitted to the project team.

15.2.2 Case Study: Paint Shop Material Handling and Model Mix Scheduling

This study involved simulation modeling and analysis of a paint shop and an adjoining automated storage and retrieval system (AS/RS) during the conceptual and detailed design phases. In addition to evaluating the design, the animation of the model was used as a visual tool to facilitate the brainstorming sessions of the design team. The objectives of this study were as follows:

During the Conceptual Design Phase:

- Evaluate the conceptual design at each iteration of the design cycle to determine the potential bottlenecks and identify alternative solution strategies.

During the Detailed Design Phase:

- Determine the throughput capability of the system.
- Assess the feasibility of the proposed shift schedules and paint booth strip sequences.
- Investigate the best stock levels of various products in the AS/RS.
- Analyze the impact of different trim line operation schedules on the number of out-of-sequence conditions.

The system consisted of the following subsystems in sequence: (1) electrocoat and phosphate, (2) sealer lines and sealer gel oven, (3) prime booth and prime oven, (4) main enamel booth and enamel oven, (5) inspection lines, (6) spot repair area, (7) second-coat paint booth and oven, (8) paper masking and repair lines, and (9) the AS/RS (see Figure 15.3). The material handling equipment in the system consisted of many two- and three-strand chain conveyors, lift tables, turntables, and power roll beds.

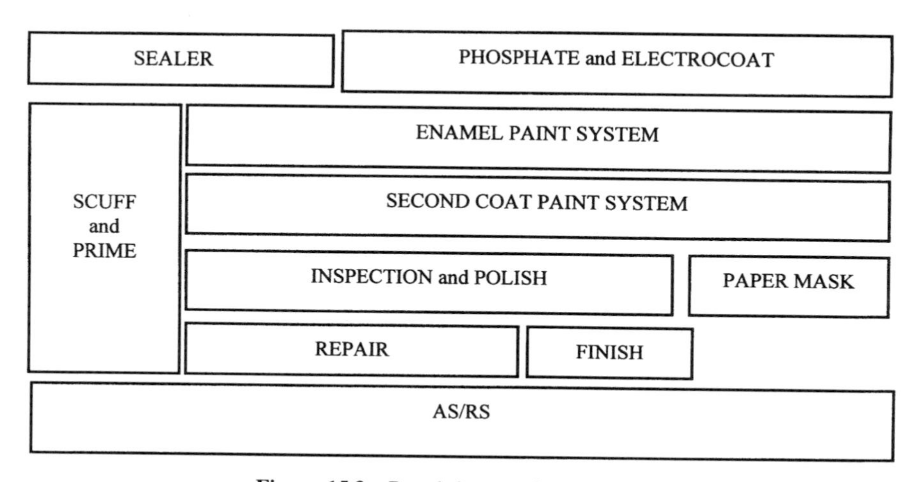

Figure 15.3 Rough layout of a paint shop.

The following parameters and variables (evaluated in what-if scenarios) were used in the simulation model: (1) conveyor speeds, spacing, and speed-up section data; (2) cycle times at repair and mask lines; (3) cycle times at spot repair area; (4) product and paint mixes; and (5) major and minor repair percentages.

The model also required a front-end scheduling routine that was customized using a programming language (e.g., FORTRAN and C). Because some of the units required several passes through paint booths, they took a longer time to be ready for delivery to the trim lines. However, since the product mix showed significant differences between shifts, and since the trim lines operated on a different shift pattern, the jobs that required long processing times were pulled ahead of their original sequence. Thus even though they would be taking more time than the other jobs, by the time they were completed, they would be able to catch their original position in the assembly sequence. Because of the randomness of the defect rate, there would be a good chance that some units would miss their sequence if they were not moved sufficiently ahead in the paint sequence. On the contrary, if they were moved too much ahead of their sequence, they would finish the paint process much earlier than the rest of the units. Therefore, to protect against such random variations in the makespan of different job types, a buffer storage bank was held in the adjoining AS/RS. This buffer would be sized to allow sufficient time for all units to catch their original sequence. The following are some of the original rules of resequencing the jobs:

- Jobs with two colors were moved ahead by 100 jobs for two product types.
- Jobs with three colors were moved ahead by 200 jobs for only one of the models.
- Pattern color jobs were moved ahead by 200 jobs.
- A job with more than one matching criterion is moved ahead by the sum of the jobs required by each criterion (e.g., a two-color job with patterns would be moved ahead by 400 jobs).

Some of the model assumptions were as follows:

- The two vehicle models, A and B, are considered. Model A vehicles have up to two coats of paint, whereas model B vehicles have up to three coats of paint.
- The model mix was known and assumed constant within a day.
- The major repair percentage is 22% and the minor repair percentage is 9% on average with random occurrences.
- Minor repair times are randomly distributed between 30 and 120 minutes and are performed at a dedicated area. Major repairs go through the second paint loop as necessary.
- Shift patterns are known and constant. The first shift is dedicated to model B and the second shift is dedicated to model A at the paint shop. The trim shop runs only one shift and makes both products.
- All conveyors run at full speed with negligible downtimes.

Analysis of the model involved an evaluation of the alternative job sequencing policies to choose one that will eliminate late jobs at the trim lines for all job types. In addition to sequencing concerns, the model was used to investigate the selectivity system (AS/RS) utilization. The runs of the model indicated that there was no reason to

TABLE 15.3 Results of Investigation

	Total Missed Jobs in a Week			
	With Resequencing		Without Resequencing	
Model A Buffer	Single Coat	Double Coat	Single Coat	Double Coat
135	120.32	16.29	0	115.76
160	14.75	4.43	0	56.45
170	1.09	4.21	0	40.32
180	0	0	0	30.20
200	0	0	0	12.29

resequence model B vehicles since the plant was planning to store a full day's worth of buffer in the AS/RS for this type. The model showed that during the second shift in the paint shop all of the longest paint jobs would be completed for the next day's production at the trim lines. Results from some of the scenarios investigated are summarized in Table 15.3 for model A only. The table depicts the results with and without resequencing. In either case, different levels of model A buffer in AS/RS were tested to find a level that will balance the buffer size and the number of missed jobs.

The results in the table indicate that a buffer size of 180 vehicles would be sufficient to avoid the missing jobs for the vehicle model A. It was also determined that with chosen buffer sizes, the utilization of the AS/RS was at a feasible level. The plant would make substantial cost savings by avoiding the reprogramming of their production monitoring system for vehicle model B.

15.2.3 Case Study: Trim and Final Assembly Lines

This simulation study was performed during the detailed design phase of a new conveyor system. An assembly plant would be making several different models of cars on one trim line. The process and flow of jobs in the system showed differences with respect to the model of cars. The conceptual design of the new system was completed following the previous version of the system. However, to accommodate the variety of the product assembly sequence, many new hardware pieces were needed. To ensure that the system could move the desired product quantities between various parts of the system, a detailed simulation model was built. An important parameter of the design was the mix of models in the target production rate. The objectives of the simulation study were:

- Verify the capability of the conveyor system to move the target number of vehicles through the trim system by considering various product mixes.
- Investigate various scenarios of assigning the size and location of empty carrier buffers by considering different product mixes.
- Analyze the impact of building a new buffer area to hold additional empty carriers.
- Determine the maximum allowable cycle times at several transfer stations by considering different product mixes.

Some of the important assumptions of the study were as follows:

- All manual operations can be completed within the given cycle time.
- All materials are always present.
- The line speed would be set up higher than the required rate so that occasional downtimes could be tolerated.
- There are three models of cars and eight possible mixes of those models.

The system consisted of a chassis buildup system, an engine delivery system, a frame buildup system, and a final trim line. All material movement were made by using power and free conveyors except for the frame buildup area where a chain conveyor was used to move the units continuously. The transfer of units and subassemblies between major areas required complicated equipment that was prone to mechanical failure. Since there was limited room for buffers, an additional storage space was designed at the mezzanine level. The size of the buffer was being questioned since there were random downtimes at major transfer points. Since there were no detailed data, the system was designed to run at a speed 12% higher than the speed required, to allow time for breakdowns and shift breaks. However, the designers wanted to confirm that the system would be capable of delivering an average number of vehicles to meet the weekly production target at a 5 to 10% downtime rate. Based on past data, only an average recovery time of 5 minutes was specified. Also, the newly designed engine assembly area required the proper number of pallets in the system to support the production of all different types of jobs. Since there could be a variety of job mixes to be produced in the system, it was necessary to determine a number of pallets that would work with all possible product mixes.

During the simulation study, first an evaluation of possible product mixes were made by using a baseline layout. This portion of the study helped to determine the maximum allowable cycle times at critical stations to accommodate a variety of product mixes. Then the study focused on evaluation of the size of the empty carrier buffer. Three different layout alternatives were investigated. The simulation runs indicated that there would be no difference between layouts with respect to the average throughput capability. However, the utilization of various subsystems would be greatly affected by allocation of the empty carriers. Table 15.4 depicts, for all three layout alternatives, the time it takes to starve various subsystems after a catastrophic breakdown at one of the critical stations. The table clearly demonstrates that the first layout alternative is significantly better than the latter two in protecting the chassis buildup system against long periods of breakdowns.

The model (see Figure 15.4 for a snapshot of a section) showed that at the transfer point from the engine build line to the engine deck area, the control logic and buffer size

TABLE 15.4 Time to Starve Subsystems After Breakdown

Subsystem	Layout 1	Layout 2	Layout 3
Chassis buildup 1 and 2	20.0	7.0	7.5
Chassis buildup 3	12.0	4.5	5.0
Chassis buildup 4	4.5	2.0	2.5
Engine load	37.0	37.0	37.0
Final line	2.0	2.0	2.0

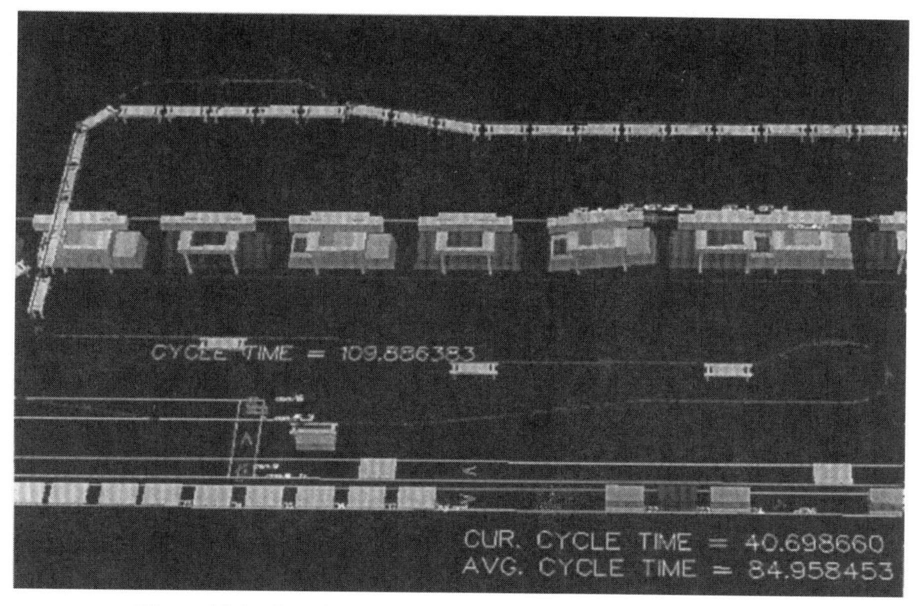

Figure 15.4 Snapshot of a section of a trim shop simulation model.

originally proposed would not support the cycle-time requirement of the engine assembly area. Also, the simulation showed that the final assembly line would be starved immediately if the downtime at the body assembly area were longer than 3 minutes. It was also determined that the buffer storage space placed in the mezzanine level would be sufficient only for breakdowns of relatively short duration. An evaluation of an alternative design showed that additional empty carrier lanes were required to support the system for a longer time should downtime occur at a body assembly point.

15.3 APPLICATIONS IN MAJOR COMPONENT PLANTS

Two of the major components in a vehicle include the sheet metal portions of the body, such as the frame, body sides, doors, roof, and hood, and the cast iron parts used to build the engine and transmission components. A typical automotive manufacturer would have a different plant for each major component. These manufacturing processes, however, could have a higher rate of production than that of cars built at an assembly plant. Consequently, it is usual that a single component plant feeds several assembly plants for several different car models. Due to the unique nature of each type of process, the nature of design problems shows variations among those component plants. In an engine plant, for example, the emphasis is on the reliability of the manufacturing machinery rather than the material handling equipment. In a stamping plant, on the contrary, the problems might be around production schedules and material movement within the plant. Two case studies are given in the following sections. Each case study is related to one type of component plant.

15.3.1 Case Study: Stamping Plant Material Handling Study

The stamping plant involved in this study had several stamping lines that made the sheet metal parts and several subassembly lines that built subassemblies to be used in the assembly plant. For a new vehicle program, a new panel assembly line was going to be put in operation. However, due to the required production volumes and complicated material flow patterns, choosing an efficient method for racking excess panels at the front end of the panel assembly line was going to be an integral part of the solution. The dynamism of the system could be captured only in a simulation model. Consequently, the study involved developing a model of the material handling system supporting the new assembly line. The main objective of the study was to find the best combination of material handling system parameters and operating rules, with an emphasis on the racking methods for excess panels.

The material handling system modeled was divided into the following four areas: (1) monorail delivery system, (2) panel assembly line, (3) empty racks load area with excess panels, and (4) full racks storage area. The simulation model also included material handling processes, job routing logic, and the operating procedures/policies. Through simulation, the system's capabilities were evaluated, resulting in improvements and modifications to the proposed system. More specifically, the study allowed plant engineers to understand product flow, identify bottlenecks, evaluate buffer space requirements, and analyze staffing, equipment, and operating plans. The results from simulation showed that one of the racking methods provided sufficient capacity to handle the racking of excess panels in a relatively inexpensive setting. Furthermore, by varying the number of forklift trucks in simulation, the best number that yielded the required throughput and minimum traffic congestion was determined.

15.3.2 Case Study: Die-Casting Cell Cycle-Time Study

In this study, a model of a die-cast cell was built to investigate the cycle time. The model included two die-cast machines, an operator, and a press machine. Parts were cast in one of the two die-cast machines that fed a single press machine to trim the cast parts. The die-cast machines were automatic needing no operator. However, an operator was required at the press. The operator was responsible for loading and unloading the parts from the press. After the parts were trimmed, they moved on to a series of operations that were not investigated in this study. The specification of a press was the decision to be made. More expensive machines had shorter cycle times. A short cycle time at the press allowed the operator a wider time window for loading and unloading.

Because of the variable nature of the manual operations and the randomness in the rate of scrap parts, a simulation model was needed to analyze the problem in more detail. The objective of the investigation was to determine the operating conditions for the system to achieve maximum throughput. To facilitate a detailed investigation, a model of the system was built to determine the press cycle time for a given range of operator utilization. The model was built so that it could be used to address the impact of the following issues on the system throughput: (1) operator utilization, (2) equipment downtimes, (3) material handling equipment cycle times, (4) individual job processing times, and (5) scrap rates.

The results from simulation analyses indicated that the two die-cast-machine approach will meet production at an acceptable rate of utilization for the operator. The study also concluded that the throughput of the system would be increased if three die-cast machines were used. The latter configuration would also improve the operator and press machine utilization if some of the flexibility issues could be resolved.

15.4 APPLICATIONS IN SMALL COMPONENTS PLANTS

All components of a vehicle other than the major components (stamped sheet metal parts, cast metal parts, and the power-train parts) can be viewed as small components for the purposes of this chapter. Manufacturing of small components is typically done in much smaller areas. However, a small-components manufacturing plant could be as big as an assembly plant, due to the variety of the components and the required production volumes. A small-components plant typically feeds many assembly plants, thus requiring a high volume of production. Many of the problems associated with such plants are due to the high volume of production. The cycle times are much smaller and the parts are smaller in size.

Many small components, such as alternators, starters, and fuel injectors, are made in separate assembly areas in a typical small-components plant. The material handling within such assembly areas is done by small but efficient and reliable conveyors. Consequently, many of the problems that constitute good opportunities for the use of simulation arise in the analysis of the part-making and assembly manufacturing processes. Reliability of machinery, scrap rates, machine setup requirements, product mix decisions, scheduling conflicts, and the efficiency of interdepartmental material handling equipment are some of the common problems with which simulation models can efficiently be of help. The following three case studies provide good examples of typical problems and their solutions using simulation.

15.4.1 Case Study: Machining Cell Design Study in an Electrical Components Assembly Plant

This study involved the study of a coil winding cell. The purpose of the study was to determine a combination of the machine configuration and operating parameters that provided the maximum possible throughput. The coil winding area consisted of robotic winding cells which (1) receive bobbins on trays, (2) insert two terminals, (3) wind bobbins, (4) trim excess wire, (5) crimp the terminals, (6) flux and solder terminals, (7) test and reject assemblies electrically, and (8) place finished coils back into the trays. Each of these operations was performed in a separate cell. An accumulating conveyor system transported trays between operations.

Numerous system parameters needed to be evaluated to determine the best configuration that would yield both satisfactory resource utilization and overall system throughput. Since the study was being performed in the very early stages of the design cycle, many of the line configuration parameters, such as the size and location of buffers and the number of machines, were variables. In addition to such higher-level variables, the length and speed of conveyors and the number of operators were lower-level variables that could be fine tuned to obtain the maximum throughput from the system.

Random downtimes and random changes in the product mix were the main contributors to the variation of the system. To represent such variation adequately and obtain meaningful performance predictions, a simulation model was needed. The model would be built to determine the operating conditions for the system to achieve maximum possible throughput. The model was built to enable a detailed study of the proposed coil winding process by analyzing the impact of the following operating issues on throughput: (1) buffer sizes and locations, (2) unscheduled equipment downtimes, (3) individual part processing times, (4) conveyor parameters, (5) scrap generated, and (6) labor involved in maintaining and repairing the equipment.

By analyzing the performance of the system under various operating conditions, the best combination of the system parameters (among those tested) was determined. The simulation showed that some of the originally proposed buffer sizes would have to be modified to improve system performance. Some other buffers, however, could be much smaller than proposed. Consequently, simulation played an important role in the design of this assembly line.

15.4.2 Case Study: Car Seat Assembly Material Handling System Study

The plant involved in this study manufactured most of the upholstered components that were put inside a vehicle. For a new vehicle program, the plant was undergoing a revamping process by installing new lines to support the production targets and cost reduction goals. The proposed seat assembly process depended on an automated guided vehicle (AGV) system for material movements within the assembly area. At distinct assembly islands within the system, front and rear seats are assembled, then joined. The general flow of the assembly process is to transport car seat parts on AGVs through the following sequence of operations: (1) pallet loading; (2) front seat parts kitting; (3) assembly islands 1, 2, and 3 (front seat assembly); (4) front seat oven heating; (5) assembly island 4 (front seat fit, finish, and inspection); (6) front seat repair; (7) rear seat parts kitting; (8) assembly islands 6 and 7 (rear seat assembly); (9) rear seat oven heating; (10) rear seat inspection and repair; (11) stretch wrapping; and (12) pallet unloading.

For a cost-effective design, several issues would have to be addressed. First, the best number of AGVs should be determined. Adding more vehicles to the system can help increase the capability of the system, but congestion might prevent an increase in the throughput. Second, the number of assembly lanes must be decided. Also, the impact of various lane assignment rules on the system performance should be evaluated. Instead of the current rule, which is relatively complicated, it was desired to simplify the operating policy for the new system. In addition, an assessment of the impact of the productivity of manual operations on overall system throughput was needed. Finally, the possibility of increasing the pace of manual operations would be weighed against more work-in-process and congestion in the system. Considering the complexity of the traffic control system, the randomness of operation cycle times, and availability of the vehicles, the simulation approach was found to be an effective analysis tool for the problem. The main objective of the study was to determine the operating parameters that would enable the seat assembly area to achieve maximum hourly throughput.

The results from simulation indicated that adding more AGVs to the current fleet, originally contemplated by the design team, would not significantly increase overall system throughput. Thus unnecessary expenditure on capital equipment was avoided. Overall system throughput increased with higher operator efficiency levels but was limited by the stretch wrapper. The throughput did not increase with further increases in worker efficiency once the bottleneck shifted to that operation. Different sequences of assigning available lanes to AGVs arriving at assembly islands did not affect overall system throughput. Opening more lanes in one assembly island did not necessarily translate to higher throughput, because of the interdependencies between consecutive assembly islands.

In this study, simulation provided a detailed picture of the behavior of the system under the assumptions of increased efficiency at various operations. Analyses clearly showed that unless some of the existing processes were improved, adding more capacity to the material handling system could not be justified. The plant engineers were able to determine the root causes of the problems, which were not evident by observation of the system or by statistical analysis of data.

15.4.3 Case Study: Instrument Panel Assembly Line

In a small-components plant, prior to installing a new instrument panel assembly line, the engineering and management teams wanted to detect potential bottlenecks that may limit capacity and to assess the utilization of resources. The new design included a conveyor system unfamiliar to the plant management and line personnel. In addition, there was a need to determine the best scheduling and loading patterns for the new system.

The system consists of a closed-loop powered roller assembly line. Each 10-ft section of conveyor has a variable-speed driver and a mechanical stop. One of two part types is loaded onto a pallet at the start of the system. The pallet enters several manual assembly stations and operations are performed on both part types. After the pallets pass through these stations, they are separated according to part type. Part A pallets are transferred across to the top of the line, where they are unloaded. Part B pallets continue on the lower line through more assembly stations and are then transferred up to the top section of the line. Then part B pallets enter one of four test cells. Following this operation, empty pallets are sent to a buffer in front of the loading station, while part B pallets merge back into the main line with part A for unloading.

The goals of this study were to (1) evaluate different line configurations identifying and correcting any problems (blockages, bottlenecks, etc.), (2) evaluate the throughput of the system using different loading patterns, and (3) determine the number of required pallets. Using the simulation model, process engineers were able to determine the buffer capacity required for empty pallets, line speeds, and control logic required at transfer and intersection points. The final system was robust enough to handle any type of product mix that could enter the system. Results showed the number of pallets required to achieve maximum system throughput to be 28, whereas the original estimate was 48, resulting in 42% savings.

15.5 NONMANUFACTURING APPLICATIONS

15.5.1 Case Study: Distribution Chain Management

A European car manufacturer was reengineering its vehicle distribution system over its North American dealership network and required a detailed study of the existing and proposed systems and recommendations on improvements to the system. Vehicles are manufactured abroad, and the distribution system uses several U.S. ports to satisfy North American demand. Dealerships in metropolitan markets get their shipments from port inventories.

The company feels that building distribution centers closer to metro markets should reduce costs and improve customer service metrics in terms of first-choice deliveries (i.e., ability to deliver to the customer immediately what she or he is asking for). There was a need for flexible analysis tools to generate and evaluate various alternatives. The objective of the study was to develop a set of models for predicting the performance of a given distribution system configuration. The performance criteria were the rate of matching customers' first choice and the total cost of installing and operating a given configuration of the distribution network.

In this study, a mathematical optimization model was developed to generate distribution center alternatives that minimize transportation-related costs per year. Once a solution is generated, the configuration was input into a simulation model that explicitly considered the probabilistic and dynamic elements in the system, and hence estimated the overall performance of the given alternative more realistically. Using an algorithm that iterated between the optimization and simulation models, a configuration was found that satisfied most of the evaluation criteria. The new design showed that it was possible to reduce transportation costs by about 25% while improving customer service.

15.5.2 Case Study: Warehousing Study

This project studied the proposed changes to a warehouse and the proposed material handling equipment. The modifications to the system were needed as a result of increased storage, shipping, and receiving volumes anticipated due to packaging changes in existing products and introduction of new product lines. It was desired to determine alternative ways of increasing both cubic storage space and material handling (shipping and receiving) capabilities. The challenge was to optimize the layout of the warehouse and select the most suitable material handling equipment to provide adequate service based on planned future volumes.

The goals of the study were to identify system constraints that could limit future space and handling requirements and to suggest potential improvements and modifications to the system design. The design alternatives were based on the following system parameters: (1) dedicated versus random storage rules, (2) original rack orientation versus rotated (perpendicular to the docks) orientation, (3) aisle width and overall storage space utilization, and (4) capacity of material handling equipment, most notably the number of lift trucks.

The results from the simulation model demonstrated that (1) randomized storage was better suited to this situation than dedicated storage, (2) the rotated orientation with the corresponding narrower aisle configuration resulted in an 85.96% increase in overall unit load storage capacity, (3) the rotated orientation also resulted in an increase from 23.57% to 41.95% in total storage space utilization, and (4) the existing number of lift trucks was sufficient to service the increased volumes.

15.6 ROLE PLAYERS AND STANDARDS

15.6.1 Role Players

Interest in determining the best use of simulation in building efficient manufacturing systems (see refs. 11 to 13) in the automotive industry continues to increase. Advances in computer hardware and software and an increasing awareness of the capabilities of simulation have helped to achieve this higher level of interest. There are several groups of people with close interests in using simulation as a productivity enhancement tool. First, almost all major manufacturers have their own consulting groups providing simulation modeling and analysis services within the corporation. These groups act as internal consultants and are closely tied to industrial engineering departments. An important function of these groups is to increase the awareness of simulation across the corporation. Working closely with process and material handling engineers, these groups help establish simulation technology as an indispensable computer-aided engineering tool. For further discussion, see ref. 14. In addition to corporatewide groups, some of the divisions of the corporation and most of the plants have access to internal personnel as simulation analysts. Another important function of these internal resources is to coordinate the acquisition of modeling services when needed. By ensuring that the models delivered are accurate and are in usable form, these internal resources also act as liaisons between simulation modeling vendors and the corporation [15].

Simulation service providers are a second group dedicated to enabling simulation technology to be used properly and to achieve its maximum benefits. Smaller companies show a high degree of variation in size and the breadth of services offered. Many simulation service providers would typically use only a few of the commercially available simulation software packages. There is only a very small number of vendors that provide expertise in all simulation software. An important contribution of simulation service providers is the expertise and focus to turn projects around at a faster rate than with the typical internal resources of a corporation. Also, simulation services providers have the ability to play a mentorship role in use of the methodology. Such mentorship programs offer an opportunity for speeding up the learning of the proper methodology. Guiding relatively novice users of the technology through complicated modeling and analysis situations is a very critical task in establishing simulation as a powerful tool in designing and operating a manufacturing system.

As indicated previously, each of the major automotive manufacturers has made it a practice to simulate new systems prior to their installation in a plant. Consequently, more and more machine tool sellers are increasing their use of simulation. Besides concurring with the requirements of the automotive companies, simulation is being used by some machine tool sellers and material handling equipment suppliers in designing, specifying, and planning the production line.

Machine tool sellers and material handling suppliers have different needs when they use simulation. From a machine tool seller's viewpoint, the objective is to evaluate the capability of the equipment to deliver the required throughput. Consequently, individual machine downtimes, machine cycle times, scrap rates, and buffer sizes become the most typical inputs to a simulation model. The typical results from a study of the machinery include the location and size of each buffer, an assessment of the line throughput, and identification of potential bottlenecks should the product mix change. A material handling equipment supplier is, however, concerned with providing the right material at the right time and at the right place. Consequently, the speed, routing and traffic logic, and capacity of the material handling equipment become predominant inputs to a simulation. Identification of congestion problems, evaluation of the speed and capacity of material handlers, and an assessment of the throughput capability of the material handling system are the typical results from a simulation study.

An examination of models built for a material handling supplier and for a machine tool seller would reveal that the simulations on each side have ignored the other even though they are ultimately for the same production line. A machine tool seller assumes that material handlers would always be present and that they would have sufficient capacity. The material handling supplier would assume that the process would always yield the target throughput. Obviously, the interactions between the two systems are seldom captured in simulation models. It is the responsibility of the buyer of the system to ensure that a complete model of the production system is built by considering both components of the manufacturing hardware. However, in a typical organization, the material handling group and the process engineers work in separate areas and, perhaps, use different simulation models. Therefore, it is seldom the case that the members of one group would gain an insight from the simulations performed by the other group. Consequently, inefficiencies, redundant effort, and even misleading conclusions are possible. Thus the management of simulation modeling for large-scale integrated design projects has some room for improvement.

15.6.2 Standards

Standardization of simulation models could have many benefits for an automotive company. A typical automotive company would have many geographically dispersed plants that are likely to be very similar in nature. For example, many of the assembly plants of the same vehicle manufacturer would look very similar. The designers of the same type of facility tend to be the same across different car programs. By keeping the same design teams on similar car programs, a vehicle manufacturer attempts to maintain and utilize the know-how generated among other programs. By the same token, building simulation models by adhering to a well-defined set of standards can reduce the costs of development. In addition, standardized animation, source codes, and reports can all help to minimize communication problems between analysts and the users of simulation results.

As discussed previously in this chapter, simulation models are being built for many reasons at many different points in a vehicle program by many different parties. Since it is spread across many functions and departments within an automotive company, the use of simulation shows a significant degree of variation.

Considering the multiplicity of users of simulation among the vendors, internal and external simulation consultants, and many facilities of a large company, structuring a set of standards for simulation modeling is a very challenging task. Furthermore, developing standards on model building, animation, analysis, and report generation has not been recognized as a problem.

Modeling tools that allow building of application templates have been an attempt at standardization on the software side of the technology. However, an important roadblock in standardization is the fact that simulation modeling is a cognitive task. Models are developed by people with different levels of education, skills, and experience. The perception of problems and understanding of the capabilities of the software tools are highly dependent on the individual. Consequently, there are differences between models of similar systems developed by different people. Furthermore, as there is a learning process and different timing requirements on different projects, even the models developed by the same person show significant differences over time.

There have been efforts toward standardizing a methodology for application of the simulation methodology. For example, refs. 16 to 20 report significant benefits from applying such methodologies. Those studies indicate that the success of a simulation study is determined primarily by how well certain guidelines on project management and model building are followed and communicated to the project team. However, at the date of this writing, there have been no published common standards of modeling and model building techniques, particularly for the automotive industry.

15.7 SIMULATION MODEL LIFE-CYCLE CONSIDERATIONS

Simulation models may have a short or long life cycle based on the use of the model through the life of the system. Short-life-cycle simulation models are developed for a single decision-making purpose at a certain point during the life of the system (e.g., conceptual design issues, detailed design issues), and once the decision is made, the model is discarded. On the other hand, long-life-cycle simulation models are built to be used at multiple points in time during the life of the system and are maintained and revalidated as conditions change in the system. Currently, about 70 to 80% of the simulation models built can be categorized as short-term models, while the rest are longer term in that they will be reused after the initial simulation study has been completed. Long-term simulation models require long-term ownership (i.e., use, control, and maintenance) of the model by a modeler and/or engineer. Long-life-cycle models are generally built for the purposes of (1) training, (2) reuse at the launch phase, and (3) reuse at the fully operational phase for changes in design and operation of the system. Training-focused long-life-cycle models are used to train engineers on the stochastic nature of the system that they are controlling and/or teaching simulation to them. Models built at the detailed design phase of a system may be used for partial personnel allocations and line segment operations in the launching phase of a new manufacturing system. The reuse of models at the fully operational phase of a system are generally for product-mix decisions as the demand for products change or as new products are introduced into the system. The successful use of long-life-cycle simulation models may require the following tasks to be accomplished in addition to the traditional steps of simulation model building [16]:

1. Construct user-friendly model input and output interfaces.
2. Determine model maintenance and training responsibility.
3. Establish data integrity and collection procedures.
4. Perform field data validation tests.

It is important that the long-life-cycle nature of the model be specified as part of the original objectives of the study because the model design is highly influenced by it. For the long-life simulation model to be effective, the simulation project team should include at least one person who is a long-term user of the model.

15.8 ROBOTICS SIMULATION

15.8.1 Overview of Robotics Simulation Technology

Robot applications are becoming more and more widely used in industries from manufacturing to health care. The most common utilization of robots is in manufacturing. In the automotive industry, primary areas of robotics applications include arc and spot welding, painting, material handling, assembly, and testing and inspection. There are many automotive assembly plants in the United States that use robots, for example, for all welding operations in body shops. Similarly, almost all new paint shops utilize robots for most paint applications. Also, many small-components manufacturing plants have robots for a significant portion of the assembly operations. In addition, in electronic component assembly lines, robots are very common. Consequently, there is a strong need for effective analysis and design tools for applying the technology successfully. With its flexibility to address a wide range of design and operational problems in robotics applications, simulation technology proves to be such a tool. Some commonly used software includes IGRIP from Deneb Robotics, ROBCAD from Technomatrix Technologies, CimStation from SILMA, and Workspace from RWT. All of these software can display a work cell in three-dimensional graphics. They also provide inversive kinematics calculations to facilitate a wide range of analyses on robotic systems, such as robot selection, robot placement, reaching capability assessment, and interference checking.

Robotics simulation applications can be categorized into four areas: (1) conceptual design and presentation applications, (2) robotics work cell design applications, (3) off-line programming, and (4) integrated simulation with ergonomics and discrete-event simulation. The first category includes applications where a proposed or existing system is modeled for demonstrating a concept, marketing, training, or documentation of different designs. A typical simulation in this category consists of machines, robots, robot tools, jigs and fixtures, material handling devices, and human operators. The main objective of these types of models is communicating the ideas and concepts through a realistic graphical representation of system operation.

The second category of robotics applications involves mostly engineering applications. Designing and evaluating the layout of a work cell, selection of robots, designing tools and fixtures, eliminating colliding motion paths, optimizing robot movements, and cycle-time assessment and task allocation are among the typical uses of robotics simulation models. Clearly, those models require a high degree of accuracy in the geometric representation of cell components.

All of the software tools mentioned above provide the capability to create robot programs in their native language (i.e., ARLA for ABB robots and KAREL for FANUC robots). Once an accurate model of a work cell is created, it is possible to develop programs for the robots. Those programs can then be downloaded to the robot controllers on the floor, eliminating the need for teaching by using a pendant. For some applications, however, there might be a need to calibrate the off-line program since there are always differences between a simulation model and real-life implementation of the work cell, due to installation errors and manufacturing tolerances. Off-line programming using robotics simulation software is explained in more detail in the following section.

Integrating a robotics simulation study with a discrete-event simulation study benefits both in various ways. The most important interaction, however, is the cycle-time determination. An assembly-line simulation can determine a time window for the cycle time, which can then be fed to a robotics simulation model for determining feasibility. On the other hand, a robotics simulation model can help determine the best and worst estimates of the cycle time in a robotics work cell. Those estimates can then constitute a basis for what-if scenarios by using a discrete-event simulation of the entire production line.

15.8.2 Off-Line Programming Using Robotics Simulation Software

The traditional "teach" method is by far the most common method of programming a robot. This method is usually satisfactory if the complexity of an application allows a relatively short programming time. With this technique, the robot program is generated by using the robot itself in its production environment. The off-line approach makes it possible to develop robot programs using a computer model. ROBCAD and IGRIP are commonly used robotics modeling and simulation software programs used for off-line programming (OLP) purposes. As exciting as it can be, off-line programming through such software could be an inefficient, expensive, and frustrating experience if precautions are not taken.

The process of generating robot programs using kinematics simulation software has three major phases: (1) preparation, (2) calibration, and (3) program development. The preparation phase is involved primarily with the generation of solid models of the related work-cell components. This phase is completed when a model of the work cell is put together using those component models. In the calibration phase, the differences between the simulated and actual environments are measured and a mathematical approximation of the actual system is constructed using techniques mostly external to the simulation software. In the third phase the actual robot programs are developed using the simulated environment. Clearly, the third phase can start immediately after the preparation phase is completed. However, the calibration phase can help determine special programming requirements that should be considered in developing final robot programs (e.g., additional work locations to compensate for deflections under heavy payloads). Consequently, those three phases follow each other in a typical OLP project.

15.8.3 Case Study: Gear Machining Cell Design

In this study, a simulation model of a proposed robotic operation that interfaced two CNC machines was developed. The operation was part of the gear-making process in an engine and transmission assembly plant. By replacing manual operations with a robotic system, it was expected that the reliability of the system would increase as the cycle time is reduced. A proposal for a robotic work cell was developed and investigated through the models built during the study.

The present process to load parts on machines was labor intensive. By using robots it would be possible to increase the efficiency and reliability of the entire operation. However, the feasibility of a robotic operation should be investigated thoroughly before taking any serious action. Among the issues that must be addressed were:

- What type of robot was required to accomplish the tasks involved in the operation?
- Where should the gantry robot and its supporting facility be placed?
- How should the robots and the existing machinery be interfaced?
- How can the robot movements be optimized to ensure that cycle-time constraints are satisfied?
- What type of gripper is required to handle both parts and containers?

To answer those questions, a model of the work cell was developed (see Figure 15.5 for a snapshot of part of the model) using a robotic simulation software tool with the following objectives:

- To develop a cell layout based on geometrical constraints, work locations, and robot work envelope.
- To develop a conceptual gripper design.
- To determine whether the robot can perform the tasks within the allowed cycle time.
- To demonstrate the initial feasibility of the robotic operation.

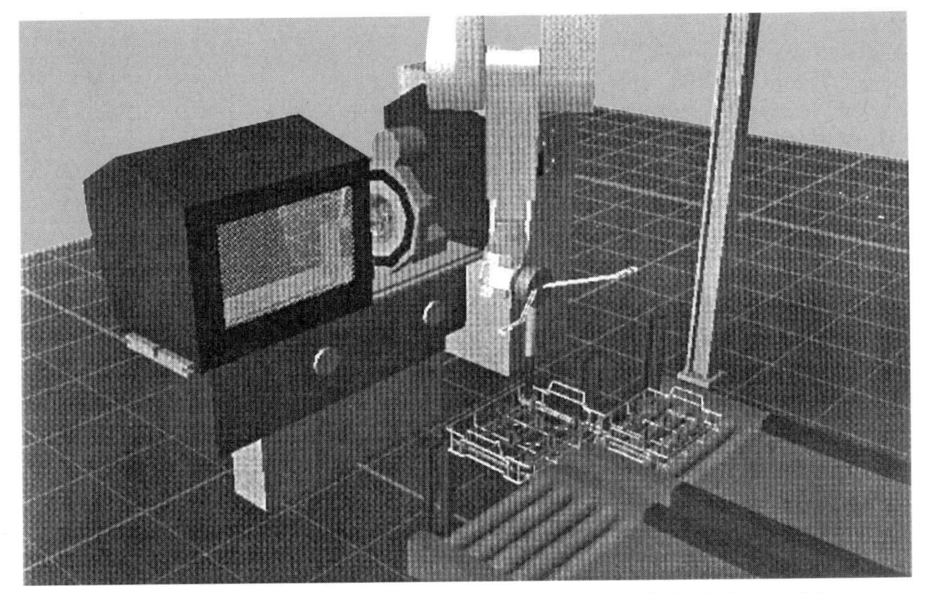

Figure 15.5 Snapshot of a section of the gear loading cell simulation model.

The work cell consisted of one gantry robot, a part container, two CNC machines, and conveyors for inbound and outbound containers. The model allowed experimentation with several parameters, including various gripper designs, and robot movement parameters, such as speed and acceleration. An arrangement for the work cell was developed by considering geometrical constraints and cycle-time limitations. The gripper design was tested and interference were eliminated. Cycle-time analysis showed that the robot could easily load two machines within the allowable cycle time. Consequently, a second gantry requiring a significant investment was avoided. The monetary savings from using simulation were much more than the cost of simulation. Also, by testing the system in a virtual environment, an efficient design was developed by testing many design alternatives in a very short amount of time. Such comparisons would take months if traditional tools of design were to be used. Finally, a videotape of the three-dimensional animation was used to demonstrate the proposed system to plant personnel.

15.9 FUTURE OF SIMULATION IN THE AUTOMOTIVE INDUSTRY

In many ways, the automotive industry was a leader in the application of discrete-event and robotics simulation in the last 35 years. Some of the earliest manufacturing simulators were developed by automotive companies in the 1970s [21] and a number of new applications of simulation are currently being tested in this industry for the first time (e.g., multiple computers running multiple manufacturing simulations in an integrated fashion, combining layout optimization with layout simulation, off-line programming of welding robots). In looking at the next 5 years, one may see the following trends in simulation in the automotive industry: (1) development of more rigorous model-building and validation procedures, (2) creation of model databases for effective model archiving and reuse, (3) development of databases for plant machinery and equipment for model input, (4) development of megamodels incorporating supply chain models with final assembly plant models, (5) use of models at various detail levels by different levels of management in a plant environment concurrently, (6) integration of kinematics models—robot, ergonomic worker with discrete-event models, (7) integration of real-time scheduling with simulation models for more effective shop floor control, and (8) expanding the virtual reality applications in simulation of manufacturing systems.

ACKNOWLEDGMENT

Edward J. Williams, Senior Simulation Analyst, Ford Motor Company, provided valuable suggestions that contributed to the clarity and organization of this chapter.

REFERENCES

1. Fox, J. G. (1991). Effective application of simulation in the life cycle of a manufacturing cell project, in *Proceedings of the 1991 Winter Simulation Conference*, B. Nelson, W. D. Kelton and G. M. Clark, eds., IEEE, Piscataway, N.J., pp. 411–418.
2. Graehl, D. (1992). Insights into carrier control: a simulation of a power and free conveyor through an automotive paint shop, in *Proceedings of the 1992 Winter Simulation Conference*, J. J. Swain, D. Goldsman, R. C. Crain, and J. R. Wilson, eds., IEEE, Piscataway, N.J., pp. 925–932.
3. Gupta, T., and S. Arasakesari (1991). Capacity planning in wood products industry using simulation, in *Proceedings of the 1991 Winter Simulation Conference*, B. L. Nelson, W. D. Kelton, and G. M. Clark, eds., IEEE, Piscataway, N.J., pp. 435–440.
4. Jeyebalan, V. J., and N. C. Otto (1992). Simulation of material delivery systems with dolly trains, in *Proceedings of the 1992 Winter Simulation Conference*, J. J. Swain, D. Goldsman, R. C. Crain, and J. R. Wilson, eds., IEEE, Piscataway, N.J., pp. 916–924.
5. Gunal, A., E. Grajo, and D. Blanck (1994). Generalization of an AS/RS model in SIMAN/Cinema, in *Proceedings of the 1994 Winter Simulation Conference*, J. D. Tew, S. Manivannan, D. A. Sadowski, and A. F. Seila, eds., IEEE, Piscataway, N.J., pp. 857–865.
6. Satyhadev, D., S. Upendram, E. Grajo, A. Gunal, O. Ulgen, and J. Shore (1995). Modeling power and free conveyor systems, in *Proceedings of the Autosimulations Symposium*, Autosimulations, Bountiful, Utah.
7. Upendram, S., and O. Ulgen (1996). Methods for implementing AGV parallel server lane selection rules in automod, paper presented at the Autosimulations Symposium.
8. Gunal, A., S. Sadakane, and J. Shore (1996). Modeling chain conveyors in automod, in *Proceedings of the Autosimulations Symposium*, Autosimulations, Bountiful, Utah, pp. 21–31.
9. Williams, E. J., and I. Ahitov (1996). Integrated use of macro and micro models within a simulation study, in *Proceedings of the AutoFact Conference*, Society of Manufacturing Engineers, Dearborn, Mich., pp. 169–179.
10. Williams, E. J. (1994). Downtime data: its collection, analysis, and importance, in *Proceedings of the 1994 Winter Simulation Conference*, J. D. Tew, S. Manivannan, D. A. Sadowski, and A. F. Seila, eds., IEEE, Piscataway, N.J., pp. 1040–1043.
11. Vasilash, G. (1995). Chrysler's CAD/CAM commitment, *Production*, November, pp. 46–50.
12. Black, J., J. Bulat, B. Colwell, R. Combs, and H. Potvin, Simulation in automotive industry:

today and the next five years, panel session chaired by O. Ulgen, *Proceedings of the Autofact Conference*, SME, Dearborn, Mich., November, pp. 947–961.

13. Berkin, D., M. Brazier, R. Klungle, A. Menawat, and H. S. Na (1995). Management of simulation technology in large companies, panel session chaired by U. Olgen, *Proceedings of the Summer Computer Simulation Conference*, T. I. Oren, L. G. Birta, eds., Society for Computer Simulation, San Diego, Calif., pp. 361–366.
14. Williams, E. J. (1996). Making simulation a corporate norm, in *Proceedings of the Summer Computer Simulation Conference*, V. W. Ingalls, J. Cynamon, and A. V. Saylor, eds., Society for Computer Simulation International, San Diego, Calif., pp. 627–632.
15. Williams, E. J. (1993). Selection of a simulation-service vendor, *Industrial Engineering*, Vol. 25, No. 11, pp. 18–19.
16. Ulgen, O. M., J. J. Black, B. Johnsonbaugh, and R. Klungle (1994). Simulation methodology in practice, Part I: Planning for the study, *International Journal of Industrial Engineering*, Vol. 1, No. 2, pp. 119–128.
17. Ulgen, O. M., J. J. Black, B. Johnsonbaugh, and R. Klungle (1994). Simulation methodology in practice, Part I: Selling the results, *International Journal of Industrial Engineering*, Vol. 1, No. 2, pp. 129–137.
18. Ulgen, O., A. Gunal, and J. Shore (1996). Pitfalls of simulation modeling and how to avoid them by using a robust simulation methodology, in *Proceedings of the Autosimulations Symposium*, Autosimulations, Bountiful, Utah, pp. 21–31.
19. Grajo, E., A. Gunal, D. Sathyadev, and O. Ulgen (1994). A uniform methodology for discrete-event and robotic simulation, in *Proceedings of the Deneb Users Group Meeting*, Deneb Robotics, Inc., Auburn Hills, Mich., pp. 17–24.
20. Ulgen, O., A. Gunal, E. Grajo, and J. Shore (1994). The role of simulation in design and operation of body and paint shops in vehicle assembly plants, in *Proceedings of the European Simulation Symposium*, Society for Computer Simulation International, San Diego, Calif., pp. 124–128.
21. Ulgen, O. (1983). Society for Computer Simulation, "GENTLE: GENeralized Transfer Line Emulation, in *Proceedings of the Conference on Simulation in Inventory and Production Control*, Society for Computer Simulation, San Diego, Calif., pp. 25–30.

CHAPTER 16

Simulation of Logistics and Transportation Systems

MANI S. MANIVANNAN
CNF Transportation, Inc.

16.1 INTRODUCTION

In highly industrialized nations, a sophisticated and widespread transportation system is an inherent need to provide both passenger and freight movements. Due to the unprecedented need for nationwide mobility, there is a requirement not only for various modes of transport but also increasingly sophisticated interfaces between customers, suppliers, and manufacturing and service industries (Wright and Ashford, 1989). Five modes account for all but a fractional percentage of all ton-mileage of freight and passenger mileage of person-travel: motor vehicles (Parsonson and Thomas, 1976; May 1990; Afshar and Azadivar, 1992; Schulze, 1993; Joshi et al., 1995), railroads (Hay, 1977; Atala et al., 1992), air transport (Ashford and Wright, 1984), water transport (Quinn, 1972; Bruun, 1981), and pipelines (Wolbert, 1979).

In this chapter we describe the application of discrete-event simulation techniques in the design, analysis, and management of logistics and transportation (L&T) systems. First, we outline the current approaches to solving various business problems and explain briefly why simulation methodology is appropriate for this industry. Next, various domains and fundamental issues that are important within L&T systems are identified. The entities, resources, and activities to be considered in the simulation model, and factors and responses (output reports) that are essential for strategic, tactical, and operational analysis of L&T systems are described in detail. Throughout the chapter we present readers with some of the common hurdles encountered during application of simulation methodology and implementation of results within the L&T domain.

Handbook of Simulation, Edited by Jerry Banks.
ISBN 0-471-13403-1

16.2 BACKGROUND

16.2.1 Direct and Indirect Resources in L&T Systems

Typically, an L&T system is built on a network composed of one or more terminals or hubs connected by a set of traffic lanes. Accordingly, the networks form hub-and-spoke arrangements and/or direct linkages between origin and destination (Lee and Fishwick, 1995). These networks and the associated topologies have evolved over long periods of time. Hence it is very expensive and often consumes enormous amounts of time and effort to make radical changes in a network. The L&T systems utilize many resources and we can classify them broadly as (1) *direct resources* used in physical transportation of freight (or physical goods) from one geographic location to another, and (2) *indirect resources* involved in sorting, storing, handling, retrieving, and consolidating at the various transit locations known as flowthrough centers, terminals, or hubs.

In a trucking system, trailers, tractors, and drivers are the *direct* (moving) resources, and the dock doors at terminals, refueling stations, fuelers, and maintenance crews are the *indirect* (stationery) resources. The terms *moving* versus *stationery* are used to distinguish the resources that move longer distances either on the road, air, or water versus the resources that tend to stay at one location (although the resource may have its mobility limited to one geographic location). Similarly, in a warehousing and distribution system, the trucks, aircrafts, and cargo ships are the moving resources; the docks, doors, forklifts, carts, storage bins, and racks inside the warehouse are the stationery resources. The moving (*direct*) resources are used to transport freight from plants to warehouses, distribution centers to customer sites, and so on. Again, in an air transportation system, the *direct* resources are aircraft, pilots, and air containers, and the scissor lifts, tug and dollies, forklifts, and hub personnel are the *indirect* resources (Horonjeff and McKelvey, 1983).

It is important that these two types of resources operate together in the most efficient manner for smooth and balanced operation of the entire L&T network. In addition, management and deployment of these resources must ensure the least amount of delay at terminals and hubs, maximum availability and utilization of resources, and on-time pickup and delivery of physical goods. A well-structured scientifically proven approach is required to accomplish these goals.

16.2.2 Challenges of L&T Systems Modeling

For the past several decades, the design, analysis, and control of transport systems were carried out mostly by field engineers (civil, structural, and traffic engineers) and operations research (OR) scientists (Ashford and Covault, 1978; Hamzawi, 1986; Ashford, 1987). A large number of L&T systems have evolved over time and become fairly huge and complex. The primary goals of an L&T business enterprise are to store, distribute and/or transport freight of varying size, form, and shape from its origin to its destination at the lowest cost in order to deliver the right quantities at the right time to its customers who are geographically dispersed; however, the underlying L&T systems that are built to guarantee on-time, damage-free, shortage-free delivery to customers have become extremely complex and often require expensive administrative, information, and decision support systems (Ashford and Clark, 1975).

Conventional L&T planning involves the development of analytical models for trip generation (Moore, 1957), trip distribution (Schneider, 1967), modal split (Wilson, 1969; Smith and Cleveland, 1976), and traffic assignment (Parsonson and Thomas, 1976).

Numerous OR models were developed and applied during the past four decades in the design and configuration of L&T systems (Schiller and Marvin, 1956; Miller 1971; Agerschou et al., 1983; Gibson et al., 1992). In recent years, descriptive modeling of L&T systems has been gaining momentum in transportation companies (Frolow and Sinnott, 1989; Hsin and Wang, 1992; Atala and Carson, 1995; Blair et al., 1995). Computer simulation models are built to evaluate a set of operation policies prior to the implementation of large and complex L&T systems (Nilson and Nicolaou, 1967; Abdus-Samad, 1977; Soros and Zador; 1977).

Major challenges face the analysts in applying simulation technologies to the L&T domain. These can be broadly listed as follows:

- L&T networks are quite complex and involve a very large number of entities and resources.
- Existing simulation software do not support all the modeling/analysis features required.
- There is unfamiliarity of simulation technology in L&T industry.
- Optimization/heuristic methods are widely applied.
- Closed-form solutions are available for many design problems.

However, there are many problem domains within L&T systems, where the simulation approach is best suited if applied properly. For instance, simulation is highly desired for evaluating alternative strategies to operate a terminal (Koh et al., 1994; Manivannan, 1996) or a warehouse. Similarly, the impact of dynamic arrival and departure times of trucks and aircraft at a central hub on time windows and expected service performance can best be understood via computer simulation (Manivannan and Zeimer, 1996).

16.2.3 L&T Problems for Simulation Modeling and Analysis

In general, L&T problems appropriate for simulation studies are divided into three major categories:

1. New design
2. Evaluation of alternative designs
3. Refinement and redesign of existing operations

Accordingly, simulation models in L&T domains are built for the following purposes:

- Models for strategic planning
- Models for tactical planning
- Models for network/traffic control
  - Off-line control
  - Real-time satellite/telecommunication control
- Models for scheduling and dispatching
  - Off-line scheduling
  - Exception handling
  - Real-time monitoring

In this section we outline a list of problems and issues that fall under each of the three categories. Although the list is not exhaustive, it provides critical issues that require effective solution strategies for an L&T business to be successful. The problem areas described in category 1 are solved, in general, using optimization/heuristic approaches. Often, the optimized new L&T designs are verified and validated using computer simulation. The problems outlined in Categories 2 and 3 can be solved using several well-known techniques; however, many L&T businesses tend to utilize simulation modeling and analysis (Kell and Fullerton, 1991; Kelsey and Bisset, 1993; Mabry and Gaudiot, 1994).

1. New Design

- Network design
  - Hub and spoke
  - Direct move
- Terminal/hub planning
  - Number of terminals
  - Location
  - Size (dock dimensions, number of doors)
- Fleet planning
- Route planning
- Least-cost transportation modes

2. Evaluation of Alternative Designs

- Transportation mode alternatives (based on the type of resources)
  - On-the-road (trucks)
    - Relay operations
    - Sleeper operations
  - Rail (trains: single/double stacked)
  - Air (planes, helicopters)
  - Ocean (ships, barges)
- Intermodal alternatives
  - Combine sleeper with rail
  - Combine relay with sleeper
  - Combine rail with relay
  - Combine trucks with air
  - Combine trucks/air with ocean
- Service performance alternatives
  - Overnight service
  - Two-day service
  - Premium service

3. Refinement and Redesign of Existing Operations

- Operational performance analysis
  - On-the-road movements
    - Linehaul, regional, and group operations
    - Terminal operations
    - Operating rules
      - Hooking, unhooking, hostling, and fueling
      - Trailer loading and trailer offloading
      - Dock operations
      - Dispatching
    - Rail movements
      - Loading strategies at the railyard
      - Train timetables
      - Capacity requirements
    - Air transportation
      - Origin terminal operations
      - End-of-line operations
      - Central/distributed hub operations

In the following sections we discuss briefly the application of simulation methodology in four L&T problem domains: (1) simulation of warehousing and distribution systems, (2) simulation of trucking operations, (3) simulation of truck docks, and (4) simulation of ramp operations in aircargo hubs. For each problem domain we describe the (1) simulation modeling issues; (2) purpose of building the model; (3) entities, resources, and critical processes that need special attention; (4) what-if scenarios; (5) input data required; and (6) outputs for statistical analysis.

16.3 SIMULATION OF WAREHOUSING AND DISTRIBUTION SYSTEMS

16.3.1 Purpose of the Simulation Model

A growing number of logistics firms utilize discrete-event simulation concepts to model the various issues of large-scale logistics networks. In one extreme, a logistics simulation model may be developed to investigate and improve the operations of a warehouse; on the other extreme, it may involve modeling and analysis of the operations of an entire supply chain. In most cases there is a common goal for developing the simulation model, which is to evaluate the performance of individual value-adding (*indirect*) resources, facilities, and operations as well as the flow of transportation entities (*direct resources*) between the plants, warehouses, and customers.

The simulation models are developed to perform a variety of what-if scenarios to accomplish the objectives of a logistics network management or its customer. These include (to name a few):

1. To evaluate strategic decisions
   - Warehouse location and allocation
   - Warehouse/distribution center designs
   - Transportation mode analysis
2. To test tactical solutions
   - Inventory management policies
   - Pull ordering between customers and plants
   - Push ordering between warehouses
   - Service levels
3. To identify operation problems on an ongoing basis
   - Changes in transportation modes
   - Changes in warehouse operation parameters
   - Changes in parts and finished products
   - Customer demand fluctuations

16.3.2 Simulation Model Development for a Logistics Application

A simulation model of a logistics network is developed to investigate the impact of the variabilities associated with production schedules, customer demand, and transportation delays. The simulation model must combine the behavior of a physical logistic network with the activities and operations of the various logistics entities within the problem domain. In general, the simulation model may emphasize the internal logistics and operations of a warehouse, or the pickup and delivery of freight within a city or a zone, or the movement of physical goods across an entire country or continent. In this section our focus is to develop a simulation model of a logistics network comprised of plants, warehouses (or distribution centers), and customers located all over the world. The simulation model can be built using several world views and/or paradigms (process interaction, event scheduling, object-oriented , etc.); however, in this section we identify a list of unique processes and activities that require special attention to adequately represent the various components of a logistics simulation model.

Often, logistics simulation models incorporate a geographic map showing the physical relationships among plants, terminals/hubs, warehouses/distribution centers, and customers. It is suggested that activities at the plants, warehouses, and customer locations are separately modeled at appropriate levels of detail. These individual models are then integrated with the underlying logistics network superimposed on a geographic map. Often, a hierarchical modeling approach is preferred to represent the logistics network as well as the operations at the individual nodes (a node may refer to a plant, a customer, or a warehouse). In this way the logistics user/designer can visualize the movement of transportation entities at the map level as well as the operations at the plant or warehouse level.

Figure 16.1 indicates the movement of orders from customers to plants and the products and parts from plants through warehouses to the customers in a logistics system. Various modes of transportation are used to move items between the origin and destination locations. Most manufacturing, retail, and service industries embrace this logistics system. Depending on the level of detail specified to generate the desired results, the

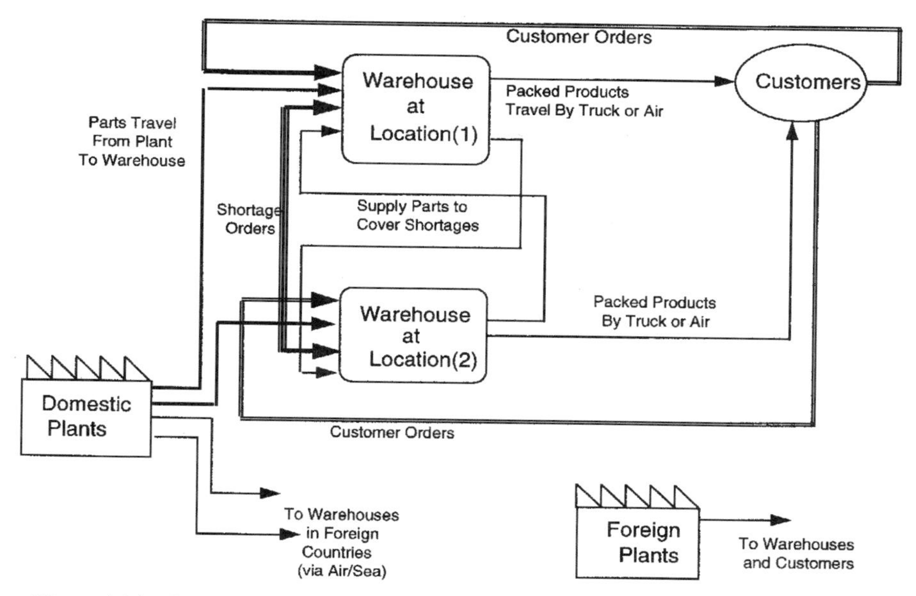

Figure 16.1 Orders and products flow between plants and customers through warehouses.

simulation modeler/analyst may represent some or all of the entities, resources, and activities in a logistics system.

An exhaustive list of processes/activities modeled and represented in a logistics simulation model follows.

- Order processing at the warehouse (manual, EDI)
  - Pull ordering system
  - Push ordering system
- Terminal operations at plants, warehouse, and customers (domestic/foreign)
- Grouping and palletizing
  - At a production plant before shipping to a warehouse
  - At a warehouse before shipping to a customer
- Ungrouping
  - At a warehouse once the parts arrive from a plant
- Transportation mode selection
  - At production plants
  - At warehouses
- Handling shortages (or surplus inventory)
  - Creation at a warehouse (domestic/foreign)
  - Send an order message to another warehouse
- Movement of parts (raw materials, semifinished)
  - From domestic plants to domestic warehouses

- From foreign plants to foreign warehouses
- From domestic plants to foreign warehouses
- From foreign plants to domestic warehouses
- Movement of finished products
  - From domestic warehouse to domestic customers
  - From foreign warehouse to customers abroad
  - Between domestic warehouses
  - Between foreign warehouses
  - Between domestic and foreign warehouses
- Customer orders
  - From U.S. customers to domestic warehouses
  - From foreign customers to warehouses abroad
- Customer locations (geo-coded)
- Transportation mode selection
  - Based on a specified service level
  - Based on the availability of resources
  - Based on shipment priorities (direct)
- Direct shipments from plant to customers
  - Items in shortage
  - Emergency items (a percentage of total shipped)
  - OEM products

16.3.3 Entities and Resources in a Warehousing and Distribution Simulation Model

Entities are physical things whose behavior change over time. Primarily, there are two sets of entities in a warehousing/distribution problem domain. Again, the level of detail depends on the goals and objectives set forth by the decision maker of logistic systems.

- Product-related entities
  - Primary and nonprimary parts
    - Produced in-house
    - Vendor supplied
  - Products (semifinished, finished)
  - Palletized (or grouped) items
    - Parts
    - Semifinished products
    - Finished products
- Information-oriented entities
  - Orders from customers
  - Orders for shortage

The following resources are encountered in the simulation model to fully represent the behavior of a logistic systems. Once again, the level of detail depends on the purpose of the simulation model. Both direct and indirect resources should be considered.

- Trucks
  - Between plants and warehouses
  - Between warehouses
  - Between warehouses and customers
- Airplanes
  - Between plants and warehouses
  - Between warehouses
  - Between warehouses and customers
- Ships and barges
  - Between domestic plants and foreign warehouses
  - Between foreign warehouses
  - Between domestic (foreign) warehouses and foreign (domestic) customers
- Internal (warehouse/plants/distribution centers/customer sites) equipment
  - Forklifts
  - Carts
  - Pallet trucks
  - Conveyors
  - AS/RS systems
  - Guided vehicles

All the static and dynamic entities, together with the resources, must be either fully or partly represented in the simulation model, depending on the level of detail desired by the logistics user(s).

16.3.4 Data Requirements for a Warehouse Simulation Model

In general, the simulation models are developed within this domain to evaluate warehouse locations and transportation modes between plants, warehouses, and customers. The input data required for these models include the following:

- Number of plants
- Number and location of warehouses
- Number of customers
- Customer demand to warehouses
- Part numbers produced at different plants
- Bill of materials
- Transportation times
- Between plants and warehouses
- Between warehouses and customers

It should be mentioned that the customer demand, transportation times, and so on, are stochastic in nature and vary over time. Accordingly, these data elements correspond to probability distributions generated using the information collected over several days, weeks, or months.

16.3.5 Simulation Outputs and Responses

Following is a list of responses that a logistics system designer or user is often interested in knowing about.

- Average utilization
  - Warehouse
  - Trucks
  - Airplanes
  - Other resources inside a warehouse
- Inventory levels
  - Production plants
  - Warehouse
- Transportation delays
  - Between the plant and a warehouse
  - Between a warehouse and customers
- Customer orders
  - Average waiting times at a warehouse
  - Number waiting at a warehouse

In summary, a good understanding of the importance and purpose of building simulation models for logistics applications, together with a list of entities, resources and critical activities that require special attention during the model-building phases, and the key data inputs and output responses required for logistics users is essential to derive valid and useful conclusions.

16.4 SIMULATION OF TRUCKING OPERATIONS

In a trucking industry, the freight may be transported in trailers either directly to its destination with no stops or transported via several intermediate terminals where freight (in trailers) is picked up and/or dropped off. Most trucking companies move freight either in LTL (less than truckload) or in DL (direct load) mode. Accordingly, freight is often transported by straight trucks or by a tractor–trailer from an origin terminal or hub or consolidation center to the destination. A straight truck is one in which the power unit (with the driver cabin) is connected to the trailer and cannot be dislodged. In the case of a tractor–trailer pair, one, two, or three loaded trailers are hooked to a tractor by the hooking personnel (hostlers) at the yard in a terminal. Truck drivers arrive at the terminal at prespecified times to drive the hooked trucks (often referred to as a schedule or truck–tractor–driver set). The number of trailers hooked to a tractor may vary depending on the amount of freight to be transported, the number of trailers available to carry freight, and most important, the geographical location of the origin and/or destination terminal. These concepts are illustrated in Figure 16.2.

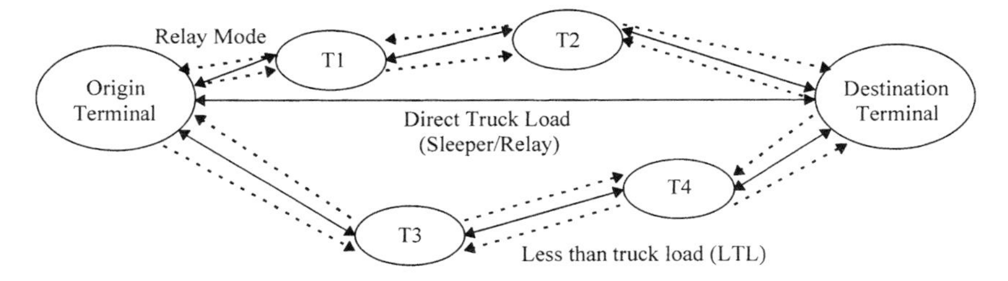

Figure 16.2 Fundamental concepts in a trucking operation.

A set number of loaded trailers are transported by relay, by sleeper, by meet-and-turn drivers, or by rail from each terminal, hub, or consolidation center. Occasionally, empty trailers may be transported, due to load imbalance. Load imbalance may occur when more freight moves in one direction than in another direction. For instance, more automobile components and parts may move from northeast toward midwest sections of the United States, whereas not much freight may move from the midwest toward the northeast, causing a load imbalance. This often leads to more loaded trucks going in one direction and returning empty in the opposite direction. In general, the driver pool is divided into three categories:

1. *Pickup and delivery drivers*, whose primary function is to stay within a city zone.
2. *Relay drivers*, who drive trucks from one terminal to the next for a shorter period (say, 8 to 10 hours) and return back to the origin terminal, which is their domicile location.
3. *Sleeper drivers*, where two drivers alternate driving and drive to farther destinations with minimum number of stops for a longer period.

In relay operations the freight tends to move through several intermediate terminals before reaching its destination. Tractor–trailer sets with LTL and straight trucks with direct loads both move between the origin and destination terminals using either relay or sleeper modes. When a truck driver is at a terminal location that is not his or her domicile point, the driver is at a foreign terminal and is referred to as a foreign driver. At foreign locations, drivers will be provided accommodation to rest for a period of time (say, 8 to 10 hours) before driving back to the domicile location. This depends on the availability of a schedule (and freight). In general, foreign drivers have higher priority over drivers domiciled at that terminal location, so as to reduce the cost of managing the foreign drivers. Once the drivers return to their domicile location, they go home and wait until their next assignment.

To reduce the cost of managing drivers when they are at foreign terminals, trucking companies often set up a pool of *meet-and-turn drivers*. These drivers are used to exchange trucks from two terminals moving toward each other when the total transportation time is within 8 to 10 hours. In such cases the drivers meet at a central point (or a predetermined point such as rest areas along U.S. highways), swap their trucks and turn back to their domicile terminals. Therefore, drivers will start from their domicile terminal and end up at their domicile location.

Once the trucks arrive at the terminal, if it is a straight truck, it will arrive at the yard and wait its turn to offload. If it is a tractor–trailer pair, the set is unhooked and waits at the yard for offloading. The unhooked tractor may be taken for maintenance checks, refueling, and so on. Each terminal has a dock with sufficient doors for the loaded trailers or trucks to arrive at a prespecified door and offload. Offloading may involve use of manual handling or forklifts or other material handling equipment. The simulation modeling and analysis of truck docks has been studied by the terminal management to evaluate the best dock procedures.

In general, truck docks are divided into areas specifically for offloading trucks (or trailers) and reloading trucks or trailers. However, there are truck docks where the offloading and reloading may be planned to occur in the same area to reduce the length of time the freight needs to be staged before it is reloaded for transportation to its final destination.

Essentially, computer simulation is applied to several areas of trucking operations. These are divided into three major topics: (1) dock simulation, (2) terminal simulation, and (3) linehaul simulation. *Dock simulation* refers to modeling and analysis of the activities of dock processes and improve the overall performance of truck docks. *Terminal simulation* refers to modeling and analysis of all activities that take place in the terminal. *Linehaul simulation* combines dock operations and terminal operations at several terminals, either along a single traffic lane or across an entire transportation network. The complexity level increases as we move from dock simulation modeling to linehaul simulation modeling.

16.4.1 Simulation Model Development for a Linehaul Trucking Operation

In this section the application of discrete-event simulation techniques to evaluate and redesign linehaul operations is discussed. This involves the (1) origin and destination terminals, (2) arrival and departure events of trucks, (3) hooking rules and driver assignment procedures, and (4) movement of trucks (tractors, trailers, and drivers) along a lane (Northern Transcon in the United States, for instance).

Fundamental Problems and Modeling Issues. The simulation model is built to perform strategic, tactical, and operational analysis and to address a variety of problems:

- Current methods of hooking trailers
- Current procedures used to assign drivers to schedules
- Load imbalances within any lane at any terminal
- Equipment needs at any terminal in the lane
- Driver needs at any terminal within the lane
- Management of these resources within the lane

This simulation model can serve not only as an operational tool to address these issues within a specific lane but can also be used to serve as an early warning system to provide dock, terminal, and linehaul management with the capability to better control and contend with exceptions that occur during day-to-day operations.

A unified, hierarchical, object-oriented, layered architecture is necessary to represent the linehaul and trucking operations. The modeling objects are designed to be more generic and flexible. Depending on the decisions to be made by the dispatchers or the operations managers, one or more simulation models may be required for an analyst or a decision maker to evaluate the impact of moving trailers by means of relay and sleeper modes, sleeper only/relay only mode, and so on. All the necessary graphical user interfaces should be part of the modeling environment so that the user can quickly modify parameter values during and at the end of simulation.

Modeling Critical Input Processes. To perform trucking simulation, the model must be built to depict the following;

1. *Trailer Closing Times at Docks.* The trailers are loaded with freight each day at each of the origin terminals according to a trailer closing process. The trailer closing times during a day are often defined in the form of a probability distribution. Most L&T companies have either centralized or distributed data processing centers that maintain on-line databases for time and freight information.
2. *Truck Arrival Process at the Docks and Terminals.* The arrival processes associated with trailers, tractors, and drivers from other terminals or hubs that are not part of the lane under study.
3. *Terminal Open Time and Close Time*
4. *Trailers Transported by a Specific Mode*

All these features must be built into the model so that the transportation user or analyst need only to enter the information for each terminal location.

Critical Processes at the Terminals. The trucks arriving at a terminal can be one of three types (Figure 16.3):

1. Trucks that originate from a terminal along a traffic lane under study and end up in a terminal along the same lane.
2. Trucks that originate from a terminal outside a traffic lane, enter the lane, utilize part of the lane, go through one or more terminal(s), and leave the lane and end up in a terminal outside the traffic lane.

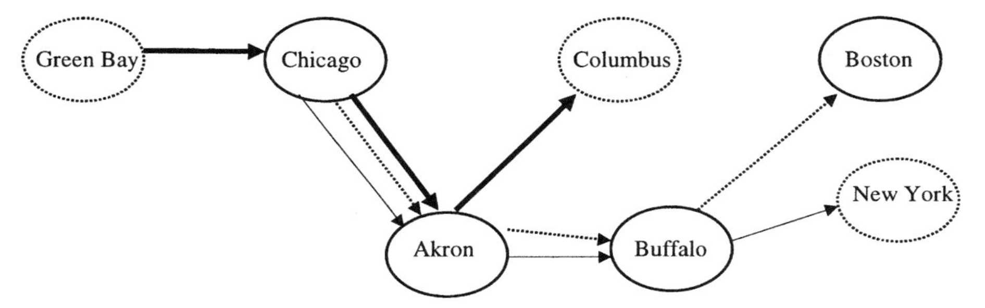

Figure 16.3 Identification of lane-specific trucks and foreign trucks.

3. Trucks that originate from a terminal inside a traffic lane, travel along the lane, utilize all or part of the lane, and leave the lane and end up in a terminal outside the traffic lane.

The terminals connected by solid and/or dashed lines form a traffic lane (under study). The trucks that originate and end up at any of the terminals along the dashed line are *type 1* trucks. Trucks moving along thick solid lines are *type 2* trucks. *Type 3* trucks are those moving along thin solid lines. A better understanding of how these types of trucks (along with drivers) are handled at various terminals is essential to represent the truck arrival and departure process in a simulation model. *Type 2* and *type 3* trucks are often called *foreign trucks*.

Modeling details at each terminal include:

- Creation of loaded trailers (dock process)
- Trailer grouping (hook assignment) rules, depending on the type of truck
- Hooking grouped trailers with a tractor
- Equipment preparation for relay or sleeper runs (hostler, hooking times associated with preparing the equipment)
- Driver assignment and management for relay, meet and turn, or sleeper runs (by employing a prespecified resource pool)
- Schedule arrival process for different types of trucks
- Unhooking trailers
- Offloading at the dock
- Schedule departure process for different types of trucks

Critical Processes at Relay Stations. The major processes and activities performed at the relay stations must be represented in the model, including:

- Driver changes (relay drivers only)
- Refueling, if needed
- Equipment preparation, if needed

Modeling Modes of Trucking Operations. Modeling details associated with the movement of trailers, tractors, and drivers include:

- Sleeper operations only (direct moves)
- Relay operations only (transport via terminals)
- Sleeper and relay operations only (transport via terminals)
- Sleeper, relay, and meet-and-turn operations (direct moves and transport via terminals)

Modeling Intermodal Issues. Often, trucking companies utilize rail to transport their loaded and empty trailers to reduce load imbalances. In this case, terminal operations are closely tied to railyard operations. The various activities related to rail operations (departure at origin and arrival at destination) should be depicted by the model. The critical processes include:

- Move rail-trailers to railyard.
- Hold trailers until train departure times.
- Split trailers at the railyard.
- Move trailers from the railyard to the terminal.
- Strip trailers and count the number of trailers.

Level of Detail Along the Traffic Lane. The simulation model must depict the flow from one or more origin terminal(s) to one or more destination terminal(s). Further, it must be easy to expand the model to handle any number of origin and destination terminals. In other words, the simulation model must encompass a variety of key processes, such as (1) truck arrival process, (2) unhooking tractors and trailers, (3) assignment of foreign and domicile drivers, (4) tractor management, (5) dock process to simulate the off loading and loading of freight from and to trailers, (6) hooking assignment, and (7) hostling and dispatch process. Each process may be parameterized individually to represent the behavior of a terminal, hub, or consolidation center.

As mentioned previously, a hierarchical modeling paradigm is preferred for simulating the trucking operations in order to represent the behavior or the various entities and resources. If object orientation is considered, the model can be built rapidly through the use of a template of objects, simply by placing the consolidation center object on a geographic map. The consolidation centers are linked together by arcs that are either (1) static and/or (2) dynamic links. The static arcs essentially used for offline analysis (i.e., during scenario playing or in strategic studies), whereas the dynamic links between terminals are used to visualize the actual movement of trucks on a lane in real time (primarily for monitoring and control purposes). Each consolidation center object placed on the map is parameterized or modified using the model-building tools to represent the specific behavior or characteristics of the terminal.

Once the simulation model is developed to represent a specific scenario, the model is tested and verified. The scenario under study is simulated for a prescribed time period, say for a period of 3 to 4 weeks of linehaul operations. The results generated is exhibited to the user in the form of visual graphs and/or tabular outputs. These are discussed in Section 16.4.3.

16.4.2 Entities, Resources, and Activities

The following entities, resources, and activities are considered in a typical simulation model. The movement of trailers, tractors, and drivers and linehaul operations at terminals or hubs or consolidation centers are focused on in this section. Sample entities, activities, and resources are provided below.

Entities

- Empty trailers at terminals
- Loaded trailers (through or closed at terminals)
  - Waiting to be grouped with other trailers
  - Waiting to be sent by relay, sleeper, or meet-and-turn (or rail) mode
- Transport operators (sleeper, relay, meet-and-turn drivers)

- Tractors (currently all tractors are considered equal)
- Trains at the railway station (if the rail mode is activated)

Resources

- Hostlers
- Dispatchers
- Fuelers

Activities/Processes That Require Special Attention. The processes that require special attention while developing the simulation model are divided into two categories.

1. If only sleeper, relay, and meet-and-turn operations are considered, the following activities are performed in a model:
   (a) Off load the loaded trailers after a trip at the dock.
   (b) Load the empty trailers at the dock.
   (c) Unhook trailers at a terminal.
   (d) Prepare (hostle) equipment at a terminal.
      - Get service from the hostler to hook/unhook trailers with tractor.
      - Get service at the fueling land or the inspection area.
   (e) Drive the truck to the next location (may be destination) via sleeper mode.
   (f) Drive the truck to the next location (may be destination terminal) via relay or meet-and-turn mode.
2. If the rail mode needs to be modeled, the following activities are performed:
   (a) Move trailers to the railyard.
   (b) Transport the trailers from the railyard to the terminal (no capacity restrictions on the pickup and deliver drivers working within city limits).
   (c) At the destination terminal, split trailers at the railyard and dispatch to the terminal.

Dynamic Processes and Associated Information. Various stochastic processes should be represented in the simulation model. These include:

- Trailer information
  - Trailers that are closed at a terminal on this lane and end up in a lane
  - Trailer closing time distribution (or)
  - Freight creation process and trailer loading time distribution
- Through trailer arrival process
  - Trailers that travel a lane but are not closed at a terminal on a lane
  - Originate from a terminal not part of a lane but end up at a terminal on a lane
  - Originate from a terminal part of a lane but do not end up at a terminal on a lane
  - Originate from a terminal not part of a lane but do not end up at a terminal on a lane

- Scheduled arrival time distribution
- Trailer information for each schedule
  - Lane destination
  - Load destination
- Empty trailer availability information
- Sleeper team arrival process at the origin terminals (call time frequency, etc.)
- Relay (or meet and turn) driver arrival process at the origin terminals
- Tractor availability information
- Other pertinent information related to any other dynamic process

16.4.3 Data Requirements and Outputs from the Model

The simulation model involves the following data for sleeper, relay, meet and turn, and rail (if included in the study) combination along a specific traffic lane under study:

Operation Data

- Hostling times (hooking or unhooking) at each terminal for tractors
- A set of intermediate locations where sleeper (or relay) trucks stop for unhooking and/or hooking and/or refueling
- Fueling/inspection times at a terminal
- Fueling/inspection times at a third-party vendor location
- Tractor service (inspection, etc.) times at terminals
- Current number of hostlers per shift at each of the terminals under consideration
- Current number of sleeper, relay, and meet-and-turn drivers
  - Standard driver roster
    - Arrival times at the terminal
    - Number of drivers
    - Assigned destination
  - Extra driver roster
    - Number of drivers
- Driver work rules and DOT regulations associated with sleepers, relay, and meet-and-turn drivers

Outputs from the Simulation Model. Using the operations data and the simulation model with features specified in Sections 16.4.1 to 16.4.2, several output statistics are collected during a simulation run. Each of these output statistics may be generated on-line as the simulation is in progress or by accumulating the results over a period of time. In the former case it is preferred that the outputs be generated in the form of visual graphs and expected arrival times (ETAs); however, in the latter case, the outputs may be created as cumulative statistics (e.g., average number and/or time, minimum and maximum values, standard deviation, etc.) in the form of tables stored in external files for further analysis and dissemination. Although the specific output reports vary depending on the purpose of the simulation study, some of the most useful outputs generated in a trucking simulation study include the following list:

- Utilization of empty trailers at each terminal
- Empty trailer buildup rate and its usage rate over a period of time
- Utilization of drivers at each terminal (both foreign and domicile) over a period of time
- Tractor utilization at each terminal and tractor buildup rate and usage rate at a terminal over a period of time
- Number of trailers transported by sleeper teams, relay drivers, or meet-and-turn drivers separately at all destination terminals for each day
- Number of trucks on the road over a period of time (both time-persistent statistics and cumulative statistics)
- Number of trailers delayed at each terminal due to the unavailability of tractors
- Delay times for trailers at each terminal due to the unavailability of tractors
- Number of drivers delayed due to the unavailability of schedules at all terminals
- Delay times for the drivers at each terminal due to unavailability of schedules
- Number of tractors delayed due to the unavailability of trailers at each terminal
- Delay times for tractors at each terminal due to unavailability of trailers
- Hostler utilization

Sample Simulation Outputs for Conducting Operational Analysis. Two kinds of outputs may be generated from a simulation model used to design, evaluate, and improve the linehaul operations. These include (1) daily, weekly, monthly behavior; and (2) snapshots at any point in time. In this section, sample output reports generated by the simulation model are presented for further analysis and communication to trucking management.

Among other reports that are essential for strategic, tactical, and operational control, some of essential reports from the simulation model include (1) traffic summary, (2) empty trailer usage, (3) driver by type usage, (4) tractor by type usage, and (5) schedule delays at terminals. The sample outputs are shown in the following tables (the values in tables are provided to illustrate the concepts and are fictitious).

Traffic Summary Report (Daily, Weekly, Biweekly, or Monthly). Table 16.1 provides the throughput information, which can be used as an effective measure to determine the operational performance of the lane being studied.

TABLE 16.1 Traffic Summary

Origin Terminal	Lane Destination Terminal	Load Destination Terminal	Total Number of Trucks Available	Average Number of Trucks in Use
Chicago	Buffalo	Akron	53	14.89
Chicago	Boston	Akron	48	23.56
Chicago	Akron	Akron	64	19.44

TABLE 16.2 Empty Trailer Usage

Terminal Location	Average Number of Trailers Waiting	Average Waiting Time (hours)	Average Percentage Trailer Utilization
Akron	4.5	2.7	89.57
Buffalo	0.5	1.3	90.01
Boston	2.5	2.2	88.76

Empty Trailer Usage Report (Daily, Weekly, Biweekly, or Monthly). This is an important statistics that assists dispatchers at various terminals along a traffic lane to learn about trailer usage at different points in time (see Table 16.2). This report also helps decision makers to identify load imbalances and potential bottlenecks along a traffic lane and take appropriate remedial actions to relieve the problems.

Driver-by-Type Usage Report (Daily, Weekly, Biweekly, or Monthly). This information, shown in Table 16.3, is highly critical to understanding the need for different types of drivers at various locations. Using this statistic the decision maker can evaluate several options to utilize the most critical resource in a linehaul trucking operation: the drivers.

Tractor Usage Report (Daily, Weekly, Biweekly, or Monthly). This information, depicted in Table 16.4, is essential to understanding the need for additional tractors at various terminal locations. Using this statistic the decision maker can evaluate several options to ensure an adequate supply of tractors, thereby keeping a smooth linehaul operation along the traffic lane.

Schedule Delays at Terminals (Daily, Weekly, Biweekly, or Monthly). This information, provided in Table 16.5, is the single most important performance measure for the decision maker to evaluate the alternative policies. The set of policies that lead to the least amount of delay time at various terminals or consolidation centers is ideal for deriving the best performance from a linehaul trucking system.

In summary, the results are generated in the form of tables and visual graphs (both off-line and real-time) to include (1) rate of change of quantities with respect to empty trailers, tractors, and drivers at each terminal due to the new arrival and movement of freight; (2) average utilization levels of empties, tractors, and drivers during a speci-

TABLE 16.3 Driver Utilization

Terminal Location	Driver by Type	Average Number of Drivers Waiting	Average Waiting Time (hours)	Average Percentage Driver Utilization
Akron	Relay only	5	2.5	85.65
Buffalo	Relay and sleeper	2	1.5	74.56
Boston	Sleeper only	3	1.0	35.00

TABLE 16.4 Tractor Usage Summary

Terminal Location	Average Number of Tractors Waiting	Average Waiting Time (hours)	Average Percentage Tractor Utilization
Akron	5	2.2	84.34
Buffalo	2	4.6	76.77
Boston	3	3.5	67.55

fied time period; (3) average number of trucks (or schedules) moved between terminals during a time period; (4) average number of trailers, tractors, and drivers delayed due to unavailability of resources; and (5) average delay times associated with the unavailability of one or more transportation resources.

16.4.4 Benefits of Trucking Simulation

By performing a series of what if scenarios and establishing a gaming environment (frequently set up by large trucking companies in a "war room" format), both terminal management and central planners can gain many benefits. We categorize these benefits as follows:

1. As a strategic tool
   (a) Evaluate and test operational strategies through an interactive, user-friendly *gaming* environment (via simulation and animation) before actually implementing the strategies.
   (b) Determine the load imbalances caused by loaded trailers, drivers, and tractors at various terminals by visual graphs.
   (c) Determine the equipment needs at various terminals, empty trailers, and tractors.
   (d) Evaluate the number of drivers (sleeper/relay/meet-and-turn drivers) and their domicile locations.
2. As a tactical tool
   (a) Estimate the resource requirements (empties, drivers, tractors) on a weekly basis based on the freight movement characteristics at different terminals.

TABLE 16.5 Cumulative Statistics to Evaluate the Overall Performance

Origin Terminal	Lane Destination Terminal	Maximum Number of Trucks Delayed	Average Number of Trucks Delayed	Average Delay Time (hours)
Chicago	Buffalo	12	1.49	3.49
Chicago	Boston	7	1.05	1.25
Chicago	Akron	4	1.66	3.20

3. As an operational tool
   (a) Manage resources, reduce the delay times at terminals, and cut down penalty pay by performing simulation experiments on a daily basis.
   (b) Handle exceptions through an early warning capability, thereby accomplishing better control. This is done by visualizing the progress of freight over time, tractor–trailer–driver movements among terminals, hostling and hooking/unhooking, and expected arrival times and departure times of schedules on one or more computer screen(s).

16.5 SIMULATION OF TRUCK DOCK OPERATIONS

Application of computer simulation techniques to model, analyze, and improve the performance of a truck dock has been gaining enormous impetus in recent years. A typical truck dock includes five major components: (1) a yard where trucks arrive and wait for an empty door; (2) doors; (3) transporters operating at the dock, such as forklifts and conveyors; (4) staging areas where the freight is temporarily stored prior to sorting and shipping; and (5) sorting systems. Trucks arrive at the yard as straight trucks or as tractor-trailer combinations. Different types of freight from the trucks are offloaded by forklifts at the doors. Depending on the freight type and its destination, they are moved either directly into another truck or staged on the dock for reloading to trucks at a later time. In most situations, a truck dock is nothing but a flowthrough center.

Primarily, forklifts move the freight from the trucks to various locations. As trucks arrive at the yard, they are assigned to doors depending on (1) the current trailers being offloaded or reloaded at the dock, and (2) the availability of dock handlers and forklifts. Only one truck is assigned to a door and offloading/loading begins immediately. The pickup and dropoff times for the forklifts depend on the freight being offloaded. As the size and shape of each type of freight vary, there is greater variation in the offloading/reloading behavior. Further, the forklift speeds, movement times, forklift turning behavior, and the level of congestion at the dock change depending on the freight type.

Several work rules are used to assign, manage, and control forklifts during the offloading or reloading process. Forklifts are assigned to a specific door or a set of door(s). During the offloading process, the forklifts may transfer loads to other trailers being loaded or to a staging area. However, they tend to stay with the doors assigned until the current truck is fully offloaded.

The staging areas are commonly located between the dock doors. These are among the most congested areas, where loose freight is offloaded from all doors by forklifts. This means that forklifts need to travel to this area more often than to other areas of the truck dock. The size, number, and exact location of staging areas play a vital role in determining the overall dock space, as do the number of doors and assignment of doors to trucks arriving from an origin terminal and departing to destination terminals. Figure 16.4 shows the facility layout of a simple truck dock in which the inbound trucks arrive at one end of the dock and depart at the opposite end. In other configurations, the inbound and outbound trucks are assigned adjacent to each other on the same side of the dock.

Often, two fundamental objectives are set forth for the simulation study: (1) studying the ways and means of increasing the throughput of a truck dock, and (2) studying the impact of changes in a facility design before implementing it, to avoid making very

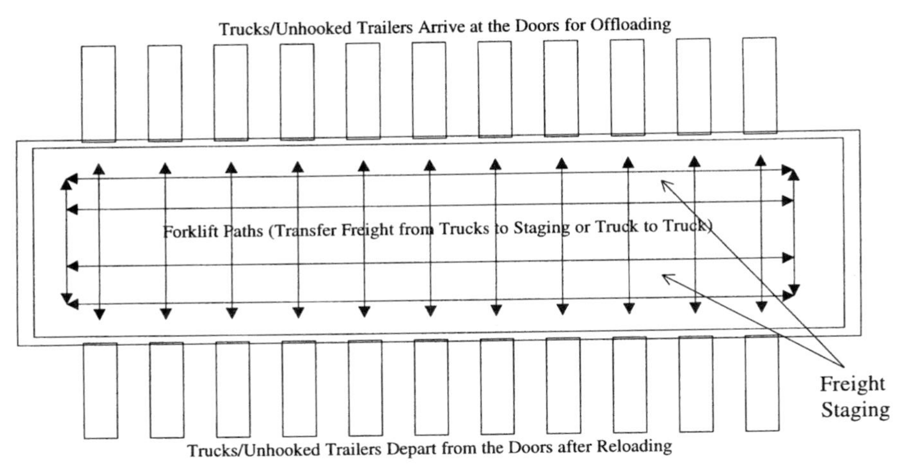

Figure 16.4 Truck dock showing the offloading and reloading processes.

costly errors. Although hundreds of what-if scenarios using a truck dock simulation model are investigated, those most frequently studied ones are:

1. To develop a simulation model that has the ability to show operation managers the existing processes, trouble spots, and changes in behavior of the existing truck dock as it undergoes changes in terms of new equipment, facility, and operations.
2. To study the impact of changes in processes and policies within each area of the truck dock.
3. To change the distances traveled and capacity/size/number of forklifts and determine the impact on performance at the truck dock area.

16.5.1 Input Data for Truckdock Simulation

For the purpose of building a truck dock simulation model, two types of input data are collected.

1. The first type involves freight types, door information, resources used, and equipment specifications.

- Door information (number of doors, door size, door locations)
- Freight types [containerized, bulk (sorted/unsorted), loose (hand-carried, top loads) freight]
- Resources [doors, forklifts, sortation conveyors (indexing)]
- Operators (forklift drivers, freight handlers, dock workers, loaders/unloaders)
- Queue locations (staging, sortation conveyors, door locations)
- Forklift specifications (number and types, length and width of forklifts by type, acceleration and deceleration, top speeds, forklifts by type, loaded speed by freight type, unloaded speed by forklift type, park locations, work assignment, turning rules, pickup and deposit rules)

- Conveyor specifications [conveyor type (indexing, roller, belt, etc.), number of segments, input station characteristics, output station characteristics, segment speed, conveyor selection rules, freight movement/stopping rules]

2. The second type of data involves rigorous data collection at the truck dock. This includes statistical distributions describing the behavior of truck dock operations, which change from day to day or night to night. These include the truck (by type) arrival process at different locations of the truck dock; pickup and dropoff times for forklifts based on freight type; operational times at the sort conveyor, staging, and so on; and the truck contents.

- Arrival process
- Type of trucks based on freight carried
- Frequency of truck arrivals (type of truck and Percentage of occurrence)
- Truck contents (freight type and Statistical distribution w/parameters)
- Truck hostling time distributions
- Door preparation time before offloading (when the truck is full)
- Door occupancy time by trucks after offloading (when the truck is empty)
- Loading/offloading times
- Forklift pickup time by freight type (at staging area, sortation conveyor, and inside trucks)
- Forklift set-down time by freight type at destination points

Both types of information are incorporated in the simulation model to describe the truck dock as well as day-to-day variabilities.

16.5.2 Controllable Factors and Performance Measures

In truck dock operations, several factors are controllable, and the decision maker is often interested in knowing about their impact on the output performance. In this section a list of factors and responses (output reports) usually generated using a truck dock simulation model are outlined. The essential factors are:

1. Number of dock doors
2. Number of forklifts
3. Number of operators
4. Number of trucks
5. Forklift operating speed

Similarly, the responses are (1) capacity, measured in number of pieces moved; (2) average number of trucks unloaded; (3) forklift utilization, measured in percentage of time the equipment is busy; and (4) average time to move a piece across a dock measured in units of time and on an hourly basis. The simulation model should be built to provide answers to many issues involving these factors. The report generator should be designed to provide all the responses listed in this section.

16.5.3 Simulating Critical Processes in a Truck Dock

Several processes and activities need to be simulated to represent the behavior of the truck dock operations. An exhaustive list of the fundamental processes that require special attention during the simulation model building phase follows.

1. Determine door assignments to trucks (based on arrival behavior) at the unloading (inbound) doors.
   - *Option 1:* Assign trucks to doors starting from left to right.
   - *Option 2:* Assign trucks based on origin or destination terminal to the preassigned door.
2. Determine door assignments to trucks (based on departure process) at the loading (outbound) doors.
   - Employ a round-robin method of door assignment, if not waiting at the yard
   - Send trucks directly to next-available door, if not waiting at the yard
   - Send trucks to doors based on origin or destination terminal
   - Send trucks to a common waiting area and to next-available door
3. Capture door arrival process.
   - Enter the dock door area for offloading.
   - Consider truck/door preparation time.
   - Read the information pertaining to the type of truck.
   - Read the information pertaining to the contents of the truck.
   - Wait for one or more forklift(s) to offload the truck contents.
4. Specify doors to forklifts
   - Assign each forklift to its door location for offloading or reloading.
5. Assign, schedule, move, and control forklifts for offloading trucks
   - Move forklift into truck to begin offloading.
   - Determine the freight picked up for transport.
   - Keep the fork forward/backward, depending on freight type.
   - Transport forklift to freight destination.
   - Drop off freight at its destination.
   - Move forklift back to its door.
   - Repeat until the entire truck is offloaded.
6. Assign, schedule, move, and control forklifts for loading trucks
   - Determine the freight picked up for loading.
   - Pick up freight from either staging area or a truck.
   - Keep the fork forward/backward, depending on freight type.
   - Transport forklift to the truck depending on freight destination.
   - Move forklift into truck to begin loading.
   - Drop off freight at its destination inside the truck.
   - Move forklift back to staging area or truck where freight is offloaded.
   - Repeat until the entire truck is offloaded.

7. Pickup and dropoff times for truck dock forklifts
   - Specify the pickup and dropoff times by freight type.
8. Select a sortation conveyor (if required)
   - Specify a list of segments for each door using one of the selection rules.
     - Nearest-neighbor rule
     - Next-available-segment rule
     - Waiting rules if no segment available
   - Select an appropriate conveyor to offload or pickup for reload.

The processes described in this section are representative of a typical truck dock and there may be some variations depending on the function, size, and purpose of the truck dock.

16.5.4 Experimental Setup to Perform What-If Scenarios

One of the fundamental reasons for developing a truck dock simulation model is to evaluate its behavior and make necessary strategic changes to facility design, work rules, and/or operational characteristics. The simulation model is usually linked with an experimental setup. These setup programs are developed to include many predesigned truck dock configurations. Occasionally, it may be necessary to make changes to both the model and the input data prior to conducting a prespecified set of simulation experiments (or scenarios). However, in the majority of cases, changes are made with the input data (and not with the model). Essentially, the experimental setup for the truck dock simulation model is developed to facilitate this function as well as to keep track of all the simulation results for further analysis. This, in essence, helps to keep the simulation model as a blackbox and make intelligent decisions. During strategic design and analysis, each experiment for a scenario is executed 10 (to 30) times and the average values for each of the responses are computed to generate statistically valid conclusions. This is mainly because the operational behavior of the truck dock varies nightly (stochastic in nature) and one replication of the simulation model is not sufficient to make useful recommendations to truck dock management.

16.5.5 Analysis Using Truck Dock Simulation Results

In this section, sample analyses and scenarios generated for improving the performance of a truck dock are discussed. As mentioned earlier, a simulation analyst may virtually perform many what-if scenarios using the truck dock model. However, a few representative case examples that are critical and often studied by a truck dock management team are described.

Impact of Eliminating Certain Truck Types at Inbound Dock. Table 16.6 shows the impact of eliminating certain types of trucks on the performance of a truck dock. For instance, the elimination of trucks carrying a certain type of freight (say, containerized freight) from the inbound doors during busy periods may affect the *average trip time* for forklifts. However, it may not affect greatly the *average number of trucks unloaded* during a certain time period. Similarly, the removal of trucks carrying bulk freight for an origin–destination pair from the doors during busy periods may affect the *average number of trucks unloaded.*

TABLE 16.6 Impact of Eliminating Certain Classes of Trucks

Description	Average Process Time at the Dock (minutes)	Average Number of Trucks
Trucks carrying no containerized freight	88.407	—
Trucks carrying containerized freight	93.082	—
Trucks carrying bulk parcels	—	29
Trucks carrying no bulk parcels	—	36

Impact of Additional Doors. Another interesting scenario involves the impact of changes in the total number of doors at the inbound and outbound ends of a truck dock. In such cases the decision maker is interested in knowing about the average increase in truck dock capacity, forklift utilization (one forklift per door), average trip time, and so on. Table 16.7 summarizes the estimated *forklift utilizations*, *average trip times*, and *average number of trucks unloaded* for a fictitious truckdock with 18-door (existing) and 40-door configurations, respectively. *New end* refers to the outputs associated with the additional 22 doors only.

Impact of Forklift Assignment Policies to Doors. The assignment of number of forklifts per door at the inbound and outbound ends of a truck dock is critical to operate and manage a truck dock effectively. This simulation study often leads to identifying the work rules on the dock, the number of forklifts and dock workers, and the best scheduling policies. Let x_1 denote the number of forklifts assigned to each door at the inbound side of the truck dock, and let x_2 denote the number of forklifts assigned to each door at the outbound side of the truck dock. A total of four scenarios are possible, depending on the number of forklifts assigned per door at the inbound and outbound ends as shown in Table 16.8.

For each of these four cases, the truck dock simulation model may be set up and executed for a prespecified number of replications. Several responses, including the average throughput, average utilization of forklifts at inbound and outbound ends, and average operation times, can be collected in each case. The results may be compared to estimate the best procedures for forklift assignment at the inbound and outbound ends of a truck dock.

In essence, the truck dock simulation models are set up to perform operation plan-

TABLE 16.7 Estimated Forklift Utilization, Average Trip Times, and Trucks Offloaded/Loaded

	Forklift Utilization (%)		Average Trip Time (minutes)		Average Number of Trucks	
Doors	18 Doors	40 Doors	18 Doors	40 Doors	18 Doors	40 Doors
Inbound	63.742	59.705	88.407	96.536	34.387	14.9
Outbound	48.740	53.867	113.506	112.142	12.00	12.0
New end	—	38.141	—	102.207	—	35.5

TABLE 16.8 Four Forklift Assignments to Inbound and Outbound Doors

Option	Number of Forklifts/Door (Inbound), x_1	Number of Forklifts/Door (Outbound), x_2
1	1.0	0.5
2	1.0	1.0
3	2.0	0.5
4	2.0	1.0

ning on an ongoing basis in a variety of logistics, trucking, and other transportation applications. Following is a list of operations planning issues studied on an ongoing basis for a given truck arrival and departure process:

- Peak and lean hours during an operating period (day, week, or month)
- Number of doors required and forklift assignment policies
- Time window to complete offloading, staging, and reloading all the trucks
- Amount of workload expected at all areas that may receive freight during a time period

16.6 SIMULATION OF RAMP OPERATIONS IN AN OVERNIGHT AIR CARGO HUB

In this section we describe the design, development, and testing of a simulation model depicting the operations of ramp operations at a central hub in an air cargo company. During the past two decades, several air cargo operations and airport terminals (Hart, 1986) have been modeled (Thompson, 1964; Stafford and Stafford, 1969; Walton and Rosenbloom, 1977), analyzed, and redesigned (Harris, 1974; Crawford, 1977; Horojeff and McKelvey, 1983; Hamzawi, 1986). However, a growing number of air cargo companies have recently begun to apply computer simulation techniques. The major benefits of an air cargo simulation model are as follows:

1. On an ongoing basis, pinpoint strategic and operational improvements on current plane offloading and reloading processes, equipment use, and facility layout due to changes in the behavior of the freight handled at the hub. This helps decision makers to improve continuously the productivity and throughput of air cargo operations.
2. Determine a priori, the appropriate levels of capital equipment, equipment needs, and a suitable facility layout due to an increase (or decrease) in the number of planes, changes in the amount and/or characteristics of freight, and unexpected exceptions. This helps decision makers to better utilize the investment dollars before spending on new equipment purchases.
3. Utilize the simulation model, on a daily (or nightly) basis, to estimate the ramp completion time for the number of planes specified and their expected times of arrival (ETAs).

Often, a simulation model built for studying the air cargo ramp operations utilizes the aircraft tail sheet information, chronological order of plane arrivals and departure data, and ATC rules that designate gates for arriving aircraft. A variety of material handling equipment, such as tug and dolly (to transport freight in containers), forklifts, and K-loaders, are used to load and offload aircraft on the ramp. The number of dollies per tug, number of tugs per plane, and total number of tugs on the ramp are design factors, and changes in these design factors affect the operation performance. The major goals for a ramp simulation model are as follows:

1. To develop a simulation model that has the ability to show ramp operations crew the problems, trouble spots, and changes in behavior of the existing ramp operations as the air cargo hub undergoes changes in terms of additional/modified K-loaders, tugs and dollies, and other auxiliary equipment.
2. To study the impact of changes in operations and policies within each area of the ramp.
3. To perform a comparative analysis: current versus new layout of aircraft parking, current versus new communication procedures, current versus new tug and dolly assignment rules, and so on.
4. To study the impact of additional aircrafts and gates on the ramp and determine the number of K-loaders and tug and dollies. Identify the changes to work rules, if any, to maximize throughput and number of planes/containers offloaded at the ramp and to minimize the average time to offload an aircraft.
5. To study the impact of changing the number of dollies per tug and number of tugs per plane on the average time required to offload a plane. Study the impact of these operation parameters on the varying number of planes that arrive at the hub during a night.

Items 1 through 5 are critical problems that require in-depth analysis before making valid recommendations to hub management. Often, an extensive statistical analysis is conducted using one or more simulation model(s) to derive the necessary conclusions.

16.6.1 Static and Dynamic Data

The ramp simulation model requires two types of input information. The first type involves the technical specifications associated with the freight, planes, gates, tugs, dollies, forklifts, and sortline conveyor. In addition, it includes material flow behavior, work rules, park location assignment rules, control rules for tugs and forklifts, and other factors. In general, a ramp simulation model includes the following representative data:

- Number of gates
- Ramp maps
- Technical specifications of sortline conveyors, tugs, dollies, and forklifts
- Plane types
- Flight information
- Tail sheet and chronological arrival/departure information
- Number of container positions that each plane type can carry
- Number of bellies

- Number of containers each belly fills up by plane type
- Number of dollies per tug
- Team makeup for offloading/reloading containers from/to a plane
- Number of tugs and dollies
- Number of forklifts designated for offloading and reloading a plane
- Sortation conveyor locations and physical characteristics
- Plane parking location rules
- Runway behavior

The second type of information involves a set of input distributions to represent the day-to-day (or hour-to-hour) variabilities found in the ramp operations. It includes the plane arrival process at different locations of the ramp, pickup times and dropoff times for tugs and forklifts based on freight type and number of containers, operation times at the conveyor, freight profile, and container contents in the plane.

16.6.2 Factors and Responses

It is important to identify a list of controllable factors and responses to evaluate alternative ramp designs. This helps immensely during the model-building stage if the simulation model has been built to represent all the necessary factors and responses. If the model does not incorporate these factors and responses, the decision maker can hardly use the model to study air cargo operations. Again, these factors and responses are decided based on a list of questions that the hub management is interested in getting answers for. For the ramp simulation model, the major factors are as follows:

- Number of planes
  - Flight profile
  - Plane arrival-time distributions
- Total number of tugs
- Number of tugs per plane
- Number of dollies per tug
- Number of K-loaders

The responses that are of interest include the following:

- Plane offload/load capacity
  - Total number of containers offloaded or loaded
  - Total number of planes offloaded or loaded
- Average time to offload/load an aircraft
- Completion time for offload/load all aircraft on the ramp

16.6.3 Simulation Model Development

A ramp simulation model should be built to interface with input data files and variables that can easily be modified to perform what-if scenarios. The model is frequently developed to represent the following critical processes that occur on a ramp:

- Plane arrival/departure process
- Gate assignment of planes
- Freight creation and movement at the aircraft
- K-loader/forklift/tug interface
- Ramp map for plane gates, container waiting, and tug parking areas
- Tug and dolly movement system
- Forklift movements
- Tracking freight by type (e.g., huts, palletized, belly freight)
- Sortation conveyors

In the simulation program, the plane arrival process is often implemented with a rule base to mimic the tower rules and parking procedures used to park the planes. The rule base utilizes the flight profile to get all pertinent information on arriving flights during a night (or day). It is important that the rule base be tested using many different flight profiles and arrival times to ensure that all arriving planes are parked at the gate locations specified.

In general, the simulation model is built with a ramp map to include the tug and dolly movement paths, the parking locations for planes and tugs, waiting areas for tugs, interaction zones for transfer forklifts to pick up containers from the tugs, and intersection blocks to prevent collision between the tugs and link blocks to keep sufficient space between tugs.

With respect to the tug and dolly system, it is preferred that the ramp simulation model be designed to depict a very explicit and in-depth representation of their behavior to perform a rigorous analysis of equipment usage. Often, a smart algorithm is designed to incorporate the plane assignment rules to tugs and dollies (belonging to a team), tugs per plane rules, tug waiting rules, offloading crew rules, collision avoidance rules, right-of-way rules, parking rules, and passing rules. In addition, the algorithm may use tug specifications such as acceleration; deceleration; loaded and unloaded speeds; and forward, reverse, and curve speeds to determine precise travel times between the aircraft and transfer areas.

The simulation logic development is one of the most important steps. The ramp model (essentially, a computer program) includes the logic and data interfaces. If the logic is incorrect or inaccurate, the output (responses) from a model cannot be utilized to make intelligent decisions (Law and Kelton, 1991). Sufficient time and effort should be spent in the design and development of the model logic to ensure quality outputs as well as flexibility to extend the model to conduct many what-if scenarios on an ongoing basis at the air cargo hub.

16.6.4 Report Generator

Often, it is suggested that a ramp simulation model be implemented with an interactive built-in output processor to create many useful on-line reports for use in making strategic and operational decisions. The results generated during each run may be stored in output files for further analysis and hub management review. The output reports from a ramp simulation model include the following information:

- The total number of containers offloaded during an hour, total number of planes offloaded during an hour, average time to offload a plane, and completion time for offloading all the planes that arrived during a night (or a day). This helps decision makers to determine exactly when the ramp completes its offloading activity.

- Offloading or reloading times for 10 planes, 20 planes, 30 planes, and so on. This provides the time at which the top side and belly are fully offloaded or reloaded. This helps the decision maker to understand the progress of ramp process during the operation window and accordingly, to determine the lean and peak periods.
- Gates for all arriving planes as created by the simulation model. This information can be utilized to park planes every day or night.
- Actual time at which each aircraft is fully offloaded or reloaded.
- Offloading time (the total time it took to offload the plane since its arrival at the park location) for each aircraft that arrived during a night (or day). Information on the aircraft reloading time for each plane is generated similarly. This information assists the decision maker to determine trouble spots during ramp operations involved in plane offloading and reloading.

16.6.5 What-if Scenarios Using Ramp Simulation Model

The ramp model can be utilized to simulate and analyze the impact of many alternative approaches to assigning tugs to aircraft and the equipment/facilities used for offloading. In addition, several critical issues can be investigated using the simulation model to study the impact of:

- Changes in number of tugs per plane
- Changes in number of dollies hooked to each tug
- Changes in total number of tugs used on the ramp
- Changes in aircraft arrival process/departure process
- Changes in number of K-loaders
- Changes in number of transfer forklifts
- Changes in work rules
- Impact of changes in the tug assignment to aircraft
- Impact of changes in the communication system used to manage tugs, K-loader, etc.

In each case the ramp simulation model can be set up to run a series of experiments, and the results discussed in Section 16.6.4 can be generated to perform in-depth statistical analyses. Based on the analyses, valid recommendations can be made to hub management to improve the operational performance of the ramp.

From the foregoing discussions it is inferred that the modeling and analysis of overnight air cargo operations pose a greater challenge to simulation experts. It is evident that a carefully designed simulation study leads to operational improvements, processing time-window reduction, and substantial cost reduction. Hence it is highly recommended that detailed simulation models combined with good experimental designs and rigorous statistical analysis procedures are set up to generate the best ramp configuration(s) and implementable operational design(s).

16.7 L&T SIMULATION SOFTWARE

Today, many commercial software packages are being employed by L&T industries, depending on the level of complexity and size of the problem investigated. These software tools range from standard linear programming packages such as LINDO, CPLEX, and OSL to special-purpose software shells such as INSIGHT, SUPERSPIN, and CAPS which are built to provide decision support in a wide range of L&T domains. With respect to commercial simulation software, a large number of vendors provide packages, built on a variety of world views and hardware platforms, that focus on modeling and analysis of simple material handling systems to complex flowthrough centers and transportation networks. These software packages provide both animation and statistical analysis capabilities for L&T domain experts to fully represent a variety of entities, resources, and critical processes. These simulation packages include Arena, AUTOMOD II, GPSS/H, MODSIM III, PROMOD, and SIMPLE ++, among many others.

16.8 CONCLUSIONS

As the degree of industrialization of an economy increases, there is a shift in preponderance from basic manufacturing industries, sometimes referred to as primary industries, to the service industries, which are secondary, tertiary, and quarternary in character. The primary industries have a greater need for freight transportation, and the existing L&T systems will continue to grow bigger and bigger and become more and more complex. To build transportation systems that are efficient, easy to operate and manage, and still cost-effective, it is crystal-clear that L&T companies will have to invest their time, money, and other resources in scientific and structured approaches for many years to come. This means that applications of mathematical modeling and numerical solution techniques such as simulation will continue to grow in L&T companies.

REFERENCES

Afshar, N., and F. Azadivar (1992). A simulation study of traffic control procedures at highway work zones, *Proceedings of the 1992 Winter Simulation Conference*, J. J. Swain, D. Goldsman, R. C. Crain, and J. R. Wilson, eds., IEEE, Piscataway, N.J.

Agerschou, H., H. Lundgren, and T. Sorensen (1983). *Planning and Design of Ports and Marine Terminals*, Wiley, New York.

Ashford, N. J. (1987). Level of service design concept for airport passengers, *Transportation Planning and Technology*, Vol. 12, No. 1.

Ashford, N. J., and J. M. Clark (1975). An overview of transport technology assessment, *Transportation Planning and Technology*, Vol. 3, No. 1.

Ashford, N. J., and D. O. Covault (1978). The mathematical form of travel time factors, *Highway Research Record 283*, Highway Research Board, Washington, D.C.

Ashford, N., and F. M. Holloway (1972). Validity of zonal trip production models over time, *Transportation Engineering Journal of ASCE*, December.

Ashford, N. J., and P. H. Wright (1984). *Airport Engineering*, 2nd ed., Wiley-Interscience, New York.

ATA (1970). *Shipper-Motor Carrier Dock Planning Manual*, Operations Council, American Trucking Associations, Washington, D.C.

Atala, O. M., and J. S. Carson (1995). A train operations simulation for Miami's SR 836 corridor, in *Proceedings of the 1995 Winter Simulation Conference*, C. Alexopoulos, K. Kang, W. R. Lilegdon, and D. Goldsman, eds., IEEE, Piscataway, N.J.

Atala, O. M., J. C. Brill, and J. S. Carson (1992). A general rapid transit simulation model with both automatic and manual train control, *Proceedings of the 1992 Winter Simulation Conference*, J. J. Swain, D. Goldsman, R. C. Crain, and J. R. Wilson, eds., IEEE, Piscataway, N.J.

Blair, E. L., F. P. Wieland, and A. E. Zukas (1995). A distributed simulation model for air traffic in the National Airspace System, in *Proceedings of the 1995 Winter Simulation Conference*, C. Alexopoulos, K. Kang, W. R. Lilegdon, and D. Goldsman, eds., IEEE, Piscataway, N.J.

Bruun, P. (1981). *Port Engineering*, 3rd ed., Gulf Publishing, Houston, Texas.

Crawford, H. R. (1977). Lake Erie Airport study, *Transportation Engineering Journal of ASCE*, Vol. 103, No. TE2.

Frolow, I., and J. H. Sinnott (1989). National Airspace System demand and capacity modeling, *Proceedings of the IEEE*, Vol. 77.

Gibson, R. R., B. C. Carpenter, and S. P. Seeburger (1992). A flexible port traffic planning model, *Proceedings of the 1992 Winter Simulation Conference*, J. J. Swain, D. Goldsman, R. C. Crain, and J. R. Wilson, eds., IEEE, Piscataway, N.J.

Hamzawi, S. G. (1986). Management and planning of airport gate capacity: a microcomputer-based gate assignment, *Transportation Planning and Technology*, Vol. 11, No. 3.

Harris, R. M. (1974). Models for runway capacity analysis, *Report MTR-4102, Rev. 2*, The Mitre Corporation, Washington, D.C., May.

Hart, W. (1986). *The Airport Passenger Terminal*, Wiley-Interscience, New York.

Hay, W. W. (1977). *An Introduction to Transportation Engineering*, 2nd ed., Wiley, New York.

Horonjeff, R., and F. X. McKelvey (1983). *Planning and Design of Airports*, 3rd ed., McGraw-Hill, New York.

Hsin, V. J. K., and P. T. R. Wang (1992). Modeling concepts or intelligent vehicle highway systems (IVHS) applications, *Proceedings of the 1992 Winter Simulation Conference*, J. J. Swain, D. Goldsman, R. C. Crain, and J. R. Wilson, eds., IEEE, Piscataway, N.J.

Joshi, S. S., A. K. Rathi, and J. D. Tew (1995). An improved response surface methodology algorithm with an application to traffic signal optimization for urban networks, in *Proceedings of the 1995 Winter Simulation Conference*, C. Alexopoulos, K. Kang, W. R. Lilegdon, and D. Goldsman, eds., IEEE, Piscataway, N.J.

Kell, J. H., and I. J. Fullerton (1991). *Manual of Traffic Signal Design*, Institute of Transportation Engineers, Washington, D.C.

Kelsey, R. L., and K. R. Bisset (1993). Simulation of traffic flow and control using fuzzy and conventional methods, in *Handbook of Fuzzy Logic and Control*, Prentice Hall, Upper Saddle River, N.J.

Koh, P. H., J. L. K. Goh, H. S. Ng, and H. C. Hg (1994). Using simulation to preview plans of a container port operations, in *Proceedings of the 1994 Winter Simulation Conference*, J. D. Tew, S. Manivannan, D. A. Sadowski, and A. F. Seila, eds., IEEE, Piscataway, N.J.

Law, A. M., and W. D. Kelton (1991). *Simulation Modeling and Analysis*, 2nd ed., McGraw-Hill, New York.

Lee, J. J., and P. A. Fishwick (1995). Simulation-based real-time decision making for route planning, in *Proceedings of the 1995 Winter Simulation Conference*, C. Alexopoulos, K. Kang, W. R. Lilegdon, and D. Goldsman, eds., IEEE, Piscataway, N.J.

Mabry, S. L., and J. L. Gaudiot (1994). Distributed parallel object-oriented environment for traffic simulation (POETS), in *Proceedings of the 1994 Winter Simulation Conference*, Orlando, Fla., December, J. D. Tew, S. Manivannan, D. A. Sadowski, and E. F. Seila, eds., IEEE, Piscataway, N.J.

Manivannan, S. (1996). Operation analysis and improvement of truckdock operations, *Proceedings of the ASI Symposium*, Salt Lake City, Utah, June.

Manivannan, S., and M. Zeimer (1996). Simulation and analysis of aircraft offloading operations, in *Proceedings of the 1996 Winter Simulation Conference*, San Diego, Calif., December, J. M. Charnes, D. J. Morrice, D. T. Brunner, and J. J. Swain, IEEE, Piscataway, N.J.

May, A. D. (1990). *Traffic Flow Fundamentals*, Prentice Hall, Upper Saddle River, N.J.

Miller, A. J. (1971). Queueing at single-berth shipping terminal, *Journal of the Waterways, Harbors, and Coastal Engineering Division of ASCE*, No. WW1.

Moore, E. F. (1957). The shortest path through a maze, *Proceedings of the International Symposium on the Theory of Switching*, Harvard University, Cambridge, Mass.

Nicolaou, S. N. (1967). Berth planning by evaluation of congestion and cost, *Proceedings of ASCE*, Vol. 93, No. WW4.

Nilson, K. O., and U. Abdus-Samad (1977). Simulation and queueing theory in port planning, *Ports '77*, American Society of Civil Engineers, New York.

Parsonson, P. S., and J. M. Thomas (1976). A case study of the effectiveness of a traffic responsive computerized traffic control system, in *Control in Transportation Systems, Proceedings of the 3rd International Symposium of IFAC/IFIP/IFORS*, International Federation of Automatic Control, Pittsburgh, Pa.

Plumlee, C. H. (1966). Optimum size seaport, *Proceedings of ASCE*, Vol. 92, No. WW3.

Quinn, A. (1972). *Design and Construction of Ports and Marine Structures*, 2nd ed., McGraw-Hill, New York.

Schiller, D. H., and M. L. Marvin (1956). The determination of requirements for warehouse dock facilities, *Operations Research*, April.

Schneider, M. (1967). Direct estimation of traffic volume at a point, *Highway Research Record 165*, Highway Research Board, Washington, D.C.

Schulze, T. (1993). Simulation of streetcar and bus traffic, *Proceedings of the 1993 Winter Simulation Conference*, Los Angeles, December, G. W. Evans, M. Mollaghasemi, E. C. Russell, and W. E. Biles, eds., IEEE, Piscataway, N.J.

Smith, R. L., and D. E. Cleveland (1976). Time stability analysis of trip generation and predistribution modal split models, *Transportation Research Record 569*, Transportation Research Board, Washington, D.C.

Soros, P., and A. T. Zador (1977). Port planning and computer simulation, in *Ports '77*, American Society of Civil Engineers, New York.

Stafford, P. H., and D. L. Stafford (1969). Space criteria for aircraft aprons, *Transportation Engineering Journal, Proceedings of ASCE*, May.

Thompson, A. W. (1964). Evolution and future of airport passenger terminals, *Journal of the Aerospace Transport Division, Proceedings of ASCE*, October.

Walton, C. M., and S. Rosenbloom (1977). Measures to reduce peak period congestion, in *Urban Transportation Efficiency*, ASCE, New York.

Wilson, A. G. (1969). The use of entropy maximizing models in the theory of trip distribution, mode split, and route split, *Journal of Transport Economics and Policy*, Vol. 3, No. 1.

Wolbert, G. S. (1979). *U.S. Oil Pipe Lines*, American Petroleum Institute, Washington, D.C.

Wright, P. H., and N. J. Ashford (1989). *Transportation Engineering: Planning and Design*, 3rd ed., Wiley, New York.

CHAPTER 17

Simulation in Healthcare

FRANK MCGUIRE
Premier, Inc.

17.1 INTRODUCTION

The spiraling increase in healthcare costs, the increasing restrictions on reimbursed payments to hospitals, the impact of managed care, and the continuing move from inpatient services to outpatient services have combined to put many hospitals in a financial bind. As hospitals came under increasing pressure to reduce costs and increase profitability, long-held beliefs (or fears) delayed the introduction of simulation. Some of those beliefs are:

1. Practices designed for manufacturing are not transferable to healthcare.
2. Efforts to increase efficiency will shortcut patient care.
3. Efforts to increase efficiency will be interpreted by the public as a reduction in the quality of medical care provided to patients.

By the mid-1990s, the resistance was greatly reduced. General acceptance of the benefits of total quality management (TQM) or continuous quality improvement (CQI) facilitated the acceptance of other efforts designed to increase efficiency in hospitals. Simulation is one of the technologies that has benefited from improvements realized from TQM/CQI efforts. Process simulation has proven to be effective as a tool used for process improvement in healthcare.

17.2 COMPARISON OF SYSTEM TYPES

17.2.1 Manufacturing

Manufacturing systems involve the transformation of a raw material into a finished product. The finished product may be designed for an end user, or it may be the raw mate-

Handbook of Simulation, Edited by Jerry Banks.
ISBN 0-471-13403-1

rial for another manufacturing system in a series of processing systems. For example, a cotton shirt is the result of four separate manufacturing processes, which often are accomplished by four different firms. The transformation from a cotton boll to some form of yarn or thread is a series of processes within a textile mill. A cloth weaving (knitting) company will then weave or knit the yarn into fabric. A finishing/dying plant will then dye and finish the fabric. An apparel manufacturer will then cut and sew the fabric into finished garments.

The efforts of industrial engineers in manufacturing have been to reduce the time variance for each task and to standardize or maximize product flow. The increasing use of robotics and automated equipment has reduced variances while reducing the human presence in manufacturing.

17.2.2 Service

A service system involves the provision of assistance to customers. Examples include food services (restaurants), financial services (banks), and healthcare services (physician offices, hospitals). While industrial engineers in the service industry also want to reduce variance (in both time and quality) in the tasks being performed, interaction with the public limits the extent to which that objective can be achieved. The main goal is to standardize the work process. Human involvement in the service industry has remained high compared with the manufacturing sector of the economy. Because of this involvement, high variability is prevalent throughout the service industry.

17.2.3 Healthcare

Healthcare is a pure service system. The entities are usually patients or are originated from patients (such as lab specimens). Intelligent entities greatly increase the variability inherent in the system. People can be moody; they can get up and walk out; they can be emotional and uncooperative; and they can be intoxicated and combative. They ask questions and argue with physicians or fail to follow a physician's orders. Patients show up late for appointments or fail to arrive at all. It can be difficult to determine the precise medical problem (diagnosis) of the patient. There are currently 495 DRGs (diagnostic related groups) and many thousands of ICD-9 codes. The DRG system is based on research efforts at Yale in which related diagnoses were grouped together as predictors of lengths of stay. A DRG is a classification of diagnoses in which patients demonstrate similar resource consumption and length-of-stay patterns. For example, DRG 127 is "heart failure and shock" [6]. ICD stands for "International Classification of Diseases for the World Health Organization." Each DRG has ICD-9 codes that make up the diagnosis group. For example, ICD-9 code 398.91 is "Failure, heart, congestive with rheumatic fever, inactive" [4]. These issues are major contributing factors to the extreme variability in healthcare systems.

17.3 RESEARCH: THE FIRST STEP

The first step an analyst must take when undertaking a simulation project is to make sure that he or she is up to date on methods being used in other hospitals to improve the process being studied. Despite the level of expertise that the analyst has in the department being studied, there is a need to refresh his/her knowledge continuously. Although reading and research will not reduce the need for direct observation and on-site interviews, it will aid in the identification of additional opportunities for improvement. The old cliché "Don't reinvent the wheel" is quite appropriate when performing a simulation project. The analyst must be able to recognize opportunities to improve processes as he or she uses a simulation model. Knowledge of how others have addressed problems in similar departments can provide the analyst with valuable ideas to explore. An analyst's credibility is also enhanced if he or she is able knowledgeably to answer questions about current trends or future directions in the department under study.

Helpful resources are often at our fingertips. Libraries usually offer ways to search periodicals for content and subject matter. Case studies are often presented in magazines and publications in the healthcare industry. Nursing and physician journals are good places to start. The American Hospital Association has many articles related to various improvement efforts by the healthcare industry. Dozens of books are published that address hospitals as a whole and the various parts thereof. These books are good starting points for those relatively unfamiliar with a particular department. However, the most current trends will not often be found in books. Journals are the best sources of information on what is new and improved in the healthcare field.

Benchmarking studies are being performed in healthcare by healthcare alliances and others. These studies are often available through the Benchmarking Clearing House or from the benchmarking team itself. Often, "best practices" are preferred when appropriate conditions exist. One of the shortcomings of some studies is that the conditions necessary for the practice to be most useful are not identified and explained. Therefore, when reading about best practices, be aware that some may not be applicable to the situation under study.

There are many databases that offer comparisons between hospitals. Some of these are compiled by states and by various entities concerned with healthcare. Some are available for purchase and some are available by being a member of an alliance. They provide information that can alert the analyst to opportunities for improvement in various departments. A profile of the various hospitals usually exists so that users of the database can have a framework for comparison.

17.4 STEPS IN A SIMULATION PROJECT

As discussed in Chapters 1 and 2, the simulation process can be broken down into a series of steps. These steps are grouped in this chapter as initial, model construction, and analysis as follows:

Initial Steps

- Identify the process to be simulated.
- Define the objective(s) of the project.
- Formulate and define model.

Model Construction

- Collect data.
- Build the model.
- Verify the model.
- Validate the model.

Analysis

- Set up alternatives for evaluation.
- Run multiple simulations on each alternative and evaluate results.
- Choose the best alternative for presentation.

The steps are sequential in nature, and "jumping the gun" often results in unnecessary additional work. Failure to plan the model structure prior to data collection often results in both inadequate and excessive data collection efforts. Proper planning will avoid these errors.

17.4.1 Initial Steps

1. Identify the Process to Be Simulated. The analyst must make sure that it is clear what aspects of the process are to be simulated. There is a lot of room for miscommunication, resulting in inappropriate expectations by stakeholders (people who will be affected by any change in the process) in the project. The starting and ending points of the process should be clearly defined and agreed upon. Knowing which process is to be simulated is the beginning of the project, but the objectives of the project must be known as well.

2. Define the Objective(s) of the Project. A simulation project is much like a TQM or CQI project, up to the point of choosing the best alternative for implementation. Simulation is an excellent tool for analysts to use during the process of a TQM project. In general, the same rules apply for both efforts.

- The objectives must be quantifiable. A unit of measure must exist for the objective in order to know when the objective has been attained.
- The number of objectives should be limited to three or less. Too many objectives keep the project from being focused and enlarge the scope of the project unnecessarily.
- The scope of the project must be narrow. The entire hospital is not a candidate for a simulation study. While many different simulation models can be combined to create one very large model, the individual models should be separate projects.
- To be useful, a project must be finished within a reasonable time.
- The results must be definitive and have a reasonable chance of being implemented.
- The size of the project has a marked inverse correlation to its ultimate success (or failure).

In defining the objectives of a simulation project, the analyst must remember that alternative results will be compared to the validated model, which will usually be compared to the historical measures of the objectives. Some exceptions are:

- A new process design that is radically different from the existing process
- A process that does not currently exist (or does not exist in that facility)
- A design for a new facility

For new facility design projects there are some advantages to collecting historical data from the old facility, should one exist. The comparison of the process flow in a new facility to the same function in an existing facility is often useful. The data can be used as a test for validation (are the results logical?) or to show where a new process is more efficient (or not). Such comparison data can be a decision variable for building of the new facility.

3. Formulate and Define the Model. Before the simulation project can proceed, the basic structure and content of the model must be designed. The type of data that need to be generated by the model (for reporting purposes) must be defined prior to the model's construction. The data that will be needed to build the model must be defined before they can be collected. If collection of some of the data is not feasible, the model will have to be redesigned accordingly.

The model must have the ability to show the relative impact on objectives of various alternatives. The model must also be designed to provide data elements that will be compared to historical data for validation efforts and to compare one alternative with another.

To emphasize an earlier point, it is helpful to restate it: The natural desire to jump right into the model building phase must be suppressed. Much time and effort will be saved if the model is planned adequately before the data collection effort begins, and especially before the actual model building starts.

17.4.2 Model Construction

1. Collect Data. Data collection is often very frustrating and time consuming. The following paragraphs cover some of the types of data needed and surrounding issues.

a. Arrivals. In healthcare, weekends are different from weekdays, and often weekdays are different from each other. Daily volumes and arrival patterns must be determined. In general, model entities are patients or patient-related items such as lab samples. Laundries, pharmacies, medical records, and x-ray film libraries are examples of departments that are less dependent on patient processing. Hospital information systems and daily log sheets are good sources of arrival information. Appendix 17.1 has an example of a data collection form used for patient chart reviews in an emergency department. Appendix 17.2 has a list of data typically needed for surgical services, and Appendix 17.3 has a similar list for emergency departments.

b. Entity Categories (Patient). As mentioned earlier, there are currently 495 DRGs (diagnostic related groups) and thousands of ICD-9 codes. The best way to plan the data collection effort is to conduct interviews with the staff prior to designing the data collection tools. The goal of the interviews will be to get an overall view of the department and to compile the list of patient categories (types). Caregivers might categorize the patients based on severity of illness (acuity) or by what procedures will be performed.

The personnel in the department under study will have a set of categories that is used on a daily basis. Communication with the department will often be easier if the analyst uses the same set of categories. Information systems sometimes present the data in too much detail. The reports are often designed for financial or utilization (resources and facilities) purposes and are not amenable to patient process analysis.

c. Flowcharts. Flowcharts will be needed for each patient type (or entity) that will be flowing through the model. These should be low-level flowcharts that include some or all of the following information:

- Percent of patients that will receive different interventions
- Percent of patients that follow a branch in the flow
- Conditions that exist before a patient may continue to the next box on the flow chart (e.g., results of previous tests must be available)
- Who performs the task
- Standard time for performance of the task

The flowchart is an important part of the data collection effort. Not only does it facilitate the construction of the simulation model, it should be an integral part of the verification process, which is discussed later in the chapter. Various companies produce quality flowcharting software packages.

d. Resources. Staff schedules are required for those models exploring staffing issues. Lunch periods and breaks should be considered. Some departments limit the number of staff by type that can be out of the department at any one time. Minimum staffing patterns may exist. Some tasks may be performed only by a specific type of resource, while others might be performed by any of several resources. For example, laboratory specimens might be drawn by a phlebotomist from the laboratory or by a technician, a nurse, or a respiratory therapist.

The time to complete tasks can be difficult to obtain. Some tasks are performed very erratically and infrequently. Trying to observe (time) 30 such tasks might be a prohibitive undertaking. Some hospitals will have standard task times that can be used. When data on time required to complete a task are not available and do not lend themselves to reasonable data collection efforts, an alternative method can be used to estimate the time available. This method involves interviewing the appropriate caregivers. Physicians and nurses are reasonably accurate when estimating the minimum, most likely, and maximum times that a task will take (for tasks less than 2 hours in duration). Those times can be used to form a triangular distribution, such as T(3,5,12). The 3 represents the minimum time in which the task can be performed, 5 represents the most likely time, and 12 is the maximum time allowed. If Unifit software is available, the three times (plus the addition of the mean—see the next paragraph for a discussion of distribution shapes) can be used to create a beta distribution.

A common mistake of beginning simulation analysts is to use a normal distribution for task times. Research by statisticians has shown that the shape of a distribution that is representing the time necessary to perform a task typically takes the form of a positively skewed distribution. In such a distribution, the mean is to the right of the median, which is to the right of the mode. The *mean* is the average of the distribution. The *median* is defined as the value above and below which lie an equal number of data observations when the data are arranged in increasing or decreasing order. The *mode* is defined as the most frequently occurring value in the data. Figure 17.1 represents a positively skewed distribution.

The shape is logical when examined closely. A caregiver will reach an average level of expertise with an average level of ability and an average amount of training. There-

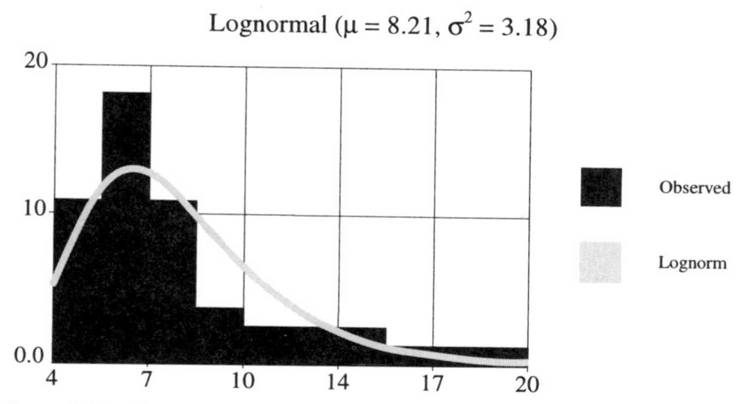

Figure 17.1 Example of a positively skewed distribution. (Courtesy of Bestfit.)

after, the caregiver will complete the task with similar times except when delayed. Any delays will cause the individual task time to take longer than the mode. The most likely time (mode) to complete the task will be less than the median time, and the median time will be less than the average time to complete the task. The most likely task time will be close to the minimum time necessary to complete the task. Although there is a discrete minimum time in which a task can be performed, there is no corresponding maximum time. Time to perform tasks is often represented by a lognormal, Weibull, or gamma distribution [1].

e. Stratification of Data. Once the flowcharts exist, demographic data will be required to define the treatment a particular diagnosis will require. Not all patients of the same type will receive the same procedures and tests. Some patients having procedures performed will have complications or delays, and others may not. Some patients will have to wait much longer for results of tests than others, based on how busy the testing department is or the availability of the clinician who will interpret the results. To incorporate the differences in waiting times and procedures, data will have to be stratified by patient type (entity type). Once stratification is complete, the analyst must summarize the data in a way that "fits" the simulation software package being used.

f. Distributions. One of the major fatal errors commonly made in simulation projects is replacing a distribution by its mean. Another error is to use the wrong probability distribution to model a data set. Distribution-fitting software is available from several vendors and is a part of most statistical packages.

2. Build Model

a. Planning. The first step in model building is to formulate and plan the model. This step was necessary before the data could be collected (see the earlier discussion). The information desired from the simulation must be evaluated in order to design the model to generate the desired statistics. Interviewing stakeholders will help identify the information that is used to measure the performance of the department/process being stud-

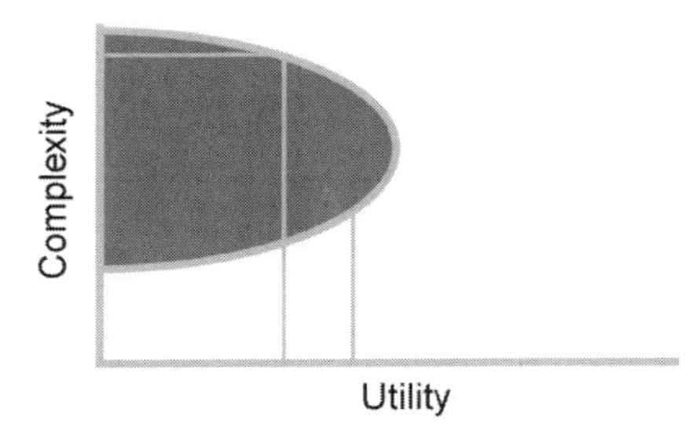

Figure 17.2 Complexity versus utility. With permission of PROMODEL Corp., Orem, Utah. Copyright © 1995 PROMODEL Corp.

ied. This information will probably be desired of the simulation model, even if it is not specifically required by the original objectives of the study. The stakeholders might well want to know the impact of an objective on their particular area of concern. The project team probably has some stakeholders represented, but it is not always practical to include on the team representatives from all stakeholders.

An efficient model would allow changes to be made to one part of the model without having to reconstruct the entire model. Novice modelers will often construct, verify, and validate a model and then start to worry about the various scenarios that will be tested. This type of shortcut will probably result in the analyst having to go back to the original (base) model and make major changes in the structure of the model in order to collect baseline data.

b. Complexity Versus Simplicity. Some of the most successful projects are very simple. Do not add complexity for complexity's sake. It is a very real temptation and one that can be difficult to avoid. Ask the question: "What additional value do I receive for this additional detail?" If the answer is little or none, do not add the complexity. If the detail is not necessary to achieve the objective, it should not be included. The ability to show great detail can be intoxicating. It can be a form of showing off. An overly complex model can confuse the issue(s). Figure 17.2 illustrates the relationship between utility and complexity.

While too much complexity is a problem, too little is also a problem. Leaving out necessary detail can cause the analyst to reach false conclusions. Figure 17.2 shows that there is a point of diminishing returns when adding complexity to a model. As complexity increases, utility also increases, at first rapidly. As complexity continues to increase, there is a point that increased utility (value) starts to level off. Addition of complexity will result in decreased value at some level.

It can be a difficult task to balance complexity and utility. Remember to ask what value will be added by additional complexity. The question will keep the analyst focused and prevent many hours of fruitless effort.

3. Model Verification

a. Comparison of Model Flow with Flowcharts. It is common to confuse verification with validation. While both are necessary before any alternatives can be tested, the two tasks are distinctly different. Verification is documenting that entity flow through the model matches the reality of entity flow in the department, while validation is a comparison of historical measures (of your objective) to data generated by the simulation model. Please see Chapter 10 for an in-depth discussion of model verification and validation.

The patient (assuming that a patient is the entity involved in the process) must arrive at the right locations, be seen by the correct caregivers for the appropriate distribution of time, receive the appropriate interventions and diagnostic tests, and proceed to the correct next locations. These steps must be compared to the flowchart from which they were constructed. When the analyst is confident that he or she has the correct process, the "expert" from the department being studied must go back over the process with the analyst to confirm that the correct process flow is being used.

b. Work with Department Staff. Verification cannot be done by the analyst without outside corroboration. The department under study has the real experts on the current process. The staff from that department must be involved in verification of the model. Their involvement is very helpful in achieving departmental acceptance of the model.

4. Model Validation. Validation methodology is discussed in detail in Chapter 10. In this section we discuss validation efforts in healthcare. This step in the project usually causes the most headaches in healthcare simulation. It is the lengthiest step of the actual simulation model process (not counting data collection). The model must reflect the correct current process (except for initial design of nonexisting processes) before validation can begin. Except for new systems, validation of a model requires that historical data be available and collected during the same period of time that the input data were collected. The following discussion assumes that the department under study currently exists and that historical data are available.

a. How to Choose Validation Parameters. The validation parameter is the measure of your objectives. It is quite common for this measure to be one related to time. A pitfall to avoid is trying to validate on multiple subsets of the decision variable. For example, if patient length of stay is the decision variable, a novice might try to validate on numerous patient types as well as the overall patient length of stay. If the patient type is a significant percentage of the total patient volume and sufficient historical data are available to provide meaningful confidence intervals, validation should include that patient type. However, some patient types may represent a small percentage of the total volume and therefore have only a few data points available. To be useful, the historical data must include enough volume to return a well-defined (narrow) confidence interval (see Chapter 7 for information on confidence intervals).

b. Validation of New Processes/Systems. When historical data do not exist, the model must still be validated. New facilities, new processes, and new technology are some reasons for the lack of historical data. If the process exists in a current facility, data can be collected from the current process and used as a baseline for validation of the simulation model. The data can provide a framework for testing whether or not the results from the simulation model are logical. In other words, do the results seem reasonable for the process under study?

The test for reasonable results is appropriate when the process is new to the organization. Benchmark reports and case studies can be researched to provide a framework for comparison. Site visits to organizations that have implemented the new process can provide information to help the analyst determine whether the results of the simulation study are logical. Once the plausibility test is completed, the results of the simulation model can then be used as the baseline for comparison of alternative scenarios.

A caveat is that the model might not yield results that accurately predict the performance of the system when implemented but will provide accurate comparison information on the alternatives tested. For example (the decision variable is patient length of stay), the simulation model might accurately predict that alternative 3 will save 15 minutes per patient compared to alternative 1. However, the model might not be accurate when predicting the total length of stay for the average patient in the new system. This limitation does not diminish the value of simulation for a new process. It is valuable to know that one alternative is superior to another alternative and by what magnitude.

c. When the Model Does Not Validate. If a model is validated the first time it is run after verification, a record has surely been set, or it is an aberration, which will later be shown to be wishful thinking. In most cases, work is still to be done. This work is one of the great values of using simulation. The processes in the model have been verified, so the analyst must search for the reason the model is not valid. The usual reasons (barring problems with the model itself) are that the processes as laid out on the flowcharts are not the only processes occurring or that the flowcharts (protocols, procedures, etc.) are not followed exactly by staff. Batching of work in progress can occur. Some examples follow:

1. Specimen collection for laboratory tests may not be performed until three requests are made.
2. Diagnostic procedures may be delayed until there are several to perform in the requesting department.
3. Lab specimens may not be picked up until there are several ready to be picked up.
4. Patient charts may not be filed until the box is full.

Other problems may exist with communication between departments and within the same department. Plans calling for cooperation among departmental staff may not be working well. Procedures that are more efficient but not part of the official plan may have been worked out by departmental staff and used daily. Some of these "illicit" procedures may make it more efficient for one area of the department but cause delays in other areas.

The analyst and the project team must find the area of the model that is causing the most deviation from the historical data and use direct observation (and other techniques) to find the offending process. It helps to have a number of intermediate data collection points, so that the process may be broken into parts. For example, patient throughput time in an emergency department might be broken into several parts, such as:

1. Elapsed time (ELAP) from arrival to triage
2. ELAP from triage to time in the treatment room
3. ELAP from arrival in treatment room until the physician sees the patient
4. ELAP from physician visit until disposition instructions
5. ELAP from arrival in treatment room until patient leaves the department

By stratifying the data, the team can focus on the section where the simulation model is most different from the historical data, thereby narrowing the scope of the effort. When the model is verified and validated, the model is ready to be used to test alternatives.

5. Choosing Alternatives for Testing. The project team should work together to identify the alternatives to be tested. These alternatives should be related directly to the project's objectives. The alternatives' results must be evaluated to make sure that any differences are statistically significant.

The results from individual alternatives are not necessarily additive. The alternatives chosen for implementation must be combined into one model and tested to determine the final impact on the objectives. Combining the alternatives into one model can also identify any problems with interdynamics (one alternative process might cause another alternative process to be less effective when the two alternatives are combined).

17.4.3 Assumptions Document

Busy analysts often neglect the assumptions document. Examples of information included in an assumption document are:

1. Flowcharts
2. Data summarization
3. Staffing levels and schedules
4. Task times and distributions
5. Staff responsible for performing tasks
6. Entity volumes and arrival information
7. Delays and delay duration
8. Equipment/location setup time and frequency

The assumptions document should be available to the project team and attached to the report as an appendix.

17.5 BARRIERS TO IMPLEMENTATION AND HOW TO DEAL WITH THEM

For any simulation project to be considered a success, the recommendations resulting from the study must be implemented. A major barrier to implementation of recommendations is the failure of the sponsoring manager to follow the studies' recommendations. The manager with direct managerial responsibility [over the department(s) included in the study] must clearly communicate what the expectations are and who must actually implement the plan (or part of the plan). Too often, a director or vice president will authorize a study and then fade into the background as his or her subordinates take over responsibility for implementation. This can work when the subordinates were the ones desiring that the study be performed and the subordinates have authority to make the necessary changes. All too often, however, an executive will insist that changes are necessary, request simulation to identify the most appropriate alternatives, and then fail to make arrangements for implementation. The expectation is that the manager over the area in question will take it from there. This does not always happen, for several reasons:

1. The manager might resent the executive's interference.
2. The manager may not have authority over all the areas that require change.
3. The manager may not be strong enough to overcome the obstacles that will be presented during the process of implementation.
4. Cooperation or help from external personnel is necessary, and the external personnel are too busy or unwilling to provide the needed assistance.
5. The manager may have to make major changes and carry out all of his or her current responsibilities as well (the manager has good intentions but not enough good-quality time available to apply to the changes).

The manager authorizing the study must follow up to see that implementation is occurring as expected. Time pressures are real and implementing significant changes take considerable time and effort. Stress levels increase, and pressure to stop short of full implementation can be intense. The manager should define steps to ensure proper implementation. These steps should include:

1. Set time frames for implementation, including incremental steps.
2. Hold periodic meetings to review progress and address any problems that exist.
3. Monitor the implementation progress along with the effectiveness of the process being put in place (the change itself).

Follow-up and guidance from the manager will not repair a faulty simulation study. However, if the study is valid, follow-up and guidance from the manager are essential to the study's implementation.

The analyst can help minimize the chance that implementation will not happen. Unless the analyst happens to be in a position to enforce implementation directly, he or she will have to settle for providing a framework for success. This would include early discussions about implementation during the project planning phases of the study. When alternatives are discussed, methods of implementation should also be discussed. The best thing that the analyst can do to help the implementation process is to make sure that the model is valid. The next most important thing is to make sure that enough people were involved in the verification stage of the simulation model. The considerations in this section are echoed in Chapters 22 and 23.

17.6 PROBLEMS UNIQUE TO HEALTHCARE DURING IMPLEMENTATION PLANNING

Implementation of alternatives in healthcare presents a special challenge. In manufacturing, decisions often revolve around implementation cost and return on investment. In healthcare, cost is an issue, but not always the major issue. Reduction in patient throughput time is often used as the main objective of a simulation study. Time has a cost factor, but that factor is often nebulous. If additional equipment or staff is necessary to reduce the throughput time, the impact on revenue or expense of the time savings can be difficult (and sometimes controversial) to determine.

Opponents to a study can question the study on an economic basis. This is, of course, a perfectly valid reason to question any recommendations being made prior to implementation. The project team should anticipate such objections and be prepared to answer them. The simulation model can be defined in such a way that factors affecting cost are captured and reported. This will allow an incremental cost analysis of each alternative.

17.7 CASE STUDY

A simulation study was performed in an emergency department using MedModel simulation software from PROMODEL Corporation. The elapsed time from initiation of the project until completion was 4 months. The model was completed as part of a hospital reengineering team's project.

The objectives are as follows:

1. Determine what actions could reduce patient length of stay in the emergency department.
2. Evaluate staffing levels and schedules
3. Determine the number of rooms needed by type of room.
4. Evaluate effect if an internal waiting room is added.
5. Evaluate effect if an observation area is added.

Following the methodology, the next step in the process is to define the model and collect the data for input into the model. This part of the project took about 4 weeks. Over 1400 patients were observed to get information about their stay in the emergency department. Tables 17.1 to 17.3 show some of the compilations from the data collection efforts (in this case the patient categories were the patient's acuity).

Table 17.1 shows the percentage of patients that arrive during each 2-hour period for each day of the week. Using MedModel's arrival editor, the patients were input into

TABLE 17.1 Patient Arrivals (%) by Period of the Day

Time Period	Sun.	Mon.	Tues.	Wed.	Thu.	Fri.	Sat.	Mean
2:00 A.M.	9.1	1.9	5.7	2.7	3.7	3.4	6.2	4.6
4:00 A.M.	3.7	0.5	1.8	0.9	3.1	3.9	4.0	2.5
6:00 A.M.	3.7	0.5	3.1	1.4	1.6	2.5	2.3	2.1
8:00 A.M.	7.0	0.0	2.6	2.7	5.2	2.5	2.8	3.2
10:00 A.M.	8.0	13.2	7.5	11.4	11.0	7.8	10.7	9.9
12:00 P.M.	12.3	16.0	19.8	12.7	16.2	12.3	18.1	15.4
2:00 P.M.	13.4	15.1	8.8	12.3	13.1	12.3	7.9	11.8
4:00 P.M.	11.8	7.1	7.5	13.6	11.5	16.2	8.5	10.9
6:00 P.M.	8.6	16.5	14.1	10.9	9.4	11.3	11.3	11.8
8:00 P.M.	9.1	8.0	11.0	11.8	8.4	10.3	7.9	9.6
10:00 P.M.	8.0	11.8	11.0	11.4	8.9	11.8	12.4	10.8
12:00 A.M.	5.3	9.4	7.0	8.2	7.9	5.9	7.9	7.4

TABLE 17.2 Ancillary Use by Patient Type and Ancillary Department

		Ancillary Use (%)							
Triage Level	Count	X-Ray	CT/US	Lab	Retest (Lab) % of lab 1	Retest (Lab) % of lab 2	Ventilation	EKG	Monitor
Adult	628	44.6	11.0	64.8	22.6	17.4	2.5	45.5	27.2
Urgent	69	21.7	5.8	30.4	19.0	25.0	1.4	15.9	4.3
Subacute	271	45.4	10.7	62.7	22.9	15.4		30.3	5.9
Acute	280	48.9	12.9	74.6	22.5	17.0	5.0	66.1	51.4
Critical	8	62.5		87.5	28.6	50.0	12.5	100.0	100.0
Pediatrics	477	9.0	1.0	15.9	13.2	20.0	0.8	2.1	1.7
Urgent	255	2.0	0.4	7.1	11.1	50.0			
Acute	214	17.3	1.9	24.8	11.3	16.7	0.9	3.7	2.8
Critical	8	12.5		62.5	40.0	0.0	25.0	25.0	25.0
Fast track	336	8.0	1.2	12.5	9.5	25.0		3.0	0.3
Overall	1441	24.3	5.4	36.4	20.2	17.9	1.4	21.2	12.5

the model with the daily and hourly rates that were observed during the data collection period. Table 17.2 summarizes the percentage of patients by patient acuity who receive diagnostic testing. Ventilation is a procedure that is performed by an ancillary department (in this case, respiratory therapy). Note that in Table 17.2 some patients require additional laboratory testing after the results of the initial tests are available. Similarly, some patients receive medications several times, which is illustrated in Table 17.3. Table 17.3 summarizes the procedures (treatments) performed for a patient during the patient's visit. The emergency department staff performs these procedures.

A floor plan was imported from an AutoCad drawing obtained from the hospital's facilities department. Figure 17.3 is a copy of the floor plan. With the floor plan in place, locations were defined. The next step in building the initial model was defining entities, resources, shift assignments, and arrival patterns. Attributes were defined to enable differentiation among patients and other entities. These attributes were assigned values of 0 or 1 (1 = true or yes) based on the probabilities of a patient receiving tests or treatments in Tables 17.2 and 17.3.

After the initial definitions were in place, the patients could be brought into the model. Figure 17.4 shows MedModel's processing editor. The entity flow is defined and controlled in this module, and the operation logic is either defined or initiated (subroutines or macros may be called).

The process window defines what entity is being processed and where the processing is taking place (e.g., patient at a location named Enter). The operation window defines the activities that take place for the patient at Enter. In this example, the patient attributes are being assigned. Attributes are used to define the entity characteristics for later activities. Examples of patient attributes are as follows:

- Patient acuity (critically ill, urgently ill, not urgently ill, etc.)
- Whether a resident physician will see the patient
- How the patient will arrive
- Whether the patient is sent home or admitted
- Which treatments and ancillary tests will be performed

TABLE 17.3 Treatments (%) by Patient Type and Treatment Type

Triage Level	Volume	Treatment (%)												
		IV	Meds	Meds 2 (% of Meds)	Meds 3 (% of Meds 2)	Suture	Bandage	NG	Foley	Ortho	Thromb	Triple	Txt1:1	Other
Adult	628	55.9	53.0	50.8	42.6	0.8	0.3	0.5	4.0	1.1	0.2	0.6	1.0	2.9
Urgent	69	18.8	33.3	26.1	50.0	2.9			2.9					
Subacute	271	51.3	44.3	49.2	32.2	0.7	0.4	1.1	3.7	2.2		0.4	0.7	2.6
Acute	280	68.6	66.1	55.1	48.0	0.4			4.3		0.4	1.1	1.4	3.6
Critical	8	87.5	62.5	40.0	50.0				12.5					12.5
Pediatrics	477	8.8	26.6	28.3	30.6	2.3	0.6	0.4	0.8	0.8	0.4	0.2	0.8	2.1
Urgent	255	2.0	19.6	8.0		0.4			0.4					0.8
Acute	214	15.9	33.2	39.4	32.1	4.7	1.4	0.5	0.9	1.9			1.4	3.3
Critical	8	37.5	75.0	66.7	50.0			12.5	12.5		25.0	12.5	12.5	12.5
Fast track	336	3.3	15.5	55.6	40.0	1.8				1.1				2.1
Overall	1441	28.0	35.5	41.8	39.7	1.5	0.8	0.3	2.0	1.0	0.2	0.3	0.7	2.4

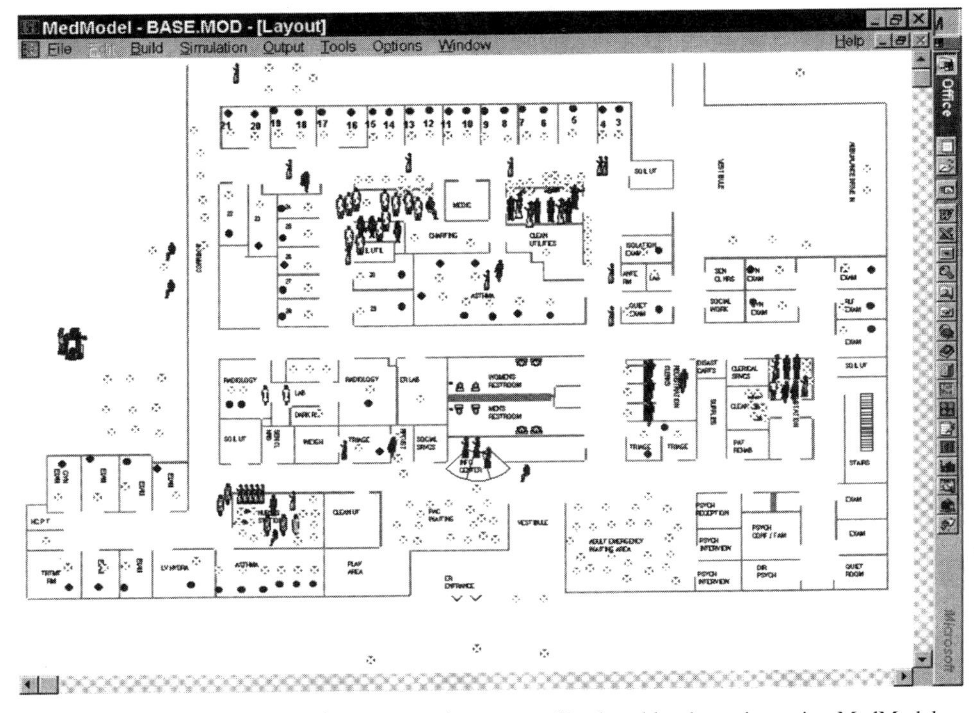

Figure 17.3 Model layout of emergency department. (Produced by the author using MedModel Software from ProModel Corporation, Orem, Utah.)

The routing window defines the next location to which the patient will be sent. In this example, a generic entity named patient was brought into the model and based on probabilities, is assigned a patient type. The patient is then renamed and routed to a location named Start.

In similar fashion the rest of the model was built. After the analyst completed the model, the initial verification process began. The arrival rates input into the model had to be checked to ensure that there were no errors. This was checked by the use of time-series plots generated by the model. These plots were compared to data observed. The daily and weekly average volumes were verified also. The numbers of procedures and tests performed in the simulation model were tracked and compared to the number of observed procedures and tests. Actual patient flow through the model was compared on a step-by-step basis to the original flowcharts by patient type. Trace files and debugging editors were used to aid the comparison.

After the analyst was comfortable that the model was accurately constructed, the same steps were performed with staff from the department. The analyst cannot complete the verification by himself or herself. Communication errors would not be caught, nor would inappropriate assumptions be identified. As errors were identified, the analyst corrected the errors and the verification was repeated to make sure that the errors were corrected.

Once verification was complete, validation began. Table 17.4 is a summary of ob-

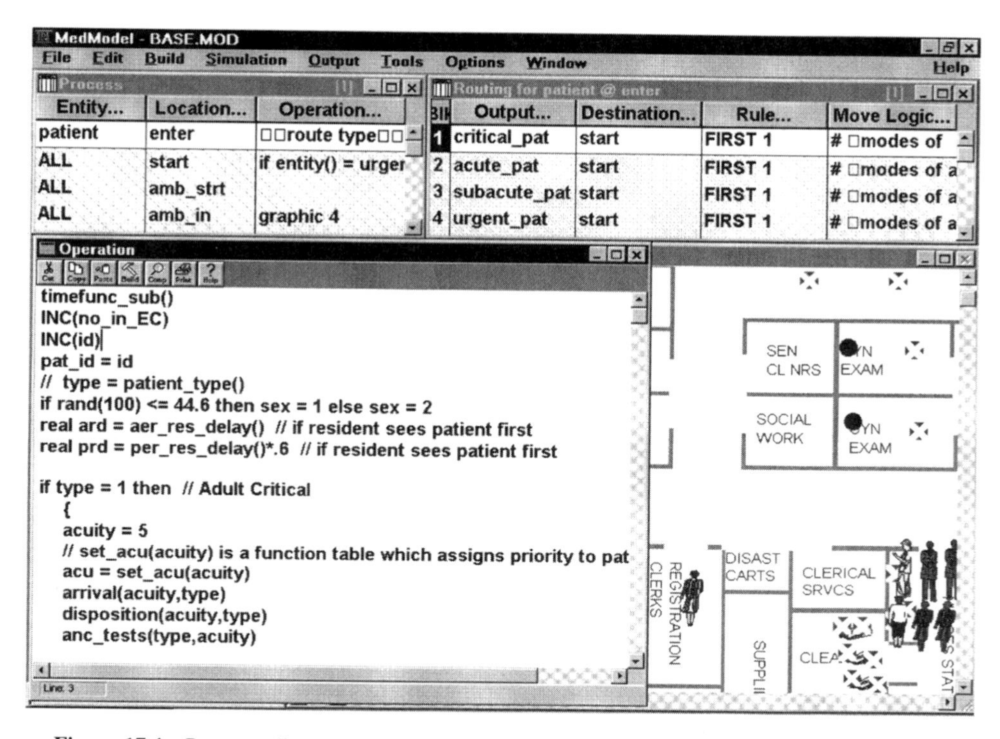

Figure 17.4 Process editor. (Produced by the author using MedModel Software from ProModel Corporation, Orem, Utah.)

served data collected for validation purposes. The data are also used for analysis purposes. The disposition order is a major point in the process. The patient will not be able to leave the treatment room until the attending physician has decided whether to admit or discharge the patient and what referrals and instructions are appropriate.

The time from the resident visit until the attending visit is that time required for a resident to reach a diagnostic decision and treatment plan. The resident will then present his or her findings to the attending physician, who will then visit the patient with the resident. The delay for discharge is measured from the time the disposition order was made until the patient actually leaves the department. LOS is an abbreviation for *length of stay*.

The simulation model was run for 8 days with a 1-day warm-up (transient) period. Thirty replications were run and the results compiled in the report editor. The observed patient's length of stay was analyzed to derive the associated confidence intervals and the simulation results compared using hypothesis testing. Initially, the simulation model's results showed patients leaving the emergency department too quickly. Additional observation of the emergency department and additional interviews were necessary to identify possible delays. Differences between the original flowcharts and observed entity flow were noted and investigated. Changes were made in the simulation model and the validation process was repeated. Two additional repetitions of the validation process were required (a total of four efforts) before the model was accepted

TABLE 17.4 Compiled Historical Data for Validation of Simulation Model

Discharge Mode	Time from Triage Until Registration	Time Patient is in room Until Resident Visit	Time from Triage Until in Room	Time from Resident Visit Until Attending Visit	Consult Delay	Time In Room Until Disposition Order	Time from Attending Visit Until Disposition Order	Delay for Discharge	LOS
All	32	12	63	82	86	220	150	129	257
Adult ER									
All	30	28	45	105	110	246	172	135	367
admit	28	45	33	112	101	280	188	198	497
discharge	32	19	54	103	110	213	155	29	284
transfer	16	28	8	15	360	141	108	218	380
Pediatric ER									
ALL	18	3	60	86	41	160	81	111	150
admit	13	8	27	147	3	240	114	178	340
discharge	18	2	63	81	52	95	54	47	134
transfer									
Fast track									
ALL	52	9	101	51	34	105	78	73	202
admit	69	8	126	15		469	491	305	827
discharge	52	9	100	51	34	70	38	12	182
transfer	13		133	30					260

TABLE 17.5 Summary of Simulation Results

	Before (minutes)	After (minutes)
All patient LOS	257	181
Adult ER LOS	367	329
Pediatric ER LOS	150	123
Fast track LOS	202	72
Average census 8 P.M.	36	28

as valid. The elapsed time was about 4 weeks. Coordination of schedules was partly responsible for the length of the process. Seven person-days were required to complete validation.

The alternatives were then tested and the results compared to each other and to the validation model's results. Table 17.5 is a short summary of the improvements made to the process as verified by simulation. The results in Table 17.5 are from a model that combined all the recommendations chosen for implementation. The alternatives tested included (but were not limited to) the following scenarios:

- Increase the number of treatment rooms in the fast track area by three.
- Revise triage criteria for routing patients to the fast track area (increase fast track volume).
- Revise patient assignment to care teams.
- Add an observation area to the emergency department.

17.8 SUMMARY

Simulation has many advantages over more traditional approaches to process improvement in healthcare. It provides an objective way to test different alternative processes. Simulation also delivers a quantified difference between the different alternatives. Simulation is not emotional and has no territorial urges. Simulation shows how a change in one area of a department will affect operations in other areas. Simulation is useful in verifying the architectural design for a new construction project. Plans can be tested and modified prior to final approval.

APPENDIX 17.1: EMERGENCY DEPARTMENT PROCESS SURVEY

*** Use Military Times with colons for all Times**

Date	
Time of Arrival	
Mode of Arrival	☐ Amb/carried o. EMS o. Wheelchair
Age	
Gender	M F
Pt. Available at time of Triage?	Yes No
Triage Time	
Meets Fast Track Guidelines?	Yes No
Pre-approval call required?	Yes No
Does pt have Family Physician?	Yes No
Did they call Family Physician?	Yes No
Registration Time?	
Pt. available for room placement?	Yes No
Time to Room	
Room #	
Time of RN assessment	
Time seen by Physician	
Consult called to come in time	
Consultant Presents to ED Time	
Grasp Classification (level)	
Physician Classification (level)	

Insurance Co.

☐ Medicare
☐ Ken-Pac
☐ Kentucky Medicaid
☐ Worker's Compensation
☐ HMO
☐ Commercial Insurance
☐ Champus/Champva
☐ Self Pay

Disposition Order Time	**Time**
Discharged **Admit**	**Time**
Transferred Air Ground	**Unit**
Expired	

Diagnosis	**Physician**

Factors contributing to increased length of stay in E.D.

X-RAY	Yes No
Portable?	Yes No
U/S	Yes No
CT	Yes No
Nuclear Med	Yes No
Time ordered	
Time to x-ray	
Time returned	
Lab	Yes No
Collected by	Lab ED
Time ordered	
Time collected	
Time results ready	
EKG	Yes No
RESPIRATORY	Yes No
Pelvic	Yes No
Cardiac Monitor	Yes No
IV/Hep lock	Yes No
Med	IV PO
IV meds type	
Suture simp/comp	Simple Complex
Bandage	Yes No
Orthopedic device	Yes No
NG Tube	Yes No
Cath for urine	Yes No
Foley catheter	Yes No
Thrombolytics	Yes No

Patient Type	**LABS**
☐ Abd. Pain	o CBC
☐ Chest Pain	o Chem 7
☐ Ent/Eye	o CPK/MB
☐ Laceration	o PT/PTT
☐ Ob/Gyn	o U/A
☐ Orthopedic	o HCG
☐ Psych/CD	o Amylase
☐ Medical/Surg	o Drug Level
☐ Minor Care	
☐ Overdose	
☐ Trauma	

APPENDIX 17.2: DATA COLLECTION CRITERIA FOR AN OPERATING ROOM PROCESS SIMULATION PROJECT

1. A floor plan (drawn to scale) of the area to be included in the simulation. Preferably an AutoCad file or other editable drawing, or a Windows bitmap file (bmp) or metafile (wmf).
2. A detailed flow-charts for each patient type. Use a different flow chart for each patient type. Depending on the project, patients may be stratified by either specialty (e.g. orthopedics, general, ENT, ob/gyn, etc.) or by surgeon. Check to see if all significant and measurable possibilities for process branching are included at each step. The most detailed sections of the flow chart should cover patient arrival-registration-preop care, and post-op care.
3. Which patient types can go to a particular OR? What conditions are placed on a patient type going to this room? Which patient types can not go to a particular room?
4. Block schedules (specifying specialty) by day of week, if used. If no block schedule, patient arrivals by day of week by specialty.
5. Arrival rate information:
   (a) Number of patients by day of week and by hour of day
   (b) Percent of patients by surgery type
6. Are rooms held open for certain types of patients at all times e.g. emergency C-Sections?
7. Number of beds in the preop area. Number of beds in the PACU.
8. Cycle times for each of the following by patient type:
   - Time from patient enters surgical room until anesthesia started/completed
   - Type of anesthesia e.g. general, spinal, MAC, local
   - Time from patient enters surgical room until incision
   - Time from incision until closure
   - Time from closure until patient leaves surgical room
   - Cleanup time
   - Setup time
9. Number of doctors, nurses, anesthesia personnel etc. needed at each cycle. If specified labor is not available and can not be pre-empted from other patients, what is the procedure to follow?
10. Shift and staffing levels by labor type. Include break and lunch schedules.
11. Percentages of patients who will receive additional work in the preop area due to incomplete charts etc. How long are these delays?
12. Is a holding area used and if so, for which patients?
13. Do outpatients and inpatients intermingle during the process or are there separate facilities e.g.
    - Preop check-in
    - Stage two recovery, for outpatients?
14. List and flow chart any additional patients who may use the PACU e.g. OP's requiring blocks by anesthesia, radiology recovery patients etc.

APPENDIX 17.3: DATA COLLECTION CRITERIA FOR AN EMERGENCY DEPARTMENT SIMULATION PROJECT

1. A floor plan (drawn to scale) of the area to be included in the simulation. Preferably an AutoCad file or other editable drawing, or a Windows bitmap file (bmp) or metafile (wmf).
2. A detailed flow-chart. Don't try to combine all types of patients on one flow chart. Instead, use a different flow chart for each patient type (e.g. trauma, orthopedic, pediatric, ob/gyn, etc.). It is helpful to first stratify the patient types into those that follow the same process flow. Check to see if all significant and measurable possibilities for process branching are included at each step.
3. Which patient types can go to a particular room? What conditions are placed on a patient type going to this room? Which patient types can not go to a particular room?
4. Which rooms are held open for certain types of patients at all times?
5. Are hallways used to hold patients or to be used as overflow treatment locations? If so, what determines whether or not the hallway is used (e.g., number of patients in waiting room).
6. Number of and type of waiting rooms and observation units (connected to the ED). What is the capacity of each waiting area?
7. The treatment patterns (caregiver(s) and time for task) for each patient type and acuity level.
8. Patient information: (a) number of patients by day of week and by hour of day; (b) percentage of patients by type and acuity; (c) does patient type and acuity vary by day of week or hour of day? (d) arrival mode; (e) admission data.
9. Staffing levels/schedules by labor type. Include break and lunch schedules.
10. Preemption practices. Which patients will nurses and/or doctors leave to go to another patient whom has a more severe acuity. How much more severe?
11. Percentage of patients that will receive ancillary services by patient type and acuity. Ancillary department turnaround times. Staff response times, etc.
12. Triage protocols; selection criteria for the fast-track area if used; which tests can be ordered by the triage RN and what criteria are used for ordering the tests.
13. Delays associated with admission (waiting for beds, attending MDs, etc.)
14. Task times for procedures. If the hospital has standard times, then these can be used. For frequent tasks, a data collection form can be used to record the start and stop times for the procedure.
15. Process flow chart for patients' charts. Where are the charts placed when waiting for test results? Where does the physician put the charts when he/she orders tests, etc.?
16. Process flow chart for lab specimens. Who draws the blood? How does the specimen get to the lab?
17. What point of care testing equipment is used?
18. Does the ED use a patient tracking grease board or software? What information is on the board? Who updates the board?

19. Who is responsible for assigning patients to rooms? Who escorts the patients to the room?
20. How are RNs assigned patients? Does the charge nurse take a patient load?

REFERENCES

1. Law, A., and W. D. Kelton (1991). *Simulation Modeling and Analysis*, 2nd ed., McGraw-Hill, New York.
2. Lorentz, E. W., and M. K. Jones (1994). *St. Anthony's DRG Guidebook, 1995*, St. Anthony Publishing, Reston, Va.
3. Wolper, L. F., and J. J. Peña (1987). *Health Care Administration: Principles and Practices*, Aspen Publishers, Rockville, Md.
4. Levy, H., and M. Ben-Horim (1984). *Statistics: Business Applications in Business and Economics*, 2nd ed., Random House, New York.
5. Law, A. M. (1994). *Five Critical Pitfalls in Simulation Input Modeling, and How Unfit II Can Help You Avoid Them*, Averill M. Law and Associates, Tucson, Ariz.

CHAPTER 18

Simulation of Service Systems

RON LAUGHERY, BETH PLOTT, and SHELLY SCOTT-NASH
Micro Analysis and Design, Inc.

18.1 INTRODUCTION

Many countries around the world are seeing a shift in the percentage of gross domestic product from manufacturing to services. As we are becoming saturated with material goods, we are relying increasingly on enhancements in the quality of our life to come from improved services. We are spending more time and money eating at restaurants, managing our finances, traveling, and engaging in various forms of recreation than ever before. There is no reason to expect that this trend will not continue.

By definition, the concept of good "service" includes consideration of the quality and timeliness of service performance. As with any economic sector, the efficiency with which a service provider can deliver timely and efficient services will determine their ability to survive in a competitive marketplace or, in the government sectors, to retain the support of the electorate.

Simulation can be used in the analysis of any system to ensure the quality, timeliness, and efficiency of stochastic, complex processes that operate in resource-constrained environments. Many service systems are exactly that—stochastic complex processes operating in resource-constrained environments. Furthermore, most service systems are fairly well defined discrete processes. As such, discrete-event computer simulation offers great potential as a means of describing, analyzing, and optimizing service systems of many types.

The following are examples of service systems that have been studied with simulation and the questions that have been addressed:

- *Banks.* What is the impact of automated teller machines on customer queue sizes and the number of tellers required? What is the impact on customer service time of better teller training and the elimination of the need to meet with bank officers for more complex transactions?
- *Food Service.* What is the impact of cash registers equipped with automatic

Handbook of Simulation, Edited by Jerry Banks.
ISBN 0-471-13403-1

price-scanning technology on customer service time and the number of cashiers needed?

- *Entertainment.* At large theme parks, what is the impact of shortening ride length and/or adding additional capacity on average and maximum customer wait time?
- *Insurance.* What is the impact on the time to process claims as a function of how the claims flow through the process and the levels at which approvals must be made for various claim types? Given a "time to process" goal, what is the impact on the number of claim adjusters required?
- *Transportation.* What is the impact on queue length of various toll booth payment strategies, and for a given rate of vehicle arrival, what is the minimum number of each type of booth required to achieve acceptable queue lengths?
- *Medical.* In an emergency room, what are the required numbers of doctors, nurses, clerks, and major and minor treatment rooms necessary to achieve acceptable patient delays? How do the delays vary during the course of normal and extraordinary days?

These are just a few examples of the types of service systems and problems that can be studied with computer simulation. If the system is stochastic, complex, resource constrained, and fairly well defined, it is a reasonable candidate for analysis with simulation.

In the remainder of this chapter we provide an overview of some of the unique aspects of modeling service systems. First, we discuss how the modeling of service systems differs from the modeling of manufacturing systems. Then we present a template of the steps that will be needed in the development of a model of a service system. Finally, we present two case studies of simulation projects for service systems. The first case study involves the analysis of a bank. This study was a one-of-a-kind analysis and, as such, represents the typical modeling project in which a simulation modeler might become involved. The second case study involves the development of a simulation "environment" specifically for the purpose of modeling maintenance systems. The focus of the second case is on the number of maintenance personnel required to achieve acceptable system performance. This case study illustrates how simulation can be used to create an environment for the "casual" user of simulation that will allow the analyst to use simulation for many specific problems.

18.2 WHAT DIFFERENTIATES SIMULATION IN THE SERVICE INDUSTRY FROM MANUFACTURING SIMULATION?

The evolution of simulation has been due largely to its use in analyzing manufacturing systems. To understand some of the unique issues in service system simulation, let us explore some of the differences between simulation in manufacturing and simulation for service systems:

1. *There is often no clearly defined set of systems and components as in manufacturing.* Consequently, the simulationist must often define system behavior without the use of hard data on the process. For example, in modeling a manufacturing process, the analyst will usually have access to drawings and specifications. Service systems will rarely have such refined documentation. The simulation modeler is often called

upon to define and understand the process in more complete ways than anyone has before.

2. *Time waiting to be served tends to have a much greater importance than throughput.* Although work in progress has an undesirable cost in production systems, it is not unreasonable to allow considerable variability in work in progress and occasionally long times in the inventory. On the other hand, people hate to wait to be served. In many service systems, time in a queue will always be the key measure of performance and excessively long waits will not be tolerated.

3. *The system's performance often depends more heavily on human beings, who tend to be more unpredictable and variable.* In most service systems, a key determinant of the system's performance is the performance of the people in the system (i.e., those who provide the services). In contrast to machines, human performance is highly variable. In some systems, a good, well-trained person can perform a task up to 10 times faster than a poor performer. Even one person's performance can be expected to vary by as much as a factor of 2. Therefore, system variability in service systems is often much higher and, therefore, of greater importance. Factors such as experience level and training recency can have a large impact on performance time and accuracy.

4. *Service systems are often short-term demand driven, and these demands can vary by day and time.* Fluctuations in demand for services will greatly affect the service system's ability to provide good service. Again, because of the variability not only of service providers but of those demanding services, it is important to understand the factors affecting demand. However, in many ways, service systems are no different from any other type of system that a simulation modeler might study. What areas are the same as those in other kinds of systems typically simulated?

5. *A process is a process is a process, and the flow of entities through a process with constrained resources is fundamentally the same as in other simulations.* Fundamentally, the same analytical issues and system modeling constructs pertain to all discrete-event processes. Therefore, the simulation tools that work with general-purpose discrete-event simulation will work well to model service systems.

6. *The focus of system analysis is usually on the resource–performance trade-off.* At the root of virtually all types of simulation, the question remains the same: How can I do more with less? Although the particular issues and emphases may differ, the basic issues are the same whether it be a manufacturing, service, or any other type of discrete system.

18.3 PROCESS OF BUILDING A SIMULATION OF A SERVICE SYSTEM

There is no absolute process for developing a service system model—or any simulation, for that matter. (See Figure 1.1 for an illustration of the simulation process.) Our experience has led us to suggest the following steps. They need not always be performed in this order and there is often iteration between the steps; however, all steps should be performed at some point in the model development and utilization process. Also, many of the steps in the simulation process are common to all types of simulation, not just service system simulation.

1. *Define the measures of system performance that are of interest.* Although this is always of importance to the analyst, it is perhaps more critical in service systems.

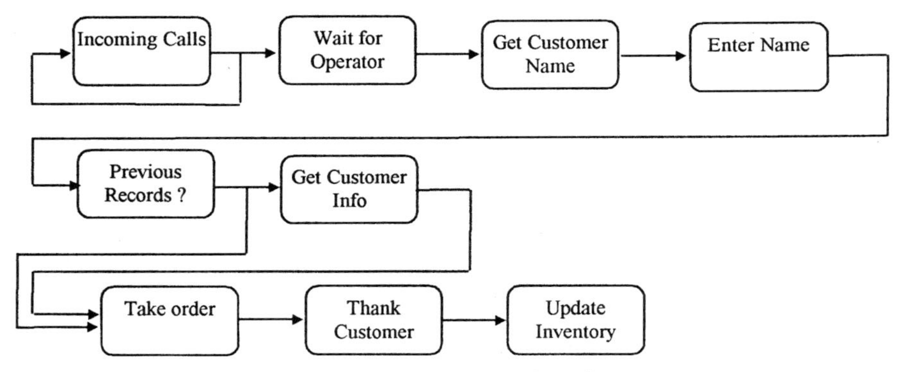

Figure 18.1 Sample process flow diagram.

Clear definitions of system performance measures (e.g., average and maximum customer waiting time) and acceptable levels of performance should be discussed from the outset. The simulation modeler can expect that these will evolve as the project progresses, but a clear definition of "what matters" will ensure proper model focus.

2. *Define the system characteristics that are to be varied to evaluate performance.* Service systems can involve hidden facets and occasional "trap doors" that influence system success. Since the service system is often not as well defined or documented as other types of systems, the data describing how the system *really* behaves are sometimes hard to find. For existing systems they can often be found only by watching the system operate and/or by interviewing service providers. For new systems, a simulation modeler should work with the process engineers or system designers. However, in many cases, the simulation modeler should expect that he or she will be the first person to define the service process rigorously and, therefore, the important system characteristics.

3. *Define general model scope.* As with other simulation projects, the model scope should be defined through joint consideration of the model goals and system characteristics, coupled with the budget that has been allocated to model development and use. It is up to the simulation modeler to ensure that the overall model design will address the desired issues while having a good chance of being performed within the budget allotted for the simulation.

4. *Develop process flow diagrams.* Through data collection techniques that are discussed in Chapter 3, the basic model is developed and represented in the form of a process flow diagram, as shown in Figure 18.1. These are also an excellent medium of communication between those knowledgeable about the system and the simulation modeler.

5. *Collect data on the process.* Once the flow of the process is determined and documented through the process flow diagram, the focus of model development becomes collecting data on model parameters such as times and error rates. As stated above, in service system models this is a particularly important phase since a large part of the service system may involve highly variable human behavior.

6. *Build the baseline process model, including resources, entities, and attributes.* This step involves entering the process flow and system data using the appropriate sim-

ulation software. The exact method used to build the model (define the tasks, resources, and attributes) will depend on the simulation software selected.

7. *Validate the baseline model.* To the extent possible, the model should be validated using the techniques discussed in Chapter 10. Because of the variable nature of service systems, validation of parts of the model will often be more feasible than validation of the overall model outputs.

8. *Run parametric studies.* Simulation is a useful tool for studying the effects of changing certain system parameters on industry performance measures. This aspect of model development is no different than when using simulation to study other discrete systems.

18.4 CASE STUDIES

Below are descriptions of two case studies where simulation was used. The first case study involves the use of simulation to study a Japanese bank and ways to improve the time spent by customers in the bank. It illustrates how simulation can be used to address a specific set of issues in a service system. The second example discusses the use of simulation to evaluate staffing issues in large aircraft maintenance organizations. In this case study, a customized tool was developed that allowed the client to examine many different alternatives for current and new aircraft designs and maintenance concepts. It illustrates how simulation can become part of the analysis process in a service system, in addition to being able to answer specific questions.

18.4.1 Case Study: Simulation in the Japanese Banking Industry

Issue. The transaction completion process in the Japanese banking industry has many areas that could be improved to better meet customer and bank needs. The current transaction completion process at Japanese banking institutions includes several tasks that are time intensive, inefficient, and may prevent tellers from keeping up with customer demand. Potential methods that could be used to decrease the time required for each transaction and improve accuracy include hardware improvements, software enhancements, and process redesign.

A provider of equipment to the Japanese banking industry determined that a cost-effective and efficient method to investigate the Japanese banking industry problems and potential solutions was simulation modeling. A recent study was performed that involved the following steps:

1. Identify the specific industry needs.
2. Obtain data on the current processes.
3. Build a simulation model that is capable of examining the effects of changing various parameters on the transaction process.

This project and the results are described in detail below.

Process of Building the Model. To examine the effects of changing various parameters of the Japanese transaction process, a base model of the current transaction tasks was developed. The steps involved in the model development were:

1. Determine what questions the model will help answer.
2. Perform a task analysis on data collected.
3. Develop the model.
4. Develop a graphical animation.

Each of these steps is discussed below.

Step 1: Determine what questions the model will help answer. One purpose of the Japanese banking project was to identify the effects on the overall transaction process of changing various parameters. In particular, it was desired to obtain results from the effects of changing system, personnel, timing, and/or resources. The output desired included time per transaction, average transaction time for each transaction type, total customer time in the bank, customer wait times, teller utilization, transaction queue times at the teller stations, and maximum queue lengths.

Step 2: Perform a task analysis on the collected data. Simulation models can reflect the real world only as accurately as the data supplied. The Japanese banking model was built using data collected at several different Japanese banking institutions. Data on the current transaction process were obtained through the use of background questionnaires, personal interviews, site analyses, videotape analyses, and information analyses. A task analysis was performed on the data collected. Through the task analysis, 14 different transaction types were identified. Then each transaction type was broken down by identifying what tasks occurred, who performed (or can perform) each task, what were the time distributions for each task, and what were the possible task sequences. The basic sequence of events for most transaction types is presented in Figure 18.2.

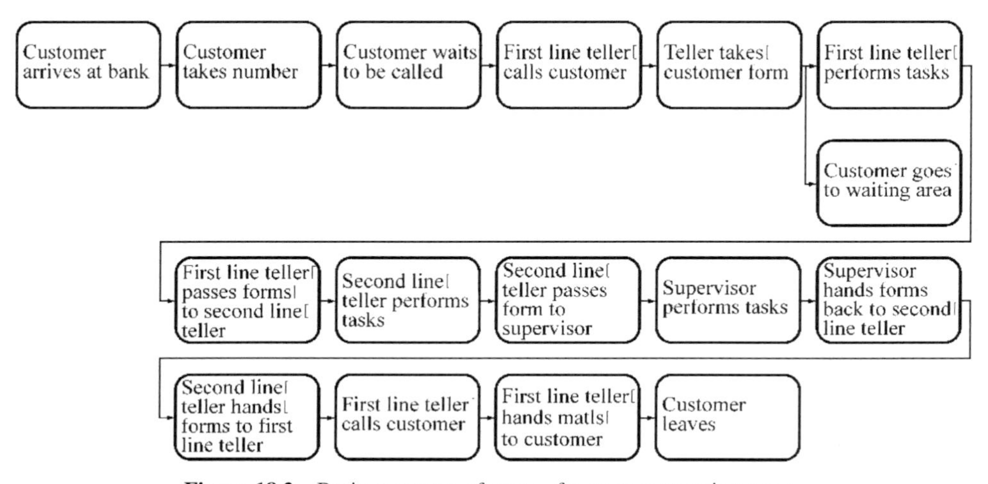

Figure 18.2 Basic sequence of events for most transaction types.

Unlike U.S. banks, where each teller works on one transaction until it has been completed, Japanese tellers may be processing multiple customers at a time. In a Japanese bank, once the first-line teller has handed one set of forms to the second-line teller, he or she may call the next customer and begin a new transaction sequence or the first-line teller may finish a transaction that has been returned to the queue. Each transaction has its own unique set of steps that are required. Some transaction types only require the first-line teller to perform tasks, others require both the first- and second-line tellers, and still others require a first-line teller, a second-line teller, and a teller supervisor.

Step 3: Develop the model. Once the task analysis was complete a model was developed for computer simulation. The model development process involved:

- Developing the task networks (i.e., the process flow diagrams)
- Entering task data
- Incorporating system logic through the use of variables
- Defining the output data
- Debugging the model

Step 4: Develop graphical animation. Graphical animation is a useful tool for debugging, demonstrating, and presenting the model. For the base Japanese banking model, a graphical animation was constructed along with the simulation to show the general layout of the bank along with various output parameters. While the model executes, the animation is updated to represent the current state of the system. The animation background is shown in Figure 18.3.

The Model Itself

Task Networks. The modeling tool *Micro Saint* was used to build the bank model. Consistent with Micro Saint's architecture, the Japanese banking model was developed in a hierarchical fashion. The top-level network (Figure 18.4) consists of standard customer initiation activities (customer arrives, customer takes service ticket, customer waits for service, teller calls the customer) followed by subnetworks for the first-line teller, second-line teller, and teller supervisor activities. Each teller network is then further decomposed into subnetworks for each of the 14 transaction types (Figure 18.5). The transaction type subnetworks (Figure 18.6) contain the individual tasks performed (verify account number, select deposit screen, insert print slip, etc.). As these example networks illustrate, the model was developed at a fairly high level of detail.

Input Parameters. The model was constructed so that certain system input parameters could easily be manipulated. The user can easily modify any of the following input parameters:

- Type of day (busy or normal)
- Customer arrival rates (based on branch and type of day, normal or busy)
- Transaction-type ratios
- Mnemonic times (mean and standard deviation)
- Number of first-line tellers

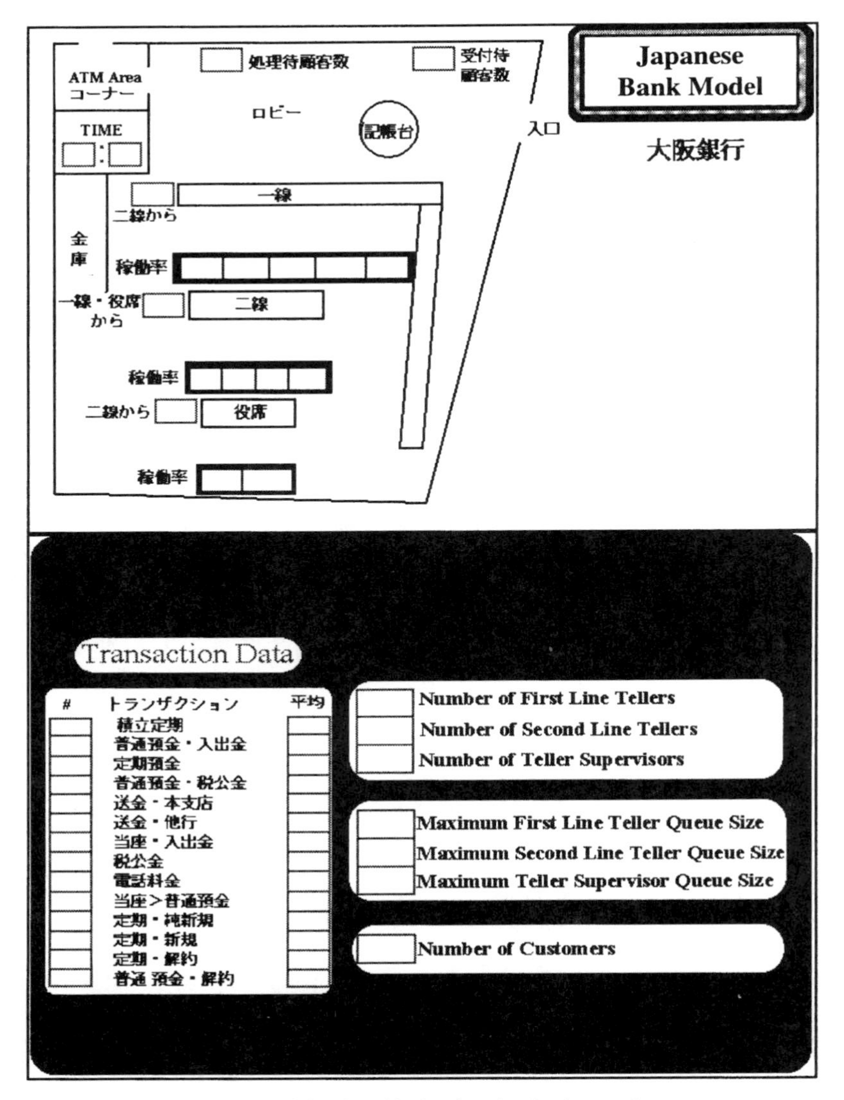

Figure 18.3 Graphical animation background.

- Number of second-line tellers
- Number of teller supervisors
- First-line teller to second-line teller assignment
- Second-line teller to supervisor assignment
- Probabilities for model decision points (% of customers who require change, % of customers who withdraw more than 1,000,000 yen, ...)

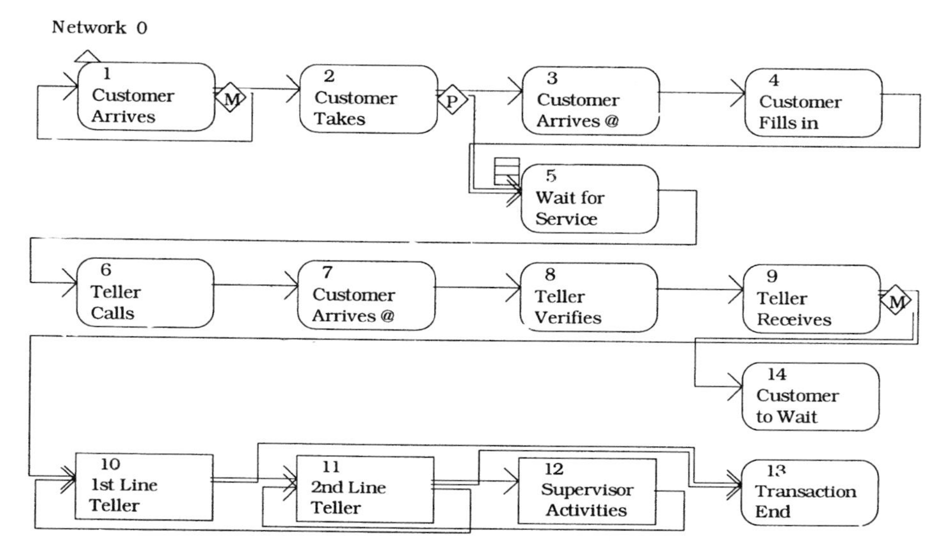

Figure 18.4 Top-level network.

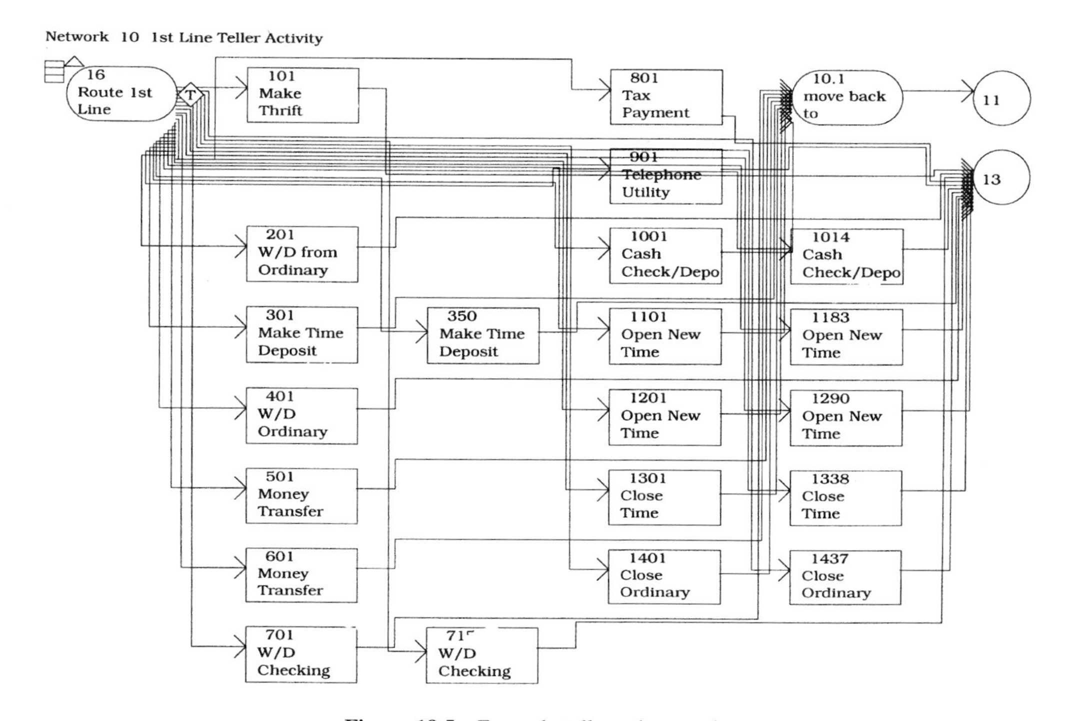

Figure 18.5 Example teller subnetwork.

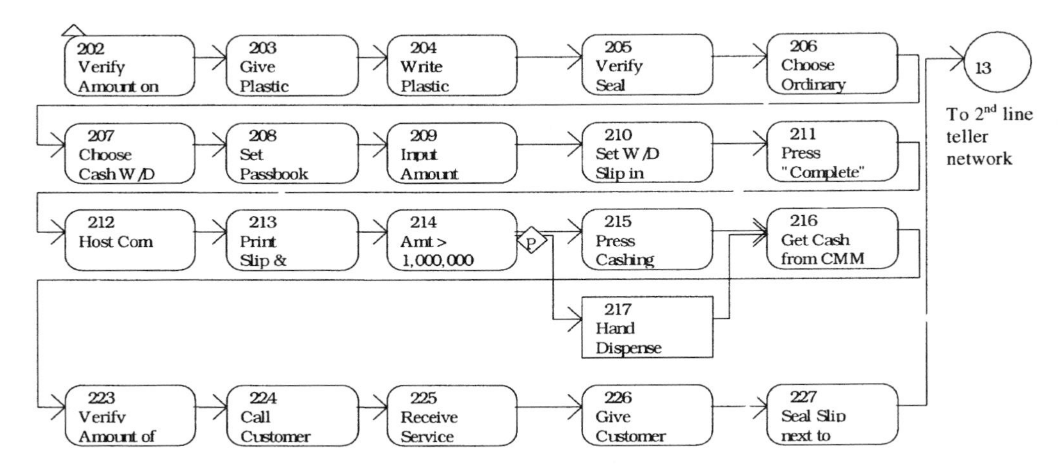

Figure 18.6 Example of a transaction subnetwork for withdrawing from an ordinary account.

Model Output. The output data collected from the model include time per transaction, average transaction time for each transaction type, total customer time in the bank, customer wait times, teller utilization, transaction queue times at the teller stations, and the maximum queue lengths.

Results and Value of the Study. As an illustration of how the model can be used, the model was run using two separate scenarios. The effect of adding one additional second line teller to the bank can be seen.

Scenario 1	Scenario 2
Branch 2, busy day	Branch 2, busy day
Three first-line tellers	Three first-line tellers
Two second-line tellers	Three second-line tellers
One teller supervisor	One teller supervisor

The data collected for each scenario are summarized in Table 18.1. These data show that by adding one additional second-line teller, the average amount of time each customer was in the bank decreased from 116.7 minutes to 9.99 minutes.

Through the use of a Japanese banking simulation model, the current and proposed transaction processes were evaluated and analyzed in order to understand, identify, and test opportunities for process improvement or reengineering. This base model is being used to evaluate different process strategies (e.g., cross training of tellers to eliminate the multitiered system) and bank automation (e.g., automated teller machines) on resource requirements (e.g., tellers) and customer throughput. In sum, simulation provided a tool to the banking industry that allows them to find ways to improve customer service while potentially reducing their operating costs.

TABLE 18.1 Summary Data

	Scenario 1		Scenario 2	
	Avg.	Std. Dev.	Avg.	Std. Dev.
Number of customers	533		533	
Average customer time in bank (min)	116.7	10.59	9.99	1.73
Maximum time customer in bank (min)	277.2	20.34	43.1	24
Average initial wait (sec)	80	7.1	83.4	8.8
Maximum initial wait (sec)	504.1	66.51	550.9	129.7
Average transaction wait (sec)	6,865.5	635.6	457.2	97
Maximum transaction wait (sec)	16,421.7	1,263	1,617	722.2
Maximum wait queue size (sec)	6.4	1.42	7	2.4
Maximum first-line teller queue size	13.9	8.9	26.1	22.9
Maximum wait for first-line teller (sec)	845.5	496	1,228	755
Maximum second-line teller queue size	186.7	12.56	17.1	3.57
Maximum wait for second line teller (sec)	16.223	1,243	806.5	121.9
Maximum supervisor queue size	2.9	0.74	4.9	1.3
Maximum supervisor wait (sec)	88.37	11.58	144.1	20.2
No. 1 First-teller utilization (%)	55.48	1.9	80.1	0.7
No. 2 First-teller utilization (%)	48.98	2.4	72.1	1.43
No. 3 First-teller utilization (%)	42.78	2.5	62.4	2.34
No. 1 Second-teller utilization (%)	99.35	0.23	82.2	1.7
No. 2 Second-teller utilization (%)	50.71	2.1	74.3	2.6
No. 3 Second-teller utilization (%)			62.12	2.7
Teller supervisor utilization (%)	41.36	2.1	59.6	1.63

18.4.2 Case Study: Using Simulation in a Government Service System for Providing Aircraft Maintenance

Governments provide services to maintain a country's infrastructure, provide basic services to the population, and support emergency operations. Several aspects of government services lend themselves well to simulation. For example, emergency operations include complex interactions between people, locations, and vehicles. Generally, these sorts of operations are more difficult to represent on paper or in a spreadsheet. Task network modeling can effectively track many tasks, people, and objects, as well as the complicated interactions between them. Declining budgets have forced constant reductions in personnel while the demand for service has remained the same. Simulation provides an excellent way to experiment with alternatives before implementation.

A particular case is that of the U.S. government maintaining a large number of aircraft, typically in excess of 500 at a time. To further complicate the problem, the demand for maintenance services could vary wildly during a high-demand period. During these periods, the right number of available aircraft was critical. Some of the available options are:

- More aircraft could be purchased.
- Aircraft could be made more reliable.
- Aircraft could be designed to be easier and faster to maintain.
- More maintainers and maintenance equipment could be utilized.

- Increased personnel training could be implemented.
- Automation could be implemented.

In fact, they could do a little of each, a lot of a few, or all of these. But what would that mean? How would these alternatives affect the number of aircraft available during a high-demand period, and what would it cost? Additionally, during periods of an emergency, the question was always posed regarding how best to deploy maintenance resources to achieve satisfactory aircraft availability.

Further complicating this issue, the questions could not just be answered once. As the maintenance organizations changed and new aircraft designs came on line, the same questions would once again need evaluation. Accordingly, the government did not need a *study* as much as they needed an *analytical tool.*

Simulation Solution. In response to this need, MANCAP II (Manpower Capabilities Analysis Tool) was developed for the Army. The tool was designed to allow those not familiar with simulation to use the technique to evaluate maintenance organization performance for different aircraft designs for a variety of fleet configurations, from platoon through division. Because of domain of application was limited (maintenance systems analysis), a simulation-based tool could be developed that was easy to use for this application, but limited only to use on this type of application.

MANCAP II consists of five separate but interrelated modules. The first module simulates the failure of each component of each weapon system in the fleet that occurs as a result of use. The second module generates component failures that result from combat. The third module is a generic model that simulates the maintenance requirements at different organizational levels given different maintenance concepts (i.e., who repairs what components). The fourth module is the personnel pool that simulates the maintenance staff hours that are available on a daily basis for each of the various types of staff. Staff members vary in the amount of training received and are qualified to repair particular subsystems of the aircraft. The last module of MANCAP II estimates the numbers of supply and support personnel that are required for a given level of maintenance activity.

The user defines a schedule of missions and how many aircraft fly each mission. As the model runs, aircraft component failures are generated based on combat hits, component utilization, and expected failure rates. When components fail, they can either cause mission aborts or simply be brought into the maintenance shop for repairs after the mission is completed. These maintenance demand drivers are represented graphically in Figure 18.7.

As a system requiring maintenance enters the shop (at whatever level of maintenance is required), the maintenance action is put into a prioritized queue and repaired when staff and parts are available. If either staff or parts are not available, maintenance is put into a queue. The concept of the delivery of maintenance is presented graphically in Figure 18.8.

MANCAP II is a software shell that incorporates a Micro Saint simulation model at its core. Each of the interrelated modules discussed above is represented by a portion of the network. The maintenance portion of the network is shown in Figure 18.9. As each aircraft enters the maintenance portion of the network, the components that need maintenance and the kind of maintenance needed are determined. Some components need to

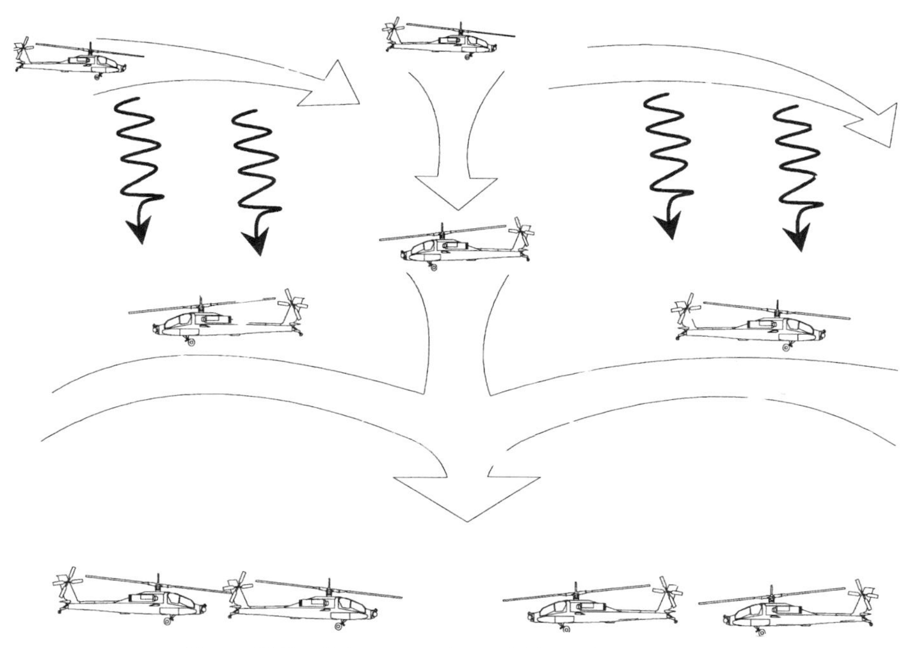

Figure 18.7 Maintenance demand drivers in MANCAP.

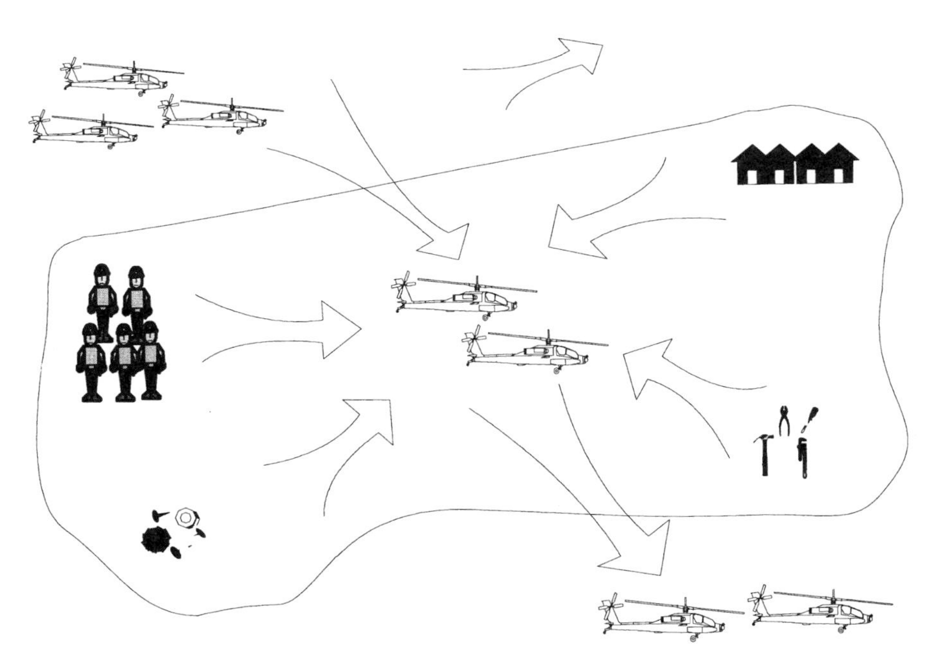

Figure 18.8 Concept of maintenance delivery.

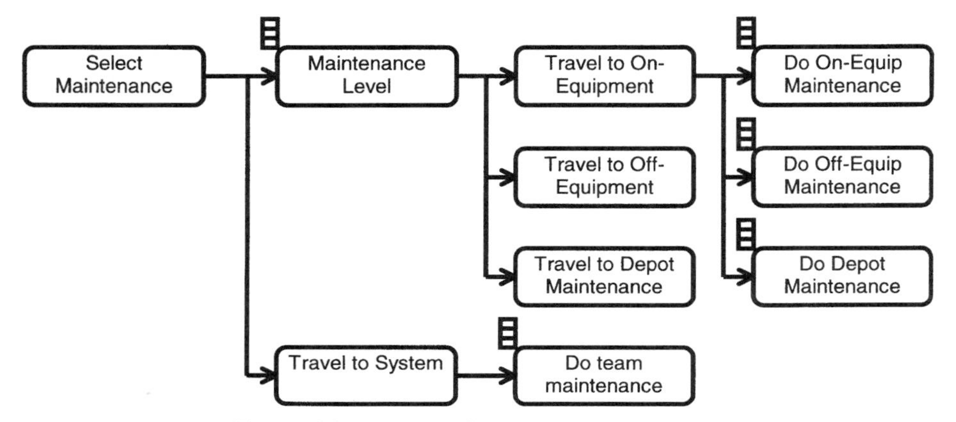

Figure 18.9 Example of maintenance task network.

be repaired without removing them from the aircraft. This is on-equipment maintenance. Others can be removed and repaired closeby (off-equipment maintenance) or sent to a remote maintenance location (depot maintenance). In some cases the maintenance team must travel into the field to repair the aircraft. Symbols that resemble ladders appearing in front of a task indicate queues.

The maintenance task network is internal within the application software. The user can not directly change the task network itself. He or she can, however, enter a wide variety of input parameters, such as "What parts does this aircraft have?", "How often do they fail?", "Who has to fix them?", and "How many of each type of maintenance person are available at the different shifts?" These parameters are input by the user via a spreadsheet-like user interface Figure 18.10 shows an example of the necessary inputs to define a mission.

MANCAP II automatically combines the input parameters with the core Micro Saint network and "runs" the model. This model generates system performance data such as scheduled and unscheduled maintenance and aircraft availability, as presented in Figure 18.11.

MANCAP II provides the user with data to diagnose system constraints such as personnel utilization, as shown in Figure 18.12. This graphic presents the number of required electrical repair specialists over a period of 30 days. These are just a few examples of the many types of output data that the MANCAP II simulation tool can collect and present to the user without the user needing to do any more than fill in a few spreadsheets describing the service system.

The U.S. government uses the MANCAP II tool routinely to study the effects of various surges in the demand for maintenance over a period of weeks or months. What they want to know is the number of available aircraft given a limited number of qualified maintainers. They also use MANCAP II to evaluate which type of aircraft they should buy based on relative costs *and* maintenance requirements. The result is that the U.S. government was able to make better-informed decisions by taking into account all the important variables as *they affected one another*. This could only be achieved through the use of simulation.

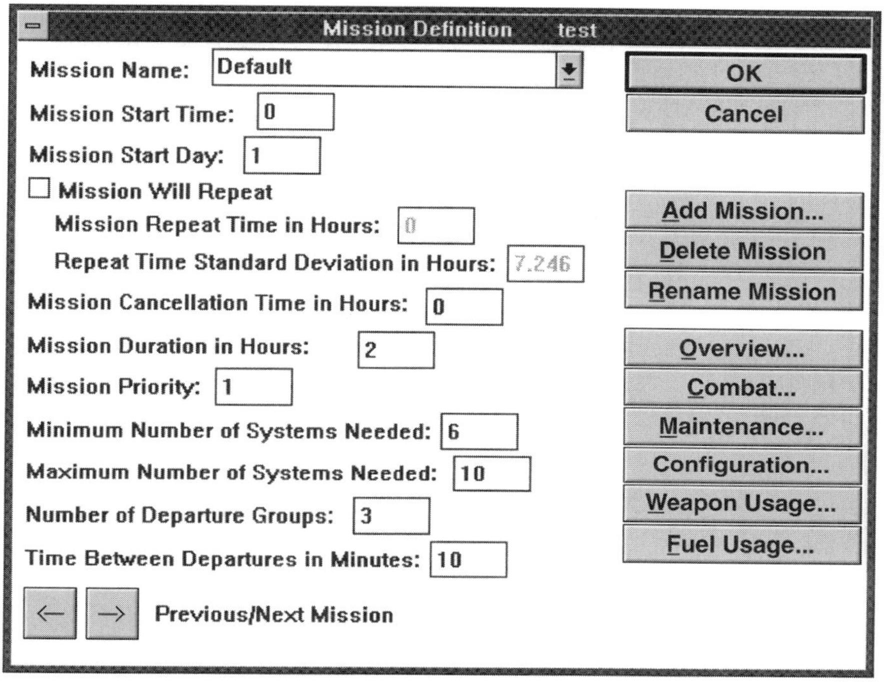

Figure 18.10 Example of the interface for entering the parameters required of the simulation model.

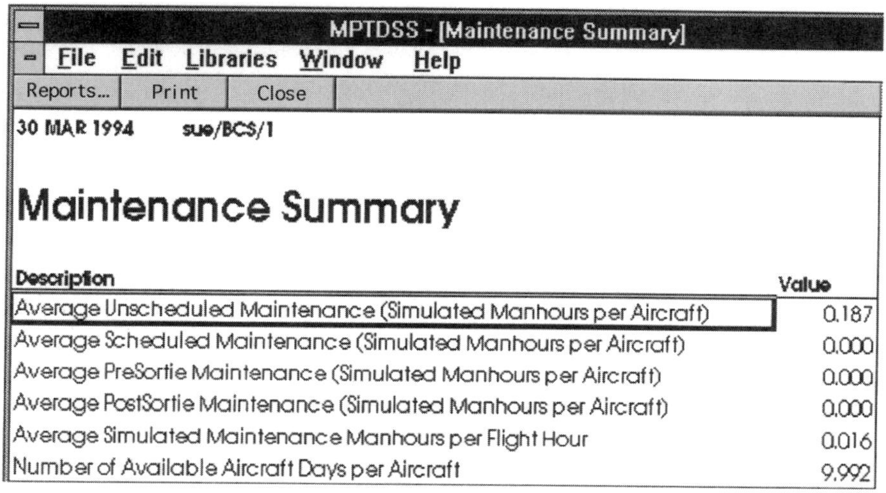

Description	Value
Average Unscheduled Maintenance (Simulated Manhours per Aircraft)	0.187
Average Scheduled Maintenance (Simulated Manhours per Aircraft)	0.000
Average PreSortie Maintenance (Simulated Manhours per Aircraft)	0.000
Average PostSortie Maintenance (Simulated Manhours per Aircraft)	0.000
Average Simulated Maintenance Manhours per Flight Hour	0.016
Number of Available Aircraft Days per Aircraft	9.992

Figure 18.11 Summary maintenance system performance data.

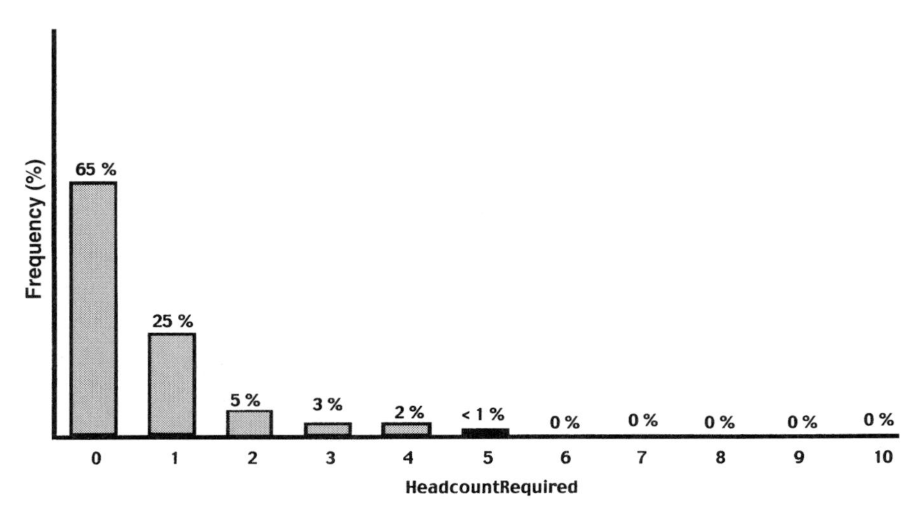

Figure 18.12 Resource utilization data on the number of required electrical repair specialists over a period of 30 days.

18.5 SUMMARY

The use of simulation to study service systems is still a fairly new concept. On the demand side, simulation will be used more as service systems strive to become more efficient and effective. The past 5 years of corporate process reengineering and downsizing have provided a significant stimulus in this regard, but we are just beginning. On the supply side, as simulation tools become more efficient and environment friendly, other types of engineers and managers will begin to use them and appreciate their value. The evolution in the usability of simulation tools over the last decade has already made major strides in opening the market for the use of simulation in service systems. Based on these trends, the analyst should look toward the use of simulation as a way to improve service systems as a great opportunity for future growth in simulation applications.

CHAPTER 19

Military Simulation

KEEBOM KANG
Naval Postgraduate School

RONALD J. ROLAND
Rolands & Associates Corporation

19.1 INTRODUCTION

Simulation has been applied extensively and successfully to a wide range of military problems, including wargaming, acquisition, logistics, and communications. For example, it has been used as a decision support tool to evaluate how a battle force should be constituted, how it might be deployed, and how the weapon systems should be acquired and maintained.

The defense area often requires domain experts since the development and use of simulation models require specialized knowledge in problems unique to the military. Most military models have been developed by highly specialized groups and used in narrowly focused user communities. For example, the Army is interested in wargaming simulation with ground forces and tanks, while the Navy is interested in battle group simulation with aircraft carriers, aircraft, and ships. The Marine Corps' interest is in amphibious operations, and the Air Force is working on space systems, strategic long-range bombing, and tactical air-to-air and air-to-ground support. Each service adopts different logistics systems for weapons systems maintenance; the Navy and the Marine Corps follow three levels (organizational level, intermediate level, and depot level) of maintenance, while the Air Force uses two levels of maintenance, and the Army has five levels.

As a result, there are many organizations and agencies involved in modeling and simulation (M&S) in different services, and each community uses its own special jargon, abbreviations, and acronyms, which makes it difficult for the various government agencies and services to communicate. [More than 150 pages of *Glossary of DoD M&S Terms* are available from Department of Defense (DoD) Directive 5000.59-M. It can be downloaded from http://www.dmso.mil/docslib/mspolicy/glossary/glossary1195.pdf.] Although the military simulation community is huge, there is a lack of systematic communication and no central resource library. Many models have been developed on a

Handbook of Simulation, Edited by Jerry Banks.
ISBN 0-471-13403-1

stand-alone, system-specific, as-needed, and as-afforded basis, which has resulted in redundant investments. Typically, more efforts are spent to develop a new simulation system that goes into building the simulation infrastructure than efforts to develop components specific to the purpose of the simulation. If the infrastructure and other simulation components could be reused, the payoff would be enormous. For this reason, object-oriented modeling and new architectures (HLA) are getting more attention. Architectures are discussed further in Section 19.4.

Although M&S has been used to investigate military problems for many years, there is very little literature available for general readers who do not specialize in military simulation. Military simulation models are different from others because (1) many of them are highly classified, with details that could not be widely disseminated; (2) weapon capabilities and use are not typically used in other M&S; (3) certain algorithms are closely controlled to avoid reverse engineering by potential adversaries; and (4) the use of certain equations (e.g., Lanchester, which is often used in wargaming simulation) is not typical of commercial M&S. The purpose of this chapter is to provide readers with an overview of military simulation and insight to its future directions. We review recent developments in military modeling, particularly in wargaming simulation. We provide sources of many DoD documents, including Internet homepage addresses, where applicable, so that the reader can retrieve the updated information after this handbook is published. Since more and more DoD documents are available electronically via the Internet, it becomes easier for nonmilitary simulationists to access information on defense simulation. There will be tremendous opportunities for expansion of simulation applications in the military.

In Section 19.2 we review background material, including the organizational structure of M&S activities in the DoD and the classification of military simulation models. In Section 19.3 we discuss a large-scale wargaming simulation model, the Joint Theater Level Simulation (JTLS), as an example. In Section 19.4 we introduce distributed interactive simulation (DIS), high-level architecture (HLA), and their current applications and future directions in military simulation. Section 19.5 covers nonwargaming simulation applications, and Section 19.6 provides concluding remarks.

19.2 BACKGROUND

19.2.1 Organizations

To understand current activities of military simulation, we briefly review several organizations that are involved. For a comprehensive list of the M&S organizations in the DoD and each service, see Chapter 3 and the appendix of ref. 14. We provide Internet homepage addresses so that the reader can retrieve up-to-date information on the organizations.

The DoD and the Joint Staff maintain their own agencies for M&S, and each service maintains M&S offices. In June 1991, the Defense Modeling and Simulation Office (DMSO) was established by the Under Secretary of Defense for Acquisition and Technology. The DMSO publishes DoD M&S policy and promotes cooperation among DoD agencies. In January 1994, the Deputy Secretary of Defense sent *DoD Directive 5000.59* to the DoD community [3]. The directive, entitled *DoD Modeling and Simulation (M&S) Management*, was a DOD-wide effort to establish policy for M&S. It was a significant step toward centrally managed DoD M&S activities.

In accordance with the DoD Modeling and Simulation Master Plan (DoD 5000.59-P, dated October 1995) [4], the DMSO is leading a DoD-wide effort to establish a common technical framework to facilitate the interoperability of all types of models and simulations among themselves and with C4I (command, control, communication, computer, and intelligence) systems, as well as to facilitate the reuse of M&S components. This common technical framework includes the high-level architecture (HLA), which represents the highest-priority effort within the DoD modeling and simulation community.

DoD Modeling and Simulation Master Plan initial definition of the M&S HLA was accomplished under the sponsorship of the Defense Advanced Research Projects Agency (DARPA) Advanced Distributed Simulation (ADS) program. It was transitioned to the DMSO in March 1995 for further development by the DoD-wide Architecture Management Group (AMG). Central to this task was the development of a set of prototypes that addressed critical issues in the HLA. In September 1996, the Under Secretary of Defense for Acquisition and Technology [USD(A&T)] approved HLA as the standard technical architecture for all DoD simulations and required that all computer simulations for military operations meet the HLA standardization requirements by FY2001. His directive also mandated that all DoD simulations failing to comply with HLA standards by a specified date be retired from service. In Section 19.4 we discuss more details on HLA. Also see http://hla.dmso.mil/ for current updates on HLA.

The Executive Council for Modeling and Simulation (EXCIMS; see http://www.dmso.mil/wrkgrps/excims/charter.txt) is a high-level advisory group on DoD M&S policy, initiatives, standards, and investments. For more details concerning DoD simulation activities, see *DoD Directive 5000.59*, available from the DMSO Internet home page.

Each service maintains its own M&S activities. The Army has long M&S history and is better organized than the rest of the services. Deputy Undersecretary of the Army for Operations Research (DUSA/OR) oversees all Army Modeling and Simulation. The Army Modeling and Simulation Office (AMSO; see http://www.amso.army.mil) is the operational activity for Army M&S. The Army maintains the modeling and simulation home page for the Army Modeling and Simulation Resource Repository (MSRR) (see http://www.msrr.army.mil). The Army National Simulation Center (see http//:leav-www.army.mil/nsc) located in Ft. Leavenworth, Kansas, supports simulation training exercises around the world. The Army currently maintains six major simulation models for training. They are Janus, VICTORS (Variable Intensity Computerized Training System), BBS (Brigade/Battalion Battle Simulation), CBS (Corps Battle Simulation), TACSIM (Tactical Simulation), and CSSTSS (Combat Service Support Training Simulation System). The details of these models and a list of other Army simulation models are available in Models & Simulations: Army Integrated Catalog (MOSAIC); see also ref. 12. A brief discussion of Janus is given in Section 19.4.

The Air Force also has a long history of M&S applications. The Commander of Air Force Agency for Modeling and Simulation (see http://www.afams.af.mil) is the single point of contact in the Air Force for policy on modeling, simulation, and analysis activities. It includes the Evaluation Support Division, Technical Support Division, Warfighting Support Division, and Air Force Studies and Analysis Agency. AWSIM (Air Warfare Simulation) is one of the Air Force unique models that are capable of tracking individual aircraft by tail number and air-to-air and surface-to-air missile (SAM) engagements.

The Navy and the Marine Corps have smaller M&S organizations than those of the Army and the Air Force. They have a Modeling and Simulation Advisory Council that guides the development of policy, coordination, and technical support and promotes the use of the Navy-wide common support services. The Navy and the Marine Corps maintain their own Modeling and Simulation Management Offices (see http://navmsmo.hq.navy.mil) and set their own M&S policies. A simulation model RESA (Research, Evaluation, and Systems Analysis) was developed as a naval warfare C3 (command, control, and communication) analysis tool for the Navy, and MTWS (Marine Air Ground Task Force Tactical Warfare Simulation) is one of the Marine Corps' tactical combat simulation models.

For more information regarding the organizations and groups involved in military M&S activities, see for example, Internet home page http://www.dmso.mil/orgs.html #OEGS. Readers will be surprised at the extent of the list.

19.2.2 Classification of Military Simulation Models

According to the Defense Science Board, military simulations are classified into three categories: *live*, *virtual*, and *constructive*. Although there is no clear-cut distinction among these categories, it is still helpful to understand the basic differences.

Live simulation involves real people and real systems. Operational test and evaluation (OT&E) and military field exercises are examples. Live simulations in support of training are conducted at the Army National Training Center (NTC) located in Ft. Irwin, California; the Navy "Strike University" in Fallon Naval Air Station, Nevada; the Air Force Red Flag Site at Nellis Air Force Base, Nevada; and the Marine Corps Air-Ground Combat Center in Twenty Nine Palms, California. Here is an example of live simulation. The NTC is a vast expanse of desert approximately the size of the state of Rhode Island. This is where the Army conducts training exercises to prepare itself for desert warfare in the Middle East. There are some 2500 soldiers permanently stationed at NTC who function as the "home team." This group pretends to be the enemy and uses all of the doctrine and tactics of the opposing army. For most of the last 30 years, they have been training to fight a war against the Soviet Union. The visiting teams arrive at NTC 12 times a year and conduct wargame-type simulations against the home team. Every move and every shot fired is monitored by a powerful laser engagement system that records all the signals from the pieces of armor and other equipment that are participating in the exercise. All of this information is fed into the computer simulation, and numerous statistics are tallied so that at the end of the exercise, both teams can be evaluated and areas of improvement can be identified.

Virtual simulation involves real people in a simulated system. This includes aircraft and tank simulators. For an example, see the Internet home page for different military flight simulator from http://www.bgm.link.com/mfs.html. This type of simulator is helpful in training and in evaluating control, decision, and communication skills. Virtual simulation has become more popular with developments in computer technology, especially computer graphics. The journal *Military Simulation & Training* (by Monch Publishing Group, Federal Republic of Germany, ISSN 0937-6348) is a good source for up-to-date information on military training simulators.

In *constructive* simulation, humans may (or may not) interact with the model, and everything is simulated. Constructive simulations of combat include wargames for training as well as for analytical tools. Constructive simulation training is usually designed for staff-level use and virtual simulation training for operator-level use. For example, JTLS is a constructive simulation that can be used for staff training as well as for planning analysis. We now describe JTLS in more detail.

19.3 EXAMPLE WARGAMING MODEL: JOINT THEATER-LEVEL SIMULATION

An important aspect of military simulation lies in wargaming. Wargames are used as an inexpensive alternative to live training exercises. They are also very useful for testing and evaluating proposed procedures, strategies, and weapons systems. As an illustration of a wargaming simulation, we present and discuss the JTLS (see http://www.rolands.com/jtls.html). The purpose of this section is to provide the reader with various aspects of wargaming simulation using JTLS as an example. The JTLS is a large-scale wargaming model including ground, air, and naval operations. More than 300 players/users can participate in one scenario simultaneously. We use the terms *player* and *user* interchangeably here.

Development of JTLS began in 1983 as a project funded by three Army organizations: the U.S. Readiness Command, the U.S. Army Concepts Analysis Agency, and the U.S. Army War College. It has had continuous functional and system upgrades since that time. Its focus is on conventional joint and combined operations and it is currently managed by the U.S. Joint Warfighting Center, Fort Monroe, Virginia.

JTLS is designed as a theater-level model for commanders and planners as a planning analysis tool, support material for education, exercise support for training, and a primary means to investigate the results of combat. It is currently used chiefly as an exercise driver where JTLS provides an environment for the dynamic interactions of intelligence, air, logistics, naval, and ground forces. This environment allows users to develop insight into the relative merits of alternative courses of action, force structures, combat systems, and procedures.

The model is, or has been, used by numerous agencies, including the Joint Warfighting Center, Warrior Preparation Center, NATO SHAPE Technical Centre, National Defense University, Army War College, Naval Postgraduate School, Combined Forces Command Korea, Australian Defense Force Warfare Centre, and the South Korean Institute for Defense Analysis. This model is being evaluated by the Ministry of Defense for Greece, the Ministry of Defense for the United Kingdom, and the Japanese Defense Agency. The Louisiana State University, MITRE Corporation, and RAND Corporation are also evaluating JTLS for application to nonwargaming environments or potential research purposes. Reference 1 discusses JTLS for military operations other than war.

JTLS is a multisided, interactive, computer-driven simulation. *Multisided* implies that there can be more than two sides, such as the blue (friendly) and red (enemy) forces used in conventional wargaming. One recent JTLS scenario includes sides called the Gulf Coalition, United Nations (UN) Forces, Israel, Iraq, and Iran. Each side in turn consists of one of more factions limited by the hardware, scenario requirements, and users' imaginations. The Gulf Coalition factions include Saudi Arabia, Kuwait, and one that represents the "civilian populace." All of the UN members are included as factions within the UN side. Factions are also included within the other sides, which permits an accurate depiction of the forces and perturbating influences within the Gulf region.

A maximum of 10 sides can be represented in JTLS. Each side can be further divided into an unlimited number of factions. A faction's side allegiance is dynamically changeable during the game (scenario). Side relationship is asymmetric and can also be changed during the game. A large number of players (over 300) can be involved in a single game simultaneously. JTLS can model coalition air, land, sea, amphibious, and special forces operations. The model supports limited nuclear and chemical effects, low-intensity conflict, and preconflict operations. The model also supports the representation of civilian and noncombatant forces within sectors of interest.

The JTLS system consists of six major software modules and numerous smaller support programs that work together to prepare the scenario, run the game, and analyze the results. Designed as a tool for use in the development and analysis of operation plans, the model is theater independent (i.e., the data for a specific scenario are stored in a database separately from the program). The database may contain highly classified information on weapons systems and military operations. However, the JTLS program itself is unclassified.

Model features include Lanchester attrition algorithms, detailed logistics modeling, and explicit air, ground, and naval force movement. In addition to the model itself, the JTLS system includes software designed to aid in scenario database preparation and verification; entering game orders; and obtaining scenario situational information from graphical map displays, messages, and status displays. The movement of forces within any combat environment is affected by the terrain. The terrain is represented as a hexagonal grid overlay on a map projection. The maximum geographic region or area used in a JTLS scenario is 2000 by 2000 nautical miles. The hexagonal overlay design is used to provide an efficient means to calculate and model force movement and to describe both terrain and human-made obstacles. Each hexagon in the database is described in terms of its relative geographic location, the terrain within the hexagon boundaries, the elevation, and the barriers on each of the six sides. Hexagon size and the number of hexagons represented in a terrain database are user-data entries. Locations of objects in the game can be displayed as a hexagonal reference, latitude/longitude, or a military grid reference. Objects can be located anywhere on the game surface and are not limited to the center of the hexagons.

JTLS does not require programming knowledge. As an interactive model, it requires human decisions to manage the processes and entities. The players receive messages and reports concerning the movement, attrition, and logistics status of their own forces, as well as intelligence summaries and capabilities of opposing forces. The player at each workstation can elect to view messages in plain language or in a special military format. Messages may be sent electronically to standard Simple Message Text Protocol (SMTP) electronic mail workstations. Electronic feeds to several military command and control systems, such as the Global Command Control System, Joint Operational Tracking System, and Joint Military Command Information System, have been demonstrated.

The players interact with the game and receive graphical feedback through the Graphics Input Aggregate Control (GIAC). They receive messages through the Message Processor Program (MPP), and status board information is presented by the Information Management Terminal (IMT). These programs obtain their data and communicate with the main simulation component, the Combat Events Program, through software modules called the G Data System, using its data server program, GENIS. A single GENIS (the primary GENIS) is connected to the Combat Events Program using the TCP/IP network protocol. A GENIS may have other GENISes or interface programs as clients. The number of clients that a single GENIS can have at one time is determined by a system parameter of the machine on which it is executing. The parameter defaults to 64 on most machines and can be modified by system maintenance personnel. A typical player's workstation has a GIAC, MPP, and IMT, all operating and connected to a GENIS.

JTLS can be operated on a single workstation, or multiple workstations, and distributed on either a local area network (LAN) or a wide area network (WAN), thus providing a distributed exercise/gaming environment. The computer system support requirements for conducting simulations or analytic excursions using the JTLS model are dependent on the specifics of the event. The purpose of one event can be quite different from another (e.g., analysis, education, contingency plan development, etc.) and could require different support systems. The computing system is a composite of resources, such as hardware devices, system software and utilities, communication lines, language compilers, and databases.

The JTLS system can be run on a workstation of very limited processing power. For very small test databases, the CEP, GENIS, and two player suites (controller and one side) can be run on a single workstation of the SPARC station 2 class, but system performance is marginal. For exercise applications, in general, each active player requires a workstation of at least SPARC station 5 capability with 32 megabytes (MB) of random access memory (RAM) to perform adequately. For medium-sized databases, the CEP and the primary GENIS each should have a SPARC station 20 level workstation with 128 MB of RAM, and each subordinate GENIS should have a workstation of at least SPARC station 20 level processing power, with 64 MB of RAM.

The JTLS source, object, and executable files occupy approximately 550 MB of disk storage. A medium-sized to large database might require another 50 MB of storage. Each checkpoint will use between two and four times as much storage as the initial database, depending largely on the intensity with which player's messages are managed. A 1.3-gigabyte disk devoted to the game directory (with tape backup) is a reasonable starting requirement.

Most of the JTLS system is written in the SIMSCRIPT II.5 programming language. It improves continuously with new technologies. Los Alamos National Laboratories has developed a graphics user interface, and ROLANDS & ASSOCIATES Corporation (R&A) (see http://www.rolands.com) has created several tools for the development of scenarios for JTLS. JTLS has been used successfully in conjunction with live simulation during exercises such as KEEN EDGE and COBRA GOLD. KEEN EDGE 95 was held at Camp Ojojihara, Japan to introduce U.S. and Japanese Ground Self-Defense Force soldiers to each other's way of doing business. COBRA GOLD is a joint exercise of U.S. and Thailand forces held in Thailand.

19.4 SIMULATION AS A TRAINING TOOL

19.4.1 Need for M&S to Support Training

An annual military exercise, called "Return of Forces to Germany," typically required approximately 97,000 troops, 7000 tracked vehicles, and 1080 tanks. The cost in 1988 was \$30.5 million, plus \$23.4 million for repayment of damage to the German countryside. In 1992 with the use of computer simulation of much of the movement of forces, the cost was reduced to \$20.7 million using only 16,500 troops, 150 tracked vehicles, and no tanks. The damage to the German countryside was limited to \$250,000 [8]. Total savings amounted to \$32.9 million. With a tank costing well over \$1 million and a single live missile costing tens of thousands of dollars, the services have no choice but to maximize the use of training simulators and simulations in lieu of live exercise.

Safety is another reason why simulation is a better vehicle for training than live exercises. Simulation training also minimizes environmental damage and overcomes battlefield space limitations. Since World War II, the amount of space required for a fighter aircraft to conduct standard aerial maneuvers has grown from 4 to 40 miles, and for an Army mechanized battalion, from 4000 to 80,000 acres [8]. Rather than expanding to meet those needs, however, training ranges are typically being reduced in size by a combination of encroaching urbanization, increased commercial air traffic, political opposition, and more stringent environmental regulation. The services have responded with ever-increasing reliance on simulation to conduct their training exercises.

19.4.2 History of Simulation as a Training Tool

The U.S. military has been using simulation as a training tool for more than 60 years since the development of the Link Flight Instrument Trainer in the late 1930s (http://www.bgm.link.com/history.html). Early simulation trainers (simulators) were designed to aid in aircraft carrier landing, in-flight emergency responses, and air-to-air combat. The early simulators used movable cameras and immense model boards. In the 1970s electronic image generators began to replace model boards, and today there are only a few model board simulators in existence. Until 1984, training simulators were used primarily for teaching and perfecting individual and crew skills. The simulators were controlled by training teams sitting at consoles throwing contrived scenarios at trainees.

Modern combat requires more than individual and crew proficiency in specific tasks. In response to this challenge, the Defense Advance Research Projects Agency (DARPA) developed SIMNET as a combined arms team training system. SIMNET is a networked, real-time simulator. Trainees in each simulator can see and interact with any and all of the other players. SIMNET consists of several hundred simulators for tanks, helicopters, fixed-wing aircraft, and one-of-a-kind experimental nodes.

After the Gulf War, a particular tank battle called *73 Easting* was thoroughly analyzed with on-the-ground measurement of tank tracks, radio traffic recordings, photographs, and interviews. The entire battle was recreated and simulated in SIMNET so that analysts could use movable viewpoints, invisible to other players, to watch from any angle during multiple replays. SIMNET and the Battle of 73 Easting has had a dramatic impact on DoD training. In the future, all distributed simulations will be required to comply with the emerging High Level Architecture (HLA) protocol. Much of the recent literature in M&S has focused on HLA and its potential applications. We discuss ADS, DIS, and HLA briefly in the next section.

19.4.3 Advanced Distributed Simulation, Distributed Interactive Simulation, and High Level Architecture

Advanced distribution simulation (ADS) technology provides various combinations of live, constructive, and interactive simulations that may be separated geographically and may involve some or even a great deal of human interaction. As mentioned at the beginning of the chapter, virtually all military simulation models were developed for a specific purpose. Up until the early 1980s, the models were built and run on a mainframe computer. Thus these models were used by a very narrow user group at the computer center. The idea of ADS is to interconnect different simulation models via high-speed communication networks, using common network interface/translation devices, so that different users, using different simulations (e.g., ground, air, navy, special operations, logistics, etc.) at different geographic locations, can communicate and interact [6, 7].

Distributed Interactive Simulation (DIS), discussed in Chapter 12, has its roots in DARPA's SIMNET, which is a virtual simulation and is an IEEE-endorsed standard for ADS. Within DIS, the environment and its participants are distributed across a local area network, a wide area network, or any combination of the two. The DIS standard supports several types of simulation entities, including live, virtual, and constructive simulation. In practice, the terms *ADS* and *DIS* are used interchangeably [11, 17].

The Synthetic Theater of War (STOW) is a good example of DIS technology applied to the joint exercise. This exercise is conducted over the Defense Simulation Internet (DSI; a set of connected site networks dedicated to support simulation projects throughout DoD), which allowed widely dispersed and different kinds of units, some actual and some synthetic, to appear together on the electronic battlefield. For example, a pilot in an F/A-18 simulator at the Naval Air Warfare Center, Patuxent River, Maryland, can fly a mission with a shipmate actually in the air over the Atlantic Ocean while both are being controlled by a cruiser in Pearl Harbor, Hawaii.

The demand for fast, flexible, high-fidelity simulation systems will grow steadily through the turn of the century. The impetus for this demand is driven by military officials seeking efficient options for honing force readiness while working under shrinking budgets. Technologies such as DIS will be at the forefront of meeting the demand for simulation systems [16].

The key to DIS is that there is no central computer required, but rather, each simulator does its own computing and is passively linked to the virtual world. This eliminates the problem of information being susceptible to traffic jams and a centralized bureaucracy. The architecture that DIS represents today is still based on the concepts of SIMNET. It consists of a large number of protocols that are sent to each station on the net. As the sizes of these distributed exercises move from the company level (approximately 150 soldiers) to a Joint Force level (a Desert Storm–size force), the bandwidth capability of the network and the ability of each simulation to process all these packets will quickly overtake the SIMNET architecture. In April 1995, an initial definition of what the next generation of DIS architecture will look like was completed. It is called the High Level Architecture (HLA). The intention is that the HLA will become the next-generation DIS. (See http://hla.dmso.mil for more details.) DoD requires that all computer simulations for military operations meet the HLA standardization requirements by FY2001. The directive also mandated that all DoD simulations failing to comply with HLA standards by a specified date be retired from service.

It is important to recognize that HLA or any other M&S technique is not a panacea guaranteeing effective training. As is true of any technology, it is only through judicious applications of this technique that its full potential will be realized. It will be up to the training and research communities to ensure that the tools and techniques of HLA-based systems function effectively in training environments.

19.4.4 Transition from Janus to JLINK Using DIS

Janus was developed as a stand-alone model for Army battalion and brigade training, and analysis of combat weapon systems. Compared to the JTLS discussed previously, it is a smaller wargaming model for Army ground operations. It was originally written in FORTRAN and is essentially still a FORTRAN-based model with some C routines that were added when Janus was ported from a VAX- to a UNIX-based system. Janus is an interactive, six-sided, closed, stochastic, ground combat simulation with detailed two-dimensional color graphics. Janus is a high-fidelity model. That is, one views the battlefield in segments of from 10 to 60 km and conducts tactical maneuvering and weapon systems analysis based on this view of the world. The high fidelity also implies that the user can evaluate the various elements at the individual entity level. In Janus this means that the user watches each entity maneuver and fire, and sees each mortar or artillery round being delivered on its target. With both the on-screen display and written output, the user can evaluate the accuracy and lethality, or effect, of the various systems and tactics.

Current research is being conducted with Janus to support various Army initiatives. This work includes the design and implementation of a networking capability, called JLINK (Janus Linked to DIS), to use Janus in a distributed mode for training in addition to analysis. JLINK was developed to support the Anti-Armor Advanced Technology Demonstration (A2ATD) directed by the Army Materiel Systems Analysis Agency. This is an excellent example of joint work by military, academia, and industry. TRAC (U.S. Army's Training and Doctrine Command Analysis Center), located in Monterey, California, had overall responsibility for building the system and for providing project management and functional area expertise. Researchers from the Naval Postgraduate School, also located in Monterey, California, developed the interface between Janus and other DIS-based simulators called the World Modeler. RAND Corporation developed the interface between Janus and the World Modeler, conducted terrain analysis, converted source code from one language to another, and modified Janus internal algorithms to make them more consistent with functionality requirements necessary for the A2ATD experiments. Rolands & Associates Corporation has demonstrated a capability to connect Janus with an Air Force F-16 simulator using DIS networking protocols. This implementation will provide the ability for several Army units, distributed over a large geographic area, to use Janus simultaneously, thus providing an integrated training environment.

The key result of the JLINK research is an advanced, more capable system with each component contributing uniquely to the resultant design. Janus brought to DIS accredited computer-generated forces and a large validated database of vehicle and weapons characteristics. DIS provided Janus with soldier-generated scenarios and a three-dimensional visualization capability for analytical insights. The combined benefits of the system include a soldier-in-the-loop capability (i.e., interactive simulation where the soldier, as a typical user, can interact directly with the model) and an integrated training and analysis tool [2, 13].

19.4.5 Other Applications of DIS Within DoD Programs

Other DIS-based simulation training programs within the DoD include the Navy/Air Force Joint Tactical Combat Training System (JTCTS), the Navy Battle Force Tactical Training (BFTT) System, and the Army Warfighters' Simulation (WARSIM) 2000.

The JTCTS is designed to provide both the Navy and the Air Force with a state-of-the-art DIS-based training system that will link live objects, such as aircraft flying on training ranges or off aircraft carriers at sea, to the synthetic battlefield. Instead of the fleet going to the range, the range can now go to the fleet. A contract for this program was awarded in March 1995 and is a combination of the Navy's Tactical Training Combat System and the Air Forces's Joint Air Combat Training System.

The BFTT is the first Navy program to design its system architecture on the standards and protocols provided by DIS. The system electronically moves real ships, crews, or individual operators located in the same or separate ports to a common synthetic theater of war that provides a realistic, interactive environment across all naval warfare areas. DIS has become the key element of the technology to achieve this transition to a common synthetic environment.

WARSIM 2000 is an ongoing project at the Army National Simulation Center. The Army envisions WASIM 2000 as its future simulation model. It will replace existing simulation models, CBS, BBS, and TACSIM, and will be designed for DIS environment for joint operation exercises. For updated information, see http://www-leav.army.mil/nsc.

19.5 OTHER ISSUES

19.5.1 Use of Simulation in Acquisition Testing and Evaluation

With the defense budget cuts in the late 1980s, weapon acquisitions have come under increased scrutiny. Reduction in available resources and the long lead time to field new technologies require that acquisition management must explore every avenue to improve the efficiency of the overall acquisition process. M&S has the potential to improve the design and effectiveness of potential systems significantly while reducing the production and testing costs.

The modern battlefield includes complex weapons systems. To ensure that the proposed system can fully integrate into the battle force, models and simulations are utilized. This integration would be costly, if not impossible, to "live" fire test. Developmental test and evaluation (DT&E) is an ongoing process. The primary benefits of models and simulations in DT&E has been the identification of risk and the reduction of development time. By identifying risky areas early, more resources can be focused on the problem, preventing time delays and cost overruns. Computerized development of modern weapons systems is standard practice now. The next huge hurdle is overcoming the resistance to the use of simulation for operational test and evaluation (OT&E).

OT&E must be completed before a weapons system can proceed past low-rate production. Title 10 of the U.S. Code (Section 138), as of 1989, precluded M&S from being included as OT&E. These tests have been restricted to "live" simulations in an operational environment. Other types of simulation can be conducted prior to, or after, these tests but cannot be the basis for proceeding to full-rate production.

In the OT&E community there is an inherent distrust for anything that is "simulated," as they view simulation as a circumvention of acceptable testing protocols. As attitudes toward current technological advances begin to be accepted, M&S will increase significantly. Although M&S will not completely replace field testing, it will improve the overall accuracy of the testing process and reduce the costs (including time) involved in completing the testing process. By integrating constructive simulations into the OT&E design process, planners will be able to examine a weapons system's entire operational spectrum, force structure, interoperating systems, and threat capabilities that can be represented in the constructive environment. Furthermore, they will be able to create more realistic developmental and operational test scenarios.

19.5.2 Nonwargaming Simulation Applications

The use of M&S is most prevalent in the areas of engineering and manufacturing. Many commercial simulation languages (e.g., AweSim, Arena, MODSIM, and Witness) are used in weapons system design, production, and maintenance. The use of M&S in manufacturing is aiming toward a future "virtual manufacturing" environment. In this approach the operational requirements identified in the synthetic battlefield environment are translated into design concepts using three-dimensional virtual simulations incorporating geometry and performance. These designs are passed along to a network of distributed manufacturing simulations that may reside throughout a vendor base to identify the manufacturing processes, facilities, and tooling requirements. This vendor base is closest to the manufacturing processes and is in the best position to develop cost and schedule estimates. These estimates may then be fed back to provide better estimates of costs and schedules to support trade-offs and system-level alternative evaluation in cost and operational effectiveness analysis [14]. The use of M&S has the potential to improve the design and effectiveness of future systems significantly while reducing the production and testing costs. Proper use of M&S will achieve cycle-time reduction, which will reduce inventory, and eventually cost.

19.5.3 Verification, Validation, and Accreditation

As military simulation models become complex it is more difficult to conduct verification, validation, and accreditation (VV&A). However, VV&A increases confidence in the models and reduces the risk of bad decisions. A comprehensive *DoD VV&A Recommended Practice Guide* [3] is currently being developed by the VV&A Technical Support Team at DMSO. (The draft can be downloaded from http://www.dmso.mil/docslib/mspolicy/vva/rpg.)

The DoD Directive 5000.59 defines VV&A as follows:

- *Verification* is the process of determining that a model implementation accurately represents the developer's conceptual description and specifications.
- *Validation* is the process of determining the degree to which a model is an accurate representation of the real world from the perspective of the intended uses of the model.
- *Accreditation* is the official certification that a model, simulation, or federation of models and simulations is acceptable for use for a specific purpose. It is a decision that a specific model or simulation can be used for specific application. Hence an accreditation does not apply to a class of models and does not apply to a number of applications. A model can receive an accreditation for use in one specific application but not be accredited for use in any other applications.

VV&A of models is intended to establish credibility or confidence in the analyses of the results from the simulation models. Since the simulation results support an important decision-making process, it is necessary that the VV&A be an integral part of the M&S. DoD's new policy is that DoD components establish VV&A policies and procedures for the M&S that they develop and manage. Also, the M&S used to support the major DoD decision-making organizations and processes should be accredited for those used by the DoD component sponsoring the application. Similarly, M&S used for joint training and joint exercises should be accredited for that purpose by the application sponsor [5]. For a general discussion of VV&A, see Chapter 10 and refs. 9, 10, and 15, among others.

19.6 CONCLUDING REMARKS

The U.S. military uses modeling and simulation extensively. The cost associated with developing applications without centralized coordination is extremely expensive. The DoD has instituted measures to coordinate and control the expenditures for M&S. The Joint Warfare Analysis and Research (JWARS) program, managed by the Joint Chiefs of Staff J-8 Office in the Pentagon, is an effort to consolidate analysis modeling. The Joint Simulation System (JSIMS) is a similar, large-scale program, to effect consolidation of training models. This JSIMS program, managed by the Joint Program Office in Orlando, Florida, is in its initial stage now but probably will provide the paradigm for future military simulations. The Army's WARSIM 2000, Air Force's NASM (National Aero Space Model), and Navy's BFTT will be linked via JSIMS for joint exercises. For updated detail of the JSIMS program, see http://www.jsims.mil/.

Every level of human behavior from the individual soldier/sailor/airman up through the highest levels of DoD has been modeled at one time or another. As distributed exercises become more complex, so does the need for more accurate models of human behavior. Consequently, the need to understand these behaviors will become increasingly important. Extensive research in this area is ongoing. For example, researchers from the Naval Postgraduate School are involved in the NPSNET project "Inserting the Human into the Networked Synthetic Environment" (http://www-npsnet.cs.nps.navy.mil/npsnet).

The availability of much of this information via the Internet provides both military and nonmilitary users with a rich source of information from which to plan and schedule M&S activities. A vast amount of information is available from the Internet home pages. Many other DoD documents are readily available electronically. M&S is an important element in military analysis and training and will continue to play an ever-increasing role as defense budgets decrease for military expenditures.

REFERENCES

1. Bolling, R. H. (1995). The joint theater level simulation in military operations other than war, in *Proceedings of the 1995 Winter Simulation Conference*, C. Alexopoulos, K. Kang, W. R. Lilegdon, and D. Goldsman, eds., IEEE, Piscataway, N.J., pp. 1134–1138.
2. Caldwell, W. J., R. Wood, and M. Pare (1995). JLINK: Janus fast movers, in *Proceedings of the 1995 Winter Simulation Conference*, C. Alexopoulos, K. Kang, W. R. Lilegdon, and D. Goldsman, eds., IEEE, Piscataway, N.J., pp. 1237–1243.
3. Department of Defense, Deputy Under Secretary of Defense (1994). DoD modeling and simulation (M&S) management, *DOD Directive 5000.59* (available from http://www.dmso.mil/docslib/directive.html).
4. Department of Defense, Under Secretary of Defense for Acquisition and Technology (1995). Modeling and simulation master plan, *DOD Directive 5000.59-P* (available from http://www.dmso.mil/mspolicy/msmp/1095msmp).
5. Department of Defense (1996). DoD modeling and simulation (M&S) verification, validation, accreditation (VV&A), *DoD Instruction 5000.61* (available from http://www.dmso.mil/mspolicy/vva).
6. Garrett, R. (1995). ADS: looking toward the future, *Phalanx*, Vol. 28, No. 2, pp. 8–10.
7. Garrett, R. (1995). A new simulation paradigm: advanced distributed simulation, *Phalanx*, Vol. 28, No. 3, pp. 25–27.

8. Anon. (1994). Virtual reality and training, *Government Executive*, June.
9. Kleijnen, J. P. C. (1993). Verification and validation of simulation models, *Technical Report 320.93.186*, Tilburg University, Tilburg, The Netherlands.
10. Law, A. M., and W. D. Kelton (1991). *Simulation Modeling and Simulation*, 2nd ed., McGraw-Hill, New York.
11. Mowbray, D. W., J. W. Wallace, A. L. Herrman, and E. S. Hirschorn (1995). An architecture for advanced distributed simulation, *Phalanx*, Vol. 28, No. 2, pp. 12–15.
12. National Simulation Center (1995). *Training with Simulations: A Handbook for Commanders and Trainers*, Combined Armed Center, Ft. Leavenworth, Kansas.
13. Pate, M. P., and G. G. Roussos (1996). JLINK: a distributed interactive Janus, *Phalanx*, Vol. 29, No. 1, pp. 12–15.
14. Piplani, L. K., J. G. Mercer, and R. O. Roop (1994). *Systems Acquisition Manager's Guide for the Use of Modeling and Simulation*, Defense Systems Management College, Ft. Belvoir, Va.
15. Sargent, R. G. (1994). Verification and validation of simulation models, in *Proceedings of the 1994 Winter Simulation Conference*, J. D. Tew, S. Manivannan, D. A. Sadowski, and A. F. Seila, eds., IEEE, Piscataway, N.J., pp. 77–87.
16. Shiflett, J. E., W. H. Lunceford, and R. P. Willis (1995). Application of distributed interactive simulation technology within the Department of Defense, *Proceedings of the IEEE*, Vol. 83, No. 8, pp. 1168–1178.
17. Sikora, J., and P. Coose (1995). What in the world is ADS? *Phalanx*, Vol. 28, No. 2, pp. 1–8.

CHAPTER 20

Discrete-Event Simulation of Computer and Communication Systems

ALFRED HARTMANN AND HERB SCHWETMAN
Mesquite Software, Inc.

20.1 INTRODUCTION

In this chapter we discuss techniques useful for constructing discrete-event simulations of computer and communications systems. While analytical methods such as queueing theory are sometimes employed (Lazowska et al., 1984) and can offer valuable insights, their scope is limited to problems for which analytical solutions methods are possible. Computer-based simulation is very broadly applicable, and with the availability of many kinds of simulation software packages and the rapid proliferation of low-cost computing power to every desktop, simulation-based analysis has moved to the fore. Today, most computer and communications system designers are familiar with the C/C++ programming language and with the notion of computing processes. This makes process-based simulation in C/C++ a natural and convenient choice for many modeling projects.

20.2 FUNDAMENTAL CONCEPTS

A computer system or communication network consists of a collection of resources. Entities (jobs, programs, tasks, transactions, messages, etc.) compete for use of these resources. Models of these systems possess, at least at some level, constructs that represent both resources and entities. Almost any simulation model of almost any kind of system will have similar constructs. In this chapter we explore the issues that make modeling computer systems and communication networks different from simulation models of other kinds of systems.

Handbook of Simulation, Edited by Jerry Banks.
ISBN 0-471-13403-1

20.2.1 Goals

The major goal of most system modeling projects is to provide estimates of system performance. With computer systems, the most important performance measure is task response time. With communication networks, the most important measure is message latency time. Thus the goal of many projects is to provide accurate estimates of these measures. Other goals include providing insight into the operation of the system and guidance for reducing the impact of performance bottlenecks.

20.2.2 Resources and Entities

As mentioned above, simulation models typically have entities (processes, customers, messages, transactions, etc.) and resources (processors, memory, buses, channels, communications links, etc.). In these models there can be several different types of entities and there can be multiple instances of each type of entity, all active at the same and at different points in simulated time. In many situations, the entities are competing for use of some of the resources of the system model. For example, in a model of a computer system, several simulated jobs can all be competing for use of the CPU resource.

Similarly, there are usually several types of resources in a system model. These can be classified into two major categories: (1) active resources, and (2) passive resources. The distinction depends on what entities do after obtaining access to an element of a resource. An element of an *active* resource is typically used or occupied for a specific interval of time and then released. An element of a *passive* resource is obtained by an entity; at this point the entity proceeds to perform other actions, including gaining access to other resources.

Examples of active resources in a computer system include CPUs, disk drives, and controllers, and examples of passive resources include main memory and buses. In a communications network, message processors and internode links might be active resources, while buffers are examples of passive resources.

The entities of these models depend on the level of the model. In a high-level model of a computer system, the entities might be programs or transactions. In a high-level model of a communications network, the entities might be messages. In a low-level model of a computer system, the entities might be individual computer instructions, I/O requests, and data transfers to and from main memory. In a low-level model of a network, the entities might be data packets or cells.

20.2.3 Workload

Every system model includes a representation of the workload for the system. Here *workload* refers to the sequence of demands by entities for the resources of the system. Obtaining an accurate characterization of the system workload is one of the most important steps toward building an accurate and useful model. With computer systems this task is, in some cases, made easier by the presence of system tools which can be used to automatically collect some of the necessary data. For example, if the entities of the model are on-line transactions, the transaction monitor may keep transaction logs that can be used to build a characterization of the input transaction stream. Similarly, in a computer network, there may be message logs at some of the nodes which can be useful in constructing profiles of the message traffic in the network.

As with most models, it is necessary to have an accurate characterization of the workload in terms of the demands for system resources. With computer systems, this may present some difficulties. For example, a critical input for a model may be the service-time distribution at a disk drive, but a system accounting package may report response times at this drive (and the mean response time at this drive would be an output of the model). In a similar vein, the accounting package might report characters per transfer to a disk drive, but it may be difficult to relate the number of characters transferred to the service time.

Typically, what is required for a model of a computer system is a profile of the kinds of jobs or tasks being modeled. This profile would consist of a stratification of the tasks into different classes, and then for each class, a summary of the demands for service for the different resources of the system. As an example, a workload for an on-line transaction processing system might be classified according to different types of transactions being submitted, and then for each type of transaction, collecting the following information:

- Number of transactions of each type
- Amount of CPU time per transaction
- Number of disk accesses per transaction
- Amount of data sent back to the requester per transaction

For a communications network, the workload characterization might consist of the following kinds of information:

- Number of messages generated at each node
- Message length distribution
- Information about the destination of each message (e.g., the distance to the destination)
- The percentage of messages requiring an acknowledgment

The end result of this workload characterization is a set of parameter values that are required to model accurately the workload in the model.

20.2.4 Output: Measures of Performance

As stated earlier, models of systems are often used to obtain estimates of the average system response time for elements of the workload. Similarly, in models of networks, estimates of the average message latencies and/or delivered bandwidth are of interest. Here *response time* means the time from when a request for service is made (starts or arrives) until the time that the request is completed (departs). *Latency* is the term used for response time in an input–output subsystem, a memory subsystem, or in a communications network. Response times (latencies) are expressed in terms of time units per request. The reciprocal of response time is throughput rate (e.g., transactions per second), while the reciprocal of latency is bandwidth (e.g., megabytes per second). All of these items are usually outputs of simulation models.

If a model is being used to gain insight into performance problems (where performance problems are often indicated by unacceptably long response times or latencies), more output data are required, to pinpoint causes of problems. These additional data usually consist of statistical summaries of times spent at individual resources of the system model or times spent in the individual activities that make up a complete workload element. In some cases it may be necessary to insert special instrumentation to collect the data necessary to pinpoint causes of performance problems.

20.2.5 Duality of Models

Process-oriented models of computer and communications systems tend to differ in their assignment of processes to simulation entities. In computer system simulations, processes tend to be assigned to the *workload* (e.g., to represent user processes or applications), and these processes acquire and release system *resources*, for which they typically compete. In communication system simulations, processes tend to be assigned to *resources* (e.g., switches, or channels), and the processes handle *workload* messages, for which they typically do not compete. These process assignments tend to keep the number of processes in the simulation to a reasonable number, since the simulation system incurs context switch overhead in dealing with multiple processes. It would normally be prohibitive to model each message as a process that competes for communication system resources, due to their vast number.

These different ways of conceptualizing models leads to a type of model duality between models of computer systems and models communication systems:

Simulation Entity	Computer System	Communication System
Workload item	Process	Token (structure)
Resource item	Structure	Process

This duality can result in a difference in perspectives between computer system simulation and communication system simulation, and perhaps a difference in choice of simulation tools.

Queue service disciplines appropriate to competitive resource acquisition in a computer system simulation may or may not satisfy the needs of message workload processing in a communication system simulation. First come, first served (FCFS) is a simple service discipline common to both computing and communications, while the common round-robin processor scheduling discipline would probably not be applicable to communications simulations. A communication system simulation may need to employ a wide variety of buffer management strategies that may not be reflected in the selection of resource allocation strategies available in a computer system simulation facility. As examples, the "leaky bucket" (Bae and Suda, 1991) buffer management strategy used for ATM network source control is unique to communications, as are combination timeout and priority-based cell disposal methods for congestion reduction in ATM networks.

20.2.6 Object-Oriented Models of Systems

Object-oriented software design methods have much to offer the simulation developer, as discussed in Chapter 11. We may consider object-oriented software design to be based on the principles of:

- *Encapsulation:* the ability to combine data and the functions that manipulate the data into abstract data types
- *Inheritance:* the ability to derive new subtypes from existing abstract types
- *Polymorphism:* the ability for functions to manipulate objects of different types that share certain type relations

The advantages that object-oriented design offers to all types of software development apply also to simulation development. Simulation entities can be encapsulated into objects, classes of simulation entities can be grouped into base object classes and differentiated using inheritance into arbitrary subclasses, and polymorphic functions can be defined appropriately as methods of the object classes. All these object-oriented techniques can serve to organize and simplify simulation model development.

Simulation development systems and programming languages provide predefined base classes for common types of simulation processes and resources. These can be inherited by newly defined subclasses in a particular simulation effort. Polymorphic functions can be predefined that perform standard manipulations on user objects that are instances of subclasses derived from the base class repertoire.

20.2.7 Quasi-parallelism in System Models

Process-oriented simulation uses multiple processes to simulate the parallel activities in the system being simulated. However, it is typical for the simulated execution of these multiple processes to be interleaved sequentially on a single processor. This interleaved execution is generally done deterministically so that results are repeatable from simulation run to simulation run. Such forms of parallelism are referred to as *quasi-parallelism*, to distinguish them from *true parallelism*, which is nonsequential and often nondeterministic. True parallelism may be exploited *underneath* the quasi-parallel simulation environment, to speed simulation, but this is transparent to the simulation design. Quasi-parallelism is not transparent to the simulation design but is an integral part of the simulation design. Building simulation models that exploit true parallelism is discussed in Chapter 12. It can be noted that some of the most successful applications of parallel simulation techniques have involved models of large communications systems.

Quasi-parallelism must provide facilities for simulated processes to *block* (i.e., waiting for simulated time to pass, for resources to become available, or for a simulated event to occur). At the appropriate point in discrete simulated time, processes must *unblock* and resume simulated execution. Multiple processes need some form of synchronized interprocess communication, which can be provided by predefined synchronization classes implementing events, signals, mailboxes, semaphores, and so on.

20.3 COMPUTER SYSTEMS

A simulation model of a computer system (MacDougall, 1987) must capture the essential parts of the real system. These essential parts consist, in broad terms, of (1) workload elements, (2) system resources, and (3) system policies (which govern allocation of resources to elements of the workload). In this section we discuss some of the issues involved in modeling these parts (Jain, 1991).

20.3.1 What Is the Workload?

Computer systems exist to process elements of the workload, where, depending on the system being modeled, a workload element could be one of the following items: (1) job, program, or task; (2) transaction or query; or (3) I/O request or request for an item in main memory. Each of these kinds of workload elements is a sequence of demands for service at the resources of the system. As an example, consider a job or task as being representative of a workload element. A job (a program) is a sequence of alternating requests for time on a CPU (CPU bursts) and input/output service (transfers of blocks of data to and from I/O devices such as disk drives). In addition, the job will request use of blocks of main memory. In most modern systems, multiple jobs are executing simultaneously, so there is competition for all of these resources. System performance (job response times) reflects the ability of the system to satisfy these conflicting requests for use of the system resources.

20.3.2 Modeling System Components

Computer systems consist of both hardware and software components. In addition, the operating system manages access to these components; it embodies the control mechanisms which are in place to guarantee correct and efficient operation of the system. In a simulation model of such a system, all of the important components, as well as the policies that control access to these components, should be represented if accurate estimates of system performance are to be achieved.

Hardware Components. The most important hardware components of a computer system are the CPU and the I/O devices. Main memory is critical to the operation of the system, but in most modern systems, main memory is quite large and is usually not a major factor in determining system performance. However, incorporating main memory into a model is usually not a difficult task. The other important and often neglected kind of component is the interconnection hardware (called a *bus*), used to tie the CPU and the I/O devices to the main memory. Modeling these hardware components is usually fairly straightforward because they can be represented by static resources. The key is to accurately model the kinds of serial or parallel accesses that are possible in each of these components.

Software Components. The software components can be more difficult to represent in a system model. Typically, a software component can embody lengthy sequences of actions and implement complex operating policies. Furthermore, some of the components are provided by third-party vendors, and their inner workings will not be disclosed to a system modeler.

One class of software component that is found in systems consists of those that offer service to user programs. An example is a database management system (DBMS). Many applications in a system could require service from the DBMS. However, there are limits to the number of requests for service that the DBMS can handle simultaneously. Thus such a server can be a limitation to the performance (job response time) for jobs that must access critical data via the DBMS. Other software components that can affect the performance of jobs in a computer system include transaction processing monitors, network access systems, and remote file servers.

If a software component is determined to be a potential performance bottleneck in a system being modeled, a special effort may be required to ensure that an accurate representation of the component is available. For example, if a commercial DBMS is an important component in a system, the system model may require a simulation model of the DBMS. To incorporate this component into the model, the modeler may have to take one of the following steps:

- Obtain additional information from the vendor or other sources.
- Perform some black-box measurement studies, so as to infer the behavioral characteristics of the DBMS.
- Obtain a prebuilt simulation model of the DBMS.

The other aspect of software components that is of importance in a model is the set of control and management policies provided by these components. These policies and rules appear in the operating system (which controls the operation of the entire system) as well as in the components mentioned above. These policies specify priorities, scheduling rules, resource constraints, and rules for dealing with congestion and overflow (e.g., maximum queue lengths at an I/O device). Depending on the level of detail in the model, these policies may be an important factor in the development of an accurate system model.

20.3.3 Asynchronous Operation

In an attempt to improve the utilization of system resources, modern systems encourage an asynchronous style of behavior. The basic principal is that a program is able to initiate I/O actions that can take place in parallel with continued use of the CPU. A model of a system must capture this parallel, asynchronous style of operation. A process-oriented simulator provides a base for implementing models in which the major components can operate asynchronously.

20.3.4 Trade-offs: Detail Versus Cost

The purpose of a system model, in the simplest terms, is to produce response intervals for workload elements. These intervals can then be used to provide the estimates that are needed to make judgments about system performance.

A response interval typically consists of one or more subintervals, in which each subinterval has a delay interval and a service interval at some resource. A very high level system model has only a few resources, and as a result, the response interval has only a few subintervals. Such a model could execute very quickly, but the resulting estimates of response times might be inaccurate.

To improve accuracy, the high-level resources can be redefined as collections of lower-level resources. When this is done, what was formerly a single service interval at a resource is now a sequence of delays and service intervals for these lower-level resources. The gain is a more accurate representation of the resource; the cost is an increased number of simulated events, which translates to a longer-running model.

As an example, in a high-level model, an input–output operation could be modeled as a single service interval. In the model there could be a single input–output resource, and programs get, in a sequential manner, service at the resource. The real input/output device and associated connection components is more complicated; capturing more of this complexity in a model might result in better estimates of the time required by a program to complete its input/output requirements. A more complex model could have a bus, a controller, and several disk drives. This more complex model could be processing multiple requests in parallel and could provide more accurate estimates of response times at the input/output resource.

All system models represent trade-offs between increased levels of detail and increased execution. Good models strike a balance between these conflicting goals.

20.3.5 Model Results

A model of a system can be used in two distinct ways [LaKe91]: (1) to provide estimates of the output values for the system operating in "steady state," and (2) to provide estimates of the output values for the system operating over a specified interval of time (e.g., for an 8-hour shift). The problems of providing accurate estimates in both of these situations are well known, as are techniques for dealing with these problems (Law and Kelton, 1991).

Models of computer systems can present some additional problems to the task of providing reliable estimates of the output variables. These issues stem from the fact that real computer systems can be large and complex. Furthermore, some of the important components are not readily visible, and producing accurate representations of their behavior may be difficult. Thus validation of the structure and operation represented in a model may be a significant issue as it is developed.

In some cases it is possible to obtain output results for the test workloads used to parameterize the model of the workload. In these cases, the results from the model can be thoroughly validated. In other cases, the model developers may be forced to review the design and implementation of the model itself, to verify the validity of the results.

20.4 COMMUNICATIONS SYSTEMS

Any project to develop a simulation model of a communications system can be divided into two main branches of work: (1) construction of the workload model, and (2) construction of the network model.

20.4.1 What Is the Workload?

A communications system workload model is, primarily, a quantitative description of the messaging traffic within a communications network. The description may be either explicit, as obtained from instrumented trace studies of a real system, or statistical, making random draws from a set of probability distributions. Regardless of the form, the workload description should accurately represent the messaging traffic for the intended network.

Workload validation is the process of ensuring appropriate accuracy in the formulation of the workload description. Even the most faithful network model will be useless if it is exercised with an invalid workload: "garbage in, garbage out." Workload construction and validation is a major portion of the effort in building any communications simulation model.

A workload driver (model) presents messages to the communications system for delivery through the network. We are concerned with the course of the message traffic through network space and time. A message may be part of a larger message stream with a given point of origination and one or more points of destination. Messages to be sent to only a single point of destination are referred to as *unicast traffic*, while messages sent to multiple destinations are *multicast traffic*, with *broadcast traffic* being intended for all destinations. A message can be sent in one piece to its destination(s), or be decomposed into smaller units, or be composed with other messages into larger units, or the pieces of multiple messages can be variously combined and separated in their course through the network.

What happens to a message *after* its entry to the network is the province of the network model. The workload model describes the kinds of messages that *enter* the network and the arrival times for these messages. In this sense the workload serves to provide the *stimulus*, while the network serves to provide the *response*. Part of the overall design of the simulation is determining the most appropriate type of stimulus that will generate the responses of interest.

Since real networks tend to be constructed in layers, the character of the workload will be greatly affected by the layer chosen for direct stimulation. For example, if only the physical links are to be studied, the stimulus can be presented at a very low level (e.g., physical data unit) in the layering, while if the performance of a distributed database operating over the network is to be studied, the stimulus is presented at a very high level (e.g., database query). The higher the layer chosen for stimulation, the more layers must be present in the network model, so the choice of the layer greatly affects the total implementation effort. If lower layers are to be treated only in the abstract, experience is necessary to judge which details to ignore and which to model.

The logical data unit may be much larger than the physical data unit transferred over the link, or the reverse may be true. For example, in Ethernet over ATM, the Ethernet packets may be kilobytes in size while the ATM cells are only a few tens of bytes. On the other hand, in ATM over SONET, the SONET frames hold a large number of ATM cells. If the number of messages can be reduced by two orders of magnitude, that is desirable if this reduction can be accomplished without loss of accuracy in the simulation. The evaluation of this trade-off may depend on which network performance attributes are important to measure; thus specific rules are difficult to give.

The workload model injects messages into the network, and each message has the following principal attributes:

- Time of origin
- Point of origin
- Point(s) of destination
- Message characteristics (size, priority, etc.)

Usually, the origin times (and possibly the destination addresses) of the injected messages are drawn from probability distributions. Or rather, the time interval between message originations at a given point of injection is drawn from a probability distribution. *Proper selection of the distribution of interorigination time intervals is a major aspect of workload validation.*

Since communications networks are frequently analyzed as queueing networks, which are most mathematically tractable with exponentially distributed time intervals, the exponential distribution is a frequent, though not necessarily valid choice. It is applicable where message origination is a Poisson (or memoryless) process, but much multimedia traffic, such as compressed voice and video streams, tends to show Markov process characteristics (or bursty behavior) (Bae and Suda, 1991). A Poisson process and a non-Poisson Markov process with the same average intermessage time interval may produce dramatically different network performance results due to traffic congestion at intermediate network nodes. Here a non-Poisson Markov process refers to a process in which some time intervals (e.g., intermessage intervals) are characterized as an exponential distribution with one of two means; the choice between these two means is determined by Markov state transitions with specified probabilities. The goal is to model a process accurately with two distinct modes of behavior.

Bae and Suda (1991) provide several examples of Markov process models for ATM network traffic of voice, video, and data content. The Markov transition probabilities typically cause state transitions that alter the message interarrival time significantly so as to produce the necessary burstiness found in the actual source. Each of the states may correspond to some simple probability distribution (such as exponential) but with different distribution parameters for each state.

All of these workload models are dependent on the types of data coding used, particularly if compression is employed, since this may alter the traffic characteristics considerably. Forward error checking (FEC) using error-correcting codes (ECC) can also alter traffic patterns by changing the incidence of re-sends of messages. Regardless of the distribution chosen, the random, uncorrelated time intervals between message originations argue for an asynchronous relation between the workload model and the network model.

To illustrate some of these issues, assume that we are developing an object-oriented simulation model in C++. Furthermore, let us propose a class named `Node` and a class named `Workload`:

```
class Node_c { . . . . };
class Workload_c { . . . . }
```

Each instance of Node could be given an instance of Workload, to supply it with new messages to send:

```
Workload_c work ( . . . );
Node_c node (work, . . . );
```

Or it might be done the other way around:

```
Node_c node ( . . . );
Workload_c work(node, . . . );
```

Either the Node_c instance, ?node, has to block until the next message origination time, or the Workload_c instance, ?work, has to block. This sort of quasi-parallel asynchrony in simulated time is not a feature of the C/C++ programming languages or of operating systems. Essentially the node and its corresponding work need to function as asynchronous coroutines in discrete simulated time, based on a simulation event queue, as discussed in the earlier section on quasi-parallelism.

20.4.2 Networks and Network Protocols

The network model is the other half of the effort, and together with the workload model constitutes a full communications system model. Networks are often represented as

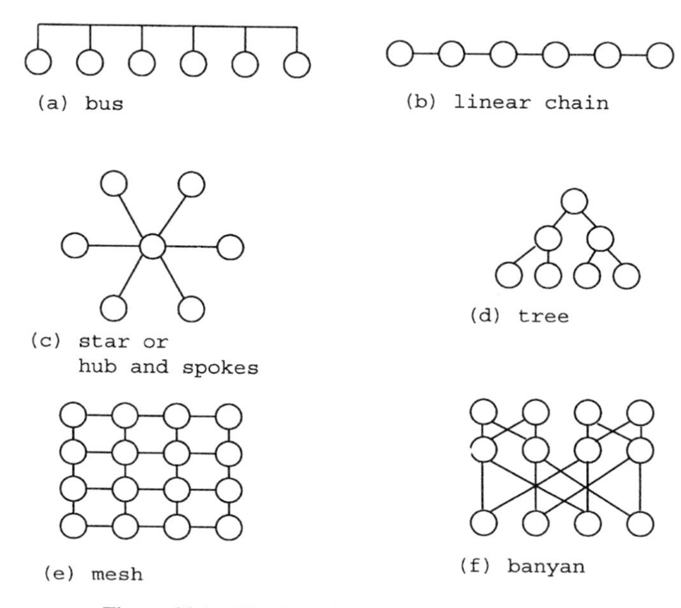

Figure 20.1 Simple and regular network topologies.

graphs, with nodes and edges (links). Multidrop links or buses can be viewed as edges in a hypergraph network representation. Usually, the terminal nodes serve as source and destination nodes, while the interior nodes serve as switch or router nodes. This graph structure becomes more elaborate when network protocols are included, especially those with stratified software layering, as in network architectures today.

The interconnection pattern of network nodes and links constitutes the *network topology*. Complex and irregular network topologies can be expensive to model, both in terms of design time and simulation run time, but cannot always be avoided. Well-known simple and regular network topologies are shown in Figure 20.1.

There need be no identity between the *logical* network topology and the *physical* network topology and the *physical* network topology. As an example we show in Figure 20.2 a logical ring network wired as either a physical ring, a linear chain (interleaved or noninterleaved), or as a star. The well-known token ring network in its usual realization is an example of the latter, since physically wiring it as a ring (Figure 20.2*c*) is often cumbersome and failure prone.

There may or may not be significance to the possible difference between logical and physical network topologies. If signal propagation delays are critical, the physical topology could be of concern in the network model. For example, in bused versions of Ethernet, the incidence of packet collisions is affected by the relative positions and separations of nodes on the bus and by the total end-to-end propagation delay.

20.4.3 Service Disciplines for Buffers, Channels, and Switches

Communications system performance can be greatly affected by choice of service discipline. Also, various service disciplines can vary in ease of implementation (depending

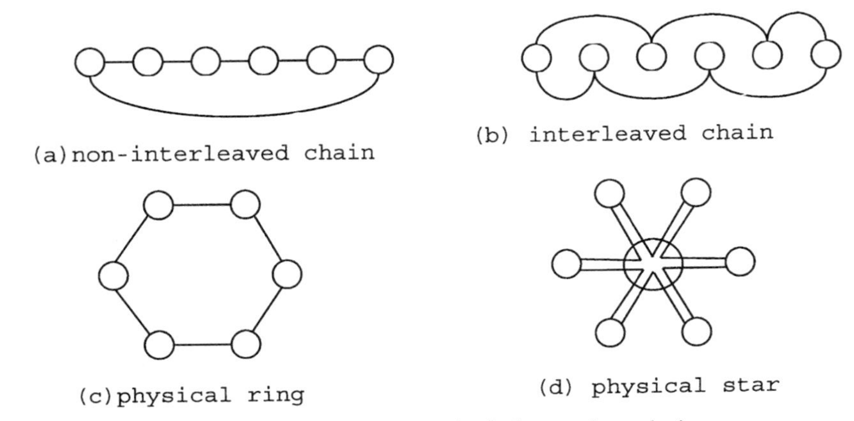

Figure 20.2 Logical versus physical network topologies.

on choice of tools and libraries) and in computational effort at simulation run time. Engineering trade-offs may argue against attempting theoretically optimal service disciplines, either in simulation or in actual network design.

Some of the more common queue service disciplines applicable to communications buffers are:

- First come, first served (FCFS), also called first in, first out (FIFO)
- Priority order, based on simple message priority or on some message attribute [e.g., time of origination, or some quality of service (QoS) parameter associated with the message or the channel or the destination]
- Random order, either purely random (uniform distribution) or skewed by some attributes of the message or time spent in the queue)

Essentially, FCFS is a special case of priority order that takes time of queue entry as the priority. Simularly, priority order can be viewed as a special case of deterministic ordering, as opposed to random ordering. The communications system architect is concerned with choosing service orders that maximize a particular figure of merit for the network. Since both network performance and network cost are multidimensional quantities, specification of an appropriate figure of merit is not simple.

In some cases there may be provably optimal service orders for a particular design goal, and frequently one desires to simulate the performance difference between optimal and nonoptimal orderings. This is especially true if the cost or complexity of implementing optimal ordering is high.

In the case of buffer management, one needs to consider (1) message entry to the buffer, (2) message retention in the buffer, and (3) message exit from the buffer (either normal or abnormal).

Message entry to the buffer may depend on the current state of the buffer (e.g., full or nonfull, priority of presently buffered messages versus incoming message priority, etc.) and the incoming message attributes. Message retention in the buffer might involve ongoing buffer content reorganization based on dynamic parameters, including the time-in-queue of the presently buffered messages. Message exit from the buffer may occur either normally as the message is serviced, or abnormally as the message is dropped for service deadline expiration, preemption by a higher-priority message, or whatever.

Frequently, buffers are used in association with switches, to store messages in the process of passing through a switch from an inbound channel to one or more outbound channels. The delays and complexities associated with message switching can have a great impact on the switch buffer management strategy. Buffers may be employed at the inputs to the switch (input-buffered switch), at the output ports from the switch (output-buffered switch), at both inputs and outputs, or at any intermediate stage within a multistage switch.

20.4.4 Accounting for Transmission Times and Propagation Delays

Depending on the scale of the simulation model in space and time, transmission times and propagation delays may or may not be significant in a given situation. A simple back-of-the-envelope analysis is usually sufficient to determine if these times need to be accounted for in the model or if they may safely be treated as zero (i.e., instantaneous transmission and/or propagation).

Because communications channel capacity and node separation distances are often fundamental in communications system simulation, it is rare that both these model parameters are ignored. Transmission speed and separation distance produce significant delays in communication that usually must be accounted for in the model.

20.4.5 Size Issues

The simplest way to manage size issues is to parameterize the model with size factors and to raise these factors judiciously as experience is gained with smaller simulations. How the computational load of simulation varies with network size or workload quantity may vary, but often may be faster than linear growth. A binary progression of 1X, 2X, 4X, 8X, and so on, can be tested and plotted against execution time to estimate practical simulation limits. Since discrete-event simulation can employ arbitrarily large amounts of computational time and expense—and still never produce perfect accuracy—seasoned judgment must prevail.

If scaling factors can be measured so that simulation results from small simulations can be extrapolated to results for large systems, significant savings can be realized. Even if such extrapolations are not trusted, the vast multidimensional design space can be explored more quickly with small-scale simulations, and the large-scale simulations be reserved for choices proven on a smaller scale.

20.4.6 Tradeoffs: Detail Versus Cost

Pragmatic engineering practice usually argues in favor of simple approaches initially, both in design of actual communications systems and in construction of their corresponding simulation models. It is often simpler and safer to converge on a final design by iterating over a sequence of successively less abstract designs. This technique of successive refinement in design and modeling can reduce total engineering effort and time to completion significantly by proceeding through a sequence of stable intermediate points.

Another valuable method is to simulate at a simple abstract level whenever possible, even if it is thought to be unrealistic, and measure the deviation from a more complex model. Only if the deviation is deemed significant should greater design or simulation effort be expended to use a more concrete model. A crude estimate in time is often far more valuable than an accurate estimate that arrives after the battle is lost. Rapid progress in communications system architecture and technology, and the swift obsolescence of existing communications systems make timeliness more of a virtue than n-digit accuracy.

20.5 EXAMPLES

20.5.1 Client–Server System

A client–server system typically is a collection of computer systems, with one system hosting a server application and the other systems hosting client processes. The systems are connected via a network (either a local area network or a wide area network). An example of this kind of system is a bank, with the server being the main accounting system and the clients being teller terminals located in the branch offices of the bank.

The clients all send requests for service to the server; the server processes each of the incoming requests and returns a reply to each requester. A major design issue for such a system is the size and processing capacity of the server, so that all of the requests for service can be handled promptly.

A process-oriented simulation model of this kind of system could be built from node objects, where each node has one or more CPUs, a collection of disk drives, main memory, and an input "mailbox" (a mailbox is a mechanism for receiving messages). One node is the server node, and the rest of the nodes are client nodes. A server process is assigned to the server node, and client processes are assigned to the client nodes.

The server process models the server application. Briefly, the server application (process) receives requests for service in its input mailbox. It decodes each request message and then assigns the request to a subprocess. The subprocess proceeds, in a parallel, asynchronous manner, to process the request and then formulate a response that is sent back to the client who originated the request. Processing this request by the server subprocess will require using a CPU for some amount of simulated time and possibly making several accesses to some of the disk drives, depending on the type of request. It is the competition for server node resources by these subprocesses that determines the time required to service requests.

Each client process is associated with a human operator (teller). This process formulates a request for service, sends it to the server's input mailbox, waits for a reply, evaluates the reply (representing a delay), formulates the next request, and so on. All of the clients operate in the same manner.

The key measure of performance is the average response time for service requests. The model can be used to configure the server node so that the average response time (or other measures of performance) meet the criteria for the system. The model can also be used to project system performance as the number of clients increases.

20.5.2 Modeling a Shared Memory Multiprocessor Computer System

A shared memory multiprocessor computer system has several CPUs, each with a private cache; each CPU (really each cache) is connected to a single bus, which in turn is connected to the main memory. This arrangement is attractive because processes execut-

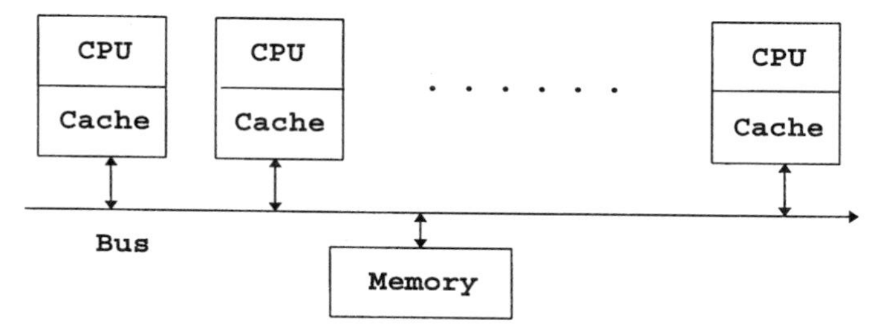

Figure 20.3 Shared memory, multiprocessor system.

ing on different CPUs (in parallel) can easily communicate with each other via common areas of the main memory (shared memory). Such a system is shown in Figure 20.3.

As could be expected, the bus between the processors and the main memory is a potential performance bottleneck. The model can be used to predict the number of processor nodes that can be accommodated on this bus. The number of processors, the instruction processing rate, the cache miss rate, and the speed of main memory all affect the performance of this system.

Another issue with such a system is the cache coherency protocol, used to make certain that all accesses to data (including modified shared data) is to the correct (latest) data values. The choice and efficient implementation of this protocol can have an impact on performance.

A simulation model was implemented using a process for each cache. The bus and the main memory are modeled as simulated resources. The cache processes decided when a cache miss occurs using a probability distribution based on the cache miss rate (an input parameter to the model). Each miss causes an access to the bus and to main memory. The competition for bus and memory accesses is modeled as competition for the simulated resources. The main performance measures are memory request latencies and bus utilization. This type of system can accommodate only a limited number of CPUs; using this kind of model can help determine this limit.

20.5.3 Modeling Ethernets and Token Rings

The two major varieties of local area networks are Ethernet (ANSI/IEEE, 1996) and token ring (ANSI/IEEE, 1995). Transmission rates are nominally 10 Mbit/s for Ethernet and 16 Mbit/s for token ring, although effective channel utilization tends to be rather low for such protocols. From a modeling standpoint the principal differences are transmission rate, network topology, and media access control (how permission to transmit is obtained).

Both the original Ethernet and token ring standards are becoming dated, and new local area network standards are being proposed for higher-bandwidth networks. The new standards are needed principally to accomplish two goals: higher transmission rates and support for multimedia traffic (mainly voice and video).

This *digital convergence* of traditional data traffic with historically analog voice and video traffic is the main impetus behind the need for higher-bandwidth and more inclusive protocols. In addition to local area networks, the telecommunications industry is supporting the Asynchronous Transfer Mode (ATM) (ATM Forum, 1995) standard as the wide area network solution for digital convergence. There is also a standards activity to bring ATM into the local area network environment as the logical layer between higher-level local area network protocols and the lower-level physical channel.

Modeling an Ethernet. While there are a number of different varieties of Ethernet, the typical transmission rate is 10 Mbit/s, or one bit every 100 ns. Access to the channel is gained by a probabilistic method of first waiting for the channel to become idle and then transmitting the message. The receiver listens during transmission, and if it should detect a collision with another sending party, a special collision signal is broadcast and then the message transmission is aborted. After a random-duration *backoff period*, the transmission is reattempted at the first idle interval following the backoff period. The random distribution from which the backoff period is drawn is adjusted according to the frequency of collisions experienced.

Collisions result in aborted message traffic and waste of channel utilization. Collisions result when two or more parties desire to send, observe the channel idle, and begin transmission within a critical interval determined by the propagation delay between senders. The greater the distance between senders, the longer the critical interval during which collision is possible. This interval is independent of transmission rate, so the cost of collisions in lost traffic only increases with increasing channel capacity.

Modeling a Token Ring. Nominal token ring transmission rates are 16 Mbit/s, or one bit every 62.5 ns. Media access is gained by acquiring a token that is passed around the logical ring topology of the token ring network. Elaborate protocols decide how the first token is created on network startup and how a new token is forged if the circulating token is lost. A sender is clear to send when the token is acquired, and the sender is guaranteed no collision (unless there is a duplicate token in the ring). It is the sender's responsibility to remove the message from the ring when it has circuited the ring back to the point of origination. MacDougell (1987) discusses simulation models of both token rings and ethernets.

20.5.4 Modeling High-Speed Networks

High-speed networks are generally considered to be networks with channel rates at or above 1 Gbit/s. At these rates, traditional methods of flow control via return messages to "slow down" are cumbersome. For example, at a signal velocity of $\frac{2}{3}c$ (about 2×10^8 m/s), Table 20.1 shows the quantity of "bits in flight" in the channel just as the leading bit hits the receiver. Thus the communications system modeler may encounter alternative flow control methods (e.g., channel capacity reservation, transmission metering, etc.) in dealing with high-performance networks.

Probably a greater concern for the network modeler confronted with simulating high-

TABLE 20.1 Channel Bit Storage

Transmission	Link Length	Bits in Flight
1 Gbit/s	1 km	5 kbit
1 Gbit/s	1000 km	5 Mbit
10 Gbit/s	1 km	50 kbit
10 Gbit/s	1000 km	500 Mbit

performance networks, though, is the tremendous simulation effort to model even a small amount of real time. While the propagation delays are unaffected by the high transmission rates, the volume of traffic in the network is increased dramatically. Other factors being equal, a 1-Gbit/s network would have 100 times the traffic in the network at any given time as a 10-Mbit/s network, assuming the same channel utilizations. If higher channel utilizations are anticipated, to utilize the costly high-performance channels more efficiently, a several-hundredfold increase in traffic could occur.

The modeler undertaking to simulate a high-performance network needs either a great deal of computing power or must be willing to wait a significant amount of time for the results. The alternative is to scale back the size and scope of the model, to compensate for the additional workload. This could reduce the simulation accuracy and workload range under study.

20.6 SUMMARY

In this chapter we have covered some of the basic techniques and issues associated with constructing and using simulation models of computer systems and communications networks. With both kinds of systems, a workable model consists of two equally important submodels: (1) the workload model, and (2) the system model.

The workload model is responsible for generating units of work that resemble the units of work for the real (modeled) system. The system model processes the stream of simulated workload units. In many cases the most important measure of system performance is the response time (latency) for these workload units as they enter and then eventually leave the system. Accurate simulation models of real systems must provide realistic estimates of these response times.

Models of computer systems and communications networks differ from other types of simulation models in some significant ways:

- *Time Units.* Most events in these types of models occur in the range of milliseconds and/or microseconds; this means that if significant periods of real time are to be spanned by the simulation time, thousands or millions of events will be simulated; this, in turn, means that simulation models could require extensive amounts of computer time.
- *Workload Models.* In many cases, existing systems can be used to generate the parameter values required to characterize the workload; also, the nature of these models means that without accurate profiles of the workload, it is difficult, if not impossible, to construct useful models of these systems.
- *System Models.* In many cases the systems being modeled cannot be readily viewed by a modeler; the real system is collection of resources and allocation policies that are controlled by software or implemented in software and are not easy to visualize. Assistance from system experts is often critical to the development of useful simulation models.

The field of simulation modeling of computer systems and communications networks covers a broad range of models, application domains, and uses. There are a number of commercially available packages that are tailored to implementing and using models in some of these domains. A person faced with the challenge of developing a model in this area should evaluate some of these packages before launching a "build from scratch" implementation project. There are also several technical conferences and workshops that include sessions addressing topics in this area. A list and/or evaluation of some of these packages and conferences is beyond the scope of this chapter.

REFERENCES

ANSI/IEEE (1995). *Standard 802.5 (Token Ring)*, IEEE, Piscataway, N.J.

ANSI/IEEE (1996). *Standard 802.3 (Ethernet)*, IEEE, Piscataway, N.J.

ATM Forum (1995). *ATM User Network Interface (UNI) Specification Version 3.1*, Professional Technical Reference Series (Paper 0-13-393928-X), Prentice Hall, Upper Saddle River, N.J.

Bae, J. J., and T. Suda (1991). Survey of traffic control schemes and protocols in ATM networks, *Proceedings of the IEEE*, February, pp. 170–189.

Jain, R. (1991). *The Art of Computer Systems Performance Analysis*, Wiley, New York.

Law, A., and D. Kelton (1991). *Simulation Modeling and Analysis*, 2nd ed., McGraw-Hill, New York.

Lazowska, E., J. Zoharjan, G. Graham, and K. Sevcik (1984). *Quantitative System Performance: Computer System Analysis Using Queueing Network Models*, Prentice Hall, Upper Saddle River, N.J.

MacDougall, M. H. (1987). *Simulating Computer Systems: Techniques and Tools*, MIT Press, Cambridge, Mass.

CHAPTER 21

Simulation and Scheduling

ALI S. KIRAN
Kiran and Associates

21.1 INTRODUCTION

In a job shop, a set of *orders* often referred to as jobs, parts, products, and so on, may require one or more *operations*, given by a process plan (job routing, etc.). The *process plan* specifies each operation and its requirements (i.e., resources and time required) as well as the *sequence of operations*. In general, there may be alternative *resources* (which can be referred to as machines, processors, workstations, and so on, for each operation. The operations themselves may be substituted by other operations based on availability and or performance considerations. Scheduling in practice usually refers to the determination of a set of orders, which will be processed by the resources during a short-term period (day, week, etc.). The selection of this period, the scheduling horizon, may be part of the decision.

A real-world scheduling problem could be stated simply as "a selection of five orders to run on Monday." In selecting the orders to be completed first, the shop supervisor uses performance measures, although in most of the cases, indirectly. For example, in the hope of reducing the number of late orders, he or she may give *priority* to orders with tighter due dates. Usually, there is more than one objective present in the practice of scheduling. In most cases the scheduling objectives cannot be even stated in terms of a quantifiable *scheduling criterion*. In addition, the concern of the scheduler is to reduce the negative impact of random events such as machine downtimes, absenteeism, scrap, and reworks.

The scheduling problem is the determination of the start and completion time for each operation of each order so that (1) no constraints are violated, and (2) some scalar function of the operation start and completion time is minimized (or maximized). The first constraint leads to the concept of a feasible schedule, whereas the second constraint defines an optimal schedule. These concepts are explored in the next section.

Scheduling problems generally include restrictive assumptions in order to be solved. A representative set is provided here for clarity [1].

Handbook of Simulation, Edited by Jerry Banks.
ISBN 0-471-13403-1

1. *Each job is an entity.* Although a job is composed of distinct operations, no two operations of the same job may be processed simultaneously. Thus we exclude from our discussion certain practical problems (e.g., those in which the components are manufactured simultaneously prior to assembly into the finished product).

2. *There may be no preemption.* Each operation, once started, must be completed before another operation may be started on that machine.

3. *Each job has m distinct operations, one on each machine.* The possibility that a job might require processing twice on the same machine is not allowed. Additionally, each job is processed on every machine; it may not skip machines. Note that the latter constraint is *not* illusory. Although it could be implied that a job which skips a machine is processed on that machine for zero time, a question remains: Where in the job's processing sequence should this null operation be placed? Because preemption is not allowed, the job could be delayed, waiting for a machine that is not needed.

4. *There may be no cancellation.* Each job must be processed to completion.

5. *The processing times are independent of the schedule.* There are two assumptions: First, each setup time is sequence independent (i.e., the time taken to adjust a machine for a job is independent of the job last processed). Second, the time to move a job between machines is negligible.

6. *In-process inventory is allowed* (i.e., jobs may wait for the next machine to be free). This is not a trivial assumption. In some problems the processing of jobs must be continuous from operation to operation. In steel mills, for example, one literally has to strike while the iron is hot.

7. *There is only one of each type of machine.* A choice of machines in the processing of a job is not allowed. This assumption eliminates the case where certain machines are duplicated to avoid bottlenecks.

8. *No machine may process more than one operation at a time.*

9. *Machines never break down and are available throughout the scheduling period.*

10. *The technological constraints are known in advance and are immutable.*

11. *There is no randomness.* In particular:

(a) The number of jobs is known and fixed.
(b) The number of machines is known and fixed.
(c) The processing times are known and fixed.
(d) The ready times are known and fixed.
(e) All the other quantities needed to define a particular problem are known and fixed.

These restrictions usually disregard practical factors such as random process and downtimes for the purpose of defining computationally tractable solutions. Therefore, as a corollary to the preceding statement, one can see that an easily solvable problem in theory may become very complex and computationally intractable as the assumptions are removed to represent real-world problems. For example, just introducing realistic shift breaks that may be different for each resource may make any multiresource scheduling problem computationally intractable.

Still, much can be gained by examining theoretical results and optimization algorithms because the main ideas behind these techniques are applicable for practical problems. For example, most optimization algorithms use the concept "keep the bottleneck machine busy" to obtain solutions. This concept is widely applied in a number of industrial scheduling problems.

Our purpose here is not to give a full description of a general job shop scheduling prolem and its solution techniques. We refer the reader to the available literature published in this area [1–4]. Our purpose is to provide a brief definition and background for exploring scheduling and simulation applications.

The chapter is organized as follows. In Section 21.2 the general job shop scheduling problem is defined. After exploring the problem parameters and basic solution approaches, in Section 21.3, the basic elements of a job shop simulation model are discussed in Section 21.4. Section 21.5 concludes with practical applications, which illustrate the use of simulation and scheduling in different real-world settings. Implementation issues are discussed in Section 21.6.

21.2 DEFINITIONS AND BACKGROUND

The general job shop problem is defined so as to determine the start/completion times for each operation of each job waiting to be processed in the shop that satisfies the following:

1. The technological constraints or processing the order for each job on all the resources
2. Optimality (i.e., minimize or maximize a given objective function) or satisfiability (i.e., a reasonably good performance with respect to one or more scheduling criteria) constraints

The term *scheduling criterion* defines a scalar value function, which measures the performance or effectiveness of a particular schedule. A performance measure is usually defined in terms of its shop or job completion characteristics and is given as a function of the job or operation completion times.

The processing of an order on a resource is called an *operation*. Each job must be processed through the machines in a particular order and may have no relation to the processing order of any other job. A special case of the general job shop scheduling problem is defined as a *flow-shop problem*, where all the jobs go through the same processing order. This special case is of particular interest because of the widespread simulation applications in flow processes, such as print shops, electronics manufacturing, and assembly operations.

Each operation requires a fixed or stochastic length of time to be completed, which is referred to as the *processing time*. In a general job shop, the processing time is assumed to be sequence or shop condition independent, a highly contested assumption. Each operation may also require a *setup time*. A setup time is required for preparing a machine for a particular operation. The setup time may or may not be required based on the current setup of a machine. Therefore, in the general problem, the setup time is assumed to be sequence dependent.

Each job may also be given a *due date* where the due date of a job may be determined by external circumstances such as seasonal demand or customer requirement dates. In some cases the due date may be internally determined, such as a required completion of components for the assembly products. Each job may be given a *release date* or the available time the first operation on a job can start. This is the time that the first

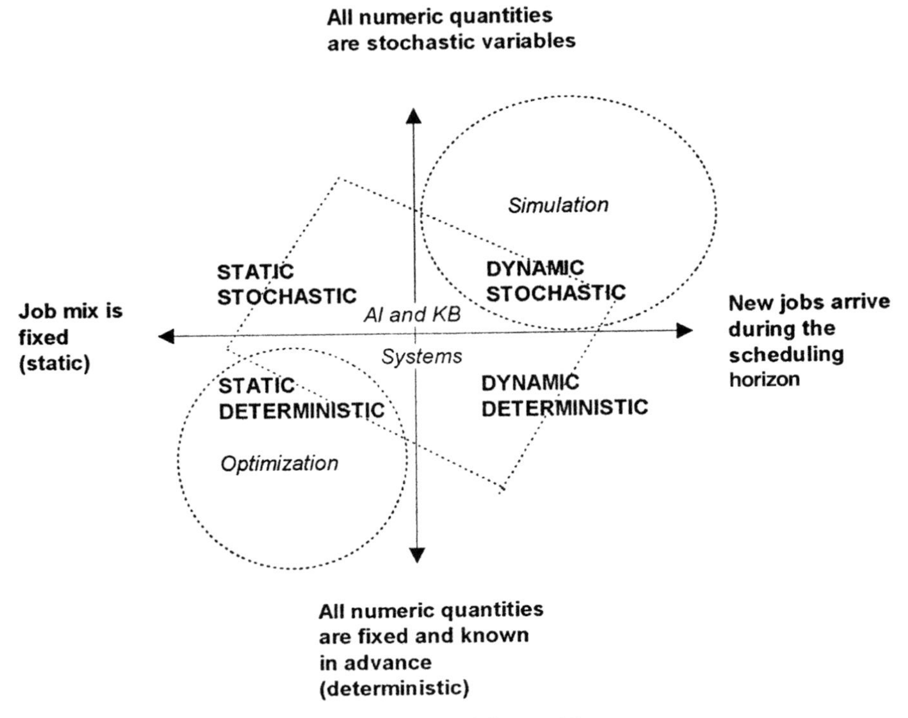

Figure 21.1 Scheduling problems.

operation of a job can start due to practical considerations such as material availability.

All numeric quantities (e.g., processing time, due dates, etc.) are assumed to be known in advance in a *deterministic* scheduling problem. In the real world, however, most numerical quantities are *stochastic*; that is, they are subject to randomness and are not known in advance.

The ready time or release time makes a considerable difference in the problem structure. If all the jobs are assumed to be available at time 0, the problem is called a *static* problem. A nonzero release or ready time for a subset of jobs defines the problem as a *dynamic* problem. Scheduling problems can fit into one of four categories as described in Figure 21.1.

Optimization-based approaches, where one tries to minimize or maximize a scalar objective function, are more suitable for deterministic–static problems where problem data are known in advance with perfect accuracy. However, as accuracy and data availability diminish, rule- and simulation-based approaches become more suitable.

21.3 SCHEDULING APPROACHES

Before one determines which scheduling approach to use, there are two important considerations: performance criteria and the availability of data. A valid scheduling approach for one set of criteria may not be valid for another. For example, an

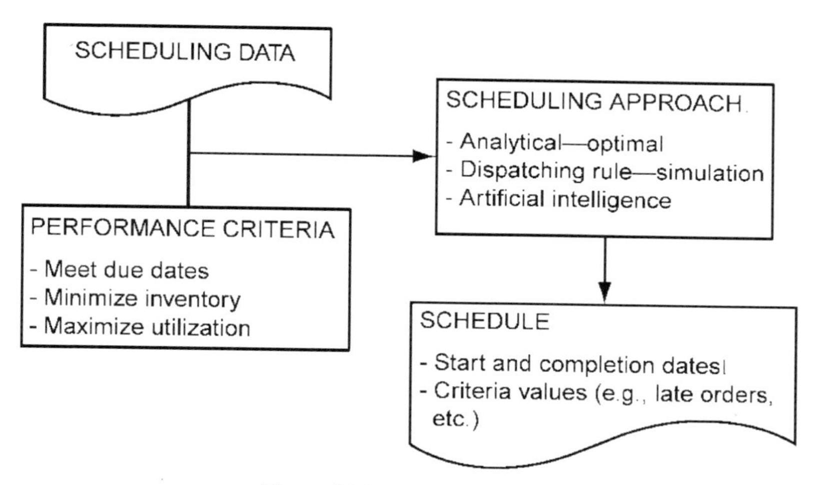

Figure 21.2 Information flow.

optimization-based approach may fail in a situation where the scheduling criterion is vague or qualitative. Similarly, a scheduling approach based on precise information about the shop floor status can be disastrous if the data are not accurate. Figure 21.2 illustrates the relation between performance criteria and that of approach and scheduling data.

Scheduling criteria may be classified into three categories:

1. *Shop Performance Based.* This set of criteria considers only time information about the start and completion dates of orders in the shop. This kind of criterion includes:

- Work in process (WIP)
- Utilization
- Completed parts per day, week, etc.

2. *Due Date Based.* Due date–based criteria also consider the customer's required due dates or internally determined shipping date information. This set of criteria may include:

- Number of late shipments
- Order lateness
- Others

Using an example, we will later show how each additional piece of information required by the scheduling solution further complicates the decision-making process. In general, due date–based criteria is harder to deal with than shop performance–based criteria.

3. *Cost Based.* On the surface, this is the most fitting criteria for industrial problems. However, the introduction of different cost components and the difficulty of obtaining accurate estimates for each of the components make cost-based criteria very complicated

for industrial scheduling problems. In most cases, costs are considered implicitly by the criteria given in the two categories above.

It has been stated [5] that the meeting of the due dates is the most important goal, followed by the minimization of setup times and the minimization of in-process inventory (in that order) in industrial scheduling. Criteria may also be classified as cost based versus time based or sequence based versus order based, and so on. Table 21.1 shows the most frequently used criteria.

There are three basic approaches to scheduling problems:

1. Dispatching rules and simulation-based approaches
2. Analytical-optimal approaches
3. Artificial intelligence, neural network, and knowledge-based systems

Non-simulation-based approaches are explored in the rest of this section as a brief review of the basic scheduling approaches. Simulation-based approaches are covered in more detail in Section 21.4.

21.3.1 Dispatching Rule–Based Approaches

The basic premise of this approach is to choose an order (job) for the available machines while moving forward or backward on a time scale. A dispatching rule is simply a rule of thumb that gives priority to one order among the many available orders at any stage. The dispatching rule does not consider the scheduling criteria explicitly. There is statistical evidence which shows that some rules perform better than others. This is sufficient enough in practice to use rules such as those given in Table 21.2.

Most of the rules given in Table 21.2 are self-explanatory. They are referred to as simple priority rules because they make decisions based on a scalar function called a *priority function*. A priority function can be calculated once (e.g., order due date) and referred to as a static rule, versus the changing of the value over time for a dynamic priority function. The function itself may be based on shop or other data, such as process times, due dates, number of operations, or setup times. It is also possible to define a scheduling rule as a combination of weighted dispatching rules and may use more than one priority function to decide which of the waiting orders will be processed next.

The following numerical examples have been constructed to illustrate the dispatching rule–based scheduling approaches.

Example 1 This example illustrates the development of a rule-based forward schedule that implicitly considers completion time– and work-in-process–based criteria. There are four orders that must be scheduled on a single available machine. The processing times for each order are given in Table 21.3.

Let us consider the shortest processing time rule for this case. The shortest processing time (SPT) rule assigns the order with the smallest run time when a resource becomes available, as shown in Table 21.4. The resulting schedule is shown on the Gantt chart in Figure 21.3.

The following criteria may be computed for this case, as shown in Table 21.5, where

$$\text{flow time} = \text{completion time} - \text{release time}$$

TABLE 21.1 Criteria Used in Job Shop Scheduling Studies

A. Criteria based on order completion times
- Average or variance of *flow time*, where flow time = completion time − release time
- Average or variance of *waiting time*, where waiting time = completion time − release time − process time
- Mean flow time per operator
- Delay factor (average waiting time/flow time)
- Variance of flow time per operation
- Distribution of flow times

B. Criteria based on in-process orders
- Number of orders in the queue
- Number of orders in the shop
- Number of orders completed
- Number of operations completed
- Number of waiting orders for more than a specified period of time
- Size of the orders waiting for more than a specified period of time
- Total work content (i.e., total processing time of all operations of all orders)
- Total work remaining (i.e., total processing time of remaining operations of all orders)
- Total processing times of the orders in the queue
- Total work completed
- Imminent operation work content
- Machine queue balance index

C. Criteria based on processor data
- Machine utilization (i.e., percentage of time machine is busy)
- Machine idle time
- Machine work balance index
- Shop work balance index
- Desired loading measure
- Total or the average setup times

D. Criteria based on due dates
- Average, maximum or distribution of Lateness where Lateness = Completion Time − Due date
- Average, maximum or distribution of tardiness where tardiness = lateness if lateness > 0 (i.e. job is late), otherwise 0.
- Average, maximum or distribution of earliness where earliness = lateness if lateness < 0 (i.e., job is early), otherwise 0.
- Fill rate (i.e., percentage of the orders meeting the due dates)
- Fraction (or %) of late orders
- Number of late orders
- Number of late orders in the queue
- Number of late operations

E. Cost based criteria
- Total cost per order
- Finished goods investment
- Storage costs of finished goods
- Average $ days of queue inventory
- Cost of carrying Work-in-process inventory
- Early (or late) completion penalty cost
- Percent of deviation in penalty cost associated with an optimal schedule
- Cost of long promises
- Setup Cost

TABLE 21.2 Simple Priority Rules

Critical ratio (CR)	This decision is based on a critical ratio index calculated as follows: The order with the minimum critical ratio is selected. CR = lead time/remaining operation time, where lead time = due date − time now.
Earliest due date (EDD)	Selects the order with the earliest due date.
Fewest operations remaining (FOPNR)	The order with the least number of operations left is selected.
First come, first served (FCFS)	Priority is given to the order that arrives into the particular queue earliest.
First in system, first served (FISFS)	Priority is given to the order that arrives into the shop earliest.
Job value (JV)	This looks at the total value of the job and selects the highest value.
Least work remaining (LWKR)	The decision is based on the total remaining time on this order.
Longest processing time (LPT)	The order with the longest processing time is selected.
Minimum operation slack time (MOST)	First, an operation slack time is calculated as follows: The job with the minimum operation slack time is preferred. Operation slack time: operation due date − remaining operation time − time now.
Most operations remaining (MOPNR)	The order with the most number of operations left is selected.
Most work remaining (MWKR)	Similar to the above. The order with the most work remaining is selected.
Number in next queue (NINQ)	This is a simple look-ahead type of rule which checks

TABLE 21.3 Data for Example 1

Orders (Jobs)	J1	J2	J3	J4
Run time	10	20	30	40

TABLE 21.4 Shortest Processing Time Schedule

Decision Time	Jobs Considered	Job Selected
0	J1, J2, J3, J4	J1
10	J2, J3, J4	J2
30	J3, J4	J3
60	J4	J4

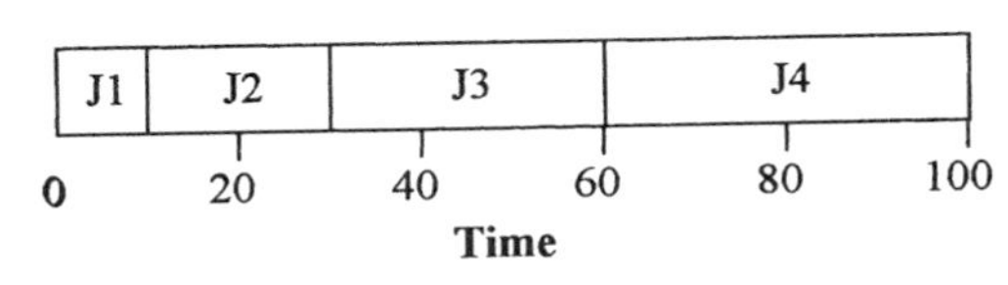

Figure 21.3 SPT schedule for Example 1.

TABLE 21.5 Flow-Time Calculations

Job	Completion Time	Release Time	Flow Time
J1	10	0	10
J2	30	0	30
J3	60	0	60
J4	100	0	100
	Total flow time = 200		

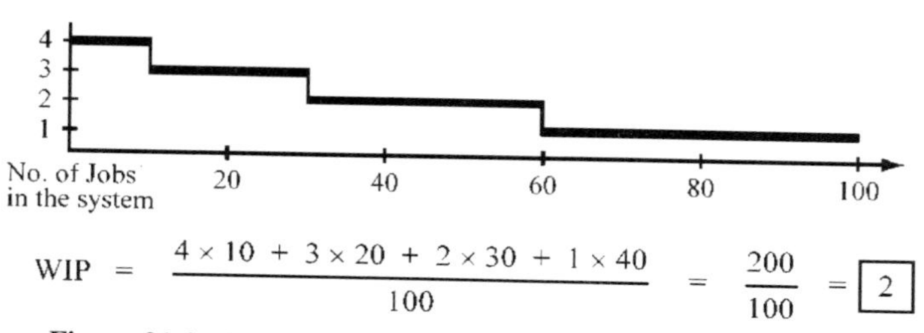

$$\text{WIP} = \frac{4 \times 10 + 3 \times 20 + 2 \times 30 + 1 \times 40}{100} = \frac{200}{100} = \boxed{2}$$

Figure 21.4 WIP inventory and job flow times for SPT schedule.

TABLE 21.6 Data for Example 2

Orders (Jobs)	J1	J2	J3	J4
Run Time	10	20	30	40
Release Time	55	20	0	25

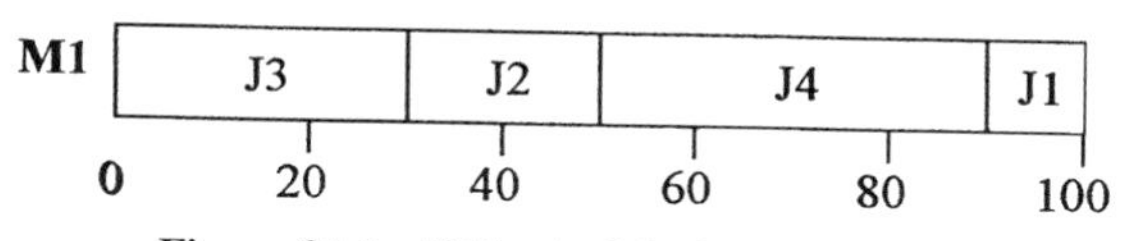

Figure 21.5 SPT schedule for Example 2.

Work in process (WIP) can be computed as follows:

$$\text{average WIP} = \frac{\text{total flow time}}{\text{length of scheduling period}}$$

The relation between the average WIP and the total flow time is shown in Figure 21.4.*

Example 2 This example illustrates additional scheduling constraints such as release times for the flow-time criterion. Consider Example 1 but assume that the jobs are now released into the shop according to Table 21.6. A simulation-based schedule using the shortest processing time rule is given in Figure 21.5 and illustrated further in Table 21.7. The criteria previously defined for this schedule are computed in Table 21.8.

*This relation, known as *Little's formula*, holds true for more general cases involving stochastic order release and processing times [22].

TABLE 21.7 Development of an SPT Schedule for Example 2

Decision Time	Jobs Considered	Job Selected
0	J3	J3
30	J2, J4	J2
50	J4	J4
90	J1	J1

TABLE 21.8 Flow-Time Schedule

Job	Completion Time	Release Time	Flow Time
J1	100	55	45
J2	50	20	30
J3	30	0	30
J4	90	25	65
		Total flow time = 170	

Compared to the results in Example 1, the total flow time is reduced. This illustrates that a controlled release of material into the shop is desirable. Actually, just-in-time (JIT) policies use this property to minimize the work-in-process (WIP) inventories and average flow times.

Example 3 This example illustrates due date information, backward scheduling, and due date–based criteria. Assume the information above with the additional due date information as given in Table 21.9. All the release times are zero. A backward schedule starts with the latest-due-date job, placed in a way that it would be completed at its due date, then moves forward in time and considers the next job. A backward schedule for the example problem is shown in Figure 21.6.

As can be seen from this schedule, one issue associated with developing backward schedules is that of unfeasibility. The situation is encountered here after the scheduling of J3 and J4. The remaining time from 0 to 20 is not enough for J1 and J2. If one proceeds with strict backward scheduling the scheduled start time of J2 would be −16,

TABLE 21.9 Data for Example 3

Orders (Jobs)	J1	J2	J3	J4
Run Time	10	20	30	40
Due Date	14	13	110	60

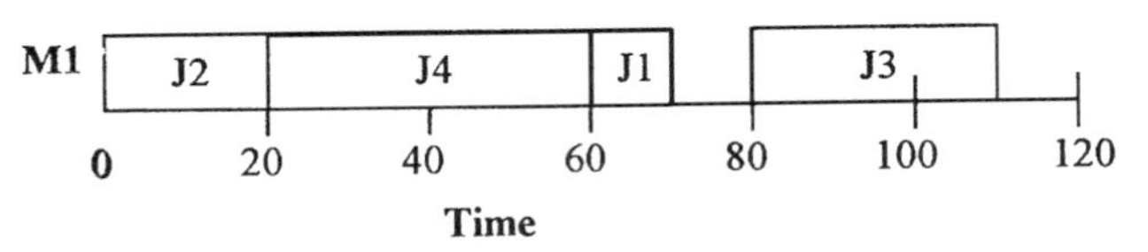

Figure 21.6 Backward schedule for Example 3.

TABLE 21.10 Computation of Due Date–Based Criteria

Job	Completion Time	Flow Time	Due Date	Tardiness
J1	70	70	14	56
J2	20	20	13	7
J3	110	110	110	0
J4	60	60	60	0
		Total flow time = 260	Total tardiness = 63	
			Number of tardy jobs = 2	

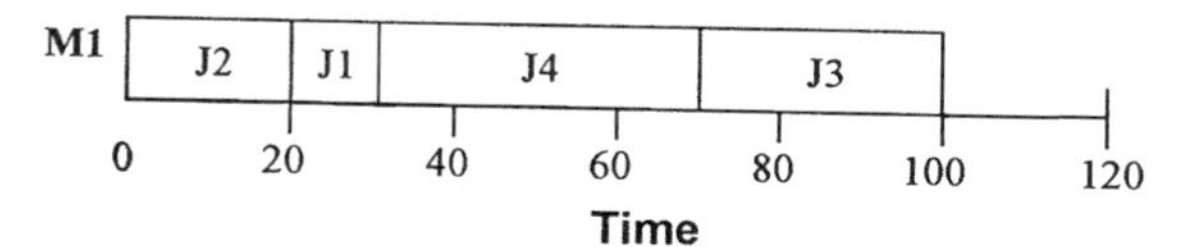

Figure 21.7 EDD schedule for Example 3.

which means that the processing of J2 should have started 16 time units earlier than the time of scheduling. Therefore, most backward scheduling procedures also define mechanisms to avoid unfeasible schedules. Here, a decision has been made to schedule J1 as a late job in order to generate a feasible schedule.

The computations of the total flow time, total tardiness, and total number of jobs are shown in Table 21.10. In contrast, a forward schedule, based on the earliest due date rule (i.e., gives priority to the order with the earliest due date), is shown in Figure 21.7. The criteria values for this schedule are computed in Table 21.11.

This example illustrates the concept of a dominant schedule with respect to criteria. If one considers the total flow time and total tardiness, the EDD schedule provides a better performance and so it dominates the backward schedule with respect to these criteria. However, if the number of tardy jobs is also important, the EDD schedule cannot be called "dominant" anymore. Because the backward schedule has two late jobs, it is therefore better than the EDD schedule for this measure of performance.

21.3.2 Optimization-Based Approaches

Optimization-based approaches are designed to develop optimal schedules to minimize or maximize a scheduling criterion. The advantage of these approaches is that

TABLE 21.11 Performance of the EDD Schedule

Job	Completion Time	Flow Time	Due Date	Tardiness
J1	30	30	14	16
J2	20	20	13	7
J3	100	100	110	0
J4	70	70	60	10
		Total flow time = 220	Total tardiness = 33	
			Number of tardy jobs = 3	

the scheduling criterion is explicitly considered during the development of a schedule. However, this requires that the quantifiable objective(s) be determined for the particular scheduling application. These approaches also require a considerable amount of solution time to obtain an optimal schedule if the number of alternative solutions is large. Therefore, some optimization-based approaches sacrifice from optimality in exchange for faster solutions. A fast solution may not consider all the possible alternatives explicitly but may choose from a subset of available alternatives and evaluate them before reaching a solution. These approaches are generally classified as *heuristics*. Some of these can also be classified as artificial intelligence– and knowledge-based approaches and they are covered in Section 21.3.3.

Optimization-based approaches covered in this section include:

1. Approaches based on optimal scheduling rules
2. Implicit enumeration techniques
3. Mathematical programming

Optimal Scheduling Rules. This approach is based on evaluating the optimality of a schedule using the mathematical properties of the problem and using the characteristics of a schedule. The schedules in question are generated using a set of rules: therefore, once optimality of the schedule is proven for most general cases of the scheduling problem being considered, the optimal scheduling can be applied to all other problems in this problem class. The following examples illustrate the various properties of the optimization-based approaches.

Example 1A Consider Example 1 as given above. Refer to Table 21.12 for the problem data. The objective is to minimize the total or average job flow times. Consider any two orders that are adjacent in any schedule. We can compare the contribution of any change in sequence to the objective function. For example, for orders J1 and J2, the calculations are as follows:

- Sequence J1–J2

$$\text{Flow time for J1} = \text{start time of J1} + 10$$
$$\text{Flow time for J2} = \text{start time of J1} + 10 + 20$$

- Sequence J2–J1

$$\text{Flow time for J2} = \text{start time of J2} + 20$$
$$\text{Flow time for J1} = \text{start time of J2} + 20 + 10$$

Since the only difference between the two schedules are the positions of J1 and J2,

TABLE 21.12 Data for Example 1A

Orders (Jobs)	J1	J2	J3	J4
Run Time	10	20	30	40

TABLE 21.13 Comparison of Alternatives for Example 1

Alternatives Compared	Difference in Objective Function Value	Alternative Selected
J1–J2 vs. J2–J1	−10	J1–J2
J1–J3 vs. J3–J1	−20	J1–J3
J1–J4 vs. J4–J1	−30	J1–J4
J2–J3 vs. J3–J2	−10	J2–J3
J2–J4 vs. J4–J2	−20	J2–J4
J3–J4 vs. J4–J3	−10	J3–J4

$$\text{start time of J1} = \text{start time of J2}$$

Then sequence J1–J2 is better than sequence J2–J1 by 10 time units. The rest of the comparisons are given in Table 21.13.

The comparisons given in Table 21.13 are applicable regardless of the positions of these orders in a sequence because of the specific objective function used in this example. Therefore, in an optimal schedule, to minimize the total flow time, J1 must be in the first position followed by J2–J3–J4, in that order. In fact, Table 21.13 can be used to prove the optimality of the shortest processing time (SPT) rule for any problem, which has the same structure as Example 1A. Hence given any problem of a similar type, an optimal schedule can be found by applying the SPT rule to minimize the average flow time.

In the example we used the *adjacent pairwise interchange technique* as the mathematical analysis technique. In this method we evaluate the alternative sequences by resequencing adjacent jobs in any given schedule and checking the objective function value. The analysis here showed that one can improve the objective function by interchanging neighboring jobs. The optimal solution is found when no other improvement is possible.

The optimization rule–based solution here has a definite advantage over total enumeration where each possible sequence must be explicitly evaluated. For example, for a 70-job problem similar to above, one has to sort 70 jobs to identify an optimum schedule. Given the state of spreadsheet software today, this is a trivial task.* By comparison, a complete enumeration-based approach would have evaluated all 70 sequences (i.e., all possible permutations of 70 jobs).† "Brute force" or complete enumeration of all the feasible schedules is therefore not possible because of the very large number of alternative solutions even for a moderate-sized industrial problem. This also leads to the use of some rules that are known from statistical evidence that they can find nearly optimal solutions. For example, Johnson's rule optimizes the maximum completion time in job shops if there are two machines in the shop. A near-optimal scheduling rule was developed using Johnson's rule for the general flow shops with any number of machines.

*Even by hand, one has to go through at most 70 × 70 (there are more efficient sorting methods available) calculations to do this task.

† For a 70-job single-machine problem, there are 70! = 10 exp(100) possible sequences. Assuming that each job sequence yields a different schedule, we leave it to the reader to calculate how long it would take to enumerate all of these schedules. (*Note:* Assume that a very fast computer is available to you and evaluate each alternative in only 1 ns.)

There are many other approaches besides the adjacent pairwise interchange technique. However, due to space limitations, we do not present any other techniques here.

Implicit Enumeration Techniques. The enumeration-based approaches use mathematical analysis (and often mathematical programming) to reduce the size of the computational task for obtaining an optimal solution. Here mathematical relations are still used to evaluate all the feasible schedules and select from the best among them. However, once it is established that some set of schedules cannot contain the optimal solution, they are eliminated from further consideration. Therefore, one can expect to eliminate enough nonpromising solutions so that the computational burden will not be impossibly large. The following example illustrates the ideas behind implicit enumeration algorithms without going into the mathematical detail.

Example 2A Refer to Example 2. The data are repeated in Table 21.14. The example illustrates cases where the adjacent pairwise interchange argument cannot be used due to the problem structure (i.e., nonzero release times). In this case an analysis that considers not only the individual pairs but also the entire schedule is necessary for finding the optimal sequence. The resulting schedule is shown in Figure 21.8. This schedule has a total flow time of 150. The improvement over the simulation-based schedule exceeds 15% in this simple example.

Like other optimization techniques, implicit enumeration requires a simultaneous evaluation of alternatives. They have the ability to "look ahead." For example, in Table 21.15, elimination of an alternative is based on the entire schedule rather than the current position in a schedule. Also, in Table 21.15 a total of eight alternatives were evaluated. A total enumeration of the alternatives would require evaluations of $4 \times 3 \times 2 \times 1 = 4! = 24$ sequences. The technique used above implicitly enumerated all of the 24 alternatives by using a mathematical argument. For evaluating all the alternative orders that can be placed in the current sequence position, the following argument has been made: If order X is placed in this position, the resulting sequence can never yield a better solution than order Y in the same sequence position, therefore do not evaluate any sequences with order X in the current position. Obviously, the ability to distinguish between the promising and the dominated sequences will increase the computational efficiency of an implicit enumeration algorithm. Such algorithms are also known by the name *branch and bound*, due to the fact that they branch to the possible solutions and they bound the possible alternatives by deleting unpromising solutions.

TABLE 21.14 Data for Example 2A

Orders (Jobs)	J1	J2	J3	J4
Run Time	10	20	30	40
Release Time	55	20	0	25

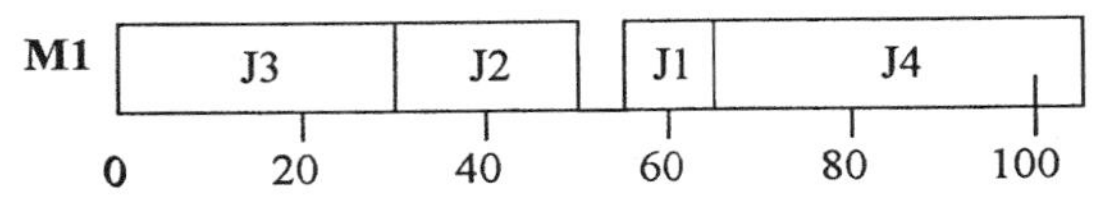

Figure 21.8 Optimal schedule for Example 2.

TABLE 21.15 Implicit Enumeration-Based Optimal Schedule

Alternative	Objective Function Value or Reasons Why Not Considered	Action Taken
J1–X–X–X	While waiting for J1, J2, and/or J3 can be completed.	Ignore all sequences starting with J1. Refer to J2-X or J3-X.
J2–X–X–X	40 – 20 = 20	Continue to develop a schedule using the partial schedule J2.
J2–J3–X–X	(40 – 20) + (70 – 0) = 90	Partial sequence J3–J2 is preferred. Ignore all sequences that include partial schedule J2–J3.
J3–X–X–X	30	Continue to develop a schedule using the partial schedule J3.
J3–J2–X–X	30 + (50 – 20) = 60	Continue to develop a schedule using the partial schedule J3–J2.
J3–J2–J1–J4	30 + (50 – 20) + (65 – 55) + (105 – 25) = 150[a]	Wait until all schedules are generated, then compare the results.
J3–J2–J4–J1	30 + (50 – 20) + (90 – 25) + (100 – 55) = 170	Wait until all schedules are generated, then compare the results.
J4–X–X–X	J3 is shorter and also available earlier.	Ignore all sequences starting with J3. Refer to J3–X.

[a] Selected.

It is also possible to use an implicit enumeration algorithm as a near-optimal procedure by limiting the search for the optimum. When the stopping criteria reaches the algorithm, it stops and presents the best solution found so far as the near-optimal solution.

Example 3A Refer to Example 3. This example illustrates the fact that the objective function itself can change the nature of the optimization seeking. The data are repeated in Table 21.16. All the release times are zero. Let us consider the criteria defined as the total tardiness. The optimum schedule is given in Figure 21.9 and the scheduling criteria defined previously are calculated for this schedule in Table 21.17.

Mathematical Programming–Based Solution Approaches. Mathematical programming refers to the methods and techniques developed for solving constrained optimization problems. Based on the mathematical structure of the relations between the problem components, mathematical programming can more specifically be referred to as linear programming (i.e., all relations can be expressed as a set of linear relations) or integer programming (some of the variables are restricted to binary or whole numbers). Most the scheduling optimization problems can be defined as integer programs as:

TABLE 21.16 Data for Example 3A

Orders (Jobs)	J1	J2	J3	J4
Run Time	10	20	30	40
Due Date	14	13	110	60

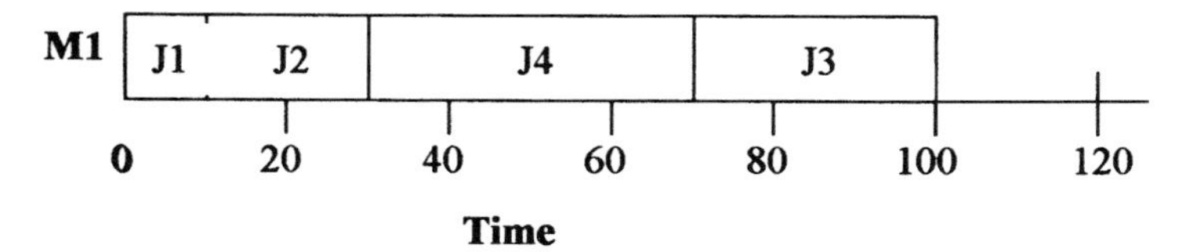

Figure 21.9 Optimal schedule for Example 3.

Minimize (or maximize)
  objective function
Subject to
  Constraints

where the objective function and the constraints are defined as mathematical functions.

This general structure is quite flexible in representing many different problem types. However, they also suffer from the fact that the solution time of an integer program may be very long even for moderately sized problems. Therefore, the use of mathematical programming for solving industrial scheduling problems is limited. Other uses of mathematical programming include the use of mathematical programming in implicit enumeration schemes and using easy-to-solve versions of programs by relaxing some of the hard constraints.

In special cases one can exploit the problem structure to obtain a solution. For example, the number of machines may be a hard constraint in a scheduling problem. But from another point of view this constraint could be removed by allowing outside help (i.e., outsourcing) for the bottleneck processes or buying more resources. Then the optimal solutions to the relaxed cases could serve as beginning solutions in forming near-optimal schedules for the realistic problem.

One of the techniques for relaxing the constraints for obtaining easily solvable versions of the problem, known as *Lagrangian relaxation*, has been used successfully for some scheduling problems. Interested readers are referred to Fisher [23] for more complete treatment of the subject.

21.3.3 Artificial Intelligence–Based Solutions

Artificial intelligence (AI)–based solutions are a developing area with much promise due to the advance of computers. The basic approach is to develop feasible schedules

TABLE 21.17 Scheduling Criteria Calculation for Schedule Given in Figure 21.9

Job	Completion Time	Flow Time	Due Date	Tardiness
J1	10	10	14	0
J2	30	30	13	17
J3	100	100	110	0
J4	70	70	60	10
		Total flow time = 210		Total tardiness = 27
				Number of tardy jobs = 2

in order to satisfy the constraints. Therefore, this is sometimes called *constraint-based scheduling*. For such a system to work, the constraints (rules) of the scheduling environment have to be clearly stated (i.e., gears must be ready before the final assembly). As a result, this poses a problem when there are many and sometimes ambiguous, scheduling rules. Second, AI-based systems are still in the developmental phase. Their solution time may not be acceptable in a situation where there are a large number of rules. The AI approaches include:

1. Rule/knowledge-based approaches (also referred to as expert systems)
2. Genetic algorithms, tabu search, and simulated annealing algorithms
3. Simulated neural network applications

Rule/Knowledge-Based Approaches. Rule/knowledge-based approaches rely on the rule that an expert would use to evaluate and develop schedules. Rule/knowledge-based systems have three components:

1. A database that stores applicable rules in a format that can be processed by the logic component.
2. A data input–output component that searches the orders to be scheduled, keeps track of the scheduled orders and objective criteria, and presents the results to the user.
3. A logic (or sometimes referred to as the *engine*) component that processes data using the rules given in the rule database.

Numerous rule-based systems have been developed with different knowledge representation and processing characteristics [6]. The basic ideas of these approaches are illustrated below by example. Refer to Examples 1 and 1A. The data are repeated in Table 21.18 for convenience. Consider the scheduling objective of minimizing the total flow time. All the jobs are assumed to be arriving at the shop at time zero: the database of an expert system at time zero with a database similar to Table 21.18.

Let us assume that the following rule is given in the expert system's rule database:

If
  processing time of job i is less than that of job j
Then
  schedule job i before job j

Using the rule, the expert system compares the jobs in the current database and determines that J1 must be scheduled first, followed by J2, J3, and J4 in that order. How this actually will be done differs from the way in which other techniques are used. For example, an expert system may choose J1 as the first job, schedule it, then may update the data table before proceeding with the selection of the second job.

TABLE 21.18 Data for Examples 1 and 1A

Orders (Jobs)	J1	J2	J3	J4
Run Time	10	20	30	40

TABLE 21.19 Data for Examples 3 and 3A

Orders (Jobs)	J1	J2	J3	J4
Run Time	10	20	30	40
Due Date	14	13	110	60

In some cases the rule database may not be sufficient to develop a schedule. Consider the problem given in Examples 3 and 3A. The data are repeated in Table 21.19. Total tardiness is selected as the scheduling criterion. Assume that in addition to the rule defined above, the expert system's rule database includes the following rule:

If
 the due date of job i is less than that of job j
Then
 schedule job i before job j

Although both rules make sense intuitively, there are conflicts for the given problem. Both rules agree that J1 and J2 should precede J3 and J4. But the rules give conflicting results if they are applied to pairs J1,J2 and J3,J4. In this case, different expert systems use different techniques. For example, between J1 and J2 one can argue that the due date difference is much smaller than the processing time difference; therefore, a measure could be developed to evaluate the magnitude of rule violations and the expert system could choose to minimize the total violation index. Another expert system could assign a weight to the different rules and then evaluate the results of each rule as a vote. For example, if the weight of a due date–based rule is 2 and the processing time–based rule is 1, J2 is selected over J1 with a vote of 2 to 1.

Genetic Algorithms. Genetic algorithms mimic the Darwinian theory of natural selection. The basic idea is to modify a set of initial alternative solutions to generate a set of alternative solutions that will hopefully include a near-optimal solution. The generation of new solutions resemble the generation of offspring by crossing the parents' genes. After enough alternatives are generated and evaluated, the best solution found is presented. A typical genetic algorithm consists of the following steps:

Step 1: Set generation counter $I = 0$.

Step 2: Create the initial population, Pop(i), by randomly generating N individuals.

Step 3: Increment to the next generation, $i = i + l$

Step 4: Create the new population, Pop(i), by selecting N individuals stochastically based on the fitness from the previous population, Pop(i–l).

Step 4a: Randomly select R parents from the new population to form the new children by application of the genetic operators (e.g., crossover and mutation).

Step 4b: Evaluate the fitness of the newly formed children by applying the objective function.

Step 5: If the stopping criterion has not been met, go on to step 3.

Step 6: Print out the best solution found.

TABLE 21.20 Data for Examples 1 and 1A

Orders (Jobs)	J1	J2	J3	J4
Run Time	10	20	30	40

Example 1B Refer to Example 1 (data are repeated in Table 21.20). Although we have an exact solution algorithm available for this problem, let us apply a simple genetic algorithm to minimize the total flow time. Assume that the initial population includes two schedules: J1–J3–J4–J2 and J4–J3–J1–J2. Let the population limit be two and the maximum number of generations be four, and that the adjacent pairwise interchange be used for generating new schedules. We also define a simple and more deterministic natural selection rule: Evaluate the performance criteria value of all the schedules generated. Select the best schedules for the development of the new generation. Drop the ones with the worst performance to keep the population in check.

Application of this simplified genetic algorithm to the problem is given below.

- *Generation 1:* The population consists of J1–J3–J4–J2 and J4–J3–J1–J2 with fitness (i.e., total flow-time values) of 230 and 290, respectively. Assume that J1–J3–J4–J2 are selected for reproduction. By interchanging J4 and J2, J1–J3–J2–J4 is generated with a fitness value of 210. Select the two schedules with the best fitness values.
- *Generation 2:* The population consists of J1–J3–J4–J2 and J1–J3–J2–J4 with fitness (i.e., total flow-time values) of 230 and 210, respectively. Assume that J1–J3–J2–J4 are selected for reproduction. By interchanging J1 and J3, J3–J1–J2–J4 is generated with a fitness value of 230. Break the tie arbitrarily between the two schedules with the same fitness value.
- *Generation 3:* The population consists of J3–J1–J2–J4 and J1–J3–J2–J4 with fitness (i.e., total flow-time values) of 230 and 210, respectively. Assume that J1–J3–J2–J4 are selected for reproduction. By interchanging J2 and J3, J1–J2–J3–J4 is generated with a fitness value of 200. Keep the schedules with the values of 200 and 210.
- *Generation 4:* The population consists of J1–J2–J3–J4 and J1–J3–J2–J4 with fitness (i.e., total flow-time values) of 200 and 210, respectively. Assume that J1–J3–J2–J4 are selected for reproduction. By interchanging J2 and J4, J1–J3–J4–J2 is generated with a fitness value of 230.

The algorithm stops with the optimum schedule of J1–J2–J3–J4 in this case. However, generally speaking, a genetic algorithm does not guarantee an optimum solution. A genetic algorithm does not scan the entire solution space (i.e., all possible schedules) implicitly or explicitly before stopping and presenting a solution. They sample only through the solutions. Although the sampling is not entirely random (it is guided by the crossing of genes), the generated sample solutions may not include the overall optimum solution. The success of the search is determined largely by the problem structure and the specific design of the genetic algorithm developed to solve the problem. This could be an advantage or a disadvantage, depending on the problem. The above-defined rudimentary algorithm illustrates some of the possible weaknesses of a genetic algorithm:

1. *A way of recognizing what changes are desirable.* The genetic algorithm could be improved by keeping track of what pairwise interchanges (called *genes*) positively affected the fitness value. This would avoid the generation of J1–J3–J4–J2 since J4–J2 is determined as a bad gene at generation 1 of the algorithm. However, evaluating a fitness function could be computationally costly if the problem size is large.
2. *Cycling (generating the same population over and over again).* This could occur, especially when the population size is very limited. Crossing could be defined to avoid this problem. The solutions can also be diversified by randomly or systematically changing the genes (i.e., mutation). Tabu search and simulated annealing specifically address some of these issues.
3. *The stopping rule.* In this example the stopping rule is arbitrary and could be improved.

A genetic algorithm was developed and used successfully [24] to solve a static job shop scheduling problem. However, the computational cost of their algorithm exceeded that of the tabu search and simulated annealing as reported in Morton and Pentico [14].

Tabu Search. The tabu search tries to avoid stopping at a local optimum point when no immediate improvement seems to be possible by searching the neighboring solutions. To avoid cycling through the same thread of solutions, all the solutions developed during the last m iterations can be kept and checked. These not-to-be generated solutions are called *tabu* and the technique is called the tabu search. The tabu search for scheduling problems relies mostly on developing a set of neighboring sequences that differ from the current solution by a pairwise interchange of two of the jobs. The tabu search is illustrated by the following example.

Example 2B Refer to Example 2. The data are repeated here in Table 21.21. Let our tabu search use an adjacent pairwise interchange to develop the neighboring schedules with $m = 5$. The tabu search algorithm then proceeds as follows:

Step 1: Start with an initial schedule.

Step 2: Develop all the neighbors of this schedule using the adjacent pairwise interchange.

Step 3: Evaluate all the neighbors using the total-flow-time measure.

Step 4: Choose, among the new schedules, the schedule with the minimum total flow time. If the number of iterations = 5, stop and present the best solution so far as the selected solution. Otherwise, use the schedule with the minimum total flow time as the new initial schedule. Repeat steps 2 to 4.

TABLE 21.21 Data for Example 2A

Orders (Jobs)	J1	J2	J3	J4
Run Time	10	20	30	40
Release Time	55	20	0	25

TABLE 21.22 Tabu Search–Based Optimal Schedule

Iteration	Initial Schedule (Value)	Neighbors Generated (Value) and Actions Taken
1	J4–J1–J2–J3 (260)	J1–J4–J2–J3 (350) J4–J1–J3–J2 (270) J4–J2–J1–J3 (270) Choose the new initial schedule (break ties arbitrarily).
2	J4–J2–J1–J3 (270)	J4–J2–J3–J1 (290) J2–J4–J1–J3 (230)[a] J4–J1–J2–J3 (tabu)
3	J2–J4–J1–J3 (230)	J4–J2–J1–J3 (tabu) J2–J1–J4–J3 (245) J2–J4–J3–J1 (250)
4	J2–J1–J4–J3 (245)	J1–J2–J4–J3 (330) J2–J4–J1–J3 (tabu) J2–J1–J3–J4 (235)
5	J2–J1–J3–J4 (235)	J1–J2–J3–J4 (320) J2–J3–J1–J4 (210) J2–J1–J4–J3 (tabu)

[a]Schedule selected by tabu search.

This algorithm is applied to Example 2B as given in Table 21.22. The tabu search did not find the optimal solution in this case. However, if the stop rule was defined as "stop after six iterations," the algorithm would have continued to search and would have found the optimal solution. The additional step that would have been performed until the optimal solution was found is given in Table 21.23.

A very simple algorithm is presented here due to the introductory nature of this section. We refer the reader to [4] and [29] for more information on this topic.

Simulated Annealing. Simulated annealing gets its name from the physical process of heating up a solid until it melts and then by cooling the melted solid until it again becomes a solid with low free energy. Application of this physical process to combinatorial optimization problems, specifically scheduling problems, is similar to that of tabu search: Instead of selecting the neighbor with the best objective function value, simulated annealing randomizes the selection of the next initial solution. The better the ob-

TABLE 21.23 Additional Steps for the Tabu Search Algorithm in Example

Iteration	Initial Schedule (Value)	Neighbors Generated (Value) and Actions Taken
6	J2–J3–J1–J4 (210)	J3–J2–J1–J4 (150)[a] J2–J1–J3–J4 (tabu) J2–J3–J4–J1 (240)

[a]Schedule selected by tabu search that is also the optimal solution.

jective value of a neighboring solution, the better chance it stands to be selected as the next starting solution. The difference in probability of the two neighbors, one with a better value of the objective function than the other, is also a function of the number of iterations (i.e., "temperature"). As the number of iterations increases (i.e., the temperature drops) the difference between the relative probability of a "good" versus "bad" schedule increases. Simulated annealing has been applied to some job shop scheduling problems successfully [25]. We refer the interested reader also to Aarts and Korst [27].

Artificial Neural Networks Applications. Artificial neural networks are information-processing systems that are motivated by the goals of reproducing the cognitive processes and organizational models of neurobiological systems [26]. The basic idea is the recognition of patterns and rules that are embedded in a good schedule, therefore quick recognition of a good schedule. Artificial neural networks accomplish this task through the use of formally defined network topography and rules for "firing neurons." This pattern of firing neurons is observed on a set of "training" problems where the artificial neural network is presented with a problem and an acceptable solution. The trained network then presented with a new problem, and based on the structure of the network and firing rules, determines a solution as the "recommended" solution.

The successful application of neural networks to scheduling includes the identification of a successful dispatching-based heuristic [26]. The authors developed and trained an artificial neural network which then recognized the most successful one-pass heuristic over 95% of the simulated problems reported in this study. The underlying approach and the details of the studies cannot be presented in sufficient detail in this section due to space constraints. We refer the interested reader to Hopfield [28].

21.4 SIMULATION OF JOB SHOPS

Simulation-based approaches are derivatives of dispatching rule–based approaches. In a simulation-based scenario, one or more dispatching rules may be used to make a decision when a resource becomes available. Simulation-based approaches are restricted mostly to a forward scheduling capability (i.e., where a schedule is constructed by starting from a reference time and then advances the simulation clock as jobs are scheduled on resources). Simulation models are able to represent the details of scheduling situations, and simulation-based approaches are useful in communicating the specific details to various levels of personnel because of the visual aids (e.g., animation) offered by simulation.

From the simulation point of view, a job shop is considered as a queuing network where an order may require several different operations by different machines and may have to wait in several different queues. If jobs arrive at the shop randomly over time, the job shop is referred to as a *dynamic* job shop. Here we present a review of the basic factors incorporated into simulation models of dynamic job shops and therefore provide the user with a basic understanding of a job shop simulation model for scheduling analysis. The second goal of this section is to provide the user with a basic understanding of scheduling rules and their performance in simulated job shops.

The most likely components of a job shop simulation model are as follows:

1. Order arrivals
2. Processing and setup times
3. Machines

4. Job routings
5. Shop load factors
6. Due dates
7. Priority rules

21.4.1 Order Arrivals

The arrival of orders is modeled in one of the following ways:

1. *Instantaneous release of orders into the shop.* In this approach the next order arrival time is defined at the time of each order arrival. The time between these arrivals is defined as a randomly generated variable.
2. *Periodic release of all available orders at the beginning of the scheduling period* (day, week, etc.). This can be modeled in two ways:
   (a) All the arriving orders as defined above accumulate at an order entry point. All of these orders are released into the shop at predefined points in time.
   (b) At each order release time, a number defined as a random variable of orders is generated.

3. *Order pooling.* This is similar to the above, except that a subset of the available orders is released into the shop at the beginning of each scheduling period. The selection of orders to be released into the shop may be based on the shop load and order characteristics.

The first approach is used in most models. The most popular arrival pattern is that of the Poisson process (i.e., the Poisson arrival rate or exponentially distributed interarrival times). If the time between arrivals is exponentially distributed, the rate (i.e., the number of arrivals per unit time) has a Poisson distribution. Therefore, the Poisson-distributed arrival rate (in orders per unit time) is translated into interarrival time of the corresponding exponential distribution. However, when periodic release or order pooling is utilized, the Poisson may directly represent the number of orders arriving in an hour, day, and so on. The Erlang distribution is the sum of exponential distributions and is also used to model order arrivals. Other distributions for the interarrival times or arrival rates are uniform, geometric, binomial, and empirical (actual shop data) distributions. Constant interarrival times are also used in investigating shop performance or the sensitivity of shop performance and the priority rules to some order and shop parameters.

The prevalence of the Poisson process is probably a result of its widespread use in queuing theory and its observed validity in some practical situations (e.g., number of calls arriving at a switchboard). The observed distribution of arrivals in actual shops, however, shows a wide variety in the arrival rate, and the Poisson distribution is not sufficient to explain or fit all the distributions observed. If the number of sources generating orders decreases, applicability of the Poisson assumption diminishes. For example, if orders to the shop are generated by a distribution center based on minimum order levels, the consolidated order patterns may be erratic, even if the customer order pattern may be a Poisson process.

Several different studies report that the arrival pattern is not important in evaluating the relative effectiveness of priority rules, although shop performance can be affected by the arrival pattern. Studies investigating periodic release and order pooling indicate similar results. In general, shop performance decreases with the increasing variance of the interarrival time distribution. Under periodic release, the following conclusions were obtained in various simulation studies:

1. The mean and variance of the inventory level are higher with an increase in the release period [7].
2. Utilization is higher than average at the beginning of a scheduling period but is lower at the end [7].
3. Fewer jobs are tardy for periodic release; however, the jobs that are tardy have longer periods of tardiness [8].
4. Due date performance is improved when period release is combined with a scheduling rule [9].

Job pooling (i.e., releasing a subset of orders at predefined intervals) is more restrictive and usually causes a decrease in shop performance [8]. However, if the subset is selected to balance the machine workloads, job pooling has a positive effect on shop and workload balance measures but has no significant effect or worse results on variance of lateness distribution and average tardiness [10].

21.4.2 Processing and Setup Times

In most job shop simulation models, the processing times are determined when an order arrives at the shop. Two approaches are possible at this point:

1. Generate the actual processing times from a specified distribution; the processing times are random variables from a distribution such as the exponential distribution.
2. Generate the estimated processing times; the estimated times are the information available for scheduling purposes. When an order is placed on a machine, a random variable called a *work rate factor* is generated and is multiplied by the estimated time to give the actual time required. The most common distributions are exponential and uniform for the estimated times and triangular, normal, or uniform for the work rate factor. This approach simulates the fact that in most cases the scheduler's knowledge of the processing times is not accurate and the process times fluctuate during day-to-day operations due to uncontrollable factors.

The specified family of processing time distributions and associated parameters affect shop performance. More important, some priority rules, such as SPT, are more sensitive to processing time distributions than are others. In general, as the variability of the order processing time decreases, performance of the non-due-date scheduling rules improves.

The setup times are included in the processing times in most models. In some models the setup time is assumed to be a function of the processing time. When sequence-dependent setup times are used, the distribution of these times must be selected along with the parameter values. As the variance increases, the desirability for minimizing the setup times increases. The relative values of the mean setup time and the mean processing time are also important considerations. When the mean setup time is large with respect to the mean processing time, there are more benefits obtainable from the scheduling rules that minimize the setup times.

In recent models, due to the increased capability and ease of modeling, a sequence-dependent setup time has been used. The presence of sequence-dependent setups changes both shop performance and the performance of the priority rules. Priority rules, which take setup times into consideration, are more successful than the others when the setup times are strongly sequence dependent (e.g., changing form J1 to J2 takes 10 minutes but changing from J3 to J2 takes 120 minutes).

21.4.3 Number of Machines

The number of machines used in simulation models varies greatly. In hypothetical models designed to evaluate the performance of scheduling rules, the number of machines range from 4 to 15. There seems to be a consensus that a four-machine shop model is large enough so that the results can be extrapolated to the more complex shops. Some comparative studies have investigated the effect of shop size on the relative effectiveness of the priority rules and conclude with similar results: that neither the size nor the configuration of the shop changes the relative effectiveness of the priority rules.

21.4.4 Job Routing

Job routing determines the required sequence of operations so as to predict the traveling pattern of orders among the machines. Diversity of order types is imported to the models via a routing matrix, which defines the transition probabilities of orders from one machine to the next. The extreme cases are the pure flow shops, where there is only one routing, and the pure job shops, where the transient probabilities between the machines are equal for subsequent operations. Pure job shop models are the most common types of models in simulation studies. This is partly because of the easier load control over the shop in pure job shop models.

Often in real systems, some or all orders may have alternative routings such that an operation may be performed on any one of a number of machines. Two approaches are possible when alternative routings exist:

1. Place the order in all feasible operation queues. Perform the operation on the first available machine; remove the order from the other queues when the order is assigned to a machine.
2. Assign the order to an idle machine that is capable of performing a feasible operation; if there is no such idle machine, place the order in a queue according to some queue selection rule (e.g., shortest queue length, least work in queue, etc.).

Alternative routing has a significant impact on shop performance and on the relative effectiveness of the priority rules: It provides better performance and reduces the difference between the priority rules.

21.4.5 Machine and Shop Utilization

The combined effects of order arrival distribution, job routing, and processing times determine machine utilization. From the standpoint of job shop simulation, machine utilization is important because it affects queue lengths. If the average queue length is too small, the scheduling rules used in the model may not be forced to make discriminating order selections. When this situation occurs, it is difficult or impossible to evaluate the effectiveness of the scheduling rule. Adverse effects also result from machine utilization that is too high. If utilization is near 1.00, transient conditions may extend over long periods and require excessive CPU time in order to obtain a steady state that permits data to be collected for comparison purposes. Machine utilization commonly found in the literature ranges from 0.85 to 0.95. Utilizations in this range usually cause queues to reach a length that permits scheduling rules to select an order from several in the queue but does not lead to very long queues.

Sequence-dependent setup times may require careful consideration due to their effects on machine utilization and hence queue lengths. When some scheduling rules yield significantly different average setup times than some other rules, the model results may be altered significantly. For example, if one rule tends to minimize setup times, it may produce small queues. On the other hand, if another scheduling rule does not consider setup times, the combined setup plus the processing times may cause a saturation of the shop and thus produce an undesirable effect.

21.4.6 Due Dates

Several different studies indicate that shop performance and the relative effectiveness of priority rules are affected by due date assignment methods as well by the tightness of due dates. The following considerations relate to due date assigment:

1. Through policy (fixed) parameters, each order is assigned a due date when it arrives at the shop. The due date assignment method may or may not use current shop information:
   (a) Static due date assignment rules consider only order data such as the arrival time, routing, and operation processing times of an order. Hence the order allowance time is a fixed amount for a given order of data and does not depend on the shop status when the order arrives.
   (b) Dynamic due date assignment methods employ order and shop data. In addition to the arrival times, operation times, and routing, other factors in due date determination may include the current shop load and the average waiting time of the orders.

2. Due date assignment and order sequencing decisions may be considered together and simultaneously as two dependent factors in the planning process. This leads to a new problem that requires determination of both the optimal set of due dates and schedules for a given situation involving the shop status and order parameters.

Table 21.24 shows the most frequently used static due date assignment rules in the literature.

21.4.7 Priority Rules

In the stochastic-dynamic scheduling literature, a variety of terms, such as *scheduling rule*, *priority rule*, *dispatching rule*, or *heuristic*, refer to the rule that selects an order from the orders waiting in a queue to be processed next by the machine. However, a distinction is possible for clarification purposes: Usually, a priority rule is defined as a

TABLE 21.24 Common Due Date Assignment Procedures

Abbreviation	Name	Definition
RND	Random	Random
N	Constant	Arrival date + constant order allowance
SLK	Slack	Arrival date + total processing time of order + constant
NOP	Number of operations	Arrival date + constant (number of operations)
TWK	Total work	Arrival date + constant (total processing time)
VTWK	Variance of TWK	Arrival date + constant (total processing time) exp(constant)

Source: Ref. 11.

method of assigning a scalar value to each order in a queue for scheduling purposes. A dispatching rule is defined similarly but implies that after the priorities are assigned, the job with the most priority will be dispatched to an available machine. A heuristic implies that more complex mathematical rules are used in determining the priorities. Finally, a scheduling rule may employ one or more priority rules and/or more complex mathematical or heuristic concepts in determining the next order to be processed. In practice, the basic term *priority* or *dispatching rule* is used for all of the above.

Priority rules may be classified according to their time dependency (static versus dynamic rules), the type of data they use (local versus global rules), or both. A static rule determines only one priority value for each operation of an order during its existence in the shop. A dynamic priority value, in contrast, changes over time; to assign the right order (i.e., the order that has the highest actual priority) to an available machine. The priority values of orders must be updated before each decision is made. Hence more calculation is involved in dynamic rules. Adam and Surkis [12] investigated the effects of updating policy and of updating the intervals on computation requirements and the effectiveness of priority rules. They concluded that due date performance is sensitive to the updating policy and to the updating intervals especially at high utilization (i.e., 92 to 96%).

Panwalkar and Iskander [13] described and categorized 113 priority rules. The functional categorization is given below.

1. (a) Simple priority rules are based on order and/or shop data such as processing times, due dates, number of operations, cost(values), setup times, arrival times, and slack.

   (b) The combinations of simple priority rules are the applications of two or more priority rules with the selection of which rule to use at a specific time being a function of the queue and order characteristics.

   (c) Weighted priority rules involve the application of rules in 1(a) and/or 1(b) combined with different weights.

2. *Heuristics* involve more complex considerations, such as the solving of static problems at the beginning of each scheduling period, look-ahead, and so on.
3. Other rules.

The body of job shop simulation research established the validity of the basic assumption that given that all the other shop parameters are the same, shop performance is strongly affected by the priority rules being used. However, there is no clear winner claimed for all the performance measures and shop and order parameters. A summary of the results is provided below.

Results for Completion Time–Based Criteria. The shortest processing time (SPT) rule is superior to the other simple priority rules for completion time–based criteria. However, some weighted priority rules have been found to be slightly more effective than the SPT rule for the average flow time and the average number of jobs in the shop. One such rule is the weighted priority rule, consisting of SPT and AWINQ (anticipated work in next queue), as indicated by Conway [3,14]. SPT could not show the same superiority for the other criteria in this group. First in system, first served (FISFS) is the best among the simple rules for the variance of flow times. SST (shortest setup time) is best for the total or average setup times. The complex heuristic rules, based on job pooling, are usually better than SPT in balancing the workload among the machines.

The simplicity of SPT has led to attempts to find rules combining SPT with FCFS (first come, first served) to yield better performance for the variance of flow times without increasing the average flow time. These attempts have included:

1. Alternate SPT and FCFS in predetermined time intervals [5].
2. Use SPT until the waiting time of a job reaches a specified limit in the queue; use FCFS for jobs with long times in queue (truncated SPT).
3. Use FCFS until the queue length reaches a specified limit; use SPT to reduce queue length (relief SPT).

The last combination effectively reduces the flow-time variance without increasing the average flow time.

Results Related to Due Dates. SPT, S/OPN, SLACK, and S/RPT are superior to the other simple priority rules in terms of all the measures of due date performance. Among these dominant simple rules, SPT and S/OPN are slightly better than SLACK and S/RPT. The consensus appears to be that SPT gives the best results for average lateness, and S/OPN is the best rule for minimizing the variance of the lateness distribution. However, the superiority of SPT and S/OPN are highly dependent on shop and job parameters for the remaining due date performance measures, such as the fraction of tardy jobs and mean tardiness. In general, high utilization, tight due dates, and due dates independent of processing times favor the superiority of SPT. S/OPN is more successful in moderate utilizations. Less tight and/or TWK due date assignments also favor S/OPN.

The relative success of SPT and S/OPN for due date–based criteria formed a basis to develop a number of combination rules, weighted priority rules, and heuristics. Although these rules incur additional computation, there is enough evidence to show that they also perform better. Improvement in performance is a function of how a rule is designed and of the shop and order parameters in which the rule has to perform. It is suggested that a specific simulation model be used to investigate which of the following approaches should be taken. The following list is used as a starting point:

1. *Weighted Rules.* The priority value is defined by a weighted total of process time and slack per operation values.
2. *Combination Rules (e.g., Truncated SPT).* The priority function changes based on shop or order parameters. For example, use SPT after an order becomes late.
3. *Sequential Application of Priority Rules.* Similar to rule 2 above; however, here the first rule is used for reducing the candidate set; then the other rules are used as sequential tie breakers. For example, a sequential rule could use slack < 5 days to eliminate the jobs with loose due dates, then could use a SPT-like rule to classify them into processing time intervals. Finally, within each group, S/OPN could be used to select the next order to be processed.
4. *Look-Ahead Heuristics.* These rules can also consider processing time, job slack, and the current load of machines. There are several studies indicating that a specifically designed look-ahead rule is usually better than a simple priority rule. Again, the question here is how much effort should be devoted to the development and testing of such a rule. Simulation is of great value in evaluating the effectiveness of such an effort as well as for the testing of heuristics.

Results for Cost-Based Criteria. The main objective in many industrial systems is to minimize the total cost. However, there are two difficulties in developing cost-based simulation models:

1. Cost parameters are difficult to obtain and their relation to schedule-dependent performance measures are hard to evaluate in real-world situations.
2. The results obtained from cost models are usually sensitive to parameter values. If any of the cost definitions or component values change, the results obtained for another set of parameter values may not be valid.

For these reasons, the results given below should be considered as thought-provoking ideas in developing one's own simulation model and not as robust results applicable to any situation.

The relation between the costs of the idle machines, of carrying work-in process, of long promises, of missed due dates, and of priority rules has long been established by the use of cost models. Again SPT and S/OPN emerge as good rules in many studies. Depending on the variance of job values and setup times, SST (shortest setup time first) and JV (job with the highest value first) have been the declared winners in some studies [8,15,16]. The relative success of these rules is a function of the variance of the setup costs and job values. For example, as the variance of job values becomes larger, the JV rule becomes more effective than the others in reducing costs if each cost component is given as a nondecreasing function of the job value.

The application of the weighted and combination rules as well as the sequential application of the simple rules is also valid for cost-based criteria. For example, Shue and Smith [17] reported a sequential application of S/OPN-JV-SST for setup, in-process inventory, and late penalty costs. They also concluded that any sequential rule performed better than its components.

Results for Some Extensions of the Basic Model. The effects of customer requested due date changes has been investigated in several studies and it has been found that those changes adversely affect the due date performance in all the priority rules, but the rules that use due date information are more responsive to due date changes.

The effects of mixing made-to-stock and made-to-customer order jobs have also been modeled. The models show that releasing made-to-stock items much earlier than their required due dates reduces their chance of being late. However, the increase in in-process and finished goods inventory costs offsets the benefits of decreasing the in-tardiness penalty.

The effects of expediting in the shop have also been investigated. The general consensus is that "shops in general are no better or worse off by using expediting" [18] on the average. However, expediting may be useful for tardiness-related criteria, especially under moderate loads and customer requested earlier due dates.

21.5 SIMULATION-BASED SCHEDULING: APPLICATION

In this section we describe the use of simulation in addressing interrelated issues such as bottleneck analysis, work center loading, due date determination, operator requirements, and shift scheduling in a complex manufacturing environment. This application also illustrates the use of simulation for some simplified scheduling needs.

The company illustrated here specializes in products that are manufactured from steel tubes. The products are a result of complex manufacturing processes with reentrant flows, batching, and process-type operations and have a high level of sensitivity with respect to due dates. There are over 100 products that are manufactured simultaneously in the shop at any given time. The specific interest of this application is a family of drawn products. Within this family there are several subfamilies of product. The first challenge of the simulation study is to organize the products into these subfamilies in such a way that the project objectives can be achieved in a short time and that the simulation models be accurate enough to represent the shop realities.

Management is particularly interested in:

1. Delivering the products on time with the existing resources
2. Minimizing the setup times and maximizing personnel and equipment utilization
3. Minimizing the work-in-process and establishing routine operating policies, such as predefined shift patterns rather than overtime.

At the beginning of the simulation project, these goals are determined at management-level meetings. After the goals are determined, data collection and transformation of goals into measurable criteria began.

21.5.1 Data Collection and Simulation Models

Although classifying modeling steps is necessary when describing model development, in reality data collection, modeling, verification, and validation go hand in hand. At the beginning of the simulation study, a preliminary data analysis showed that the products in question could be divided into eight major classes and that one other class described all the other products. For each of these eight major classes, routings were identified and developed using spreadsheets. Engineering and production personnel verified these routings. While this activity was taking place, modeling started. The first model took two products of families that could easily be identified and whose data were relatively complete. The simulation model included the following components, which are found in most simulation models:

1. Reentrant or looping operations
2. Yield factors that may be a function of the operation
3. Batching and unbatching operations
4. Alternative routings
5. Various shift patterns for equipment and different operators

The model also included some specific characteristics of the manufacturing process. For example, a somewhat unique first article inspection was included. After verification, the models were run against the real operations. Also, a walk-through was performed to validate the models. Validation of the pilot models was finalized at a top-level management meeting. The project goals were again summarized, along with the model assumptions, and a "results versus reality" discussion took place.

Of particular interest were the criteria that established the success of the project. These was defined as:

1. The model's ability to predict the due date performance of the shop
2. The model's ability to show the key equipment and personnel utilizations

After verification and validation of the pilot models, the data analysis continued with respect to the other subfamilies. Clustering techniques were used to analyze the process routings for the various subfamilies and to classify them. For better time management, the simulation models were built incrementally, adding one product family at a time, verifying and validating the model after each major milestone. This allowed for a better understanding of the models by company personnel as well as for quick completion of the model development. At the end of the model development, another formal top-level validation meeting was held, which provided a green light for the rest of the study.

21.5.2 Simulation Experimentation and Results

Experimentation included running the models under different shift and order schedule scenarios. We chose a simulation length of 6 months to 1 year. Instead of a warm-up period, a realistic work-in-process status in the shop was used. This was achieved through a file downloaded from the company's tracking system. The ASCII file was manipulated on a spreadsheet program for proper formatting before it could be read by the simulation model.* This file contained information regarding the order number, order quantity, product type (subfamily name), order due date, current location of the order in the shop, and other attributes that were used in the simulation model. An example of this file is shown in Table 21.25.

*ProModel for Windows simulation software was used in this study. However, there is other simulation software that can fulfill modeling needs. Due to very rapid changes in the simulation software industry, we refrain from providing a list of possible alternatives here. Frequent reviews appear in related periodicals, such as *IIE Solutions* [publisher: Institute of Industrial Engineers (IIE)], *APICS* [publisher: American Production and Inventory Control Society (APICS)], *OR/MS Today* [publisher: Institute for Operations Research and Management Science (INFORMS)], *Simulation* [publisher: Society for Computer Simulation (SCS)], and *Manufacturing Engineering* [publisher: Society of Manufacturing Engineers (SME)].

TABLE 21.25 Sample Order File for ProModel Input

Entity	Location	First Arrival	Order Size	Job Number	Number of Draw Cycles
Small_Getter_Tube	Weld_in	984 hr	100	3933	1
Larger_Getter_Tube	Weld_in	264 hr	100	3935	1
Round_2_B4C	QARI	2280 hr	400	420603	2
Round_2_B4C	Clean	0 hr	500	420601	2
Round_2_B4C	QARI	240 hr	600	424701	2
Round_2_B4C	QARI	720 hr	400	420602	2
Round_2_B4C	QARI	888 hr	500	424702	2
Round_2_B4C	QARI	1656 hr	700	424703	2
Round_2_B4C	QARI	2618 hr	600	424704	2
Round_2_B4C	Draw_in	2448 hr	500	424705	2
Round_2_B4C	QARI	3456 hr	600	424706	2
Race_wire	Straighten_in	0	46	3780	1
Race_wire	Straighten_in	0	32	3781	1

The simulation model's external file read and write capability was very useful in performing the experimentation. The output of the model included the standard output, such as:

1. Equipment utilization
2. Operator utilization
3. Average work in process and queue sizes
4. Statistical output with respect to user-defined criteria such as order lateness.

The output of the model also included two files:

1. An ASCII file that was used to assess order tardiness. This file was read directly onto a spreadsheet program for computing the planned shipment versus the simulated completion of each order (Table 21.26).
2. An order tracking file in the shop that was utilized to develop the Gantt charts.

21.5.3 Continuous Use of Simulation Models for Shop Loading

The successful completion of the simulation project included evaluation of the shop performance. Therefore, the project included the development of models for continuous use. This required not only the read and write capability mentioned above, but graphical tools that would provide quick output for the user to analyze the shop performance. For this purpose, ProS Scheduler software from MMS Soft Corporation was used.

ProS read the files (called *duration files*) that indicated the operation start and completion times for each operation. This information was obtained via explicit write statements in the simulation model. It was, therefore, user-definable. Figure 21.10 shows the information flow among the various components of the system. Upon reading the duration file, ProS initialized the start of the scheduling horizon and placed the operations on a Gantt chart, as shown in Figure 21.11.

The user could also manipulate the results by changing the shift patterns. This allowed the user to see the impact of adding new shifts or of changing the existing

TABLE 21.26 Sample Output for Order Lateness

Job Number	Order Release Date	Planned Ship Date	Simulated Ship Date
424705	8/9/96	10/1/96	10/5/96
3780	8/19/96	10/11/96	10/26/96
3781	8/20/96	10/15/96	11/4/96
420601	9/8/96	11/1/96	10/28/96
424701	9/15/96	11/8/96	11/19/96
3933	9/19/96	11/15/96	12/14/96
3935	10/17/96	12/6/96	12/30/96
420602	11/12/96	1/3/96	12/31/96
424702	11/19/96	1/10/97	1/14/97
424703	11/26/96	2/7/97	2/8/97
420603	12/31/96	3/7/97	3/4/97
424704	1/15/97	3/18/97	3/15/97
424706	1/28/97	3/31/97	3/28/97

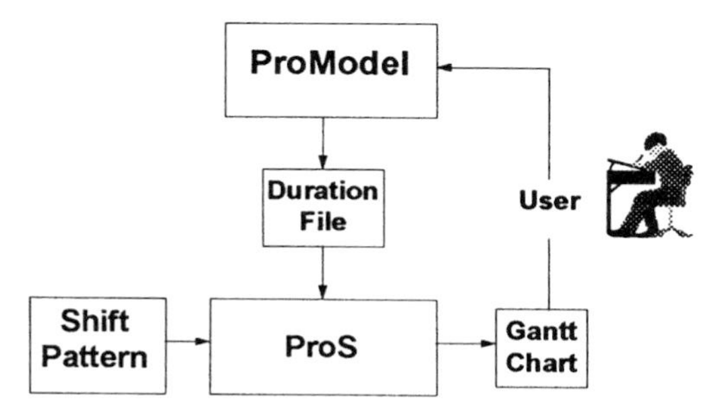

Figure 21.10 Information flow.

shifts on the schedule completion of orders. The Gantt charting program also provided an order-based or resource-based reporting capability. These reports showed the operation completion and start times for each order on the resources. Therefore, it could be used for analyzing the performance and also could be used as a dispatch list to distribute the plan to the shop floor.

21.5.4 Case Study: Heatcraft, Inc. Scheduling Success

Mixed Model Scheduler (MMS),* a finite capacity scheduling software, was installed to schedule the production of heat transfer coils for the ADP business segment of Heatcraft, Inc., replacing a very time consuming manual process [21]. Each morning the Master

*MMS by MMS Soft Corporation is used here. There is other scheduling software that could possibly fulfill the requirements. Due to very rapid changes in the software industry, we refrain from providing a list of possible alternatives here. However, frequent reviews appear in related periodicals such as *IIE Solutions* [publisher: Institute of Industrial Engineers (IIE)], *APICS* [publisher: American Production and Inventory Control Society (APICS)], *OR/MS Today* [publisher: Institute for Operations Research and Management Science (INFORMS)], and *Manufacturing Engineering* [publisher: Society of Manufacturing Engineers (SME)].

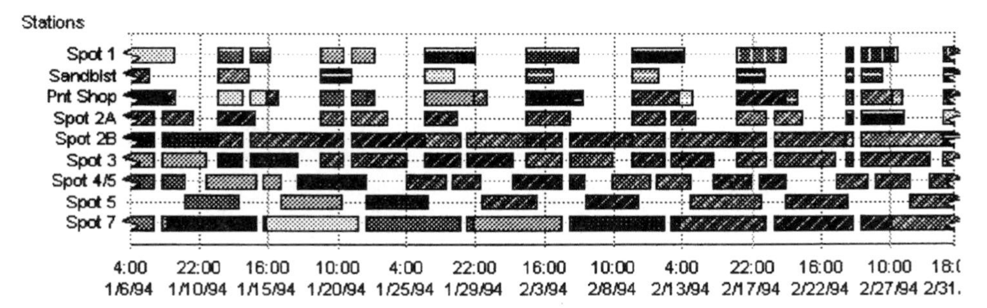

Figure 21.11 Sample Gantt chart.

Scheduler had to create a manual schedule for the next day. It took about 4 hours a day to create the next day's schedule. Updating the schedule due to a machine breakdown took about another 4 hours. This downtime was very costly to the ADP. Moving an order up or down in the schedule or changing assembly lines also took about 4 hours. Flexibility of the scheduling process was a major problem. Also, visibility of the purchased parts was limited. Often, an order would reach the assembly line before ADP personnel discovered that a purchased part needed was not available for the order. Therefore, units would have to be set aside to wait for the purchased part. Before the MMS implementation, the ADP had an average of 75 orders per week to schedule. This took about 4 hours a day. Today, with MMS, ADP has an average of 140 orders per week to schedule and it takes about 5 minutes. The scheduling of these many orders manually would take at least 7 hours a day. The time saved allows the Master Scheduler to look more closely at the purchased parts inventory. This has eliminated the problem of running out of purchased parts for an order when it has already reached the assembly line. In this application the following issues were very important to the success of the project.

Data Downloads and Uploads. Making sure that the downloaded information was accurate and in the required format took several iterations. The interface between the scheduling system and the MRP and/or shop floor data collection systems had to take into consideration possibilities such as material availability, orders on hold, and quality-related issues. A database management system shell was used to interface between the scheduling system and the company-wide MRP and management information systems. This should be fully automated and require only a routine inquiry to be run by the scheduling personnel for data downloads and scheduling uploads.

Variety of Components. Typically, there are multiple components that must be completed before initiating the final assembly. Each component has its own operation steps, setup base grouping, and resource availability issues (Figure 21.12). The scheduling systems must include constraints coming from the component part completions.

Sequence-Dependent Setup Times. Sequence-dependent setup times combined with product mix may cause unexpected bottlenecks at the resources. In some cases, these resources are dynamic bottlenecks. The current mix of products and the running sequence at each resource determine the bottleneck status. The scheduling system must account for the sequence-dependent setup times and the current availability of the various order types. A resource with enough available time does not need to be optimally

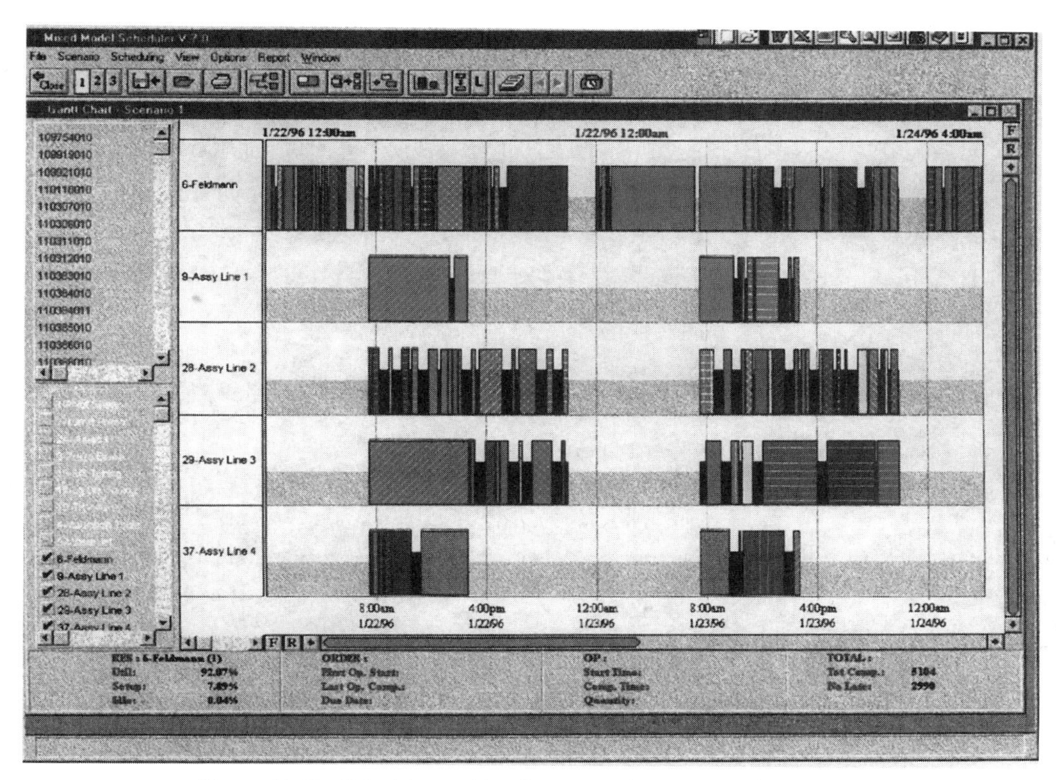

Figure 21.12 Schedule resulting from operation, setup, and resources.

scheduled. However, running an arbitrarily bad sequence on a nonbottleneck resource with a given mix can turn a nonbottleneck resource into a temporary bottleneck. This situation is called a *dynamic bottleneck* and cannot be addressed by static tools such as spreadsheets or simple sorting-based algorithms. It is also difficult to resolve using simulation-based approaches because of the large number of possible alternatives. A FCS system such as the Mixed Model Scheduler with its setup optimization features accommodates dynamic bottlenecks. Upon determining the bottlenecks, the scheduling system is capable of sequencing jobs to minimize the setup times. The user should also be able to determine the set of noncritical resources, which can be scheduled in such a way as to provide the bottleneck with enough work at all times.

Resource Utilization Issues. The scheduling system should keep track of the various grades of labor and other resources utilized at the various machine and assembly line resources. The labor and work centers are usually the constraining resources when creating a schedule, with the work centers typically having a larger influence. Due to this, labor requirements will fluctuate throughout the schedule. It is important for the FCS system to recognize these requirements and inform the scheduler of such. These numbers can be used for scheduling additional shifts or overtime.

Work-in-Process and Waiting Times. Schedulers should be able to evaluate the effects of the schedules produced on the work-in-process (WIP) levels. As mentioned earlier, dynamic bottlenecking can play havoc in scheduling the shop floor. Large WIP inventory may move around on the shop floor, depending on the schedule run. It is important to be able to evaluate the WIP levels created by the schedule and alter the schedule appropriately to minimize this inventory. Also, knowing the average time a job spends in queue prior to a resource is beneficial in determining the potential bottlenecks.

Dynamic Updates. Although not very common, it is possible for a bottleneck resource to go down. This causes a major change in the availability of certain components and may require complete rescheduling of orders for the final assembly as well as at the component levels. Changing the resource availability on the fly and rescheduling based on the new resource availability is a definite advantage within the scheduling software.

An estimated downtime savings of 2 hours per month was attributed to the MMS implementation because of the speed of generating a new schedule based on an unavailable resource. This represented a savings of $360,000 per year. The average number of units per day increased 40%. Most of this increase was attributed to converting to a demand flow approach to manufacturing that was supported by the tailored algorithm used in the MMS software.

21.6 IMPLEMENTATION ISSUES

For most organizations, the question is not whether to use simulation and/or scheduling, but how to solve an existing or future problem related to capacity planning and scheduling. Although are guidelines are provided, there is no foolproof recipe that guarantees success.

In practice, the desire to use a software tool to solve a problem usually arises in a crisis situation, such as in times of rapid expansion, change of business climate, or physical movement of business location. At such times, the internal (i.e., usually top management) or external (i.e., consultants) sources often order personnel to "implement scheduling" or "use simulation" to help with the crisis. Forming a committee to identify and evaluate the software usually follows the recommendation. After several vendor visits and evaluations, software is finally chosen and implementation starts. At this point the success of the project is still more of a chance event than a sure result if the following issues are not observed:

1. Start with an end in mind with respect to the desired solution and the objectives. Do not start with software or general methodology predetermined.
2. Investigate the problem and possible solution techniques. Use a team building approach, not a "software evaluation committee." Use the diversity of the expertise in the team to approach the problem from different viewpoints. For example, the information system (IS) specialist point of view could be very different from that of a production supervisor. Use all points of view to arrive at solutions and do not pit one against the other. Remember that there are no opposing objectives in an organization—only a single goal: to become a more profitable organization (short and long term).

3. Rely on both theoretical expertise and common sense. If any solution does not make common sense, do not buy the argument. Any truth, however complex, can be explained in a layperson's terms. If this is not possible, there is a fundamental fault in the proposed theoretical solution.
4. Implement the solution properly and completely without hesitation once it is started. Allocate the proper resources and time for its success. Most implementations fail because of improper allocation of resources (budget constraints) or time (due date constraints).

21.6.1 When to Use Scheduling or Simulation

Although they are not mutually exclusive, finite capacity scheduling and simulation tools address difficult problems. Traditionally, simulation is a tool for medium- to long-run capacity planning, whereas scheduling is used for day-to-day detailed order sequencing. However, technological advances have made it possible, so that these tools can be used somewhat interchangeably. Although their academic validity may be questionable, the following observations offer some insight based on the practice of scheduling [20].

1. *Finite-capacity scheduling is mainly a deterministic tool, whereas simulation is a stochastic tool.* Most finite-capacity scheduling models or systems do not consider randomness explicitly. The rework and reject loops and the probabilistic routings between the operation steps are not considered in finite-capacity scheduling (FCS). The FCS models use deterministic process times, setup times, and transport times, whereas simulation almost invariably uses probability distributions for the process, setup, and transport times. Some of this is due to the aggregation of different products or different units into an "entity" in simulation models. Most of the variability, however, is valid even for one product. For example, assembly times in an assembly line may be different from one assembly to another.

2. *Finite-capacity scheduling addresses shorter horizons, whereas simulation addresses longer planning horizons.* Current use suggests that simulation is a valuable tool for addressing long-term design and planning problems, whereas finite-capacity scheduling is an excellent tool for addressing short-term and on-line execution problems. This is partially related to the uncertainty issue above. Finite-capacity scheduling, being a deterministic approach, is more successful when the level of accuracy is high. Usually, the shorter the scheduling interval, the higher the accuracy. Therefore, most current applications suggest a shorter time horizon for scheduling applications.

3. *The degree of detail is different in finite-capacity scheduling than in simulation.* This is not to suggest that finite-capacity scheduling models are more detailed than simulation, but the degree of detail is different because the purpose of these modeling tools is different. Scheduling systems have a more detailed representation of the real part numbers, product types, and routing steps, whereas simulation models are much more representative in probabilistic routings and in modeling secondary and additional resources.

21.6.2 Step-by-Step Implementation Plan for Scheduling

Following a sound scheduling methodology will help to ensure that implementation of the scheduling software will be successful. The basic steps of this methodology are:

1. *Team Building.* Develop an implementation plan that draws expertise from all areas of the facility.
2. *Definition of Performance Metrics.* Determine how you will measure the success of the scheduling project.
3. *Selection of a Generalized Approach.* Evaluate if the problem is a candidate for optimization.
4. *Design of Required Reports.* Ensure that the software chosen is capable of producing all of the output reports required.
5. *Determination of Data Sources.* Determine where data are required for the creation of a schedule [e.g., Material Requirements Planning (MRP) and/or shop floor data collection systems].
6. *Analysis and Refinement of Specific Approaches.* Use computer simulation to test some of the basic ideas for the scheduling solution. This process will also help to cost justify the scheduling implementation.
7. *Selection of the Implementation Method.* Since most of the preplanning work has been completed, many finite-capacity scheduling (FCS) packages can be eliminated at this point. The user should now be comfortable in fitting the correct software to the scheduling problem.
8. *Deployment.* Choose a prototype site within the organization. Once the solution has been successfully implemented at that site, expand the solution to the rest of the facility.

By following these steps, the scheduling implementation should progress relatively smoothly through a successful completion.

21.6.3 Role of Simulation in Scheduling

Using simulation as a precursor to the implementation of a finite-capacity scheduling system not only saves time and money in the long run but will also allow for testing of some of the basic scheduling rules. Avoidance of design and operational mistakes, as well as the minimization of cycle times, are just a few of the areas that can be studied in detail prior to scheduling. Simulation can be used to determine the medium- to long-term resource levels. It can be used as a strategic tool in determining the number of required machines, for example. The scheduling system specifically should be used as a tactical tool to determine how jobs should be sequenced through the machines in the next scheduling period.

Simulation allows the user to analyze a system without disrupting its operation. The capacity of the system can be evaluated before a new product is added to the production process. Instead of buying additional capital equipment at the start, simulation can help to justify the purchase or prove that the existing equipment is capable of handling the increased workload. This method takes into account variability and randomness such as machine breakdowns that occur on the shop floor. Simulation is also a good tool for identifying the constraints or bottlenecks on the shop floor. Once these have been identified, the scheduler knows where to focus.

It is important to remember that striving for high efficiencies at all the operations does not guarantee a higher overall system throughput. Having high efficiencies at all the operations may lead to increases in WIP at the bottleneck as well as other operations upstream. It is quite possible that starving or blocking nonbottleneck operations will have no impact on the overall system throughput. This is easily shown through simulation by studying the impact of the varying efficiencies (i.e., evaluating different downtime, repair time, and cycle time distributions) on the operations. By doing this sensitivity analysis, the engineer will have a better understanding of the impact of varying parameters in the system and how the bottlenecks or constraints move around the system.

Through better management of the constraints, throughput in the facility can be improved. Some basic scheduling algorithms can be utilized in the simulation model to see their effects on the overall system. By showing improvements in the system through simulation, the cost justification of a more detailed scheduling solution, such as a FCS system, will be easier.

21.7 CONCLUDING REMARKS

Scheduling problems and the solution approaches were introduced in this chapter. First, the scheduling problems are defined and classified. The non-simulation-based solution approaches were discussed as a background with an emphasis on various solution techniques. Each major technique is illustrated with a numerical example. Simulation of job shops is covered, with an emphasis on the effects of priority rules on various scheduling performance measures and other components of simulation models which also affect scheduling performance.

The research and the practical applications suggest that simulation could be a very useful tool in approaching scheduling problems, specifically in the areas of system design, algorithm testing, and system integration. Simulation provides the ability to investigate various what if's that may require questioning of some of the constraints. For example, instead of accepting a given number of machines and trying to optimize their utilization, a properly constructed simulation model could justify acquiring more machines, therefore increasing the scheduling performance. The simulation-based approaches are also very useful in identifying and solving implementation issues.

As suggested by the cases presented in this chapter, scheduling and simulation are complementary approaches. So the question that arises in many situations: "Do I need to use simulation or do I need a tool for scheduling?" should not be the focus. The recommended approach is to use the right tool at the right time. So in this short chapter we have tried to give the reader the main ideas that he or she could use for a decision on when to use available simulation and scheduling models and techniques.

REFERENCES

1. French, S. (1982). *Sequencing and Scheduling*, Ellis Horwood, Chichester, West Sussex, England.
2. Baker, K. R. (1974). *Introduction to Sequencing Scheduling*, Wiley, New York.
3. Conway, R. W., W. L. Maxwell, and L. W. Miller (1967). *Theory of Scheduling*, Addison-Wesley, Reading, Mass.
4. Morton, T. E., and D. W. Pentico (1993). *Heuristic Scheduling Systems*, Wiley, New York.
5. Panwalkar, S. S., R. A. Dudek, and M. L. Smith (1973). Sequencing research and the indus-

trial scheduling problem, in *Proceedings of the Symposium on the Theory of Scheduling and Its Applications*, Springer-Verlag, Berlin, pp. 29–38.

6. Kusiak, Andrew, ed. (1989). Knowledge-Based Systems in Manufacturing, New York: Taylor & Francis Ltd.
7. Panwalkar, S. S., M. L. Smith, and R. A. Dudek (1976). Scheduling with periodic release of orders for production, presented at ORSA/TIMS Special Interest Conference on Theory and Application of Scheduling, Orlando, Fla.
8. Ulgen, O. (1979). Application of system methodologies to scheduling, unpublished Ph.D. dissertation, Texas Tech University, Lubbock, Texas.
9. Holloway, C. A., and R. T. Nelson (1974). Job shop scheduling with due dates and variable processing times, *Management Science*, Vol. 20, No. 9, pp. 1264–1275.
10. Irastorza, J. C., and R. H. Deane (1974). A loading and balancing methodology for job shop control, *AIIE Transactions*, Vol. 6, pp. 302–307.
11. Kiran, A. S., and M. L. Smith (1984). Simulation studies in job shop scheduling, I: A survey. *Computers and Industrial Engineering*, Vol. 8, No. 2, pp. 87–93.
12. Adam, N. R., and J. Surkis (1980). Priority update intervals and anomalies in dynamic ratio type job shop scheduling rules, *Management Science*, Vol. 26, No. 12, pp. 1227–1237.
13. Panwalkar, S. S., and W. Iskander (1977). A survey of scheduling rules, *Operations Research*, Vol. 25, No. 1, pp. 45–61.
14. Conway, R. W. (1964). An experimental investigation of priority assignment in a job shop, *Memo. RM-3789-PR*, February, RAND Corporation, Santa Monica, CA.
15. Aggarwal, S. C., and B. A. McCarl (1974). The development and evaluation of a cost-based composite scheduling rule, *Naval Research Logistics Quarterly*, Vol. 12, No. 1, pp. 155–169.
16. Jones, C. H. (1973). An economic evaluation of job shop dispatching rules, *Management Science*, Vol. 20, No. 3, pp. 293–307.
17. Shue, L., and M. L. Smith (1978). Sequential approach in job shop scheduling, *Journal of the Chinese Institute of Engineers*, Vol. 1, pp. 75–80.
18. Hottenstein, M. P. (1970). Expediting in job-order-control systems: a simulation study, *AIIE Transactions*, Vol. 2, pp. 46–54.
19. Kiran, A. S., and M. L. Smith (1984). Simulation studies in job shop scheduling, II: Performance of priority rules, *Computers and Industrial Engineering*, Vol. 8, No. 2, pp. 95–105.
20. Kiran, A. S., and T. Unal (1993). Simulation and scheduling in a TQM environment, Presented at *IIE Conference*.
21. Kiran, A. S., and J. Green (1996). Manufacturers meet global market demands with FCS software, *IIE Solutions*, August.
22. Little, J. D. C. (1989). A proof for the queueing formula: $L = \lambda W$, in *Stochastic Modeling and the Theory of Queues*, Ronald W. Wolff, ed., Prentice Hall, Upper Saddle River, N.J., pp. 383–387.
23. Fisher, J. L. (1993). An applications oriented guide to Lagrangian relaxation, in *Heuristic Scheduling Systems with Applications to Production Systems and Project Management*, T. E. Morton and D. W. Pentico, eds., Wiley, New York, pp. 10–21.
24. Della Croce, F., R. Tadei, and G. Volta (1992). A genetic algorithm for the job shop problem, D.A.I., Politecnico de Torino, Italy. Also, *Heuristic Scheduling Systems with Applications to Production Systems and Project Management*, T. E. Morton and D. W. Pentico, eds., Wiley, New York.
25. van Laarhoven, P. J. M., E. H. L. Aarts, and J. K. Lenstra (1988). Solving job shop scheduling problems by simulated annealing, preprint, Philips Research Laboratories, Eindhoven.
26. Rabelo, L., S. Alptekin, and A. S. Kiran (1989). Synergy of artificial neural networks and knowledge-based expert systems for intelligent FMS scheduling, in *Proceedings of the 3rd*

ORSA/TIMS Conference on Flexible Manufacturing Systems: Operations Research Models and Applications, Cambridge, Mass., Elsevier Science Publishers, Amsterdam, pp. 361–366.

27. Aarts, E., and J. Korst (1989). *Simulated Annealing and Boltzmann Machines: A Stochastic Approach to Combinatorial Optimization and Neural Computing*, Wiley, New York.
28. Hopfield, J. J. (1988). Neural networks and physical systems with emergent collective computational abilities (Neural Systems Incorporated), *Neural Networks—Theory and Application*, August, pp. 2554–2558.
29. Glover, F. (1990). Tabu search: a tutorial, *Interfaces*, Vol. 20, No. 4, pp. 74–94.

PART V

PRACTICE OF SIMULATION

CHAPTER 22

Guidelines for Success

KENNETH J. MUSSELMAN
Pritsker Corporation

22.1 INTRODUCTION

Few things in this world are static. This is particularly true of simulation projects. They seek continually to redefine themselves. As a project develops, discoveries are made. Some of these discoveries reinforce the project's original direction, while others force a change in heading. Most of these heading corrections result in only slight changes. Others, however, force major adjustments, possibly even reversing general understandings held at the start of the project. Beyond these recurring discoveries, interests change. What first drives a customer to pursue a problem may not be what sustains his or her interest over the course of the project. As knowledge is gained, some issues take on less importance, while others begin to dominate the customer's attention. Indeed, the customer can easily become engrossed with questions that were not even considered when the project began.

Success in such a dynamic environment requires working to a set of principles. Without well-founded principles, a simulation project can drift aimlessly, increasing its chance of failure. With these principles, a project remains true to course. Reactionary moves and radical changes in direction are far less prevalent. Issues are judged with consistency and forethought. All this helps to keep the project moving in the right direction and away from troubled waters.

The principles presented here serve as navigational guidelines for conducting a successful simulation project. They highlight the underlying managerial skills needed in each step of the process, such as directing the project, controlling model development and analysis, and improving customer relations. Technical fundamentals are not emphasized. This is not to minimize their importance, for technical knowledge and competence are necessary to ensure valid results. Instead, the purpose is to emphasize the role practical management plays in every successful project. By following these principles, the technical aspects of a project continue with fewer distractions and disruptions.

Handbook of Simulation, Edited by Jerry Banks.
ISBN 0-471-13403-1

The result is improved project performance and increased likelihood of the project's recommendations being implemented.

22.2 PROJECT STEPS

A simulation project is a process of interpretive, developmental, and analytical steps [1–3]. Table 22.1 lists these steps, which are intrinsic to all simulation projects. The process outlined provides a framework within which many diverse yet interdependent activities relate in an orderly sequence.

Admittedly, not all simulation projects follow this exact sequence. Often, the project team is challenged into rethinking the model late in the process because of knowledge gained along the way. This, in turn, forces revisitation of earlier steps, such as model conceptualization, data collection, and model building. Recurrent processing of these steps is characteristic of the vast majority of simulation projects. In the end, though, all steps are visited, for each is important to the process of delivering a valued and defensible result.

22.3 GUIDELINES

In this section we present specific guidelines for conducting a simulation project. Each step in the process is repeated below, followed by assorted principles that support the

TABLE 22.1 Simulation Project Steps

Step	Type	Description
1. Problem formulation	Interpretive	Define the problem to be studied, including a written statement of the problem-solving objective
2. Model conceptualization	Analytical	Abstract the system into a model described by the elements of the system, their characteristics, and their interactions, all according to the problem formulation.
3. Data collection	Developmental	Identify, specify, and gather data in support of the model.
4. Model building	Developmental	Capture the conceptualized model using the constructs of a simulation language or system.
5. Verification and validation	Analytical	Establish that the model executes as intended and that the desired accuracy or correspondence exists between the model and the system.
6. Analysis	Analytical	Analyze the simulation outputs to draw inferences and make recommendations for problem resolutions.
7. Documentation	Interpretive	Supply supportive or evidential information for a specific purpose.
8. Implementation	Developmental	Fulfill the decisions resulting from the simulation.

Source: Refs. 4 and 5.

activities at that step. The positioning of each principle reflects where it should be given particular attention in the process. In acting on a principle, the project team is more likely to avoid a usual pitfall, reduce risk, or maximize the activity's effect on the project's outcome. Keep in mind, however, there are no definitive rules about when to employ a given principle. Some principles may need to be initiated well ahead of where they are positioned here and engaged repeatedly. Furthermore, these principles are directed at the project team. This team may consist of people internal to the customer's organization, external to it, or a combination of both. Whatever the team's composition, the principles remain equally relevant.

22.3.1 Problem Formulation

Problem formulation is the most important step in the process, for it guides all other project activities. It is here that the team begins to establish the central issues and project scope. Guidelines that direct this activity include:

1. Start off on the right foot.
2. Work on the right problem.
3. Manage expectations.
4. Question skillfully.
5. Listen without judgment.
6. Communicate openly.
7. Predict the solution.

Start Off on the Right Foot. First impressions are important. The customer is entrusting the project team with solving a significant and probably costly problem. From the start, give the customer assurance that this team is right for the task. Establish confidence in this election to move forward with the project by beginning with a clear and effective start.

To get off to a good start, consider holding a kickoff meeting. This meeting should be held after some preliminary investigations have been done but before any substantial work has begun. Its purpose is to lay a foundation for success. In the meeting, introduce all the team members, restate and possibly refine the project's objectives, lay out the proposed schedule with milestones, define areas of responsibility and accountability, establish lines of communication, and begin to develop an understanding—of both the simulation process (for the benefit of the customer) and the system to be examined (for the benefit of the project team). Explain why the project is being conducted, what the specific approach is, and what opportunities exist. Gauge the customer's understanding of the project and the simulation process. Early on, work to surface everyone's needs. Honor all views, but be careful not to let those who invent or reinforce obstacles dominate the meeting. Acknowledge their concerns, but work to mitigate them. Focus on benefits and functionality more than technology. Getting a good start like this can make a big difference in how well the project eventually finishes.

As with all meetings, control the flow. Have respect for the customer's time. First, break the agenda down into small topics that can be clearly stated and managed. Then communicate this agenda in advance. People contribute more if they know what is expected. Invite key organizational members, both support staff and upper-level management, but keep the meeting as small as possible, opening it only to contributing members. Allowing someone to attend who will not add to the discussion or be responsible for an aspect of the project can affect the meeting negatively. Agree at the start on the topics to be discussed and the time allotted for each topic. This further signals to those gathered what the most important topics are and the level of discussion expected. Take notes on what is discussed. This not only provides a log of what was said, but also makes the participants feel as though their ideas are being heard, inspiring them to contribute even more. Then use their ideas to target the solution approach. Handle all objections with finesse. There are likely to be people in the meeting with some strongly held beliefs, which is probably why they are there. Welcome their objections. View them as requests for more information. Never discard them, for they might just hold the key to the project. Importantly, stay on time. If discussion on a topic is to last an hour, hold to that. Finally, summarize the meeting before adjournment. This 5-minute activity brings proper closure to this critical early step.

During the kickoff meeting and afterward, be observant. Watch team member interactions. These are the people who will be instrumental in seeing the project to a successful conclusion. Look for signs of leadership among them. Success requires a project champion: someone at the customer's site who is a consistent advocate and at a high enough level in the organization to exercise influence. When the project needs a push, this is the person who will make it happen. But most of all, be mindful of the fact that the project ultimately depends on how the team members collectively perform. Without their total support, success is unlikely.

Maxim A strong finish begins with an effective start.

Work on the Right Problem. Nothing is less productive than finding the right solution to the wrong problem. Although this is never the team's intention, it happens more frequently than it should. Often, this is due to misunderstood or poorly stated objectives. For this reason, establishing sound objectives is critically important. Obscure objectives make it difficult to succeed. Unless the project team members know what they need to accomplish, they have little chance of doing it. Accordingly, construct objectives that are precise, reasonable, understandable, measurable, and action oriented to convey a proper sense of direction and to distinguish between primary and subordinate issues. Write them down. Then continue to have the team refer to and be guided by them throughout the project. The goal is to satisfy these objectives, not simply to complete the project. Having these objectives clearly defined goes a long way to keeping the project focused and, in the end, measuring its degree of success.

Maxim Fuzzy objectives lead to unclear successes.

Manage Expectations. The customer's expectations must be properly set and continually managed. Allowing them to go unchecked can easily force a project off course. Expectations must always agree with the project's objectives. Otherwise, project team members can find themselves obligated to work on tasks only to satisfy an expectation, not an objective.

Start by setting the correct expectations upfront. Be enthusiastic, but be careful not to oversell. Be realistic about what can be accomplished and how much time it will take. Make sure that the customer understands what issues the model will and will not address. There are limits to what technology can do. It is far better to discuss these limitations from the beginning than to state them for the first time when the results are presented.

Once set, expectations must be managed continually. Quickly throttle any unrealistic hopes or ideas. Do not allow the customer to be misled into assuming that the team is delivering capability it cannot or does not intend to deliver. The team needs to be clear about what it plans to do. Not only is this the right thing to do; it is the smart thing to do. If expectations go unchecked, both parties come away disappointed from the misunderstanding. Therefore, work with the customer to understand his or her viewpoint and come to a mutually acceptable resolution.

As issues arise, do not rush to agreement. Timing can be important. Every project has its ebb and flow. Knowing when to draw the line or adjust a position can be critical to keeping the project moving forward. If agreement cannot be reached immediately, look for a way to work around the problem. For technical issues, consider using the model to explore options or to bound the problem to see if the issue really matters. For nontechnical issues, usually the passage of time can bring both parties to a reasonable and balanced compromise. Learn to practice patience. Often, it is time well spent.

Maxim It is easier to correct an expectation now than to change a belief later.

Question Skillfully. Questions wield tremendous power. They can cause customers to reveal, often for the first time, key pieces of information or cause them to keep silent, refusing to cooperate. They can lead people toward joint discoveries or divide them against each other. They can help bring about a quick solution or complicate the problem to such an extent that the project becomes a worthless, endless journey. Therefore, project team members should be as attentive to question selection as they are to the customer's answers.

Skillful questioning requires forethought. Work to create a safe and open forum for gathering information and exploring new ideas. To accomplish this, keep the following in mind. First, probe with proactive questions. Begin to force new thinking on the part of the customer. For example, consider the following questions: What if one releases the work later? Earlier? Suppose that one carries more work-in-process inventory? Less? How about eliminating it altogether? Try compressing, adding, subtracting, multiplying, substituting, combining, and reversing the process. Lack of system knowledge is an advantage here. Look at the system and explore how it could run. Encourage the customer to look at it in the same light. This not only helps clarify how the system currently works, but also opens the customer's mind to what is possible.

Second, learn to ask open-ended questions. Simple "yes" or "no" questions stifle customer participation. Open-ended questions, on the other hand, invite explanation and clarification. Phrase the question so that it leads to a more informative response. By way of example, consider the following:

Poor: If you were to rerun these orders, would you sequence them the same way?
Better: If you could change the order sequence, what would you do?
Poor: Is your system functioning the way you want?
Better: What aspects of the system need attention, and why?

Third, avoid putting the customer on the defense. Many people are not comfortable with having to justify their actions. Be less threatening. Avoid letting the customer feel as if he or she is the problem. Try, instead, to make the customer feel a part of the solution process. For example:

Poor: Why did you select that order to run next?
Better: How do you decide order sequence?

Finally, be forward focused. Direct people's energies toward what should be done rather than to what is wrong and who is to blame. For example:

Poor: Why do you route these parts to the other side of the plant when everything could be done here?
Better: If you could redesign your flow process, what changes would you make?

Maxim The right solution starts with the right questions.

Listen Without Judgment. Successful simulationists are often viewed as gifted modelers. They appear to capture with ease and creativity the essence of any system. Upon further study, though, one learns that as good as they are at modeling, they are even better at listening. They continually give the customer a chance to change their way of thinking. Their statements are few; their questions are replete; and their listening is intent. They remember that the goal is to solve the customer's problem, not theirs.

With this in mind, work on suspending judgment until the system and the situation are better understood. Project team members should not be too anxious to showcase their knowledge. Quickly sizing up the problem and then leaping to what is believed to be the cause can easily backfire. The team members think they are proving their worth, while the customer feels that he or she is not being heard. Instead, take time to know who the customer is, what is being asked, and why. Concentrate on what is, as well as what is not, being said. Draw out the facts and encourage further clarification. Do not let vague answers or unfamiliar terminology curb the need to know. In short, fully understand the problem first. Next, isolate it. Then, and only then, venture out to solve it.

Also, be more circumspect. Part of finding a solution involves understanding its implications. Be continually sensitive to the customer's needs, values, beliefs, and attitudes. Watch for clues on how the customer, individually as well as collectively, views the problem and the project. Then couch all comments, reports, documents, findings, and presentations accordingly.

Maxim Don't look for a solution without first listening to the problem.

Communicate Openly. Poor communication is the single biggest reason that projects fail. There is simply no substitute for good communication. Start by reaching an understanding with the customer about the project. Settle on objectives, key questions, performance measures, scope, assumptions, and model input. Then put these findings in writing and get it approved by the customer. At this stage of the process, everyone needs to understand clearly the project team's collective knowledge and intent.

Next, orient the customer by establishing a project plan. Gain the customer's support by explaining how the project will proceed and what to expect. Break the project into phases. Emphasize the benefits associated with doing each phase as planned and prepare the customer for potential problems. In short, give the customer a "road map" of the project. By knowing what is coming and why, the customer is in a better position to lend support. Without this knowledge, the customer may unintentionally work against the project.

Finally, keep the customer informed. People like knowing where they are. Have plenty of "sign posts" along the way. Easily identifiable deliverables are excellent for this. Moreover, keep in mind that when it comes to projects, people do not like surprises. When a problem arises, tell the customer. By reporting it early, the customer has more options available and more time to deal with it.

Maxim Keep people informed, for the journey is more valuable than the solution.

Predict the Solution. In the beginning of a project, people often miss an excellent opportunity to set the stage for success. They are too eager to get started. As a result, they fail to do a simple, yet effective exercise.

At project initiation, ask the customer to conduct a quick, even crude, analysis of the problem. This accomplishes several things. First, it gets the thought process started. The customer begins to concentrate on the problem more than the model. This provides early insight into the issues ahead and possible solutions. Caution is advisable, however. Preconceived ideas can restrict thinking. Do not let this happen. Keep creativity alive. This exercise should strengthen the thought process, not stop it.

Second, this starting solution provides a comparison base. If the project's results turn out differently, interest is aroused. This leads to exploration into why these differences exist and, eventually, to an even deeper understanding of system behavior.

Finally, this solution defines the customer's thinking at the beginning of the project. This is important because, as stated earlier, as the project unfolds, knowledge is gained. Without this beginning reference point, it is difficult to establish how the project advanced everyone's understanding. Accordingly, the project's true value is unfairly discounted.

Maxim Only by knowing where you started can you judge how far you've come.

22.3.2 Model Conceptualization

Following problem formulation, the task of conceiving the model begins (see Chapter 2). This involves dividing the actual or proposed system into its salient elements and then synthesizing these elements into a simulation model. Guidelines include:

1. Direct the model.
2. Take calculated risks.

Direct the Model. With the project's objectives and scope now set, one is tempted to begin model building immediately. Yet, before this activity can start, there is an important next step: model formulation. Leaping into model building without first properly setting its direction is a wasteful venture.

So how should the model be formulated? Is it better to go for a direct, one-to-one translation between the system and the model or to work at a more abstract level? The answer lies in what the team aims to get out of the process. If they seek, for example, better intercellular coordination in the plant, a broader, higher-level model may be appropriate. On the other hand, if accurate order sequencing within a cell is most important, it may be better to formulate a more detailed model which differentiates among the many intricate routes within this cell. In short, the model needs to reflect its pur-

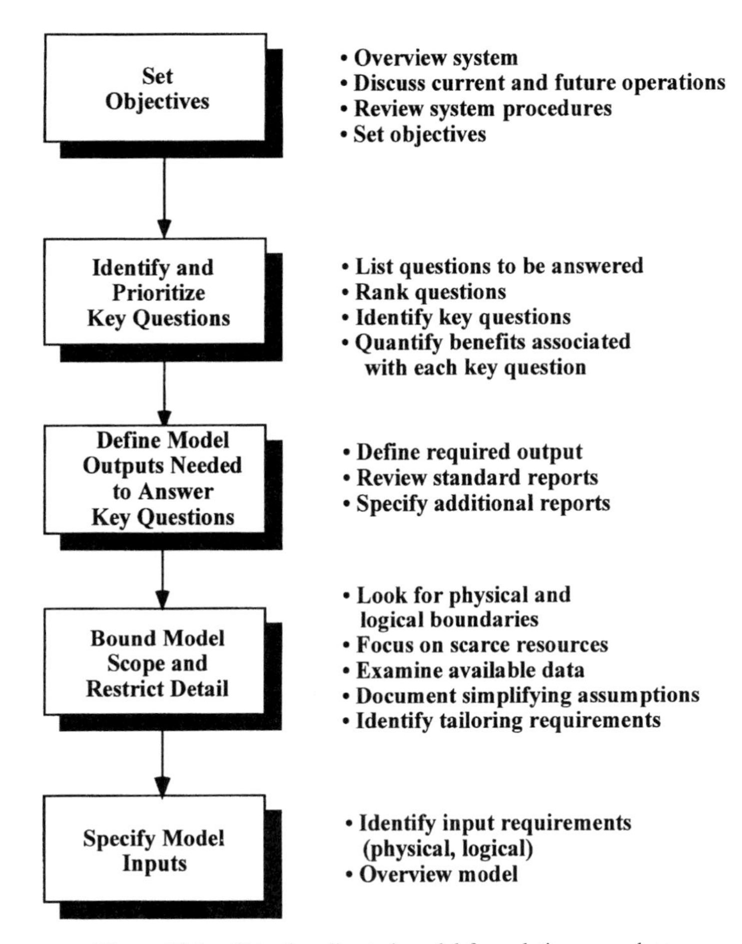

Figure 22.1 Objective-directed model formulation procedure.

pose. That is why the best guide for formulating a simulation model is a well-defined set of objectives. This is also why these objectives must be set upfront. Through them, the criteria against which all modeling decisions should be judged is established. This results in a more directed model that better fits the needs of the project.

The most effective way to incorporate these objectives into the formulation process is through a backward pass. Figure 22.1 displays the steps involved. By working from the objectives, first generate a list of key questions. These questions serve to support the objectives and identify those areas of the system that need to be modeled. For example, suppose that an objective is to improve system throughput. Representative questions in support of this include the following: What are the bottleneck work centers? Do any work centers have excess capacity? How should orders be sequenced at the mill to do the best job of getting the work out on time? As another example, suppose that an objective is to reduce operational costs. Questions reflective of this objective include: What is the expected backlog at each work center over the next several months? If we change our current scheduling policy, can we reduce overtime without incurring any additional late orders? How much maintenance in fabrication can be done on-shift before it starts affecting delivery dates? Listing all questions, ranking them in importance, and selecting the key ones helps to further direct the modeling process.

At this point, the team should also work to quantify benefits. Establishing the value of an answer *before* knowing what the answer is increases the value of the benefits associated with that answer. By settling on these benefits early, the value of the solution is less apt to be as deeply discounted once the results are in.

Next, decide what output measures are necessary to answer these questions. By focusing on these measures and only these measures, the model becomes even more narrowly defined. For example, an objective regarding work center performance could lead to the need for various measures of machine utilization. At this general level of interest, dispatch lists for each machine in the work center are not justified. To answer a question regarding utilization, a dispatch list provides more detailed information than necessary. The objective here is to better understand the work center's load, not to schedule it. By including only those system elements that support the key measures of interest, the model is reduced in size without sacrificing value.

After the output measures are set, begin to establish the model's boundaries, both physical and logical. These should be based on how the measures are defined. If, for example, an output measure calls for line utilization, modeling the detailed differences among the stations in that line may not be necessary. Conversely, knowing the size of a particular station's queue in the line could prove helpful in routing upstream work. Orders could be diverted based on the workload at this station. If this level of logic is required to answer a key question, a more detailed representation of this line and possibly other associated lines is warranted.

In turn, model boundaries drive input requirements. Consider system resources, for example. Since only scarce resources set a system's flow rate, noncritical resources need not be modeled explicitly. Instead, activities involving these elements may be left out, represented simply as time delays, or modeled at a more generic level, not requiring the express capture and release of specific resources within a group. Again, in this way the model is fitted to its purpose.

Finally, in formulating the model, work to keep each new element at a level commensurate with other elements. Broad characterizations in any one area can easily offset exactness in other areas. Seek a balanced representation across the model. For example, setting an activity time based on the entity types that transit this activity is pointless unless entity attributes can be characterized properly. Poor data characterization can easily undermine the value otherwise offered by the logic that is dependent on these data.

The objective-directed formulation approach described above benefits the simulation process in many ways. First, it streamlines model development. The resultant model, which is usually cleaner and leaner by following this approach, is easier to understand, build, verify, and maintain. One is not constantly wrestling with needless modeling detail. This saves both time and money. Second, it supports quick feedback. Early initial results, albeit preliminary in nature, can pay big dividends, especially in those situations where the team is led in directions not considered previously. Third, the model is easier to adjust. If the model is too cumbersome to adapt easily to change requests, the project team can find itself arguing against a change only because of the effort involved—a weak argument against a valid request. Fourth, the project's degree of success improves because the model remains consistent with the objectives originally set forth. The project is less apt to drift aimlessly during model formulation and development. Fifth, by reminding everyone at the end of the project of the benefits established earlier, the value delivered is further reinforced. Finally, this approach better serves the users' needs since their key questions are what drive the model's scope, level of detail, and reporting. Also, by being involved in the process, the users better appreciate the model's strengths and have greater confidence in its predictions.

Maxim Advance the model by formulating it backward.

Take Calculated Risks. Growth requires stretching. By reaching beyond one's limits, learning takes place. Consequently, constantly challenge the status quo. Be bold! Begin to color outside the lines. Strive to try something new on every project. Maybe it is approaching the problem in a new way, using an unfamiliar section of the simulation language, or experimenting with a new analysis technique. Whatever it is, make a commitment to break with convention, provided that the risks do not outweigh the rewards. For when one does, one frees oneself to see other possibilities for the first time.

Taking calculated risks also means experiencing failure. If the project team continually pushes the envelope, it is impossible to avoid this condition—despite the team's best efforts. Things can and do go wrong. Very few projects beat a direct path to success. Even the most successful projects experience setbacks. The crucial issue is how the team deals with them. Failure should influence, not dictate, direction. The team must learn to use failure as a steppingstone to new ideas. Moreover, by being open to failure, the team can take on additional risk, which, in turn, can lead to even better ideas. Taken in this way, failure simply becomes a part of the process, not the end of it.

What about reporting failure? Should the team report the problem immediately and risk unnecessarily alarming the customer, or try to recover quickly and not notify the customer unless the problem persists? The answer depends on many factors. As a rule, however, no one appreciates being the last to know. Keeping the customer informed, be it good news or bad, is always prudent. Customers are usually much more accepting of a message, whatever the content, provided that it is timely. The customer looks for continual assurance that somebody is standing watch over the project. Having an early warning system, although it may be flawed, is much better than having no warning system at all.

Maxim If you never fail, you limit the extent to which you can succeed.

22.3.3 Data Collection

All simulation models require data (see Chapter 3). Collecting system information, or estimating it if the system does not exist, is a necessity. Furthermore, this activity is often repeated, for additional data are frequently requested as the system becomes better understood. Guidelines here include:

1. Question the data.
2. Make assumptions.

Question the Data. Challenge all data collected. Do a quick audit. Consider the source, what was collected, when it was collected, and how it was collected. Does it make sense? Is it at an appropriate level of detail? Is it within the scope of the project? How is it going to bias the results? Good data are critically important. If the data are limited in some way, so are the results. Moreover, if the data appear in error, inconsistent, or irrelevant, it undermines the customer's trust in the results. Without this trust, no action is taken. Therefore, support a strong defense of the results by making sure that all input data are statistically representative of the process being modeled.

Strive to get all the necessary data, but be open to compromise. A model need not be so rigid as to force an extensive data collection effort. Requiring more data can easily delay a project. The more prudent approach is to start with the data that are available. Then request additional information once the need can be verified. Working in this way, model conceptualization and data collection proceed in parallel, with each influencing the other activity.

For example, suppose that the data collection activity grouped truck interarrival times at a distribution center. That is, the time between arrivals of a particular truck type are confounded with other truck types. If truck type is important in processing work at the distribution center, this information is also important to the model and should be a distinguishable attribute of arriving trucks. One could require recollection of the interarrival times by truck type, but this would probably delay model development. To avoid this delay, one could use instead the confounded interarrival times together with an estimate of the relative frequency of each truck type. In this way, progress on the model could continue. Then, by testing the sensitivity of the model's outputs to the full range of input possibilities, in this case the truck type frequency, one can judge if better data are necessary. If sensitivity is shown, good reason exists to get the additional information.

Keep in mind that data easily affect what a model can address. Expecting a model to handle issues that cannot be supported with the requisite data is wishful thinking. Suppose, for example, a project team seeks to build a scheduling model, yet the manufacturer's data collection system only records closed operations, not partial completions. If operation times were of significant length, not knowing the time remaining on all partially completed operations would adversely affect schedule quality. Although this level of imprecision jeopardizes the use of the model for detailed scheduling, it has less of an effect on planning. The requirement for precise startup conditions diminishes as the horizon of interest lengthens. Since better scheduling begins with better planning, perhaps by formulating a planning model first the project would be better served with the data that are available. This better planning model could then lead the way to an improved scheduling environment.

Maxim Don't take data for granted.

Make Assumptions. Data collection is inevitably on the critical path. Do not let the project be held captive by a lack of information. Be willing to make assumptions to keep the project moving forward. Have the customer take an educated guess at what is not known. Then use the model to judge the sensitivity of the system to this information. It may be of little consequence.

As the project develops, regularly review and adjust, as necessary, all modeling assumptions. Assuming something at the start of the project does not mean that the team must hold to it. Encourage flexibility. The team must learn to revise its thinking and assumptions throughout a project. Making an unconfirmed assumption is a problem only if it is treated as a fact. Information is being gathered continually. Expect this to force reconsideration of certain system assumptions. As stated earlier, this rethinking of the model may force revisitation of earlier steps. This should be anticipated and embraced. Then, before presenting results, have the team reexamine and revalidate all assumptions. If the project's results are to be challenged, the argument will ultimately be directed at the assumptions.

Maxim Be willing to make assumptions.

22.3.4 Model Building

A simulation model is constructed using the conceptual blueprint established earlier. Yet, just as with other applications, substantially different models can be built from the same blueprint. To help further direct the model building task, consider the following guidelines:

1. Focus on the problem.
2. Start simple.
3. Curb complexity.
4. Maintain momentum.
5. Review.

Focus on the Problem. Many simulation projects inappropriately concentrate on model building more than problem solving. Getting the model up and running becomes the overriding objective. Understanding the problem and deriving possible solutions become subordinate. Consider, instead, spending more time experimenting and less time building. Develop the initial model in less than half of the total project time. Then work with the remaining time to explore various what-if questions. This encourages use of the model to generate ideas. By having the time and freedom to explore beyond current thinking, the team gains additional insight into the problem and possible solutions. Remember, building the model is not the primary task; finding the right solution is.

Maxim Focus on the problem more than the model.

Start Simple. Use the model to *reduce* the problem. Building a detailed model can easily undermine the team's ability to solve the problem. Too much time is wasted managing the model, verifying it, and then trying to understand it. The model can become more complex than the problem it is trying to solve. Avoid this by starting with a simple model, one that captures just the essence of the system. This may be some portion of the original system or a simplified view of it.

For example, when modeling a manufacturing system, start with just the basic production process, void of any complex flows, decision logic, or disruptions. Work to recognize the cause-and-effect relationships in this simplified model. This establishes a foundation on which to build an in-depth understanding of the system's more complex behavior. Then, following verification and validation of this basic model, begin to expand the model both down and out. In this example, this may involve adding complex routings, rework, planned maintenance, material handling systems, shifts, or random downtimes. At each new stage of development, the model needs to be reverified and revalidated. Here again, the simulation process loops back on itself—a convolution of data collection, model building, verification and validation, and analysis.

By incrementally developing the model in this way, the team can better see and understand why things happen as they do. Complex interactions are seen in stages, basic and secondary effects are easier to discern, and the verification and validation process becomes a less daunting task.

Simple models also help maintain modeling freedom. If a model is too cumbersome to work with, the effort involved in changing it becomes a significant factor in deciding if the change should be made. As detail is added, the model begins to impose its own limitations. It is much better to maintain modeling freedom as long as possible. This is done by considering, at each new stage of development, not only the value gained by adding the detail, but also the freedom that may be lost as a result.

Maxim Add detail; don't start with it.

Curb Complexity. Complexity can easily creep into a model. With perfect information available, a model's control logic, for example, can become so sophisticated that it allows the model to outperform the actual system. Alternative routes may be chosen based on instantaneous knowledge of where all the other work currently resides. This level of control in the model can conceal fundamental system design flaws. Be careful not to let the model cloud actual behavior. Its purpose is to expose these flaws, not hide them.

Moreover, do not lose sight of what is reasonable to implement. Consider, for example, a situation where one is deciding the logic to use in selecting the next-best job to run on a machine. By including both waiting and incoming jobs in this search, system performance is likely to improve. Yet how intelligent should the model be? Can this logic be implemented on the shop floor? Be ever mindful of what is practical. It can save precious modeling time and effort.

Maxim Don't let the model become so sophisticated that it compensates for a bad design, or so complex that it goes beyond the customer's ability to implement the solution.

Maintain Momentum. A simulation project is a journey. Along the way, it is important to update the customer on how the project is coming. By showing progress and having the customer acknowledge this progress, enthusiasm for the project is kept high.

One means of showing progress is through frequent deliverables. These need not be major pieces of work. The best deliverables are quick to accomplish, hold value for the project, and are clearly identifiable. Examples include a model specification, prototype demonstrations, model delivery, animations, training, analyses, model documentation, progress reports, presentations, and a final report. Timing these deliverables judiciously over the project breaks the project into manageable tasks, gives the customer a reliable measure of progress, and reinforces confidence in the team's approach.

Maxim It is better to work with many intermediate milestones than with one absolute deadline.

Review. Review the project while it is ongoing. Set aside time to sit back and reflect, stepping outside the details of the model for a moment. Evaluate the significance of what the team is now doing. If it is not advancing the project, stop! Reconsider. Direct all energies toward the project's objectives.

Formal project reviews are helpful for this. Periodically, examine how far the project has come, what needs to be done next, and how much more there is to do. The timing of these reviews can vary. Some projects require frequent and regular interaction since they tend to lose sight of their objectives. Others, conversely, develop along predictable lines and can be effectively managed with fewer, irregularly scheduled reviews, such as when project milestones are met. As a general guideline, though, monthly reviews are sensible. At this frequency, enough has happened on the project to warrant another exchange. More frequent reviews than this tend to oversteer the project, while less frequent reviews cause important opportunities to be missed.

Maxim Take time to realign the project.

22.3.5 Verification and Validation

For verification, the model's actions are tested (see Chapter 10). This entails examining if the data are read in correctly and if the entities flow as specified. Once verified, model validation begins. Here the model is calibrated for reasonable agreement between simulated and actual performance or, in the case of proposed systems, simulated and expected performance. Guidelines for this phase include:

1. Control changes.
2. Be mindful of the customer's perceptions.

Control Changes. Change requests are inevitable. As a project progresses, newly acquired information forces reexamination of the model. This provokes the customer into calling for change requests. Therefore, expect them, plan for them, and most of all, control them.

Begin by being smart from the start. Do not accept a small change just because the project is going well now. Argue against any change that is not properly aligned with the project's objectives and scope. Admittedly, it can be difficult to say no to a customer. But when circumstances justify it, the project team has no choice. Naively agreeing even to a minor change can be disastrous, for the true implications of a change usually only become evident much later in the project: during verification, validation, and implementation. Costly delays easily result.

Agree to a change only if necessary. Be aware that accepting a change can lead to other changes, with possibly even bigger consequences. Therefore, proceed with caution. Remember also that a change can delay deliverables that others are expecting. Consequently, be sure to inform all affected parties of the implications of each change, especially as to the added time and costs.

Consider delaying changes by moving them to a later phase of the project. Get baseline results out first. Then look to incorporate these deferred changes. This allows progress to continue under the original schedule while still acknowledging and eventually satisfying the change requests.

Include the customer in all change request meetings. This gives the customer another chance to become even more involved. Further involvement strengthens the customer's advocacy for the project. It also helps him or her understand, in light of the project's objectives and scope, the reasonableness of these change requests, and their impact on the project.

Get all change requests in writing. A record of these requests and their judgments fosters good customer relations. If no record exists of what was agreed to or, more important, what was not, the project team is more likely to consent to the change.

Be especially alert to changes in project personnel. Most successful projects have, throughout their lifetime, maintained the same team members. Therefore, when circumstances justify it, argue against a switch in personnel. Understandably, this is not always possible. When there is no choice, hold a project meeting immediately. Review the team's accomplishments, what tasks remain, and who needs to be involved. Be cognizant that anyone who enters the project in midstream does not have the benefit of earlier discussions and decisions. Take the time necessary to properly transition all technical and administrative information. Make sure that each team member knows what this means to him or her personally. Poorly executed transitions can quickly cause an otherwise successful project to derail. Do not miss this opportunity to get the project back on track. At the end of the project, it is too late to take the time to do it right.

Maxim Verbal agreements aren't worth the paper they're printed on.

Be Mindful of the Customer's Perceptions. Through experience, one develops a way of doing things that works well. The process becomes almost rote. What is important to establish early and what can be left for later is understood and proven to work well.

Customers, on the other hand, have their own experience base to draw on and for that reason may have a different perspective. What is currently unimportant to a project team, based on where the project is in the process, may be very important to a customer. A known, minor programming error, for example, if left unattended, can rankle a customer, causing a lack of confidence in the model and the project. This, in turn, can cause the verification and validation effort to extend well beyond what would otherwise be considered reasonable. What would have been a simple correction turns out to be a major disruption. Therefore, quickly correct any mistakes the customer perceives as a problem.

Maxim Customer perceptions require attention.

22.3.6 Analysis

Experiments are conducted with the simulation model to draw inferences. This involves the application of analytical techniques, such as estimation procedures, variance reduction techniques, and design of experiments (see Chapters 6 to 9). The inferences drawn relate directly to the objectives established earlier for the project. In performing this step, the following practical guidelines are offered:

1. Work the model.
2. Question the output.
3. Understand the model's limits.
4. Know when to stop.
5. Present a choice.
6. Sell success.

Work the Model. A simulation model is built to stimulate thinking. With it, the project team begins to see, often as never before, the system working as a dependent set of subsystems. Given the opportunity, the model can lead the team to fascinating discoveries, each revealing valuable insights about the system's various cause-and-effect relationships.

These insights come from the team exercising the model, not from the model itself. The model is the medium through which perceptive observations are made. Think of it as a window to the system. Through it, the system can be studied from various perspectives. At times, the picture may be blurred, but by using the team's collective judgment skills, the details can be filled in. At other times, the picture may be especially clear, offering a telling view of dependencies and unexplored relationships. By continually repositioning the window, the model becomes an invaluable tool to generate and test ideas, observe the system's sensitivity to changes, and examine its underlying behavior.

Maxim Let the model work for you.

Question the Output. Challenge the model's outputs. Can they be explained? Can they be defended without getting into technical details? Do they make intuitive sense? If not, check the work. Something is wrong! Examine the assumptions. Reverify and revalidate the model. There must be a rational explanation.

At the end of the project, compare the results with the crude analysis the customer did at the beginning of the project. Are they different? If so, why? If not, evaluate whether the project should have been done. What value did it offer? How did it change the customer's thinking? In the future it may not be necessary to simulate a system that exhibits such predictable behavior.

Maxim If it doesn't make sense, check it out.

Understand the Model's Limits. At best, a model is less than reality. By its very nature, it is an abstraction of the system, one that is roughly right. This means a solution for the model is not necessarily a solution for the system. A degree of interpretation must accompany each analysis. Caution is advised here. Be careful not to stray beyond reasonable limits. Extrapolating results outside the model's inference space is risky. To judge what is reasonable, revisit the model's objectives, assumptions, and inputs.

For example, suppose that a model does a superior job of evaluating policy decisions over a range of demand patterns. Predicting the relative performance of these policies is within the scope of the model. However, it does not necessarily follow that this model is also a good predictor of future workload. Besides setting the policy, this would require identifying *the* representative demand pattern of the future. Absolute results such as this require another level of model validation. Carrying this further, assume that these policies are on a national level. Claiming regional behavior in the output could be valid only if regional effects were characterized in the input. Without this, any regional findings are groundless.

Models are not meant to replace individual thought. Their purpose is to support the decision-making process, not to supplant it. Successful projects depend on valid models, sound statistical analyses, and *cogent reasoning*. This is why it is so important to keep the customer involved. Ultimately, it is the customer who must drive the value of a solution. The model cannot do this.

Maxim People decide; models don't.

Know When to Stop. More can always be done on a project. The model can be expanded, the data refined, and the output reformatted. Analysis can also continue. Perhaps it would be helpful to test the sensitivity of the results to boundary conditions, conduct a more rigorous statistical evaluation, or examine several more alternatives. And, if this is not enough, surely more training and better documentation are in order. The list is never ending.

Prepare the customer for this eventuality. During the project, work with the customer to define a suitable stopping point. This is a judgment call, but one with which the customer must feel comfortable. Transition the customer into owning the continuation of the project. This could involve teaching the customer how to modify the model, decipher errors, and interpret output. The project is deemed far more successful if progress can continue following the project team's involvement.

Maxim Ultimate truth is not affordable.

Present a Choice. The customer may ask for a solution but really wants a choice. Narrowing the solution set to one option is often too confining. It limits the extent to which other reasoning can enter the picture. Moreover, if this solution is rejected for any reason, the project has no place to go. The team risks having the entire project rejected. Alternatively, by presenting a range of possibilities, the customer has more freedom, naturally becomes more involved in the evaluation process, and gains a better understanding of why the best is best. The knowledge gained here by the customer can be more valuable than the solution itself. Besides, with a choice, the customer has a harder time finding fault. It is much easier to dispute one solution than a set of solutions.

Maxim People don't resist their own discoveries.

Sell Success. Selling the value of a project continues throughout its life ... and beyond. It starts with the proposal and continues past implementation. Seize every opportunity to reinforce the original reasons for doing the project and trumpet the value—from the customer's perspective—of what has been done. Pursue these opportunities aggressively. Success does not come to the project; the team needs to go after it.

Learn to accentuate the positive. The team will have enough support in exposing the negative. Moreover, sell success, not underachievement. Compare the project's progress more to where it was than to where it is going. It will always be short of its objectives until the end. Continue to remind everyone where the project started and how far it has come. At every opportunity, make known what *has* been accomplished.

Maxim Report successes early and often.

22.3.7 Documentation

Simulation models are built to be used. Through use, they evolve. Having complete documentation of the model and the project provides the requisite information to support continued use of the model and to extend, if necessary, the model building and analysis efforts. Guidelines here include:

1. Adhere to standards.
2. Report progress.

Adhere to Standards. Following good programming and documentation standards is always prudent. While some simulation models are built to examine a current issue and are never used again, others are built to study a full range of issues, both current and future. The latter models can last decades. The importance of maintaining good programming and documentation practices is quite evident in those cases where the model is projected to have an extended life. As for the short-lived models, on the other hand, the need is not as evident. This is why these "disposable" models are not often held to the same set of standards. Yet model longevity cannot be predicted with certainty. Many so-called quick-and-dirty models have resurfaced years later. With little documentation to go by, the effort to reverify the model can be substantial. Moreover, if the need arises to revise or extend the model, the team is severely handicapped. Consequently, good programming and documentation practices should be applied consistently whatever the model's life expectancy.

The value of establishing and working to a set of standards is manifested in many ways. Communication among team members is improved, debugging time is minimized, and model maintenance is less onerous. Although working to standards is usually not recognized and seldom appreciated by the customer, it nevertheless is a worthwhile activity. This is especially true if the customer becomes involved with debugging, modifying, or maintaining the model.

Maxim Conform for clarity.

Report Progress. Progress reports provide an important, written history of a project. They give a chronology of work done, decisions made, what worked, and what didn't. This can prove to be invaluable as one strives to keep the project on course.

Reporting should occur at least monthly. In this way, people who are not directly involved in the project's day-to-day activities can stay involved. By regularly reporting the project's status and plans, these people have an extended opportunity to advance the project and its chance of success. Regular reporting also surfaces misunderstandings early in the process, when problems are most easily resolved and changes are less costly. The project team cannot afford to have problems fester. By reporting on them early, their impact is minimized.

Keeping a project log is also important. The log should provide a comprehensive record of accomplishments, noteworthy problems, change requests, key decisions, ideas for follow-on work, and anything else of major or even minor importance. This can be indispensable when developing a historical record of the project, especially one in which objectives, scope, schedules, or personnel have changed.

Maxim Document, document, document!

22.3.8 Implementation

A simulation project needs to lead to some concrete action by the customer, for ultimately this is how a project is judged. If system performance improves as a result of taking this action, the project is considered successful. To get to this all-important point of action, the following guidelines are offered:

1. Inspire trust.
2. Have a winning attitude.
3. Foster teamwork.
4. Involve key influencers.
5. Structure presentations.
6. Advocate improvement.
7. Follow up.

Inspire Trust. If there is any quality that is vitally important in the team's relationship with the customer, it is trust. This is the root of the relationship. It is this firm belief in the integrity and reliability of the project team that, in the end, gives the customer the necessary confidence to take action. Having to live with the consequences of any project action, the customer remains keenly aware of these qualities in the team. Without this confidence, no action results.

The project team must constantly work to win this trust. This is not easily done, for trust is something that is earned—not through words but through actions. It is built by maintaining high principles and keeping one's word. Once earned, trust must never be taken for granted. One thoughtless moment can quickly erase what the team has worked so hard to accomplish. Once lost, it is all but impossible to recover.

Maxim Only promise what can be delivered and always deliver what was promised.

Have a Winning Attitude. Failure is far too common—failure to understand, failure to try, failure to take action, failure to perform. On the other hand, success, which is unquestionably more pleasing, is not common enough. So how can the team improve the project's chance of success? Attitude is a great equalizer.

Any project has a much better chance of succeeding if the team's attitude is aligned accordingly. Setting a positive tone creates a successful climate. Stay upbeat. Exhibit positive, optimistic thinking. Expect success and keep it in the forefront of everyone's mind. Truly own the attitude. Feel it, show it, say it, repeat it, and picture it. Help others to see and feel the same way. The team's attitude goes a long way toward influencing the results that can be achieved.

Keep in mind that wanting to succeed is not the same as expecting to succeed. Everyone on the team wants to be associated with a successful project, but not everyone expects to be. Some expect the approach to the problem to fail. Some expect interest in the project to wane. Others expect the results to be inconclusive. Still others expect no action to be taken by the customer. Work to create positive thinking at the beginning and throughout the project. Dispel any doubt and disbelief. As the team gets accustomed to thinking the project will succeed, it will.

Maxim Success is an attitude.

Foster Teamwork. A simulation project is more than building a model or managing a process; it is working with people. The team must work well together if the project is to succeed. This means having trust and respect for each other. It means taking full advantage of the team's collective strengths, with each member adding to the team's understanding based on his or her distinct knowledge and experience. It means working as a unified body, with all members being aware of their role and importance to the general outcome of the project. It means avoiding pointless conflicts. And it means being open to sharing crucial information that will help advance the project. By having the team specify, build, search, discover, and conclude together, the entire project is enriched. To the degree the team succeeds, the project succeeds; to the degree the team fails, the project fails.

Recognize that some people could have a self-interest in seeing the project fail, for they may feel threatened by its success. This could be for many reasons, including job security, exposure, or power. Identify these people early and move quickly to resolve the conflict. People working in harmony get a lot more done than those who are working at cross purposes with each other.

Organizations are inherently political. It follows that implementation, which involves working with people at all levels of an organization, is necessarily a political process. To succeed requires skill in practical politics. Observe the customer closely and make an effort to understand what is occurring not only in his or her department, but the company in general. Use this knowledge to put the right spin on the project, whether it is in presenting the results or in asking for an extension. Be continually armed with accurate information about the company and the industry in which it competes. Then, in moving the implementation forward, use this information to rally the project team and work the organization. In other words, go beyond just solving the problem; move the customer to action.

Maxim Focus on possibilities, not personalities.

Involve Key Influencers. Serve the needs of both the key decision makers and the process owners to avoid project meltdown. If the key decision makers are not already directly involved, meet with them upfront to agree on objectives, scope, and approach. Get their commitment, incorporate their ideas, and allay their concerns. Then meet with them periodically to inform them of the team's progress and to reaffirm their commitment. As for the process owners, they are the ones who will be most affected by the project's findings. For this reason, they have a vested interest in the results. Work diligently to gain their support. Their involvement is particularly important during problem formulation, model conceptualization, validation, analysis, and implementation. By getting both the key decision makers and the process owners involved from the start, neither camp is likely to disrupt progress in the middle of the project by resetting expectations. This avoids what could be a costly and demoralizing development.

Watch for signs of the project not being a priority, such as data-gathering delays, end-user unavailability, management dissociation, decision postponements, deadline slippages, and apathetic or even hostile attitudes. When these signs appear, call on upper management to reset the priorities.

As stated earlier, changes in project personnel can be very harmful to a project. To help prevent these changes, enlist the support of departmental management. Make sure that they understand the commitment required of their people and the project's value to their department. By getting their early support and then keeping them informed of the project's progress, the project team is less likely to be forced to replace or reassign people.

Maxim Pull in key influencers to prevent the project from being pushed out.

Structure Presentations. Too often project teams lose a great opportunity to make a difference because they fail to make a persuasive presentation. Their techniques obscure the message. While good techniques cannot save poor results, poor techniques can certainly weaken good results.

Take time to present the message in a clear manner by focusing on organization, editing, preparation, and delivery. Organize the presentation to be clear, interesting, and relevant. Have an unmistakable purpose, direction, and strategy. Anticipate questions, preanswering them in the presentation material. Strive for a logical flow throughout, appealing to reason and building toward a believable and convincing close.

Next, edit the presentation for punch. Know what the project team wants from the audience and go directly after it. Brevity is an asset. Show more than write. Illustrate more than explain. Stress benefits more than details or features. Focus on qualitative insights more than quantitative results. Use visuals to hold attention, emphasize points, improve understanding, and clarify the message. But mainly, get to the point. A presentation is a compendium, not a thesis. The customer can get the full story later.

Once the message is set, prepare the delivery. Become familiar with the material, committing to memory the important points and their sequence. Memorize a strong introduction and a concise close. Learn the presentation point by point, not word for word. Then rehearse it to correct potential problems.

Finally, deliver the message. Flavor the presentation using voice pitch and tone to emphasize particular points. Use pauses to attract attention. When presenting visuals, inform the audience of what they should see. Explain meanings and implications more than the visual itself. Carefully choose words and phrases to get the desired results. Avoid excuses, for they deflate the message. Use gestures to add interest, variety, and impact. Through careful use of voice and body, one can inform, convince, and motivate the customer into action.

Maxim Presentation is as important as content.

Advocate Improvement. A simulation project usually results in system improvements. These improvements are the result of change, such as a new operating policy, a different cell configuration, or a new order allocation scheme. Whatever the change, expect resistance and take action to overcome it. Enlist management support, educate those affected as to the value of the change, sell the team on the importance of being proactive and enthusiastic about the change, and educate the customer as to the benefits associated with carrying out the change. By involving everyone in the change process, resistance is mitigated.

As stated earlier, getting customers to make a change is heavily dependent on their confidence to take action. Giving a customer a sense of control over the project helps instill this confidence. Learn, therefore, to guide the simulation process without usurping the customer's control. With loss of control, the customer becomes either angered or uninterested. In either case, the project falls into disfavor, and any attempt to implement constructive change is unlikely.

Maxim Be a change agent; have a bias for action.

Follow Up. At the end of the project, know that it is not over. Keep the line to the customer taut by taking the time to stay in touch. Be proactive in making sure that the objectives are satisfied. Were the recommended actions taken? How were they received?

Have any new issues been raised? How has performance changed? Are the benefits being realized? Focusing on the objectives throughout the project naturally leads to a genuine interest in how these objectives are eventually fulfilled.

After the project, continue to nourish the relationship with the customer. The effort put forward to establish this valued connection in the first place should not be wasted. Periodically, renew contact. Keep the relationship fresh. Find out what the customer's current needs are and offer suggestions. Even though the project went well, further contact and encouragement are required to maintain a lifelong partnership. By occasionally investing the little time necessary to make another positive impression, one continues to build a strong and healthy relationship with the customer.

Maxim Projects never end; they just transition to another state.

22.4 CONCLUSIONS

Everyone's project experience is different. What may hold true for one simulationist may not hold for another. Yet, as one's experience grows, so does one's belief in certain fundamental truths. The specifics of the simulation process invariably change from project to project, but the basic activities associated with this process do not. Accordingly, a set of principles can be developed.

This is not to suggest there is only one way to be successful or only one set of principles by which to conduct a simulation project. The complex combination of factors that affect a project is too bewildering for this. However, working from a well-founded set of principles can make a difference. Projects proceed in a more directed manner with fewer distractions and delays. The result is an increased likelihood of success.

The guidelines presented here are straightforward actions that have repeatedly proven their value. While each is exceedingly important, one fundamental truth transcends them all: customer value governed by responsible action. By accepting and following an ethic that places the welfare of the customer first, the more temporal goals of building a valid model, performing an exhaustive analysis, or completing the project on time and within budget become overshadowed. By putting the customer first, in the end and long after these other goals have faded, one is left with the deep satisfaction that comes with having succeeded for the customer.

ACKNOWLEDGMENTS

The contents of this chapter are taken primarily from two papers [6, 7] originally published in the *Proceedings of the Winter Simulation Conference*. Both papers have been revised and updated for this chapter.

REFERENCES

1. Banks, J., J. S. Carson II, and B. C. Nelson (1996). *Discrete-Event System Simulation*, 2nd ed., Prentice Hall, Upper Saddle River, N.J.
2. Law, A. M., and W. D. Kelton (1991). *Simulation Modeling and Analysis*, 2nd ed., McGraw-Hill, New York.

3. Pritsker, A. A. B., C. E. Elliott Sigal, and R. D. J. Hammesfahr (1989). *SLAM II Network Models for Decision Support*, Prentice Hall, Upper Saddle River, N.J.
4. Pritsker, A. A. B. (1995). *Introduction to Simulation and SLAM II*, 4th ed., Wiley, New York.
5. Pritsker Corporation (1993). *Simulation: A Decision Support Tool*, Pritsker Corporation, West Lafayette, Ind.
6. Musselman, K. J. (1992). Conducting a successful simulation project, in *Proceedings of the 1992 Winter Simulation Conference*, J. J. Swain, D. Goldsman, R. C. Crain, and J. R. Wilson, eds., IEEE, Piscataway, N.J., pp. 115–121.
7. Musselman, K. J. (1993). Guidelines for simulation project success, in *Proceedings of the 1993 Winter Simulation Conference*, G. W. Evans, M. Mollaghasemi, E. G. Russell, and W. E. Biles, eds., IEEE, Piscataway, N.J., pp. 58–64.

CHAPTER 23

Managing the Simulation Project

VAN NORMAN
AutoSimulations, Inc.

JERRY BANKS
Georgia Institute of Technology

23.1 INTRODUCTION

In this chapter we discuss management of the simulation project. It begins with the motivations for simulation. Then we take a reversed role and discuss situations when simulation is not appropriate. Next, the manager's role as a consumer of simulation is discussed. Then we describe the steps in starting a simulation project and the roles of the manager after the startup is completed. Next we suggest ways to avoid problems or pitfalls in the simulation process. Finally, we discuss ways that the manager can get information and training about simulation.

23.2 MOTIVATIONS FOR SIMULATION

There are many possible motivations for simulation. We have discussed this material in other settings (Banks and Norman, 1995) as "justifying simulation to managers." Our concern then was for the simulation analyst who needed to convince management that simulation was the appropriate technology. In this setting we present the material for the same reason as previously, but also for the case where the manager must also convince others once the decision has been made to use simulation.

23.2.1 Will the System Accomplish Its Goals?

Perhaps the most frequent question asked of the simulation analyst is: Will the system accomplish its goals? A system is designed to achieve a specified performance level or throughput, and the designer wants to know if the intended throughput will be achieved.

Handbook of Simulation, Edited by Jerry Banks.
ISBN 0-471-13403-1

The simulation model to answer this question can possibly be less detailed than when more specific questions are asked, such as those about the waiting times of entities at different parts of the system.

23.2.2 The Current System Won't Accomplish its Goals; Now What?

The flip side of the first question is the second question: The current system won't accomplish its goals—now what? In this case, the system already exists, but it is not functioning at its designed level, or the requirements have changed. In this situation the simulation analyst will examine many possible scenarios. These scenarios are posed by the client. In an unusual situation, the client, out of desperation, asked the simulation analyst to develop a scenario. By the time the situation had reached this state (many scenarios had been examined, but none accomplished the goals of the system), the simulation analyst had become so familiar with the interactions within the system that a scenario was posed that was better than any suggested by the client (although the system still did not accomplish its goals).

23.2.3 Need for Incremental or Continuous Improvement

This situation arises in several ways. In some instances there is excess capacity, underutilization of resources, and rightly or wrongly, it is desired to increase utilization. So the simulation analyst is asked to see how much additional throughput is possible. In the other case, an ultimatum might be given, such as: "Give us a 5% increase in output, or economics dictate that this unit is going to have to be closed." This statement was made to a large producer of an important commodity that became a major user of simulation. Simulation led to improvements, much greater than 5%, and the production facility is still going strong many years after the ultimatum was given.

23.2.4 Resolve Disputes

An actual case led to the inclusion of this as a reason to simulate. The client wanted to increase throughput substantially in an existing facility, as it was staggering under the existing load. A systems consultant made a claim to the client that the current facility could handle the existing load without any expansion required. The client challenged the systems consultant. After some failed attempts to prove the point, the system consultant called a simulation consultant to help them prove that their intuition was accurate. The simulation consultant showed that indeed the system consultant was right with a reorganization of the current system. The client saved a great deal of money.

23.2.5 Solve a Specific Problem

The client recognizes that a bottleneck exists. The bottleneck is a symptom rather than a cause. Two such situations, both in the field of rapid transit, are provided. A rapid transit system in a major U.S. city was very concerned about the impact on system performance related to providing legally mandated handicapped access. When a handicapped person wanted to enter the train, the loading time at a station was much greater than usual. This affected the entire system of trains on the line. But handicapped access occurred only on an occasional basis. The transit system wanted to better understand the systemic effects of handicapped access, and they wanted to try different control strategies in the event of handicapped access. Simulation was very instrumental in establishing policies to operate under the new laws.

In another case, a major city used parallel lines for their rapid transit system (like a highway with two lanes). A major repair of the northbound line was planned between two stations. The rapid transit planners wanted to know what was going to happen to the system when both northbound and southbound trains had to travel on the same track. Control algorithms were simulated to help understand how the system could be operated while the repairs were under way.

23.2.6 Create a Specification or Plan of Action

In this instance the model algorithm becomes the plan of action. There are at least two situations where this might occur. Consider the situation where a new manufacturing system is being designed. A machine is needed, but the designer doesn't know what the machine specifications should be. Simulation might be used with different machine capabilities inserted for the unknown machine. When the system operates as intended (in the simulation), the (simulated) machine specifications are provided to the machinery supplier as design requirements.

We are seeing an increased use of simulation to create a plan for implementation (the second situation). Thus the simulation model is constructed and fine tuned. Then the simulation is followed as the implementation plan. An example is a new method of assembly-line sewing in which workers move forward to the next operation (and physical location) with the lot that they are sewing if there is no worker currently at the next operation. There are many variations of this concept.

23.2.7 Sell an Idea

With the advent of simulation animation, managers (and customers) can readily understand a new concept or proposal. After all, many managers immediately say "No!" to any new idea. It's easier to say "No" than "Yes." But with simulation, much more information is transmitted to the manager than with a thick report, lots of tables, and numerous graphs. The new simulation analyst will be quite pleased to see the smile on a manager's face and that "Aha!" body posture when the manager or customer gets the new idea via animation. Even the old simulation analyst gets a positive charge from this.

A real example illustrates how simulation was used to sell an idea. However, the name of the firm has been changed. The first day of the model review, there were 12 Apex Products people in the meeting from various parts of the organization. Some were at the VP level of operations, some were floor managers, and others ran the shipping docks. For most of them it was their first exposure to simulation. Most were skeptical, and there was not much participation with what was going on in the model. We were projecting the model 6 ft by 6 ft on the wall. The first day there were a lot of arms folded, and we weren't sure whether they would come back for the second day of review and experimentation.

The second day, each person who was responsible for a different part of the plant took a turn in front of the screen where we were projecting. Each seemed to take a sense of ownership in the area that had been redesigned. The simulation became the focal point of the discussions, and each person anxiously awaited the results of the next run. We found problems and fixed them on the fly. We did what-ifs to satisfy the curiosity of several of the area managers. We pushed the system to see where it would break. At the end of the day, all seemed to be satisfied that we had tested the system adequately.

What happened between the first and second days? There were several factors. First, the model was not quite verified on the first day. We were down to the wire getting the model up and running for the review meeting. Within 20 minutes that lack of verification was apparent to all. With animation there is no hiding what is really happening, and those who knew their operation were not amused. Also, because of the (Apex Products) project team's unfamiliarity with simulation, there was a certain "decompression" period required before they were comfortable that simulation wasn't just a toy, and that serious answers could be derived from the model.

Even though animation was very important in the example, simulation analysts realize that animation is just pictures. It is useful to have animation for selling ideas and for verifying and validating models. But the important aspect is the understanding that comes from the simulation.

23.2.8 Direct Order

Fortunately for the simulation community, the direct order is becoming more prevalent. In days past, selling simulation was an uphill battle. Now, more managers are aware of the technology and are insisting that it be used.

An example is given by the following: A well-known sporting goods firm had a single distribution center in the United States. This distribution center had been designed without the use of simulation. The distribution center never achieved its goals, regardless of what was done: adding shifts, adding overtime, and so on. The distribution center manager called for a massive redesign of the system but insisted that it be simulated first. At the outset it was stated that there are many reasons for simulation. Eight of them were given here, and many more could be given, each with examples from the real world.

23.3 WHEN SIMULATION SHOULD NOT BE USED

Simulation modeling has become an essential tool for many types of industrial systems: for analyzing anticipated performance, for validating a design, for demonstrating and visualizing the operation, for testing hypotheses, and many other uses. It is the preferred tool in a variety of industries. In some industries, simulation is now even required prior to decision making for all major capital investments. A question that is often overlooked should be asked: Is simulation modeling the right tool for the problem? In this section we discuss cases in which simulation modeling is inappropriate. This section is based on an article by Banks and Gibson (1997).

In the past, simulation modeling was reserved for very large or specialized projects that required one or more programmers or analysts with specialized training and much experience. The recent proliferation of simulation software has lead to a significant increase in applications—many by users without appropriate training or experience. It has also lead to an increasing dependence on simulation to solve a variety of problems. Although many of these projects are successful, the tool can be—and sometimes is—misapplied. We're concerned that this can lead to unsuccessful projects and that simulation modeling—or the simulation software—can be mistakenly held at fault. An awareness of when "quantitative" problem requirements or "qualitative" project dynamics indicate that simulation may not be appropriate should help avoid this mistake.

In this section we present some guidelines to consider before selecting simulation as an analysis tool and 10 rules for determining when the technique is not appropriate or may not lead to a successful outcome.

23.3.1 The Problem Can Be Solved Using Commonsense Analysis

Consider an example such as the following: An automobile tag facility is being designed. Customers arrive at random to purchase their automobile tags at a rate of 100 per hour. The time for a clerk to serve a customer varies but averages 5 minutes. What is the minimum number of clerks required?

The utilization rate, ρ, is given by

$$\rho = \frac{\lambda}{c\mu}$$

where λ is the arrival rate (100 per hour), μ is the service rate (12 per hour), and c represents the servers (the unknown quantity). To avoid an explosive condition, $\rho < 1$. Thus

$$\frac{\lambda}{c\mu} < 1$$

Multiplying across gives

$$\lambda < c\mu$$

Solving for c gives

$$c > \frac{\lambda}{\mu}$$

So

$$c > \frac{100}{12} = 8.33$$

Thus, to avoid an explosive situation, at least nine clerks will be needed. The more clerks, the shorter will be the average time waiting. This problem could have been analyzed by simulation, but that is unnecessary and would take much longer to program and run than the solution above.

23.3.2 The Problem Can Be Solved Analytically

There are steady-state queuing models, probabilistic inventory models, and others that can be solved using equations (i.e., in a closed form), and this is much less expensive then simulation would be. In the license tag example above, assume that all the times are exponentially distributed. How long do the customers wait in the queue, w_Q, on average, if there are 10 clerks? This is called an M/M/c queueing model, where the first M indicates Markovian arrivals, the second M indicates Markovian servers, and c is the number of parallel servers. Markovian is another way of saying that the time values are exponentially distributed.

An equation can be used to determine the probability that the system is empty, from which the average number in the system can be determined. A graph was developed by Hillier and Lieberman (1995) to accomplish the same result. Using that graph, the average number in the system, L, is 10.8.

Little's equation relates L and w, the time in the system, as follows:

$$L = \lambda w$$

so

$$w = \frac{L}{\lambda} = \frac{10.8}{100} = 0.108 \text{ hour}$$

Customers spend their time either waiting in queue or being served. Thus

$$w_Q = w - \frac{1}{\mu}$$

where $1/\mu$ is just the average service time, or $\frac{1}{12}$ hour. Then

$$w_Q = 0.108 - 0.083 = 0.025 \text{ hour}$$

This is certainly a much faster analysis than using simulation.

23.3.3 It Is Easier to Change or Perform Direct Experiments on the Real System

That it is easier to change or perform direct experiments on the real system might generally seem obvious, but not always. Consider the case where a detailed model of a drive-through fast-food restaurant was constructed and used to test improvements on customer service time by adding a second drive-up window. The model took weeks to complete. A competitor tested the same concept by staging a second person with a remote hand-held terminal and voice communication along the drive-up line and completed the entire study in a matter of days.

The rule of thumb here is: If the problem involves an existing system which can be perturbed or measured without undue consequences, look first for a direct experiment to answer the questions. In addition, a direct experiment avoids all questions relating to whether the model was detailed enough, was properly validated, and so on.

23.3.4 The Cost of the Simulation Exceeds Possible Savings

Although almost every simulation project has many "qualitative" benefits, the expense of the model, data collection, and analysis is usually justified by the expected quantitative payback. It is possible that a model may actually cost more than the potential savings at stake. Accurately estimating the total costs of a simulation project requires some experience, and many factors must be considered, including:

- Project planning, problem definition, and process documentation
- Model development and testing
- Data collection, review, and formatting
- Model validation
- Experimentation and analysis
- Possible updates or enhancements to the model
- Project documentation and presentations

Also to be considered are the costs of the simulation software (if not readily available) and computer resources. Simulation of a complex problem can easily run into the tens of thousands of dollars. Models of large facilities with complex operating procedures and control logic (such as a large distribution center) or the need to use real (historical) data or actual product location and quantities can raise the cost even higher.

Generally, simulation project costs are compared to potential savings or cost avoidance. If the potential savings are not clearly greater than the estimated simulation costs, the model may not be justified. On the other hand, some simulation projects are undertaken because of perceived risk—for systems that are too complex to understand otherwise. These models provide a level of insurance to understand if and where possible problems lurk.

23.3.5 Proper Resources Are Not Available for the Project

Primary resources required to complete a successful simulation project include people, software, and money. *The* most critical component in any successful simulation project is people—an experienced analyst who understands the problem, selects the proper level of detail, translates it into a simulation model requirement, programs the model, and so on. It is surprising how often there are attempts to solve important problems with simulation by a person with little or no training and without proper experience.

The advanced simulation software now widely available certainly helps, but it is not a substitute for the proper people resources for a project. If a properly trained simulation modeler is not available for a project, it might be best (and less risky) to look for outside help. Remember that a poorly constructed model is worse than no model at all, because the flawed results may be used anyway.

Proper funding for a project is also critical. Assume that you are to manage a project, that you have properly trained people, and that you have the appropriate software, but the project cost estimate is twice the available project funding. How to proceed? Our recommendation is: *Don't* simulate. The project objectives will probably have to be compromised and corners cut in the model design and planned analysis experiments in order to come close to the budget. This will put the project goals at risk in that the resulting model may not be capable of providing the required results. Simulation, or the software selected, or both, will mistakenly be held at fault. Often they are not at fault—they are just misapplied.

23.3.6 There Isn't Enough Time for the Model Results to Be Useful

A time constraint is usually caused by one of two reasons: (1) the project schedule is too short, or (2) model development and testing takes too long. This is a very frustrating but not uncommon problem. You've worked hard to complete a model, carefully verified and validated it, and are in the middle of running experiments when you're told: "The decision has already been made to proceed with the facility—we didn't have time to wait for the simulation results." Sometimes a project has a decision "window" when results are needed. Earlier results may be ignored or may miss critical data or assumptions not yet available. Late results may arrive after the window has been closed—the decisions are already made. Timing the results when they are needed can be a challenge!

Simulation studies tend to be commissioned at the last minute, often as a "final check." Often, the schedule is unrealistic to begin with. If there isn't sufficient time to conduct a proper project, the analyst must make coarser assumptions, skip details, or otherwise cut corners in an attempt to meet the schedule. How do you know if critical detail was left out and the results are not meaningful? No textbook can define where the level of detail or complexity should be set—this is based on experience, on a project-by-project basis.

A simulation model should be detailed enough so that the questions posed can be answered, but not too detailed. A typical error for an inexperienced user to make is to start with too much detail in the model, which invariably takes longer to develop and test than was initially estimated and scheduled.

If not enough time has been allowed in the overall project schedule to produce results and put them to use, it may be better if you don't use simulation. This means allowing time to change the system design and resimulate if needed!

23.3.7 There Are No Data—Not Even Estimates

During the design phase of a simulation project, one of the most critical tasks is to determine if the data required to meet project expectations and support the level of detail planned for the model are available, and if not, how they can be obtained. In some cases the data may not be available, and either impossible, impractical, or too expensive to collect. Don't fall into the trap of committing to a project and building the model before checking to see if the necessary data are available. The temptation will be to proceed with the analysis anyway, since you've already expended the effort to build the model, and people may be counting on the results.

It is possible to perform sensitivity testing using estimates of the data values, but this still requires estimates about the range of values for critical data items! Guesses can lead to inaccurate conclusions; remember, a model is only as accurate as the quality of its input data.

23.3.8 The Model Cannot Be Verified or Validated

An inability to verify or validate is usually the result of a lack of one of three critical ingredients; people, data, and time.

1. The project analyst may not understand how to verify the model properly (lacks sufficient training and/or experience).

2. There may not be useful performance data for comparing the model results against test scenarios in order to validate the model.
3. The project schedule doesn't allow for sufficient testing and/or validation activities.

As indicated in Chapter 10, these are many test procedures that help to build confidence in the model, but you must still question if those that are used are sufficient to support the decisions that may be made based on simulation results. The topics of verification and validation are mentioned again in Sections 23.6.6 and 23.6.7. If the model is not properly verified and validated, results will be questioned and may not be accepted (probably shouldn't be accepted!).

23.3.9 Project Expectations Cannot Be Met

Nine times out of 10, the failure to meet project expectations is due to a lack of properly educating management decision makers about what is realistic and possible when solving the problem with a simulation model. Management may have unreasonable expectations—usually, they expect too much too fast. When it can't be delivered, they may mistakenly blame simulation technology or the analyst. People with no experience in simulation often conclude that once a system is modeled, the model will be capable of answering any question that they may wish to ask of it. It can be difficult to explain that models are only capable of answering explicit questions that they were designed to address.

In the remaining one of 10 times, the analyst overestimated either his or her own capability or the capability of the software being used. Here's an example: The problem is to simulate a large beverage plant, with production in the 100,000 cases per day range. Questions to be answered require a very detailed model. The analyst decides to model each individual can or bottle through the entire process. With a massive number of active entities flowing through a detailed simulation, the model runs in something close to real time, and the animation proves useless because of an unreasonably high demand on graphics. The analyst struggles to validate the model but can produce only one experiment run a day; management quickly loses interest. If the project expectations are unreasonable and cannot be modified or controlled, the project will be very difficult to complete successfully and may be a "don't simulate" candidate.

23.3.10 System Behavior Is Too Complex or Cannot Be Defined

The system to be simulated must be thoroughly understood and properly documented before modeling begins. If not, there is no way to create an accurate model. The analyst will be forced to guess or "be creative." Some systems are so complex that building an accurate model (within an acceptable schedule and budget) is not possible. This is often the case where (complex) human behavior is a critical part of the simulated system.

For example, modern automated distribution centers are complex, and because of this are frequently simulated prior to implementation or modification. Most are driven by computerized warehouse management system (WMS) software, which select and combine orders to process. Almost all of the actual order processing (picking) is performed manually—people run the facility, even in automated environments. Typically, the scenario simulated is an average (in some cases, a peak) day, and the model results can be quite accurate. But in a real facility when an unusual event occurs and the orders start falling behind schedule, people will change their normal behavior or activities to "find a way around" the system constraints in an attempt to meet the schedule. This behavior can be quite varied and virtually impossible to describe and simulate completely for all possible scenarios. Model results for these "crash case" scenarios almost never match what occurs in the real system and are simply unreliable.

23.4 MANAGER AS CONSUMER OF A SIMULATION PROJECT

In this section we examine the role of the manager as a consumer of simulation. In this role the manager has determined that simulation is necessary, then works with the simulation analyst either internally or externally. We provide some common problems from our experience as simulation consultants in working with our clients. We have observed both effective and ineffective ways that managers work with simulation analysts. Some of the important issues for managers are presented in the following paragraphs.

23.4.1 Decision-Making Window

Most decisions about systems have a window during which information is useful. But after the window is closed, the information is virtually useless. Simulation studies need to be planned and scheduled to provide information that can be used effectively, in a timely fashion, by the decision maker.

23.4.2 Time and Resources

To meet the decision-making window, resources must be available. We estimate that the costs for conducting a simulation are less than 1% of the total project cost. We also estimate that it costs 10 times as much to make a change in a system than doing it right the first time. There is also a cost of lost productivity during the correction period. With this information and some idea of the subjective probability of making a mistake, a decision can be made as to whether and how much resources should be expended for simulation.

With respect to the cost of doing simulation, in 1998 dollars, hiring a simulation consultant costs about $5000 per week, and an average simulation project lasts around 5 or 6 weeks. Thus the average project costs in the $25,000 to $30,000 range. If simulation software is being purchased, its cost is additional.

23.4.3 Internal Versus External Simulation Analysts

Given the information in the preceding sections, the decision as to whether to have an internal or an external team conduct a simulation study is in order. The first consideration is the timeliness. The decision-making window must be considered. If an internal team is used, learning or relearning the software must take place unless the team is constantly performing simulation projects. If simulation is going to be a continuous activity of the firm, it may be wise to have an internal capability. The analyst cost would probably be less, the simulation team would be in place and able to schedule the project, and the simulation team would be more familiar with the system.

We have had success in combining forces between the internal group and the external simulation consultant. The consultant spends several intensive days at the startup, then provides general direction to the internal group. Consultants also work for firms that have internal teams. This is often the case when the internal team is fully occupied with other simulation work, but a project must be accomplished by the time the window closes.

23.4.4 Experience and Qualifications

What type of people make good simulation analysts? First and foremost, simulation analysts need to be system thinkers. They have to see the whole as the sum of its parts and to see the relationships and interactions between the components of a system. Simulation analysts need to have a knowledge of engineering statistics (i.e., hypothesis testing and regression) as a minimum. Experience in model building and analysis is necessary for success as a simulation analyst. We suggest 3 to 6 months of training on small projects, or assisting in segments of large projects, before an analyst does any modeling at all. Whether an internal team is being constructed or an external team is being considered, the same experience and qualifications are required. More on this topic is given by Rohrer and Banks (1998).

23.4.5 Questions That Simulation Can Answer

There may be incorrect assumptions about what simulation can do. Generally, simulation is principally descriptive rather than prescriptive. Given a design, the simulation analyst can model the system and determine how the system will operate.

Simulation models are run rather than solved. In a system with stochastic behavior, sufficient history is generated to develop confidence intervals on measures of performance (average WIP, for example).

Several simulation software packages have recently added a capability to use heuristic procedures such as evolutionary algorithms and scatter search to converge on a solution that approximates a global optimum. This topic is discussed in Chapter 25. Combining simulation with this type of optimization is an active area of research as improvements in efficiency and speed are sought.

23.4.6 Data Requirements

Simulation requires data—sometimes, lots of data. A problem occurs when the data do not exist. Then they need to be collected, summarized, estimated, or guessed. The simulation modeling can proceed, even in the absence of data. There are cases where an estimate is sufficient. For example, the model may be insensitive to the input.

23.4.7 Suggestions

The simulation process in Chapter 1 requires managerial involvement. For a successful simulation project, the involvement of the manager is a must. This point is made repeatedly here. The manager should also ascertain that the model is being verified and validated as the modeling progresses. Failure to do so until the modeling is complete will result in a veritable mess. The model will just be too big to verify and validate.

While preparing training materials for a simulation software recently, one of the authors remarked how difficult it had been to verify and validate a rather small (say, 40 lines of code) simulation model. Two people had been performing verification and validation for about 2 hours, and more time was required. Those who begin to perform verification and validation after the model is built (say with a minimum of 2000 lines of code) face a monstrous problem. Start simply, grow the model, and verify and validate at each step in the process.

23.5 EXTENDED ROLE OF THE MANAGER IN THE SIMULATION PROCESS

This section is for the manager of a simulation team, perhaps new to the task. It provides a set of activities that might be useful in starting a simulation project, at least to the point where the technical activity associated with the simulation actually begins. Then we discuss four roles of the manager during the technical aspects of the simulation process.

23.5.1 Starting a Project

Banks and Gibson (1996) present 12 steps in starting a simulation project:

1. Define the problem.
2. Understand the system.
3. Determine your goals and objectives.
4. Learn the basics of simulation.
5. Confirm that simulation is the right tool for your problem.
6. Attain support from management.
7. Learn about software for simulation.
8. Determine the data that are needed and what is available.
9. Develop a set of assumptions concerning the problem.
10. Determine the outputs needed to solve the problem.
11. Determine whether the simulation will be done internally or externally.
12. Kick off the project.

These steps are intended for the simulation analyst, but in many cases the manager needs to be involved, particularly when the analyst has limited experience. For example in step 3 it is important to have agreement from all that are involved, and this includes management. Even though step 5 is partially technical, as explained in Section 23.3, there are still many questions that a manager should address.

Step 6 mentions the manager directly. Without managerial support continuing throughout the simulation study process, the potential for failure is very high. Step 7 can also be accomplished by managers. Managers can be helpful in the selection process by opining on the importance of the various criteria.

Step 11 should involve the manager. It is suggested that first-time users and those who are infrequent users of simulation use external consultants or work in combination with external consultants. The learning curve for proficiency can be rather high for simulation analysts. It is also time consuming to relearn a software package that has not been used for a year, which is now represented on the shelf by a newer version.

Finally, in step 12, the manager should be involved in the kick off of the project. This will give the simulation credibility and provide for a strong beginning.

23.5.2 During Simulation Process

There are four aspects in which the manager can significantly improve the simulation process: promoting quality, technical facilitation, corrective action, and continuing support. These are explored below.

1. *The manager needs to promote quality.* The manager improves quality by ensuring that no mistakes are made during any step of the simulation process. Additionally, this is accomplished by requesting frequent reports and by challenging statements and actions that may be unfounded.

2. *The manager needs to provide technical facilitation.* This does not necessarily mean that the manager provides the technical capability. But the manager should recognize when, and if, the simulation analysts are "in over their heads" and be willing to call for external or additional assistance.

3. *The manager needs to provide for corrective action.* If the manager observes that the analysts are off course, the manager should suggest corrective action. It is possible that the analyst is so immersed in the technical aspects of the simulation that he or she does not see what is happening from a larger perspective. The manager should be the questioner of reasonableness. As an example, a group of military simulation analysts were examining a scenario using some new strategy and tactics. Their simulation presentation was being made to a general. The simulation analysts made a statement that the forward progress of the force was something on the order of 60 km per day. The general (manager) indicated that this was much higher than the advance of the Israelis in the Six-Day War, and that this was unreasonable and sent the simulation analysts back to their computers. The simulation analysts found that they had made some faulty assumptions.

4. *The manager needs to provide support throughout the simulation process.* Managerial support is required for success. Here is an example from the consulting area: A large distribution center was being simulated. The question to be answered was: How much buffer space is needed in front of each picking station? The consultants were well received by the distribution center manager. The project got off to a good start. But the distribution center manager was called to New York for a major reorganization study. This manager was at the distribution center for only one day per week. It was next to impossible to get an audience with the manager during that one day. The project was completed, the consultants were paid, but the results were never implemented (no management support).

23.6 MANAGER'S ROLE IN ENSURING QUALITY AND SUCCESS

This section is an aid to ensuring that the entire simulation process is successful. We have returned to the steps in the simulation process mentioned in Chapter 1, and we show how these steps can go awry (Banks, 1994). By understanding the possible pitfalls in the simulation process, the manager can assure a higher-quality result.

23.6.1 During Problem Formulation

The manager should understand thoroughly the problem and the environment that created the need for a simulation. It is possible to nod one's head "yes" but to really have no conception of what is being discussed. This could be dangerous, as the manager certainly will need to ask questions about the simulation project as it proceeds. If the problem is not understood, it will be difficult to ask intelligent questions.

It is also possible that a type III error will be committed if the problem is not formulated accurately. A type III error occurs when the wrong problem is solved. The simulation might be technically marvelous in this case, but it just doesn't solve management's problem.

During problem formulation, a set of assumptions about the system needs to be developed. These assumptions are an interpretation of the system to be modeled. The manager should be certain that these assumptions are valid and that all assumptions are included. At the end of the simulation project, the modelers will fall back on the assumptions if there is a question as to whether the real system is represented by the simulated system. If there is a difference between the two, and the difference comes from the list of assumptions, additional costs may be incurred to revise the model so that it represents the real system.

A major problem can occur if the manager is not totally committed to the simulation study. Involvement of management throughout the study process is imperative in assuring that the simulation study results in a successful implementation.

During the problem formulation, the manager needs to develop the questions that are to be answered by the simulation. The simulation model can then be constructed with these questions in mind. Asking the questions after the model is constructed almost always results in added costs and time.

The manager can also propose a set of scenarios to be investigated. This might be accomplished in concert with the modelers. Scenarios are the different possible implementations of the system. The current system, or base case, is the first scenario. For example, we might consider adding another machine X as a second scenario.

23.6.2 During the Setting of Objectives and Overall Project Plan

During this step the manager should help in stating the important measures of system performance. There are many, many possible measures of how a system is performing but generally only a few that are important in each case. Average time in system may be important in one case but unimportant in another case. The manager should work with the simulation modelers in deciding the appropriate measures.

Also, during this step, the manager, with the help of the simulation modelers, will want to specify exactly what the study excludes and includes. The more detail included, the longer and more expensive the study. Yet it is possible to exclude certain aspects, to narrow the boundary, without sacrificing very much in information gained.

Simulation is not free. Its cost was discussed in Section 23.4.2. The manager is warned not to allow a project to start with fewer resources than anticipated to complete a simulation study. This produces an unsatisfactory result and can be harmful to future simulation opportunities.

Similar to the preceding, the manager should not promise the results of a simulation sooner than they can realistically be completed. There are those that want to be agreeable and make promises that are likely to be broken. But rushing a simulation study only results in failing at some aspect of the project.

The person new to simulation may wonder how it is possible that the time required to complete a study can be estimated with any accuracy. There are at least two approaches to these estimates. First, a manager could allocate days for each step in the simulation process, and sum those. Another technique is based on experience. After the project formulation, the complexity of the problem is related to similar projects that have been solved. For a problem of similar complexity, x days were required, so the current project should also take approximately x days. The authors' experience indicates that experienced modelers can regularly estimate the days required within plus or minus 20%.

23.6.3 During Model Conceptualization

A major concern during this step is to start the model simply and add complexity incrementally. It is virtually impossible to prepare the entire model of a realistic system and then begin verification and validation on the total model. As indicated previously, we strongly advise the manager to insist that verification and validation be accomplished as the model grows. For example, in a manufacturing and material handling environment, first model the manufacturing processes, then add the conveyors, then add the AGVs, then add the downtimes, then add the shift schedules, and so on, verifying and validating the model after each addition.

Your simulation modelers may ask whether to model system symptoms or causes. Sometimes it is acceptable to model symptoms and sometimes it is better to model causes. For example, if a complex machine is breaking down, it is usually more appropriate simply to model the time between failures as some statistical distribution rather than model all the interactions between the components inside the machine (modeling the symptoms). On the other hand, if bottlenecks occur, it is not acceptable to model their occurrence using a statistical distribution. Rather, you should insist on modeling their causes.

23.6.4 During Data Collection

Request only necessary data from the client. It simply infuriates a client to ask why you need the data and have you answer "I don't know, but we might need it later." Another problem with data collection includes using summary measures instead of individual values. Simulation is usually based on random inputs. These are obtained, hopefully, from historic data. However, using only the mean value is not using the distribution of all the values, and these mean values are not as helpful in simulating the real system. Also, be concerned about using past data when conditions have changed. This may be necessary if there is nothing else available. But it is risky.

23.6.5 During Model Translation

Buy the most powerful and flexible software that you can afford. The price differential among the popular software packages is not great. When the cost of a simulation analyst is about $5000 per week, saving a few thousand dollars between software packages may rapidly be offset by substantially decreased productivity. For example, software can vary greatly in the implementation of a material handling construct. Some implementations are simplified, others are much more realistic. If there is no competition for the resource, treating it simply may be acceptable. But this may not be the case.

Consider a situation in which there are two cranes that are operating along the same bridge. Having simulation software that can adequately model bridge cranes can save a great deal of time, as this is a rather tricky device to model (they compete for space on the bridge). So the manager needs to help in the selection of the right tools for accomplishing the simulation. Power, speed, and accuracy of the tools are important. But also consider other factors, such as technical support provided by the software vendor, the quality of the documentation, and the frequency of updates and enhancements to the software. The topic of simulation software selection is discussed further in Chapter 25.

23.6.6 During Verification

The major problem that occurs is when no planning for verification takes place. Resources have not been allocated. If verification occurs at all, it is an afterthought rather than something conducted from commencement of the modeling.

Even if verification is conducted, it may not exercise the model fully. For example, have all possible combinations (routes that an entity can take through the system) been tested? There are many techniques that can be used, as indicated in Chapter 10. A very easy verification technique, often overlooked, is to check the output for reasonableness.

Virtually all of the major simulation software have an interactive run controller (IRC); many call it a *debugger*. Using this feature, the progress of entities through the model can be followed event by event. The IRC might have a feature in which it is turned on whenever an entity enters a particular section of the model. Or perhaps the IRC turns on whenever a particular situation occurs (e.g., whenever a particular queue has 20 or more entities, turn on the IRC). The IRC can be tedious, but also very helpful.

Another verification tool that should be pushed by the manager is documentation. Don't allow the simulation analyst to write 10,000 lines of code and then begin to document what was done. Documentation should take place while the code is being written.

Often, a model is built for the use of someone other than the developers of the model. This might be called a *turnkey* situation. Thus your group builds the model, but another group, perhaps totally unfamiliar with simulation, will exercise the model. A big concern is data corruption by turnkey users. There are several ways to avoid such corruption. One way is to build an interface that allows for changes in the model data, but only through the interface. This will keep the turnkey user out of the model logic. Another method is to use a run-time-only version of the model. This version might allow for limited data changes but prohibit logical changes.

23.6.7 During Validation

The biggest problem is the manager who fails to insist on any validation whatsoever. Many studies are reported in the literature that don't mention the word *validation*. Whether validation was conducted or not is known only by the author. The definition of *simulation* in Chapter 1 refers to the ability to draw inferences from the model. Validation provides that ability as a positive evaluation, saying in effect: "Yes, the simulation model can be substituted for reality."

If validation is conducted, many stop it far too soon. They say, "yep, looks OK to me." But look at the many techniques that can used, as indicated in Chapter 10. The ultimate technique is where the simulation replicates the real environment with statistical confidence. We only wish that it were that easy. Frequently, the environment that we are simulating doesn't exist—it's only a concept in design.

23.6.8 During Experimental Design

Problems can occur in experimental design, including the failure to account for a warm-up period or too few replications. Perhaps the manager is not a statistician, or studied experimental design too long ago. Then it would be useful to become reacquainted with the topic by reading Chapter 6. Also, some of the software products have built-in statistical capability, including some basic experimental designs. The problem is that using this capability blindly can result in some embarrassing moments, especially when the manager is called upon to explain the technique or to explain how such statistical results could occur.

23.6.9 During Production Runs and Analysis

Don't provide point estimates unless you are forced to do so; and then, do so only with the caveat that it is a point estimate. This will save the manager from potential problems when the actual value of the implemented system is different from that point estimate. Even though many software packages have built-in statistical capability, pushing a button but not really understanding what is happening can be a very bad idea.

Improper statistical interpretation can be a serious problem. Some experience in statistics is needed. As indicated in Chapter 7, this subject is not to be taken lightly.

23.6.10 During More Runs

A concern to the manager at this step in the process is the amount of resources required to conduct the appropriate number of replications and what options are being considered. For example, if the variance is high and the required precision of the simulation is also high, the number of replications can be very large. Let us say that the result of the combination is $n = 240$, meaning that 240 replications are needed. Say that each replication takes 30 minutes. Then 120 hours of computing are going to be required to simulate this particular scenario. What if you don't have 120 hours of time to give to this scenario? What if there are five scenarios and each takes about 120 hours? There are several possibilities. One is to reduce the precision. Often, people ask for far too much precision. They say, "I want the result to be within 1% of the true value of the performance measure with 99% confidence." Perhaps it would be satisfactory to say, "I want the result to be within 5% of the true value of the performance measure with 90% confidence." Rather than 240 (or whatever) replications being required, 20 might do the job. Users need to evaluate the real requirements.

In consulting we are often faced with the dilemma of constrained time for analysis. We would never perform just one replication of a simulation for a system with randomness. Set a minimum number of replications, say five, and report the resulting confidence interval.

23.6.11 During Documentation and Reporting

A serious matter is reporting what the client wants to hear rather than the actual model results. Think of simulation as science, and the simulation analyst as a scientist, a provider of information. Documentation should be thorough. It should describe all the assumptions and methodology used in the simulation. Practice your presentation. We are very concerned that simulation be understood, perceived positively, and used more widely. Documentation of the progress of the simulation is important also. Chapter 22 has much more to say about this.

23.6.12 During Implementation

Our position is that the simulation analyst be a reporter of the findings but not an advocate. This is different from the position taken in Chapter 22. You decide which is best for you. Also, the manager should anticipate continued support requirements (if you turned the model over to another group). You need to plan on helping your client understand the model and fixing any problems for an agreed-upon length of time, say 60 days.

23.7 SIMULATION TRAINING FOR THE MANAGER

In this section we discuss ways in which the simulation manager can enhance his or her knowledge of simulation. We have participated in all the activities mentioned and understand the benefits that can be derived. Let us assume that the manager has a technical background but that this background does not include simulation. The realistic opportunities for the manager to become familiar with simulation are as follows:

- Winter Simulation Conference
- Simulation vendor–sponsored user group meetings
- Short courses
  - Simulation in general
  - Software specific
- Mentoring services provided by simulation consulting firms
- Other conferences

23.7.1 Winter Simulation Conference

The Winter Simulation Conference (WSC) is the premier conference for discrete-event simulation. It occurs in early December of each year; every third year in Washington, D.C. WSC is sponsored by six technical societies and the National Institute of Standards and Technology (NIST). Approximately 750 attend WSC, including academicians, practitioners, and software vendors. Surveys indicate that about one-third of the attendees have never been to a WSC. Thus many attendees may be new to simulation.

The technical program includes "tracks" that are of interest to newcomers to simulation. One of these tracks consists entirely of introductory tutorials on all aspects of simulation, presented by eminent persons in the field. These introductory tutorials are well attended by approximately 80 people in each session. Another track of interest to newcomers is on simulation software. The vendors, and sometimes others that use the software, describe the latest release and give live demonstrations. Also of interest to newcomers are the various applications tracks. Here newcomers can learn how others solved a problem using simulation.

In addition to these formal sessions, there are three other features of WSC that are important to newcomers. There is an exhibits area populated by some 30 booths. Here the software vendors meet potential customers on a one-to-one basis. Another feature is the reception, at which attendees mingle and discuss their favorite subject, simulation. Third, the major software vendors have user group meetings at which they discuss how they have addressed the challenges of the past year and explain features that are upcoming in the next year.

The WSC is a boon to newcomers to simulation. Much information is exchanged. The top people in simulation are in attendance and discussing their craft.

23.7.2 Simulation Vendor–Sponsored User Group Meetings

Generally, the major software vendors have an annual conference lasting from 3 to 5 days at a site near their principal location. These conferences include presentations on applications of the software and training from basic to advanced on using the software. These conferences provide an excellent opportunity for newcomers to see the entire team that a vendor can field. It is an opportunity to talk to the vendor's support staff, their training team, the software developers, and so on. Based on these interactions, potential purchasers of simulation software can determine if there is a good fit between their needs and what the vendor can provide. Usually, the vendor has a 'free for all' in which they explain what improvements have been made in the software and what enhancements are planned for the near future. Many of the attendees are extremely familiar with the vendor's software and are not afraid to vocalize their concerns. This is an opportunity for newcomers to glean how supportive vendors are to reasonable requests by users. *Warning:* Some requests are unreasonable and newcomers may have difficulty in separating these from reasonable requests.

23.7.3 Short Courses

There are two types of short courses: simulation in general and software specific. General simulation courses are of two types, public and private. Public courses are offered by an educational institution or training group. These courses last from 1 to 3 days. This is an opportunity to receive a vast amount of information in a short time.

Private courses may be more economical, depending on the number of people that are to attend. Instead of sending five people to a public course for 3 days, it may be less expensive to have the presenter come to your site for 3 days (one person travels for 4 days instead of six people traveling for 4 days). Also, the course can be tailored to meet your specific needs. For example, your business may be heavily involved with material handling. You may need a presenter who can talk about the simulation of power-and-free conveyors and overhead cranes. Alternatively, your business may be services; you may not need a discussion of material handling simulation at all.

23.7.4 Mentoring Services

Some consulting firms are willing to mentor those who need to learn more about the simulation process. The visiting person works with simulation managers and consultants on a full- or part-time basis on a specific project, usually related to the visitor's problem. Much can be learned in a short time by "living the process."

23.7.5 Other Conferences

Recently, other conferences have been organized to spread the word of simulation. For example, the Society for Manufacturing Engineers (SME) has become quite involved with simulation of manufacturing systems as has the Institute of Industrial Engineers (IIE). Other groups that are supporting simulation presentations as part of a larger show or conference include AUTOFACT (sponsored by SME) and PROMAT or NAMH (both sponsored by the Material Handling Industry of America).

23.8 CONCLUSIONS

We have provided some suggestions for the manager of a simulation project. This manager might be involved directly as manager of a simulation team, or indirectly as a consumer of simulation, having ordered the service from a team not necessarily under the manager's technical control. In either case, the suggestions made in this chapter should help in completing the project successfully. This chapter is not a complete treatise on the matter; it represents only what we have experienced over many years of simulation consulting.

ACKNOWLEDGMENTS

The authors would like to thank Matt Rohrer of AutoSimulations Inc., and Randall R. Gibson of Automation Associates Inc. for their suggestions. Material in this chapter is adapted from Banks and Norman (1995), Banks and Gibson (1996), and Banks and Gibson (1997), with the permission of the Institute of Industrial Engineers, 25 Technology Park, Norcross, GA 30092, 770-449-0491, Copyright © 1995, 1996, 1997.

REFERENCES

Banks, J. (1994). Pitfalls in the simulation process, *Proceedings: New Directions in Simulation for Manufacturing and Communication*, S. Morito, H. Sakasegawa, K. Yonedo, M. Fushimi, K. Nakano, eds., Tokyo, August 1–2.

Banks, J., and R. R. Gibson (1996). The 12 steps for getting started with simulation modeling, *IIE Solutions*, November.

Banks, J., and R. Gibson (1997). Don't simulate when: 10 rules for determining when simulation is not appropriate, *IIE Solutions*, September.

Banks, J., and V. Norman (1995). Justifying simulation in today's manufacturing environment, *IIE Solutions*, November.

Hillier, F. S., and G. J. Lieberman (1995). *Introduction to Operations Research*, 6th ed., McGraw-Hill, New York.

Rohrer, M., and J. Banks (1998). Required skills of a simulation analyst, *IIE Solutions*, May.

CHAPTER 24

How Discrete-Event Simulation Software Works

THOMAS J. SCHRIBER
The University of Michigan

DANIEL T. BRUNNER
Systemflow Simulations, Inc.

24.1 INTRODUCTION

A black-box approach is often taken in teaching and learning discrete-event simulation software. The external characteristics of the software are studied, but the foundation on which the software is based is ignored or touched on only briefly. The correspondence between the foundation and its implementation in the software might not be studied at all and related to step-by-step model execution. The modeler therefore might not be able to think things through to develop good approaches for modeling complicated situations, might not be able to use interactive tools effectively to come to an understanding of error conditions arising during model development, and might not be able to use interactive tools to verify that complex system logic has been accurately captured in a model. The objective of this chapter is to bring about a better understanding of the particulars of discrete-event simulation and to motivate modelers to study the implementation of these particulars in the simulation software they use. The result will be to improve the effectiveness with which modelers can build, verify, and use discrete-event simulation models.

The approach taken in the chapter is to develop a generalized view of the logical foundations of discrete-event simulation, introducing generic vocabulary and constructs to support this view. Three instances of commercial simulation software (SIMAN, which is the simulation language within ARENA; ProModel; and GPSS/H) are then discussed in terms of the generalized view. Differences among the three implementations in several modeling situations are described to highlight the need to understand the characteristics of simulation software.

The chapter begins with a discussion in Section 24.2 of the transaction-flow world

Handbook of Simulation, Edited by Jerry Banks.
ISBN 0-471-13403-1

view and the nature of discrete-event simulation, including units of traffic, events, and identical event times. The discussion continues in Section 24.3 with entities, resources, control elements, and operations, and goes on to a summary of model execution in Section 24.4. The generic treatment concludes with the topics of entity states and entity management structures in Sections 24.5 and 24.6. The mechanisms used in SIMAN, ProModel, and GPSS/H to satisfy the logical requirements of discrete-event simulation are described in Section 24.7. The chapter concludes with examples in Section 24.8 of how the differences in implementation of the Section 24.7 software lead to differing outcomes in several modeling situations.

Terms used generically in the chapter are not given special emphasis but terms used by SIMAN, ProModel, and GPSS/H are capitalized. Tables relating generic terms to their software-specific equivalents are included to help distinguish between the generic and the specific.

24.2 ASPECTS OF DISCRETE-EVENT SIMULATION

24.2.1 Transaction-Flow World View

The *transaction-flow world view* often provides the basis for discrete-event simulation.* In this view, a system consists of discrete units of traffic that compete with each other for the use of limited resources while moving ("flowing") from point to point in the system. The units of traffic are sometimes called *transactions*, resulting in the phrase *transaction flow*.

A simple example of units of traffic competing for use of a limited resource is the one-line, one-server system of Figure 24.1, where units of traffic are shown as circles, the square is the server ("resource"), and the circle within the square is a unit of traffic being served. The line of traffic waiting for service is called a *queue*. The combination of the server, the unit of traffic being served, and those waiting to be served is termed a *queuing system*. In practice, the units of traffic might be work in process and the server might be a machine, or the traffic might be print jobs and the server might be a printer, or the traffic might be patients and the server might be a physician, and so on.

Another example of units of traffic making use of limited resources is the one-line, multiple-server system of Figure 24.2. Traffic waiting for service forms one line. The unit at the head of the line goes to the next idle server. Such a system might consist of incoming phone calls and a pool of phone operators at a phone-order business (e.g., J. Crew, L. L. Bean, Lands End), or of customers and a set of bank tellers, or of travelers and check-in clerks at an airline counter.

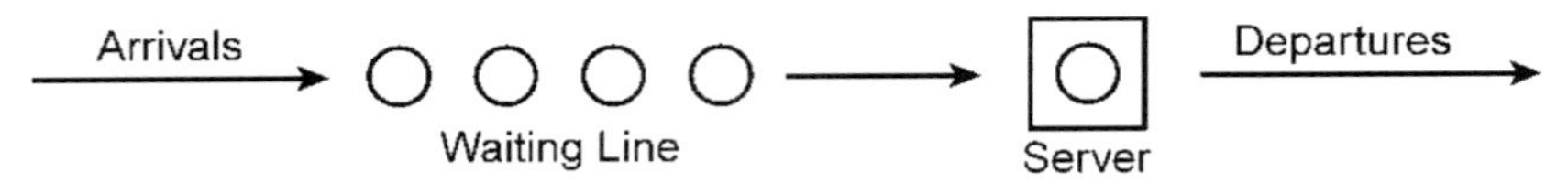

Figure 24.1 One-line, one-server queuing system.

*We estimate that 80% to 90% of current commercial discrete-event simulation software is based on the transaction-flow worldview. Various alternative worldviews are discussed in Balci 1988.

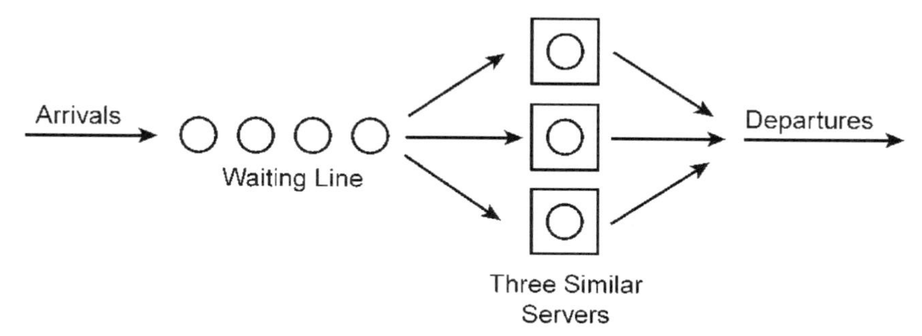

Figure 24.2 One-line, multiple-server queuing system.

Another type of transaction-flow system is the multiple-line, multiple-server system of Figure 24.3. Here there are multiple waiting lines, one per server. A unit waits to be served by the server at the head of its line. Systems of this design include toll-collection points on toll roads, checkout systems in supermarkets, and airports providing two or more takeoff runways (where the units of traffic are planes and the runways are servers).

The types of systems in Figures 24.1 to 24.3 are building blocks for more complex systems. For example, consider the simple harbor system of Figure 24.4. Ships (units of traffic) come to the harbor to load and unload cargo. There are two types of ships: type A and type B. There are three types of servers: tugboats, type A berths, and type B berths. Ships of type A and B can only use berths of type A and B, respectively. Ships use tugboats to move nonstop into the harbor and into a berth, and later use tugboats again to move nonstop out of berth and out of the harbor. The number of tugboats needed by a ship depends on the type of ship and whether the ship is inbound or outbound. This system is characterized by multiple traffic types, multiple resource types, traffic needs that depend on traffic type, static servers (berths), mobile servers (tugboats), and the need for simultaneous control of more than one resource type by units of traffic (e.g., a ship needs to control a berth *and* a tugboat or tugboats before it can move nonstop into the harbor and into a berth). Try to visualize how the building blocks of Figures 24.1 to 24.3 can come into play in the system of Figure 24.4.

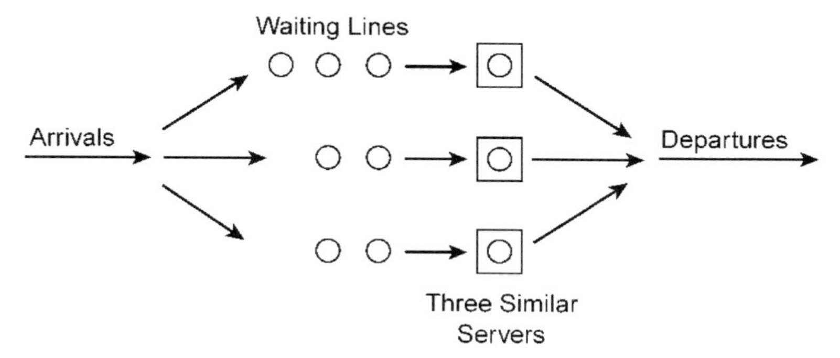

Figure 24.3 Multiple-line, multiple-server system.

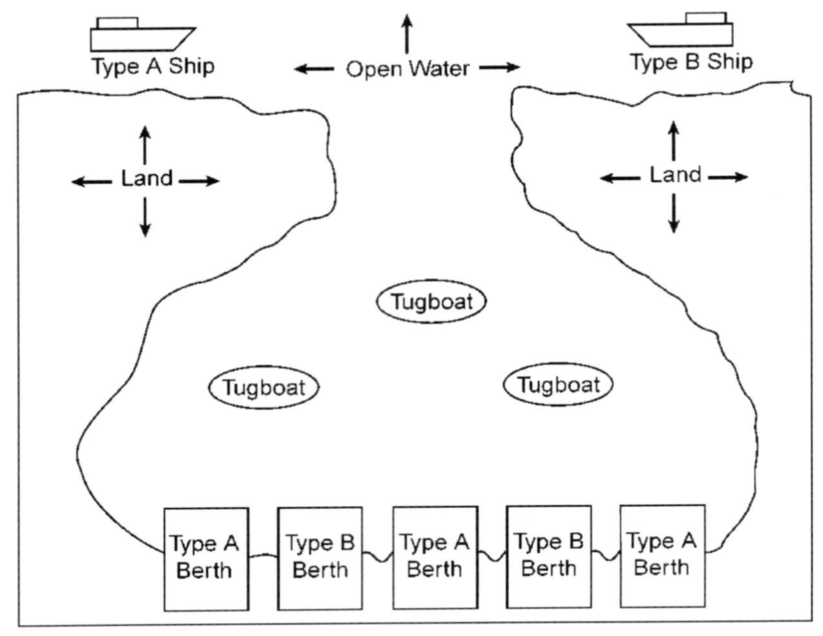

Figure 24.4 Harbor system.

Space is often a limited resource. In one interpretation of the multiple-line, multiple-server system of Figure 24.3, for example, planes are traffic and the servers are runways (space). In the Figure 24.4 harbor system, nothing is said about space. It is implied that the harbor mouth is wide enough, for example, to have multiple ships move through it simultaneously. Suppose, however, that the mouth is wide enough for only one ship at a time. Then space is a limited resource, too, as shown in Figure 24.5. (Space interior to the harbor is probably a limited system resource, too, but this is not shown in Figure 24.5.)

Numerous systems are subject to a transaction-flow interpretation. Included are many manufacturing, health care, transportation, civil, communication, defense and information processing systems, and queuing systems in general.

24.2.2 Nature of Discrete-Event Simulation

A discrete-event system is one in which the state of the system changes at only a discrete, but possibly random, set of time points, known as *event times*. An *event* is a change in system state. For example, suppose that a type A ship arrives outside the harbor mouth in the Figure 24.5 system. The arrival is an event. It occurs at a point in time and changes the state of the system. (The number of type A ships outside the mouth of the harbor increases by one.) Similarly, suppose that a type B ship takes control of ("captures") a tugboat. The capturing of the tugboat is an event, taking place at a point in time and changing the capture status of the tugboat from "idle" to "captured."

A simulated clock records the time points at which events occur in a discrete-event

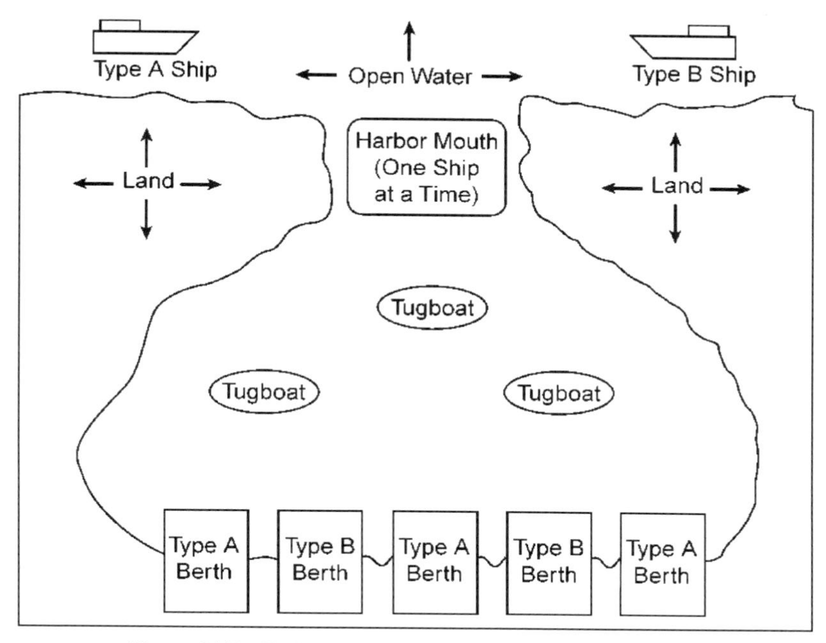

Figure 24.5 Harbor system with space as a limited resource.

simulation. Such a clock is provided by discrete-event simulation software and its value is managed automatically by the software. The clock's value advances during a simulation, only registering the discrete time points at which events occur.

Some aspects of the state of some systems change continuously over time, instead of changing at discrete points in time. For example, suppose that the process of loading cargo onto a ship begins at 1:30 P.M. and continues for four hours, ending at 5:30 P.M. Then the degree of loading completion (which is a state of the system) changes continuously during the 4-hour loading process. The loading process can be modeled in discrete-event terms, however, by focusing on the two discrete events corresponding to the *initiation* and the later *completion* of the process, interposing a simulated time lag (representing the duration of the loading process) between these two events. Using this approach, it is possible to model many continuous state changes in discrete-event terms.

Discussion here is limited to systems in which all changes in system state can be modeled as discrete events.

24.2.3 Units of Traffic, Events, and Identical Event Times

Units of traffic act when system conditions permit or require it. Such action results in one or more changes in system state (events). In a harbor system, an *arrival event* occurs when a ship arrives outside the mouth of the harbor. A *capture event* occurs when a ship captures a tugboat. A *service-initiation event* occurs when the ship initiates the process of being tugged into a berth.

Two or more events often take place at the same time point; that is, they have identical event times. For instance, two events occur at the same time point if a ship captures a tugboat and immediately initiates the process of getting tugged into a berth. Here, action taken by *one* unit of traffic results in a sequence of two events at one time point. The simulation clock remains fixed in value while events with identical event times are carried out one after the other. Real (wall-clock) time goes by while the computer works to update the state of the model at the simulated time point in question.

Some sequences of events occur at increasingly later time points. For example, suppose that the time between consecutive arrivals of ships at a harbor varies at random and always exceeds zero. When a ship arrives, its successor will not arrive until a later simulated time. As another example, if a ship initiates a loading process at one time point, it will not complete the loading process until a later time point has been reached.

Two units of traffic can be involved in multiple events at the same time point. In the one-line, one-server system of Figure 24.1, for example, suppose that a unit of traffic causes a service-completion event when the waiting line is not empty. This sets the stage for another unit of traffic (the next to be served) to cause a capture event at that time and a service-initiation event, too. Here, occurrence of an event caused by one unit of traffic sets the stage for the occurrence of two follow-on events involving another unit of traffic at the same time point.

Now consider a situation in which *three* units of traffic cause multiple events with identical event times. Suppose that a ship in the Figure 24.4 harbor system is using two tugboats to move out of the harbor. When the ship finishes this process it causes a service-completion event, changing the capture status of the two tugboats from "captured" to "idle." If two other ships are each waiting to capture one tugboat, three units of traffic can cause multiple events with identical event times (service completion; then a tugboat capture, perhaps immediately followed by a service initiation; then another tugboat capture, perhaps also immediately followed by a service initiation).

The simulated clock remains fixed in value while events with identical event times are carried out one after the other. The first event occurs, then the second event occurs, and so on. The fact that real time goes by at a fixed simulated time while the computer pays attention to multiple units of traffic, one by one, and carries out multiple events, one after the other, is illustrated in Figure 24.6. The figure corresponds to the scenario described in the preceding paragraph, where three units of traffic cause multiple events with identical event times.

The real-time order in which two or more events occur at a fixed time point is sometimes dictated by logical dependencies. For example, the next in line cannot capture a server until the preceding user has put the server into an "idle" state, so the "free the server" event precedes the next "capture the server" event. Similarly, a ship cannot initiate a berthing process until it has captured a tugboat. Here, logic dictates the event sequences.

The real-time order of events with identical event times is not always dictated by logic. For example, when a ship puts two tugboats into an "idle" state and two ships are each waiting to capture one tugboat, logic does not dictate the real-time order in which the two capture events are to occur. Could the real-time order matter? Yes, it could. Suppose that the two tugboats differ in type (e.g., one is more powerful and faster than the other) and both ships prefer to capture this tugboat. The first ship to act will capture the preferred tugboat, leaving the lesser tugboat to the other ship.

Another example in which two or more events can occur at a fixed time point, but in

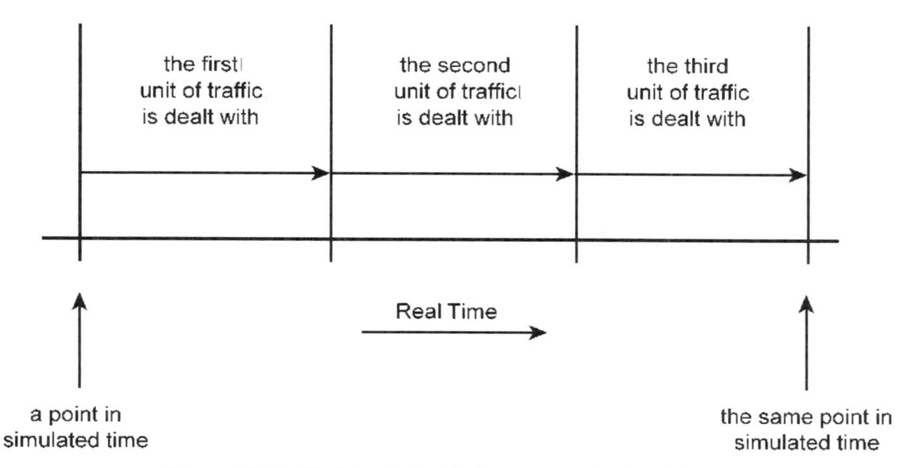

Figure 24.6 Real (wall-clock) time versus simulated time.

an arbitrary real-time order, involves a global change in system state. In the Figure 24.5 harbor system, for example, suppose that a storm is in progress, so that no ships can move out of the harbor. Eventually the storm subsides, so ships can now safely move out of the harbor. If two ships are waiting to leave but only one at a time can move through the harbor mouth, the real-time order in which the two ships take action will determine which moves out of the harbor first and which must wait its turn to move out later.

The preceding discussion involves situations in which sequences of *dependent* events occur at the same time point. It is also possible for *independent* events to occur at the same time point. In the Figure 24.4 harbor, for example, a type A ship might arrive at the harbor at the same time a Type B ship completes a loading process. If the time between arrivals of type A ships varies at random, and if the duration of a loading process varies at random, the probability that the arrival and service-completion events have identical event times is small. If the event times are identical, however, the real-time order in which units of traffic try to take action can matter. In the example at hand, suppose that the type A ship needs one tugboat to move into the harbor and the type B ship needs one tugboat to move out of the berth and harbor. Also suppose that exactly one tugboat is in an "idle" state. If the type A ship acts first and captures the tugboat, the type B ship must wait. Alternatively, if the type B ship acts first and captures the tugboat, the type A ship must wait. Which is it to be? It can be left to chance in the model, or the modeler can determine how the real system operates in this case and then build the model to imitate the real-system behavior accordingly.

The fact that multiple events can occur at a common time point can lead to logical complexities in discrete-event simulation. These complexities must be understood and taken into account both by the *model* designer and, at a higher level, by the *language* designer. The model designer must take the complexities into account in specific modeling contexts, whereas the language designer must do so in a generalized way. Choices and trade-offs exist, especially for the language designer. As a result, although discrete-event simulation languages are similar in broad terms, they can differ in subtle and important particulars.

24.3 ENTITIES, RESOURCES, CONTROL ELEMENTS, AND OPERATIONS

Systems consist in part of entities, resources, control elements, and operations. The following subsections provide particulars.

24.3.1 Entities

The term *entity* is used here as the generic name of a unit of traffic (a "transaction"). Entities model such things as ships in a harbor system, work in process in a manufacturing system, shoppers in a supermarket, planes at an airport, phone calls in a communication system, and so on. As we have seen, entities take actions that change the state of a system.

Modeling languages provide tools used to create and destroy entities during a simulation. Entities come into a model from time to time, just as ships come to a harbor from time to time. Similarly, entities leave a model from time to time, just as served ships leave a harbor. The number of entities in a model usually varies at random during a simulation.

Entities can have attributes. Attributes of a ship at a harbor include arrival time, ship type, the number of tugboats needed for berthing, a distribution of loading times, the number of tugboats needed for deberthing, and so on.

It is useful to distinguish between two types of entities, here referred to as external entities and internal entities. *External entities* are those whose creation and actions are explicitly visualized and arranged for by the modeler. Entities of the types mentioned above (e.g., ships, work in process, phone calls) are examples of external entities.

There often are two or more conceptual classes of external entities in a model, each with its own characteristics and types of attributes. In a manufacturing system, for example, there might be an entity class for orders of a certain kind, an entity class for workers of a certain kind, and so on. (An order might have due date as an attribute, but a worker would not. A worker might have skills as attributes, but an order would not.) In terms of modeling, distinctions among different entity classes might only exist in the mind of the modeler, who then would build a model that is logically consistent with these conceptual distinctions. In contrast, formal definition of various classes of external entities might be required in a model. Whether such formal definition is required would depend on the modeling software being used.

In contrast to external entities, which are highly visible conceptually, *internal entities* are "behind the scenes" entities that some modeling languages use to support various needs in discrete-event modeling. Internal entities are created and manipulated implicitly by the simulation software itself, and in this sense are invisible to the modeler. The model designer might not even be aware that internal entities are at work in a model.

For example, internal entities are used in some modeling languages to trigger machine breakdowns and the later return of the broken-down machines to a working-order state. (The specifications for breakdowns and repair have to be supplied by the model designer, of course, but the designer does not have to provide the logic for implementing the breakdowns if internal entities are used for this purpose.) In contrast, some modeling languages do not use internal entities to model machine breakdowns. In such languages, the model designer works with external entities and explicitly provides the logic needed to implement such breakdowns.

As another example, an internal entity is used in some languages to stop a simulation. A model designer might state that a simulation is to stop at the end of the eighth simulated hour, for example, and the modeling software provides an internal entity to make this happen. (If internal entities are not provided for this purpose, the modeler works with external entities to achieve the effect desired.)

24.3.2 Resources

A *resource* is a system element that provides service. Resources in a harbor include tugboats and berths. Resources in a manufacturing system include machines, machine operators, transportation devices (such as automated guided vehicles and conveyors), and space for temporary storage of work in process and finished goods. Among the resources at an airport are parking spaces for cars, shuttle buses, redcaps, ticket agents, security equipment, walkways, check-in counters, jetways, planes, and runways.

Some resources can only serve one user at a time. For example, a parking space can hold only one car at a time, and a jetway can connect only one plane at a time to a terminal. In some cases, however, a resource can serve two or more users at a time. An automated guided vehicle might be able to move three units of work in process from point A to point B at the same time, and a shuttle bus can move multiple people from a parking lot to an airport terminal.

Resources are limited in number. At a harbor there might be three tugboats, two berths of type A and three berths of type B. An airport might have three runways. There might be 250 spaces in a parking garage. There might be four automated guided vehicles (AGVs) in a manufacturing system.

The users of resources are usually entities. A ship-entity captures a berth and then a tugboat so it can get pulled into the berth. A work-in-process entity captures space in an input buffer feeding the next machine it will use, and then captures an AGV so that it can be moved from its current position to the input buffer. An airline passenger might sequentially use a shuttle bus, a redcap, security equipment, a series of walkways, a check-in counter, a seat, and a jetway.

The fact that resources are limited means that entities must sometimes wait their turn to use resources. When a unit of work in process asks for an AGV, it might have to wait for its request to be filled. When the work in process eventually has been put into the input buffer feeding its next machine, it might have to wait to use the machine.

Modeling languages have constructs that are used to control direct access to resources by entities. Such constructs provide for the automatic recording of the resource's capture status ("idle" or "captured") and operating condition ("in working order" or "broken down"). When an entity tries to capture a resource, its capture status and operating condition can be tested by the software to determine if the attempted capture can take place.

24.3.3 Control Elements

In addition to resource constructs, modeling languages provide other constructs to support various control-related aspects of a system's state. The term *control element* is used here for such constructs. A switch is an example of a control element. A switch is a two-state variable (on or off). A switch might be used in a model of a harbor system, for example, to signal whether a storm is currently in progress. (If it is storming, ships might be forced to wait in the harbor until the storm is over.) In a banking context, a switch might be used to indicate whether the doors into the bank are locked.

Counters are another type of control element. A counter might be used to count the number of engine blocks that have had holes drilled in them since the last time the drill bits were changed in a drilling machine. The policy might be to replace the drill bits after 100 uses. Implementing this policy requires that a count be kept. The counter is used to help control this aspect of system behavior.

Arithmetic expressions can be the basis for control elements. Consider a supermarket that uses a multiple-line, multiple-server system for its checkout lanes. There might be 12 checkout lanes, but only several might be open at a given time. Suppose that if the average number of customers waiting in checkout lanes is five or more, another checkout lane will be opened. The modeler can introduce an arithmetic expression to compute the average number of customers waiting in checkout lanes. An entity used to simulate the "lane manager" can monitor the value of this expression to determine if conditions require opening up another checkout lane. Here an arithmetic expression is used to control the behavior of the lane manager.

Boolean expressions (truth-valued expressions composed with boolean operators such as *and*, *or*, and *not*) can also be used as control elements. A ship might not be able to leave a harbor, for example, until it can capture either one powerful tugboat *or* two less powerful tugboats, *and* there is no storm in progress.

Like resources, control elements can force entities to wait, delaying their movement through a system. The ways in which modeling software manages delayed entities are discussed in Section 24.6.

24.3.4 Operations

An *operation* is a step or action carried out by or on an entity during its movement through a system. Examples of operations include the arrival of an order in an order-processing system, the capturing of an AGV by a unit of work in process, the x-raying of a patient's broken arm, and the transfer of a unit of finished goods to finished-goods inventory.

An ordered set of operations is a sequence of steps or actions taken or experienced by an entity while it moves from point to point in a system. An integrated sequence of operations is sometimes called *operation logic*. For example, this might be the operation logic for movement of a ship through a harbor: arrive outside the harbor; capture a berth; capture two tugboats; use the tugboats to get pulled nonstop into the harbor and into the berth; free the tugboats; use the berth to load or unload cargo; capture one tugboat; use the tugboat to get pulled nonstop out of the berth and out of the harbor; depart.

24.4 OVERVIEW OF MODEL EXECUTION

24.4.1 Projects, Experiments, and Replications

Conducting a simulation project involves carrying out one or more experiments and, within each experiment, performing one or more replications (trials). This pattern is shown in Figure 24.7, where m experiments are indicated, each consisting of n replications. Experiments are differentiated by alternatives in a model's logic and/or data. Replications are (usually) differentiated by using different sets of random numbers from replication to replication and across experiments.

For example, suppose that a simulation project is conducted for the harbor system

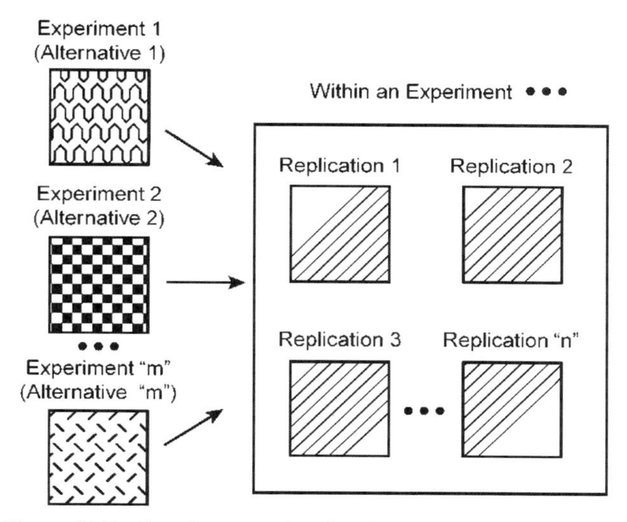

Figure 24.7 Experiments and replications in a simulation project.

of Figure 24.5. The purpose of the project, let us assume, is to study alternatives for decreasing the delay experienced by ships at the existing harbor. Assume that the system shown in Figure 24.5 describes the harbor "as is" (e.g., three tugboats, three type A berths, two type B berths, and a harbor mouth only wide enough to accommodate one ship at a time), and operation data are available (e.g., distributions of interarrival times, berthing and deberthing times, and loading and unloading times for type A and type B ships). Suppose that the service order for use of tugboats and berths is first come, first served. Experiment 1 in Figure 24.7 might be with a model of the harbor as is. (The purpose of experiment 1 might be to validate the model by comparing its characteristics to characteristics observed in the real system. For example, the distributions of harbor residence times for type A and type B ships in the model might be compared to those in the real system.)

Experiment 2 of Figure 24.7 might study the effect (on the distributions of harbor residence times) of widening the harbor mouth to accommodate two ships at a time. Experiment 3 might study the effect of providing one additional tugboat. Experiment 4 might study the combined effect of widening the harbor mouth *and* providing one additional tugboat. Experiment 5 might study the effect of giving type A ships higher priority for tugboat use than type B ships. Experiment 6 might study the effect of replacing the three tugboats with three higher-speed tugboats. And so on.

As suggested in Figure 24.7, each experiment consists of one or more *replications* (*runs*). A replication is execution of a simulation model that incorporates the model logic and data for the experiment but uses a set of random numbers unique to that replication. A replication produces statistical results differing from those produced by other replications. The statistical results can then be analyzed across the set of replications.

In some experimental designs involving variance reduction, the set of random numbers used in a replication might be deliberately correlated with those used in a companion replication. Furthermore, in some designs only one relatively long replication might be performed.

24.4.2 Anatomy of a Replication

The phases within a replication are discussed in this subsection. For the sake of concreteness, the broad considerations are interpreted in the context of the harbor system of Figure 24.4.

Initialization Phase. The Figure 24.8 flowchart shows the several phases making up a replication. The replication begins with an initialization phase (box 1), during which the simulation clock has (we assume) an initial value of 0.0. Simulated times will then be expressed relative to this starting value. For example, time 0.0 might correspond to real-system time 9:00 A.M. on the first day in a series of simulated days making up a replication.

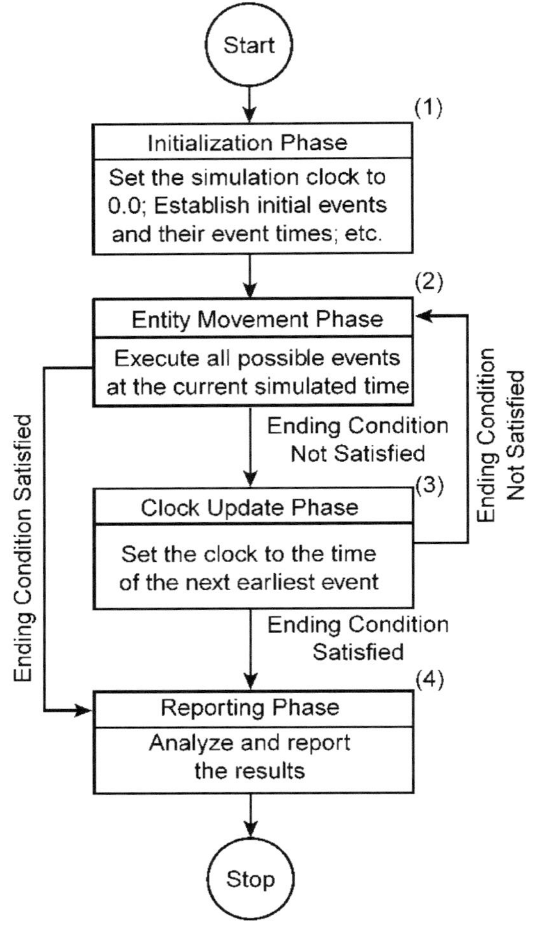

Figure 24.8 Anatomy of a replication.

At the start of the initialization phase, no entities yet exist. During initialization, one or more external entities are created and their immediate and/or eventual arrival at the model is planned, as indicated in Figure 24.3 (box 1). In the harbor system of Figure 24.4, for example, the initial ship of type A to arrive at the harbor would be created during the initialization phase and its eventual time of arrival, such as simulated time 23.5 minutes, would be determined. The determination is made by drawing a sample from the interarrival time distribution of type A ships. Relative to the initial clock value of 0.0, this sampled value is the simulated time at which the first type A ship will eventually arrive. (For convenience, we assume that the clock units are minutes, but the unit of time can be whatever the modeler wants it to be.) During initialization the initial type B ship to arrive at the harbor would also be created and its eventual time of arrival (let us say at simulated time 18.2 minutes) determined.

The preceding discussion is for the initialization-phase creation of *external* entities. If the model involves *internal* entities too, any needed initialization takes place for them during the initialization phase, too. For example, suppose that the Figure 24.4 harbor model is built with software that uses an internal entity to control the duration of a simulation and the replication is to run until time 43,200 (the number of minutes in thirty 24-hour days). Then an internal entity would be created during the initialization phase accordingly.

Entity Movement Phases. As shown in Figure 24.8, an entity movement phase follows the initialization phase (from box 1 to box 2). Additional entity movement phases then also take place eventually, after clock update phases (from box 3 back to box 2). The purpose of an entity movement phase is to have all qualifying entities take whatever action they can at the current simulated time. An entity qualifies to try to take action if its planned action time equals the current simulated time.

The word *movement* is used in "*entity movement phase*" because in the transaction-flow world view, units of traffic are visualized as moving from point to point in a system. It is during the entity movement phase that entities carry out actions, some of which simulate the movement of units of traffic in a system. For example, a type B ship, having arrived at a harbor, acts to capture a tugboat (either immediately, or as soon as possible) and then moves into the harbor. Action and movement are closely coupled. Such phrases as *action time*, *move time*, and *event time* are often used interchangeably.

Initial Entity Movement Phase. The entity movement phase is initially entered from the initialization phase at simulated time 0.0. If no initialization entities (either external or internal) have a planned action time of 0.0, no actions take place during this EMP. In the harbor model, for example, we have assumed that the arrival times of the initial type A and type B ships are 23.5 and 18.2 minutes, respectively. In this example there are no actions to be taken by external entities during the initial entity movement phase.

More generally, it might be necessary to take action during the initial EMP. For example, if a bank opens for business at 9:00 A.M. (simulated time 0.0), there might be three customer-entities waiting for the door into the bank to be unlocked. These three entities would then each act one by one during the initial entity movement phase.

Subsequent Entity Movement Phases. Action always takes place during the entity movement phases that follow clock update phases (from box 3 back to box 2, Figure 24.8). This is because a clock update phase (see below) always brings the state of the model to the next earliest simulated time at which at least one entity is scheduled to act.

In the harbor system we have assumed that the initial type B ship arrives at time 18.2. After the entity movement phase at time 0.0 (during which there is no action), the clock update phase sets the time to 18.2 (the next earliest time for which action has been scheduled; the initial type B ship has been scheduled to arrive later, at time 23.5). Then the subsequent entity movement phase takes place and the initial type B ship acts to capture an idle berth, capture an idle tugboat, and initiate the process of being moved by the tugboat into the berth.

Clock Update Phases. After all possible actions have been carried out during an entity movement phase, a clock update phase (CUP) takes place (from box 2 to box 3, Figure 24.8). The purpose of the CUP is to set the clock to the next earliest simulated time at which one or more actions have been scheduled. This *next earliest simulated time* might be the current simulated time, depending on whether two or more entities have been scheduled to act at the same simulated time *and* depending also on implementation choices made by language designers. Alternatively, this next earliest simulated time might be a *later* point in simulated time. (The simulated clock never *decreases* in value. Discrete-event simulations are not designed to make it possible to go *backward* in simulated time.)

After a clock update phase is finished, the entity movement phase is performed again (from box 3 back to box 2 in Figure 24.8) to give all qualifying entities the opportunity to act at the current (newly established) simulated time. Then the clock update phase takes place again, then the next entity movement phase takes place, and so on.

The heart of a replication is this alternating execution of the entity movement and clock update phases. It is during the entity movement phase that simulated time remains fixed and real time elapses while the state of the model is updated (see Figure 24.6). An understanding of the complexities of the entity movement phase supports the modeler in expressing subtle system logic correctly and making effective use of tools provided by software to help troubleshoot and verify models.

Statistics Gathering. A goal in discrete-event simulation is to collect statistical observations about the behavior of the system being simulated. Discrete-event simulation software is designed to gather many types of statistics automatically during a replication. For resources, such things as capture counts, average holding time per capture, and utilization might be measured automatically. For waiting lines, such things as average content, average time in line, and maximum line length might be measured automatically. In addition to gathering such statistics automatically, simulation software also typically provides tools the modeler can use to gather customized statistics.

The observing and recording of statistics typically takes place during entity movement phases (box 2 in Figure 24.8), both for statistics gathered automatically by the software and for the gathering of customized statistics specified by the user. The resulting information is then used to produce reports at the end of the replication.

End of a Replication. When a replication takes place, a run-ending condition eventually occurs during either an entity movement phase or a clock update phase. The ending condition might be time-based (e.g., 24 hours of harbor operation have been simulated) or count-based (e.g., 500 units of product have been manufactured), or can be more complex (e.g., the doors into a bank have been locked at the end of the afternoon *and* all customers already in the bank when the doors were locked have been served). The ending condition can occur during either an entity movement phase (box 2 to box 4, Figure 24.8) or a clock update phase (box 3 to box 4).

A reporting phase is then carried out (box 4), completing the replication. The state of the model at the conclusion of the replication might be described, indicating such things as the value of the simulated clock, the number of entities of various types that were brought into the model, and the number of entities currently in the model. Statistical aspects of the replication are typically summarized in the form of reports, including resource and waiting line statistics.

24.5 ENTITY STATES

An entity is created at a point in simulated time, works its way through a model while simulated time advances, and then is destroyed. During its life cycle the entity migrates from state to state, usually passing through various states multiple times before it is destroyed. (There is no requirement, however, that an entity must eventually be destroyed; some entities might loop through part of a model repeatedly, but with intermittent pauses, as part of the model design.)

There are five alternative entity states. These states, termed the *ready state*, *active state*, *time-delayed state*, *condition-delayed state*, and *dormant state*, are shown in Figure 24.9. Also shown in the figure are points of entity creation and destruction, as well as the paths along which entities migrate from state to state. The paths are numbered to support the following discussion. Figure 24.9 will be discussed "inside out." That is, the discussion begins with the active state, then goes on to the ready state and the time- and condition-delayed states and the dormant state. Entity creation and destruction are also discussed.

24.5.1 Active State

There can only be one moving entity at any moment of wall-clock time. The active state is the state of the currently moving entity. The active-state entity (active entity) moves nonstop until it is delayed, or destroyed, or chooses to yield the active state to some other entity before reentering the active state itself at the same simulated time. No simulated time goes by while an entity is in the active state.

If the active entity is delayed, it migrates from the active state to one of three alternative delay states: the time-delayed state (path 5 in Figure 24.9); the condition-delayed state (path 7), or the dormant state (path 9). The roles played by these delay states are discussed in Sections 24.5.3, 24.5.4, and 24.5.5. If the active entity makes a move that results in its destruction (path 11 in Figure 24.9), it will be removed immediately from the model and ceases to exist.

In some models the active entity might choose to leave the active state temporarily, with the understanding that it will later reenter the active state at the same simulated time. The objective in such a case is to let one or more *other* entities pass through the active state to accomplish one or more tasks before the initiating entity again becomes active itself at the same simulated time. The initiating entity accomplishes this temporary yielding of the active state by migrating from the active state back to the ready state (path 4, Figure 24.9).

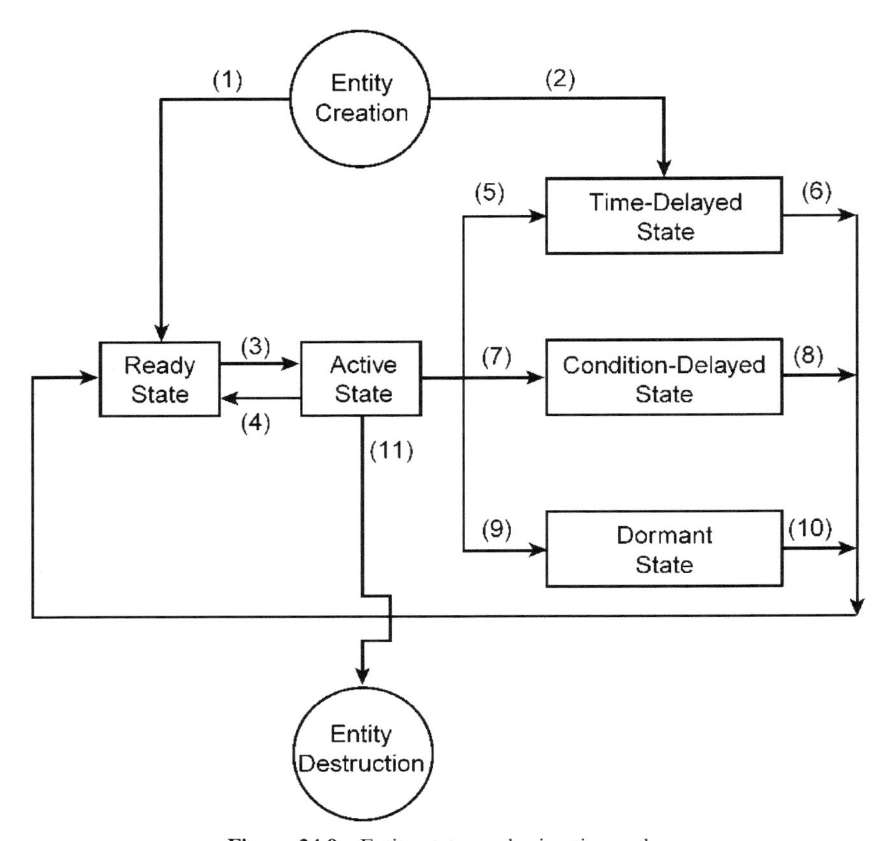

Figure 24.9 Entity states and migration paths.

24.5.2 Ready State

At a given simulated time there might be more than one entity ready to move, but only one entity at a time can be in the active state. If two or more entities want to move at a freshly established simulated time, all but one must wait their turn to enter the active state. The ready state is the state of those entities waiting to enter the active state at the current simulated time. No simulated time goes by while an entity is in the ready state.

There are several paths in Figure 24.9 along which entities can migrate into the ready state. Migration from the time-delayed, condition-delayed, and dormant states is always into the ready state (paths 6, 8, and 10, respectively), for example. Among these possibilities, migration from the time- and condition-delayed states to the ready state occurs most frequently.

The path to the ready state from the dormant state (path 10) is less frequently traveled, because models do not always make use of the dormant state. The path from entity creation to the ready state (path 1) is followed only by entities that are ready to move at the simulated time of their creation (rather than first being ready to move at some later simulated time). The path from the active state to the ready state (path 4) is least frequently followed. Most models do not require use of this migration path. (Some discrete-event simulation software does not even provide this path.)

24.5.3 Time-Delayed State

The time-delayed state is the state of entities waiting for a *known* future simulated time to be reached so that they can then (re)enter the ready state as a prelude to moving again. Entities migrate into the time-delayed state from the active state (path 5, Figure 24.9) and from the point of entity creation (path 2). Later they migrate from the time-delayed state to the ready state (path 6), when the known simulated time for which they have been waiting is reached.

Consider an example of the interplay among the ready, active, and time-delayed states. Suppose that an entity simulates a unit of work in process that needs to have a hole drilled in it, and suppose that when the entity becomes ready for this operation, the drill is idle. At the simulated time in question, the entity migrates from the ready state to the active state (path 3, Figure 24.9), captures the drill, and then migrates to the time-delayed state (path 5), remaining in that state while waiting for the known future simulated time at which the drilling operation will end. (The entity "knows" the applicable future simulated time because it samples the drilling time from the drilling-time distribution when it migrates into the time-delayed state.)

Extending the example, now suppose that the simulation has reached the simulated time when the drilling operation ends. The entity then migrates from the time-delayed state to the ready state (path 6, Figure 24.9) as a prelude to moving. The entity then migrates from the ready state to the active state and acts to change the capture status of the drill from "captured" to "idle." Suppose that the entity next needs to be transferred by an AGV to its next destination. Also suppose that the AGV is idle. Continuing nonstop in the active state, the entity captures the AGV and then migrates back into the time-delayed state (path 5) while waiting for the known (sampled) future simulated time when the transport operation will end.

When an entity is created, it starts its existence in the ready state or in the time-delayed state, depending on whether the simulated time when it will first move equals or exceeds the simulated time of its creation. If it first moves at the time of its creation, the entity starts in the ready state (path 1, Figure 24.9); otherwise, it starts in the time-delayed state (path 2). Entities usually start their existence in the time-delayed state.

Consider an example of entity creation. Assume that the time between arrivals of phone calls to an information operator is exponentially distributed, and an entity simulates a call. When a call arrives, the future time the *next* call will arrive is determined by sampling from the exponential intercall-time distribution. A corresponding call-entity is created and put into the time-delayed state (path 2, Figure 24.9), waiting for the known future simulated time at which it will first move ("arrive," causing the phone to ring).

For a second example of entity creation, suppose that when a bank opens at 9:00 A.M. (simulated time 0.0), three customers are waiting for the door to be unlocked. Each of these customers will move into the bank when the door is unlocked. (In reality, several seconds would go by while the customers enter the bank, one by one; but as a first approximation, let us assume that all three enter instantaneously.) At simulated time 0.0, the first customer-entity is created and put into the ready state (path 1, Figure 24.9). The second customer-entity will also be created at time 0.0 and put into the ready state. Finally, the third customer-entity will also be created at time 0.0 and put into the ready state. During the entity movement phase at time 0.0, each of these customers will migrate serially from the ready state into the active state and then take action.

24.5.4 Condition-Delayed State

The condition-delayed state is the state of entities whose movement is currently hindered because of some condition. Such entities wait in the condition-delayed state until their delay condition has been resolved. While condition-delayed, entities do not know *when* the delay condition will be resolved, so they are waiting for an *unknown* simulated time to be reached. (Contrast the condition-delayed state with that of the time-delayed state, in which entities wait for a *known* simulated time to be reached.)

As an example of the condition-delayed state, consider a unit of work in process (WIP) that needs to have a hole drilled in it. Assume that when the WIP-entity is ready for this operation, the drill is not idle. (The drill is being used by some other unit of work in process.) The active WIP-entity attempts to capture the drill but finds itself hindered. Assuming that it has no other choice, it then migrates to the condition-delayed state (path 7, Figure 24.9) and starts to wait its turn to use the drill. (There might be other WIP-entities waiting to use the drill, too.) The WIP-entity remains in the condition-delayed state until the drill becomes idle *and* it is this WIP-entity's turn to use the drill. At that time it will migrate to the ready state (path 8), then into the active state (path 3), and then will capture the drill and migrate into the time-delayed state (path 5).

As suggested by the preceding example, a single event might be all that is needed to resolve delay in some cases. These are cases of *simple* delay. If one or more units of work in process are delayed, waiting their turn to use a drill, then when the current drill-user puts the drill into an "idle" state, this single event resolves the delay for the next in line. It is easy to detect the resolution of simple delay. Migration from the condition-delayed state to the ready state can easily be related to the delay-resolving event in such cases.

For an example of a more complex condition-delayed state, consider a ship-entity that having loaded cargo at a harbor, is ready to move out of the harbor. Suppose that two conditions must be in effect for the ship to move out of the harbor: (1) the ship needs a tugboat, and (2) the weather must be calm. Still in the active state after causing the loading-completion event, the ship-entity checks to see if there is an idle tugboat *and* if the weather is calm. (An on-or-off switch might be used as a control element to signal the state of the weather. A boolean expression might be used as a control element to test for an idle tugboat *and* calm weather.) If this compound condition is satisfied, the ship-entity migrates from the active state to the time-delayed state (path 5, Figure 24.9) for the known (sampled) interval of simulated time required to move out of the harbor; otherwise, it migrates from the active state to the condition-delayed state (path 7) and starts to wait for the unknown simulated time at which the two conditions will be satisfied simultaneously.

In this case we have an example of *complex* delay. Resolution of complex delay cannot be related to the occurrence of a single event. When a tugboat becomes idle, there is no guarantee that the weather will be calm. When the weather becomes calm, there is no guarantee there will be an idle tugboat. It is more challenging to detect the resolution of complex delay than of simple delay.

The mechanisms used in discrete-event simulation software to cause an entity to migrate from the condition-delayed state to the ready state will be considered in Section 24.6.5. For now it is enough to point out that like migration from the time-delayed state, migration from the condition-delayed state takes place *automatically* during a simulation; that is, discrete-event simulation software is designed to be intelligent enough to make this migration take place when conditions permit.

24.5.5. Dormant State

Like the time-delayed and condition-delayed states, the dormant state is one in which entities are put into suspension for an interval of simulated time. Entities in the dormant state are *managed by the modeler*, however, instead of being *managed automatically by the software*. Particulars follow.

The ready, active, time-delayed and condition-delayed states are all managed according to rules built into modeling software. For example, qualifying entities are automatically transferred at the right simulated time into the time-delayed state and then into the ready state.

Similarly, qualifying entities are automatically transferred into the condition-delayed state and then eventually from there to the ready state when model conditions permit. The modeler is provided with some flexibility in this regard, but only on a limited basis. Consider a situation, for example, in which five entities are in a condition-delayed state, waiting for a machine. When the machine becomes idle, *which* of the five waiting entities will be the next to use it? That is, what service order is in effect? The modeler can typically pick from a short list of alternative service orders in this regard (e.g., simple first come, first served; or first come, first served based on a prioritized ranking; or last come, first served).

Sometimes more flexibility than provided by the condition-delayed state is required to model complicated situations. In these cases the dormant state often can be used to achieve the required entity behavior in a model. The key to the usefulness of the dormant state is that the *modeler* specifies *when*, *which*, and *how many* entities will migrate from the dormant state to the ready state (path 10, Figure 24.9) at appropriate times in a simulation.

For an example of dormant-state use, consider a manufacturing system in which "least remaining slack" is the service order used to determine which waiting job will be the next to use a machine. Remaining slack is a measure based on the relationship among a job's due date, the time now (i.e., the time at which remaining slack is being computed), and the job's remaining operation time. (Remaining operation time is the estimated remaining total processing time required by the job before it will leave the system as a finished job.) Remaining slack is computed this way:

$$\text{remaining slack} = (\text{due date}) - (\text{time now}) - (\text{remaining operation time})$$

The smaller a job's remaining slack, the more urgent it is to give the job preferential treatment in an attempt to complete the job by its due date. Note that remaining slack changes (decreases) while simulated time (time now) goes by. A job's remaining slack therefore is not computed when a job *arrives* at a machine; instead, the remaining slack is computed *later*, when the time has come (time now) for the machine to start another job. While waiting for that time to come, delayed jobs can be kept in a dormant state. When the time comes for the machine to start another job, the remaining slack for each waiting dormant-state job can be computed and the job with the least remaining slack can be transferred from the dormant state to the ready state, poised to capture the machine.

TABLE 24.1 Alternative Entity States

State	Description and Comments
Active	The active state is the state of the currently moving entity.
Ready	The ready state is the state of entities waiting to enter the active state at the current simulated time.
Time-delayed	The time-delayed state is the state of entities waiting for a *known* simulated time to be reached so that they can then move into the ready state.
Condition-delayed	The condition-delayed state is the state of entities waiting for the *unknown* simulated time when the condition causing their delay will be resolved. When conditions permit, such entities will be transferred *automatically* from the condition-delayed state to the ready state *using rules built into the software.*
Dormant	The dormant state is the state of entities waiting for the *unknown* simulated time when the condition causing their delay will be resolved. When conditions permit, such entities will be transferred from the dormant state to the ready state *using logic provided by the modeler.*

24.5.6 Summary of the Entity States

The various entity states are summarized with a brief description and comments in Table 24.1.

24.6 ENTITY MANAGEMENT

Simulation software manages entities by organizing them in linear lists. Each list corresponds to an entity state and is ordered, with some entity at the top of the list, another entity behind it, and so on, down to the bottom of the list. A generalized entity list is pictured in Figure 24.10, where each rectangle represents an entity.

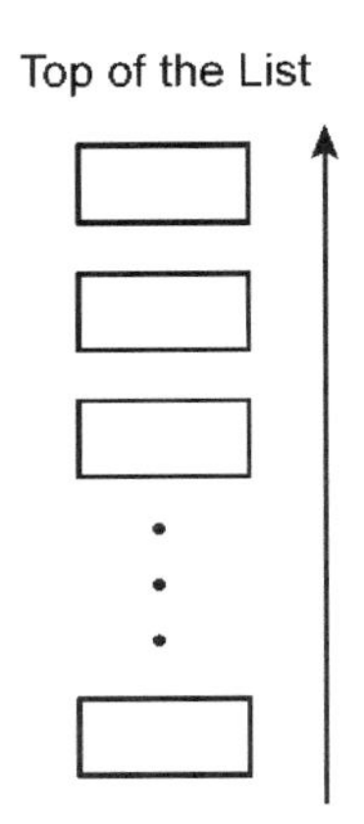

Figure 24.10 Entity list.

The ordering of entity lists raises questions about the rules used to determine the order and the role order plays in list management. When an entity is inserted into a list, how is its insertion point determined? When an entity is removed from a list, from which position is the entity taken? Is it possible for entities to initiate changes in their relative position in a list? These and other details of the five types of entity lists are described in the following subsections.

24.6.1 Active-Entity List

There can only be one active-state entity, so this entity occupies a list of length one. This "list" is not given a name here. The active entity moves nonstop at the current simulated time until it migrates to another state (transfers to another list) or is destroyed.

24.6.2 Current Events List

Entities in the ready state form a single list named the current events list (CEL). This name reflects that ready-state entities are poised to move at the current simulated time and that their movement will cause events (such as arrivals, resource captures, and departures) to occur in the model.

Various rules can be used to determine the insertion point for entities being put into the current events list. In some cases, entities migrating to the CEL might be put at the bottom of the list. In other cases, entities might be arranged on the CEL in order of decreasing priority, where priority is an entity attribute. Priority ties might be resolved by inserting the newcomer into the list below those with matching priority. In yet other cases, entities might be put at the top of the CEL. These alternative approaches can influence the real-time order in which ready-state entities will become active, and so have important implications for the designer of simulation models. When the time comes to transfer an entity from the ready state to the active state, the entity is typically taken from the top of the current events list.

24.6.3 Future Events List

Time-delayed entities form a single list called the *future events list* (FEL). This name reflects that these entities will not try to move again until some (known) future simulated time is reached. The future events list is typically ranked top-down in order of increasing entity move time. (An entity's move time is the simulated time at which the entity will attempt to make its next move or set of moves in a model.) When an entity is inserted into the FEL, its move time is calculated by adding the simulated clock time to the known (sampled) duration of the time-based delay. Move-time ties are typically resolved "first in, first out."

When a clock update phase occurs, the next earliest simulated time at which new movement will take place in the model equals the move time of the entity at the top of the future events list. The clock update phase sets the simulation time to this entity's move-time value, then removes the entity from the FEL and inserts it into the current events list. When a clock update phase starts, two or more entities at the top of the future events list might have identical move times. The response of modeling software to such move-time ties is implementation dependent. One approach is to transfer each time-tied entity from the FEL to the current events list during a single clock update phase.

Another approach is to take a "one entity transfer per clock update phase" approach. (This means that in case of time ties, two or more consecutive clock update phases and entity movement phases will take place at the same point in simulated time.)

Languages that work with internal entities often use the future events list to support the timing requirements of these entities. The FEL is then typically composed both of external and internal entities in such languages. During a discrete-event simulation, the software often inserts entities into the future events list. If the list is long, finding the insertion point can take relatively large amounts of computer time. This motivates language designers to develop efficient algorithms to find insertion points in lists. As a result, some modeling software works with future events lists that are not linear, but instead, involve other types of data structures. Nevertheless, for all practical purposes we can visualize the future events list as being linear.

24.6.4 Delay Lists

Delay lists are composed of entities in the condition-delayed state. These entities are waiting for delay-inducing conditions to be resolved so that they can then be transferred automatically into ready state on the current events list. There can be many delay lists in a model, one (or more) for each delay-inducing condition. This contrasts with the existence of only one active-state "list," current events list, and future events list. In the harbor system of Figure 24.4, for example, there could be a delay list consisting of type A ships waiting for type A berths; another list of type B ships waiting for type B berths; and another list composed of type A and type B ships waiting for a tugboat or tugboats.

When an entity is inserted into a delay list, its list position is determined using a ranking criterion chosen by the model builder. There are usually several choices in this regard. Delay lists can typically be ranked first in, first out; or ranked last in, first out; or ranked ascending or descending on a user-specified entity attribute or arithmetic expression. Each alternative ranking has implications, of course, for the order in which waiting entities will eventually be removed from the delay list and put into ready state on the current events list.

When a delay-inducing condition is resolved, the most highly ranked entity on the applicable delay list will be transferred automatically by the software into ready state on the current events list. For example, when a type A berth in the Figure 24.4 harbor system becomes idle, the type A ship-entity at the top of the associated delay list will be removed and put into ready state on the current events list as a prelude to its taking control of the type A berth.

Not all entities on a delay list are necessarily candidates at all times for transfer from the condition-delayed state to the ready state. Whether they are is implementation dependent. A language designer might decide, for example, that the only current contender for transfer from a delay list is the *queue leader*, that is, the entity at the top of the delay list. In such cases, other entities on the delay list will not be considered for transfer from the list until they have become the queue leader. In contrast, a language designer might decide that all entities on a delay list are candidates at all times for transfer from the condition-delayed state to the ready state. The approach taken can lead to different outcomes in cases in which delay has not yet been resolved for the queue leader, but has been resolved for one or more other entities behind the queue leader on the delay list. (See Section 24.8.3 for a specific example.)

24.6.5 Related Waiting and Polled Waiting

Two approaches can be used by simulation software to provide for the automatic removal of entities from delay lists. If delay can easily be related to a single event that eliminates the delay, a *related waiting* approach can be used to manage the delay list. For example, suppose that a machine's capture status changes from "captured" to "idle." In direct response to this change in machine status, the software can remove the highest-ranked waiting entity from the applicable delay list and put it in ready state on the current events list. (This entity will be the next to use the machine.)

In some circumstances, the resolution of delay might require that two or more particular events occur in a model. The occurrence of one of these events doesn't necessarily mean that the delay has been resolved. The conditions needed to justify transfer of an entity from a delay list to the ready state therefore cannot simply be related to a single-event occurrence. In these cases a *polled waiting* approach can be used to manage delayed entities. In polled waiting, an entity isn't transferred from a delay list in *direct* response to the occurrence of a single event; instead, the software *eventually* checks (at a later wall-clock time, but at the same simulated time) to see if a combination of circumstances has come about that justifies the transfer of one or more delayed entities from the delayed state to ready state on the current events list.

For an example of polled waiting, consider a harbor in which a ship can't move into a berth until a berth is idle and a tugboat is idle as well. (Assume that the harbor is managed in such a way that the ship doesn't first claim an idle berth and then ask for a tugboat, but instead, waits until both of these resources are simultaneously idle before claiming either of them.) Note that a boolean condition is implied here: "wait until a berth is idle *and* a tugboat is idle." Change of a berth's status to idle doesn't necessarily eliminate delay. Change of a tugboat's status to idle doesn't necessarily eliminate delay, either. Resolution of delay can't be related exclusively to either of these single status changes. Polled waiting can be used, however, to determine whether the delay has been resolved. The polling can be carried out routinely and automatically by the software at some point during each entity movement phase (such as when the ongoing entity movement phase is coming to an end).

24.6.6 User-Managed Lists

User-managed lists are composed of entities in the dormant state. Like delay lists, there can be many user-managed lists in a model. In contrast to delay lists and the current and future events lists, however, which are created and managed automatically by the software, the modeler himself or herself must arrange for the creation of user-managed lists and supply the logic needed to insert entities into and remove entities from these lists.

Entities decide whether to put themselves onto a user-managed list. They conduct user-designed tests to make this decision, conducting the tests while they are in the active state. In the context of the "least remaining slack" service order discussed in Section 24.5.5, for example, when a job entity's next operation is to use the machine, it can conduct a test to determine if the machine is idle. The job entity can capture the machine if it is idle or can migrate from the active state to the dormant state in a user-managed list if the machine is in a state of capture.

The modeler typically has choices when specifying an insertion point for a user-managed list. The choices are implementation dependent. Alternatives might include

TABLE 24.2 Entity States and the Associated Entity Lists

Entity State	Generic Name of Entity List(s)	Comments
Active	None	There is a maximum of one active entity.
Ready	Current events list	There is only one such list.
Time-delayed	Future events list	There is only one such list.
Condition-delayed	Delay list	There are potentially many such lists.
Dormant	User-managed list	There are potentially many such lists.

inserting an entity at the bottom of the list, at the top of the list, or into the list ranked ascending or descending on a user-specified entity attribute.

Entities on a user-managed list cannot cause themselves to be removed from the list, and the software will not automatically remove entities from the list, either. Removal of an entity from a user-managed list takes place when *some other entity* acts to make this happen. The action is based on a modeler-supplied test this other entity conducts to determine if it is appropriate to transfer one or more entities from a specified user-managed list to ready state on the current events list.

The modeler has flexibility in choosing which entity or entities to remove from a user-managed list. The simplest choices are to remove entities from the top or the bottom of the list. A more elaborate choice is to scan the list top down, repeatedly evaluating a user-defined boolean expression, entity by entity, removing the entity or entities for which the boolean expression is true. The value of the boolean expression would depend on one or more attributes of the entity being evaluated. The choices a modeler has for imposing removal conditions on entities in the dormant state are implementation dependent.

24.6.7 Summary of the Lists Used for Entity Management

The various entity states and the generic names given here to the lists used to manage entities in those states are summarized with brief comments in Table 24.2.

24.7 IMPLEMENTATION IN THREE INSTANCES OF SIMULATION SOFTWARE

The discrete-event simulation software whose implementation particulars will be reviewed here are Systems Modeling Corporation's SIMAN V, which is the simulation language within ARENA (Banks et al., 1995b; Pegden et al., 1995); ProModel Corporation's ProModel version 3.0 (Benson 1997; ProModel Corporation, 1996; and Wolverine Software Corporation's GPSS/H, release 3 (Banks et al., 1995a; Crain, 1997; Henriksen and Crain, 1998; Schriber, 1991). SIMAN and GPSS/H are general purpose, whereas ProModel is oriented toward manufacturing applications.

SIMAN, ProModel, and GPSS/H are among about 50 tools for discrete-event simulation reported in a 1997 survey (Swain, 1997). Some of the other tools might be better suited than any of these three for particular modeling activities. The choice of these three is based on the belief that they are reasonably representative. Furthermore, there are some interesting contrasts in the underlying approaches used by these three tools to manage entities and control entity movement.

TABLE 24.3 SIMAN Terminology

Generic Term or Phrase	SIMAN Term
External entity	Entity
Internal entity	No special term is used
Resource	Resource, Conveyor, Transporter
Control element	Blockage
Operation	Block or Blocks
Current events list	Current Events Chain
Future events list	Future Events Heap
Delay list	Attached Queue, Internal Queue
User-managed list	Detached Queue
Entity movement phase	No special phrase is used
Clock update phase	No special phrase is used

It is not necessary, of course, to use discrete-event simulation software to build a discrete-event simulation model. Those interested in possibilities for implementing discrete-event simulation models in a high-level *programming* language such as C or C++ are referred to Balci (1988). Those who might be interested specifically in C++ are referred to Joines and Roberts (1997).

24.7.1 SIMAN

In the following sections SIMAN terminology is summarized, particulars of the SIMAN entity movement phase and clock update phase and their interplay with the SIMAN equivalents of the current and future events lists are described, and SIMAN's delay lists and user-managed lists are discussed.

SIMAN Terminology. SIMAN V equivalents for many of the generic terms or phrases presented earlier are given in Table 24.3. More details are provided in the following sections.

Entities, Resources, Control Elements, and Operations. External entities are called Entities in SIMAN. Consistent with the generic discussion of entities, SIMAN Entities are used to model objects, whether animate or inanimate, that move through a system and cause changes in the state of the system. Entities can have attributes (e.g., a unit of work in process can have an arrival time, a due date, a customer identification number, etc.). Various classes of external entities are not formally defined in SIMAN.

Internal entities are used by SIMAN to manage such things as the beginning and ending of downtime periods for resources and for stopping a simulation when a modeler-specified simulated time has been reached. They have no special name in SIMAN. (Internal entities result from elements specified by the modeler in the SIMAN experiment file.) SIMAN Resources model objects that provide service on behalf of Entities. Conveyors and Transporters are special-purpose resources used to model the movement (transportation) of objects along fixed and free paths, respectively, in systems.

In SIMAN, Blocks are used to describe operations carried out by or on Entities. Blocks are arranged in sequences in the order of operations and are connected by paths. Entities move along these paths from Block to Block, triggering Block execution when they move. Each type of Block has a key word and operands whose values particularize instances of the Block type. For example, the CREATE Block is used to create Entities; the distribution followed by the intercreation-time random variable is specified in one of its operands. The WAIT Block is used to put an Entity into the time-delayed state; it has an operand to describe the distribution of time delays. The SEIZE Block is used by an Entity to request one or more units of a Resource; it has operands to identify the particular Resource and the number of units being requested. And so on. SIMAN provides more than 40 types of Blocks.

Current Events Chain. The SIMAN Current Events Chain has the characteristics of the current events list discussed generically. Ready-state Entities are put onto the Current Events Chain by the clock update phase, by cloning during the entity movement phase, and by resolution of related and polled waiting during the entity movement phase. At the end of an entity movement phase, the Current Events Chain is empty.

Entity Movement Phase. The steps followed to carry out SIMAN's entity movement phase are displayed in the flowchart of Figure 24.11, with various components numbered to support discussion. SIMAN does not have a special name for the entity movement phase.

At the start of the SIMAN EMP, a test is conducted (box 1, Figure 24.11) to see if the Current Events Chain contains ready-state Entities, or is empty. Normally, one or more ready-state Entities *will* be on the CEC when an entity movement phase begins, but the CEC *could* be empty because the presence of internal entities on the future events list can trigger entity movement phases to support polling, even if there are no ready-state entities to be processed. (The case of an initially empty CEC and polling is discussed below with the clock update phase.)

Assume an initially nonempty CEC. After the box 1 test, SIMAN puts the Entity at the top of the CEC into the active state (box 2) and then moves the Entity through as many Blocks (operations) as possible (box 3) until the Entity migrates from the active state into the time-delayed state (is transferred to the FEC), or condition-delayed state (is transferred to an Attached or Internal Queue), or dormant state (is transferred to a Detached Queue).

After an Entity has migrated out of the active state, the CEC is tested (box 4, Figure 24.11) to see if it is now empty or if there are still one or more ready-state Entities on it. If the CEC is not yet empty, and entity movement phase continues by putting the ready-state Entity at the top of the CEC into the active state (from box 4 back to box 2), and so on. The looping process (consisting of boxes 2, 3, and 4) continues until the CEC is empty.

The active Entity might execute a Block with the effect of creating one or more clones (copies of itself). In this case, the clone or clones are put immediately into ready state at the top of the CEC (ahead of any other CEC Entities) in first-in, first-out clone order. For example, if the active Entity produces one clone, this clone will be put at the top of the CEC and will be the next ready-state Entity to enter the active state.

After the CEC has been emptied (by the box 2–3–4 looping), a test is conducted (box 5) for the presence of SCAN Blocks in the model. SCAN Blocks are used by SIMAN to implement polled waiting. When a SCAN Block is executed, an associated delay list is checked (box 6, Figure 24.11) to see if the delay-inducing condition has been resolved for the Entity at the top of the list (the queue leader). If it has, this Entity is put into ready state on the CEC (box 7). After *each* SCAN Block has been executed, the entity movement phase continues by processing any resulting ready-state Entities (boxes 2, 3,

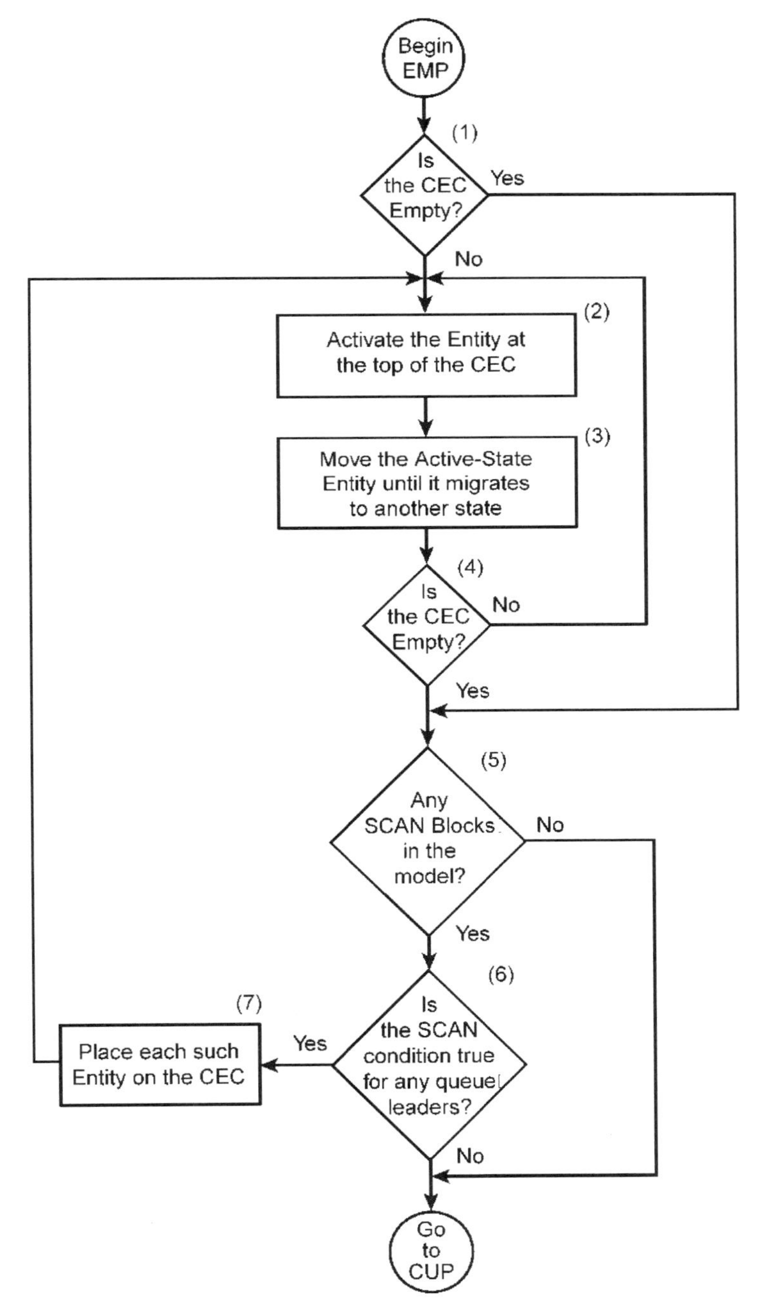

Figure 24.11 SIMAN entity movement phase.

and 4). The entity movement phase continues in this way until the CEC is empty and no more Entities are put into ready state on the CEC by polling. The SIMAN model has then been updated completely at the current simulated time. The next step is to execute the next clock update phase ("no" exit from box 6 in Figure 24.11).

Future Events Heap. In SIMAN, time-delayed entities (both external and internal) are located in a structure named the Future Events Heap (FEH). This structure behaves like a list ranked on increasing move time. The entity with the smallest move time is the next one removed from the FEH when a clock update phase occurs. If there are move-time ties, the order in which entities are removed from the Future Events Heap will not necessarily match their order of insertion into the heap.

Clock Update Phase. The clock update phase advances the simulation clock to the move time of the entity at the top of the Future Events Heap. What happens next depends on whether the entity is external or internal. The alternatives are outlined in Figure 24.12, which shows a flowchart for SIMAN's clock update phase (CUP) and numbers the various components to support discussion.

After the simulation clock has been set equal to the move time of the entity at the top of the Future Events Heap (box 1, Figure 24.12), the entity is removed from the heap (box 2) and is tested (box 3) to determine if it is internal or external. If internal, the entity is processed immediately (box 4). If external, the Entity is inserted at the top (last in, first out) of the Current Events Chain (box 5).

Either way, removal of the entity from the FEH means that some other entity is now at the top of the FEH. To check for potential time ties, the next step (box 6) is to test the top FEH entity's move time to see if it matches the current simulated time. If there is a match, this entity is dealt with appropriately (boxes 2 and 3, and then box 4 or 5). Then another test is conducted (box 6) to determine whether yet another FEH entity might be involved in a time tie. After time ties have been taken into account, SIMAN executes the entity movement phase ("no" path from box 6) to update the state of the model at the freshly established simulation time.

Study of Figure 24.12 shows that a SIMAN clock update phase does not necessarily put one or more external entities on the Current Events Chain. Nevertheless, SIMAN follows each clock update phase with an entity movement phase, whether or not there are external entities on the CEC when the entity movement phase begins. This ensures that a timely check of polled waiting conditions will be made. (Inspection of the Figure 24.11 flowchart for the entity movement phase shows that in this case the entity movement phase immediately takes the path from box 1 to the start of the polled-waiting logic at box 5.)

Attached and Internal Queues. Attached and Internal Queues are the two types of SIMAN lists composed of Entities in a condition-delayed state. These lists have the potential to be formed with Blocks known as Hold Blocks. For example, SEIZE, the Block used by an Entity to capture a Resource, is a Hold Block. If an active Entity tries to execute a SEIZE Block and discovers that it must wait its turn to use the requested Resource, it will be transferred from the Current Events Chain into a condition-delayed state on either an Attached or an Internal Queue, where it will wait for the Hold condition to be resolved.

If a QUEUE Block immediately precedes a Hold Block, an Attached Queue results. An Attached Queue is a named list of Entities waiting to execute the associated Hold Block. Entities are put into an Attached Queue first in, first out, or last in, first out,

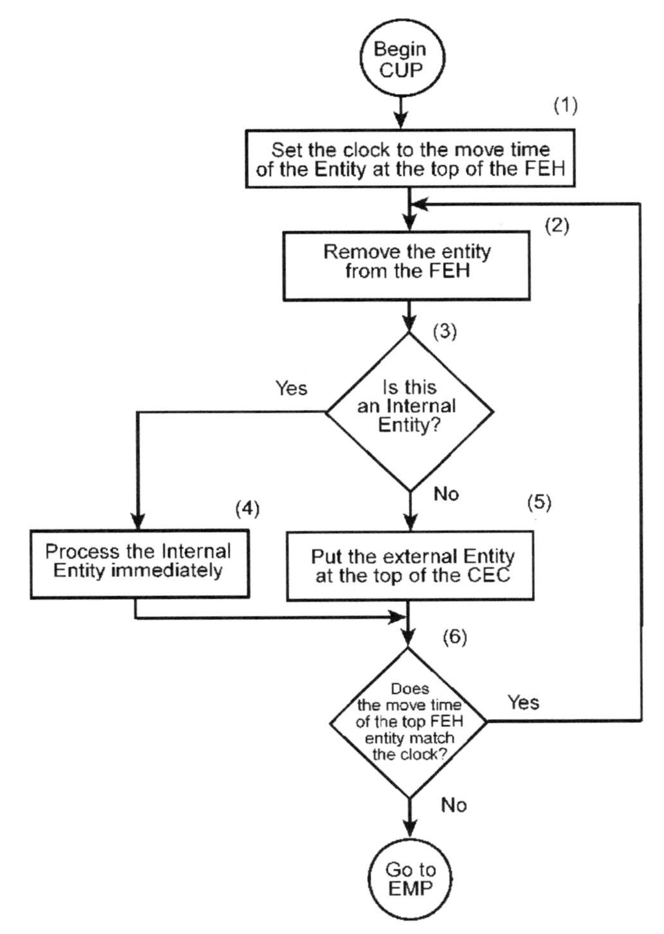

Figure 24.12 SIMAN clock update phase.

or are inserted into the Queue ranked on the value of a modeler-supplied arithmetic expression. If no QUEUE Block precedes a Hold Block, SIMAN provides an Internal Queue for that Hold. An Internal Queue is an unnamed Queue maintained in first-in, first-out order.

Sometimes it is convenient to use identical Hold Blocks at multiple points in a model (e.g., to use a "SEIZE DRILL" Block two or more places in a model). The modeler can associate a separate Attached Queue with each such Hold Block. The resulting Attached Queues are *unshared*, because Entities in two or more Queues are then waiting in separate delay lists for the same Resource. (Hold-Block priority is used to determine which Attached Queue will supply the next Entity to get the Resource.) Alternatively, the modeler can put Entities delayed at two or more identical Hold Blocks into the same Attached Queue, a *shared* Attached Queue. The highest-ranked Entity in this Queue will be the next to get the Resource.

The related-waiting approach is generally used to manage the transfer of Entities from Attached and Internal Queues to ready state on the Current Events Chain. An exception is made for the type of Hold Block known as a SCAN Block. A SCAN Block delays Entities until a modeler-supplied arithmetic expression (typically involving system-state information and/or data values) becomes true. Attached and Internal Queues that form at SCAN Blocks are polled toward the end of each entity movement phase, as shown in the Figuer 24.11 flowchart for the entity movement phase (boxes 5, 6, and 7).

Detached Queues. Detached Queues are Entity lists used by SIMAN to implement the dormant state. Entities are put into Detached Queues when they execute QUEUE Blocks with the DETACHED modifier specified. Such Entities are later transferred (as the result of action taken by other Entities) from the dormant state to the ready state. (These other Entities use either SEARCH and REMOVE or QPICK and MATCH Blocks to transfer Entities from Detached queues to the CEC.)

24.7.2 ProModel

In the following sections ProModel terminology is summarized, particulars of the ProModel entity movement phase and clock update phase are described, and the approach used in ProModel to handle delayed entities is discussed.

ProModel Terminology. ProModel equivalents for many of the generic terms presented earlier are given in Table 24.4. More details are provided in the following subsections.

Entities. External entities are called Entities in ProModel. Various classes of Entities must be formally defined when modeling in ProModel. These entity classes are called Entity Types. The various Entity Types are each given a unique name supplied by the modeler. ProModel uses internal entities, calling them Internal Events. The role played by Internal Events is discussed further below.

Resources. Locations are one of the ProModel constructs that correspond to the generic term *resource*. As the name suggests, Locations represent fixed places. They

TABLE 24.4 ProModel Terminology

Generic Term or Phrase	ProModel Equivalent
External entity	Entity
Internal entity	Internal Event
Resource	Location, Resource, Node
Control element	Variable
Operation	Process Step
Current events list	Action List
Future events list	Future Events List
Delay list	Waiting List
User-managed list	There is no generalized equivalent
Entity movement Phase	No special phrase is used
Clock update Phase	No special phrase is used

are used to model such things as waiting areas (space), workstations, and immobile machines. Locations are also used to model accumulating and nonaccumulating conveyors. Resources are another ProModel construct corresponding to the generic term *resource*. Resources can be used to model such things as people and equipment. Resources can be static, modeling for example an immobile worker assigned to operate a fixed-position machine. Alternatively, Resources can be dynamic (mobile), modeling for example forklift trucks or workers who move from point to point in a system.

Dynamic Resources move about in networks composed of paths (path networks). There are several types of path networks. For example, there is a type in which dynamic Resources must move single file, and there is another type in which dynamic Resources can pass each other. Nodes are another ProModel *resource* construct. Nodes are the beginning and ending points of the paths making up the networks within which dynamic Resources move. Resources themselves can compete with each other for the use of Nodes, moving through a network in search of something to pick up, for example, or for a place to be idle.

Because dynamic Resources move, the timing of their movement comes into play; and because they compete for Nodes, the possibility of delay also comes into play for them. As a result, dynamic Resources have much in common with Entities. (Whereas Entities are units of traffic that move *through* systems and are service consumers, dynamic Resources are units of traffic that move *within* systems and are service providers.) Like Entities, dynamic Resources can be put into a time-delayed state while they wait for the known future time when they will reach their destination. They can also migrate from the time-delayed to the ready state, and then the active state. Because of Node competition, they are candidates for the condition-delayed state, too. ProModel uses its Future Events List, Action List, and Waiting Lists not only to manage Entities but also to manage dynamic Resources. (The Future Events and Action Lists are also used to manage Internal Events.)

Control Elements. In ProModel, a Variable is a general-purpose data element whose value can be the object of a ProModel WAIT UNTIL. Like all control elements, a Variable (used in conjunction with a WAIT UNTIL) has the potential for causing Entities to be delayed.

Operations. The transaction-flow part of ProModel is specified by the modeler as an ordered collection of Process Steps contained in a Process Table. Every Process Step specifies the name of an Entity Type (or All) and the name of a Location (or All). An Entity "flows" from one Process Step to the next in a Process Table by jumping to the next Process Step that matches its Type and Location (looping back to the top of the Process Table if necessary). These Process Steps determine what this Entity Type is to do at this Location.

Operation Logic and Routing Logic are the components from which Process Steps are formed. A Process Step can consist of zero or more of each of these components. Competition among Entities for nontransportation Resources is spelled out in the Operation Logic. Competition among Entities for Locations and transportation Resources is spelled out in the Routing Logic. Routing Logic is applied after Operation Logic has been executed.

Many model-definition constructs in ProModel have optional user-defined Logic fields (e.g., Downtime Logic and Location Exit Logic). Logic is a collection of statements that are executed automatically and whose execution has the potential of putting Internal Events on the Future Events List. When processed, these Internal Events might go back onto the Future Events List or onto a delay list. They might also cause Entities or dynamic Resources to be put on the current events list.

Action List. Action List is the ProModel name for the current events list. The Action List can contain Entities, dynamic Resources, and Internal Events. (The term *element* will be used from this point forward to designate any one of Entity, dynamic Resource, or Internal Event.) The Action List is empty when an entity movement phase begins. That is, there are no elements in the ready state. An Entity, dynamic Resource, or Internal Event is already in the active state at this point, however. Entities, dynamic Resources, and Internal Events can be put on the Action List while the active-state element is being moved. Any such list newcomers are put at the top of the list. When the active element migrates from the active state, the element at the top of the Action List is the next to enter the active state. This results in last-in, first-out Action-List management.

Entity Movement Phase. ProModel's entity movement phase is shown in Figure 24.13. One element is in the active state when the entity movement phase begins (as explained below in the discussion of the clock update phase), and the Action List is empty. If the active element is an Entity or a dynamic Resource (box 1, Figure 24.13), it executes operations until it is forced to migrate from the active state (box 2); otherwise, it is an Internal Event and all the corresponding logic is acted upon (box 3).

Either way, one or more new ready-state elements might be created as a result and, if so, will be put on the Action List last in, first out. (For example, an Entity might create one or more copies of itself.) If this is the case ("no" path from box 4, Figure 24.13), the element at the top of the Action List is made active (box 5) and the entity movement phase continues from there. Otherwise, the Action List is still empty and a clock update phase comes next ("yes" path from box 4).

Future Events List. ProModel's Future Events List is composed of elements (Entities; dynamic Resources; Internal Events) in the wait-delayed state. The elements are arranged in the list top down in order of increasing move times. ProModel always removes exactly one element from the top of the list per clock update phase. (This means that in case of time ties there will be two or more consecutive clock update phases and entity movement phases at the same point in simulated time.)

Clock Update Phase. ProModel's clock update phase is summarized in Figure 24.14. The phase begins by setting the simulation clock equal to the move time of the element at the top of the Future Events List (box 1). This element is put immediately into the active state (box 2). This concludes the clock update phase, and the entity movement phase begins. In the ProModel approach, then, there are no ready-state entities when an EMP begins. (The Action List is empty.) Furthermore, if two or more elements at the top of the Future Events List have identical move times, two or more consecutive clock update phases and entity movement phases will take place at the same point in simulated time.

Waiting Lists. Waiting Lists are ProModel delay lists. There are Waiting Lists for Entities and for dynamic Resources. A Waiting List for Entities is attached to each Location, to each static Resource, to each dynamic Resource, and to each Variable that is subject to a WAIT UNTIL. A Waiting List for dynamic Resources is attached to each Node.

A single Entity (or dynamic Resource) can be represented simultaneously in many delay lists. As a result, ProModel does not use polled waiting to manage the resolution

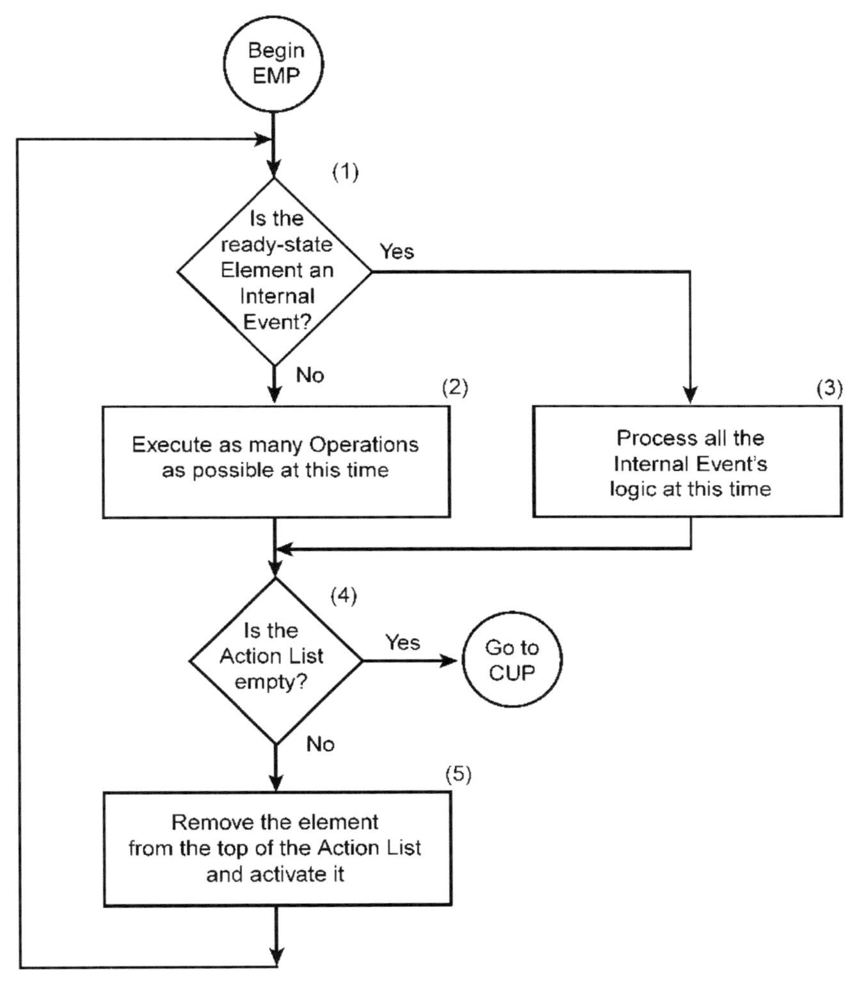

Figure 24.13 ProModel Entity Movement Phase.

of complex delay. Instead, related waiting is used to manage Waiting Lists, with the provision that ProModel removes the (representation of the) Entity (or dynamic Resource) from all relevant delay lists as soon as the related-waiting mechanism transfers the Entity (or dynamic Resource) from any one of them to the Action List.

ProModel has no lists corresponding to the generically described user-managed lists. However, JOIN, LOAD, and SEND are all Routing Logic options that place Entities on special Location-specific lists where they await a corresponding JOIN, LOAD, or SEND Operation Statement to be executed by another Entity at the destination Location. This explicit Entity-based triggering makes these lists resemble user-managed lists. Because the lists are specific to Locations, they are considered here to be in the category of delay lists.

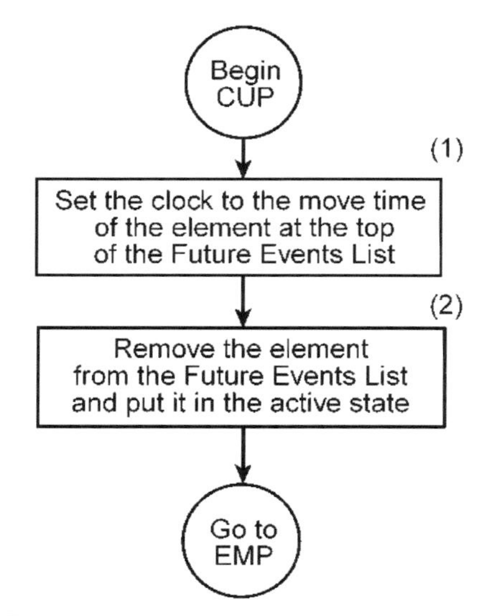

Figure 24.14 ProModel Clock Update Phase.

24.7.3 GPSS/H

In the following sections GPSS/H terminology is summarized, particulars of the GPSS/H entity movement phase and clock update phase are described, and user-managed lists are discussed.

GPSS/H Terminology. GPSS/H equivalents for many of the generic terms presented earlier are given in Table 24.5.

Entities, Resources, Control Elements, and Operations. External entities are called Transactions (Xacts for short) in GPSS/H. Various classes of Transactions are not

Table 24.5 GPSS/H Terminology

Generic Term or Phrase	GPSS/H Equivalent
External entity	Transaction
Internal entity	System Transaction
Resource	Facility; Storage
Control element	Logic Switch, Arithmetic Expression, Boolean Expression
Operation	Block
Current events list	Current Events Chain
Future events list	Future Events Chain
Delay list	Current Events Chain
User-managed list	User Chain
Entity movement phase	Scan Phase
Clock update phase	Clock Update Phase

formally defined. Transactions do have numeric-valued attributes whose role is determined by the modeler. A specialized numeric-valued Transaction attribute known as Priority Level is used to determine a Transaction's rank when it is placed on the current events list.

GPSS/H makes some use of internal entities. These are called System Transactions. System Transactions are used to initialize arrival processes, to support making certain types of statistical observations, and to support the list-processing algorithms employed to manage the current and future events lists. GPSS/H uses the future events list in some cases to manage the timing of System Transactions. GPSS/H does not use internal entities to model resource downtimes or simulation stop-times. Instead, downtimes and stop-times are modeled with external entities (Transactions) that execute Blocks (discussed below) to implement downtime and stop-time logic.

Facilities and Storages model resources. Facilities model unit-capacity resources; that is, a Facility is a resource that can only serve one Transaction at a time. Storages model resources having any user-specified capacity. Storages are capable of serving multiple Transactions simultaneously.

Logic Switches are on-or-off variables used for control purposes in GPSS/H. Transactions can be forced to wait until a specified Logic Switch is in an indicated state, or can alternatively move either sequentially or nonsequentially, depending on whether a specified Logic Switch is or is not in an indicated state.

Arithmetic and Boolean Expressions can also be used as control elements in GPSS/H. Transactions can be forced to wait until a specified Arithmetic Expression satisfies a specified relational condition (or until a Boolean Expression has a specified truth value), or can alternatively follow a nonsequential path conditioned on the value of an Arithmetic or Boolean Expression.

In GPSS/H, blocks are used to describe operations carried out by or on Transactions. Blocks are arranged in sequences in the order of operations, and are connected by paths. Transactions move along these paths from Block to Block, triggering Block execution when they move. Each type of Block has a key word and operands whose values particularize instances of the Block type. For example, the GENERATE Block is used to create Transactions; the distribution followed by the intercreation-time random variable is specified in one of its operands. The ADVANCE Block is used to put a Transaction into the time-delayed state; it has an operand to describe the distribution of time delays. The SEIZE Block is used by a Transaction to request a Facility; it has an operand to identify the particular Facility. And so on. GPSS/H provides about 65 types of Blocks.

Current Events Chain. The GPSS/H Current Events Chain holds Transactions that are in the ready state. It also holds the active-state Transaction. The Current Events Chain additionally serves as a single global GPSS/H delay list, holding condition-delayed Transactions in a model. These condition-delayed Transactions are commingled with the active-state Transaction and ready-state Transactions on the CEC.

Transactions have a specialized numeric-valued attribute known as Priority Level. A Transaction's Priority Level is assigned when the Transaction is created and can be changed dynamically any number of times during the Transaction's life cycle. Transactions on the CEC are ranked first in, first out within Priority Level. This results in a "first come, first served, within Priority Level" service order for condition-delayed Transactions. (User Chains, discussed below, are used to model other service orders.)

GPSS/H has a Queue construct and a QUEUE Block, but unlike SIMAN's QUEUE

Block, the GPSS/H QUEUE Block, does not perform a list management function. GPSS/H queues are used for statistics-gathering purposes only, consistent with the fact that except for the CEC (and some CEC-supporting internal delay lists discussed below), there are no additional delay lists in GPSS/H. Because of its delay-list role, the GPSS/H CEC frequently is not empty when an entity movement phase ends.

Scan Phase. The entity movement phase is known as the Scan Phase in GPSS/H. The Scan Phase must take into account the presence of commingled ready-state and condition-delayed Transactions on the Current Events Chain. Transactions cannot always simply be examined only one time during a Scan Phase, from the top of the CEC to the bottom. This is because when the active-state Transaction executes a Block that resolves a delay, the affected condition-delayed Transaction(s) might be somewhere above the active Transaction on the CEC. The affected Transaction(s) are switched from the condition-delayed state to the ready state and must then eventually be put into the active state and moved before the simulation clock is advanced. This forces the Scan Phase to redirect its CEC focus upward after the active Transaction leaves the active state. The Scan Phase does this by returning to the Transaction at the top of the CEC. From there it again starts to examine Transactions top-down, guaranteeing that it comes across any Transactions that might have been switched from the condition-delayed state to the ready state.

The logic of the GPSS/H entity movement phase (Scan Phase) is shown in the form of a flowchart in Figure 24.15, where various components have been numbered to support discussion. As indicated in Figure 24.15, GPSS/H starts (or restarts) a Scan Phase with a test (box 1) to see if there is a Transaction at the front of the CEC. (The CEC might be empty if the Scan Phase is taking place at simulated time 0.0, or if the scan is being restarted.) If the CEC is empty, the Clock Update Phase comes next ("no" path from box 1).

Assume that there is a Transaction at the top of the CEC. If the Transaction is in the ready state (box 2 test), the Transaction is made active and moved (box 3) until it is forced to migrate from the active state. A check is then made (box 4) to determine if the actions just taken might have resolved one or more delay-inducing conditions. If delay might have been resolved, the scan of the CEC restarts at the top of the CEC ("yes" path from box 4 back up to box 1); otherwise, the scan continues to the sequential CEC Transaction, if any ("no" path from box 4 to box 5). If there is no sequential Transaction, the last Transaction on the CEC has been dealt with and the Clock Update Phase comes next ("no" path from box 5).

If the Transaction being dealt with at box 2 in Figure 24.15 is not in the ready state, it is condition-delayed. The "no" path is taken from box 2 to box 5 in such cases. The effect is to skip over condition-delayed Transactions during the Scan Phase. The GPSS/H approach of keeping condition-delayed Transactions on the CEC and examining them one or more times during the Scan Phase to see if they have been put into the ready state means that all of these Transactions are fundamentally in a polled-waiting condition.

More About Restarting the Scan. The scan restarts discussed above result in extra processing demands while GPSS/H reexamines condition-delayed Transactions on the CEC. To offset this, each Transaction has an on-or-off switch called the Scan Skip Indicator that signals whether the Transaction is waiting for the resolution of a simple delay condition. When the scan examines a CEC Transaction, its Scan Skip indicator is

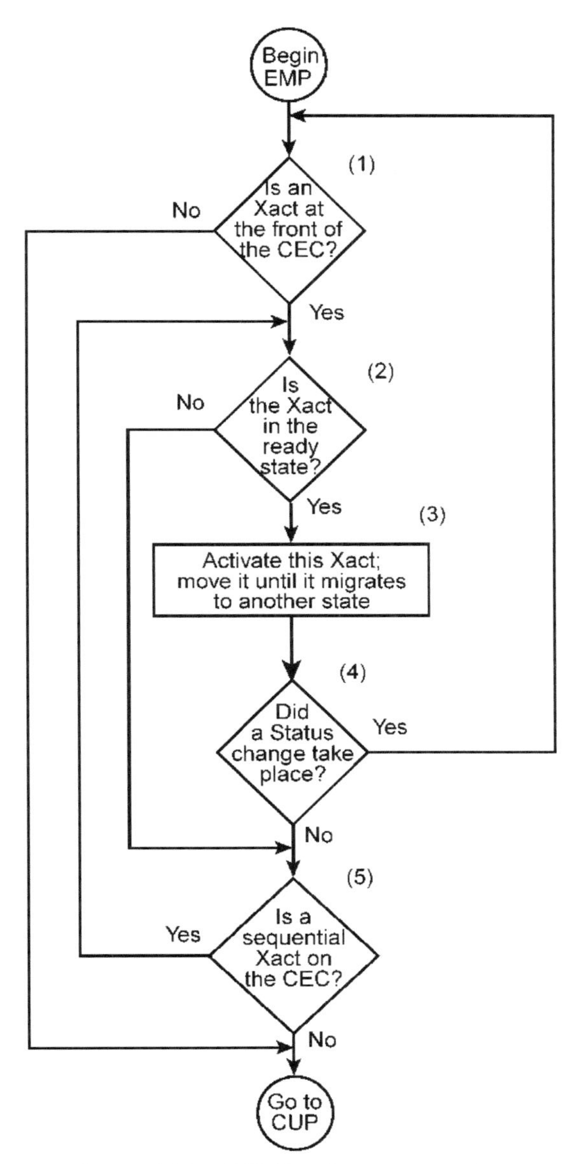

Figure 24.15 GPSS/H entity movement phase.

tested to determine if the Transaction is condition-delayed. If it is, the scan skips over the Transaction and goes to the sequential CEC Transaction. (See the "no" path from box 2 to box 5 in Figure 24.15.)

A condition-delayed Transaction's Scan Skip Indicator is switched off at the moment the delay-inducing condition is resolved, putting the Transaction into the ready state.

Internal delay lists are used to track which Transactions should have their Scan Skip Indicators switched off when the associated delay-inducing condition is resolved. Because these internal delay lists are related to the delay-inducing conditions, the GPSS/H Scan Phase is really a hybrid of the polled-waiting and related-waiting approaches.

GPSS/H has an internal Status Change Flag that GPSS/H switches on whenever a delay-inducing condition has just been resolved. (GPSS/H also puts corresponding condition-delayed Transactions into the ready state as well by switching off their Scan Skip Indicator.) When the active Transaction migrates out of the active state, the Status Change Flag is tested (box 4, Figure 24.15) to determine whether to restart the scan. The need for scan restarts under qualified conditions complicates the logic of the Scan Phase. (Full-blown particulars of the logic are provided in Schriber, 1991.) Nevertheless, the use of Scan Skip Indicators, internal delay lists, and the Status Change Flag to support these scan restarts results in extremely fast execution of GPSS/H models.

There is one additional aspect of the Scan Phase and scan restarts to be mentioned. This involves the ability of the active Transaction to call for an *immediate* restart of the CEC scan. The active Transaction achieves this by executing the GPSS/H YIELD Block, with the effect of returning itself temporarily to the ready state (corresponding to path 4 in the Figuer 24.9 entity-state diagram) and forcing an immediate scan restart. (This possibility could be shown in the Figure 24.15 flowchart of the entity movement phase with a conditional path from box 3 back up to box 1.) The restarted scan will eventually reexamine the YIELDing Transaction on the CEC at the same simulated time, which will then become the active Transaction again. This ability of a Transaction to yield control deliberately but only temporarily to one or more other Transactions can be useful from time to time in discrete-event modeling.

Future Events Chain. The future events list is called the Future Events Chain (FEC) in GPSS/H. Transactions are arranged on the FEC in order of increasing move time, with move-time ties resolved first in, first out. The GPSS/H Clock Update Phase removes multiple Transactions from the Future Events Chain in top-down order if they are tied for the earliest move time, inserting them one by one (ranked by Priority Level) into the Current Events Chain.

Clock Update Phase. The logic of the Clock Update Phase is shown in Figure 24.16. The simulated time is set equal to the move time of the Transaction at the top of the Future Events Chain (box 1); then this Xact is transferred from the Future to the Current Events Chain (box 2). A check is then made (box 3) to see whether the move time of the Transaction now at the top of the Future Events Chain matches the simulated time. If a match is made (move-time tie), this Transaction will also be transferred from the top of the FEC to the CEC (see the "yes" path from box 3 back up to box 2). This looping process is repeated until the Transaction at the top of the Future Events Chain has a move time exceeding the current simulated time. The clock Update Phase is then complete, and the Scan Phase comes next ("no" path from box 3).

User Chains. GPSS/H implements the dormant state with User Chains, which are user-managed lists of Transactions. Transactions can be inserted into a User Chain at the top, at the bottom, or ranked ascending on a modeler-specified Transaction attribute. After a Transaction puts itself onto a User Chain, it can only be removed by an active-state Transaction. The order of Transaction removal from a User Chain can be top-

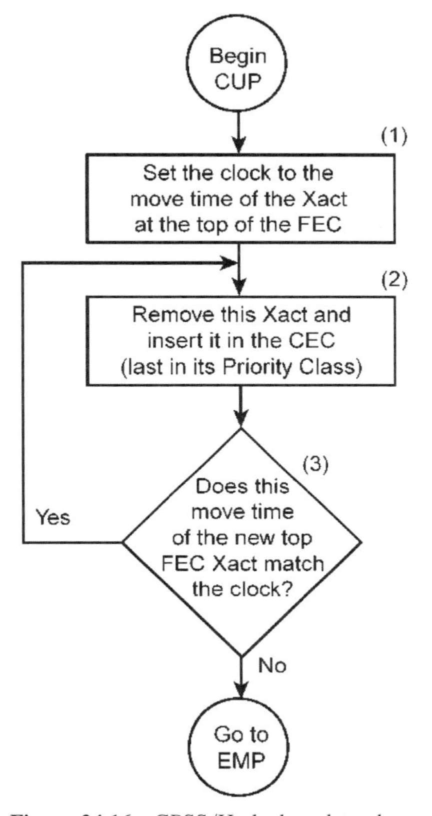

Figure 24.16 GPSS/H clock update phase.

down, or bottom-up, or can be based on the truth of a user-supplied Boolean Expression. Removed Transactions are put into ready state on the Current Events Chain and the Status Change Flag is switched on (see above). As a result, the scan of the CEC will be restarted ("yes" path from box 4 back up to box 1 in Figure 24.15) after the active Xact migrates from the active state. This ensures that the ready-state newcomers to the Current Events Chain will be put into the active state and move before the next Clock Update Phase takes place.

24.8 WHY IT MATTERS

24.8.1 Overview

Three scenarios will now be described that reveal some of the practical differences in the design and implementation of discrete-event simulation software. These differences are illustrated for SIMAN, ProModel, and GPSS/H, but there are underlying implications that extend to whatever software is being used for discrete-event simulation modeling. Furthermore, these scenarios are only a few of many that could be presented.

The section concludes with comments on how knowledge of the logical foundations of discrete-event simulation and implementation of logical considerations internal to modeling software is needed to make effective use of software tools provided for troubleshooting and verifying models.

24.8.2 Trying to Recapture a Resource Immediately

Suppose that a part releases a machine, then the part immediately recompetes for use of the machine. The objective is to let a more highly qualified waiting part (that might have come along in the meantime) be the next to use the machine. It is possible, however, that the releasing part itself might be most highly qualified, in which case it is to recapture the machine without delay. Waiting parts might be ranked on due date, for example (the nearer the due date, the higher the rank), and the question is whether the releasing part is the most highly qualified to use this same machine for the next operation to be performed on the part. Suppose further that the modeler thinks this situation can be modeled in straightforward fashion, using fundamental techniques provided in the languages under discussion. Then the question is whether the modeler achieves the intended effect.

Of interest here is the order in which events take place following the giving up of a resource. There are at least three alternatives that are based on the *timing* used by the software to determine which entity will be next to use the machine that has just been made idle. These are the alternatives:

1. Coupled with the giving up of the resource is the *immediate* choosing of the next resource owner from the associated delay list. In this case the releasing entity is still the active-state entity when the next owner is chosen; the releasing entity is not on the delay list and so is not a contender for the resource.
2. The choosing of the next resource owner is deferred until the releasing entity has migrated out of the active state; this gives the releasing entity the opportunity to be put on the associated delay list (even though the resource is idle) and therefore to be a contender for the resource.
3. "Neither of the above"; that is, the releasing entity, without paying heed to the possible presence of other contenders, continues to be the active entity and recaptures the idle resource immediately, even though a more highly qualified entity might be waiting for the resource.

Each of these three alternatives is in effect in one or another of the three instances of simulation software being discussed here, reflecting differing implementation choices made by the software designers. In particular, SIMAN, ProModel, and GPSS/H, respectively, implement the first, second, and third alternatives outlined above, as summarized in Table 24.6.

None of the three Table 24.6 alternatives is intrinsically either "right" or "wrong." The point is that the modeler must be aware that alternative design choices exist for the *language* designer, and must know which design choice has been made in the software the modeler is using. Otherwise, if the modeler is unaware that this is potentially an issue, it is quite possible to model a given situation with an unintended effect and then through inadequate model verification not even become aware of this fact.

TABLE 24.6 Alternative Outcomes When Recompeting for a Resource

Alternative	When is the Successor Chosen?	Is the Releaser a Contender at That Time?	Which Language (and How)?
1	At the time the resource is released	No	SIMAN (RELEASE followed by SEIZE)
2	After the releaser has become a member of the delay list	Yes	ProModel (FREE or USE followed by GET or USE)
3	When the active entity tries to capture it	Yes	GPSS/H (RELEASE followed by SEIZE)

24.8.3 The First In Line Is Still Delayed

Suppose that two or more condition-delayed entities are waiting in a list because no units of a particular resource are idle. Assume that the entity at the top of the list needs two units of the resource, whereas the next entity in the list only needs one unit. (For example, the waiting entities might be ships, and the resource might be tugboats. The ship at the top of the list needs two tugboats, but the next ship in the list only needs one tugboat.) Now suppose that one unit of the resource becomes idle. The needs of the entity at the top of the list cannot yet be satisfied, but the needs of the next list entity can. What will happen?

There are at least three possible alternatives: (1) Neither entity claims the idle resource unit; instead, both entities continue to be condition-delayed; (2) the first entity claims the one idle resource unit and waits in a condition-delayed state for a second unit; or (3) the first entity's needs cannot be met but the second entity's can, so the second entity claims the idle resource unit and migrates from the condition-delayed state to the ready state.

If the modeler takes a fundamental modeling approach for entities requesting resources in the languages being discussed (SEIZE in SIMAN; GET or USE in ProModel; ENTER in GPSS/H), each of the three alternatives described above will be achieved, one per language. The outcomes are summarized in Table 24.7.

Alternative 1 in Table 24.7 is in effect in SIMAN because only the entity at the head of the associated delay list (the queue leader) is a contender for the resource; the next

TABLE 24.7 Alternative Outcomes When the First in Line is Still Delayed

Alternative	Behavior	Which Language (and How)?
1	Neither entity claims the idle resource unit; both continue to be condition-delayed	SIMAN (SEIZE)
2	The first entity claims the idle resource unit and continues in the condition-delayed state, waiting for a second unit	ProModel (GET or USE)
3	The second entity claims the idle resource unit and migrates to the ready state	GPSS/H (ENTER)

entity in the delay list isn't yet in active contention for the resource because it is not yet the queue leader.

In ProModel, alternative 2 in Table 24.7 is in effect because the entity at the head of a ProModel delay list is permitted to make a partial capture. The entity captures the one idle resource unit and remains in a delayed state, waiting to capture another resource unit.

In GPSS/H, all condition-delayed Transactions waiting for a resource are active contenders for the resource, and partial captures are not permitted. (The approach taken in GPSS/H when a Transaction requests multiple units of a resource is "all or nothing at all.") Both of the affected condition-delayed Transactions are switched from the condition-delayed state to ready state on the Current Events Chain when the one resource unit is made idle. The Transaction requesting two units becomes active first but is switched back to the condition-delayed state because there aren't two idle resource units for it. The Transaction requesting one unit then becomes active and captures the one idle resource unit, resulting in alternative 3 in Table 24.7. The comment made at the conclusion of Section 24.8.2 applies here, too: the modeler needs to know that software behavior depends on choices made by the *language* designer and must understand what the implications are for *model* design.

24.8.4 Yielding Control Temporarily

Suppose that the active entity wants to give control to one or more ready-state entities, but then wants to become the active entity again before the simulated clock has been advanced. This might be useful, for example, if the active entity has opened a gate permitting a group of other entities to move past a point in the model in zero simulated time, and then needs to reclose the gate after the other entities have moved through it. (Perhaps a group of identically flavored cartons of ice cream is to be transferred from an accumulation point to a conveyor leading to a one-flavor-per-box packing operation. After the group has moved past the accumulation point, conditions must be reset so that the next group can begin to accumulate there.)

There are a number of ways to accomplish this effect, with the particulars dependent on the language being used. The scenario described here is more complicated than either of those in Sections 24.8.2 and 24.8.3, so a modeler without much experience might not be able to model this scenario "instinctively," but might have to give it some thought. The modeler might evolve a plan like this:

1. A "watchdog-entity" is used to monitor the group of other entities, waiting until the group has reached the predetermined size (e.g., the number of cartons of ice cream per packing box). Meanwhile, the other entities are arriving over time at the accumulation point and are being put into a condition-delayed state because the gate is closed.
2. When the watchdog detects that the required group size has been reached, it opens the gate, then "stands aside" while the other entities move through the open gate, one after the other. This movement consumes very little time in the real system, so let us assume that the movement is modeled as taking zero simulated time.
3. The watchdog then takes action again at the same simulated time, reclosing the gate to force subsequent entities arriving there to wait until the next group being formed has reached the predetermined size, and so on.

This leads to the question of whether and how the modeler can achieve the effect of having the watchdog take action early in an entity movement phase (to open the gate), then "stand aside" (leave the active state) while the other entities move, and then reenter the active state during the same entity movement phase to take more action (to reclose the gate). Referring to the entity-state diagam in Figure 24.9, the watchdog-entity wants to go from the active state to the ready state (path 4) and then eventually (but at the same simulated time) return to the active state (path 5).

The effect can be accomplished, either approximately or exactly, in each of the three modeling languages being discussed, as follows:

1. In SIMAN, the effect can be accomplished approximately by putting the watchdog entity into a time-delayed state (DELAY Block) for a very short simulated time. The watchdog will then enter the active state again after the next clock update phase and can reclose the gate then. (Presumably no additional entities have come along in the meantime. See the further discussion below.)
2. In ProModel, the effect can be accomplished exactly by putting the watchdog entity into a time-delayed state for *zero* units of simulated time ("WAIT 0"). (This can't be done in either SIMAN or GPSS/H; if the active entity executes a "DELAY 0" Block in SIMAN or "ADVANCE 0" in GPSS/H, the entity is not put onto the future events list but remains active and immediately tries to execute the sequential Block.) At the conclusion of the ongoing entity movement phase, ProModel will carry out the next clock update phase, putting the watchdog into the active state without changing the simulated clock. (In other words, the watchdog will traverse the 5–6–3 path in the Figure 24.9 entity-state diagram in zero simulated time.) The watchdog can then move to reclose the gate at precisely the same simulated time at which it had been opened.
3. In GPSS/H, the effect can be accomplished exactly by having the watchdog execute a YIELD Block right after it opens the gate. This Block execution causes the watchdog to migrate immediately from the active state back to the ready state, with the entity movement phase then refocusing on the entity at the top of the Current Events Chain. (The logic of YIELD-Block execution is not shown in the Figure 24.15 flowchart of the GPSS/H Scan Phase. It could be shown there with a line leading from box 3 to box 1.) Qualifying entities above the watchdog on the CEC will then become active, one by one (with ice-cream entities, for example, now able to move through the open gate). As the entity movement phase continues, the watchdog will be reencountered by the scan and then will be made active again. It can then reclose the gate at precisely the same simulated time at which it had been opened.

Would experienced modelers take the watchdog approach described above to model this scenario? Not necessarily. Other approaches are possible, making it unnecessary, for example, to settle for approximate effects or to assume that each entity in the group moves through the conceptual gate in zero simulated time. The important point is that description of the approach presented here, and language-specific discussion of its particulars, would be difficult or impossible without appealing to an understanding of the foundations of discrete-event simulation. Modelers who do not have this understanding are at a disadvantage relative to those who do.

24.8.5 Interactive Troubleshooting and Verification of Models

The modeler is faced with several types of challenges when building discrete-event simulation models. One challenge is to develop correct approaches for modeling logically complicated situations. A second challenge is to come to a careful understanding of model behavior so that execution errors occurring during model development can be diagnosed and remedied. A third challenge is to verify that a model accurately captures the logic intended for it. (There are other challenges too in a simulation project, but these three bear especially on the subject of this chapter.)

A modeler should be able to handle each of these challenges, and to handle them smoothly and quickly. For this purpose, modeling software can provide an arsenal of tools to support interactive probing of model behavior. These tools make it possible to put a magnifying glass on a simulation model while it executes in slow motion. The modeler can follow the active entity step by step, watch the hand-off of the active state from one entity to the next, display the current events list at the start of an entity movement phase and at intermediate points during the phase, and display the future events lists and the delay and user-managed lists at will. The status of resources can also be displayed, as well as statistical reports summarizing various aspects of the model.

Here are examples of some of the types of things a modeler might be able to do when interactively *troubleshooting* a model:

1. Suppose that during a trial batch-mode run being made during model development, an error message is issued at a specified simulated time saying that a particular entity is trying to make a forbidden move (such as freeing up a resource that it doesn't have under its control; or such as requesting its own destruction while it still has resources under its control). The question then arises: How did this particular entity erroneously reach the point in the model at which it was attempting such a move? To probe for the answer to this question, the modeler can rerun the model in interactive mode. (Because pseudorandom numbers are used, the previous run will be reproduced exactly.) Before the simulation starts, the modeler can "set a trap" on the simulation clock, then indicate that the simulation is to continue until reaching the beginning of the entity movement phase at the simulated time the error is known to occur. When this time comes, the software will then interrupt the simulation and the modeler will be put in control. Next, the modeler can "set a trap" on the offending entity, so that when this entity next becomes active, the simulation will be interrupted again. When the entity does become active, the modeler can display its whereabouts in the model and single-step it along its path, right up to the point that the execution error occurs. By this time the modeler has probably gained enough insights to realize why the model is misbehaving.
2. Suppose now that in the preceding scenario, the offending entity has already gotten off its proper path sometime *before* the simulated time at which the error occurs. In the preceding approach, the modeler might have started to monitor the entity's behavior too late, that is, not until *after* the entity has already made its first improper move in the model. How can this circumstance be handled? In this case, at the *start* of the simulation the modeler can set a trap on the offending entity (even though it probably has not yet been created), then let the simulation continue. When the offending entity has been created and becomes active for the first time, the simulation will be interrupted and the user will take control. The

modeler can then look at the entity's point of entry to the model to see which path it is on initially. The entity can then be followed step by step whenever it becomes active during the ongoing simulation. In the process it will probably occur to the modeler why the model is misbehaving.

Here are some of the types of things that might be done when interactively *verifying* segments of a model, or an entire model:

1. A modeler might be wrestling with possibilities for modeling one system complexity or another (perhaps such as the scenario described in Section 24.8.4). Suppose that the modeler has developed a proposed approach but doesn't know whether the approach will work correctly. The modeler can then build a small model designed to investigate the behavior of the approach, then use interactive tools to determine whether or not the approach is sound.
2. The modeler might want to be assured that in the setting of a full-scale model, one or another logical complexity has been modeled correctly. In interactive mode, the modeler could set a trap on one or more model blocks (or statements) that are entity entry points to the model segment in question. The simulation would then continue until an entity or entities interact with such a block (or statement). The modeler could then set a trap on one or more of these entities and monitor their behavior and experiences while they move through the model segment, watching carefully to see whether the entities are making the right moves under the model conditions in effect then.
3. For overall verification of a model, the modeler might set traps on a number of randomly chosen entities, then monitor the behavior and experiences of these entities while they work their way through the model. (This is analogous to putting radioactive tracers into a biological system.) Does each of these entities make the right moves at the right times under the model conditions in effect when the moves are being made? The process of trying to explain and understand the moves of entities under a range of model circumstances forces the modeler to think clearly about the model and can bring model flaws into focus as a result.

In more general terms, interactive model execution can also be used for self-education in the characterisics of a modeling language itself. A language might be relatively new to a modeler, and it might not be clear what the characteristics are of this or that aspect of the language. For example, in the scenario of Section 24.8.3, the modeler might not know what the effect will be in the software of choice, and it might not be possible to find the answer either with on-line help or in a reference manual. Then the modeler could quickly build a small test-case model and make an interactive run to determine how the software behaves.

The types of activities described above yield valuable insights into model and language behavior for the modeler who knows the relevant concepts. These insights can lead to a rapid understanding of why error conditions are coming about, and whether a model segment or an entire model is behaving in the way intended for it. Without understanding the concepts and knowing the vocabulary the software uses to issue interactive messages and display information, the modeler might not take full advantage of the interactive tools provided, and might even avoid using the tools entirely.

24.9 CONCLUSIONS

The transaction-flow world view of discrete-event simulation has been described and illustrated through discussion of a series of examples of queuing systems. The complexities of modeling large-scale systems of this type and orchestrating the movement of units of traffic in such systems have been indicated. Entity states and the structures and procedures used to manage entities and bring about their migration from state to state during a simulation have been described. Details of the implementation of these generic concepts in three instances of discrete-event simulation software have been provided. Finally, examples have been given to illustrate how important it is for the modeler to be knowledgeable about the foundations of discrete-event simulation software and the implementation particulars for the modeler's software of choice.

ACKNOWLEDGMENTS

Information in this chapter has been derived in part from conversations with software-vendor personnel, some of whom then reviewed the written material. The support provided by David T. Sturrock and C. Dennis Pegden of Systems Modeling Corporation, Charles Harrell and Eric Du of ProModel Corporation, and Robert C. Crain and James O. Henriksen of Wolverine Software Corporation is acknowledged with thanks. This chapter has been expanded from a tutorial on this topic given at several Winter Simulation Conferences (Schriber and Brunner, 1997). The contributions of those attending the tutorial and asking insightful questions are acknowledged with thanks. Finally, thanks to handbook editor Jerry Banks for his comments, encouragement, and leadership, and to two fellow handbook chapter authors who reviewed the chapter and made useful suggestions for improving it.

REFERENCES

Balci, O. (1988). The implementation of four conceptual frameworks for simulation modeling in high-level languages, in *Proceedings of the 1988 Winter Simulation Conference*, M. Abrams, P. Haig, and J. Comfort, eds., Society for Computer Simulation, La Jolla, Calif., pp. 287–295.

Banks, J., J. S. Carson, and J. N. Sy (1995a). *Getting Started with GPSS/H*, 2nd ed., Wolverine Software Corporation, Annandale, Va.

Banks, J., B. Burnette, H. Kozloski, and J. Rose (1995b) *Introduction to SIMAN V and Cinema V*, Wiley, New York.

Benson, D. (1997) Simulation Modeling and Optimization using Promodel, in *Proceedings of the 1997 Winter Simulation Conference*, Society for Computer Simulation. La Jolla, Calif. pp: 587–593.

Crain, R. C. (1997). Simulation using GPSS/H, in *Proceedings of the 1997 Winter Simulation Conference*, J. M. Clarnes, D. J. Morrice, D. T. Brunner, and J. J. Swain, eds., Society for Computer Simulation, La Jolla, Calif., pp. 567–573.

Henriksen, J. O., and R. C. Crain (1998). *GPSS/H Reference Manual*, 4th ed., Wolverine Software Corporation. Annandale, Va.

Joines, J. A., and S. D. Roberts. (1997). An Introduction to Object-Oriented Simulation in C++, In *Proceedings of the 1997 Winter Simulation Conference*, Society for Computer Simulation. La Jolla, Calif. pp: 78–85.

Pegden, C. D., R. E. Shannon, and R. P. Sadowski (1995). *Introduction to Simulation Using SIMAN*, 2nd ed., McGraw-Hill, New York.

ProModel Corporation (1996). *ProModel Version 3 User's Guide*, ProModel Corporation, Orem, Utah.

Schriber, T. J. (1991). *An Introduction to Simulation Using GPSS/H*, Wiley, New York.

Schriber, T. J., and D. T. Brunner (1997). Inside Simulation Software: How It Works and Why It Matters, in *Proceedings of the 1997 Winter Simulation Conference*, Society for Computer Simulation, La Jolla, Calif., pp: 14–22.

Swain, J. J. (1997). Simulation Goes Mainstream. *OR/MS Today*, October, 1997, 35–46.

CHAPTER 25

Software for Simulation

JERRY BANKS
Georgia Institute of Technology

25.1 INTRODUCTION

In this chapter we discuss the principles of simulation software selection. Then we describe computer simulation software. Finally, we provide a list of software vendors and how they may be contacted.

25.2 SOFTWARE SELECTION

Making a choice from the vast amount of software that is available for discrete-event simulation is bewildering for the newcomer to the field. The 1998 simulation Buyer's Guide in *IIE Solutions* [1] showed 53 entries. Of these, 42 indicated that they were useful in manufacturing. Some 39 of the entries provide animation. There were 30 vendors represented. An extremely careful software selection decision can take as many as 6 months, although that may be the extreme case. This section is based on an article by Banks and Gibson [2].

Possible features to consider in software selection include the following:

- Input
- Processing
- Output
- Environment
- Vendor
- Cost

Prior to beginning a discussion of these features, a few warnings are given.

Handbook of Simulation, Edited by Jerry Banks.
ISBN 0-471-13403-1

- ***Warning 1:*** *Know which features are appropriate*! A particular feature may or may not have relevance to your situation. For example, a feature may be the ability to model bridge cranes. Your firm may not have any bridge cranes and never will have any bridge cranes. Thus this feature is not important to your situation.
- ***Warning 2:*** *Don't make judgments on the basis of "yes" and "no"*. Assume that the software that is being considered has the ability to model conveyors. But can your conveyors be modeled by the software? A simplification of your conveyors may be the limits of the software that is being considered. If more detail is needed, this simplification can lead to problems downstream. For example, both the length and width of the conveyor may be relevant. However, the software that is being considered is concerned only with the length of the conveyor. A discussion of the use of check marks (yes or no indications) in selecting simulation software is given by Banks [3].
- ***Warning 3:*** *Consider the avoidance of a feature.* Assume that one of the features of the software under consideration is that an interface with C code is possible. However, simulation software should be judged on its ability to avoid C code. Thus powerful constructs within the software can prove to be very useful.

25.2.1 Input Considerations

These considerations include the following:

- Point-and-Click Capability. Simulation software users expect and desire the familiar environment for their software applications. Most simulation software has migrated to this environment.
- CAD Translation. If there exists a CAD drawing of the static background, there is no reason to redraw this for purposes of the simulation. A CAD translator will take a CAD drawing and convert it into the drawing system used by the simulation. A CAD translation feature can save a great deal of time when laying out large or complex conveyor models, for example.
- Importing a File. This feature provides the capability to import a data file for use in the simulation. There are numerous situations in which this can be important. The database can be very large, and generated and updated electronically without human intervention. Obviously, having to rekey such information would be a waste of time and would probably introduce errors.
- Exporting a File. Perhaps the output file will be used as input to a spreadsheet for drawing business graphics beyond or different from those generated by the simulation software.
- Syntax. The syntax should be easily understood, consistent, and unambiguous. There are many counterexamples to this. For instance, in several of the popular simulation languages, a QUEUE block is not necessary to have a waiting line; the QUEUE block just initiates data collection. This is just one of many ambiguous notions that long-term users have come to accept. Chapter 24 discusses the difference in interpretation among several simulation software packages.
- Interactive Run Controller (IRC). Some call this feature by the name *debugger*. Even the best of simulation analysts makes mistakes or commits logical errors

when building a model. The IRC assists in finding and correcting those errors in the following ways (although not every IRC will perform the following, and some will do even more).

1. The simulation can be monitored as it progresses. This can be accomplished by advancing the simulation until a desired time has elapsed, then displaying model information at that time. Another possibility is to advance the simulation until a particular condition is in effect and then display information.
2. Attention can be focused on a particular area of the simulation or a particular entity. For instance, every time an entity enters a specified area, the simulation will pause so that information can be gathered. As another example, every time that a specified entity becomes active, the simulation will pause.
3. Values of selected model components can be observed. When the simulation has paused, the current value or status of variables, attributes, queues, resources, counters, and so on, can be observed.
4. The simulation can be temporarily suspended, or paused, not only to view information, but also to reassign values or redirect entities.

- *Interface to Other Language.* This is the ability to drop into another language for additional detail. Earlier this particular capability was questioned. However, if it cannot be avoided, its use can be of great importance. Banks and Gibson [4] discuss this topic in much detail.
- *Input Data Analysis Capability.* This feature provides the ability to determine whether input data can be described by a statistical or mathematical distribution. Many software vendors have included the ability of third-party software to provide this capability.

25.2.2 Processing Considerations

- *Powerful Constructs.* For the situation, powerful capabilities may be needed. For example, modeling bridge cranes with interference requires powerful constructs.
- *Speed.* When there are many entities in a system, the software speed should not degrade to the point of slow motion.
- *Run-Time Flexibility.* An example of this feature is batching, the ability to stack up a series of runs, and the necessary data collection in an orderly fashion, without intervention. Another example is scenario generation. In this case, with some prodding, the simulation software will automatically generate alternative possibilities for simulation. An example of scenario generation allows input data to vary over a range (e.g., the cycle time of a machine can range from 4 to 8 seconds).
- *Random Variate Generator.* There are about 12 statistical distributions that are commonly used in simulation. Most, but not all simulation software has the ability to generate random variates using these 12 distributions.
- *Reset.* For steady-state analysis, it is important to have the ability to reset the statistics that have been collected to zero. This reset is accomplished without clearing the entities that are currently in the system.
- *Independent Replications.* Multiple replications using different sets of random numbers should be possible. Otherwise, the same results would occur repeatedly.
- *Attributes and Global Variables.* Attributes are local values available to the entity

possessing that attribute, and global variables are available to all entities. The question is how many of each of these is available. Real simulation models can use large numbers of attributes and global variables.

- *Programming.* This feature is also called *custom logic representation*. The ability to mimic custom logic accurately to any desired degree of detail usually requires some type of internal programming capability or underlying language. It is an absolute necessity for modeling complex problems or systems in order to build high-fidelity models (ones that, like a good audio system, can reproduce the sound of the original system). Consider whether the software allows you to represent a problem to any desired level of detail or complexity or if it forces you to choose canned representations or simplifications.
- *Portability.* This feature enables the software to be run on various classes of computer without changes in the software.

25.2.3 Output Considerations

These are many output considerations when selecting simulation software. A few of the many are given in the following paragraphs:

- *Standardized Reports.* Examples of standardized output measures are the average number in queue, average time in queue, and throughput. The software can produce these and other values automatically or upon request.
- *Customized Reports.* These are tailored presentations such as those that would be shown to a manager. The format can be set by the simulation analyst.
- *Business Graphics.* The software can have the ability to generate bar charts, pie charts, and histograms that are of such high quality that they can be shown to managers and included in reports.
- *Database Maintenance.* One possibility is to collect stacks of paper output representing the replications from each scenario. However, this could become rather a large stack of paper. At the other extreme is a database that contains all these outputs in an organized fashion.
- *Collection of Mathematical Expressions Desired.* This feature allows the specification of measures of interest to the modeler. For example, suppose that each type 1 output completed is to be multiplied by a royalty of $1.53 and each type 2 output is to be multiplied by a royalty of $1.86. These royalties are to be reported by shift and week, as well as their sum.
- *Custom Performance Measures.* Does the software allow the analyst to define and create new or custom measures of performance for a model? Or is the analyst constrained to a predefined set of performance measures?
- *Write to a File.* Does the software allow data, events, or system variables to be written to a file whenever desired? This feature allows the analyst later to import the file into a spreadsheet or database program for further customized analysis or manipulation.

25.2.4 Environment Considerations

There are many environment considerations, some more important than others. Examples include the following:

insurance company or an injection molding machine, the object will know how what to do.

SIMPLE++'s open architecture allows it to communicate or be integrated with a variety of other software packages [13]. For example, SIMPLE++ can communicate with databases to allow on-line real-time scheduling in the automotive and silicon wafer manufacturing industries. It has also been linked to expert systems for training and diagnosis, spreadsheets, and other off-the-shelf software, work measurement, and virtual reality packages. In addition, SIMPLE++ has been embedded within software for layout optimization, Business Process Re-engineering (BPR), and ergonomic workstation design. By doing this, a dynamic analysis is possible in tools otherwise not able to do this.

Extend. Entend, from Imagine That, Inc., is a visual, interactive simulation tool [14]. Extend contains a built-in development system that allows the user to construct components and build custom user interfaces. Models are constructed graphically by dragging and dropping blocks (high-level model components) from library windows onto the model worksheet. Data can be entered directly into block dialogs, interactively using controls, or read from files as the simulation runs. Output is in the form of plots, histograms, tables, and customizable reports. The block development environment includes a full-featured, compiled, C-based language that allows simulation modelers to add custom functionality. Also included are hooks to external languages such as C or Fortran. Additional features include integrated animation that evolves as the simulation progresses, multiple scenario analysis, unlimited hierarchical decomposition, model-building wizards, hot links, drag and drop reports, work areas for centralized model control and reporting, over 300 built-in functions, 32-bit computing, and compatibility across platforms.

25.3.2 Manufacturing-Oriented Software

The software discussed in this section is limited to seven within the category, including ProModel, AutoMod, Taylor II, WITNESS, AIM, Arena, and Extend+Manufacturing. For the most part, this software is used for the simulation of manufacturing and material handling systems. References for these software packages include the following: Benson [15] for ProModel, Rohrer [16] for AutoMod, Markt and Mayer [17] for WITNESS, and King [18] for Taylor II. For AIM, the reference is Ehrlich and Lilegdon [19]. For Arena, the reference is Takus and Profozich [20]. The reference for Extend+Manufacturing is Rivera [14].

ProModel. ProModel, from PROMODEL Corporation, has programming features within the environment and the capability to add C or Pascal subroutines to a program. Some of the features of ProModel, Release 4, are as follows:

1. Models are created using a point-and-click approach. Intuitive interfaces, interactive dialog, and on-line help are provided. An autobuild feature guides the user through the model building process. An on-line trainer is available.
2. The software operates in the Windows environment, as a 32-bit application, taking advantage of memory management techniques, synchronized windowing, and data exchange. Windows fonts, printer drivers, cooperative multiasking, and the Dynamic Link Library are available.

3. Virtually unlimited model size is offered.
4. The simulator offers a two-dimensional graphics editor with scaling, rotating, and so on. Icons can be defined using either vector-based or pixel graphics. These icons are saved as bitmaps at run time for fast animation during the simulation.
5. CAD drawings as clip art can be imported as well as process information and schedules. Customized output reports and spreadsheet files can be produced. If the data are another Windows application, cutting and pasting can be accomplished.
6. The static and dynamic elements of the animation are developed while defining the model. That is, the simulation model and animation are integrated.
7. Business output graphics are provided automatically and may be printed in color.
8. Only standard hardware is required (IBM or compatible with VGA graphics). No special graphics cards, monitors, or math coprocessor chip is needed.
9. Preprogrammed constructs are provided. This allows for fast modeling of multiunit and multicapacity locations, shared and mobile resources, downtime, shifts, and so on.
10. Automatic statistics are available.
11. Submodels allow the creation of a library of templates of work steps, activities, or subprocesses that can be reused. This allows for model construction to be accomplished by a team with later merger of submodels into one model.
12. A free run-time, multiple-scenario capability is provided. Statistics from multiple scenarios can be viewed simultaneously.
13. Multiple views can be saved for easy navigation and graphic layout.
14. Cranes can be added with multiple bridges.
15. Models can be encrypted for protection of data.
16. Interactive subroutine can be added to allow changes to model parameters during model execution.
17. Extensive modeling support for continuous flow processes, tanks, levels, and transfers.

A model is constructed by defining a route for a part or parts, defining the capacities of each of the locations along the route, defining additional resources such as operators or fixtures, defining the mobile/shared resources, scheduling the part arrivals, and specifying the simulation parameters. The software then prompts the user to define the layout and the dynamic elements in the simulation.

AutoMod. AutoMod, software from AutoSimulations, Inc., has general model-building features, including the specification of processes, resources, loads, queues, and variables. Processes are specified in terms of traffic limits, input and output connections, and itineraries. Resources are specified in terms of their capacity, processing time, MTBF, MTTR, and cycles for shifts and preventive maintenance. Loads are defined by their shape and size, their attributes, generation rates, generation limits, and start times, as well as their priority. Users can define any number of load attributes.

AutoMod software is very powerful in its description of material handling systems. AGVs and other path-guided transporters, conveyors, bridge cranes, AS/RSs, and power and free devices can be defined rapidly. The range of definition is extensive. For example, an AGV can be defined in terms of the following: multiple vehicle types, multiple-capacity vehicles, path options (unidirectional or bidirectional), variable-speed paths, control points, flexible control and scheduling rules, arbitrary blocking geometries, automatic shortest-distance routing, and vehicle procedures.

Numerous control statements are available in the AutoMod language. For example, process control statements include If-Then-Else, While-Do, Do-Until, and Wait For. Load control, resource control, and other statements are also available. C functions may be defined by the user, although they are not usually required because of the flexibility of the AutoMod language.

The animation capabilities include true-to-scale three-dimensional graphics, rotation, and tilting, to mention a few. A CAD-like drawing utility is used to construct the model. Business graphics can be generated. In 1993, AutoSimulations added the Simulator to AutoMod. Features of the Simulator include its spreadsheet interface. The spreadsheet interface allows the user to enter complex logic without programming.

A separate utility option is AutoStat. It provides simulation warm-up capability, scenario management, confidence interval generation, and design of experiments capability. AutoView is a postprocessor that provides three-dimensional walkthrough capability for presentation-quality animations. Users can configure a walkthrough with smooth pans and zooms and can even mount a camera on board a moving object such as a load or vehicle.

Taylor II. Taylor II is a product marketed by F&H Simulations. It runs under Windows. Working with Taylor II starts with building a model. All model building is menu driven. A model in Taylor II consists of four fundamental entities: elements, jobs, routings, and products. The element types are inout, machine, buffer, conveyor, transport, path, aid, warehouse, and reservoir. One or more operations can take place at an element. The three basic operations are processing, transport, and storage. Defining a layout is the first step when building a model. Layouts consist of element types. By selecting the elements in sequence, the product path or routing is defined. Routing descriptions may be provided from external files.

The next step is detailing the model. In this step the parameters are provided. In addition to a number of default values, Taylor II uses a macro language called TLI (Taylor Language Interface). TLI is a programming language that permits modifications of model behavior in combination with simulation-specific predefined and user-definable variables. TLI can also be used interactively during a simulation run to make queries and updates. Interface to C, Basic, and Pascal is also possible. Local and global attributes are available.

During simulation, zoom, pan, rotate, and pause are options. Modifications can be made on the fly. The time representation is fully user definable (hours, days, seconds, and so on, can be mixed). Output analysis possibilities include predefined graphics, user-defined graphics, predefined tabular reports, and user-defined reports. Examples of predefined graphics are queue histograms and utilization pies. User-defined outputs include bar graphs, stacked bars, and other business graphics. Predefined tables include job, element, and cost reports.

Animation capabilities include both two and three dimensions. The three-dimensional animation can be shaded. Standard indicators can be shown for elements. Icon libraries for both two- and three-dimensional animation are provided. Each of these libraries contains more than 50 icons. Additional features include 650 pages of on-line, context-sensitive help with index and page-search capability. Educational support materials are available.

WITNESS. WITNESS, a Windows application from the Lanner Group, contains many elements for discrete-part manufacturing. For example, machines can be single, batch, production, assembly, multistation, or multicycle. Conveyors can be accumulating or nonaccumulating. Options exist for labor, vehicles, tracks, and shifts. WITNESS also contains elements for continuous processing, including processors, tanks, and pipes.

Variables and attributes may be specified. Parts can be smart (having their routing attached) or dumb (elements of the process decide the appropriate routing). Distributions and functions can be used for specifying operation times and for other purposes. Machine downtime can be scheduled on the basis of operations, busy time, or available time. Labor is a resource that can be preempted, use a priority system, and be scheduled based on current model conditions.

Track and vehicle logic allow requests for certain types of jobs, vehicle acceleration and deceleration, park when idle, and change destinations dynamically. Many types of routing logic are possible in addition to the standard push and pull. For example, If-Then-Else conditions may be specified. Simulation actions, performed at the beginning and end of simulation events, may employ programming constructs such as For-Next, While-End, and GoTo-Label. The user can look at an element at any time and determine the status of a part.

Reporting capabilities include dynamic on-screen information about all modeled elements. Reports may be exported to spreadsheet software. C-LINKS allows detailed programming and subroutines to be attached to WITNESS. Data inputs can be numeric, variables, distributions, or a user-defined equation. Debugging or brainstorming can be accomplished by stopping the model, changing desired parameters, and continuing with the model from the same point in simulation time.

An animation is constructed along with the model definition. This animation and statistical feedback can be turned on or off during any run. Many changes to the model may be made at any time. Built-in experimentation capabilities are available from the menu bar. The results of the experiments are output to a CSV file by default, or other file types by user choice. The CSV file is in a format that allows the internal statistics package to create confidence intervals.

The capabilities of Witness, version 7.3, include the following:

1. Unlimited number of variables or attributes. Variables can be arrayed.
2. Up to 1000 distinct random number streams.
3. Bitmap import/export with icon sizes increased to 256 × 256 pixels.
4. Module element for hierarchical modeling. One icon represents the detail existing in another submodel.
5. Double-click on an object to open a menu for filling fields with rules, distributions, built-in functions, and so on, for defining elements "on the fly."
6. Utilizes WIN32S to run in 32-bit mode.
7. OLE2 compatibility.

FACTOR/AIM. FACTOR/AIM, from Pritsker Corporation, is a manufacturing operations simulation system for discrete and discrete batch manufacturing. AIM integrates with other FACTOR product family applications for operations planning, scheduling, order promise, and release applications. AIM models are Windows '95, NT, or OS/2-based and built graphically with icons that represent machines, operators, conveyors, and so on, placed directly on the screen. The animations are created in a virtual window.

The current release of AIM is version 7.0. During a simulation, the simulator can be stopped to check model status or add components, then continue the simulation. Performance data are dynamically displayed and updated while the simulation is running. AIM is database oriented, with all model inputs and outputs stored in an Access database. It includes hundreds of built-in reports and graphs. Custom reports and graphs can be created using features of the Access database program.

Features of AIM include the following:

1. *Manufacturing Representation.* Manufacturing specific modeling components can represent a variety of discrete and discrete manufacturing processes. Standard rules provide choices that are common in many manufacturing environments. Custom rules may be written to extend the logic of a model.
2. *Manufacturing Data.* AIM is built around a relational database that stores the manufacturing operation and simulation output. Part descriptions, process plans, order release schedules, machine locations and schedules, shift schedules, and so on, can be transferred from other data sources to the AIM database.
3. *Animation Support.* AIM models are built graphically and are animated automatically during model construction. Typical AIM models are built from CAD facility drawings. AIM uses a scaling parameter to determine the material handling movement distances directly from these layouts.
4. *Interactive Model Building and Simulation.* Components are located on a scaled facility background. Intelligent defaults are provided for all components. Completing forms customizes components. During execution, the modeler can change the status of a component and observe the simulated impact on the manufacturing system.
5. *Comparison of Alternatives.* The AIM project framework organizes all aspects of a manufacturing simulation project. Alternative models of the manufacturing process are stored in the AIM database. Comparison reports and graphs show model performance data to identify differences between alternatives.
6. *Gantt Charts.* AIM automatically creates Gantt charts for faster verification and validation of models. Load charts show the flow of individual loads through the system, verifying model logic. Resource charts aid in reviewing model performance with operations personnel.
7. *Detailed Cost Modeling.* AIM provides for the modeling of cost allocations and financial analysis. Direct cost models can be created to allocate costs to each load as it utilizes resources. AIM has the ability to predict accurately activity-based costs (ABC). ABC cost drivers and methodology can be tested on a specific system to allocate indirect costs in a second-pass simulation.

Arena. Arena, a product of Systems Modeling Corporation, is an extendible simulation and animation software package. It provides a complete simulation environment that supports all steps in a simulation study. Arena combines the modeling power and flexibility of the SIMAN simulation language, while offering the ease of use of the Microsoft Windows and Microsoft NT environments. Arena is a graphical modeling/animation system that is based on hierarchical modeling concepts. It allows users to create new modeling objects called modules, which are the building blocks of model creation. All aspects of a particular process—logic, data, animation, and statistics collection—can be modules to represent the process through which entities flow.

Arena offers Application Solution templates that can be used to tailor the software to a specific animation. Templates are groups of modules that have been designed to capture the entities, processes, and terminology of a specific application. Application Solution Templates in areas such as business process reengineering, call centers, high-speed manufacturing, semiconductor wafer fabrication, and others currently exist or are under development. The Arena system includes the Input Analyzer, designed to give users the ability to read raw input data, and the Output Analyzer, for simulation data viewing and analysis. Arena also features Microsoft Visual Basic for applications enabling users to build interfaces between their Arena models and other applications such as Microsoft Excel.

Extend+Manufacturing. Extend+Manufacturing is a product of Imagine That, Inc. It has all the features of Extend, mentioned previously, plus the following:

1. Blocks that represent machines, stations, labor buffers, bins, fixtures, and so on
2. Reneging, preemption, and interruptible processes
3. Ability to specify sequences for merging and routing streams of entities
4. Scheduled and unscheduled downtimes
5. Material handling constructs
6. Automatic statistical reporting
7. Built-in costing

25.3.3 Business Process Reengineering

Several vendors have developed software to facilitate the reengineering process. These include BP$im, ProcessModel, SIMPROCESS, TimeMachine, and Extend+BPR.

BP$im. BP$im, a product of Systems Modeling Corporation and Technology Economics International, is an Application Solution Template under Arena. Model building is focused on the activities that comprise a business process. Models are created by placing activity modules in the Arena work space and providing information about each activity. The software can be used for activity-based costing by defining busy and idle costs for any resource. BP$im also provides constructs for modeling human and technological resources. Costs, schedules, downtimes, efficiencies, and other resource attributes can be specified. An interface to Lotus and Excel is provided. The software is capable of reading pertinent process data from existing IDEF0 models. BP$im also interfaces with Visio to create models based on flowcharts.

ProcessModel. ProcessModel, a product of PROMODEL Corporation, interfaces with the ABC Graphics Suite [21]. Flowcharts are developed using ABC FlowCharter. Modeling capabilities include the following:

1. Staff scheduling and shift planning
2. Task allocation prioritization and interruption
3. Use of multiple or alternative resources

4. Individual or group processing
5. Matching of orders to correct customers
6. Document splitting and tracking
7. Customers dropping out of line
8. Activity-based costing
9. Resource scheduling for breaks and downtimes
10. Ability to change graphics during simulation

SIMPROCESS. SIMPROCESS, a product of CACI Products Company, integrates icon-based process flowcharting, hierarchical event-driven simulation, activity-based costing, and data analysis capabilities into a tool for business process reengineering. Other features include reusable templates and advanced modeling constructs, such as If-Then-Else logic. SIMPROCESS is based on CACI's MODSIM, an object-oriented architecture for simulation. It runs under both Windows and UNIX operating systems.

Time Machine. Time Machine is a product of F&H Simulations. It is a Windows-based dynamic simulation worksheet used for flowcharting, process modeling, what-if analysis, concept validation, and the communication of ideas. It is built on Taylor II, described previously.

Extend+BPR. Extend+BPR, a product of Imagine That, Inc., is based on the Extend software discussed previously. It has the following features:

1. Blocks that represent operations, transactions, work-flow stacks, labor resources, decisions, and so on
2. Ability to have entities renege
3. Specification of rules for the flow of work throughout the operation
4. Additional blocks for cost reporting
5. Automatic statistical reporting

25.3.4 Simulation-Based Scheduling

Several of the many simulation-based scheduling software currently available are Tempo, AutoSched, and FACTOR. These are discussed briefly in the following paragraphs.

Tempo. Tempo, from Systems Modeling Corporation, is a visually interactive finite-capacity scheduling system. Tempo features include automatic sequencing using forward, backward, and bidirectional rules, job selection by priority, due date, or first come, first served. Tempo offers users the ability to generate algorithmic or simulation-based schedules and includes the ability to create user-defined rules. A variety of output reports are available. Comprehensive data import and export capabilities for integration with other manufacturing technologies (e.g., ERO, MRP, MES) are included.

AutoSched. AutoSched, from AutoSimulations, Inc., is based on AutoMod, a simulation system that was described previously. The following data/information is handled by AutoSched for capacity analysis and scheduling [22]:

1. Production resources consisting of workstations, storage locations, operators, and tools
2. Products consisting of parts to be manufactured along with their routings
3. Production requirements consisting of orders and lots
4. Operating rules, including how tasks are selected, and calendars that specify when workstations are unavailable

Much of the information required by AutoSched may already exist in a database such as an MRP. If it does exist, AutoSched can import it. AutoSched allows multi-dimensional task selection rules for each resource. More than one rule (logic filter) is allowed. Significant scheduling improvements can be achieved through the application of these sophisticated rules.

AutoSched provides simulation and detailed dispatching of the shop floor according to the task selection rules provided by the user. Two outputs are provided, graphical and statistical. Business graphs can be created to track any statistic(s). These graphs are updated dynamically. AutoSched also contains an interactive Gantt chart from which an event can be selected for detailing to include the quantity of orders in a workstation's queue when an order was selected, the quantity of orders in the next workstation's queue, and so on. Statistical or historical reports are also available. These include the master schedule file, performance report, workstation report, and many user-defined reports.

FACTOR. FACTOR, from Pritsker Corporation, is an integrated software system that provides capacity management applications to help manufacturers meet customer demand [23]. FACTOR includes finite-capacity scheduling, operations planning and loading, and order promising capabilities. Since its introduction, the software has been enhanced to include relational database capabilities, client–server installations, interactive schedule adjustment, and automatic scheduling capability.

To use FACTOR, a model is built by combining components consisting of order characteristics, shop floor status, production calendar, shift schedules, resources, functional resource groupings, preventive maintenance plans, parts definitions with the associated routings, material availability, tooling requirements, and operator capability. Any of these model components may be defined directly or taken from preexisting manufacturing system files. A simulation of selected alternatives is conducted for various scheduling strategies, with results stored in a database for analysis and eventual shop distribution.

In practice, FACTOR is used to schedule operations at regular intervals (e.g., shifts) or upon demand (when a significant event occurs). In either case, up-to-date status information is transferred into the FACTOR database and the FACTOR scheduling engine generates a recommended schedule based on the objective (e.g., maximize throughput, minimize order lateness). The user reviews a summary of the performance anticipated from the schedule and may then accept it, revise the objective and generate another schedule, or interactively make manual adjustments to the schedule.

25.3.5 Animators

Most simulation animators are integrated with the software. However, this is not always the case, and the introduction of general-purpose animation packages allows the use of custom made environments.

Proof Animation. Proof Animation is a product of Wolverine Software Corporation [24]. Any software that can write ASCII data to a file can drive Proof Animation. Thus BASIC, C, FORTRAN, GPSS/H, Arena, and SIMSCRIPT II.5, among others, can serve as drivers. Animation is accomplished by using a static background, the layout file, and a trace file that contains dynamic events. Some of the features of the software are as follows:

1. Graphics are vector based, similar to CAD programs.
2. AutoCad-compatible layouts can be imported and exported.
3. Zoom in, zoom out, and multiwindow animations are supported. Maximum resolution is assured at any scale.
4. Drawing takes place on a coordinate grid using mouse-driven primitives.
5. Moving objects are defined internally by their geometry.
6. Statistics, graphs, and plots can be displayed dynamically.
7. Animation occurs in a postprocessing mode.
8. Motion is smooth on VGA PCs.
9. There is a steady ratio of animated (simulation) time to viewing (wall clock) time. This ratio may be varied while the animation is running.
10. Top view can be changed to isometric and back to top view instantly.
11. An option allows the construction of a demo disk.

25.3.6 Simulation Support Software

Two products, among many that are available, are discussed in this section. The first is ExpertFit from Averill M. Law and Associates, for input data modeling [25]. The second is Stat: :Fit from Geer Mountain Software Corporation also for input data analysis [26].

ExpertFit. ExpertFit is a software package for selecting what probability distribution to use in a simulation model. It will automatically and accurately determine what distribution best represents a data set. It will also determine whether the distribution is appropriate in an absolute sense, or whether an empirical distribution should be used. In either case, the distribution will be put into the proper format for direct input to 34 different simulation software products. ExpertFit provides access to 39 probability distributions by providing special constructs that can be used with simulation software. There more than 30 two- and three-dimensional graphical plots available. The software helps in the choice of a distribution in the absence of data and determines if data sets are similar.

Stat: :Fit. Stat: :Fit statistially fits data to one of 21 distributions. The Auto: :Fit function automatically fits continuous distributions, provides relative comparisons between distribution types, and an absolute measure of each distribution's acceptability. The Export function translates the fitted distribution into specific forms for simulation software. Features include descriptive statistics, parameter estimates, goodness of fit tests, graphical analysis, random variate generation, and printed reports.

25.4 OPTIMIZATION

A rigorous treatment of optimization in simulation is discussed in Chapter 9. However, practical applications are not claimed. In this section, optimization capabilities based on heuristic approaches that are used in three software products are described. Additionally, implementation of one of the techniques in simulation software discussed in Chapter 8, selection of the best, is discussed. Heretofore, simulation analysts have relied on intuition or experience to direct their search for a good solution. That is, they varied the factors, watching the response as they do so, and have headed in the direction leading toward the optimal objective. This is a complicated matter when the number of factors is even modest. Furthermore, the solution using this procedure may be a local optima rather than a global optimum.

Another possibility is complete enumeration. For example, a job shop may have two parallel machining processes. Let us say that there may be anywhere from 1 to 10 machines in each process. Then there are 100 cases that must be examined. If each of these cases can be examined in 10 seconds, the simulation will be complete in less than 20 minutes. But if there are five parallel machining processes, each with 1 to 10 machines, then 100,000 simulations must be conducted using complete enumeration. At 10 seconds per simulation, the time required approaches 12 days. Such a time requirement is probably excessive even for this modest problem.

Recent developments in optimization have led to intelligent search procedures that have the capability to find optimal or near-optimal solutions by exploring a small fraction of the possible alternatives. These procedures are used by SimRunner an add-on to ProModel, OptQuest in Micro Saint, and OPTIMIZER in WITNESS. These software operate differently as explained in this section. AutoStat, an extension to AutoMod, has a ranking and selection feature that selects the best from a small number of alternatives. This capability is also discussed in this section.

SimRunner. SimRunner, Release 2.0, an add-on to ProModel, is discussed generally by Benson [15]. The analyst indicates the input factors to change and how to measure performance. SimRunner will then run an optimization analysis or a simulation optimization. An optimization analysis is a factorial experiment that indicates whether changing a specific input factor affects the objective function and by how much. A simulation optimization tries various combinations of input factors to arrive at the combination that provides the best value of the objective function. SimRunner features a third module, statistical advantage, that assists the analyst in preparing models that produce statistically significant results (e.g., helping to determine the warm-up period, run length, and number of runs).

Bowden [27] indicates that the optimization procedure used in SimRunner evolves a population of solutions to the problem such that each solution's survival is dependent on its performance in the simulated environment. The population of solutions evolves for a number of generations, at which time the search is terminated and the best surviving solution is forwarded. Bowden also shows how the algorithm avoids being trapped by local optima using two examples; the first deterministic and the second probabilistic. Akbay [28] gives three examples of the application of SimRunner. IBM's microelectronics plant in Vermont had many factors to test in their models. Simulation experimentation was becoming very time consuming. SimRunner was applied and IBM's capacity-planning capabilities were improved.

Sverdup Fcilities, Incorporated and GPR Planners Collaborative were contracted to determine if a building addition would improve transporter operations in a hospital setting. SimRunner was used to determine the optimum number of transporters required by each department with large reductions in analysis time. In another medical setting, simulation was used to test various staffing plans for a new clinic. Then the various proposed configurations were optimized using SimRunner.

OptQuest. OptQuest in Micro Saint is discussed generally by Barnes and Laughery [29]. The optimization procedure in OptQuest is discussed by Glover et al. [30]. The technology used is called *scatter search.* It operates on a set of reference points that constitute good solutions obtained from previous solutions. Linear combinations of these reference points guide the search. These linear combinations are formed on the basis of tabu search. Essentially, tabu search prevents the reinvestigation of solutions that have already been evaluated.

In using OptQuest, the optimization problem is described in terms of the factors selected and an objective function. The problem can include constraints. For example, the total number of A and B machines must be less than or equal to 12. Once the problem is specified, OptQuest repeatedly calls Micro Saint to perform simulation runs during the search for an optimal solution.

OPTIMIZER. OPTIMIZER in WITNESS is discussed by Markt and Mayer [17]. It operates by first specifying an objective function defined as a WITNESS function. An example objective function might be:

$$\text{value of throughput} - \text{cost of machines} - \text{cost of staff},$$

During the optimization process, different aspects of the model are varied, and the resulting value for the objective function will be compared to previous values to determine if improvements have occurred. Any number of constraints can be imposed to reduce the possible number of combinations to be investigated. An example constraint might be:

$$\text{staff1} + \text{staff2} \leq 10,$$

Several optimization methods are provided ranging from simply running all possible combinations to more complex algorithms. An example of a more complex algorithm in the OPTIMIZER is simulated annealing (SA). Users are allowed to adjust the parameters if using SA.

AutoStat. AutoStat, discussed by Carson [31], offers a "select the best" ranking and selection procedure, discussed in Chapter 8. The procedure assumes that output for each is normally distributed and IID (independent and identically distributed). The procedure selects the single best system, or a subset containing the best system. The user specifies the probability of correct selection and the indifference amount. An example problem statement might be the following: I want the best two of six alternative systems with the probability of being correct greater than or equal to 0.95 if the difference between the average time in system for the two selected systems and the next highest of the remaining four systems happens to be at least 2.5 minutes. Pilot runs, say 10, are conducted. Then AutoStat tells the analyst how much computer time will be required to achieve the objective. If the time is excessive, the analyst may need to be somewhat flexible in the problem statement (reducing the probability of a correct selection, increasing the indifference amount, or both). Additionally, there may be more than one response of interest. This also increases the number of replications and may require additional flexibility from the analyst (e.g., reduce the number of responses).

25.5 VENDORS OF SOFTWARE

The products mentioned in Section 25.3 are listed below with vendor contact information indicated.

ARENA, BP$im, Tempo
Systems Modeling Corporation
504 Beaver Street
Sewickley, PA 15143

AutoMod, AutoSched, AutoStat
AutoSimulations, Inc.
655 Medical Drive
Bountiful, UT 84010

AweSim, AIM, FACTOR
Symix
8910 Purdue Road, Suite 600
Indianapolis, IN 46268

ExpertFit
Averill M. Law & Associates
P.O. Box 40996
Tucson, AZ 85717

Extend, Extend+Manufacturing, Extend+BPR
Imagine That, Inc.
6830 Via del Oro, Suite 230

GPSS/H, SLX, PROOF Animation
Wolverine Software Corporation
Little River Turnpike, Suite 900
Annandale, VA 22003-2603

Micro Saint
Micro Analysis and Design, Inc.
4900 Pearl East Circle, Suite 201 E
Boulder, CO 80301

ProModel, ProcessModel
PROMODEL Corporation
1875 State Street, Suite 3400
Orem, UT 84058

Simple++
Aesop Corporation
223 South Wacker Drive
Sears Tower, Suite 9604
Chicago, IL 60606

SIMSCRIPT II.5, SIMANIMATION, SIMPROCESS
CACI Products Company
3333 North Torrey Pines Court
La Jolla, Ca 92037

Stat: :Fit
Geer Mountain Software Corporation
104 Geer Mountain Road
South Kent, CT 06785

Taylor II, Time Machine
F&H Simulations, Inc.
P.O. Box 658
Orem, UT 84059-0658

Witness
Lanner Group, Inc.
11000 Richmond Ave., Ste. 680
Houston, TX 77042

ACKNOWLEDGMENTS

The material in Section 25.2 is adapted from Banks and Gibson (1997) with the permission of the Institute of Industrial Engineers, 25 Technology Park, Norcross, GA 30092, 770-449-0491. Copyright © 1997.

REFERENCES

1. Anon. (1998). Simulation software buyer's guide, *IIE Solutions*, May, pp. 48–54.
2. Banks, J., and R. G. Gibson (1997). Selection of simulation software, *IIE Solutions*, May, pp. 30–32.
3. Banks, J. (1996). Interpreting simulation software checklists, *OR/MS Today*, June, pp. 74–78.
4. Banks, J., and R. G. Gibson (1997). Caution: some programming required, *IIE Solutions*, February, p. 26 ff.
5. Banks, J. (1995). Semantics of simulation software, *OR/MS Today*, December, pp. 38–40.
6. Banks, J. (1996). Software for simulation, in *Proceedings of the 1996 Winter Simulation Conference*, J. Charnes, D. Morrice, D. Brunner, and J. Swain, eds., Association for Computing Machinery, New York, pp. 31–38.
7. Crain, R. C. (1996). Simulation Using GPSS/H, in *Proceedings of the 1997 Winter Simulation Conference*, S. Andradóttir, K. J. Healy, D. H. Withers, and B. L. Nelson, eds., Association for Computing Machinery, New York, pp. 567–573.
8. Henriksen, J. O. (1997). An introduction to SLX^{TM}, in *Proceedings of the 1997 Winter Simulation Conference*, S. Andradóttir, K. J. Healy, D. H. Withers, and B. L. Nelson, eds., Association for Computing Machinery, New York, pp. 559–566.
9. Russell, E. C. (1993). SIMSCRIPT II.5 and SIMGRAPHICS tutorial, in *Proceedings of the 1993 Winter Simulation Conference*, G. W. Evans, M. Mollaghasemi, E. C. Russell, and W. E. Biles, eds., Association for Computing Machinery, New York, pp. 223–227.
10. Pritsker, A. A. B., and J. J. O'Reilly (1997). AweSim: the integrated simulation system,

in *Proceedings of the 1997 Winter Simulation Conference*, S. Andradóttir, K. J. Healy, D. H. Withers, and B. L. Nelson, eds., Association for Computing Machinery, New York, pp. 545–551.

11. Geuder, D. (1995). Object-oriented modeling with Simple++, in *Proceedings of the 1995 Winter Simulation Conference*, C. Alexopoulos, K. Kang, W. R. Lilegdon, and D. Goldsman, eds., Association for Computing Machinery, New York, pp. 534–540.
12. Levasseur, G. A. (1996). The case for object-oriented simulation modeling, *OR/MS Today*, August, pp. 65–67.
13. Kalasky, D. R., and G. A. Levasseur (1997). Using SIMPLE++ for improved modeling efficiencies and extending model life cycles, in *Proceedings of the 1997 Winter Simulation Conference*, S. Andradóttir, K. J. Healy, D. H. Withers and B. L. Nelson, eds., Association for Computing Machinery, New York, pp. 611–618.
14. Rivera, J. (1997). Modeling with Extend™, in *Proceedings of the 1997 Winter Simulation Conference*, S. Andradóttir, K. J. Healy, D. H. Withers, and B. L. Nelson, eds., Association for Computing Machinery, New York, pp. 674–679.
15. Benson, D. (1997). Simulation modeling and optimization using ProModel, in *Proceedings of the 1997 Winter Simulation Conference*, S. Andradóttir, K. J. Healy, D. H. Withers, and B. L. Nelson, eds., Association for Computing Machinery, New York, pp. 587–593.
16. Rohrer, M. (1997). AutoMod tutorial, in *Proceedings of the 1997 Winter Simulation Conference*, S. Andradóttir, K. J. Healy, D. H. Withers, and B. L. Nelson, eds., Association for Computing Machinery, New York, pp. 657–662.
17. Markt, P. L., and M. H. Mayer (1997). WITNESS simulation software: a flexible suite of simulation tools, in *Proceedings of the 1997 Winter Simulation Conference*, S. Andradóttir, K. J. Healy, D. H. Withers, and B. L. Nelson, Association for Computing Machinery, New York, pp. 711–717.
18. King, C. B. (1996). Taylor II manufacturing simulation software, in *Proceedings of the 1996 Winter Simulation Conference*, J. M. Charnes, D. J. Morrice, D. T. Brunner, and J. J. Swain, eds., Association for Computing Machinery, New York, pp. 569–573.
19. Erlich, J. N. and W. R. Lilegdon (1997). Making better manufacturing decisions with AIM, in *Proceedings of the 1997 Winter Simulation Conference*, S. Andradóttir, K. J. Healy, D. H. Withers, and B. L. Nelson, eds., Association for Computing Machinery, New York, pp. 552–558.
20. Takus, D. A., and D. M. Profozich (1997). Arena software tutorial, in *Proceedings of the 1997 Winter Simulation Conference*, S. Andradóttir, K. J. Healy, D. H. Withers, and B. L. Nelson, eds., Association for Computing Machinery, New York, pp. 541–544.
21. Gladwin, B. D., and C. Harrell (1997). Introduction to ProcessModel and ProcessModel 9000, in *Proceedings of the 1997 Winter Simulation Conference*, S. Andradóttir, K. J. Healy, D. H. Withers, and B. L. Nelson, eds., Association for Computing Machinery, New York, pp. 594–600.
22. Lindler, B. (1997). AutoSched tutorial, in *Proceedings of the 1997 Winter Simulation Conference*, S. Andradóttir, K. J. Healy, D. H. Withers, and B. L. Nelson, Association for Computing Machinery, New York, pp. 663–667.
23. Lilegdon, W. R. (1993). Manufacturing decision making with FACTOR, in *Proceedings of the 1993 Winter Simulation Conference*, G. W. Evans, M. Mollaghasemi, E. C. Russell, and W. E. Biles, eds., Association for Computing Machinery, New York, pp. 159–164.
24. Henriksen, J. O. (1997). The power and performance of PROOF animation, in *Proceedings of the 1997 Winter Simulation Conference*, S. Andradóttir, K. J. Healy, D. H. Withers, and B. L. Nelson, eds., Association for Computing Machinery, New York, pp. 574–580.
25. Law, A. M., and M. G. McComas (1997). ExpertFit: total support for simulation input modeling, in *Proceedings of the 1997 Winter Simulation Conference*, S. Andradóttir, K. J. Healy,

D. H. Withers, and B. L. Nelson, Association for Computing Machinery, New York, pp. 668–673.

26. Blaisdell, W. E., and J. Haddock (1993). Simulation analysis using SIMSTAT 2.0, in *Proceedings of the 1993 Winter Simulation Conference*, G. W. Evans, M. Mollaghasemi, E. C. Russell, and W. E. Biles, eds., Association for Computing Machinery, New York, pp. 213–217.
27. Bateman, R., R. Bowden, T. Gogg, C. Harrell, and J. Mott (1997). *Improvement Using Simulation*, 5th edition, PROMODEL Corporation, Orem, Utah.
28. Akbay, K. S. (1996). Using simulation optimization to find the best solution, *IIE Solutions*, May, pp. 24–29.
29. Barnes, C. D., and H. R. Laughery, Jr. (1997). Advanced uses for Micro Saint Simulation Software, in *Proceedings of the 1997 Winter Simulation Conference*, S. Andradóttir, K. J. Healy, D. H. Withes, and B. L. Nelson, eds., Association for Computing Machinery, New York, pp. 680–686.
30. Glover, F., J. P. Kelly, and M. Laguna (1996). New advances and applications of combining simulation and optimization, in *Proceedings of the 1996 Winter Simulation Conference*, J. M. Charnes, D. J. Morrice, D. T. Brunner, and J. J. Swain, eds., Association for Computing Machinery, New York, pp. 144–152.
31. Carson, J. S., II (1997). AutoStat^{TM}: output statistical analysis for AutoMod users, in *Proceedings of the 1997 Winter Simulation Conference*, S. Andradóttir, K. J. Healy, D. H. Withers, and B. L. Nelson, eds., Association for Computing Machinery, New York, pp. 649–656.

INDEX